Microbial Food Safety in Animal Agriculture

CURRENT TOPICS

Microbial Food Safety in Animal Agriculture

CURRENT TOPICS

Mary E. Torrence

Richard E. Isaacson

Iowa State Press
A Blackwell Publishing Company

Mary E. Torrence, DVM, PhD, DACVPM, is the National Program Leader for Food Safety at the USDA's Cooperative State Research Education and Extension Service, and serves on various federal and national food safety committees. Dr. Torrence has held positions as a small animal veterinarian, State Veterinarian, Associate Director at Purdue University, and at the Food and Drug Administration. Prior to her current position, she was the Chief of the Epidemiology Branch, Division of Surveillance and Biometrics, Center for Devices and Radiological Health, FDA.

Richard E. Isaacson, PhD, is Professor and Chair of the Department of Veterinary PathoBiology at the University of Minnesota. Dr. Isaacson has held positions at the National Animal Disease Center; was Assistant Professor in the Department of Epidemiology, School of Public Health at the University of Michigan; and was the Manager of Immunology and Infectious Diseases at Pfizer Central Research. Prior to his position at the University of Minnesota, he was Professor in the Department of Veterinary Pathobiology at the University of Illinois.

© 2003 Iowa State Press
A Blackwell Publishing Company
All rights reserved

Any opinions, findings, conclusions, or recommendations expressed in this publication are those of the authors and do not necessarily reflect the views of the U.S. Department of Agriculture.

Iowa State Press
2121 State Avenue, Ames, Iowa 50014

Orders:	1-800-862-6657
Office:	1-515-292-0140
Fax:	1-515-292-3348
Web site:	www.iowastatepress.com

Authorization to photocopy items for internal or personal use, or the internal or personal use of specific clients, is granted by Iowa State Press, provided that the base fee of $.10 per copy is paid directly to the Copyright Clearance Center, 222 Rosewood Drive, Danvers, MA 01923. For those organizations that have been granted a photocopy license by CCC, a separate system of payments has been arranged. The fee code for users of the Transactional Reporting Service is 0-8138-1495-2/2003 $.10.

⊗ Printed on acid-free paper in the United States of America

First edition, 2003

Library of Congress Cataloging-in-Publication Data

Microbial food safety in animal agriculture : current topics / Mary E.
Torrence and Richard E. Isaacson, editors.
      p. cm.
Includes bibliographical references and index.
   ISBN 0-8138-1495-2 (alk. paper)
   1. Food—Microbiology. 2. Food—Safety measures. 3. Agricultural
microbiology. I. Torrence, Mary E. II. Isaacson, Richard E., 1947-
   QR115.M4564 2003
   664′.001′579—dc21   2002156431

The last digit is the print number: 9 8 7 6 5 4 3 2 1

CONTENTS

CONTRIBUTORS

MARY E. TORRENCE
U.S. Department of Agriculture
Cooperative State Research Education and
Extension Service
AgBox 2220
1400 Independence Ave. SW
Washington, D.C. 20250

RICHARD E. ISAACSON
University of Minnesota
Department of Veterinary PathoBiology
1971 Commonwealth Ave.
St. Paul, Minnesota 55108

ALWYNELLE S. AHL
College of Veterinary Medicine
Food Animal Production Building
Tuskegee University
Tuskegee, Alabama 36088

CRAIG ALTIER
College of Veterinary Medicine
North Carolina State University
4700 Hillsborough St.
Raleigh, North Carolina 27606

SHEILA M. ANDREW
University of Connecticut
Department of Animal Sciences
Storrs, Connecticut 06269

ART BAKER
U.S. Department of Agriculture
Food Safety and Inspection Service
Aerospace Bldg.
1400 Independence Ave. SW
Washington, D.C. 20250-3700

MARY J. BARTHOLOMEW
Center for Veterinary Medicine
U.S. Food and Drug Administration
7519 Standish Place
Rockville, Maryland 20855

ANDREAS J. BÄUMLER
Department of Medical Microbiology and
Immunology
College of Medicine
Texas A&M University System Health Science Center
407 Reynolds Medical Building
College Station, Texas 77843-1114

ANDREW BENSON
University of Nebraska
Department of Food Science and Technology
Lincoln, Nebraska 68583

THOMAS E. BESSER
Department of Veterinary Microbiology and
Pathology
Washington State University
Pullman, Washington 99164-7040

MONICA BORUCKI
U.S. Department of Agriculture
Agricultural Research Service
Animal Disease Research Unit
Pullman, Washington 99164

DANIEL M. BYRD, III
Life Sciences Research Organization
Bethesda, Maryland 20815

DOUGLAS R. CALL
College of Veterinary Medicine
Washington State University
Pullman, Washington 99164

STEVE A. CARLSON
U.S. Department of Agriculture
Agricultural Research Service
National Animal Disease Center
Pre-harvest Food Safety and Enteric
Disease Research Unit
Ames, Iowa 50010

JOSHUA T. COHEN
Harvard Center for Risk Analysis
Harvard School of Public Health
Boston, Massachusetts 02115

M. COLEMAN
Office of Public Health and Sciences
U.S. Department of Agriculture
Food Safety and Inspection Service
Maildrop 334 Aerospace Center
1400 Independence Ave. SW
Washington, D.C. 20250-3700

PETER COWEN
Department of Farm Animal Health and
Resource Management
College of Veterinary Medicine
North Carolina State University
4700 Hillsborough St.
Raleigh, North Carolina 27606-1499

ANNE K. COURTNEY
Office of Public Health and Science
U.S. Department of Agriculture
Food Safety and Inspection Service
Aerospace Building
1400 Independence Ave. SW
Washington, D.C. 20250

ROBERT H. DAVIES
Veterinary Laboratories Agencies (Weybridge)
New Haw
Addlestone
Surrey KT15 3NB
United Kingdom

HELEN C. DAVISON
Veterinary Laboratories Agency
New Haw, Addlestone, Surrey, KT15 3NB
United Kingdom

A. B. DESSAI
College of Veterinary Medicine
Food Animal Production Building
Tuskegee University
Tuskegee, Alabama 36088

J. P. DUBEY
Parasite Biology, Epidemiology, and
Systematics Laboratory
Animal and Natural Resources Institute
Agricultural Research Service
U.S. Department of Agriculture
Building 1044, BARC-East
10300 Baltimore Avenue
Beltsville, Maryland 20705-2350

ERIC EBEL
Office of Public Health and Sciences
Food Safety and Inspection Service
U.S. Department of Agriculture
555 S. Howes St.
Fort Collins, Colorado 80521

WILLIAM T. FLYNN
Center for Veterinary Medicine
U.S. Food and Drug Administration
7519 Standish Place
Rockville, Maryland 20855

TIMOTHY S. FRANA
Iowa State University
College of Veterinary Medicine
Department of Veterinary Microbiology and
Preventive Medicine
Ames, Iowa 50011

DANIEL L. GALLAGHER
Virginia Tech University
Department of Civil and Environmental Engineering
418 Durham Hall
Blacksburg, Virginia 24061

IAN GARDNER
Department of Medicine and Epidemiology
University of California, Davis
Davis, California 95616

GEORGE M. GRAY
Harvard Center for Risk Analysis
Harvard School of Public Health
Boston, Massachusetts 02115

YRJO T. GROHN
Department of Population Medicine and
Diagnostic Sciences
Cornell University
Ithaca, New York 14853

JEAN GUARD-PETTER
U.S. Department of Agriculture
Agricultural Research Service
Southeast Poultry Research Laboratory
934 College Station Road
Athens, Georgia 30605

M. GUO
Departments of Surgery and Medical Physiology
Cardiovascular Research Institute
Texas A & M University System Health Science
Center
702 Southwest H.K. Dodgen Loop
Temple, Texas 76504

DALE D. HANCOCK
Field Disease Investigation Unit
Department of Veterinary Clinical Sciences
Washington State University
Pullman, Washington 99164-6610

DONALD E. HANSEN
College of Veterinary Medicine
Oregon State University
Corvallis, Oregon 97331

DOLORES E. HILL
Parasite Biology, Epidemiology, and
Systematics Laboratory
Animal and Natural Resources Institute
Agricultural Research Service
U.S. Department of Agriculture
Building 1044, BARC-East
10300 Baltimore Avenue
Beltsville, Maryland 20705-2350

CHARLES L. HOFACRE
Department of Avian Medicine
The University of Georgia
Athens, Georgia 30602

MARCIA L. HEADRICK
Center for Veterinary Medicine
U.S. Food and Drug Administration
USDA/ARRS/RRRC, Rm 207
950 College Station Rd
Athens, Georgia 30605

ALLAN T. HOGUE
U.S. Department of Agriculture
Food Safety and Inspection Service
1400 Independence Avenue SW
Room 389, Aerospace Bldg.
Washington, D.C. 20250-3700

KATHERINE HOLLINGER
Food and Drug Administration
Office of Women's Health
5600 Fishers Lane
Rockville, Maryland 20855

B. K. HOPE
Oregon Department of Environmental Quality
811 SW Sixth Ave.
Portland, Oregon 97204-1390

WILLIAM D. HUESTON
Center for Animal Health and Food Safety
College of Veterinary Medicine
University of Minnesota
1971 Commonwealth Ave.
St. Paul, Minnesota 55108

T. J. HUMPHREY
University of Bristol
Division of Food Animal Science
Langford, Bristol BS40 5DU
United Kingdom

LEE-ANN JAYKUS
Department of Food Science
North Carolina State University
Box 7624
Raleigh, North Carolina 27695-7624

LYNN A. JOENS
University of Arizona
Departments of Veterinary Science and
Microbiology
Tucson, Arizona 85721

F. JORGENSEN
University of Bristol
Division of Food Animal Science
Langford, Bristol BS40 5DU
United Kingdom

JOHN B. KANEENE
The Population Medicine Center
Department of Epidemiology
Michigan State University
East Lansing, Michigan 48824

SOPHIA KATHARIOU
North Carolina State University
Department of Food Science and
Graduate Program in Genomic Sciences
339 Schaub Hall, Box 7624
Raleigh, North Carolina 27695

J. KAUSE
Office of Public Health and Sciences
U.S. Department of Agriculture
Food Safety and Inspection Service
Maildrop 334 Aerospace Center
1400 Independence Ave. SW
Washington, D.C. 20250-3700

KATHLEEN KEYES
Department of Medical Microbiology and
Parasitology
College of Veterinary Medicine
The University of Georgia
Athens, Georgia 30602

JOHN D. KLENA
Washington State University
School of Molecular Biosciences,
Pullman, Washington 99164

MICHAEL E. KONKEL
Washington State University
School of Molecular Biosciences,
Pullman, Washington 99164

SILVIA KREINDEL
Harvard Center for Risk Analysis
Harvard School of Public Health
Boston, Massachusetts 02115

ALECIA LAREW-NAUGLE
The Ohio State University
1900 Coffey Rd., 251 Sisson
Columbus, Ohio 43210

DAVID LARSON
Iowa State University
Veterinary Diagnostic Laboratory
Ames, Iowa 50011

MARGIE D. LEE
Department of Medical Microbiology and
Parasitology
College of Veterinary Medicine
The University of Georgia
Athens, Georgia 30602

JEFFERY T. LEJEUNE
Food Animal Health Research Program, OARDC
The Ohio State University
1680 Madison Ave
Wooster, Ohio 44691-4096

JAY F. LEVINE
Department of Farm Animal Health and
Resource Management
College of Veterinary Medicine
North Carolina State University
4700 Hillsborough St.
Raleigh, North Carolina 27606-1499

ERNESTO LIEBANA
Veterinary Laboratory Agency-Weybridge
Department of Bacterial Diseases
Woodham Lane
Addlestone
Surrey, England KT15 3NB
United Kingdom

CAROL W. MADDOX
University of Illinois
Department of Veterinary Pathobiology
2001 S. Lincoln Ave.
Urbana, Illinois 61802

S. MALCOLM
Department of Food and Resource Economics
210 Townsend Hall
University of Delaware
Newark, Delaware 19717

SCOTT MARTIN
University of Illinois
Department of Food Science and Human Nutrition
486 Animal Sciences Laboratory
Urbana, Illinois 61801

JOHN J. MAURER
Department of Avian Medicine
College of Veterinary Medicine
The University of Georgia
Athens, Georgia 30602

PATRICK F. McDERMOTT
Office of Research
Center for Veterinary Medicine
U.S. Food and Drug Administration
Laurel, Maryland 20708

R. McDOWELL
Army Medical Surveillance Activity
Bldg T-20 Room 213 (mchb-ts-edm)
6835 16th Street NW
Washington, D.C. 20307-5100

ROBERTA MORALES
Research Triangle Institute
Raleigh, North Carolina 27709

MARSHALL R. MONTEVILLE
Washington State University
School of Molecular Biosciences
Pullman, Washington 99164

TERESA Y. MORISHITA
Department of Veterinary Preventive Medicine
The Ohio State University
Columbus, Ohio 44691

RODNEY A. MOXLEY
University of Nebraska-Lincoln
Institute of Agriculture and Natural Resources
Department of Veterinary and Biomedical Sciences
Lincoln, Nebraska 68583-0905

C. NARROD
Animal Production and Health Division
Food and Agriculture Organization, Room C539
Rome, Italy

DIANE G. NEWELL
Veterinary Laboratories Agency
New Haw, Addlestone, Surrey, KT15 3NB
United Kingdom

EDWARD J. NOGA
College of Veterinary Medicine
North Carolina State University
4700 Hillsborough St.
Raleigh, North Carolina 27606

KATHLEEN ORLOSKI
Office of Public Health and Sciences
Food Safety and Inspection Service
U.S. Department of Agriculture
555 S. Howes St.
Fort Collins, Colorado 80521

SHEILA PATTERSON
Department of Veterinary Pathobiology
University of Illinois
2001 S. Lincoln Ave.
Urbana, Illinois 61802

MARY PORRETTA
U.S. Department of Agriculture
Office of Public Health and Science
Food Safety and Inspection Service
Aerospace Building
1400 Independence Ave., SW
Washington, D.C. 20250

RACHEL CHURCH POTTER
The Population Medicine Center
Department of Epidemiology
Michigan State University
East Lansing, Michigan 48824

MARK POWELL
Office of Risk Assessment and
Cost Benefit Analysis
U.S. Department of Agriculture
Room 5248
1400 Independence Ave. SW
Washington, D.C. 20250

DAVID G. PYBURN
U.S. Department of Agriculture
Animal and Plant Health Inspection Service
Veterinary Service
Des Moines, Iowa 50010

WOLFGANG RABSCH
National Reference Center for Salmonellae
and Other Enteric Pathogens
Robert Koch-Institut
Burgstr. 37
D-38855 Wernigerode
Germany

DAN RICE
Field Disease Investigation Unit
Department of Veterinary Clinical Sciences
Washington State University
Pullman, Washington 99164-6610

TANYA ROBERTS
Economic Research Service
U.S. Department of Agriculture
Room N4069, 1800 M street NW
Washington, D.C. 20036-5831

ORHAN SAHIN
Food Animal Health Research Program
Department of Veterinary Preventive Medicine
Ohio Agricultural Research and Development Center
The Ohio State University
1180 Madison Ave.
Wooster, Ohio 44691

L. J. SAIF
Food Animal Health Research Program
Ohio Agricultural Research and
Developmental Center
The Ohio State University
1180 Madison Avenue
Wooster, Ohio 44691

MO D. SALMAN
Departments of Environmental Health
and Clinical Sciences
College of Veterinary Medicine and
Biomedical Sciences
Colorado State University
Ft. Collins, Colorado 80523

JAN M. SARGEANT
Food Animal Health and Management Center
College of Veterinary Medicine
Kansas State University
Manhattan, Kansas 66506

WAYNE SCHLOSSER
Office of Public Health and Sciences
U.S. Department of Agriculture
Food Safety and Inspection Service
2700 South Earl Rudder Pkwy., Suite 3000
College Station, Texas 77845

LINDA SCHROEDER-TUCKER
U.S. Department of Agriculture
Animal and Plant Health Inspection Service
National Veterinary Services Laboratory
Ames, Iowa 50010

YNTE H. SCHUKKEN
Department of Population Medicine and Diagnostic
Sciences
Cornell University
Ithaca, New York 14853

WILLIAM P. SHULAW
The Ohio State University
1900 Coffey Rd., 251 Sisson
Columbus, Ohio 43210

RANDALL S. SINGER
Department of Veterinary Pathobiology
College of Veterinary Medicine
University of Illinois
2001 S. Lincoln Ave.
Urbana, Illinois 61802

DAVID R. SMITH
Department of Veterinary and Biomedical Science
Institute of Agriculture and Natural Resources
University of Nebraska–Lincoln
Lincoln, Nebraska 68583

LINDA TOLLEFSON
Deputy Director
Center for Veterinary Medicine
U.S. Food and Drug Administration
7519 Standish Place
Rockville, Maryland 20855

HELMUT TSCHÄPE
National Reference Center for Salmonellae and
Other Enteric Pathogens
Robert Koch-Institut
Burgstr. 37
D-38855 Wernigerode
Germany

J. VINJÉ
Virology Laboratory
Department of Environmental Sciences and
Engineering
CB # 7400 McGavran-Greenberg Hall, Room 3206
University of North Carolina
Chapel Hill, North Carolina 27599

LYLE VOGEL
American Veterinary Medical Association
1931 N. Meacham Rd.
Schaumburg, Ilinois 60173

DAVID VOSE
David Vose Consultancy
Les Leches
France

BRUCE WAGNER
Departments of Environmental Health
and Clinical Sciences
College of Veterinary Medicine and
Biomedical Sciences
Colorado State University
Ft. Collins, Colorado 80523

ROBERT. D. WALKER
Office of Research
Center for Veterinary Medicine
U.S. Food and Drug Administration
Laurel, Maryland 20708

IRENE V. WESLEY
U.S. Department of Agriculture
Animal and Plant Health Inspection Service
National Animal Disease Center
Ames, Iowa 50010

DAVID G. WHITE
Office of Research
Center for Veterinary Medicine
U.S. Food and Drug Administration
Laurel, Maryland 20708

RICHARD WHITING
Food and Drug Administration
Center for Food Safety and Nutrition (CFSAN)
200 C Street SW
Washington, D.C. 20204

MARTIN WIEDMANN
Department of Food Science
Cornell University
Ithaca, New York 14853

A. GEBREYES WONDWOSSEN
Department of Farm Animal Health and
Resource Management
College of Veterinary Medicine
North Carolina State University
4700 Hillsborough St.
Raleigh, North Carolina 27606-1499

CLIFFORD WRAY
Greendale Veterinary Laboratory
Lansbury Estate
Knaphill
Woking GU21 2EW
United Kingdom

MAX T. WU
U.S. Department of Agriculture
Agricultural Research Service
National Animal Disease Center
Pre-harvest Food Safety and Enteric Disease
Research Unit
Ames, Iowa 50010

QIJING ZHANG
Food Animal Health Research Program
Department of Veterinary Preventive Medicine
Ohio Agricultural Research and Development Center
The Ohio State University
1180 Madison Ave.
Wooster, Ohio 44691

INTRODUCTION

Concerns about and advances in food safety have been evolving over the last fifty years. The graphic details of food processing presented in Upton Sinclair's book *The Jungle* raised enough food safety awareness that new regulations, standards, and laws were created and remain in effect to this day. The next major impetus for change in food safety measures was prompted by a large foodborne outbreak in the 1990s caused by *Escherichia coli* resulting from improperly cooked hamburger. Because of the media attention and the fact that deaths of children occurred, this outbreak created a significant influx of resources and funding toward the improvement of food safety. Robin Cook's book *Toxin*, and Eric Schlosser's book *Fast Food Nation*, re-emphasized the role of agriculture and animal processing in food safety.

Food production consists of a highly complex set of systems that are defined by the specific food product. Consequently, although the phrase "food safety" conjures up various connotations, no standardized definition exists. The definition in the National Academy of Science report "Ensuring Food Safety" denotes food safety as the "avoidance of food borne pathogens, chemical toxicants, and physical hazards, but also includes issues of nutrition, food quality, and education." The current focus is on "microbial, chemical, or physical hazards from substances that can cause adverse consequences." The 1997 Presidential Food Safety plan includes public health concerns arising from both traditional and novel (that is, genetic modification) methods of food production, processing, and preparation, and covers domestic and imported foods. Consequently, approaches to understand the etiology and pathogenesis of foodborne illness and to the prevention and control of foodborne outbreaks are as varied as the definition of "food safety" itself. Adding to the complexity of the issues are the multiple and interwoven methods that comprise food production. For example, within the poultry industry, the production process for laying chickens is quite different from that of broiler chickens. Likewise, poultry production is dissimilar to pork and beef production. In addition, the production of an animal or plant is just one phase of a multi-stepped process of food production. The product is transported to a processing plant, after which it is packaged and shipped to retail stores, eventually reaching a consumer. Each phase of this process influences the next.

Food safety issues embrace both public health and animal health. Efforts geared toward good management practices, new intervention strategies, and the understanding of the pathogenesis of disease can lead to reduced prevalence of microbial organisms at the farm level, healthier animals, and fewer costs to the producers. However, improvements and advances in one phase of food production may not translate to differences at the end of the chain. Each phase along the "farm to fork continuum" may affect the other. Improved food safety really is a shared responsibility.

Because of the expansiveness of food safety, solutions require a multidisciplinary approach. Integration of microbiology, epidemiology, animal science, veterinary medicine, food science, virology, and many other disciplines are vital for the success of research and educational outcomes. Efforts focused on the food safety area include research, education, extension, and the change in or development of new regulations and policies.

This book focuses primarily on research. Education and extension efforts focusing on the most recent research and evolving as a result of continuing research findings will have increasing impact on the food safety area. In addition, understanding the pathogenesis of food-related illness may provide better scientific data for regulatory and policy decisions as well as future risk assessments. The priorities in food safety research continue to evolve and change, particularly as technology advances.

The advancing field of genomics certainly will enhance food safety as well as animal health. Learning to map the particular genes responsible for antimicrobial resistance, for example, will help researchers to understand how antimicrobial resistance develops and becomes transferred among foodborne pathogens. Advances in genomics also may lead to better detection or diagnostic methods and even intervention strategies. New molecular tools, such as microarrays or biosensors, may lead to more rapid detection of foodborne pathogens and permit quicker interventions or control strategies. Microbial risk assessments continue to evolve and will improve. Increased funding and the recognition of better and more risk assessments in food safety will also help to advance the science. Significant events in U.S. history have motivated a change of emphasis toward food security issues. Luckily, these issues are closely related to current food safety areas, and research can continue to build on recent successes, such as detection methods.

Because of the expansive nature of food safety, this book concentrates only on the pre-harvest aspects of food safety and related microbial organisms. Although educational advances in food safety are mentioned, this book focuses primarily on the most recent research areas. We have attempted to provide the most current information on a variety of topics to give a broad overview of pre-harvest microbial food safety. The information presented here is by no means comprehensive, but the authors have provided detailed reference sections with each chapter.

We hope that you find this book useful. However, even within the relatively short time that will have elapsed between manuscript production and printed book, new information about pathogens and production systems will likely have been found and a new organism or disease may have emerged as the latest concern dominating the food safety arena.

Microbial Food Safety in Animal Agriculture

CURRENT TOPICS

PART I

OVERVIEW OF FOOD SAFETY

1

U.S. FEDERAL ACTIVITIES, INITIATIVES, AND RESEARCH IN FOOD SAFETY

MARY E. TORRENCE

INTRODUCTION. Although foodborne illnesses have always been a public health concern, the *E. coli* 0157:H7 outbreak in the early 1990s gained media and consumer attention. This heightened concern and awareness about food safety prompted demands for action by the federal government in solving this problem. This chapter provides a broad overview of federal food safety activities and research. The Web site, http://www.foodsafety.gov, provides a gateway to governmental food safety information.

HISTORY. In response to increased concerns about foodborne illnesses, President Clinton in 1997 introduced the Food Safety Initiative (FSI) (White House 1997a). FSI was formalized by a report entitled, "Food Safety from Farm to Table: A National Food Safety Initiative—A Report to the President" (Anon. 1997). This report was requested by the President and submitted to him by the Environmental Protection Agency (EPA), the Department of Health and Human Services (DHHS), and the Department of Agriculture (USDA). This report consequently stimulated the formation of numerous task forces, committees, initiatives, and funding incentives. The initial focus and goal of FSI

was to reduce the number of illnesses caused by microbial contamination of food and water. It was subsequently expanded to other foodborne hazards such as chemicals. An early activity within FSI was an announcement in the fall of 1997 of the Produce Safety Initiative to specifically help ensure the safety of domestically grown and imported fruits and vegetables (White House 1997b). This new initiative called for the development of good agricultural and manufacturing practices for fruits and vegetables. Consequently, DHHS's Food and Drug Administration (FDA) in October 1998 published a "Guide to Minimize Microbial Food Safety Hazards for Fresh Fruits and Vegetables" (Anon. 1998a). This guidance was science based, voluntary, and addressed specific microbial food safety hazards. The guide promoted good agricultural and management practices used in all phases of production of most fresh fruits and vegetables. This document and the subsequent educational training were a major collaborative effort among government, academia, and industry.

About the same time that these initiatives were unveiled, Congress requested the National Academy of Sciences (NAS) to study the current food safety system. The report of the Academy's findings was released

3

in 1998 and entitled "Ensuring Safe Food from Production to Consumption" (Institute of Medicine 1998). In this report, the narrow definition of food safety was expanded from microbial contaminants to "the avoidance of foodborne pathogens, chemical toxicants, and physical hazards, and other issues such as nutrition, food quality, labeling, and education" (Institute of Medicine 1998). The NAS report recommended that a comprehensive National Food Safety Strategic Plan be developed and emphasized the need for integrated federal, state, and local activities. It also emphasized an increase in research and surveillance activities and an increase in other efforts to ensure the safety of imported foods.

In response to the NAS's recommendations, the President issued Executive Order 13100 establishing the President's Council on Food Safety (FSC) (White House 1998a). The Food Safety Council consisted of the Secretary of Agriculture, the Secretary of Commerce, the Secretary of Health and Human Services, the Administrator of EPA, the Director of the Office of Management and Budget, the Assistant to the President for Science and Technology, the Assistant to the President for Domestic Policy, and the Director of the National Partnership for Re-Inventing Government. The Secretaries of Agriculture and DHHS and the Assistant to the President for Science and Policy were co-chairs.

The charge to the FSC was to improve food safety and to protect public health. This was to be accomplished through science-based regulations, coordinated inspection and enforcement programs, and research and educational programs. In addition, the FSC was charged to develop a five-year federal strategic plan for food safety that would include recommendations for implementation, advice to federal agencies on priorities, and the enhancement of coordination among all government and private sectors. To accomplish all these charges, the FSC established an Interagency Strategic Planning Task Force. The co-chairs of the Task Force were the FDA Commissioner and the USDA Under Secretary for Food Safety. Other representatives included the USDA's Food Safety Inspection Service (FSIS) Administrator, the Assistant Director for FSI in the Centers for Disease Control and Prevention (CDC), the USDA's Under Secretary for Marketing and Regulatory Programs, the Office of Management and Budget, the Domestic Policy Council, the Office of Science Technology and Policy, the Acting Director for Policy in FDA, the USDA Deputy Under Secretary for Research Education and Economics, the Director for Center for Food Safety and Nutrition (CFSAN), the Director of Seafood Inspection Service, the National Marine Fisheries Service in the Department of Commerce, the Deputy Assistant Administrator of the EPA, and the Acting Assistant Administrator for Prevention Pesticides and Toxic Substances in EPA. The primary goal for this task force was to develop the strategic plan. The FSC also was charged to oversee the Joint Institute for Food Safety Research (JIFSR) that

was being formed (White House 1998b). JIFSR was established in an attempt to help coordinate all the federal food safety activities and research. An Executive Research Committee was established to provide leadership for JIFSR; the committee included representatives from the Office of Science and Technology Policy in the White House, DHHS, and USDA. In addition, policy and budget committees were formed to be responsible for coordinating budget and planning food safety activities for the federal agencies. The first Executive Director was appointed in June 2000 with a two-year appointment.

The primary mission of JIFSR was to coordinate the planning and setting of research priorities for food safety research. This agency was formed in response to the criticism in the National Academy's report that more than twenty agencies were involved in food safety with no oversight or coordination among them. JIFSR was to optimize the current funding and infrastructure in food safety research and to help facilitate better communication among federal agencies. As part of this goal, JIFSR activities included enhancing intramural and extramural research programs and enhancing partnerships among government agencies, academia, and industry. To date one significant report was published on food safety research conducted by federal agencies in 2000. As of June 2002 the formal structure of JIFSR no longer exists.

The Risk Assessment Consortium (RAC) also was established in 1997 (Anon. 1997). Its role was to promote the advancement of the science of risk assessment and to encourage the development of new predictive models and tools (http://www.foodsafety.gov/~dms/fsreport.html). The Consortium helps to promote the use of scientific research in risk assessments and helps coordination and communication among federal agencies involved in risk assessments. Initial goals for the RAC were to identify data gaps and critical research needs, and to identify and catalogue research and scientific methods, models, and data sets needed for risk assessments targeted at food safety issues. The Consortium serves as a technical resource for federal agencies and provides advice to member agencies. CFSAN (FDA) was appointed the lead agency with a designated representative as chair. The RAC is made up of at least twenty federal agencies and holds a public meeting annually (http://www.foodriskclearinghouse.umd.edu). The current RAC Web site and its clearinghouse are maintained by the Joint Institute for Food Safety and Nutrition (JIFSAN), University of Maryland, with guidance from the RAC (http://www.foodriskclearinghouse.umd.edu/risk_assessment_consortium.html). JIFSAN was formed as an academic-government partnership and is discussed in more detail in Chapter 2. In 2002, the RAC, JIFSAN, and others sponsored the First International Conference on Microbial Risk Assessment: Foodborne Hazards at the University of Maryland. Proceedings are available online at: http://www.foodriskclearinghouse.umd.edu/RACConferencehome.html.

Another interagency coordinating group that was formed in 1998 was the Foodborne Outbreak Response Coordinating Group (FORC-G) (Anon. 1997; Anon. 1998b). FORC-G is co-chaired by the Under Secretary for Food Safety and the Assistant Secretary for Health and Human Services. The primary task of FORC-G was to improve intergovernmental coordination and communication among federal agencies (DHHS, USDA, EPA) involved in the response to interstate foodborne outbreaks (http://www.foodsafety.gov/~dms/forcgpro.html). The goal for this committee was to develop a coordinated national response system that would provide efficient use of resources and expertise during an emergency. Organizations participating in FORC-G were federal agencies and associations representing local and state health departments, laboratory directors, and state departments of agriculture. Similarly the National Food Safety System (NFSS) was established in 1998 after participants at a fifty-state meeting of federal, state, and local officials in health and agriculture recognized the need for coordination in response to foodborne outbreaks and other food safety issues (http://www.cfsan.fda.gov/~dms/fs-50ltr.html). NFSS has a steering committee that oversees five workgroups which generate activities and ideas that will help improve the coordination among federal, state, and local agencies during foodborne outbreaks, and help to improve other issues such as standardization of laboratory methods. A major result from NFSS was the development of eLEXNET (Electronic Laboratory Exchange Network). This is the first Internet-based data exchange food safety system that provides a repository of pathogenic findings and enhances communication and collaboration among federal, state, and local laboratories. A pilot study with eLEXNET participants linked *E. coli* 0157:H7 data of eight laboratories. Future goals include the evaluation of four pathogens and fifty additional laboratories

As food safety activities increased, the need for more food safety research, education, and surveillance was heightened. Although CDC had started FoodNet (Foodborne Diseases Active Surveillance Network) in 1996, its growth has continued after the influx of interest and funding from FSI (http://www.cdc.gov/foodnet). FoodNet is a collaborative effort among federal agencies (FSIS, FDA, and CDC), the state health departments, and local health department investigators to collect data that allows better tracking of the incidence of foodborne illness. This system helps provide better monitoring and detection of outbreaks as well as monitoring of the effectiveness of food safety programs. Currently, nine FoodNet locations exist, which represent more than 10 percent of the U.S. population. Another CDC initiative is PulseNet (National Molecular Subtyping Network for Foodborne Disease Surveillance), which was established in 1998 as a national computer network of public health laboratories to help rapidly identify and stop episodes of foodborne illness (White House 1998c). In PulseNet, bacterial subtypes are determined at state and local laboratories, using pulse field gel electrophoresis, and transmitted digitally to a central computer at CDC. The computer then matches newly submitted pulse field gels to those in the databank and confirms whether different outbreaks are connected to a common source. PulseNet includes all fifty state public health laboratories and four local public health laboratories as well as seven FDA laboratories and a USDA-FSIS laboratory. The first organism tracked in PulseNet was *E.coli* 0157:H7. Since then, PulseNet has expanded to cover *Salmonella, Shigella,* and *Listeria.* PulseNet allows epidemiologists in different parts of the country to respond and identify sources more quickly (http://www.cdc.gov/pulsenet). These networks may provide useful models for planned surveillance and laboratory networks related to counter-terrorism.

In 1998, USDA implemented the 1996 Pathogen Reduction and HACCP Rule (FSIS 1996). This rule was passed to help reduce microbial pathogens in processing plants and to clarify federal and industry roles in food safety. This rule required the development of standard operating procedures for sanitation, process control validation, pathogen-reducing performance standards for *Salmonella,* and the establishment of preventative controls. Under HACCP, the federal government was responsible for setting appropriate food safety performance standards, maintaining inspection oversight, and operating a strong enforcement program. In January 1998, HACCP was implemented for three hundred of the largest meat and poultry plants; in 1999 it was implemented in federally and state-inspected small meat and poultry plants; and in 2000 it was implemented in very small plants.

Finally the first National Food Safety Strategic Plan was signed by the President in January 2000 (http://www.foodsafety.gov/~fsg/cstrpl-4.html). The broad goal of the strategic plan was "the protection of public health by significantly reducing the prevalence of foodborne hazards through science-based and coordinated regulations, surveillance, inspection, enforcement, research, and education programs." Other specific targets included limiting increases in antimicrobial resistance, improving food safety practices, and expanding uses of safer pesticides. To provide some measure of the impact of the plan, an outcome measurement was established. The goal by 2004 is to reduce by 25 percent the incidence of the most common foodborne illnesses and reduce by 50 percent the amount of residues of carcinogenic and neurotoxic pesticides. The strategic plan is organized into three interrelated goals: 1) Science and Risk Assessment; 2) Risk Management; and 3) Risk Communication. Within each goal, specific priority areas are to be addressed in the following years (Table 1.1).

The emphasis and importance of food safety continues as the current Administration has expanded on the original Food Safety Council. In June 2002, President Bush drafted a new Executive Order that establishes the President's Council on Food Safety and Security. This council will have the same co-chairs as the previous

Table 1.1—List of Three Goals and Priority Areas Identified in the 2000 National Strategic Food Safety Plan

Goal 1 - Science and Risk Assessment
    Develop needed scientific skills and methods
    Strengthen link between research funding and risk assessment
    Begin work on a comparative risk assessment
    Perform priority risk assessments

Goal 2 - Risk Management
    Enhance public health, food products, and marketplace surveillance and improve
        the response to public health emergencies
    Set rigorous food safety program standards
    Establish food safety inspection priorities
    Improve import inspections
    Accelerate technology development and implementation

Goal 3 - Risk Communication
    Develop an effective national information network
    Promote public-private partnerships for communication, education, and training.

FSC and will have slightly different federal agency membership. This Council will continue to emphasize food safety issues but will expand the scope to food security areas as described in the Public Health Security and Bioterrorism Preparedness and Response Act of 2000, Food Safety and Security Strategy, Public Law 107–88 (http://www.cfsan.fda.gov/~dms/sec-ltr.html). For example, the foodsafety.gov Web site will remain active, as will such groups as the RAC and NFSS. FORCE-G will be subsumed by the new Council. Relative to new food security concerns, the new Executive Order specifically mentions the development of tests and sampling technologies for adulterants in foods and for tests and sampling technologies that will provide accurate and rapid detection, particularly for food inspections at ports of entry. This new Order also mentions the development of a crisis communication and education strategy with respect to bioterrorist threats to the food supply, such as technologies and procedures for securing food processing, manufacturing facilities, and modes of transportation.

REGULATORY AND SUPPORTING FOOD SAFETY AGENCIES. Federal agencies involved in food safety can be loosely categorized as regulatory or supporting. Regulatory agencies (EPA, FSIS, FDA) are responsible for enforcing legal statutes. Supporting agencies are involved in other areas of food safety such as research, education, surveillance, or extension. These activities support the regulatory agencies.

Regulatory. Three federal agencies have major regulatory responsibilities for food and food safety: EPA, FDA (DHHS), and FSIS (USDA). These three agencies administer the major statutes that are relevant to food safety regulation (http://www.foodsafety.gov/~fsg/fssyst2.html). FSIS ensures the safety of all domestic and imported meat, poultry, and some egg products in interstate commerce. FDA, through its Center for Food Safety and Nutrition and its Center for Veterinary Medicine, ensures the safety of all domestic and imported foods that are marketed in interstate commerce (except meat, poultry, and some egg products) and the safety of

game meat, food additives, animal feeds, and veterinary drugs. EPA establishes the maximum allowable limits or tolerances for pesticide residues in food and animal feed while FDA and FSIS enforce these tolerances.

Supporting. CDC leads federal efforts to specifically collect data on foodborne illness and investigate illnesses and outbreaks. CDC helps identify preventive strategies and monitor the effectiveness of those preventive and control efforts. Grants have been available from CDC to state and local health departments to help them increase their laboratory and epidemiology capacities. Also within DHHS, the National Institutes of Health provides grants and conducts food safety research. Most is focused on basic laboratory research and consequential product development. In 2002, the National Institutes received a large increase in funding for bioterrorism research. As part of these granting programs, increased funding will be available for food and waterborne organisms and other areas related to food safety (http://www.niaid.nih.gov/dmid/bioterrorism).

Within the USDA, the Agricultural Research Service (ARS), the Cooperative State Research Education and Extension Service (CSREES), and the Economic Research Service (ERS) conduct various types of food safety research. The Agricultural Marketing Service provides voluntary poultry, dairy, and fruit and vegetable inspection and grading programs to ensure quality and safety. The Animal and Plant Health Inspection Service is responsible for the surveillance of zoonotic disease, tracebacks of affected animals to herds of origin, and most recently for the protection of the food animal supply against possible deliberate acts of contamination.

Within the Department of Commerce, the National Marine Fisheries Service conducts a voluntary seafood grading and inspection program to help ensure commercial seafood quality and safety.

FEDERAL FOOD SAFETY RESEARCH. As federal initiatives and activities increased, so did funding for food safety research. The funding for related FSI activities has increased every year since 1997 (Table 1.2). Numerous federal agencies are conducting food safety research in a diverse range of areas. This in itself presents a problem because of the potential for overlapping research with little direction. The ability to coordinate research and identify common priorities has not been easy. Several small successes have occurred. For example, in April 2002, JIFSR released a comprehensive report entitled "2000 Federally-funded Food Safety Research Portfolio." This research was organized into thirteen categories, such as behavior, education, detection, control, pathogenicity, epidemiology, risk assessment, food handling, and others. Funding was listed for most of the federal agencies.

In 2000, a new food safety inventory database called the Food Safety Research Information Office (FSRIO) was established by the National Agricultural Library (USDA) (http://www.nal.usda.gov/fsrio). This database is continually updated and provides information on food safety research projects from government agencies, educational institutions, and other organizations as well as information such as food safety reports, conference proceedings, and Congressional research reports. FSRIO also has direct links to online bibliographical databases to search for scientific publications and more than twenty food-related scientific journals.

Relative to pre-harvest issues, USDA is the major food safety research agency. Each agency within USDA seems to fill a necessary, specific niche of research to help provide a broad overlapping coverage of the food safety area. For example, ERS conducts economic research and provides analyses of economic issues related to food safety and the food supply (http://www.ers.usda.gov/emphases/safefood). In a report in 2000, ERS estimated the human illness costs of foodborne disease at $6.9 billion per year for the five major foodborne organisms (Crutchfield and Roberts 2000). This analysis is continually updated and expanded by ERS. ERS also provides benefit/cost analyses of programs for food safety improvements. Publications by ERS can be obtained from its Web site (http://www.ers.usda.gov).

ARS is the primary intramural research agency for USDA with more than 2,200 scientists in one hundred locations. A close relationship with industry and other stakeholders provides the opportunity to transfer newly developed methodologies and technologies where needed in the field. Microbial food safety remains a major emphasis. Research includes the development of methodologies to detect and quantify pathogens as well as the development of technologies for pathogen reduction in both pre-harvest and post-harvest. A major objective of these programs is to provide the needed knowledge and technologies to help implementation of HACCP programs. In recent years, ARS has developed partnerships with various universities and has funded research grants (http://www.nps.ars.usda.gov). ARS also has established collaborations with the Institute of Food Research in Norwich, United Kingdom, and has begun to identify specific program areas for collaboration between ARS and scientists from the European Union.

CSREES is the primary extramural research agency with a strong partnership with the land grant university system so that CSREES may provide leadership in research, education, and extension programs. Through its competitive food safety grant programs (the National Research Initiative, the National Integrated Food Safety Initiative, and the Initiative for the Future of Agriculture and Food Systems) as well as other special grants, CSREES provides needed funding and direction for food safety research (http://www.reeusda. gov/1700/programs/programs.htm). The vision of the

Table 1.2—Food Safety Funding for Selected Agencies 1997–2002

Federal Agency	1997	1998	1999	2000	2001	2002
	Dollars in Thousands					
USDA:						
Agricultural Research Service	$49,647	$54,949	$69,868	$81,168	$91,074	$92,318
Cooperative State Research, Education, & Extension Service	6,234	8,765	24,765	27,400	29,570	29,967
Agricultural Marketing Service	0	0	112	112	5,995	6,279
Food Safety & Inspection Service	1,000	2,065	18,532	21,432	751,693*	790,605*
Economic Research Service	485	485	938	1,391	1,391	1,391
Office of the Chief Economist	89	98	196	196	436	436
National Agricultural Statistics Service	0	0	0	2,500	2,500	2,600
Food & Nutrition Service	0	0	2,000	2,000	1,998	1,998
Subtotal USDA	**57,455**	**66,362**	**116,411**	**136,199**	**884,657**	**925,594**
DHHS:						
Food & Drug Administration	109,335	133,282	158,300	187,300	216,674	483,204
Centers for Disease Control	4,500	14,500	19,476	29,476	39,476	
Subtotal	**113,835**	**147,782**	**177,776**	**216,776**	**256,150**	

*This reflects FSIS's total budget. In 2001, $703,120 was appropriated for increased inspections—not necessarily food safety. In 2002, $728,281.

Source: OMB and individuals within involved agencies. The amounts listed may vary slightly depending on the budget coding and definition of food-safety related activities within each Agency.

various grant programs is to provide researchers with funding opportunities for research that is integrated and that can follow the food production continuum. The National Research Initiative (NRI) is a major competitive granting program of CSREES. The NRI's Ensuring Food Safety Grant Program provides funds (up to $300,000) primarily for basic laboratory research including molecular research or biotechnology. A strong emphasis has been on method development, such as biosensors, for the detection of foodborne pathogens. To provide the opportunity for more applied population-type studies, the NRI's Epidemiologic Approaches for Food Safety Grant Program was established in 1999. This program provides larger grants (up to $1.5 million) for epidemiologic (population-type) studies. The National Integrated Food Safety Initiative (formerly the Food Safety and Quality National Education Initiative) provides researchers with an opportunity to link basic or applied research with an educational or extension program. This competitive grant program funds proposals (up to $600,000) that integrate research, education, and extension components into a multifunctional program. The Initiative for Future Agriculture and Food Systems was mandated by Congress to provide science-based solutions to critical emerging issues in priority mission areas. One of these areas in 2000 was food safety with an emphasis on functional foods, nutrition, and the reduction of microbial hazards on raw agricultural commodities such as fruits and vegetables. The emphasis of funding changes each year and funding depends on Congressional appropriations; in 2002, for example, no funding was provided (http://www.reeusda.gov/1700/programs/IFAFS/IFAFS.htm).

FSIS (USDA) is primarily a regulatory agency. Yet to make science-based decisions, FSIS needs to have access to cutting-edge research. Through a Congressional mandate, ARS is to provide such research. Additionally, the Animal and Egg Production Food Safety (FSIS) staff has entered into cooperative agreements with states to facilitate training and education of food safety. The design of the program was to enhance the knowledge of food animal producers about food safety and quality assurance practices and to enhance the ability of processing plant packers to implement HACCP. These state-level partnerships with animal health, public health, industry, and academic representatives will help deliver food safety information, education, and quality assurance outreach. To date, approximately twenty-five states are involved (http://www.fsis.usda.gov/OPPDE/animalprod/partners/statesow.htm). In addition, the Office of Public Health and Science (FSIS) uses a risk-assessment approach to focus programs on the potential human health outcome of consumption of meat, poultry, and egg products. For example, the Food Animal Science Division monitors and evaluates public health hazards associated with animal populations from the farm through processing. Field laboratories in various regions coordinate and conduct laboratory analysis in support of FSIS. The Office of

Risk Assessment and Cost Benefit Analysis is the major risk assessment agency within USDA and has provided impetus for food safety–related risk assessments, advice on risk analysis policy, and organized monthly seminars on risk assessment issues (http://www.usda.gov/agency/oce/oracba/index.htm). The Animal and Plant Health Inspection Agency (USDA) is primarily responsible for animal health issues and conducts National Animal Monitoring System Studies on different animal species each year. Although the studies are focused on animal health issues, these national surveys can provide useful preharvest food safety information such as management practices and demographic data (http://www.aphis.usda.gov).

Within DHHS, the Center for Veterinary Medicine (FDA) has used FSI funding for intramural and extramural research. The extramural research is limited but involves several cooperative agreements with universities for specified food safety research studies relevant to their mission. A major component of funding is to continue and expand the National Antimicrobial Resistance Monitoring System, which is described in more detail in Chapter 7. Intramural research is coordinated through the Center's Office of Research. There are three areas of priority: antimicrobial resistance in foodborne pathogens that are zoonotic; antimicrobial resistance in both pre- and post-approval production; and the microbial quality of animal feeds. The Office of Research is responsible for developing specific detection and analytical methods, particularly in residue chemistry such as detection of fluoroquinolone residues in eggs and compounds in animal feed. The Office also provides basic and applied animal research to support regulatory issues. Recently, the Office has expanded its research in antimicrobial resistance, specifically in the isolation and identification of pathogens in retail foods (http://www.fda.gov/cvm/fsi/fsior/fsior.htm). CFSAN (FDA) provides leadership through intramural and extramural research in methodology development, for example, analytical and quantitative methods and dose response. CFSAN is also active in risk assessment, particularly post-harvest (http://www.fda.gov/cfsan).

SUMMARY AND CONCLUSIONS. As this brief description reveals, funding and initiatives in the area of food safety have produced a wide variety of programs and research as well as results. More effort is still needed in providing direction for new food safety research areas and the integration of activities and research priorities. Despite the influx of funding in recent years, large population studies to evaluate the movement of pathogens in the food chain and sophisticated new molecular detection techniques are needed. These studies will require larger resources. New food safety issues will continue to emerge, whether they involve a newly emerging or re-emerging pathogen or the possibility of the deliberate contamination of the food supply. The current progress in research and in federal activities has provided a flexible framework that can adapt and respond to future challenges.

REFERENCES

Anon. 1997. Food safety from farm to table: a national food-safety initiative—a report to the President. U.S. Department of Agriculture, U.S. Department of Health and Human Services. May. Available at http://vm.cfsan./fda/gov/~dms/fsreport.html.

Anon. 1998a. Guide to minimize microbial food safety hazards for fresh fruits and vegetables. U.S. Department of Health and Human Services, Food and Drug Administration. Washington, D.C. October 26.

Anon. 1998b. Memorandum of understanding among U.S. Department of Agriculture, U.S. Department of Health and Human Services, and the Environmental Protection Agency. May. Available at: http://www.foodsafety.gov/~dms/forcgmon.html.

Crutchfield, S.R., and T. Roberts. 2000. Food safety efforts accelerate in the 1990s. *Food Review* 23(3):44–49.

Food Safety and Inspection Service, USDA. 1996. Pathogen Reduction: Hazard Analysis and Critical Control Point (HACCP) systems, final rule. *Federal Register* 61(144):38806–38989. July 25. Available at: http://www.fsis.usda.gov/OA/fr/haccp_rule.htm.

Institute of Medicine and National Research Council. 1998. Ensuring safe food from production to consumption. National Academy Press, Washington, D.C. August.

White House. 1997a. Office of the Press Secretary. "Radio address of the President of the Nation". President Clinton announces new measures to reduce foodborne illness. January 25.

White House. 1997b. "Memorandum for the Secretary of Health and Human Services, The Secretary Agriculture." Initiative to ensure the safety of imported and domestic fruits and vegetables. October 2. Available at: http://www.foodsafety.go/~dms/fs-wh2.html.

White House. 1998a. Office of the Press Secretary. President's Council on Food Safety. August 25. Executive Order 13100. Available at http://www.foodsafety.gov/~dms/fs-wh12.html.

White House. 1998b. Memorandum for the Secretary of Health and Human Services, The Secretary of Agriculture. Joint Institute for Food Safety Research. July 3. Available at: http://www.foodsafety.gov/~dms/fs-wh1.html.

White House. 1998c. Office of Vice President. National molecular subtyping network for foodborne disease surveillance. Vice President Gore launches computer network to fight foodborne illness. May.

2

ACADEMIC ACTIVITIES IN FOOD SAFETY: CENTERS, CONSORTIA, AND INITIATIVES

LEE-ANN JAYKUS

INTRODUCTION. Foodborne disease is a major cause of morbidity and mortality worldwide. In the U.S. alone, it has been estimated that foodborne disease may be responsible for as many as 76 million illnesses, 325,000 hospitalizations, and 5,000 deaths annually (Mead et al. 1999). These illnesses cost up to $9 billion a year in medical expenses and lost productivity (Council for Agricultural Science and Technology 1994). The situation is far more grave in developing countries, where food and waterborne diseases are an important cause of infant death and for which there are few statistics and less control. Furthermore, it is very clear that food safety issues are viewed as increasingly complex. For instance, although scientists recognize the inherent ability of pathogens to evolve, relatively little is known about how they do it. And because the pathogens do not exist in a vacuum, the environment to which a pathogen is exposed during food production and processing, as well as upon entry in the host, may further modify its ability to survive and cause disease. Scientists frequently refer to a "trinity" of factors that influence the occurrence of food borne disease: the pathogen, the host, and the environment (Institute of Food Technologists 2002). All three are important and must be better understood if we are to adequately address microbial food safety.

Because of the complexity of food safety issues and the perceived need to protect an increasingly global food supply, experts from academia, industry, and government alike cite the need for a farm-to-table systems approach, a multidisciplinary perspective, and a larger food safety work force (National Research Council 1996). Traditionally, the land grant university, through its roles of teaching, research, and extension, has provided food safety education (Jaykus and Ward 1999). In large part, research and academic functions have been mostly discipline specific and frequently provided almost exclusively by Food Science departments. As scientists recognize the need for multidisciplinary research and education programs, many of these teaching and research functions have evolved from single discipline projects and training to multidisciplinary, collaborative efforts (Jaykus and Ward 1999). Some of these efforts have a long history, beginning long before the current national interest in food safety that began in the late 1990s; others are more recent efforts that have been motivated by increased funding and the demand for a larger and more diverse food safety workforce.

Regardless, the need to produce a new generation of food safety professionals who are knowledgeable about the entire food safety continuum from farm to table and have the requisite skills to understand and address issues wherever they occur along that continuum is essential (Jaykus and Ward 1999). Training these professionals is perhaps the most important function of the university community.

The purpose of this chapter is to describe the historical and more current efforts in microbial food safety as provided by academic institutions and their consortia. Although this book focuses on pre-harvest food safety topics as related to animal agriculture, most of the initiatives described in this chapter have adopted the farm-to-table approach. These initiatives are broad based, with pre-harvest food safety serving as only one component of a more collective approach.

RESEARCH AND OUTREACH INITIATIVE. Historically, only a few model research and outreach initiatives have existed in food safety, some limited to a single institution and others involving large, collaborative efforts. As the national interest in food safety has increased over the last five years, these initiatives have also grown, with several newer consortia arising. In most of the earlier centers, research collaborations were somewhat limited to scientists from the agricultural disciplines such as food science, agricultural engineering, animal sciences, and agricultural economics. More recently, these scientists can be found collaborating with individuals in the veterinary medical sciences, human and animal epidemiology, and even human medicine. Some of these multi-institutional collaborations involve several land-grant universities. Others combine the efforts of academic institutions and agency scientists such as those representing the FDA, USDA, or CDC. Funding sources are diverse and include industry resources, state and national budget appropriations, extramural contracts and grants, cooperative agreements, and other creative means. A few of the larger initiatives are described in the pages that follow.

Multi-Institutional Food Safety Centers and Consortia. The concept of multi-institutional food safety centers began in the 1980s, with two of these centers being particularly successful over the last decade or so.

Others established in the late 1990s promise to provide creative and integrated approaches to solving emerging food safety issues.

THE FOOD SAFETY CONSORTIUM (FSC) (HTTP://WWW.UARK.EDU/DEPTS/FSC/). Established by Congress in 1988 by a Special State Cooperative Research grant, the Food Safety Consortium is a multi-institutional group whose charge is to conduct extensive investigation into all areas of poultry, beef, and pork meat production, from farm to table. This is a true consortium in that it involves a cooperative arrangement among institutions with no one central facility in which research is conducted. The congressionally mandated research areas include: (1) development of technologies for the rapid identification of infectious agents and toxins; (2) development of a statistical framework necessary to evaluate potential health risks; (3) determination of the most effective intervention strategies to control microbiological and chemical hazards; and (4) development of risk-monitoring technologies to detect hazards in the distribution chain. Since this mandate, the FSC has expanded its mission to include risk assessment, HACCP, and technology transfer. Each of the three participating universities is charged with the primary function of performing research associated with the specific animal species for which that university is uniquely qualified; that is, the University of Arkansas (poultry), Iowa State University (pork), and Kansas State University (beef). Each institution has multiple faculty members from different departments involved in the Consortium. At the University of Arkansas, many of the Consortium members come from the Food Science, Biological and Agricultural Engineering, and Poultry Science departments, as well as the College of Veterinary Medicine. Iowa State members hail from multiple departments in the Colleges of Agriculture and Veterinary Medicine, whereas Kansas State participants are largely from the Department of Animal Science and Industry. Other groups, including the USDA Agricultural Research Service (ARS) National Animal Disease Center, participate in some projects. Preharvest food safety research has focused on production practices and on-farm sanitation. Although the ultimate mission of the FSC is research, some outreach has been done in the form of newsletters and industry roundtables.

THE NATIONAL CENTER FOR FOOD SAFETY AND TOXICOLOGY (NCFST) (HTTP://WWW.IIT.EDU/ ~NCFS/). The National Center for Food Safety and Toxicology was established in the late 1980s by a cooperative agreement between the U.S. Food and Drug Administration and the Illinois Institute of Technology (IIT). It is a research consortium composed of scientists from academia, FDA, and food-related industries, providing a neutral ground where these scientists can pool their scientific expertise and institutional perspectives in a program that consists of both research and outreach components. Funding comes from the cooperative agreement, industry membership fees, and external grant monies, with governance provided by an oversight advisory committee consisting of academic, industry, and government representatives. A physical facility consisting of multiple buildings exists on the Moffett campus of IIT in the western suburbs of Chicago. The Center goals include the following: (1) to foster scientific and technical exchange among diverse segments of the food science community; (2) to better understand the science and engineering behind food safety; (3) to conduct research promoting the safety and quality of the U.S. food supply; and (4) to conduct research to answer regulatory questions regarding food safety. Almost all funded projects focus on post-harvest food safety, as do most of the extension programs. NCFST has, however, taken a lead role in educating the industry about food security and risk management.

THE JOINT INSTITUTE OF FOOD SAFETY AND APPLIED NUTRITION (JIFSAN) (HTTP://WWW. JIFSAN.UMD.EDU). The Joint Institute of Food Safety and Applied Nutrition was established in 1996 through the signing of a Memorandum of Understanding (MOU) that established a joint venture between the University of Maryland and the U.S. Food and Drug Administration (FDA). This is a jointly administered, multidisciplinary research and outreach program that is directed by delegates from the University and representatives from FDA's Center for Food Safety and Applied Nutrition (CFSAN) and Center for Veterinary Medicine (CVM). The goals for this collaboration are to expand food safety research and education programs necessary to provide the FDA with expertise and knowledge necessary to recognize and effectively handle emerging food safety issues as well as to enhance regulatory review capabilities. Areas of collaborative research include microbial pathogens and toxins, food constituents and applied nutrition, animal health sciences, food safety, and risk assessment. The JIFSAN office is located on the University of Maryland campus and the MOU established a set of relationships that closely link the University with CFSAN and CVM by committing to shared space, personnel, and intellectual resources, which eventually will include future shared facilities. The proximity of JIFSAN to FDA research facilities, as well as to other governmental offices and Washington-area national and international organizations, provides the Institute with opportunities for establishing critical food safety partnerships. Although the organization is in its infancy, it has provided a small amount of competitive funding for collaborative projects in food safety, including pre-harvest efforts aimed at surveillance of avian species for carriage of multidrug-resistant *Enterococcus* species and multiple fluoroquinoline–resistant *E. coli*. JIFSAN is perhaps best known for the Food Safety Risk Analysis Clearinghouse (http://www.foodriskclearinghouse.umd.edu), which it maintains in collaboration with the Risk

Assessment Consortium. The Institute has already established a presence in outreach, including organization of seminars and symposia covering pre-harvest food safety issues such as transmissible spongiform encephalopathies and good agricultural practices for fresh produce production.

THE NATIONAL ALLIANCE FOR FOOD SAFETY (NAFS) (HTTP://NAFS.TAMU.EDU). The National Alliance for Food Safety was established in 1998 with 24 member universities. The purpose of NAFS was to formally network research and educational expertise in food safety within the national university system and USDA ARS. The NAFS developed a full-fledged partnership with ARS through a congressional funding line focused on *L. monocytogenes* and *E. coli* O157:H7 research. To date, about $1.7 million has been awarded in competitively funded NAFS-ARS partnership projects, with continued ARS support anticipated to be about $1 million per year. Pre-harvest projects have focused on addressing the following: the prevalence of contamination, dissemination, and antibiotic susceptibility profiles of *E. coli* O157:H7 in cattle; the fate of pathogens in animal manures; and control/decontamination of pathogens in fresh produce. The NAFS is probably the largest and most diverse of the food safety consortia. It is overseen by a Board of Directors (one voting delegate per participating institution) and is managed by a Secretariat and an Operations Committee, which is a subset of the Board of Directors. It is truly a "virtual" organization in that it has no single location, although daily activities are managed through the office of the Secretariat at his or her university. Currently, more than 100 scientists are affiliated with the NAFS. In the process of organization, the Alliance established six disciplinary-oriented virtual centers (Detection and Typing Methods; Education and Outreach; Food Toxicology; Microbial Physiology and Ecology; Pathogen Control; and Risk Analysis and Policy) and six commodity-oriented virtual centers (Beef; Dairy; Plant Products; Pork; Poultry; and Seafood/Aquaculture) that serve as institutional focal points for the major dimensions of the science of food safety. Center directors are elected by participating faculty; these directors are responsible for leading their group in developing multi-institutional proposals for competitively awarded grants as well as for fostering collaboration and cooperation among the member scientists. Although these centers are focused on specific needs and science, the overall Alliance is in the process of positioning itself to provide a coherent focal point for university-based expertise that will broadly address the most pressing issues facing the U.S. agencies responsible for food safety.

Single-Institution Food Safety Centers and Consortia. Historically, there have been relative few single-institution food safety centers. However, with increased national interest in food safety, as well as enhanced funding opportunities, many institutions have invested time and resources over the last decade into establishing centers.

THE FOOD RESEARCH INSTITUTE (FRI) (HTTP://WWW.WISC.EDU/FRI/). Probably the oldest food safety center is the Food Research Institute, which was founded in 1946 at the University of Chicago and later moved to the University of Wisconsin at Madison in 1966. The FRI is located in its own building at the University of Wisconsin. It serves as both a research institute and holds standing as an academic department, the Department of Food Microbiology and Toxicology within the College of Agriculture and Life Sciences. Its mission is to maintain a leadership role in identifying and addressing food safety issues to meet community, industry, and government needs; to interact with industry, regulators, academia, and consumers on food safety issues and provide accurate, useful information and expertise; to facilitate and coordinate food safety research at the University; and to deliver quality food safety education and training. Currently the Institute (and department) boast six principal faculty members and several others holding affiliate positions. Pre-harvest food safety research has focused predominantly on farm ecology, the evolution of *E. coli* O157:H7, and bacterial stress response. Funding sources are mostly extramural, with outreach activities taking the form of periodic newsletters, research summaries, and annual meetings.

THE CENTER FOR FOOD SAFETY (CFS) (HTTP://WWW.GRIFFIN.PEACHNET.EDU/CFS). Formerly called the Center for Food Safety and Quality Enhancement, this Center was established in 1993 and is affiliated with the University of Georgia but is located 50 miles off campus in Griffin, Georgia. The CFS has a stand-alone facility and houses about six faculty members and 65 other employees, including graduate students and post-doctoral-research associates. Other faculty members from the Athens, Georgia campus are also members, and employees of local USDA ARS facilities and the CDC are participants in the activities of the Center. The CFS was created to work closely with industry, and numerous food companies contribute financially and serve as board members. Although industry funds a few projects, most of the research funding comes from extramural sources. The Center's research programs address food safety issues in three targeted areas: (1) developing and evaluating tools for detecting foodborne pathogens; (2) elucidating and characterizing conditions required to prevent the contamination of foods with pathogens; and (3) identification and evaluation of treatments to eliminate pathogens from foods and animal reservoirs. Examples of pertinent pre-harvest food safety research include investigation of the following: competitive exclusion cultures to reduce the carriage of *C. jejuni* and *Salmonella* in broilers; antibiotic resistance in *E. coli* O157:H7 and *C. jejuni*; and optimization of conditions to inactivate *E. coli* O157:H7 in manure. In recent years, the CFS has been very active in pre- and post-harvest food safety of fresh produce. The primary goal of the CFS remains food safety research, although some outreach is done in the form of newsletters and meetings.

THE NATIONAL FOOD SAFETY AND TOXICOLOGY CENTER (NFSTC) (HTTP://WWW.FOODSAFE.MSU. EDU). The National Food Safety and Toxicology Center is located on the campus of Michigan State University (MSU). Dedicated in 1997, the Center is housed in a large building that was completed in 1998, the funding for which was provided by USDA, MSU, and the MSU Foundation. The mission of the NFSTC is to conduct research that will increase the understanding of chemical and microbial hazards in foods, and to use this knowledge to develop a safer food supply, a well-founded public policy, and an increased public understanding of food safety issues. This mission is to be accomplished using the combined expertise of MSU faculty members with specialization in a variety of disciplines impacting food safety. The Center boasts close to 30 faculty members representing six colleges and 14 departments, with especially close ties to Toxicology, Epidemiology, Food Science, Sociology, and Microbiology and Immunology. Not all faculty members are located in the Center building, but the building does house well over 100 faculty, students, and staff. Research areas of particular interest have included: (1) food toxicology, particularly mechanisms of toxicity, dietary exposure risks, and public policy implications; (2) emerging and re-emerging pathogens, including microbial evolution and pathogenesis; (3) advanced analytical methods for detection of contaminants such as pesticides and mycotoxins in food and water; (4) collaborative work between epidemiologists, microbiologists, agriculturalists, and the medical community to determine the origin and extent of foodborne disease hazards; (5) research on the social dimensions of risk and food safety, including changing agricultural practices and individual/organizational food safety behaviors; and (6) provision of science-based information and education on food safety for the public and food professionals. Risk analysis is a common theme underlying much of this work. Pre-harvest food safety projects have included investigation of the ecology and evolution of *E. coli* O157:H7 and a wide variety of toxicological studies. Extension and outreach activities are considerable and include some high-profile conferences organized jointly with groups such as the National Food Processors Association (NFPA), National Alliance for Food Safety, USDA, FDA, and CDC. This Center is unique amongst those presented thus far because it also has invested considerably in educational initiatives (see below).

THE UT FOOD SAFETY CENTER OF EXCELLENCE (HTTP://WWW.FOODSAFE.TENNESSEE.EDU). The Food Safety Center of Excellence was established by the University of Tennessee (UT) in December 2000 as one of nine UT Centers of Excellence. As part of this initiative, UT will fund the Center in the amount of $5 million over the next five years. The rest of the research funding for the Center comes from a wide variety of sources including extramural contracts and grants, cooperative agreements, and targeted projects for industry. As with most centers, the focus is food safety from farm to table; the Center boasts faculty members from a diverse group of colleges and departments, as well as collaborations with federal research units. Research priorities to date include the following: (1) development of methods and technology for rapid detection and control of foodborne pathogens; (2) identification of mechanisms associated with the development and transfer of antibiotic resistance in foodborne pathogens, with the goal of devising strategies to reduce the occurrence of resistant pathogens; (3) development of non-antibiotic approaches for the prevention and control of diseases in food-producing animals that impact food yield and quality; (4) development of strategies to reduce environmental contamination with foodborne pathogens during animal production and food processing; and (5) development and dissemination of instructional food safety information. A number of projects in pre-harvest food safety are under way, and others continue to be funded. The Center is likely to produce some significant findings over the next several years.

THE AUBURN UNIVERSITY DETECTION AND FOOD SAFETY CENTER (AUDFS) (HTTP://AUDFS.ENG. AUBURN.EDU/). The Auburn University Detection and Food Safety Center was established in 1999 and designated as a University Peak of Excellence. With this designation came funding from the State of Alabama to initiate a systems engineering approach to food safety research. As with other centers, additional funding comes through extramural sources such as grants from federal agencies, industry projects, and industrial partner membership. The mission of the Center is to improve the safety of the U.S. food system by developing science and engineering required to rapidly identify, pinpoint, and characterize problems that arise in the food supply chain through the integration of sensor and information systems technology. Researchers from five Auburn University colleges (Agriculture, Engineering, Human Sciences, Sciences and Mathematics, and Veterinary Medicine) form the core faculty. Major achievements to date include development of phage-based acoustic wave sensors (AWS) for the detection of foodborne pathogens and identification of biomarkers to detect meat and bone meal contamination of livestock feeds.

Other Research and Outreach Centers. The list of research and outreach centers described above is far from complete. It can safely be said that almost every land grant institution in the U.S., as well as some private institutions, have some sort of initiative in food safety. These may be either loosely or tightly defined and associated, may cover the entire food chain from farm to table, or may focus on only a single aspect of that chain. All are at various stages of organization. A common theme for these consortia remains multidisciplinary collaboration in food safety research. A few additional initiatives are listed in Table 2.1.

GRADUATE AND CONTINUING EDUCATION INITIATIVES.

As is the case for research, a multidisciplinary perspective is essential for graduate and continuing education initiatives. It is now widely recognized that future food safety professionals must be thoroughly trained in food science, but also must be exposed to the basic sciences such as microbiology and toxicology. They must have the methodological background to be able to apply techniques that are increasingly essential tools in our current and future food safety research, including those provided by disciplines such as molecular biology, immunology, epidemiology, and mathematical modeling. In order to see the "big" picture of food safety, these future professionals must have some familiarity with the agronomic, animal, and veterinary medical sciences, as well as environmental and public health sciences. And because food safety policy and international trade are increasingly important issues, public policy training will be key to the success of these individuals as well. As food safety educational programs grow, future professionals may also need specific training in adult education methods and program efficacy evaluation. All these primary disciplines must be integrated into coursework that relates specifically to both historical and emerging food safety issues (Jaykus and Ward 1999). Many institutions recognize the need for such training programs, and some have provided model programs.

Graduate Education Programs. Graduate education is central to the mission of many institutions of higher learning. In food safety, graduate education programs provide future professionals with targeted training, as well as providing laborer resources necessary for research. The benefit to the student lies in the skills obtained through participation in these types of training programs.

NORTH CAROLINA STATE UNIVERSITY (NCSU) GRADUATE MINOR IN FOOD SAFETY (HTTP://WWW.CALS.NCSU.EDU:8050/NCFSA/GRADMINOR.HTML). The purpose of the graduate food safety minor at North Carolina State University is to prepare science profes-

sionals with the breadth of training necessary to understand and address food safety challenges from farm to table. The program attracts graduate students from a wide variety of disciplines such as veterinary medical sciences, toxicology, microbiology, genetics and biotechnology, epidemiology, and food science, to name a few. The students receive basic science training in their major discipline and formal interdisciplinary training in food safety. The minor includes participation from departments in the Colleges of Agriculture and Life Sciences and Veterinary Medicine (NCSU) as well as the School of Public Health at the University of North Carolina at Chapel Hill (UNC-CH). Participating graduate students are required to have, or to develop during the early part of their training, appropriate knowledge in the basic scientific disciplines of chemistry, biochemistry, and microbiology. It is also highly desirable that formal course training in genetics and statistics become evident in each student's academic program. Students in a masters or doctoral program are required to complete ten credit hours of core courses to earn the food safety minor. These core courses include: (1) Pre-harvest Food Safety, in which students study all the major food production systems in animals and plants; (2) Post-harvest Food Safety, which covers the relationship of post-harvest handling of agricultural commodities and food products to food safety; and (3) Food Safety and Public Health, a course that addresses issues and developments related to the relationship between food safety and public health, including emerging foodborne pathogens, virulence and pathogenicity, foodborne toxins, epidemiological techniques used in the investigation of foodborne disease, rapid detection methods, and quantitative microbial risk assessment. The final course in the sequence is Professional Development and Ethics in Food Safety, a seminar series that focuses specifically on the development of effective written and oral communication skills, as well as ethical issues related to the management of food safety issues. Ancillary courses in related disciplines provide additional training in areas of special interest for the individual student. Most graduate students seeking the minor supplement their program with a specific research project in a food safety–related area. Since the program's inception in the fall of 2000, consistent interest has been shown on the part of graduate students and faculty alike. To my knowledge, this is the first such program of its kind in the U.S.

Table 2.1—Additional Food Safety Initiatives

Center	University	Website
Farm-to-Table Comprehensive Initiatives		
Food Safety Institute	Clemson	http://www.clemson.edu/scg/food/index.htm
NC Food Safety Alliance	NC State Univ.	http://www.cals.ncsu.edu:8050/ncfsa/index.html
Center for Food Safety	Texas A&M	http://ifse.tamu.edu/Centers/Food%20Safety/foodsafetymain.html
Targeted Initiatives		
National Food Safety Database	Univ. Florida	http://foodsafety.ifas.ufl.edu/index.htm
Food Safety Engineering Project	Purdue	http://foodsafety.agad.purdue.edu/
Dairy Food Safety Laboratory	Univ. CA-Davis	http://www.vetmed.ucdavis.edu/DFSL/dfslintro.html

CORNELL UNIVERSITY INTERDISCIPLINARY FOOD SAFETY GRADUATE TRAINING PROGRAM (HTTP://WWW.FOODSCIENCE.CORNELL.EDU/FWS/TRAINING.HTML). The Cornell University Colleges of Agriculture and Life Sciences and Veterinary Medicine have initiated an interdisciplinary Ph.D. program in Food and Water Safety. The training program allows Ph.D. candidates to pursue training in one of three graduate fields: Food Science and Technology, Microbiology, or Veterinary Medicine. Included in the program are the following: (1) a core food safety curriculum consisting of courses in pathogenic bacteriology, toxicology, epidemiology, bacterial genetics, advanced food microbiology, statistics, and food safety assurance; (2) a multidisciplinary dissertation research program that includes the opportunity to perform lab rotations; (3) a journal club and seminar series; and (4) opportunities for internships with industry, regulatory agencies, or both. A relatively new initiative, this training program boasts close to fifteen faculty member trainers and is currently recruiting Ph.D. candidates.

Distance Education Initiatives. Distance education is certainly a trend of the twenty-first century. Some institutions are capitalizing on the ability to use computer technology to reach a broader and, in some instances, more diverse student population.

MICHIGAN STATE UNIVERSITY (HTTP://WWW.FOODSAFE.MSU.EDU). Michigan State University's National Food Safety and Toxicology Center (NFSTC) has announced the development of an online Professional Master of Science (proMS) Program in Food Safety, with an anticipated start date of summer 2002. The proMS program will operate through MSU's existing "virtual university" and consists of a core curriculum with a choice of multiple tracks of major emphasis. Courses will be presented through Web-based instructional modules with a traditional length of ten to fifteen weeks. The core curriculum is preceded by an introductory food safety and professional development course and then consists of four three-credit-hour courses, which include: Food Law; Food Toxicology; Evolution and Ecology of Foodborne Pathogens; and Public Health Impact and Risk Assessment of Foodborne Diseases. Students then select nine credit hours of elective distance learning courses. They conclude the thirty-credit-hour degree program with six hours dedicated to an on-site, applied special project. There will be opportunities for internships as well. Although in its infancy, this program should offer an innovative means by which to serve the educational needs of industry.

NORTH CAROLINA STATE UNIVERSITY. North Carolina State University (NCSU) has focused its distance learning efforts on educational programs for the food industry designed to train food processing personnel to identify and control hazards that are unique to each food processing plant, using the HACCP model. The program seeks to broadly train employees with varying degrees of educational background in the concepts of food safety control. The curriculum is a two-tiered system whose successful completion of each tier is rewarded by a specific certificate.

Completion of Tier I results in the HACCP Coordinator Certificate. The objective is to provide in-depth HACCP training as well as training in regulatory programs considered "prerequisite" to HACCP using a distance education delivery system. The course curriculum will consist of the following three courses: (1) Hazard Analysis and Critical Control Point–HACCP; (2) Good Manufacturing Practices (GMPs); and (3) Sanitation Standard Operating Procedures (SSOPs). The HACCP course has been developed to meet the accreditation requirements of the International HACCP Alliance, although it goes beyond the actual requirements and provides a more in-depth instructional experience to enhance the students' understanding. Students successfully completing the HACCP course will receive a certificate bearing the seal of the International HACCP Alliance, and a HACCP Coordinator Certificate is awarded to each student who successfully completes the three courses in tier I. These first-tier courses have already been designed and the certificate program is being marketed.

Completion of the second tier results in a Food Safety Management Certificate. The courses in this curriculum are intended to expand on the educational base established in tier I and focus on the scientific principles associated with the food safety sciences. The course curriculum consists of the following three courses: (1) Microbiology/Microbial Hazards, covering the basic principles of food microbiology with specific emphasis on foodborne pathogens and their control; (2) Sanitation, presenting a discussion of hygienic practices, requirements for sanitation programs, and modern sanitation practices in food processing facilities; and (3) Hazard Analysis/Risk Assessment. Upon successful completion of the second tier, a Food Safety Managers Certificate will be awarded to each student. Because of the higher level of understanding that the Food Safety Managers Certificate signifies, this certificate is not awarded until the student successfully completes the tier I requirements for the HACCP Coordinators Certificate. The tier II certification program is expected to be available in early 2003.

Other Educational Initiatives. As is the case with research initiatives, the preceding list is incomplete with respect to educational initiatives in food safety. For example, many of the centers described previously also have educational initiatives in preparation. Suffice it to say that this is a fruitful area of curricular development at many universities, and we can anticipate numerous innovative programs to be launched in the coming years.

SUMMARY AND CONCLUSIONS. Clearly, the "golden age" of food safety has arrived. Because of increasing national and international concerns about food bioterrorism and an increasingly global food supply, this golden age seems likely to persist for some time. An important benefit to society of increased food safety emphasis will be seen through research initiatives aimed at improving many aspects of the food chain from farm to table. From the perspective of preharvest strategies, we can expect the following: improved methods to detect pathogens during production; better understanding of the transmission of pathogens and their ecology on the farm; and improved methods to control pathogen colonization in the food animal. Success in these research areas will continue to require multidisciplinary, multi-institutional collaborations providing the technical expertise required to solve highly complex food safety problems. The food safety centers and consortia described here are well positioned to be leaders in these future research endeavors.

Likewise, integrated food safety education offers an opportunity for a wide array of seemingly unrelated disciplines to come together in a common educational theme (Jaykus and Ward 1999). With the advent of distance education technologies, there are increasingly varied options for delivering the educational message and designing creative curricula. Ultimately the greatest benefit associated with an integrated approach to food safety education will be from the students graduating from such programs, who will be more diverse, better qualified, and able to extend and model this collaborative teaching and research theme throughout their careers (Jaykus and Ward 1999). This is perhaps the highest calling for the academic institution: to provide a qualified, creative labor force that will make important contributions to food safety in the years to come.

REFERENCES

Council for Agricultural Science and Technology. 1994. Foodborne pathogens: Risks and consequences. Task Force Report No. 122. Council for Agricultural Science and Technology, Ames, IA.

Institute of Food Technologists. 2002. Emerging microbiological food safety issues: Implications for control in the 21st century. IFT Expert Report, Institute of Food Technologists, Chicago, IL.

Jaykus, L., and D.R. Ward. 1999. An integrated approach: The future of graduate food safety education. *Dairy, Food, and Environ. Sanit.* 19:14–17.

Mead, P.S., L. Slutsker, V. Dietz, L.F. McCaig, J.S. Bresee, C. Shapiro, P.M. Griffin, and R.V. Tauxe. 1999. Food-related illness and death in the United States. *Emerg. Infect. Dis.* 5:607–625.

National Research Council. 1996. *Colleges of agriculture at the land grant universities: Public service and public policy.* Washington, D.C.: National Academy Press.

FOOD ANIMAL INDUSTRY ACTIVITIES IN FOOD SAFETY

PETER COWEN, DONALD E. HANSEN, CHARLES L. HOFACRE,
EDWARD J. NOGA, DAVID G. PYBURN, AND A. GEBREYES
WONDWOSSEN

INTRODUCTION. Public concerns and new research about food safety and contamination of meat, milk, and egg products prompted the U.S. Department of Agriculture's (USDA) Food Safety Inspection Service (FSIS) to issue regulations for Pathogen Reduction; Hazard Analysis of Critical Control Point (HACCP) systems (Federal Register 1996). This system was to verify that the pathogen reduction systems of the industry processing plants were effective in controlling the contamination of foodborne organisms. To test the effectiveness, *Salmonella* performance standards testing was done to evaluate the prevalence of *Salmonella* levels at the majority of meat and poultry processing plants for a three-year period (http://www.usda.gov). The reported success of this program stimulated thoughts about prevention of further contamination of food products by initiating actions at the farm level. Industry groups from all major commodities began to initiate and implement new food safety activities.

POULTRY INDUSTRY. The poultry industry in the United States is a fully integrated system of animal agriculture. This integration results in each company having control over all fiscal decisions and bird husbandry aspects of production from the day-old parent breeder to the marketing and distribution of the final products. The poultry industry actually consists of three different industries: commercial table egg layer chickens, broiler chickens, and turkeys. Each of the three segments has unique bird husbandry conditions that result in uniquely different food safety issues. For example, the concern for the table egg layer industry is primarily *Salmonella enterica* serotype *enteritidis* (S.E.), whereas all *Salmonella enterica* serotypes and *Campylobacter* spp. are issues for the broiler chicken and turkey producers.

Vertical integration allows each company the advantage of control over its operation's aspects. Consequently, when consumers, retailers, or specific restaurant chains set food safety standards for poultry meat and eggs, the poultry company can readily adopt programs on the farm that can meet these standards.

Table Egg Layers. In 1999, the USDA's Animal and Plant Health Inspection Service (USDA-APHIS) performed a survey, as part of the NAHMS studies, of the U.S. table egg layer industry (NAHMS 2000). The study accounted for more than 75 percent of the table egg layers in the U.S. The study found that 56.1 percent of the farm sites participating in the survey had an S.E. quality assurance program, with the most common being a company-sponsored program (40.3 percent).

The 1999 layer study also demonstrated S.E. in 7.1 percent of the surveyed layer houses (NAHMS 2000). This study highlighted many of the risk factors within a table egg layer farm that increases the likelihood of having a positive flock. For example, houses with S.E. will have twice the rodent index (rodents trapped in twelve traps per week) of houses that are negative for S.E.

The first on-farm table egg layer program began in 1989 by addressing the egg transmission of S.E. from the primary breeder company to the egg layers. It was an extension of the successful *Salmonella pullorum* eradication program, a voluntary industry program, with USDA-APHIS oversight, called the National Poultry Improvement Plan (NPIP 2001). The U.S. *Salmonella enteritidis* Monitored Program was adopted in 1989 as a component of NPIP (NPIP 2001). This program is intended to be the basis from which the table egg layer breeding-hatching industry can conduct a program for the prevention and control of S.E. A flock of table egg layer breeders is eligible for the classification of U.S. *Salmonella enteritidis* Monitored Program if it meets the following criteria:

1. Originated from a U.S. *S. enteritidis* Monitored flock or meconium from chick boxes and a sample of chicks that died within seven days after hatching are cultured S.E. negative at an authorized laboratory.
2. All feed either is pelleted or contains no animal protein other than protein produced under an Animal Protein Products Industry (APPI) Salmonella Education/Reduction Program.
3. Environmental samples are to be taken for bacteriological examination for S.E. when the flock is two to four weeks of age and then every thirty days thereafter.
4. A federally licensed S.E. bacterin may be used in flocks that demonstrated to be culture negative for S.E.
5. Hatching eggs are collected as quickly as possible and are sanitized or fumigated in approved methods.
6. A flock found to be positive for an environmental culture for S.E. is required to have sixty live birds

cultured for S.E. at an authorized laboratory. If no S.E. is recovered from any of the bird specimens, the flock is eligible for the S.E. Monitored classification.

This S.E. Monitored program of the NPIP is the foundation for all the industry, state, and company S.E. quality assurance programs. The 1999 Layers study found that 40.3 percent of the U.S. layers participated in a company-sponsored quality assurance program; 28.4 percent participated in commodity group programs, such as the United Egg Producers (UEP 2000); and 22.7 percent participated in a state program (NAHMS 2000). Many of the state or company programs were modeled after the commodity group programs.

EGG QUALITY ASSURANCE PLANS. All the state and commodity egg quality assurance programs are voluntary and require specific bird husbandry/management practices in conjunction with testing or validation, often carried out by state or federal officials. The basic requirements found in most of the programs are the following:

1. Purchase chicks or pullets from hatcheries participating in the NPIP's U.S. S.E. Monitored Program.
2. Clean and disinfect poultry houses prior to arrival of new birds.
3. Obtain feed from mills that follow accepted feed industry "Good Manufacturing Practices" as well as the "Recommended Salmonella Control for Processors of Livestock and Poultry Feeds."
4. Designate an employee as the quality control supervisor.
5. Maintain a flock health program.
6. Maintain a farm rodent monitoring and reduction program.
7. Develop and implement a biosecurity plan and train employees on the proper procedures to execute the program.
8. Follow proper egg processing procedures such as egg washing (90°F) and refrigerating eggs (45°F)
9. Maintain records.
10. Have program validated by environmental and/or egg culturing as well as by a third-party review (such as a representative of extension service, a State Veterinarian, USDA, or other qualified party.)

Broiler Chickens and Turkeys. The U.S. Salmonella Monitored and U.S. *S. enteritidis* Clean Programs, components of the NPIP, form the basis of the broiler breeding industry's program of prevention and control for primary breeders (NPIP 2001). The U.S. Sanitation Monitored is the name of the NPIP Program for the turkey industry (NPIP 2001). The requirements for all three NPIP programs are very similar to one another:

1. Chicks/turkey poults must originate from tested flocks or have hatchery debris samples cultured.

2. The poultry house for these young future candidate breeders must be cleaned and disinfected.
3. Feed must be pelleted, and if any feed contains animal protein, the protein products should originate from an APPI participant.
4. Feed will be stored and transported in a manner to prevent possible contamination.
5. Environmental cultures will be taken starting at four months and then every thirty days thereafter for broiler chickens. Turkeys require the first environmental sample to be taken at twelve to twenty weeks and again at thirty-five to fifty weeks of age.
6. Broiler breeder flocks may be vaccinated with a Salmonella vaccine.
7. Egg sanitation practices will be followed.
8. If a broiler breeder flock under the U.S. S.E. Clean Program has a positive environmental culture, samples from twenty-five birds are cultured.
9. In turkeys, the flocks enrolled in the U.S. Sanitation Monitored Program require hatchery debris to be cultured from each hatch of this flock as a means of evaluating the effectiveness of the program.

ON-FARM FOOD SAFETY PROGRAMS OF BROILER CHICKENS AND TURKEYS. The broiler industry has addressed the issue of on-farm food safety by developing guidelines for Good Manufacturing Practices–Fresh Broiler Products (NCC 2000). This outline has served as the basis for more than 90 percent of the individual broiler companies' quality assurance programs from day-old breeders through the processing plant (Pretanik 2000). The program includes biosecurity, breeder management for food safety, and controls to ensure the absence of pharmaceutical, microbiological, and chemical residues in feed.

The turkey industry's program entitled Food Safety–Best Management Practices (FS-BMP) is a comprehensive program that begins with the requirement to obtain breeders from hatcheries participating in the U.S. Sanitation Monitored Program of the NPIP (NTF 2000). There are five modules that cover foundation and multiplier breeding, commercial turkey hatching, meat bird production, live-haul and feed manufacturing, and delivery. Each module contains sixteen control points and nine monitoring feedback areas. The control points emphasize such areas as biosecurity, cleaning and disinfection of facilities, vector (rodent/insect) control, feed management, drinking water management, and disease diagnosis and control. The microbial monitoring is primarily culturing the environment to assess the effectiveness of the cleaning and disinfecting program.

Neither the broiler nor the turkey best management program requires a third-party auditor. Because the product of the on-farm food safety program is the company's own processing plant, the success of the on-farm FS-BMP is assessed by the FSIS *Salmonella* culture tests in the processing plants from the carcass rinses.

CAMPYLOBACTER. The ecology of *Campylobacter* spp. on broiler chicken and turkey farms is poorly understood. Therefore it has been difficult to determine whether the best management programs designed primarily for control of *Salmonella* spp. will have an impact on *Campylobacter* spp., especially because large numbers of *Campylobacter* foodborne cases are associated with consumption of poultry. Some industry professionals believe that *Salmonella* initiatives on the farm will lower the level of *Campylobacter* spp. at the processing plant (Pretanik 2000). Even though the ecological niche for *Campylobacter* spp. within the intestines of the bird is different from *Salmonella* spp., *Campylobacter* spp. is still spread primarily via the fecal-oral route. Yet to be determined are what the reservoirs within the poultry farm are and whether egg transmission of *Campylobacter* occurs.

CATTLE INDUSTRY. By 1990, all cattle industry leaders recognized the importance of developing and applying safety and quality assurance standards for meat and milk products. They concluded that the best approach included participation of veterinarians, producers, and processors for complete success. The means used to gain participation differed between the dairy and beef groups.

Beef Cattle. The beef industry, under the combined leadership of the National Cattlemen's Beef Association (NCBA), the American Veterinary Medical Association (AVMA), and the American Association of Bovine Practitioners (AABP), followed the automobile industry's concept of bottom-up team building and developed ways to promote producer-driven quality assurance programs for beef production systems. These leaders encouraged adoption of these programs by state associations for local implementation. By 1994, virtually all states had some form of a voluntary beef quality assurance program in operation. Quality assurance programs for beef cattle continue to be voluntary and controlled by state and local associations rather than the NCBA. Currently, all state beef quality assurance programs include methods to avoid illegal drug and chemical residues. They also include a spectrum of preventive practices that address potential defects in the finished carcass. Some of these include reduction of injection-site damage in back and rump muscles, reduction of tissue damage from bruises, and reduction of excessive fat trim. A number of quality deficits in market cows and bulls, such as advanced lameness, down cows, advanced cancer eye, inadequate muscling in cows, heavy live-weights in bulls, and low dressing percentages, are also included in many state programs.

ILLEGAL RESIDUE AVOIDANCE. Violative residue avoidance is a basic focal point in quality assurance programs for all food animal species. For most harvested cattle, contamination from illegal drug or chemical residues may be moot because the annual incidence of violative residues in this group remains at or near zero. However the incidence of violative residues in harvested veal, bob veal calves, and market cows continues to be problematic.

The first step in avoiding an illegal drug residue of meat commodities is to prevent occurrence of diseases that require use of drugs considered harmful to the consumer. Designing and implementing an effective disease-prevention plan is the basic first step in all quality assurance programs. However, when prevention fails and disease occurs or when drugs and chemicals are needed for disease prevention, the following guidelines are recommended by all states for cow/calf producers to minimize the risk of violative drug residue in animals at harvest:

1. Follow label directions for dosage, delivery method, (for example, intramuscular or subcutaneous), and withdrawal time.
2. Use prescribed products only as and when directed by a veterinarian with a valid veterinarian-client-patient relationship.
3. Permanently identify and track treated animals.
4. Keep treated animals until withdrawal time has been met.
5. Keep records of treated animals that include animal identification, product name, dosage, dates used, administration route, and withdrawal time.
6. All individual animal treatments will strictly follow Federal guidelines.
7. Treatment procedures will comply with label directions or those prescribed by a veterinarian with a valid veterinarian-client-patient relationship, and these procedures will observe required withdrawal times.
8. Treatments will be recorded on a group/pen basis if given to an animal preweaning, or on an individual basis if given to an animal postweaning. Records will consist of date, pen/individual identification, product used, amount given, route and location given, and withdrawal time.
9. All treatment records will be kept for at least two years, made available to cattle owners on request, and transferred with the cattle as they move from the operation.
10. All cattle shipped will be checked to verify that withdrawal times have been met. This verification will be documented by the veterinarian's dated signature, and this information will be sent with those cattle.
11. A veterinarian can evaluate any question about withdrawal periods being met by comparing the treatment history against information provided by the Food Animal Residue Avoidance Databank. The animal(s) in question may be required to pass an on-site residue screening test.

NATIONAL CERTIFICATION GUIDELINES FOR STATE BEEF PROGRAMS. The NCBA developed the following guidelines for states to adopt for their quality assurance certification programs. To date, all states that offer quality assurance certification have adopted these guidelines. The guidelines promote a strong education program directed toward producers and their veterinarians. The elements include training for Beef Quality Assurance Practices, testing for understanding by participants, agreement to follow the assurance practices, and maintaining records of practices where applicable. These elements are detailed in the following sections.

TRAINING

1. Training must be offered to all cattle producers/employees and/or production segments.
2. Training must include NCBA National Beef Quality Assurance (BQA) Guidelines.
3. The training program can meet or exceed standards set forth in the BQA guidelines.
4. The training program must follow FDA/USDA/EPA Regulations.
5. The format will be set by individual state program.

TESTING

1. Each participant is required to satisfactorily complete the program requirements.
2. The format can be open book or oral.
3. The format will be established by individual state program.

AGREEMENT

1. An agreement is to be maintained between the program administrators and the program's participants.
2. The agreement should require producers to follow all guidelines and program procedures.
3. The agreement may include a clause to inspect records or verify operating procedures.
4. The format will be established by individual state program.

CERTIFICATION

1. Participants maintain a record of certification information.
2. The certification process must be available to all production sectors.
3. The certification process must provide for continuing education, recertification, or both.
4. The format will be determined by individual state program.

Dairy Cattle. After a one-year voluntary trial, the dairy industry embraced top-down regulatory methods to assure quality in the milk supply by expanding the regulations of the Pasteurized Milk Order (PMO). The PMO ultimately governs on-farm aspects of producing grade-A milk by regulating standards for all shipments of milk. The PMO was changed in 1992 to control and restrict antibiotic storage, labeling, and use in U.S. dairies, and it required preprocess testing of bulk milk for antibiotics.

ILLEGAL DRUG RESIDUES. The dairy industry also developed a voluntary program aimed at drug residue avoidance in dairy-beef animals that has been widely distributed since 1991. The following critical control points are recommended for dairy farmers. They encompass the PMO and voluntary Dairy-beef Quality Assurance Program.

1. Practice healthy herd management.
2. Establish a valid veterinarian-client-patient relationship.
3. Use only FDA-approved over-the-counter or prescription drugs with a veterinarian's guidance.
4. Make sure all drugs used have labels that comply with state and/or federal labeling requirements.
5. Store all drugs correctly (drugs for lactating animals must be kept separate from drugs for nonlactating animals).
6. Administer all drugs properly and identify all treated animals.
7. Maintain and use proper records on all treated animals. Records include animal ID, product name, treatment date, dosage, route of administration, withdrawal time for meat and milk, and name of the person administering the product.
8. Use drug residue screening tests.
9. Implement employee/family awareness of proper drug use to avoid marketing adulterated products.

Cattle Feeders. The section of the industry that is responsible for feeding cattle in feedlots and pasture grazing systems adopted a more comprehensive set of guidelines for its operations and imposed a voluntary system of certification. For feedlot operators who are interested, most states offer a feedlot certification program that requires strict adherence to specific management practices and significant record keeping. The following is a summary of feedlot certification requirements adopted by most participating states:

1. A quality feed control program will be maintained for all incoming ingredients. This program will analyze any suspect contamination and eliminate any products contaminated by molds, mycotoxins, or chemicals.
2. Only FDA-approved medicated feed additives will be used in rations. They will be used in accordance with the FDA label, including administration procedure, dosage, and withdrawal time.
3. Extra Label Drug Use of feed additives will not be used at any time or for any reason.

4. In compliance with the FDA regulations on prohibited ruminant proteins, such illegal feed components will not be stored or fed.
5. Feed records will be maintained for at least two years and will contain the batches of feed produced that contain the additive, date run, ration number or name, and amount produced.

Foodborne Pathogen Reduction for All Cattle Enterprises. In the late 1990s, efforts were directed toward enhancing food safety and reducing the number and prevalence of foodborne pathogens on finished carcasses. These efforts were sparked by outbreaks of foodborne illness involving specific pathogens found in hamburger and other meat products such as *E. coli* O157:H7. Although Federal law currently regulates control at processing facilities, the cattle industry will need to initiate efforts directed toward reducing and controlling specific pathogens in animals at production facilities such as feedlots and cow/calf operations. Current methods for pathogen reduction at cattle production centers are not well developed or are still theoretical. National cattle organizations such as the NCBA continue to invest in and support research efforts to discover methods to reduce the pathogen load at the herd level. As research continues, practical prevention and reduction methods may be developed that can be applied at various production levels in the cattle industry.

PORK INDUSTRY. By the middle 1980s, the pork industry recognized that the future involved integrated intensive production practices, the development of global markets and innovative products, and pork safety and quality. The pork producers recognized the value of forming partnerships with government and universities to implement pork safety and quality programs. The Trichinae Certification program and the Pork Quality Assurance (PQA) program are two good examples of innovative programs affecting pork safety and quality.

Pork Quality Assurance (PQA) Program. Since 1989, pork producers have utilized the PQA Program to ensure that the pork produced from their animals was as safe as possible (National Pork Board 2001). The primary goals of this voluntary educational program are to prevent antimicrobial residues in pork and to improve on-farm herd health. Improved herd health could result in lower overall usage of antimicrobials in the animals and theoretically fewer microbes transferred along the pork production chain. This program also helped to raise the pork producer's awareness of their on-farm food safety responsibilities.

The PQA Program outlines ten Good Production Practices (GPPs) focusing on medication usage, swine health, and feed management for the industry. These GPPs instruct producers on pig management practices that result in higher quality and safer pork. The PQA Program has been so successful that now many meat packers accept pigs only from producers who are certified in the PQA Program. This program has become part of their in-plant HACCP monitoring plan. The Good Production Practices recommended in the PQA program are the following:

1. Identify and track all treated animals.
2. Properly store, label, and account for all drug products and medicated feeds.
3. Educate all employees and family members on proper administration techniques.
4. Establish and implement an efficient and effective herd health management plan.
5. Follow appropriate on-farm feed and commercial feed processor procedures.
6. Maintain medication and treatment records.
7. Use a valid veterinarian-client-patient relationship as the basis for medication decision making.
8. Use drug residue tests when appropriate.
9. Provide proper swine care.
10. Complete the Quality Assurance Checklist every year and the education card every two years.

Certification of On-Farm Good Production Practices. There is a need for a verified on-farm pork safety program that will reassure consumers and fulfill the needs of an HACCP-based pork safety system that permeates the production system, including on-farm, in-plant, and retail. In response to this need, on-farm pork safety certification programs are beginning to be developed and implemented in the industry. As it exists today, the PQA Program is still an education program and within it producers are "certified" as being *educated* on the Good Production Practices (Lautner 2001). They are not presently certified as *following* the practices that are detailed within the program.

As these certification programs are implemented on farms, a third party will assure that the specified GPPs are implemented and adhered to by the pork producer. Third-party, on-farm certification by USDA-APHIS veterinarians was a major component addressed by the Swine Futures Project, which was a 1999 policy initiative jointly developed by the pork industry and USDA.

Trichinae Certification Program. In the mid 1980s, the convergence of three factors provided a powerful rationale for the development of industry-supported programs to improve food safety related to Trichinae. First, the presence of trichinosis in U.S. swine declined to such a low level that a disease-free status could be envisioned (Cowen 1990; Hugh-Jones 1985; Schad 1985). Second, pork industry leaders recognized that international markets were denied to U.S. producers because of a perception that U.S.-produced pork had a comparatively high risk of being infected with Trichinae. Finally, the development of a rapid, ELISA (Enzyme Linked Immunosorbent Assay)-based diagnostic test provided a relatively inexpensive tool that could be utilized in a control program (Gamble 1983). A precursor to current Trichinae Certification Programs was the 1987 evaluation of a prototype Trichinae-Safe Program in North Carolina, which aimed to label pork as Trichinae-free based on slaughterhouse testing of carcasses with an ELISA test. Even though the sensitivity and specificity of the test were 93 percent and 92 percent, respectively, the prevalence of seropositive swine was so low (0.38 percent) that many false positive carcasses were removed from the production line at the slaughterhouse (Cowen 1990). This program did, however, underscore the need for a combined on-farm herd certification program with initial slaughterhouse testing to provide a comprehensive picture of Trichinae status. These principles were incorporated in the present certification program that is being developed by USDA. The Trichinae Certification Program is based on scientific knowledge of the epidemiology of *Trichinella spiralis* and numerous studies demonstrating how specific GPPs can prevent exposure of pigs to this zoonotic parasite. The Trichinae Certification Program is a pork safety program requiring documentation of pork production management practices that minimize the risk of exposure of pigs to *Trichinella spiralis*. Although the prevalence of this parasite is extremely low and the number of human cases due to domestic pork consumption is rare, trichinosis remains a perception and trade issue for U.S. pork (McAuley 1991; Kim 1983). This program is currently being tested on a pilot basis on pork production sites and in midwestern U.S. pork processing facilities. The completion of the pilot phase in 2003 will lead to implementation of the federally regulated certification program throughout the U.S. pork industry.

In the pilot program, pork production sites are audited by veterinarians who have been qualified and accredited by the USDA. The purpose of the audits, both scheduled and random, is to observe and collect information about the production site, pig sources, feed sources, feed storage methods, rodent and wildlife control, carcass disposal procedures, and facility hygiene. Information is collected on USDA-approved official program audit forms. The USDA regulates the audits to ensure that the program standards are met, and the USDA certifies that the specified Good Production Practices are in place and maintained on the audited pork production site. USDA maintains a database containing program records for each certified pork production site.

For pigs originating from certified sites to be sold into commerce, the slaughter facility must have in place a procedure by which pigs from certified sites, as well as edible pork products derived from pigs from certified sites, are segregated from pigs and edible pork products originating from noncertified sites. This process is verified by USDA's FSIS. Swine slaughter facilities processing pigs from certified sites are responsible for conducting testing to confirm the Trichinae-free status of those pigs originating from certified production sites. On a regular basis, statistically valid samples of pigs from certified herds are tested at slaughter for Trichinae. This process verification testing is performed using a USDA-approved tissue or blood-based postmortem test and is regulated by the USDA Agricultural Marketing Service.

The Trichinae Certification Program will document the safety of pork produced under scientifically proven methods for raising pigs free from risk of exposure to *Trichinella spiralis*. More important, the development and implementation of the Trichinae Certification Program will provide an infrastructure for addressing more complex on-farm quality assurance and food safety issues. It represents a structured yet innovative partnership between producers, other industry segments, and the government to institute science-based risk-management strategies for improved food safety that open and expand markets previously constricted by food safety concerns.

Pork Safety Research. The pork industry, through the National Pork Board, routinely dedicates a significant portion of its resources to the advancement of pork safety science through the funding of pork safety research projects (Wagstrom 2002). In the past three years, this research initiative has resulted in the funding of more than $1.4 million of pork safety research projects in a number of areas. Some of the areas covered by this research are as follows: (1) antimicrobial resistance and the development of alternatives to antimicrobials; (2) improved diagnostics for zoonotic pathogens; (3) on-farm prevalence of pathogens of foodborne significance; (4) on-farm reduction or elimination strategies for pathogens of foodborne significance; (5) and on-farm and in-plant prevention of physical hazards in pork. Every two years, the industry then holds a Quality and Safety Summit to transfer the new technology gained through this research to those involved along the entire production chain.

FSIS mandated the movement of the food animal industry to an HACCP-based inspection system in 1996 (Federal Register 1996). All segments of the animal agriculture industry responded with programs not only in the processing plants but also on the farms. These on-farm HACCP programs were primarily voluntary to join. However, after a producer/farmer chooses to be a member of one of these programs, certain standards must be met, with third-party verification.

SUMMARY AND CONCLUSIONS. Each of the different food animal industries has unique food safety issues. For example, the table egg layer industry is principally concerned with egg transmission of *Salmonella enteritidis* with the U.S. *Salmonella enteritidis* Monitored Program, whereas the pork industry has addressed *Trichinella spiralis* through the Trichinae Certification Program. The less integrated production systems of the beef and dairy industries make it more difficult to begin food safety initiatives at the national level; therefore, many programs begin at the producer level, with state associations working to encourage participation. Despite the differences among the food animal industries, each group has realized the importance of food safety and has taken a role in its improvement. The success of early HACCP programs is leading to a reduction in the level of foodborne pathogens to consumers (http://www.usda.gov). Consumers likely will continue to pressure the food animal industry to continue this downward trend, resulting in further enhancement of the on-farm food safety programs.

REFERENCES

Cowen, P., S. Li, and T.J. McGinn. 1990. Survey of trichinosis in breeding and cull swine, using an enzyme-linked immunosorbent assay. *Am J Vet Res* 51(6):924–928.

Federal Register. 1996. Pathogen Reduction; Hazard Analysis and Critical Control Point (HACCP) systems; Final Rule. *Federal Register* 61(144):38805–38989.

Gamble, H.R., W.R. Anderson, C.E. Graham. 1983. Diagnosis of swine trichinosis by enzyme-linked immunosorbent assay (ELISA) using an excretory-secretory antigen. *Vet Parasitol* 13:349–361.

Hansen, D.E.. 1999. The bovine practitioner's role in beef quality assurance. In *Current Veterinary Therapy Food Animal Practice*. Eds. L.J. Howard and R.A. Smith. Philadelphia: WB Saunders, 121–127.

Hansen, D.E., and W.A. Zollinger. 2000. Steadying up for the 21st century. Corvallis, OR: Oregon State University, 1–10.

Hentschl, A., V. Green, and G. Hoffsis. 1991. Milk and dairy beef residue prevention protocol. *Agri-Education* 1–38.

Hugh-Jones, M.E., T.B. Stewart, C. Raby, J.E. Morrison, R.S. Isenstein, and T.R. Lei. 1985. Prevalence of trichinosis in southern Louisiana swine. *Am J Vet Res* 46:463–465.

Kim, C.W. 1983. Epidemiology II. Geographic Distribution and Prevalence. In *Trichinella and trichinosis*. Ed. W.C. Campbell. London and New York: Plenum Press, 445–500.

Lautner, E. 2002. U.S. Pork Producers Role in Providing Safe Pork Products. Personal communication.

Lautner, E. 2002. Systems Approach to Pork Safety. Personal communication.

McAuley, J.B., M.K.Michelson, and P.M. Schantz. 1991. Trichinosis Surveillance, United States, 1978–1990. In CDC Surveillance Summaries, Dec. 1991. *MMWR* 40 (No. SS-3):35–42.

McKean, J.D. 2001. Farm to Table Quality Assurance Programs for the World Marketplace. Unpublished.

McKean, J.D. 2001. Good Production Practices: Vehicle for Enhanced Pork Safety. *Proceedings of the 32nd Annual Meeting of the American Association of Swine Veterinarians.* 279–315.

National Animal Health Monitoring System (NAHMS). 2000. Layers '99, *Salmonella enteritica* serotype *Enteritidis* in table egg layers in the U.S. Centers for Epidemiology and Animal Health USDA: APHIS: VS. Available at http://www.aphis.usda.gov/vs/ceah/cahm/Poultry/lay99SE.htm

National Chicken Council (NCC). 2000. Good manufacturing practices, fresh broiler products. National Chicken Council publication, Washington, D.C.

National Pork Board. 2001. Pork Quality Assurance: A Program of America's Pork Producers. Edition Level III.

National Poultry Improvement Plan (NPIP). 2001. Subchapter G-Poultry Improvement. Code of Federal Regulations 9(part 145):725–764.

National Turkey Federation (NTF). 2000. Food safety best management practices for the production of turkeys. National Turkey Federation Publication FS-BMP Manual, 2nd ed. Washington, D.C.

Pretanik, S. 2000. Industry updates on quality assurance activities-broilers. In *Proceedings: National Conference on Animal Production Food Safety*. Eds. J.R. Ragan and A.Thaler, 44–46.

Pyburn, D.G., H.R.Gamble, L.A. Anderson, and L.E. Miller. 2001. Verification of Good Production Practices Which Reduce the Risk of Exposure of Pigs to Trichinella. Submitted to *Preventative Veterinary Medicine*.

Schad, G.A., D.A. Leiby, and C.H. Duffy. 1985. Swine trichinosis in New England slaughterhouse. *Am J Vet Res* 46:2008–2010.

United Egg Producers (UEP). 2002. UEP "5-Star" total quality assurance program. Atlanta, Georgia: United Egg Producers Publication.

Wagstrom, L. 2002. National Pork Board Pork Safety Projects Funded for 2000–2002. Personal communication.

PART

ANTIMICROBIAL RESISTANCE IN FOODBORNE ORGANISMS

4

EPIDEMIOLOGY AND ECOLOGY OF ANTIBIOTIC RESISTANCE

RANDALL S. SINGER

INTRODUCTION. The increasing rate of development of bacterial resistance to antimicrobials has been well documented (Levin et al. 1998; Levy 1995; Levy 1997; Levy 1998; Salyers and Amabile-Cuevas 1997; Davis et al. 1999). This development of resistance has resulted in human and animal bacterial infections that are refractory to many forms of treatment currently available. Bacterial pathogens that have been effectively treated for years may soon give cause for concern, especially in children, the elderly, and in other immunocompromised individuals. This decreasing bacterial susceptibility to antimicrobials will also have an impact in animals. The emergence and spread of antimicrobial resistance among bacterial populations will have major health and economic consequences in both human and animal populations. However, the reasons for including a section on antimicrobial resistance in a book on food safety may not be immediately obvious.

In the recent scientific literature, there have been several key manuscripts that demonstrate the connection between antimicrobial resistance and food safety. Three good examples appeared in the New England Journal of Medicine. In 1999, Smith et al. investigated human *Campylobacter* infections in Minnesota from 1992–1998 and specifically aimed to assess trends in and risk factors for quinolone resistant *Campylobacter* infections (Smith et al. 1999). The authors used a variety of techniques, including statistical analyses and molecular typing of *Campylobacter* isolates, to conclude that the increases in quinolone-resistant infections acquired domestically result largely from the acquisition of resistant strains from poultry, and that this increase in resistance may be caused by the use of the antibiotic in poultry. In a second study, Molbak et al. (1999) documented the transmission of an antibiotic resistant strain of *Salmonella* (*Salmonella enterica* serotype typhimurium DT104) from pigs to people, and following the human infection, there was increased morbidity and mortality because of the antibiotic resistance. Finally, Fey et al. (2000) used molecular techniques to relate a ceftriaxone-resistant strain of *Salmonella* that infected a child to *Salmonella* isolates obtained from cattle. The authors concluded that the *Salmonella* isolate from the child was likely obtained from cattle and that the resistant nature of the isolate increased the morbidity of the child's illness.

These examples highlight the potential risk to human health posed by a resistant bacterial isolate that is transferred through the food chain. Although the studies lack the data to support the notion that antibiotic use in these animals caused the bacterial isolate in question to become resistant, it is clear that any use of antimicrobials will exert a selective pressure that can lead to the creation, amplification, dissemination, or persistence of antibiotic resistance mechanisms. This effect of antimicrobials occurs in many different populations and settings, including animals, plants, humans, and the environment. All of these will add to the selective pressures exerted on bacteria; therefore, discerning the cause of an organism to be resistant to an antibiotic or tracing the origin of an antibiotic resistance mechanism is extremely difficult. Regardless, many different ways exist for antibiotic resistance in agricultural animals to have an impact on human and animal health, in addition to the obvious linear path of a resistant isolate moving through the food chain. Some of these pathways are discussed in this chapter.

The risks of antibiotic resistance and their link to food safety issues are becoming increasingly recognized and addressed, as evidenced by numerous educational campaigns and research funding opportunities. The educational campaigns are diverse and have been directed at many different groups of people. For example, efforts have been targeted at veterinary and medical practitioners through groups such as the American Veterinary Medical Association (AVMA), the American Medical Association (AMA), the Center for Veterinary Medicine at the Food and Drug Administration (FDA), and the Alliance for the Prudent Use of Antibiotics (APUA) by their adoption of judicious use guidelines of antibiotics. More detailed information can be found in Chapter 8. A considerable amount of research funding has been made available by the United States Department of Agriculture (USDA), the National Institutes of Health (NIH), the Centers for Disease Control and Prevention (CDC), FDA, and other organizations in order to investigate this issue. In particular, many ongoing studies in animal agriculture are attempting to determine the origins of specific antibiotic resistance mechanisms, the risk factors for the creation, dissemination, and persistence of antibiotic resistance mechanisms and resistant organisms, and the risks that these resistance genes and resistant organisms have on human and animal health. Understanding the epidemiology and ecology of antibiotic resistance and finding solutions to counter this problem will be difficult, primarily because of the complexity of the issue.

COMPLEXITIES OF THE PROBLEM. For those readers familiar with the principles of epidemiology and ecology, the title of this chapter might be problematic. Both epidemiology and ecology are population-based disciplines. A standard definition of epidemiology would include the study of the distribution of health or disease in populations and the factors that influence this distribution. Ecology, although related to epidemiology, might focus more on the distribution and abundance of organisms within populations and the interactions among these populations. When either of these terms is juxtaposed with antibiotic resistance, we are left asking ourselves what the population is that we are attempting to study.

In studies of antibiotic resistance, the final outcome of interest, resistance, is challenging to define and can possess a variety of forms. Antibiotic resistance might be defined clinically as the inability to treat a specific bacterial infection. For example, the interest might be the expected proportion of chickens in a flock that will develop colibacillosis that is resistant to treatment with antibiotics. Antibiotic resistance might be defined phenotypically in an *in vitro* system. This type of study might focus on factors that influence isolates of *Streptococcus* spp. from bovine mastitis cases to possess an MIC above a specific "resistant" threshold. Finally, antibiotic resistance might be defined genotypically as the presence of a resistance gene within the bacterium that is capable of rendering an antibiotic useless. In this case, the study could have objectives such as estimating the prevalence and distribution of *Escherichia coli* that possess a certain gene that confers resistance to third-generation cephalosporins.

The challenge with antibiotic resistance is that it is a concept and often an abstract one. It is not a defined entity such as the presence of *Salmonella* in feces or the prevalence of individuals with antibodies against a specific pathogen. Antibiotic resistance is a relative term whose definition and determination is variable and can change with time. Even after a definition of resistance is delineated, the study and monitoring of antibiotic resistance are far from being simple tasks. One basic challenge is the determination of which population to study. If antibiotic resistance is being studied in the context of health (animal, human, or both), then we must decide on which outcomes we wish to measure this impact. If we are interested in describing the patterns of antibiotic resistance across spatial and temporal scales, we must decide on which environments, hosts, organisms, and genes to study. Another complexity with antibiotic resistance studies is the relationship between laboratory findings and a real, *in vivo* effect. Finally, studies of antibiotic resistance require a solid background in evolutionary theory to understand the selective pressures involved in the emergence, dissemination, and persistence of antibiotic resistance mechanisms. This chapter attempts to explain these challenges to further the understanding of the epidemiology and ecology of antibiotic resistance.

SELECTION OF THE POPULATION. In any epidemiological or ecological study, the target population to which inferences will be made must be clearly defined. Within the framework of antibiotic resistance, the potential units of analysis have a hierarchical structure, which is demonstrated in the flow diagram that follows. The ways in which we design our studies and the types of inferences we can make from our studies will vary depending on the unit of analysis.

Gene(s) → Bacterium → Host → Population

One type of study would use the gene as the unit of analysis. Within an epidemiological framework, this type of study might assess the prevalence and distribution of specific resistance genes. This type of study might also examine the genetic variability and cellular location of the resistance gene and place this information into an evolutionary framework. For example, Winokur et al. (2001) conducted a study in which the presence of CMY-2, a gene that produces an AmpC-like-lactamase that can confer resistance to extended-spectrum cephalosporins, was assessed in *Salmonella* and *E. coli* isolates from cattle, swine, and humans. The study estimated the prevalence of isolates that possessed the gene. The authors found little clonal relatedness of the *E. coli* isolates in animals and humans. However, similar plasmids containing the CMY-2 gene were identified in *Salmonella* and *E. coli* isolates from humans and animals. The authors concluded that although the dissemination of the CMY-2 gene does not appear to be clonal, ample evidence exists of plasmid transfer between *Salmonella* and *E. coli* and possible transmission between humans and animals. Assuming that the appropriate data are collected, studies such as this could proceed beyond the descriptive and could assess the factors that might influence the probability of the bacteria possessing this gene or of a host possessing bacteria with this gene. Within an ecological framework, research on resistance genes could focus on the environmental distribution of the resistance gene by studying the relationships between different communities of bacteria. PCR-typing methods were used to investigate the environmental distribution of tetracycline resistance genes in waste lagoons and groundwater near two swine farms (Chee-Sanford et al. 2001). The genes were found in DNA extracted from many different water samples as well as in animal bacteria and normal soil bacteria, thus suggesting an environmental reservoir of resistance genes.

A factor that dramatically influences the way in which these genetic studies are designed, conducted, and interpreted is the means by which the resistance genotype is spread. The two primary mechanisms for the acquisition of resistance include the transfer of genetic material among bacteria and the mutation of specific genes in the bacterium. For resistance genes that are mobile, such as genes contained on plasmids, transposons, or integrons, inferences about factors that influence the distribution and persistence of these genes become more complex. The most accurate way to assess the factors influencing the distribution of these genes would be to quantify the frequency with

which these genes can be transferred among bacteria, including transfer events among very different bacteria. In the study by Chee-Sanford et al. (2001), the tetracycline resistance genes were amplified in a diversity of bacteria. Although it is difficult to prove, even with sequence data, that a direct horizontal transfer of these genes occurred from the animal bacteria to the soil bacteria, the investigators were required to cultivate bacteria from heterogeneous environments to understand the full ecological picture. Conversely, for those resistance mechanisms that are based on chromosomal mutations and are not likely to be transferred to other bacteria, the frequency at which these mutations result in viable resistant organisms must be estimated. This type of genetic mechanism of resistance becomes easier to study because the spread of the gene is tied to the clonal spread of the bacterium and thus becomes practically synonymous with using the bacterium as the unit of analysis. Fluoroquinolone resistance in *Campylobacter* is predominantly mediated by a point mutation in the DNA gyrase gene, *gyr*A. Because of the limited number of point mutations that will confer resistance and allow the organism to survive, the use of DNA fingerprinting techniques on fluoroquinolone resistant *Campylobacter* isolates will afford almost the same resolution of identification as would the addition of the sequencing of the *gyr*A gene to determine the exact mutation.

A more common and feasible level of analysis in antibiotic resistance studies is the focus on specific bacteria that possess phenotypic resistance, genotypic resistance, or both to a specific antibiotic. Studies of this type could investigate the diversity of bacteria that possess phenotypic resistance to a specific antibiotic. Bischoff et al. (2002) studied neonatal swine with diarrhea and isolated beta-hemolytic *E. coli* with decreased susceptibility to chloramphenicol in order to determine whether these isolates were clonal. The investigators found a diversity of *E. coli* genotypes, many possessing a common resistance gene, *cml*A, indicating a broad dissemination of the gene and potentially explaining the persistence of chloramphenicol resistance in the absence of selection pressure. When studying the bacterium as the unit of analysis, the focus might be on the categorization of resistance using some predefined cutpoints, such as those delineated by National Committee for Clinical Laboratory Standards (NCCLS). Conversely, studies of this nature may choose to analyze the quantitative level of resistance, expressed through an MIC or a zone of inhibition, and focus on questions related to changes in levels of susceptibility. Studies that analyze quantitative levels of resistance are much less common than those that categorize an isolate into susceptible or resistant classes. Studies that categorize resistance data are typified by temporal surveys of specific organisms, such as a study by Davis et al. (1999) in which the antimicrobial resistance patterns in *Salmonella enterica* serovar typhimurium isolates from humans and cattle were analyzed over a sixteen-year time period. This study categorized the *Salmonella* isolates into resistant and susceptible groups in order to assess changes in the prevalence of resistance over time. The study could also have analyzed the raw quantitative data, such as the median MIC or MIC range, to make inferences about the changing temporal dynamic of antimicrobial susceptibility. Finally, studies that focus on the bacterium as the unit of analysis might investigate the relationship between the genotypes of the bacteria and the genotypes of the antibiotic resistance determinant in order to make inferences about patterns of transmission. In the study of resistance to extended-spectrum cephalosporins by Winokur et al. (2001); the finding of identical CMY-2 genes on very similar plasmids in *Salmonella* and *E. coli* of humans and animals suggests transfer events between different bacterial species and may indicate transmission of resistance between animals and humans.

Another potential unit of analysis is the animal or animal population. This unit is more typical of the population-based approach with which epidemiologists and ecologists are most familiar. A type of study that would fit in this framework would include an assessment of the factors that influence the likelihood of an individual or a herd possessing bacteria that are either phenotypically resistant or that possess a specific resistance gene. For example, a randomized field trial investigated potential risk factors for the development of penicillin-resistant *Staphylococcus aureus* mastitis in early-lactation dairy cows (Osteras et al. 1999). These types of studies are useful in identifying the practices at the animal or herd level that influence the development, acquisition, or maintenance of antibiotic-resistant bacteria. They are especially useful when assessing the distribution and spread of new resistance mechanisms or in the monitoring of the emergence of resistance to a new compound. In a national survey of ceftriaxone-resistant *Salmonella*, none of the ceftriaxone-resistant *Salmonella typhimurium* had identical PFGE fingerprints, but most had a similar AmpC plasmid-mediated resistance that has since been investigated more thoroughly (Dunne et al. 2000).

MEASUREMENT OF TRENDS. Implicit in the methods of epidemiology and ecology is the ability to describe distributions and patterns. Even in the classic and apparently simple example of describing the prevalence of animals with antibodies to a specific pathogen, the situation is made more complex when attributes of the diagnostic test, age of the animals, and prevalence of cross-reacting pathogens are taken into account. The task of describing patterns of antibiotic resistance adds further levels of complexity, and some of the reasons for this have already been mentioned.

The most common way of describing temporal and spatial trends is by using the percentage of isolates that are "resistant" to the antibiotic. How do we decide what constitutes "resistant?" Although the more effec-

tive and accurate method may be to study changes in the levels of susceptibility over time (for example, by using the raw MIC data), a cutoff between "susceptible" and "resistant" is more commonly utilized. On the surface, this choice of a cutoff might sound similar to a cutoff on a serological assay, but these methods are inherently different. In the serological assay, the cutoff is based on the value that provides a desired level of accuracy for detecting known positives and known negatives correctly. In the case of the antibiotic resistance cutoff, the value is based on the theoretical concentration of antibiotic that would effectively inhibit or kill the bacterium within the host at a specific anatomic location. When testing these bacteria in an *in vitro* system, these values become much less interpretable.

For several reasons, spatial and temporal trends of antibiotic resistance can be difficult to interpret. Often the desire is to link these trends to specific antibiotic use practices. Several difficulties arise with making these associations, thus caution must be used when making inferences from these types of studies or surveys, especially when attempting to assess cause-effect relationships between antibiotic use practices and changes in antibiotic resistance prevalence. A good review of this topic has been previously published (Lipsitch 2001). One problem with interpreting changes in the distribution or temporal trend in resistance is the difficulty in determining whether the fluctuations are caused by real changes in the prevalence of resistance mechanisms or simply by changes in the prevalence of a single resistant bacterial clone. For example, studies of the resistance of *Salmonella enterica* serovar typhimurium have documented temporal increases in resistance (Davis et al. 1999). More detailed investigations have shown that these changes are predominantly caused by the widespread clonal dissemination of the multidrug-resistant *Salmonella* definitive phage type 104. The study of chloramphenicol-resistant *E. coli* in neonatal swine described previously (Bischoff et al. 2002) revealed a large genetic diversity of isolates, suggesting that observed changes in resistance prevalence were caused by the transfer of resistance determinants.

A second difficulty in assessing trends in antibiotic resistance relates to the issue of genetic linkages. The goal of resistance monitoring is often to associate changes with antibiotic use practices. If the selection pressure imposed by a certain antibiotic is removed, one might expect the prevalence of resistance to that antibiotic to decline. However, if the resistance gene is genetically linked to another gene that is still under some selection pressure, then the prevalence of resistance to the antibiotic may not decline and, in fact, may actually increase. Furthermore, no reason exists to assume that simply because the antibiotic is removed, the resistance should immediately decline. If the presence of the resistance gene does not impose a "fitness cost" on the transmissibility and survival of the bacterium, then the resistance gene might persist for extended periods. This effect has been studied *in vitro* (Lenski 1997) and through mathematical models (Levin et al. 1997; Stewart et al. 1998).

When measuring trends in antibiotic resistance, we might expect to find spatial and temporal clustering of antibiotic-resistant isolates, and perhaps of resistance genes as well. Agricultural premises that are located in geographic proximity are often served by a single veterinarian or veterinary group. Consequently, a similar treatment program may be administered to these farms for similar infections. In addition, farms in proximity might be more likely to share common pathogens. With similar antibiotic pressures and the potential for movement among farms, one might expect specific bacterial clones or specific resistance genes to be clustered on a local or regional scale. For example, Singer et al. (1998) investigated the spatial and temporal distribution of ampicillin and tetracycline resistance in *Pasteurella multocida* and *Mannheimia hemolytica* (previously called *P. haemolytica*) in bovine pneumonia cases in California from 1991 to 1996. Throughout the state were spatial clusters of resistant isolates that suggested a potential effect of regional antibiotic usage practices as well as the unfortunate potential outcome of regional treatment failures.

Most of the investigations of antibiotic resistance that have been conducted in agricultural settings have focused on the animals in those settings, and occasionally on some aspects of the surroundings of the animals. The farm has thus been treated as its own ecosystem, and some ecological principles have been utilized in studying the dynamic of antibiotic-resistant bacteria and antibiotic resistance genes on these premises. However, few studies have extended this notion of the ecosystem to the broader relationship between the farm and the environment outside the farm's boundaries. There are many ways in which microorganisms can flow to and from agricultural settings, including animal movements, water, manure distribution on fields, wildlife, and humans. This view of a broader ecosystem further complicates the antibiotic resistance problem and our ability to assess trends. When considering the dissemination and persistence of antibiotic resistance genes, the possibility of gene transfer events between animal bacteria and environmental bacteria must be addressed. These environmental bacteria might thus become effective long-term reservoirs of antibiotic resistance genes, as was postulated in the study in which tetracycline resistance genes were found in a diversity of bacteria in the ground water and soil near swine facilities (Chee-Sanford et al. 2001).

In addition to the dissemination of antibiotic-resistant organisms and resistance genes, the possibility also exists for the flow of antibiotics and antibiotic residues. Any active ingredient in these compounds can potentially create a selective pressure on the environmental bacteria, thereby increasing the rates of gene transfer events, mutations, and pressures for the long-term persistence of resistance genes (Jorgensen and Halling-Sorensen 2000). These events would further complicate the interpretation of resistance trends and distributions.

When discussing antibiotic resistance trends, the topic that most commonly comes to mind is the effect that a change in antibiotic usage has on the prevalence or persistence of antibiotic-resistant organisms or genes. Intuitively, one would guess that a reduction or cessation in antibiotic usage would ultimately result in the disappearance of resistant organisms and the relevant resistance genes. Very few studies in animal agriculture have addressed this issue. One large study monitored the changes in resistance in *Enterococcus faecium* and *E. faecalis* following the abolishment of in-feed antibiotics in Denmark (Aarestrup et al. 2001). Declines in the percentage of resistance to several antibiotics were documented; therefore, the authors concluded that the elimination of the agricultural selective pressure was responsible for the observed decline in the prevalence of resistant organisms. This study may have documented a real cause-and-effect relationship; the magnitude of the perceived relationship, however, may have been biased for the various reasons discussed in this chapter. Assigning a causal link to an intervention and an apparent response in antibiotic resistance profiles should be done with caution because of the plethora of factors that can influence the perceived measurements.

RISK ASSESSMENT AND MATHEMATICAL MODELING. Quantitative risk assessment generally involves the development of a model to estimate the probability of an outcome given a set of input parameters. Using this model, it is then possible to estimate the excess risk associated with specific values of the input parameters or values that represent specific conditions in the real world. A question that has been raised lately is whether a risk assessment on antibiotic resistance can and should be performed. Specifically, many people have debated whether it is possible to create a model of antibiotic resistance that has sufficient accuracy and validity to make the inferences valuable and thereby enable policy changes and risk-reduction strategies to be based on the model.

Issues related to antibiotic resistance, some of which have already been discussed, provide a very difficult framework for assessing risk. First, many possible outcome measurements can be assessed. This is primarily due to the fact that many possible questions can be asked relative to antibiotic resistance. Second, risk models that relate to antibiotic resistance will inevitably have high uncertainty and high variability. Uncertainty reflects a lack of knowledge and understanding of components of the model and therefore an inability to estimate the parameters with precision and accuracy. Variability reflects the natural random error associated with measuring and quantifying parameters of the model. In a risk assessment it is critical to separate uncertainty from variability, and in the case of modeling the dynamics of antibiotic resistance, both will be high. Finally, as previously discussed, the various possible levels of analysis of the antibiotic resistance problem further complicate the risk assessment effort.

One of the first tasks that must be undertaken when developing a risk assessment model is determining what outcomes should be modeled; this is especially true in antibiotic resistance models. Most of the risk assessment models discussed to date have involved the risks to human health. However, it should not be forgotten that the risks associated with antibiotic resistance can affect both human and animal populations. Some of the animal risks that may be of interest include the rate of treatment failure, the number of excess mortalities caused by treatment failure, the amount of lost production because of inefficacious treatment, or the economics of treatment failure. Some of the risks to human health and well being may include the rate of treatment failure, the number of excess days of morbidity, the number of days of work lost, the mortality rate associated with treatment failure, and the economics of treatment failure (for example, increased medical costs). Finally, it might also be possible to model how these risks to human health and well being affect the public perception of agriculture.

After the outcome has been defined, specifying the way in which antibiotic resistance will be quantified is then necessary. This aspect is the most troublesome because of the difficulty in deciding what the unit of analysis should be, as discussed earlier. At the most basic level, the assessment of risk at the level of the gene might be chosen. Thus, models would be developed that use the presence of specific resistance genes on farms, in hospitals, or in populations, as the unit of concern. Genes that confer resistance would thus be important regardless of the bacteria in which they are contained. Several mathematical models have been developed to demonstrate the emergence, dissemination, and persistence of resistance genes (Austin et al. 1997; Austin and Anderson 1999; Levin et al. 1998; Stewart et al. 1998). Most of these models relate to selection pressures specific for human hosts, but models that are specific for different animal agricultural ecosystems are being developed. Models such as these can be used to make predictions as well as identify parameters that might be important in the development of the risk assessment model. Using the gene as the unit of analysis may prove to be the most useful for predicting the risk of antibiotic resistance; it is the gene that is ultimately responsible for the phenotypic manifestation of resistance.

Another type of risk model would use specific bacteria possessing phenotypic resistance, genotypic resistance, or both as the units of analysis in estimating the risk of antibiotic resistance. At this level, the interest might be in quantifying the risk that a specific pathogen that possesses phenotypic resistance would be transferred from animals to humans and subsequently cause additional morbidity because of the resistance. In a model such as this, the prevalence of the specific pathogens that possess specific genetic resistance determinants must be quantified. Again, though, if the genetic determinant can move from bacterium to bacterium, the prevalence of resistant bacteria could change rapidly. The probability of genetic transfer of the resistance mechanisms, not only between bacteria of the same genus and species but also among any bacteria that possess the same genetic determinant and that are capable of transferring genetic material, would have to be estimated. Thus, one of the main questions to be addressed in the development of this type of model is the decision of which animal host(s) and environments need to be sampled and included in the model.

The use of the animal or animal population as the unit of analysis could provide a more population-based approach to the assessment of risk. However, this unit presents some additional problems. Again, the familiar question of which unit of analysis to incorporate in this population-based approach must be addressed. A herd could be sampled and the unit of risk specified as follows: (1) the presence of resistance genes in the herd; (2) the presence of resistant pathogens in the herd; (3) the number of animals in the herd with resistance genes; (4) the number of animals in the herd with resistant pathogens; or (5) the number of animals in the herd with specific pathogens possessing specific resistance mechanisms. Finally, if the goal of the risk assessment is to quantify the risk to human health, then the probability that this unit of resistance can be transferred between the animal and human populations and create some hazard to human health must now be determined.

Very few risk assessments have been conducted on issues related to antibiotic resistance, and based on the preceding discussion, the reasons for this should be obvious. The FDA's Center for Veterinary Medicine (CVM) recently completed a risk assessment that can be used as a case study for dealing with the problems just described. Details of this risk assessment model can be found in Chapter 30.

Although the assumptions and interpretations of the CVM model have been actively debated, the model was a good beginning in the effort to develop a risk assessment model of antibiotic resistance. However, for several reasons, risk assessments involving antibiotic resistance will not get any easier than the scenario modeled by CVM. The system that was modeled, fluoroquinolone-resistant *Campylobacter*, has many simplifying attributes. These include: (1) fluoroquinolone resistant *Campylobacter* infecting humans is a one-organism problem (a necessary cause); (2) the model assumes one host transmission to humans (chickens); (3) the model assumes one major vehicle of transmission to humans (consumption of chicken meat); and (4) a chromosomal point mutation confers resistance to fluoroquinolones and therefore requires clonal spread of the resistant bacteria. The model did not have to consider the presence of the resistant organisms in multiple hosts, the possibility of stepwise changes in decreasing susceptibility, nor the existence of other mechanisms of resistance, such as efflux pumps or integrons. As soon as these complexities are added to the antibiotic resistance risk assessment, the number of assumptions and uncertainty of the model will rise dramatically.

SUMMARY AND CONCLUSIONS. Many of the issues discussed in this chapter have been qualified by statements about the lack of information regarding their importance and dynamic in animal agricultural settings. Many areas of the antibiotic resistance problem are in need of additional research, especially because antibiotic resistance involves animal populations. It is also important to remember that antibiotic resistance is an ecosystem problem; consequently, additional research into the interactions of soil, water, plant, and animal bacteria is essential.

If resistance is studied at the genetic or the bacterium level, probabilities associated with the transfer of resistance genes will need to be incorporated. Unfortunately, the little research that has been conducted in this area has largely been conducted in the laboratory. Some have addressed this topic in human population settings, such as in a hospital, and others have utilized mathematic models to estimate these effects. The probability of gene transfer needs to be assessed in agricultural environments, but that raises another question: Where does gene transfer occur? Should the frequency of gene transfer be estimated in the gastrointestinal tract (or other site) of the animal, in the environment, or within the gastrointestinal tract (or other site) of the human host? Perhaps all play a role in the spread of antibiotic resistance mechanisms; consequently, if the antibiotic resistance mechanism that is being assessed is mobile, then all these modes of spread must be evaluated. Additional research needs to be conducted in agricultural ecosystems in a longitudinal fashion to quantify the frequency of these events.

There is a need to incorporate the concepts and tools of evolutionary genetics into the study of antibiotic resistance and its transfer. Often the goal of resistance studies is to infer a directionality of transmission (such as from animals to humans), but in the realm of food safety, the use of genetic data to infer directionality (gene flow) has not been widely practiced. The appropriate and accurate use of phylogenetic analyses must be increased in order to infer the source of the resistance mechanism, the rate of spread of the resistance mechanism, and the future distribution of the resistance mechanism.

Epidemiological methods need to be refined or developed to improve the scope of antibiotic resistance studies. For example, epidemiologists are trained in adjusting rates, and these techniques can and should be applied to the study of antibiotic resistance to control for environmental, host, and other factors that affect the apparent prevalence and distribution of resistance as well as the frequency of resistance gene transfer. Epidemiologists also are trained to estimate effective contact rates. These rates can be used to determine the frequency of bacterium-bacterium and host-host contacts that result in transfer of resistance genes or resistant organisms. Epidemiologists spend a considerable amount of time on sampling strategies, and these methods need to be further adapted to the study of antibiotic resistance in an agricultural environment. Depending on the unit of analysis, the strategy will vary considerably. The spatial and temporal dynamics of resistance will need to be incorporated, and as with any sampling strategy, the way in which variance is partitioned will have to be accounted for. For example, if the animals within a farm share the same resistance genes but a considerable diversity of resistance genes exists among farms, then a monitoring effort would be best to study a diversity of herds with fewer individuals per herd.

Finally, considerable attention needs to be given to the inferences that are derived from trend and distribution data. At this time, resistance prevalence trends that decrease are viewed as a good thing, whereas trends of increasing resistance in the face of decreased antibiotic use are viewed as failures or are rationalized by the likely existence of genetic linkages. Inherent in epidemiological theory is the identification and control of the impact that confounding variables and effect modifiers have on study inferences. These types of variables can change the magnitude of trends or the appearance of spatial patterns and, in some cases, can even change the direction of the trend entirely. Research into the identification of these variables, the effect they have on perceived outcomes, and the means by which they can be controlled are essential if we are to make accurate inferences about the dynamics of antibiotic resistance.

REFERENCES

Aarestrup, F.M., A.M. Seyfarth, H.D. Emborg, K. Pedersen, R.S. Hendriksen, and F. Bager. 2001. Effect of abolishment of the use of antimicrobial agents for growth promotion on occurrence of antimicrobial resistance in fecal enterococci from food animals in Denmark. *Antimicrob. Agents and Chemother* 45:2054–2059.

Austin, D.J., and R.M. Anderson. 1999. Studies of antibiotic resistance within the patient, hospitals and the community using simple mathematical models. *Philos. Tran. R. Soc. Lond. B. Biol. Sci.* 354(1384) 721–38.

Austin, D.J., M. Kakehashi, and R.M. Anderson. 1997. The transmission dynamics of antibiotic-resistant bacteria: the relationship between resistance in commensal organisms and antibiotic consumption. *Proc. R. Soc. Lond. B. Biol. Sci.* 264(1388) 1629–38.

Bischoff, K.M., D.G. White, P.F. McDermott, S. Zhao, S. Gaines, J.J. Maurer, and D.J. Nisbet. 2002. Characterization of chloramphenicol resistance in beta-hemolytic *Escherichia coli* associated with diarrhea in neonatal swine. *J. Clin. Microbiol.* 40:389–394.

Chee-Sanford, J.C., R.I. Aminov, I.J. Krapac, N. Garrigues-Jeanjean, and R.I. Mackie. 2001. Occurrence and diversity of tetracycline resistance genes in lagoons and groundwater underlying two swine production facilities. *Appl. Environ. Microbiol* 67:1494–1502.

Davis, M.A., D.D. Hancock, T.E. Besser, D.H. Rice, J.M. Gay, C. Gay, L. Gearhart, and R. DiGiacomo. 1999. Changes in antimicrobial resistance among *Salmonella enterica* serovar typhimurium isolates from humans and cattle in the Northwestern United States, 1982–1997. *Emerg. Inf. Dis* 5:802–806.

Dunne, E.F., P.D. Fey, P. Kludt, R. Reporter, F. Mostashari, P. Shillam, J. Wicklund, C. Miller, B. Holland, K. Stamey, T.J. Barrett, J.K. Rasheed, F.C. Tenover, E.M. Ribot, and F.J. Angulo. 2000. Emergence of domestically acquired ceftriaxone-resistant Salmonella infections associated with AmpC beta-lactamase. *JAMA* 284:3151–3156.

Fey, P.D., T.J. Safranek, M.E. Rupp, E.F. Dunne, E. Ribot, P.C. Iwen, P.A. Bradford, F.J. Angulo, and S.H. Hinrichs. 2000. Ceftriaxone-resistant salmonella infection acquired by a child from cattle. *N. Engl. J. Med* 342:1242–1249.

Jorgensen, S.E., and B. Halling-Sorensen. 2000. Drugs in the environment. *Chemosphere* 40:691–699.

Lenski, R.E. 1997. The cost of antibiotic resistance—from the perspective of a bacterium. *Ciba Foundation Symposium* 207:131–140.

Levin, B.R., R. Antia, E. Berliner, P. Bloland, S. Bonhoeffer, M. Cohen, T. DeRouin, P.I. Fields, H. Jafari, D.Jernigan, M. Lipsitch, J.E. McGowan Jr., P. Mead, M. Nowak, T. Porco, P. Sykora, L. Simonsen, J. Spitznagel, R. Tauxe, and F. Tenover. 1998. Resistance to antimicrobial chemotherapy: a prescription for research and action. *Amer. J. Med. Sci* 315:87–94.

Levin, B.R., M. Lipsitch, V. Perrot, S. Schrag, R. Antia, L. Simonsen, N.M. Walker, and F.M.Stewart.1997. The population genetics of antibiotic resistance. *Clin. Inf. Dis.* 24: Suppl 1, S9–16.

Levy, S.B. 1995. Antimicrobial resistance: a global perspective. *Adv. in Experimental Med. and Biol* 390:1–13.

Levy, S.B. 1997. Antibiotic resistance: an ecological imbalance. *Ciba Foundation Symposium* 207: 1–9.

Levy, S.B. 1998. Multidrug resistance—a sign of the times. *N. Engl. J. Med* 338:1376–1378.

Lipsitch, M. 2001.The rise and fall of antimicrobial resistance. *Trends in Microbiology* 9:438–444.

Molbak, K., D.L. Baggesen, F.M. Aarestrup, J.M. Ebbesen, J. Engberg, K. Frydendahl, P. Gerner-Smidt, A.M. Petersen, and H.C. Wegener. 1999. An outbreak of multidrug-resistant, quinolone-resistant *Salmonella enterica* serotype typhimurium DT104. *N. Engl. J. Med* 341:1420–1425.

Osteras, O., S.W. Martin, and V.L.Edge. 1999. Possible risk factors associated with penicillin-resistant strains of *Staphylococcus aureus* from bovine subclinical mastitis in early lactation. *J. Dairy Sci* 82:927–938.

Salyers, A.A., and C.F. Amabile-Cuevas. 1997. Why are antibiotic resistance genes so resistant to elimination? *Antimicrob. Agents Chemother.* 41:2321–2325.

Singer, R.S., J.T. Case, T.E. Carpenter, R.L. Walker, and D.C. Hirsh. 1998. Assessment of spatial and temporal clustering of ampicillin- and tetracycline-resistant strains of *Pasteurella multocida* and *P. haemolytica* isolated from cattle in California. *J.A.V.M.A.* 212:1001–1005.

Smith, K.E., J.M. Besser, C.W. Hedberg. F.T. Leano, J.B. Bender, J.H. Wicklund, B.P. Johnson, K.A. Moore, and M.T. Osterholm. 1999. Quinolone-resistant *Campylobacter jejuni* infections in Minnesota, 1992–1998. Investigation Team. *N. Engl. J. Med* 340:1525–1532.

Stewart, F.M., R. Antia, B.R. Levin, M. Lipsitch, and J.E. Mittler. 1998. The population genetics of antibiotic resistance. II: Analytic theory for sustained populations of bacteria in a community of hosts. *Theor. Pop. Biol* 53:152–165.

Winokur, P.L., D.L. Vonstein, L.J. Hoffman, E.K. Uhlenhopp, and G.V. Doern. 2001. Evidence for transfer of CMY-2 AmpC beta-lactamase plasmids between *Escherichia coli* and *Salmonella* isolates from food animals and humans. *Antimicrob. Agents Chemother.* 45:2716–2722.

5 ANTIMICROBIAL SUSCEPTIBILITY TESTING METHODOLOGIES

DAVID G. WHITE, PATRICK F. McDERMOTT, AND ROBERT D. WALKER

INTRODUCTION. The advent of antimicrobials heralded the golden age of medicine. Beginning with the clinical use of penicillin in the 1940s, antimicrobials have significantly reduced morbidity and mortality associated with infectious diseases. However, the global rise of microorganisms that are resistant to antimicrobials poses a major threat to human health. The development of bacterial antimicrobial resistance is neither an unexpected nor a new phenomenon. It is, however, an increasingly troublesome situation because of the frequency with which new emerging resistance phenotypes are occurring among many bacterial pathogens. After sixty years of successful antimicrobial use, we are seeing the emergence of multiresistant bacterial pathogens, which are less responsive to therapy. The emergence of antimicrobial resistance is increasing the overall mortality, morbidity, and health care costs associated with treating bacterial infections caused by resistant organisms worldwide (Holmberg et al. 1987; Williams and Heymann 1998). Thus, addressing the issue of antimicrobial resistance is one of the most urgent priorities in the field of infectious disease today.

Historically, medical practitioners and veterinarians selected antimicrobials to treat bacterial infectious diseases based primarily on past clinical experiences. However, with the increase in bacterial resistance to traditionally used antimicrobials, it has become more difficult for clinicians to empirically select an appropriate antimicrobial agent (Hubert et al. 1998; Walker 2000). As a result, *in vitro* antimicrobial susceptibility testing (AST) of the relevant bacterial pathogens, from properly collected specimens, is now standard procedure (Gould 2000; Hubert et al. 1998; Murray et al. 1999; Woods 2002). Antimicrobial susceptibility testing (AST) was initiated in many countries throughout the world soon after the introduction of antimicrobials for treatment of bacterial diseases (Greenwood 2000). In fact, one of the earliest antimicrobial sensitivity tests was developed by Alexander Fleming and consisted of a crude version of the broth dilution method (Fleming 1929). Other methods soon followed, including the creation of a plate diffusion method in 1941 (Greenwood 2000). This plate diffusion method attempted to quantify the effects of antimicrobial compounds according to inhibition zone size criteria (Wheat 2001). Paper disc diffusion methods were developed in the 1940s, followed by the development of modern-day agar dilu-

tion AST techniques as initially described by Schmith and Reymann (Wheat 2001). However, the majority of early methods for measuring bacterial susceptibility to antimicrobial agents *in vitro* were developed with little direction or coordinated effort (Barry 1989). When a need arose, most laboratory workers simply devised a method that would solve their immediate problem. Unfortunately, this approach resulted in the creation of an array of testing procedures that were being used in different laboratories, which usually lacked any type of quality control measures to ensure accurate and reproducible data (Barry 1989).

As the science of AST has progressed, a greater understanding of the multiple factors that could affect the overall outcome of susceptibility testing became clearer. In 1961, the World Health Organization (WHO) released an expert report addressing the need for method standardization when conducting microbic sensitivity testing (WHO 1961). A working group was created under the auspices of the WHO to initiate an international collaborative study to standardize AST methods. This study was followed by the release of the International Collaborative Study (ICS) working group recommendations for susceptibility testing in 1971 (Ericsson and Sherris 1971). The ICS report was the catalyst for establishing a series of recommendations for standardizing the performance of AST methods that are in use today. During this same time frame, Bauer et al. described a novel disk diffusion technique. This method rapidly gained acceptance and was instrumental in the early standardization of AST methods (Bauer et al. 1966; Wheat 2001). In 1975, this method became the basis for disk diffusion standards adopted by the National Committee for Clinical Laboratory Standards (NCCLS) in the United States.

ANTIMICROBIAL SUSCEPTIBILITY TESTING (AST) METHODOLOGIES. Although a variety of methods exist, the goal of *in vitro* antimicrobial susceptibility testing is the same: to provide a reliable predictor of how an organism is likely to respond to antimicrobial therapy in the infected host. This type of information aids the clinician in selecting the appropriate antimicrobial agent, provides data for surveillance, and aids in developing antimicrobial use policy. Some bacterial pathogens remain predictably susceptible to certain antimicrobial agents (continued penicillin susceptibility of *Arcanobacterium pyogenes*), thus AST is seldom performed. However, this is more the exception then the rule. Therefore, selecting the appropriate antimicrobials to test can be difficult given the vast numbers of agents available. Because some organisms can be intrinsically resistant to particular classes of compounds, it is unnecessary and misleading to test certain agents for activity *in vitro*. In addition, antimicrobials in the same class may have similar *in vitro* activities against select bacterial pathogens. In these cases, a representative antimicrobial is selected that predicts susceptibility to other members of the same molecular class.

In vitro antimicrobial susceptibility testing methods are performed by exposing a known concentration of a pure bacterial culture to increasing concentrations of a select antimicrobial drug. The endpoint measurement, based on the inhibition of bacterial growth, is reported either qualitatively (susceptible, intermediate, resistant) or quantitatively as the minimal inhibitory concentration (MIC, usually expressed in micrograms per milliliter or milligrams per liter). A number of AST methods are available to determine bacterial susceptibility to antimicrobials. The selection of a method is based on many factors such as practicality, flexibility, automation, cost, reproducibility, accuracy, and individual preference. All these AST methods are based on one of three approaches: broth dilution, agar dilution and agar diffusion (Figure 5.1).

Dilution. AST techniques based on dilution consist of bacterial growth assays using serial two-fold dilutions of antimicrobials. Using serial dilutions results in an endpoint MIC for the drug being tested. The quantitative data provided by dilution methods enables both the clinician to determine optimal pharmacokinetic and pharmocodynamic ratios to enhance clinical efficacy as well as provides information for surveillance programs that monitor changes in susceptibility over time. Note that the MIC does not represent an absolute value. Strictly speaking, the MIC lies somewhere between the observed endpoint and the next lower dilution (NCCLS 1999). It is important that the antimicrobial ranges assayed span both the clinically relevant breakpoints (susceptible, intermediate, and resistant) and quality control ranges for at least one of the reference organisms.

Broth dilution is a technique in which a standardized suspension of bacteria is tested against incremental concentrations of an antimicrobial agent in a standardized liquid medium. The broth dilution method can be performed either in tubes containing a minimum volume of 2 ml (macrodilution) or in small volumes (50–100 µl) (microliter) using microtitration plates (microdilution) (NCCLS 1999). Microtiter plates, which can be made in-house or purchased commercially, contain prediluted antibiotics in the wells. Commercial systems are frequently accompanied by automated plate readers, enhancing the high throughput analyses required by most clinical and reference laboratories. However, commercial systems are less flexible than agar dilution or disk diffusion methods in terms of modifying drugs to test, and such systems can be quite costly.

Agar dilution is similar to broth dilution, except that the antimicrobial agent is incorporated into an agar medium, and the bacterial inoculum is applied to the surface of that medium, usually by using a multipin replicating apparatus. The MIC is determined by unaided visual inspection of the agar surface. Using agar dilution methods has several advantages. The medium is very flexible, allowing one to extend the antibiotic concentration range as far as necessary. It allows the testing of multiple bacteria (32–36 isolates) simultaneously on the same set of agar plates. It is the only recommended NCCLS standardized antimicrobial susceptibility testing method for many fastidious organisms, such as many of the anaerobes, *Helicobacter* and *Campylobacter* species (Jorgensen and Ferraro 2000). The NCCLS considers agar dilution to be the reference method for *in vitro* AST; however, it requires extensive training of personnel and may be more labor intensive than other testing methods.

Diffusion. Diffusion methods generate an endpoint based on the diffusion of an antimicrobial agent from a solid carrier (e.g., paper disk or gradient strip) into a culture medium seeded with a known bacterial inoculum. This gradient is formed by diffusion of drug over time into a semi-solid agar medium. The two most widely used variations of this method, disk diffusion (Kirby-Bauer method) and the Etest, are described.

DISK DIFFUSION. In disk diffusion, a known amount of drug is incorporated into a paper disk, which is placed on the agar surface. After incubation, the diameter of the zone of inhibition is measured and used to predict drug susceptibility. It is important that the method for seeding the plate, the growth of the organism being tested, the time of incubation, the depth of the agar, and the diffusion of drug from the solid phase be consistent. Regardless, zone sizes can be difficult to measure because of partial inhibition at the border. Thus, other elements must be consistent, such as the intensity and angle of the light source used to examine the test plates and the measuring device used. As long as standardized methodologies are employed and zone interpretive standards are in place, reliable results can be obtained with disk diffusion tests. Note, however, that disk diffusion tests based solely on the presence or absence of a zone of inhibition, without regard to the size of the zone, are not generally acceptable.

Disk diffusion is technically straightforward to perform, reproducible, and does not require expensive equipment. The main advantages of the disk diffusion method are the low cost and the ease in modifying test formats when needed. On the other hand, the time required to manually measure endpoints may make this method impractical for those laboratories that perform large numbers of tests and record zone sizes (Andrews

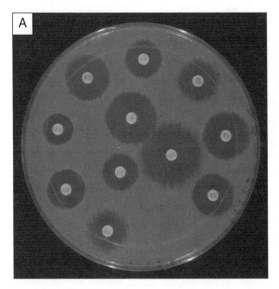

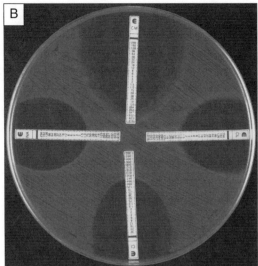

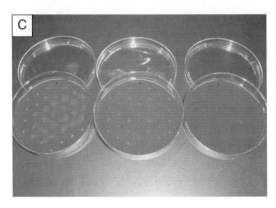

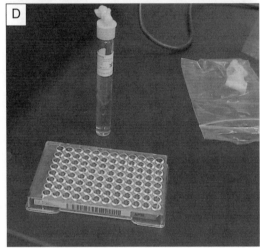

FIG. 5.1—Antimicrobial Susceptibility Testing Methods Methods. A. Disk diffusion (Kirby-Bauer Procedure). B. Antimicrobial Gradient Method E-test. C. Agar Dilution. D. Broth Microdilution.

et al. 2000). Automated zone reading devices, which can interface with laboratory reporting and data handling systems, are available, (Andrews et al. 2000; Hubert et al. 1998). Last, it is important to remember that no more than twelve disks should be placed on one 150 mm agar plate, nor should more than five disks be placed on a 100 mm plate, and disks should be evenly distributed and no closer than 24 mm from center to center (NCCLS 1999).

E-TEST (EPSILOMETER). The E-test is a quantitative agar diffusion method. With this AST method, a concentration gradient of an antimicrobial drug is deposited on one side of a plastic strip, which is placed on a seeded agar surface (Brown and Brown 1991). After incubation, an ellipse is formed, intersecting the strip at a point in the gradient indicating the MIC. Because E-test comprises a continuous gradient, MIC values between two-fold dilutions can be obtained. In performing the E-test, special attention must be paid to carefully placing the strips on the agar surface. In addition, users must be aware of particular interpretive guidelines for different bacterium/antimicrobial combinations. Peculiar zone patterns can lead to MIC discrepancies for some organisms when compared with agar dilution results (Brown and Brown 1991). The relatively high cost of the strips makes it most useful for a limited number of drugs in special circumstances.

In vitro AST methods have been standardized mostly for aerobic and facultative bacteria with antimicrobial agents that are intended for systemic use. Methods have not been standardized, and in some cases are not recommended, for some uncommon or fastidious bacteria. The appropriate method will ultimately depend on the growth characteristics of the test bacterium (Table 5.1). For example, disk diffusion is not practical for testing slow-growing bacteria such as the Mycobacteria. Similarly, disk diffusion should not be used for *Campylobacter*, anaerobes, or other bacteria with highly variable growth rates (Jorgensen and Ferraro 2000; Woods 2002). Regardless of the method used, procedures must be standardized to ensure accurate and reproducible results. Standardization, among other things, requires the identification of a stable quality control organism(s) suitable to the growth requirements of the test population. It is important that this quality control strain be tested routinely when AST is performed to confirm day-to-day reproducibility of the testing procedure.

In certain situations, it may be clinically advantageous to test only for the presence of a particular resistance phenotype, rather than susceptibility endpoints for multiple drugs for some pathogens. For example, certain beta-lactamase–producing bacteria are best detected using a direct nitrocephin-based chromogenic test (NCCLS 1999). Extended-spectrum beta-lactamase activity in certain bacteria can also be detected by using standard disk diffusion susceptibility test methods utilizing specific cephalosporins (cefotaxime and ceftazidime) in combination with a beta-lactamase inhibitor (clavulanic acid) and measuring the resulting zones of inhibition (Threlfall et al. 1999). In addition, chloramphenicol resistance attributed to production of chloramphenicol acetyl transferase can be detected in some bacteria via rapid tube or filter paper tests within one to two hours (NCCLS 1999).

INTERPRETATION OF ANTIMICROBIAL SUS-CEPTIBILITY TESTING RESULTS. The qualitative designation of an organism as susceptible (S), intermediate (I), or resistant (R) to a particular antimicrobial is arrived at by determining *in vitro* breakpoints. Breakpoints are based both on obtainable serum concentrations of the antimicrobial agent administered at therapeutic doses and on clinical trials (Murray et al. 1999; Walker 2000). *Susceptible* implies that the antimicrobial agent should be successful in treating the bacterial infection when administered at approved doses. *Intermediate* indicates that the antimicrobial agent may be successful in treating the bacterial infection if high levels of the agent can be achieved at the site of infection, for example, beta-lactams for treating urinary tract infections. Intermediate may also serve as an interpretation buffer zone preventing small, uncontrolled technical factors from causing discrepancies in interpretation, for example, preventing resistant organisms from being called susceptible or a susceptible isolate from being called resistant (NCCLS 1999). *Resistant* implies that the bacterium would not be inhibited *in vivo* by usually achievable systemic concentrations of the antimicrobial agent administered at approved doses. An organism may also be deemed resistant if it is known to possess a specific resistance mechanism, for example, methicillin-resistant *Staphylococcus aureus* (MRSA), which should be reported as resistant to all beta-lactams (NCCLS 1999).

There are two primary factors that contribute to the accuracy of AST in determining whether a bacterium is susceptible or resistant to an antimicrobial agent. The first factor is the use of quality control (QC) organisms with established QC ranges. This quality control is essential for validating the specific AST method used. Incorrect testing methods frequently involve using some combination of incorrect QC organisms, growth media, incubation conditions, and tested antimicrobial concentrations. When performing AST for a particular bacterium/antimicrobial agent combination, the person performing the test must use the medium, incubation conditions, and drug concentrations indicated by the appropriate testing organization (such as NCCLS, BSAC, SIR, CAFSM, or other). The person performing

Table 5.1—Suggested Modifications of Standard Methods for Susceptibility Testing of Some Fastidious and Special Problem Veterinary Pathogens

Organism	Method	Medium	Incubation	Comments
Staphylococcus hyicus	Broth Microdilution	CAMHB + thymidine phosphorylase (0.2 IU/mL); for sulfonamides and trimethoprim only	35°C/18-24 h	Wegener et al.
Haemophilus somnus	Agar Dilution with Disk Diffusion	Chocolate Mueller-Hinton agar	35°C/5-7% CO_2/20-24 h	
Actinobacillus pleuropneumoniae	Broth Microdilution	Veterinary Fastidious Medium (VFM)	35°C/5-7% CO_2/20-24 h	
Enterococci	Broth Microdilution	CAMHB	35°C/16-24 h	Accurate detection of vancomycin resistance requires that the plates be incubated for a full 24 h.
Listeria spp.	Broth Microdilution	CAMHB + LHB (2-5% v/v)	35°C/18 h	

Parts reproduced with permission from NCCLS publication M31-A2 Performance standards for antimicrobial disk and dilution susceptibility tests for bacteria isolated from animals; Approved standard-2nd edition (ISBN 1-56238-000-0).

the test also should use the QC appropriate for the particular assay. Failure to follow these standardized methods may result in the generation and of reporting erroneous results.

This has been especially challenging for AST of *Campylobacter* species. In performing AST of *Campylobacter* species, several testing methods, including disk diffusion, broth microdilution, agar dilution, and E-test (AB Biodisk, Sweden) have been used to determine the *in vitro* susceptibilities of this organism to antimicrobial agents (Fernandez et al. 2000; Gaudreau and Gilbert 1997). However, until recently, none of these methods had been standardized on either a national or international scale. Thus, the large number of testing variables as well as the different interpretive criteria used make comparing results from various studies difficult. Tentative MIC susceptible and resistant breakpoints for *Campylobacter* spp. to only erythromycin and ciprofloxacin have been proposed by the British Society for Antimicrobial Chemotherapy (BSAC) (23). The BSAC-proposed resistant MIC breakpoint for ciprofloxacin for *Campylobacter* spp. is ≥4 µg/ml, and for erythromycin ≥2 µg/ml (King 2001). The Comité de L'antibiogramme de la Société Française de Microbiologie (http://www.sfm.asso.fr/) has proposed *Campylobacter* interpretive criteria for fourteen antimicrobials for organisms belonging to this genus. To date, the NCCLS has not approved interpretive criteria for *Campylobacter*. Recently, the NCCLS subcommittee on Veterinary Antimicrobial Susceptibility Testing approved the agar dilution test as a standard susceptibility testing method for *Campylobacter*, and *C. jejuni* ATCC 33560 as the quality control (QC) organism. The antimicrobial agents for which quality control testing has been established for *Campylobacter jejuni* are indicated in Table 5.2. *In vitro* susceptibility testing of this organism with other agents is not recommended until appropriate quality control parameters have been established.

When *in vitro* antimicrobial susceptibility tests are performed without the use of appropriate QC organisms, comparing data between laboratories and between countries is difficult. Alternatively, appropriate QC organisms may be used, but the dilution scheme does not encompass the ranges for the select QC organism. As alluded to previously, when using a dilution testing method, the dilution scheme should encompass

Table 5.2—Tentative Quality Control Ranges for Agar Dilution Testing of *Campylobacter jejuni* (ATCC® 33560)

Antimicrobial Agent	MIC (µg/mL)
Ciprofloxacin	0.12–1
Doxycycline	0.5–2
Erythromycin	1–8
Gentamicin	0.5–4
Meropenem	0.004–0.015
Nalidixic Acid	8–32
Tetracycline	1–4

the entire QC range of the QC organism. For example, the NCCLS QC range for *Escherichia coli* ATCC 25922, when tested against ciprofloxacin, is 0.004 µg/ml to 0.16 µg/ml. If a laboratory uses a ciprofloxacin dilution scheme from 0.16 µg/ml to 4 µg/ml, that laboratory's data may not be in proper control. One could argue that if the MIC for the *E. coli* 25922 is 0.16 µg/ml, then its data is properly controlled. But because the MIC is the first well in which no visible growth occurs, there is no way to know whether growth would have occurred at the next lowest dilution, that is, 0.008 µg/ml, because that concentration was not tested. Ultimately, if *E. coli* 25922 was the only QC organism used, and dilutions below 0.16 µg/ml were not tested, the data would not have passed quality control and thus should be considered invalid.

On the other hand, other NCCLS QC organisms would be more appropriate for this particular testing scheme, namely, *Staphylococcus aureus* 29213 (QC range 0.12 µg/ml to 0.5 µg/ml), *Enterococcus faecalis* 29212 (QC range 0.25 µg/ml to 2 µg/ml), or *Pseudomonas aeruginosa* 27853 (QC range 0.25 µg/ml to 1 µg/ml). Alternatively, the laboratory could change its dilution scheme to include *E. coli*'s QC range. To do so, the laboratory would need to test dilutions down to 0.002 g/ml to ensure that the entire QC range was controlled. These types of considerations are often overlooked but are essential to generating reliable results.

After the QC ranges for the QC microorganisms have been established, the second major component needed to interpret susceptibility results is the determination of interpretive criteria (IC). The determination of IC differs among the many professional societies and regulatory agencies around the world. Yet, most share similar data requirements for generating IC. The NCCLS uses population distributions of relevant microorganisms, pharmacokinetic/pharmocodynamic parameters of the antimicrobial agent, and clinical efficacy data (Acar and Goldstein 1995; Walker 2000). For the population distribution data, the MICs of 300 to 600 clinical isolates that are most likely to be tested by a laboratory in the anticipated use of the antimicrobial agent are plotted against the zone of inhibition generated for the same isolates. The MIC and zone diameters are then presented as a scattergram. This is then analyzed by linear regression analysis and error rate-bounding to establish potential interpretive breakpoints. The use of the appropriate QC organisms for the generation of the data is essential to ensure the accuracy of the IC from this analysis.

The second set of data that is used to determine the IC is the pharmacokinetic/pharmacodynamic parameters of the antimicrobial agent in the target animal species. Data should be relevant to the method of drug administration used in clinical efficacy trials such as the route of administration, dose, and frequency of dosing. This information might also include the length of time that the serum or plasma concentration exceeds the susceptible breakpoint, the ratio between the peak serum or plasma concentrations to the susceptible breakpoint MIC, and ratio of the area-under-the-curve (AUC) to the susceptible breakpoint MIC. If possible, the concentration of the drug and its active metabolites at the site of infection are considered. In addition, the drug's sponsor may provide the NCCLS subcommittee with such information as postantibiotic effect, concentration-dependent killing effect, and other details.

The last set of data required by NCCLS to determine the IC of an antimicrobial agent is the correlation between the *in vitro* susceptibilities of bacterial isolates from animals enrolled in clinical trials and the therapeutic outcome. Where possible, samples are collected pre- and post-treatment, and quantitative susceptibility test results are submitted on individual isolates. Ideally, the *in vitro* susceptibility data submitted includes both MICs and zone of inhibition diameters, including the appropriate quality control data. The efficacy of the antimicrobial agent is then evaluated using pharmacokinetic/pharmacodynamic data (based on the same dose, route of administration, treatment regimen, and so on), antibacterial activity, and therapeutic outcome. Antimicrobial susceptibility breakpoints derived by professional societies or regulatory agencies in various countries are often very similar to each other. However, different countries can have notable breakpoint differences for the same antimicrobial agent/bacterial pathogen combination. These differences may be caused by many factors such as variation in technical AST factors (inoculum density, test media, and test method) and the fact that different countries use different dosages or administration intervals for some antimicrobials. Some countries also are more conservative in setting interpretive criteria for specific antimicrobials. Additionally, it is important to remember that IC developed for human clinical medicine are not always relevant for veterinary use, as pharmacokinetics and relevant infectious agents may differ significantly (NCCLS 1999).

STANDARDIZATION AND HARMONIZATION OF AST METHODOLOGIES. The most effective approach for national and international surveillance of antimicrobial resistance would be for all participating laboratories to use a common AST method, including similar quality control reference organisms and ranges. Currently, there is no consensus on one AST procedure, and no one existing method has been adopted on a global scale (Cotter and Adley 2001). The many variations in methodologies, techniques, and interpretive criteria currently being used globally make adopting one method a difficult task.

A number of standards and guidelines currently are available for antimicrobial susceptibility testing and subsequent interpretive criteria. These include standards and guidelines published by NCCLS, BSAC, Japan Society for Chemotherapy (JSC), Swedish Reference Group for Antibiotics (SIR), Deutshes Institute für Normung (DIN), Comité de L'antibiogramme de la Société Française de Microbiologie (CASFM), Werkgroep richtlijnen gevoeligheidsbepalingen (WRG system, NL) and others (1997; 2002; Comité de L'antibiogramme de la Société Française de Microbiologie [CA-SFM] 2002; Deutsch Industrie Norm-Medizinsche Mikrobiologie 1994; NCCLS 1999; Working Party on Antibiotic Sensitivity Testing of the British Society for Antimicrobial Chemotherapy 1991). The variations in diffusion and dilution AST methods (such as choice of agar medium, inoculum size, growth conditions) and the differing IC (including breakpoints) among the many countries, makes comparing susceptibility data from one system with another difficult. Not surprisingly, therefore, several member countries in the European Union are contemplating coordination of AST methods to facilitate resistance surveillance (Cotter and Adley 2001; White et al. 2001).

In both veterinary and human medicine, antimicrobial resistance data is being shared among a number of laboratories through the creation of antimicrobial resistance surveillance networks. Some of these networks are linked internationally. These networks have helped to facilitate the standardization and harmonization of AST methods among participating laboratories, which adhere to the appropriate published guidelines and standards of AST and quality control monitoring to ensure accuracy and comparability of the data. Examples of international and national surveillance systems employing standardized methods include the European antimicrobial resistance surveillance system (EARSS), Alexander project for respiratory pathogens, Antibiotic resistance in bacteria of animal origin (ARBAO), SENTRY, The surveillance network (TSN, now known as Focus Technologies), DANMAP, WHONET, NARMS, and Enter-Net (Bager 2000; Livermore et al. 1998; Marano et al. 2000; Threlfall et al. 1999; Trevino 2000). Data generated from these surveillance and monitoring programs will eventually play a key role in the development of national and perhaps international polices concerning the use of antimicrobials in animals and possible linkages to public health. The continuing success of these surveillance and monitoring programs suggests that standardization and harmonization of AST methods are both conceivable and progressing globally.

However, to determine the comparability of results originating from different surveillance systems around the globe, antimicrobial susceptibility test results must be reported quantitatively, including information on the methods, quality control organisms, and ranges tested (Walker and Thornsberry 1998). Thus, minimum inhibitory concentration (MIC) values or zone diameters (rather than just S, I, R) should be reported so that shifts in antimicrobial susceptibility among the target bacterial pathogens can be monitored and compared. These MIC values or zone diameters can be achieved by either broth or agar dilution methods, or by statistical transformation of the zone of inhibition diameters obtained by disk diffusion methods to MICs (Acar and Goldstein 1995).

Quality Control and Quality Assurance of AST. The implementation of quality control in laboratories that perform AST is imperative to help monitor the precision and accuracy of the AST procedure, the performance of the appropriate reagents, and the personnel involved (NCCLS 1999). Stringent adherence to standardized AST methods in conjunction with quality control of media and reagents is necessary for the collection of reliable and reproducible antimicrobial susceptibility data. Preferably, records will be kept regarding lot numbers and expiration dates of all appropriate materials and reagents used in AST. It is critical that the appropriate quality control reference bacteria be appropriately maintained in the laboratory and tested frequently to ensure standardization regardless of the AST method used. Reference bacterial strains to be used for quality control should be catalogued and characterized with stable, defined antimicrobial susceptibility phenotypes such as those obtained from the American Type Culture Collection (NCCLS 1999). Recognized quality control strains should be tested every time a new batch of medium or plate lot is used and on a regular basis in parallel with the bacterial strains to be assayed (Table 5.3).

The preferred method for analyzing the overall performance of each laboratory is to test the appropriate quality control bacterial strains each day that susceptibility tests are performed (NCCLS 1999). Because using this method may not always be practical or economical, the frequency of such quality control tests may be reduced if the laboratory can demonstrate that its procedures are reproducible. After a laboratory doc-

uments the reproducibility of its susceptibility testing methods, testing may be performed on a weekly basis (NCCLS 1999). If quality control errors emerge, the laboratory is responsible for determining the causes and repeating the tests. If the laboratory cannot determine the source of errors, then the quality control testing should be re-initiated on a daily basis. Whenever a QC error is discovered, careful consideration should be given as to whether to report the data, especially if the source of the error may have potentially affected the interpretation of the AST results (NCCLS 1999).

External proficiency testing (that is, third-party testing) of participating laboratories should be mandatory for major bacterial species included in national surveillance systems. Designated reference laboratories should be appointed to coordinate quality assurance for the participating laboratories. The responsibilities of the reference laboratory may include developing a set of reference bacterial strains with varying antimicrobial susceptibilities to be tested at the participating laboratories. Each laboratory will test these strains under the same AST conditions that it normally employs. Proficiency testing on a regular basis helps to ensure that reported susceptibility data is accurate and without question.

SUMMARY AND CONCLUSIONS. The use of genotypic approaches for detection of antimicrobial resistance genes has been promoted as a way to increase the rapidity and accuracy of susceptibility testing. Numerous DNA-based assays are being developed

Table 5.3—Antimicrobial Susceptibility Testing Quality Control Organisms used Worldwide

	Strain Number for the Indicated Collections:					
Organism	ATCC[1]	DSM[2]	UKNCC[3]	JCM[4]	CIP[5]	CCUG[6]
Escherichia coli	25922	1103	12241	5491	7624	17620
Escherichia coli	35218	5923	11954		102181	30600
Pseudomonas aeruginosa	27853	1117	10896		76110	17619
Staphylococcus aureus	29213	2569		2874	103429	15915
Staphylococcus aureus	25923	1104	12702	2413	7625	17621
Staphylococcus aureus	43300					
Enterococcus faecalis	29212	2570	12697	2875	103214	9997
Enterococcus faecalis	33186		12756		100750	
Enterococcus faecalis	51299					
Actinobacillus pleuropneumoniae	27090				100916 T	
Haemophilus somnus	700025					
Campylobacter jejuni	33560		11351		70.2 T	
Streptococcus pneumoniae	49619					
Klebsiella pneumoniae	700603					

[1]American Type Culture Collection (www.atcc.org)
[2]Deutsche Sammlung von Mikroorganismen und Zellkulturen (www.dsmz.de)
[3]United Kingdom National Culture Collections (includes National Collection of Type Cultures [NCTC] and National Collection of Industrial and Marine Bacteria [NCIMB]) (www.ukncc.co.uk)
[4]Japan Collection of Microorganisms (www.jcm.riken.go.jp)
[5]Collection of Bacterial Strain of Institut Pasteur
[6]Swedish Reference Group for Antibiotics (www.srga.org)
Reproduced with permission, from NCCLS publication M31-A2 Performance standards for antimicrobial disk and dilution susceptibility tests for bacteria isolated from animals; Approved standard-2nd edition (ISBN 1-56238-000-0).

to detect bacterial antibiotic resistance at the genetic level. This in turn is changing the nature of laboratory methods from susceptibility testing to resistance testing. The newest and perhaps most state-of-the-art approach is to predict antimicrobial resistance phenotypes via identification and characterization of the known genes that encode specific resistance mechanisms. Methods that employ the use of comparative genomics, genetic probes, microarrays, nucleic acid amplification techniques (for example, polymerase chain reaction [PCR]), and DNA sequencing offer the promise of increased sensitivity, specificity, and speed in the detection of specific known resistance genes (Jorgensen and Ferraro 2000; Paulsen et al. 2001; Strizhkov et al. 2000). This approach is especially useful for organisms that are difficult to culture, such as *M. tuberculosis*. Molecular methods to identify *M. tuberculosis* in clinical specimens, and to simultaneously detect resistance mutations (such as *rpoB*), can provide physicians with important clinical information in a much shorter time (Crawford 1994). Genotypic methods have been successfully applied to supplement traditional AST phenotypic methods for other organisms as well. These include the verification of methicillin resistance in staphylococci, vancomycin resistance in enterococci, and detection of fluoroquinolone resistance mutations (Brakstad et al. 1993; Dutka-Malen et al. 1995; Jorgensen and Ferraro 2000; Walker et al. 2000). PCR methods have also been described for beta-lactamases, aminoglycoside inactivating enzymes, and tetracycline efflux genes, to name a few (Bradford 2001; Chopra and Roberts 2001; Frana et al. 2001).

Technological innovations in DNA-based diagnostics should allow for the detection of multiple resistance genes, variants, or both during the same test. The development of rapid diagnostic identification methods and genotypic resistance testing should help reduce the emergence of antimicrobial resistance by enabling the use of the most appropriate antimicrobial when therapy is initiated. Additionally, new technological advances may facilitate the ability to probe bacterial species for large numbers of antimicrobial resistance genes quickly and cheaply, thereby providing additional relevant data into surveillance and monitoring programs. However, despite the new influx of genotypic tests, standardized phenotypic AST methods will still be required in the near future to detect emerging resistance mechanisms among bacterial pathogens.

REFERENCES

Acar, J. and F.W. Goldstein. 1995. Disk susceptibility test. In *Antibiotics in laboratory medicine*. V. Lorian, ed. Baltimore, MD: Williams & Wilkins, 1–51.

Andrews, J.M., F.J. Boswell, and R. Wise. 2000. Evaluation of the Oxoid Aura image system for measuring zones of inhibition with the disc diffusion technique. *J. Antimicrob. Chemother.* 46:535–540.

Bager, F. 2000. DANMAP: monitoring antimicrobial resistance in Denmark. *Int. J. Antimicrob. Agents.* 14:271–274.

Barry, A.L. 1989. Standardization of antimicrobial susceptibility testing. *Clin. Lab Med.* 9:203–219.

Bauer, A.W., W.M. Kirby, J.C. Sherris, and M. Turck. 1966. Antibiotic susceptibility testing by a standardized single disk method. *Am. J. Clin. Pathol.* 45:493–496.

Bradford, P.A. 2001. Extended-spectrum beta-lactamases in the 21st century: characterization, epidemiology, and detection of this important resistance threat. *Clin. Microbiol. Rev.* 14:933–51, table.

Brakstad, O.G., J.A. Maeland, and Y.Tveten. 1993. Multiplex polymerase chain reaction for detection of genes for Staphylococcus aureus thermonuclease and methicillin resistance and correlation with oxacillin resistance. *APMIS.* 101:681–688.

Brown, D.F. and L. Brown. 1991. Evaluation of the E test, a novel method of quantifying antimicrobial activity. *J. Antimicrob. Chemother.* 27:185–190.

Chopra, I. and M. Roberts. 2001. Tetracycline antibiotics: mode of action, applications, molecular biology, and epidemiology of bacterial resistance. *Microbiol. Mol. Biol. Rev.* 65:232–260.

Cars, O. Antimicrobial susceptibility testing in Sweden. 1997. *Scand. J. Infect. Dis.*, Suppl. 105.

Comité de L'antibiogramme de la Société Française de Microbiologie (CA-SFM). 2002. Définition des catégories thérapeutiques et Méthode de détermination de la Concentration minimale inhibitrice en milieu solide pour les bactéries aérobies à croissance rapide. *Bull Soc Fr Microbiol.* 8:156–166.

Cotter, G. and C.C. Adley. 2001. Comparison and evaluation of antimicrobial susceptibility testing of enterococci performed in accordance with six national committee standardized disk diffusion procedures. *J. Clin. Microbiol.* 39:3753–3756.

Courvalin, P., and C.J. Soussy. 2002. Report of the Comité de l'antibiogramme de la Société Française de Microbiologie. . *Clin. Microbiol. and Infect.* Suppl. 1.

Crawford, J.T. 1994. New technologies in the diagnosis of tuberculosis. *Semin. Respir. Infect.* 9:62–70.

Deutsch Industrie Norm-Medizinische Mikrobiologie. 1994. Methoden zur Empfind-lichkeitsprunfung von bakteriellen Krankheitserregen (ausser Mykobakterien) gegen Chemotherapeutika, Geschaftsstelle des NAMeds im DIN. *Deutsche Industrie Norm Medizinsche Mikrobiologie.* 58:940.

Dutka-Malen, S., S. Evers, and P. Courvalin. 1995. Detection of glycopeptide resistance genotypes and identification to the species level of clinically relevant enterococci by PCR. *J. Clin. Microbiol.* 33:24–27.

Ericsson, H.M. and J.C. Sherris. 1971. Antibiotic sensitivity testing. Report of an international collaborative study. *Acta Pathol Microbiol. Scand.[B] Microbiol. Immunol.* 217:Suppl.

Fernandez, H., M. Mansilla, and V. Gonzalez. 2000. Antimicrobial susceptibility of Campylobacter jejuni subsp. jejuni assessed by E-test and double dilution agar method in Southern Chile. *Mem. Inst. Oswaldo Cruz.* 95:247–249.

Fleming, A. 1929. On the antibacterial action of cultures of a penicillum with special reference to their use in the isolation of B. influenzae. *Br. J. Exp. Pathol.* 10:226–236.

Frana, T.S., S.A. Carlson, and R.W. Griffith. 2001. Relative distribution and conservation of genes encoding aminoglycoside-modifying enzymes in Salmonella enterica serotype typhimurium phage type DT104. *Appl. Environ. Microbiol.* 67:445–448.

Gaudreau, C. and H. Gilbert. 1997. Comparison of disc diffusion and agar dilution methods for antibiotic susceptibility testing of Campylobacter jejuni subsp. jejuni and Campylobacter coli. *J. Antimicrob. Chemother.* 39:707–712.

Gould, I.M. 2000. Towards a common susceptibility testing method? *J. Antimicrob. Chemother.* 45:757–762.

Greenwood, D. 2000. Detection of antibiotic resistance in vitro. *Int. J. Antimicrob. Agents.* 14:303–306.

Holmberg, S.D., S.L. Solomon, and P.A. Blake. 1987. Health and economic impacts of antimicrobial resistance. *Rev. Infect. Dis.* 9:1065–1078.

Hubert, S.K., P.D. Nguyen, and R.D. Walker. 1998. Evaluation of a computerized antimicrobial susceptibility system with bacteria isolated from animals. *J. Vet. Diagn. Invest.* 10:164–168.

Jorgensen, J.H. and M.J. Ferraro. 2000. Antimicrobial susceptibility testing: special needs for fastidious organisms and difficult-to-detect resistance mechanisms. *Clin. Infect. Dis.* 30:799–808.

King, A. 2001. Recommendations for susceptibility tests on fastidious organisms and those requiring special handling. *J. Antimicrob. Chemother.* 48 Suppl 1:77–80.

Livermore, D.M., A.P. Macgowan, and M.C. Wale. 1998. Surveillance of antimicrobial resistance. Centralised surveys to validate routine data offer a practical approach. *BMJ.* 317:614–615.

Marano, N.N., S. Rossiter, K. Stamey, K. Joyce, T.J. Barrett, L.K. Tollefson, and F.J. Angulo. 2000. The National antimicrobial resistance monitoring system (NARMS) for enteric bacteria, 1996–1999: surveillance for action. *J. Am. Vet. Med. Assoc.* 217:1829–1830.

Murray, P.R., E.J. Baron, M.A. Pfaller, F.C. Tenover, and Y.H.Yolken. 1999. Antimicrobial agents and susceptibility testing. *Manual of Clinical Microbiology.* Washington, DC.: American Society for Microbiology.

NCCLS. 1999. Performance standards for antimicrobial disk and dilution susceptibility tests for bacteria isolated from animals; approved standard. *NCCLS document M31-A.* Wayne, PA: National Committee for Clinical Laboratory Standards.

Paulsen, I.T., J. Chen, K.E. Nelson, and M.H. Saier, Jr. 2001. Comparative genomics of microbial drug efflux systems. *J. Mol. Microbiol Biotechnol.* 3:145–150.

Schwartz, B., D.M. Bell, and J.M. Hughes. 1997. Preventing the emergence of antimicrobial resistance. A call for action by clinicians, public health officials, and patients. *JAMA.* 278:944–945.

Strizhkov, B.N., A.L. Drobyshev, V.M. Mikhailovich, and A.D. Mirzabekov. 2000. PCR amplification on a microarray of gel-immobilized oligonucleotides: detection of bacterial toxin- and drug-resistant genes and their mutations. *Biotechniques.* 29:844–2, 854.

Threlfall, E.J., I.S. Fisher, L.R. Ward, H. Tschape, and P. Gerner-Smidt. 1999. Harmonization of antibiotic susceptibility testing for Salmonella: results of a study by 18 national reference laboratories within the European Union-funded Enter-net group. *Microb. Drug Resist.* 5:195–200.

Trevino, S. 2000. Antibiotic resistance monitoring: a laboratory perspective. *Mil. Med.* 165:40–42.

Walker, R.A., A.J. Lawson, E.A. Lindsay, L.R. Ward, P.A. Wright, F.J. Bolton, D.R. Wareing, J.D. Corkish, R.H. Davies, and E.J. Threlfall. 2000. Decreased susceptibility to ciprofloxacin in outbreak-associated multiresistant Salmonella typhimurium DT104. *Vet. Rec.* 147:395–396.

Walker, R.D. 2000. Antimicrobial susceptibility testing and interpretation of results. In *Antimicrobial Therapy in Veterinary Medicine.* J.F. Prescott, J.D. Baggot, and R.D. Walker, eds. Ames, IA: Iowa State University Press, 12–26.

Walker, R.D. and C. Thornsberry. 1998. Decrease in antibiotic susceptibility or increase in resistance? *J. Antimicrob. Chemother.* 41:1–4.

Wegener, H.C., J.L. Watts, S.A. Salmon, and R.J. Yancey, Jr. 1994. Antimicrobial susceptibility of Staphylococcus hyicus isolated from exudative epidermitis in pigs. *J. Clin. Microbiol.* 32:793–795.

Wheat, P.F. 2001. History and development of antimicrobial susceptibility testing methodology. *J. Antimicrob. Chemother.* 48 Suppl. 1:1–4.

White, D.G., J. Acar, F. Anthony, A. Franklin, R. Gupta, T. Nicholls, Y. Tamura, S. Thompson, E.J. Threlfall, D. Vose, M. van Vuuren, H.C. Wegener, and M.L. Costarrica. 2001. Antimicrobial resistance: standardisation and harmonisation of laboratory methodologies for the detection and quantification of antimicrobial resistance. *Rev. Sci. Tech.* 20:849–858.

WHO. Second report of the expert committee on antibiotics. *Standardization of methods for conducting microbic sensitivity tests.* 201:1–24. 1961. World Health Organization Technical Report Series.

Williams, R.J. and D.L. Heymann. 1998. Containment of antibiotic resistance. *Science* 279:1153–1154.

Woods, G.L. 2002. In vitro testing of antimicrobial agents. *Antibacterial therapy: in vitro testing, pharmacodynamics, pharmacology, new agents.* Philadelphia, PA: WB Saunders Co., 463–495.

Working Party on Antibiotic Sensitivity Testing of the British Society for Antimicrobial Chemotherapy. 1991. A guide to sensitivity testing. *J. Antimicrob. Chemother.* 27:1–50.

6

ANTIBIOTICS: MODE OF ACTION, MECHANISMS OF RESISTANCE, AND TRANSFER

KATHLEEN KEYES, MARGIE D. LEE, AND JOHN J. MAURER

INTRODUCTION. Man's search for remedies against aches, pain, and illness is probably as old as mankind itself, but before the actual etiological agents of infectious disease were identified, treatment relied for the most part on the vagaries of chance and empirical observation. Much of ancient folk medicine was also based on observations that natural substances could alleviate the symptoms of disease. Ancient Egyptian texts from the middle of the sixteenth century B.C., such as the Great Medical Papyrus, describe treatment of a pustular scalp ailment by rubbing the crumbs of moldy, wheaten loaf on it (Ebell 1937). Inflammations of the bladder and urinary tract were also treated with moldy, wheaten preparations, which enjoyed the reputation of "soothing the pipes" (Bottcher 1964). The Egyptian documents also describe dispensation of a "medicinal earth," which possessed curative properties and carried with it the recommendation "prepared by Schow for Re" (Ebell 1937; Bottcher 1964)—and what more could one want from a remedy than one prepared by one god for another! The ancient Egyptians had worked out a system of antibiotically active concoctions: the moldy bread no doubt provided a goodly dose of penicillin and the "medicinal" dirt offered its accumulation of secondary metabolites or natural antibiotics from the soil organisms.

In the second half of the nineteenth century, Robert Koch observed that some microorganisms could destroy others (Koch 1876–1877). This phenomenon was confirmed by Louis Pasteur (Pasteur 1880), who believed it might be utilized in medicine. By the end of that century, the German bacteriologist Paul Ehrlich had begun his quest to systematically seek new, antimicrobial compounds (Himmelweit 1960). When he found a compound that showed at least limited antimicrobial activity, he would set about trying to modify the molecule to produce derivatives or closely related compounds in order to find more effective ones, an approach still used by drug chemists today. In 1909, Paul Ehrlich discovered the first chemical "cure" for a disease, the arsenical compound, arsphenamine (marketed as Salvarsan), that was selectively toxic for *Treponema pallidum* (Ehrlich and Hata 1910). The medical profession dubbed this compound the magic bullet because it killed the specific "germs" that caused syphilis.

Over the next few, short decades, from the late 1920s to the 1960s, came the true era of wonder drugs known as antibiotics. In 1935 the German bacteriologist and pharmacologist Gerhard Domagk discovered that the dye protonsil red used to tint cloth was effective in the treatment of streptococcal infections in mice (Domagk 1935). When his own young daughter was close to dying with a streptococcal infection, he in desperation injected this dye into the girl (Schadewaldt 1985). Her fever dropped almost immediately, with a recovery termed at that time nothing short of miraculous. A Swiss-borne chemist, Daniel Bovert, then showed that the antibacterial activity of protonsil red was caused by the sulfanilamide component of the dye. This compound already lay in the public domain when its antimicrobial effects were discovered; by 1940, sulfanilamide was available under at least thirty-three different trade names and a start had been made on producing the numerous sulfanilamide derivatives that subsequently became available (Magner 1992).

It was in early September of 1928 that Alexander Fleming, returning from holiday to his laboratory at St. Mary's Hospital in London, made his famous observation on an old, uncovered culture plate of bacteria. Along with the expected staphylococci on the petri plate, he noticed a blue-green mold and something in the mold that was attacking the bacteria. He identified the mold as *Penicillium notatum,* cultured it in nutrient broth, filtered it, and discovered in the filtrate a substance that ravaged bacteria. He named the discovery penicillin (Fleming 1929). Fleming was unable to purify the penicillin himself and was also unable to arouse much interest in his discovery. It took the economic and political (as well as medical) pressures of disease-ridden soldiers and civilians of World War II to awaken pharmaceutical companies' interest in penicillin. The British pathologist/bacteriologist Howard Florey and German chemist Ernst Chain extracted the first real sample of penicillin in the spring of 1942, which was a million times more powerful than Fleming's original filtrate. Penicillin became the starting point for modern antibiotic therapy. Mass production of penicillin began in 1941 in the United States, shortly after America became involved in the war, for use by the military to treat sick and wounded soldiers. In 1945 Fleming, Florey, and Chain were awarded the Nobel Prize for Medicine (Hoel and Williams 1997).

THE ANTIBIOTICS: AN OVERVIEW. The term *antibiotic,* from the Greek derivation that literally means *against life*, first appeared in the scientific literature in 1942 (Waksman and Woodruff 1942; Waksman et al. 1942). Coined by Selman Waksman to embrace those newly discovered antimicrobial substances, such as pyocyanin, penicillin, actimycin, and others, Waksman used this term to describe and define those substances of microbial origin that specifically inhibit the growth of other microorganisms. The usage for this term has now been extended to include any low-molecular-weight compound, whether a microbial or other living organism's metabolite or synthetic compound, that at low concentrations will kill or inhibit the growth of other microorganisms (O'Grady et al. 1997).

Antibiotic agents affect the growth of bacteria in three general ways. Those that kill bacterial cells outright, but where lysis or cell rupture does not occur, are known as *bacteriocidal*. Bacteriocidal antibiotics generally bind tightly to their cellular targets and are not removed by dilution (Lancini et al. 1995). *Bacteriolytic* agents induce killing by cell lysis and include those antibiotics that inhibit cell wall synthesis and those that damage the cytoplasmic membrane (Lancini et al. 1995). Antibiotics that do not kill but inhibit growth are known as *bacteriostatic*. Bacteriostatic antibiotics are frequently inhibitors of protein synthesis and act by binding to ribosomes. The binding is not tight, and when the concentration of the antibiotic is lowered, it becomes free from the ribosome and resumes growth (Lancini et al. 1995).

The sensitivity of microorganisms to antibiotics varies. Gram-positive bacteria are usually more sensitive to antibiotics than gram-negative bacteria, although some antibiotics act only on gram-negative species (Greenwood 1997). An antibiotic that kills or inhibits the growth of many types of bacteria is called a *broad-spectrum* antibiotic. A broad-spectrum antibiotic will generally find wider medical usage than a *narrow-spectrum* antibiotic, which is one that acts only on a single group of organisms.

The majority of antibiotics in human and veterinary use as antibacterial agents are "natural products," elaborated as secondary metabolites by living organisms, primarily bacteria and fungi (Bennet and Bentley 1989; Davies 1990; Vining 1990). Most antibiotics are the products of the secondary metabolism of three main groups of microorganisms: actinomycetes, eubacteria, and filamentous fungi (Greenwood 1995b). The actinomycetes produce the largest number and greatest variety of known antibiotics, with more than six thousand active substances isolated from them.

TARGETS AND MECHANISMS OF ACTION OF ANTIBIOTICS.

Most antibiotics are directed against some target that is peculiar to bacteria, interfering with some structure or process that is essential to bacterial growth, survival, or both, while causing little or no harm to the eukaryotic host harboring the infecting bacteria (Betina 1983). Antibiotics block the growth of sensitive bacteria by inhibiting the action of a molecule, usually a macromolecule, such as an enzyme or a nucleic acid, essential for cell multiplication. At the molecular level, this means that the antibiotic molecule is able to bind to a specific site on the target macromolecule, forming a molecular complex that is no longer able to accomplish its original function (Lancini et al. 1995).

Determining the mechanism of action of an antibiotic requires identification of the target molecule and its function. It is usually easier to identify the *function* that is blocked rather than the particular macromolecule involved. The four proven principle targets for the main classes of antibiotics are as follows: (1) bacterial cell-wall biosynthesis (peptidoglycan); (2) bacterial protein synthesis (bacterial ribosomes); (3) bacterial DNA replication and repair (bacterial enzymes involved in DNA supercoiling); and (4) cytoplasmic membrane function (Walsh 2000). A few antibiotics are antimetabolites, acting as competitive inhibitors and mimicking important growth factors needed in cell metabolism (O'Grady et al. 1997).

Various schemes for classification of antibiotics have been proposed, although none have been universally adapted. In general, antibiotics and other chemotherapeutic agents are grouped together based either on mechanism of action or, more usually, chemical structure (Lancini et al. 1995). This classification is useful in practice, because the components of one class usually share many biological properties. The major classes of antibiotics include: (1) β-lactam antibiotics (penicillins and cephalosporins), which inhibit cell wall biosynthesis; (2) glycopeptides, which inhibit cell wall biosynthesis; (3) tetracyclines, which inhibit protein synthesis by binding to the 30S ribosomal subunit; (4) aminoglycosides, which inhibit protein synthesis by binding to the 30S ribosomal subunit, (5) macrolides and lincosamides, which inhibit protein synthesis by binding to the 50S ribosomal subunit, (6) quinolones, which inhibit DNA replication; and (7) miscellaneous drugs, such as the antitubercular drug rifampin, which inhibits DNA-directed RNA polymerase, and chloramphenicol and analogues, which inhibit protein synthesis by binding to the 50S ribosomal subunit (Greenwood 1995b; Lancini et al. 1995; O'Grady et al. 1997).

ANTIBIOTIC RESISTANCE.

Microbes display a truly amazing versatility in terms of their ability to avoid, withstand, or repel the antibiotic onslaught (Jacoby and Archer 1991; Neu 1992; Courvalin 1996). Often, the use of antibiotics disturbs the delicate bacterial ecology within the body of humans as well as animals, allowing the proliferation of resistant species and sometimes initiating new infections that are worse that the one originally treated (Levy 1997). The historical cycle that we have witnessed in the last fifty years is that drugs were discovered, and diseases were supposedly conquered, only to re-emerge resistant to chemotherapeutic treatment. Antibiotic drug resistance allowed diseases such as cholera, bacterial meningitis, tuberculosis, pneumonia, and even plague to spring back with a renewed vengeance (Fraimow and Abrutyn 1995; Caputo et al. 1993; Doern et al. 1996; Neu 1992). Among hospital- and community-acquired infections from organisms including *Pseudomonas aeruginosa, Acinetobacter baumanii,* and vancomycin-resistant enterococci (VRE), there are now multidrug-resistant isolates, which means that some individuals may contract infections for which only one drug treatment, or in some cases none, is available (Levy 1998b).

Antibiotic resistance in clinical bacterial isolates has been recorded since these agents first came into use. The first antibiotic resistance mechanism was identified in 1940 when Abraham and Chain described the presence of a penicillinase, an enzyme that inactivates penicillin, in resistant *Escherichia coli* (Abraham and Chain 1940). Close on the heels of that mechanism, a similar penicillin resistance was reported in 1944 in an isolate of *Staphylococcus aureus* (Kirby 1944). Although the enormity and complexity of the current problem of antibiotic resistance could hardly have been foretold in the early days of antibiotic use, even Alexander Fleming was able to recognize the threat and factors that would promote its appearance (Fleming 1929).

The term *antibiotic resistance* is often used in a general sense to signify the lack of effect of an antibiotic agent on a bacterial cell. A commonly accepted definition states that a bacterial strain derived from a species that is susceptible to an antibiotic is said to be *resistant* when it is no longer inhibited by the minimal concentration of the antibiotic that inhibits the growth of typical strains of that species (Greenwood 1995). The resulting resistance is dependent on the relationship among (1) a given bacterial strain, (2) the particular antibiotic, and (3) the concentration of that antibiotic (Lancini et al. 1995). However, in practice, the decision to label a given bacterial isolate *sensitive* or *resistant* depends ultimately on the likelihood of a infection responding to treatment with a given antibiotic. A bacterial strain is considered truly resistant to a given antibiotic only if it can grow in the presence of a concentration equal to or greater than that which the antibiotic can reach in the serum or tissues (Towner 1995).

THE GENETIC BASIS OF ANTIBIOTIC RESISTANCE. Bacterial resistance to antibiotics is essentially of two main types: *intrinsic resistance,* a natural property of the bacteria, or *acquired resistance,* which results from mutation or acquisition of new genetic material.

Intrinsic resistance refers to bacteria that are insensitive, in their natural state, to an antibiotic without acquiring resistance factors. Not all bacterial species are intrinsically sensitive to all antibiotics. The most obvious determinant of intrinsic resistance is the absence or inaccessibility of the target for the drug's action. For example, the lipopolysaccharide (LPS) outer envelope of Gram-negative bacteria is important in determining resistance patterns, because numerous, hydrophobic antibiotics cannot penetrate this barrier to reach their intracellular target (Brody et al 1994).

Acquired resistance in bacteria occurs when a bacterium that has been sensitive to antibiotics develops resistance. This may happen by mutation or by acquisition of new DNA. Mutation is a spontaneous event that occurs regardless of whether an antibiotic is present. Changes in only a few base pairs, causing substitution of one or a few amino acids in a crucial target (for example, enzyme, cell structure, or cell wall) may lead to new resistant strains of bacteria (Sanders and Wiedemann 1988). In any large population of bacterial cells a few individual cells may spontaneously become resistant. Such "resistant" cells have no particular advantage in the absence of an antibiotic, but following the introduction of treatment with an antimicrobial agent, all sensitive bacterial cells will be killed, so that the (initially) very few resistant cells can proliferate until they form a completely resistant population. Many antibiotics have been shown to select, both *in vitro* and *in vivo,* for this type of acquired resistance in many different bacterial species (Towner 1995).

By far, the much more prevalent type of acquired resistance occurs when genes conferring antibiotic resistance transfer from a resistant bacterial cell to a sensitive one. Bacteria can acquire these foreign genes by three main mechanisms: transformation, transduction, and conjugation (Mazodier and Davies 1991). These foreign genes can provide a resistant version of the normal cellular target, an additional drug efflux pump, or enzymes that modify the antibiotic, rendering it inactive (Salyers and Amabile-Cuevas 1997). Many of the antibiotic resistance genes are carried on plasmids, transposons, or integrons that can act as vectors that transfer these genes to other members of the species, as well as to another genus or species of bacteria (Ochman et al. 2000).

Bacteria exploit a variety of mechanisms to combat antibiotics. These strategies include the following: limiting the intracellular concentrations of the antibiotic by decreased influx or increased efflux of the drug; modification or neutralization of the antibiotic by enzymes that reversibly or irreversibly inactivate the drug; alteration of the "target" of the antibiotic so that the drug no longer interferes with it; and elimination of the target altogether by the utilization of different metabolic pathways. Bacteria may use or combine multiple mechanisms against a single agent or class of agents, or a single change may result in development of resistance to several different agents.

MECHANISMS OF GENE TRANSFER BETWEEN BACTERIA. Probably the most common way resistance genes are picked up is by conjugation, which forms a "bridge" between two bacteria through which genetic material can be transferred from one to another by cell-to-cell contact. A number of conjugative elements and types of mobile DNA, such as plasmids, transposons, and integrons, can be transferred in this fashion. Conjugation was first described in 1946 by Lederberg and Tatum, based on studies showing that the intestinal bacteria *E. coli* uses a process resembling sex to exchange the circular, extrachromosomal elements, now known as plasmids (Lederberg and Tatum 1946). Plasmids were not linked to transferable drug resistance (R-factors) until the early 1960s in a retrospective study describing the emergence of multiple antibiotic resistance in clinical isolates of *Shigella* from 1952–1955 (Wantanbe and Fukasawa 1961). The same resistance patterns were observed in many strains of different serotypes of *Shigella,* as well as in *E. coli* and *Citrobacter freundii* strains isolated from the same patients, ruling out the possibility of clonal selection. The resistance could also be transferred to sensitive strains by cell-to-cell contact, but not when a filter (permeable to DNA and bacteriophages) separated the bacteria. Conjugative plasmids were present in pathogenic clinical isolates of bacteria, such as *Shigella*, *E. coli, Klebsiella,* and *Proteus* (from the Murray collection) prior to the use of antibiotics, but they did not carry resistance genes (Hughes and Datta 1983). The evolution of R-plasmids, some of which carry as many as seven different resistance determinants, is a recent evolutionary event, selected for by the intensive use of antibiotics in clinical therapy (Davies 1997).

VECTORS OF ANTIBIOTIC RESISTANCE.
Development of antibiotic resistance can occur through the acquisition of foreign genes that reside on mobile genetic elements.

Plasmids. Plasmids are one of the key players in the team of mobile genetic elements that fuel bacterial adaptability and diversity, and plasmids are capable of promoting the rapid spread of antibiotic resistance genes. By definition, a plasmid is a unit of extra-chromosomal inheritance (Thomas 2000). Most plasmids are self-replicating, circular molecules of DNA that are maintained in a bacterial cell in a stable manner. Some plasmids also have alternative lifestyles. They can exist in the autonomous extra-chromosomal state or they can be inserted into the bacterial chromosome and then be carried as part of it.

Natural plasmids vary considerably in size, copy number, host range, and the various genetic traits conferred to their hosts (resistance to antibiotics and heavy metals, virulence factors, and so on) (Top et al. 2000). Whatever the size of a plasmid, however, its genes and the sequences required for its replication and control are usually clustered in a small region called the replicon (Espinosa et al. 2000). Generally, basic replicons do not exceed 3 kb and consist of: (1) an origin of replication (*ori*); (2) *cop/inc* genes(s) involved in regulating the initiation of replication; and in most cases (3) rep gene(s), encoding Rep proteins required for replication and that often participate in the replicon's control (Espinosa et al. 2000).

Plasmids, as self-replicating and mobile genetic elements that are separate from the chromosome, generally provide accessory, but not essential, functions to their hosts. In particular, traits that confer adaptations to locally restrictive conditions tend to cluster on plasmids (Eberhard 1989). Therefore—and in spite of any energetic drain imposed upon host cell metabolism—plasmids can be viewed as desirable elements for host cells. They provide a mechanism for carriage of functions that can be required for survival under conditions of environmental stress, but that are dispensable at other times. From the plasmid's selfish point of view, it makes sense to provide selectable functions to potential host cells, because doing so can enhance host competitiveness and thus host and plasmid survivability in cases of selective pressure in the environment (van Elsas et al. 2000).

Often it has been convenient to divide plasmids into two main groups: narrow and broad host range or promiscuous plasmids. Broad host range plasmids are defined as those for which replication is not restricted to one particular species of bacteria, or more typically, a plasmid that can replicate in many of a selected group of host species (Valla 1998; Szpirer et al. 1999). Broad host range plasmids may further be subdivided into two subgroups: conjugative or self-transmissible, and non-conjugative but mobile. These self-transmissible, broad host range plasmids are possibly the most active vehicles for a potential "horizontal gene pool" of antibiotic resistance genes that are available to many bacteria, of many species and even families (Thomas 2000).

In many cases, when a gene, such as one encoding antibiotic resistance, moves onto a plasmid, its copy number per bacterium rises, increasing the chance for mutations to show up in one of the copies. Plasmids that are self-transmissible or mobilizable increase the chance of antibiotic resistance genes moving between bacteria, and they remove the need for an antibiotic resistance gene to integrate into the bacterial chromosome in order to become established in its new host. Furthermore, most plasmids can replicate within the species of at least one genus so that they can easily spread between the species of that genus.

Transposons. Antibiotic resistance genes on plasmids or in the bacterial chromosome are often located on *transposons,* genetic elements capable of moving from one DNA molecule to another, independent from the normal *recA*-dependent type of recombination. Following the discovery of R-factors, many antibiotic resistance genes were found to be located on small pieces of DNA that can independently move around. The original concept of a DNA segment capable of physical movement inside a genome dates back to the early 1970s, when Hedges and Jacob (1971) proposed that the gene associated with resistance to ampicillin was part of a mobile genetic entity they named a transposon. This was, in fact, the end of a long story that started with the obscure, and ridiculed, work of Bar-

bara McClintock on the variegation of maize kernels (McClintock 1951).

Transposons are linear pieces of DNA that range in size from 2.5 to 23 kilobase pairs (kb) and contain two identical insertion sequences (IS elements), one at each end of the molecule. The insertion sequences act as a point of insertion into a new molecule, as well as contain the transposase enzyme required for transposition (Kleckner 1981). Different genes, such as toxin genes and antibiotic resistance genes, can be located between the IS elements. The IS sequences of some transposons (such as Tn5) are themselves capable of independent movement, not carrying with them the antibiotic resistance genes (Berg and Berg 1983). These IS elements therefore are sometimes found not only flanking resistance genes but also inserted into the bacterial chromosome at random sites. This finding suggests that transposons, such as Tn5, may have evolved by the movement of two IS elements to flank an antibiotic resistance gene, thus forming a composite unit (Berg et al. 1984). A *composite transposon* is a modular unit consisting of an antibiotic resistance gene(s) flanked by two IS elements.

Transposons appear to be ubiquitous in nature and have been identified in many types of organisms, including plants and animals, as well as bacteria (Berg and Howe 1989; Sherrat 1989). There are several classes of transposons, grouped according to their general structural and functional organization, using features such as size, conserved DNA regions, number of open reading frames (ORFs), presence of host genes, and particularly the transpositional pathway (Mahillon 1998). Some transposons, such as *Tn10*, which encodes resistance to tetracycline, move in a "conservative" manner, without replication (Kleckner 1989). Other transposons, however, copy themselves to a new location while maintaining the original location; the transposition includes replication of the transposon (Lupski 1987). Conjugative transposons, found in many gram-positive and a few gram-negative bacteria, are integrated DNA segments that excise from the chromosome to form a circular intermediate. The circular intermediate then transfers itself to a recipient, where it integrates once again into the chromosome (Salyers and Shoemaker 1997). Conjugative transposons resemble conventional transposons in that they integrate into DNA, but the mechanism of integration is completely different from that used by "regular" transposons. Unlike regular transposons, conjugative transposons regenerate the DNA segment from which they are excising and form a circular intermediate that is self-transmissible by conjugation. Conjugative transposons resemble plasmids in that they are transferred by conjugation and have a circular intermediate, but the circular intermediate does not replicate (Salyers and Shoemaker 1997).

Conjugative transposons, for example, *Tn916*, transfer not only among species within the gram-positive or within the gram-negative group but also between gram-positive and gram-negative bacteria (Bettram et al. 1991; Scott et al. 1988; Scott and Churchward 1995). As can self-transmissible plasmids, conjugative transposons can mediate the transfer of other DNA. They can mobilize co-resident plasmids, and some can mediate the transfer of linked segments of chromosomal DNA (Salyers et al 1999). Unlike phages and some plasmids, most conjugative transposons do not exclude closely related elements unless they have a single site that is no longer available after it is filled, and allows a bacterial strain to acquire more than one conjugative transposon. Some evidence also exists that related conjugative transposons increase each other's transposition frequencies when they are present in the same strain (Salyers and Shoemaker 1997).

Integrons. Although plasmids and transposons play a significant role as vectors in the acquisition and dissemination of antibiotic resistance genes, we now know that other routes are available for the incorporation of resistance genes into the bacterial genome. Detailed genetic analysis revealed that the regions that flank antibiotic resistance genes in many transposons and several broad host range plasmids of gram-negative bacteria often show a high degree of similarity (Ouellete and Roy 1987; Sundstrom et al. 1988; Cameron et al. 1986; Wiedermann et al. 1987; Hall and Vockler 1987). This realization led to the discovery of a new mobile genetic element, dubbed the *integron* (Stokes and Hall 1989). The term *integron* was originally coined to describe the group of apparently mobile elements that contain one or more antibiotic resistance genes located at a specific site, as well as contain the determinants of the site-specific recombination system responsible for insertion of the resistance genes (Stokes and Hall 1989).

Integrons are "natural" vectors, gene-capture and expression systems (Ploy et al. 2000c), for exogenous DNA that incorporate open reading frames and convert them into functional genes. The integron vector codes for an integrase gene (*intI*) that mediates recombination between a proximal primary recombination site (*attI*) and a secondary target called an attC site (or 59 base pair element). The *attC* site is normally found associated with a single open reading frame (ORF), and the *attC*-ORF structure is termed a gene cassette (Hall and Collis 1995). Insertion of the cassette at the *attI* site, which is located downstream of a resident promoter internal to the *intI* gene, drives expression of the encoded proteins (Levesque et al. 1994). Most of the *attC* sites of the integron's cassettes identified to date are unique. Their length and sequence vary considerably (from 57 to 141 base pair, or bp), with similarities primarily restricted to their boundaries. Two consensus sequences of 7 bp frame the boundaries of each *attC* site and are designated the "core site" and "inverse core site." These are the actual targets of the recombination process (Hall et al. 1991).

More than sixty different antibiotic resistance genes, covering many of the antimicrobial drugs presently in use, have been characterized in cassette structures (Mazel and Davies 1998; Mazel et al. 2000; Nield et al. 2001), as well as genes conferring resistance to disinfectants and heavy metals, such as mercury. Open reading frames encoding unknown functions, but associated with *attC* sites in the form of cassettes, have also been described among several integrons (Recchia and Hall 1995).

At least eight distinct classes of chromosomal and plasmid-borne integrons have been identified, based on sequence differences between the integrases they encode (Nield et al. 2001). Classes 1, 2, and 3 are the best studied and all are implicated in antibiotic resistance (Recchia and Hall 1995).

Class 1 includes the majority of integrons found in clinical isolates to date (Stokes and Hall 1989). These elements constitute the most intensively studied integrons and are the only group for which gene cassette movement has been demonstrated experimentally (Hall and Collis 1995). The integrase of class 1 integrons, *intI1*, is a 337 amino acid protein (Ploy et al. 2000a, 2000c). The structure of the class 1 integrons includes a 5′ and 3′ conserved segment and a variable region (Levesque and Roy 1993; Liebert et al. 1999; Paulsen et al. 1993; Stokes and Hall 1989). The 5′ conserved segment consists of the integrase gene, *IntI1*, and a promoter region for expression of the inserted gene cassette(s) (Stokes and Hall 1989). Most class 1 integrons have a 3′ end region containing three open reading frames. The first, *qacE*Δ1, is a truncated derivative of the *qacE* gene, conferring resistance to quaternary ammonium compounds (Paulsen et al. 1993). The second open reading frame is the *sul1* gene, which encodes resistance to sulfonamides (Stokes and Hall 1989). The third open reading frame, ORF5, does not code for any known function. The variable region, located between the two conserved segments, is the site for the insertion of the antibiotic gene cassettes (Gravel et al. 1998; Hansson et al. 1997; Stokes et al. 1997).

Class 2 integrons are present on transposon Tn7 and related elements. The *intI2* integrase gene encodes a protein exhibiting 46 percent homology with *intI1*, but truncated by 12 amino acids (Ploy et al. 2000b). The 3′ end region of class 2 integrons is composed of genes involved in the transposition mechanism of Tn7 (Recchia and Hall 1995). Class 2 integrons contain a dihydrofolate reductase gene cassette (Fling and Richards 1983).

A third class of integrons, whose integrase *IntI3* exhibits 61 percent identity with *IntI1*, was described in 1995 (Arakawa et al. 1995) in *Serratia marcescens*. Only one integron of class 3 has been described (Arakawa et al. 1995) in a transposon-like element with several gene rearrangements (Shibata et al. 1999). This integron carried the *bla*$_{imp}$ gene, which encodes broad-spectrum β-lactam antibiotic resistance.

A fourth class of integron, dubbed the "super-integron", has been identified in several *Vibrio* species, including *V. cholerae* (Mazel et al. 1998) and *V. metschnikovii* (Rowe-Magnus et al. 1999). This chromosomal, super-structure harbors hundreds of gene cassettes, and the encoded functions, when identifiable,

are linked to adaptations extending beyond antibiotic resistance and pathogenicity (Mazel et al. 1998; Heidelberg et al. 2000). The cassette-associated *attC* sites of class 4 integrons, termed VCRs, for *Vibrio cholerae* repeats, display a high degree of sequence relatedness (Rowe-Magnus et al. 1999) and the activity of the associated integrase is identical to that of the class 1 integrase, *intI1*. The integrase enzymes found in both these species of *Vibrio*, *IntI4* and *IntI6* respectively, apparently insert repeated sequences of genes in clusters that mirror the gene cassette arrays typically found in other integrons (Rowe-Magnus et al. 1999; Mazel et al. 1998). Equivalent integron superstructures have now been identified in nine distinct bacterial genera (Rowe-Magnus et al. 2001). The partial sequence for a fifth *intI* gene, *intI5*, associated with *Vibrio mimicus*, has been reported as having 75 percent identity with *intI4* (Clark et al. 2000). It is unclear whether this represents a new integron class or another example of a super-integron.

Finally, a very recent study recovered integron sequences from environmental DNA samples (Nield et al. 2001). The sequence diversity of nearly complete integron sequences in these samples was sufficient to classify them as belonging to three previously undescribed classes of integrons (Nield et al. 2001).

The integron/gene cassette system provides a simple mechanism for the acquisition and dissemination of new antibiotic resistance genes by existing bacterial genomes. Although integrons themselves are defective for self-transposition (Brown et al. 1996; Rowe-Magnus and Mazel 1999), they are often found associated with insertion sequences, transposons, and/or conjugative plasmids that can serve as vehicles for their intra- and interspecies transmission (Rowe-Magnus et al. 1999; Liebert et al. 1999). The potency of such a highly efficient gene-capture and expression system, in combination with broad host range mobility, is obvious. Indeed, the strength of this partnership is confirmed by the wide spread occurrence of integrons in nature, in human, animal, and environmental isolates (Nield et al. 2001; Goldstein et al. 2001; Bass et al. 1999; Chang et al. 1997, Schnabel and Jones 1999; Daly et al. 2000; Lucey et al. 2000). The fact that horizontal and vertical transfer readily occurs is demonstrated by the presence of many common gene cassettes among the *Enterobacteriaceae* and *Pseudomonas*, as well as the marked differences in codon usage among cassettes in the same integron, indicating that the antibiotic resistance determinants are of diverse origin (Rowe-Magnus and Mazel 1999).

The integron/gene cassette system also allows bacteria to stockpile exogenous DNA, such as antibiotic resistance genes, and researchers believe that it has played an important role in the development of multiple drug resistance. The finding of super-integrons with gene cassettes coding for other determinants, such as biochemical functions and virulence factors (Rowe-Magnus et al. 1999), implies a role for these structures in bacterial genome evolution before the antibiotic era.

Integrons and super-integrons may represent a means for rapid adaptation to unpredictable changes in the environment by allowing bacteria to scavenge foreign genes that may ultimately endow them with increased fitness.

Selection for Antibiotic Resistance: The Role of Antibiotics. Bacteria become resistant to antibiotics either by mutations or by acquisition of a "pre-existing" resistance gene (or genes) by horizontal transfer from another organism. Prior to 1950, scientists were divided into two camps: those who believed resistance to be the result of post-treatment adaptation by the microbes, that is, that bacteria were "trained" to withstand the "poisonous" antibiotics (Hinshelwood 1946); and those who believed that random mutations were responsible for the development of all types of microbial diversity. The matter was solved in 1952 by a publication that introduced the technique of *replica-plating*, which allowed the study of mutations independent from selection pressures (Lederberg and Lederberg 1952). The experiment in this study proved that streptomycin-resistant *E. coli* could be isolated and propagated without the selection pressure from streptomycin. This result led to the important realization that mutations, as well as gene transfer of resistance determinants, happens independent of antibiotic use.

After a mutation exists in, or an antibiotic resistance gene is acquired by, a bacterium in the population and antibiotics are then introduced, the condition exists for resistant bacterium to become predominant in the population. Emerging clones of resistant bacteria actually are the result of selection by antibiotics (Levy 1992b, 1998a). Antibiotic use selects, and promotes the evolution and growth of, bacteria that are resistant to that drug. The selection process is fairly straightforward and the scenario unfolds as follows. When bacteria are exposed to an antibiotic, bacterial cells that are susceptible to the drug will die. But cells that have some resistance from the start, or that acquire it later (through mutation or gene exchange), may survive and grow, especially if too little drug is given to overwhelm every last cell that is present. Those cells, facing reduced competition from the susceptible bacteria, will then go on to proliferate. When confronted with the antibiotic, the most resistant cells in a group will inevitably out-compete all others. Over time, such antibiotic-resistant populations can dominate a localized environment, be it the gut of an animal or human. Recognizing that antibiotics select for the growth and dissemination of otherwise rare, resistant microorganisms, this ability of antibiotics has been termed "the antibiotic paradox" (Levy 1992b): that antibiotics meant to overwhelm bacteria in a clinical infection can actually select for the survivors that then thwart their efficacy.

It has also been argued that resistance cannot be fully explained by antibiotics selecting antibiotic-resistant organisms, because they are in part (the bacteria, that is) only the side effects of the evolution of subcellular entities that infect microorganisms and spread resistance genes (Heinemann 1998). This is akin to the "self-ish gene" theory (Dawkins 1976): the selection is not necessarily for resistant bacteria, but rather for the vectors that carry the resistance genes. Most resistance genes are mobile, moving from organism to organism as part of horizontally mobile elements, conjugative plasmids, transposons, and integrons, or flow between organisms through processes such as natural transformation (Lorenz and Wackernagel 1994; Baquero et al. 1998; Souza and Eguiarte 1997). The evolution of these horizontally mobile elements, because of their autonomous reproduction, can be very different from the "cellular life" carrying such elements (Souza and Eguiarte 1997).

Antibiotics also have been shown to specifically effect the development and evolution of resistance in bacteria in numerous ways. Some antibiotics are known to enhance the antibiotic resistance gene transfer between bacteria in humans and animals (Salyers and Shoemaker 1996) and influence the frequency of such events *in vitro* (Davies and Wright 1997). Antibiotics have been shown to increase gene transfer frequency in the laboratory by reducing the effectiveness of the cell surface as a barrier to the release and uptake of genetic material or by making the bacteria susceptible to fusion with other bacteria and vesicles. For example, antibiotics have been shown to influence DNA uptake by weakening the bacterial cell wall and making cells more permeable to DNA, and/or decreasing the concentration of periplasmic nucleases (Vaara 1992; Dreiseikelmann 1994). β-lactam antibiotics have been shown to increase the frequency of interspecies DNA transmission between combinations of *E. coli*, *Staphylococcus aureus*, *Listeria moncytogenes*, *Streptococcus faecalis* and *Bacillus anthracis* (Ivins et al. 1988; Trieu-Cuot 1993).

Some antibiotics induce the genes controlling horizontal transfer by acting on the vectors themselves, and consequently, on their antibiotic resistance genes. For example, levels of the antibiotic regulate transmission of some conjugative transposons. Tetracycline stimulates transmission of a transposon that carries tetracycline resistance and is native to the gram-negative *Bacterioides*, by 100–1,000 fold (Salyers et al. 1995). The transmission of Tn925, a conjugative transposon that is native to many gram-positive bacteria and which, like the *Bacteriodes* transposon, encodes tetracycline resistance, is also responsive to the drug. The transmission of this transposon to antibiotic-sensitive bacteria was enhanced 5–100 fold following culture of the resistant host in tetracycline (Torres et al. 1991).

Although antibiotics may effectively halt bacterial reproduction, they rarely inhibit the metabolism necessary for accepting and distributing the genes on plasmids. Such cells, called *dead vectors* because the antibiotic prevents the cells from dividing, nonetheless may sometimes remain active in the process of conjugation via plasmid-mediated transfer of genes (Heinemann 1998). For example, bacteria harboring plasmids that contain antibiotic resistance genes may continue to transfer the plasmid long after the "donor" bacterium has been killed. Mitomycin-C at some doses irreversibly damages DNA, but bacteria killed by these agents can still receive plasmids and then redistribute them (Heinemann 2000). Thus, it is possible that an antibiotic may convert a bacterial pathogen into a vector that is undetectable by conventional culture assays, but may persist in the patient or any other local environment.

Some antibiotics are able to induce resistance to other drugs. For example, induction or constitutive expression in multiple-antibiotic resistance operator (*marO*) or repressor (*marR*) mutants confers cross-resistance upon cells to dissimilar agents by decreasing intracellular concentrations (Miller and Sulavik 1996). The *marO* and *marR* mutants, selected on tetracycline or chloramphenicol, were 1,000 times more likely also to acquire resistance to the structurally unrelated fluoroquinolones, compared with populations exposed first to the fluoroquinolone itself (Cohen et al. 1989).

Some bacteria can adapt their physiology to new environments following the reception of a signal (usually from the new environment) to switch physiological states (Heinemann 1999). By mimicking signals that induce alternative physiological states, some antibiotics appear to induce phenotypic resistance to themselves and other drugs. For example, researchers found that exposing *Pseudomonas aeruginosa* to the aminoglycoside antibiotic gentamycin for brief periods of time induces resistance to the drug among those few that survive the initial exposures (Karlowsky et al. 1997). Depriving the bacteria of Mg2+ can also induce the resistance phenotype. Gentamycin-resistant cells concomitantly display resistance to other toxic agents, such as netilmicin, tobramycin, neomycin, kanamycin, streptomycin, polymyxin, and EDTA (Karlowsky et al. 1997). That this is a physiological rather than mutational adaptation is inferred from the uniform appearance of the phenotype, after induction by either Mg2++ starvation or gentamycin, and then its uniform reversal in descendent generations. Although the mechanism of physiological adaptation is not known, it may be associated with a secondary effect that gentamycin has on the pathway for drug uptake.

Antibiotics can also act to increase recombination and mutation rates. Drugs that cause DNA damage, such mitomycin-C, can directly elevate the frequency of mutation (Lorenz and Wackernagel 1994). Antibiotics that affect translational fidelity can also boost the mutation rate in bacteria. Mutants arose more frequently during culture of wild-type strains of *E. coli* in the presence of streptomycin, which decreases translational fidelity, than during culture in the absence of streptomycin (Boe 1992). The antibiotic was not a mutagen *per se*, because the mutation rate of strains resistant to streptomycin was independent of culture conditions. Theoretically, by stimulating the basal mutation rate, antibiotics increase the probability that a new resistance determinant will arise in a population. If the mutant gene is on some type of horizontally transmissible vector, the gene has the potential to be disseminated even independently of the survival of the host cell.

As a final footnote, antibiotics may be (or may have been) involved in the development of antibiotic resistance by virtue of actually being the source of antibiotic resistance genes. A study by Webb and Davies (1993) found that a number of commercial antibiotic preparations were contaminated with DNA sequences containing antibiotic resistance genes from the organism used in their production. Genes encoding resistance to antibiotics have inadvertently been co-administered with antibiotics for years and under the simultaneous selection pressure of the antibiotic, uptake of one or more resistance genes by a bacterium in the host could lead to antibiotic resistant organisms being constructed by natural "genetic engineering." Subsequent inter- and intraspecific genetic transfers would permit other bacteria to become resistant to the antibiotics also.

THE COST OF RESISTANCE. Conventional wisdom in the older evolutionary and microbiological literature assumed that bacteria (and other microorganisms) must pay a metabolic or physiologic price for the acquisition of antibiotic resistance, be it by mutation or carriage of an accessory element, such as a plasmid. In an environment that contains an antibiotic, possession of a corresponding resistance gene is clearly beneficial to a bacterium and worth the cost. It was thought, however, that in the absence of the antibiotic, resistant genotypes may have reduced growth rates and be at a competitive disadvantage compared to their sensitive counterparts. It was believed that the frequency and rates of ascent and dissemination of antibiotic resistance in bacterial populations was directly related to the volume of antibiotic use and inversely related to the cost that resistance imposes on the fitness of the bacteria (Andersson and Levin 1999).

A number of older studies have indeed shown that resistant genotypes are less fit than their sensitive progenitors in an antibiotic-free medium, indicating a cost to the mechanisms of resistance. The principle element of these studies involved competition between sensitive and resistant genotypes that are otherwise isogenic (Lenski 1997). Some of these studies demonstrated costs associated with the carriage of plasmids and expression of plasmid-encoded resistance functions (Zund and Lebek 1980), whereas other studies have demonstrated the side effects of mutations that impair growth (Jin and Gross 1989).

These older studies were performed, however, by putting an antibiotic resistance gene (either a plasmid-encoded function or a chromosomal mutation) into "naïve" bacteria, which have had no evolutionary history of association with the resistance genes (Lenski 1997). Evolutionary theory suggests that bacteria might overcome the cost of resistance by evolving adaptations that counteract the harmful side effects of resistance genes.

Several recent laboratory studies have now shown that although most resistance genes do engender some fitness cost, at least initially, the cost of resistance may be substantially diminished, even eliminated, by evolutionary changes in bacteria over rather short periods of time.

Bouma and Lenski (1988) found that growth of *E. coli* containing the drug resistance plasmid pACYC184 for 500 generations, with selection for maintenance of the plasmid, resulted in adaptation of the host to eliminate the cost of plasmid carriage. The "adapted" *E. coli* host containing pACYC184 had a competitive advantage over the original plasmid-free *E. coli* host, even when grown in the absence of the antibiotic. Selection for increased fitness during long-term growth resulted in a strain that actually grows better with pACYC184 than without it. A similar study examined the co-evolution of *E. coli* and a derivative plasmid pBR322 that encodes resistance to ampicillin and tetracycline (Modi et al. 1991). After approximately 800 generations, Modi et al. found that the cost of plasmid carriage to the bacterial host had been significantly reduced. Bacteria that inhabit an animal host also evolve to compensate for the cost of resistance. A study by Bjorkman et al. (2000) showed that not only do bacteria in animal hosts adapt to the cost of resistance through secondary mutations that compensate for the loss of fitness, but that in Salmonella, the process of adaptation to the costs of resistance are different, depending on whether the bacteria "grew" in mice or in culture broth.

SUMMARY AND CONCLUSIONS. Antibiotics are the wonder drugs of the 20th century. These drugs, which were first used on a large-scale basis in 1944, continue to lose their efficacy in treating bacterial diseases because of emerging resistance in bacteria. Bacteria have adapted various mechanisms to combat antibiotics including alteration or destruction of the drug, modification of drug's target, or efflux of the drug. Microorganisms develop resistance through antibiotic selection of microbes that have acquired resistance either through spontaneous mutation(s) or through the acquisition of foreign, antibiotic resistance gene(s). Physiological changes also can alter the organism's susceptibility to these drugs. The widespread dissemination of antibiotic resistance among diverse microorganisms is attributed to mobile genetic elements, plasmids, integrons, and transposons. Certain plasmids and transposons are self-transmissible, genetic elements with narrow to broad-host range and are an important factor in the transmission of antibiotic resistances. Together, plasmids, integrons and transposons are key factors in the evolution, rapid development and spread of antibiotic resistance to diverse microorganisms. Once believed to reduce the organism's fitness, there is evidence to support the idea that bacteria overcome the cost of resistance through compensatory mutations and may explain why antibiotic resistance persists in the absence of drug selection pressure.

REFERENCES

Abraham, E.P., and E. Chain. 1940. An enzyme from bacteria able to destroy penicillin. *Nature* 146:837.

Andersson, D.I., and B.R. Levin. 1999. The biological cost of antibiotic resistance. *Curr. Opin. Microbiol.* 2:489–43.

Arakawa, Y., M. Murakami, K. Suzuki, H. Ito, R. Wacharotayankun, S. Ohsuka, N. Kato, and M. Ohta. 1995. A novel integron element carrying the metallo-beta-lactamase gene bla_{IMP}. *Antimicrob. Agents Chemother.* 39:1612–1615.

Baquero, F., M.C. Negri, M.I. Morosini, and J. Blazquez. 1998. Antibiotic-selective environments. *Clin. Infect. Dis.* 27 Suppl. 1:S5–S11.

Bass, L., C.A. Liebert, M.D. Lee, A.O. Summers, D.G.T. White, S.G., and J.J. Maurer. 1999. Incidence and characterization of integrons, genetic elements mediating multiple-drug resistance, in avian *Escherichia coli*. *Antimicrob. Agents Chemother.* 43:2925–2929.

Bennet, J., and R. Bentley. 1989. What is a name? Microbial secondary metabolism. *Adv. Appl. Microbiol.* 34:1–5.

Berg, D.E., and C.M. Berg. 1983. The prokaryotic transposable element Tn5. *Biotechnology* 1:417–435.

Berg, D.E., C.M. Berg, and C. Sashawa. 1984. Bacterial transposon Tn5: evolutionary inferences. *Mol. Biol. Evol.* 1:411–422.

Berg, D.E., and M.M. Howe. 1989. *Mobile DNA*. Washington, D.C.: American Society for Microbiology, , 972.

Beringer, J.E. 1998. Concluding remarks. *APMIS Suppl.* 84:85–87.

Betina, V. 1983. *The Chemistry and Biology of Antibiotics*. Amsterdam: Elsevier.

Bettram, J., H. Stratz, and P. Durre. 1991. Natural transfer of conjugative transposon Tn916 between Gram-positive and Gram-negative bacteria. *J. Bacteriol.* 173:443–448.

Bjorkman, J., I. Nagaev, O.G. Berg, D. Hughes, and D.I. Andersson. 2000. Effects of environment on compensatory mutations to ameliorate costs of antibiotic resistance. *Science* 287:1479–1482.

Boe, L. 1992. Translational errors as the cause of mutations in *Escherichia coli*. *Mol. Gen. Genet.* 231:469–471.

Bottcher, H. 1964. *Wonder Drugs: A History of Antibiotics*. New York: J.B. Lippincott Company.

Bouma, J.E., and R.E. Lenski. 1988. *Evolution of a Nature bacteria/plasmid association*. i335:351–352.

Brody, T.M., J. Larner, K.P. Minnerman, and H.C. Neu. 1994. *Human pharmacology*. St. Louis, MO: Mosby-Year Book, Inc..

Brown, H.J., H.W. Stokes, and R.M. Hall. 1996. The integrons In0, In2, and In5 are defective transposon derivatives. *J. Bacteriol.* 178:4429–4437.

Cameron, F.H., D.J.G. Obbink, V.P. Ackerman, and R.M. Hall. 1986. Nucleotide sequence AAD (2″) aminoglycoside adenyltransferase determinant aadB. Evolutionary relationship of this region with those surrounding aadA in R538-1 and dhfrII in R388. *Nucl Acids Res.* 14:8625–8635.

Caputo, G.M., P.C. Applebaum, and H.H. Liu. 1993. Infections due to penicillin-resistant pneumococci: Clinical, epidemiologic, and microbiologic features. *Arch. Intern. Med.* 153:1301–1310.

Chang, L.L., J.C. Chang, C.Y. Chang, S.F. Chang, and W.J. Wu. 1997. Genetic localization of the type I trimethoprim resistance gene and its dissemination in urinary tract isolates in Taiwan. Kaohsiung *J. Med. Sci.* 13:525–533.

Clark, C.A., L. Purins, L. Kaewrakon, T. Focareta, and P.A. Manning. 2000. The Vibrio cholerae 01 chromosomal integron. *Microbiol* 146:2605–2612.

Cohen, S.P., L.M. McMurry, D.C. Hooper, J.S. Wolfson, and

S.B. Levy. 1989. Cross-resistance to fluoroquinolones in multiple-antibiotic-resistant (Mar) *Escherichia coli* selected by tetracycline or chloramphenicol: decreased drug accumulation associated with membrane changes in addition to OmpF reduction. *Antimicrob. Agents Chemother.* 33:1318–1325.

Courvalin, P. 1996. Evasion of antibiotic action by bacteria. *J. Antimicrob. Chemother.* 37:855–869.

Daly, M., J. Buckley, E. Power, C. O'Hare, M. Cormican, B. Cryan, P.G. Wall, and S. Fanning. 2000. Molecular characterization of Irish *Salmonella enterica* serotype Typhimurium: detection of class I integrons and assessment of genetic relationships by DNA amplification fingerprinting. *Appl. Environ. Microbiol.* 66:614–619.

Davies, J. 1990. What are antibiotics? Archaic functions for modern activities. *Mol. Microbiol.* 4:1227–1232.

Davies, J., and G.D. Wright. 1997. Bacterial resistance to aminoglycoside antibiotics. *Trends Microbiol.* 5:234–240.

Davies, J.E. 1997. Origins, acquisition, and dissemination of antibiotic resistance determinants. *Ciba Foundation Symposium* 207:15–35.

Dawkins, R. 1976. *The Selfish Gene.* New York: Oxford University Press.

Doern, G.V., A. Brueggemann, H.P.J. Holley, and A.M. Rauch. 1996. Antimicrobial resistance of *Streptococcus pneumoniae* recovered from outpatients in the United States during the winter months of 1994–1995: results of a 30-center national surveillance study. *Antimicrob. Agents Chemother.* 40: 1208–1213.

Domagk, G. 1935. Ein beitrag zur chemotherapie der bakteriellen infektionen. *Dt. Med. Wochenschrift* 61:250–253.

Dreiseikelmann, B. 1994. Translocation of DNA across bacterial membranes. *Microbiol. Rev.* 58:293–316.

Ebell, B. 1937. *The Papyrus Ebers, The Greatest Egyptian Medical Document.* Levin and Munksgaard, Copenhagen.

Eberhard, W.G. 1989. Why do bacterial plasmids carry some genes and not others? *Plasmid* 21:167–174.

Ehrlich, P., and S. Hata. 1910. *Die experimentelle chemotherapie der Spirillosen,* Berlin.

Espinosa, M., S. Cohen, M. Couturier, G. del Solar, R. Diaz-Orehas, R. Giraldo, L. Janniere, C. Miller, M. Osborn, and C.M. Thomas. 2000. Plasmid replication and copy number control. In C.M. Thomas, ed. *The Horizontal Gene Pool.* Amsterdam: Harwood Academic Publishers, 1–47.

Fleming, A. 1929. On antimicrobial activity of Penicillium cultures. *Br. J. Exper. Path.* 10:226.

Fling, M.E., and C. Richards. 1983. The nucleotide sequence of the trimethoprim-resistant dihydrofolate reductase gene harbored by Tn7. *Nucl Acids Res.* 11:5147–5158.

Goldstein, C., M.D. Lee, S. Sanchez, C. Hudson, A.O. Summers, D.E. White, and J.J. Maurer. 2001. Incidence of Class 1 and 2 integrons in clinical and commensal bacteria from livestock, companion animals, and exotics. *Antimicrob. Agents Chemother.* 45:723–726.

Gravel, A., B. Fournier, and P.H. Roy. 1998. DNA complexes obtained with the integron integrase IntI1 at the attI1site. *Nucl Acids Res.* 26:4347–4355.

Greenwood, D. 1995a. *Antimicrobial Chemotherapy.* New York: Oxford University Press.

Greenwood, D. 1995b. Historical introduction. In D. Greenwood, ed. *Antimicrobial Chemotherapy.* New York: Oxford University Press, 1–10.

Greenwood, D. 1997. Mode of action in antibiotic and chemotherapy. In F. O'Grady, H. P. Lambert, R. G. Finch and D. Greenwood, eds. *Antibiotic and Chemotherapy.* New York: Churchill-Livingstone.

Hall, R.M., D.E. Brookes, and H.W. Stokes. 1991. Site-specific insertion of genes into integrons: role of the 59-base element and determination of the recombination cross-over point. *Mol. Microbiol.* 5:1941–1959.

Hall, R.M., and C.M. Collis. 1995. Mobile gene cassettes and integrons: capture and spread of genes by site-specific recombination. *Mol. Microbiol.* 15:593–600.

Hall, R.M., and C. Vockler. 1987. The region of the IncN plasmid R46 coding for resistance to β-lactam antibiotics, streptomycin/spectinomycin and sulphonamides is closely related to antibiotic resistance segments found in Inc W plasmids and Tn21-like transposons. *Nucl Acids Res.* 15:7491–7501.

Hansson, K., O. Skold, and L. Sundstrom. 1997. Non-palindromic *attI* sites of integrons are capable of site-specific recombination with one another and with secondary targets. *Mol. Microbiol.* 26:441–453.

Hedges, R.W., and A.E. Jacob. 1971. Transposition of ampicillin resistance from RP4 to other replicons. *Mol. Gen. Genet.* 132:31–40.

Heidelberg, J.F., J.A. Eisen, W.C. Nelson, R.A. Clayton, M.L. Gwinn, R.J. Dodson, D.H. Haft, E.K. Hickey, J.D. Peterson, L. Umayam, S.R. Gill, K.E. Nelson, T.D. Read, H. Tettelin, R. D., M.D. Ermolaeva, J. Vamathevan, S. Bass, H. Qin, I. Dragoi, P. Sellers, L. McDonald, T. Utterback, R.D. Fleishmann, W.C. Nierman, and O. White. 2000. DNA sequence of both chromosomes of the cholera pathogen Vibrio cholerae. *Nature* 406:477–483.

Heinemann, J.A. 1998. Looking sideways at the evolution of replicons. In C. Kado and M. Syvanen, eds. *Horizontal Gene Transfer.* New York: Chapman and Hall, 11–24.

Heinemann, J.A. 1999. How antibiotics cause antibiotic resistance. *Drug Discov. Today* 4:72–79.

Heinemann, J.A., R.G. Ankenbauer, and C.F. Amabile-Cuevas. 2000. Do antibiotics maintain antibiotic resistance? *Drug Discov. Today* 5:195–204.

Heinemann, J.A., and P.D. Roughan. 2000. New hypotheses on the material nature of horizontally mobile genes. *N.Y. Acad. Sci.* 906:169–186.

Himmelweit, F. 1960. *The Collected Papers* Paul Ehrlich. New York: Pergamon Press.

Hinshelwood, C.N. 1946. The Chemical Kinetics of Bacterial Cells. Oxford: The Clarendon Press.

Hoel, D., and D.N. Williams. 1997. Antibiotics: Past, present, and future. *Postgrad. Med.* 101:114–122.

Hughes, V.M., and N. Datta. 1983. Conjugative plasmids in bacteria of the pre-antibiotic era. *Nature* 302:725–726.

Ivins, B.E., S.L. Welkos, G.B. Knudson, and D.J. Leblane. 1988. Transposon Tn916 mutagenesis in *Bacillus anthracis. Infect. Immun.* 56:176–181.

Jacoby, G.A., and G.L. Archer. 1991. New mechanisms of bacterial resistance to antimicrobial agents. *New Engl. J. Med.* 324:601–612.

Jin, D.J., and C.A. Gross. 1989. Characterization of the pleiotropic phenotypes of rifampin-resistant rpoB-mutants of *Escherichia coli. J. Bacteriol.* 172:5229–5231.

Karlowsky, J.A., S.A. Zelenitsky, and G.G. Zhanel. 1997. Aminoglycoside adaptive resistance. *Pharmacotherapy* 17:549–555.

Kirby, W.M.M. 1944. Extraction of a highly potent penicillin inactivator from penicillin resistant *staphylococci. Science* 99:452–453.

Kleckner, N. 1981. Transposable elements in procaryotes. *Annu. Rev. Genet.* 15:341–404.

Kleckner, N. 1989. Transposon Tn10. In D. E. Berg and M. M. Howe, eds. *Mobile DNA.* Washington, D.C.: American Association for Microbiology, 227–268.

Koch, R. 1876–1877. Cohn's Beit. z. Pflanzen, Breslau.

Lancini, G., F. Parenti, and G. Gualberto. 1995. *Antibiotics: a multidisciplinary approach.* New York: Plenum Press.

Lederberg, J., and E.M. Lederberg. 1952. Replica plating and indirect selection of bacterial mutants. *J. Bacteriol.* 63:399–406.

Lederberg, J., and E.L. Tatum. 1946. Gene recombination in *Escherichia coli. Nature* 158:558.

Lenski, R.E. 1997. The cost of antibiotic resistance—from the perspective of a bacterium. *Antibiotic Resistance: origin, evolution, and spread.* New York: Wiley,131–151.

Levesque, C., L. Piche, C. Larose, and P.H. Roy. 1994. PCR mapping of integrons reveals several novel combinations of resistance genes. *Antimicrob. Agents Chemother.* 39:185–191.

Levesque, C., and P.H. Roy. 1993. PCR analysis of integrons. In D. H. Persing, T. F. Smith, F. C. Tenover and T. J. White, eds. *Diagnostic Molecular Microbiology.* Washington, D.C.: American Society of Microbiology, 590–594.

Levy, S.B. 1992b. *The Antibiotic Paradox: How Miracle Drugs are Destroying the Miracle.* New York: Plenum Press.

Levy, S.B. 1994. Balancing the drug-resistance equation. *Trends Microbiol.* 2:341–342.

Levy, S.B. 1997. Antibiotic resistance: an ecological perspective. In D. J. Chadwick and J. Goode, eds. *Antibiotic resistance: origins, evolution, and spread.* New York: Wiley, 1–14.

Levy, S.B. 1998a. The challenge of antibiotic resistance. *Sci. Am.* 275:46–53.

Levy, S.B. 1998b. Multidrug resistance—a sign of the times. *N. Engl. J. Med.* 338:1376–1378.

Liebert, C.A., R.M. Hall, and A.O. Summers. 1999. Transposon Tn21, flagship of the floating genome. *Microbiol. Mol. Biol. Rev.* 63:507–522.

Lorenz, M.G., and W. Wackernagel. 1994. Bacterial gene transfer by natural genetic transformation in the environment. *Microbiol. Rev.* 58:563–602.

Lucey, B., D. Crowley, P. Moloney, B. Cryan, M. Daly, F. O'Halloran, E.J. Threlfall, and S. Fanning. 2000. Integronlike structures in *Campylobacter* spp. of human and animal origin. *Emerg. Infect. Dis.* 6:50–55.

Lupski, J.R. 1987. Molecular mechanisms for transposition of drug-resistance genes and other movable genetic elements. *Rev. Infect. Dis.* 9:357–368.

Magner, L. 1992. *A History of Medicine.* New York: Marcel Dekkor.

Mahillon, J. 1998. Transposons as gene haulers. *APMIS Suppl* 84:29–36.

Mazel, D., and J. Davies. 1998. Antibiotic resistance. The big picture. *Adv. Exp. Med. Biol.* 456:1–6.

Mazel, D., B. Dychinco, V.A. Webb, and J. Davies. 1998. A distinctive class of integron in the Vibrio cholerae genome. *Science* 280:605–608.

Mazel, D., B. Dychinco, V.A. Webb, and J. Davies. 2000. Antibiotic resistance in the ECOR collection: integrons and identification of a novel aad gene. *Antimicrob. Agents Chemother.* 44:1568–1574.

Mazodier, P., and J. Davies. 1991. Gene transfer between distantly related bacteria. *Annu. Rev. Genet.* 25:147–171.

McClintock, B. 1951. Chromosome organization and genetic expression. *Cold Spring Harbor Symp. Quant. Biol.* 16:13–47.

Miller, P.F., and M.C. Sulavik. 1996. Overlaps and parallels in the regulation of intrinsic multiple-antibiotic resistance in *Escherichia coli. Mol. Microbiol.* 21:441–448.

Mitsuhashi, S, K. Inoue, and M. Inoue. 1977. Nonconjugative plasmids encoding sulfanilamide resistance. *Antimicrob. Agents Chemother.* 12:418–422.

Modi, R.I., C.M. Wilke, R.F. Rosenzweig, and J. Adams. 1991. Plasmid macro-evolution: selection of deletions during adaptation in a nutrient-limited environment. *Genetica* 84:195–202.

Neu, H.C. 1992. The crisis in antibiotic resistance. *Science* 257:1064–1073.

Nield, B.S., A.J. Holmes, M.R. Gillings, G.D. Recchia, B.C. Mabbutt, K.M. Nevalainen, and H.W. Stokes. 2001. Recovery of new integron classes from environmental DNA. *FEMS Microbiol. Lett.* 195:59–65.

Ochman, H., J.G. Lawrence, and E.A. Groisman. 2000. Lateral gene transfer and the nature of bacterial innovation. *Nature* 405:299–304.

O'Grady, F., H.P. Lambert, R.G. Finch, and D. Greenwood. 1997. *Antibiotic and Chemotherapy: Anti-infective Agents and Their Use in Therapy.* Pp. 987. Churchhill-Livingstone, New York.

Ouellette, M., and P.H. Roy. 1987. Homology of ORFs from Tn2603 and from R46 to site-specific recombinases. *Nucl Acids Res.* 15:10055.

Pasteur, L. 1880. Compt. Rend. Acad. d. sc. Paris xci:86, 455,697.

Paulsen, I.T., T.G. Littlejohn, P. Radstrom, L. Sundstrom, O. Skold, G. Swedberg, and R.A. Skurry. 1993. The 3' conserved segment of integrons contains a gene associated with multidrug resistance to antiseptics and disinfectants. *Antimicrob. Agents Chemother.* 37:761–768.

Ploy, M., T. Lambert, A. Gassama, and F. Denis. 2000a. The role of integrons in dissemination of antibiotic resistance. *Ann. Biol. Clin.* (Paris) 58:439–444.

Ploy, M.C., T. Lambert, J.P. Couty, and F. Denis. 2000b. Integrons: an antibiotic resistance gene capture and expression system. *Clin. Chem. Lab. Med.* 38:483–487.

Recchia, G.D., and R.M. Hall. 1995. Gene cassettes: a new class of mobile element. *Microbiol.* 141:3015–3027.

Rowe-Magnus, D.A., A.M. Guerot, and D. Mazel. 1999. Super-integrons. *Res. Microbiol.* 150:641–651.

Rowe-Magnus, D.A., and D. Mazel. 1999. Resistance gene capture. *Curr. Opin. Microbiol.* 2:483–488.

Salyers, A., N. Shoemaker, G. Bonheyo, and J. Frias. 1999. Conjugative transposons: transmissible resistance islands. In J. P. Kaper and J. Hacker, eds. *Pathogenicity Islands and Other Mobile Virulence Elements.* Washington, D.C.: ASM Press, 331–346.

Salyers, A.A., and C.F. Amabile-Cuevas. 1997. Why are antibiotic resistance genes so resistant to elimination? *Antimicrob. Agents Chemother.* 41:2321–2325.

Salyers, A.A., and N.B. Shoemaker. 1996. Resistance gene transfer in anaerobes: new insights, new problems. *Clin. Infect. Dis.* 23 Suppl. 1:S36–S43.

Salyers, A.A., and N.B. Shoemaker. 1997. Conjugative transposons. *Genet. Eng.* 19:89–99.

Salyers, A.A., N.B. Shoemaker, A.M. Stevens, and L.Y. Li. 1995. Conjugative transposons: an unusual and diverse set of integrated gene transfer elements. *Microbiol. Rev.* 59:579–590.

Sanders, C.C., and B. Wiedemann. 1988. Conference summary. *Rev. Infect. Dis.* 10:679–680.

Schadewaldt, H. 1985a. 50 Jahre Sulfonamide. *Dt. Med. Wochenschrift* 110:306–317.

Schnabel, E.L., and A.L. Jones. 1999. Distribution of tetracycline resistance genes and transposons among phylloplane bacteria in Michigan apple orchards. *Appl. Environ. Microbiol.* 65:4898–4907.

Schrag, S.J., and V. Perrot. 1996. Reducing antibiotic resistance. *Nature* 381:120–121.

Schrag, S.J., V. Perrot, and B.R. Levin. 1997. Adaptation to the fitness costs of antibiotic resistance in *Escherichia coli. Proc. R. Soc. Lond. B. Biol. Sci.* 264:1287–1291.

Scott, J.R., and G.G. Churchward. 1995. Conjugative transposition. *Annu. Rev. Microbiol.* 49:367–397.

Scott, J.R., P.A. Kirchman, and M.G. Caparon. 1988. An intermediate in transposition of the conjugative transposon Tn916. *Proc. Natl. Acad. Sci. USA* 85:4809–4813.

Sherrat, D. 1989. Tn3 and related transposable elements. In D. E. Berg and M. M. Howe, eds. *Mobile DNA*. Washington, D.C.: American Society for Microbiology, 163-184.

Shibata, N., H. Kurokawa, T. Yagi, and Y. Arakawa. 1999. A class 3 integron carrying the IMP-1 metallo-beta-lactamase gene found in Japan. *39th Interscience Conference on Antimicrobial Agents and Chemotherapy*, San Francisco.

Souza, V., and L.E. Eguiarte. 1997. Bacteria gone native vs. bacteria gone awry?: plasmidic transfer and bacterial evolution. *Proc. Natl. Acad. Sci.* U.S.A. 94:5501–5503.

Stokes, H.W., and R.M. Hall. 1989. A novel family of potentially mobile DNA elements encoding site-specific gene integration functions: integrons. *Mol. Microbiol.* 3:1669–1683.

Stokes, H.W., D.B. O'Gorman, G.D. Recchia, M. Parsekhian, and R.M. Hall. 1997. Structure and function of the 59-base element recombination sites associated with mobile gene cassettes. *Mol. Microbiol.* 26:731–745.

Sundstrom, L., P. Radstrom, G. Swedberg, and O. Skold. 1988. Site-specific recombination promotes linkage between trimethoprim and sulphonamide resistance genes. Sequence characterization of dhfrV and sulI and a recombination active locus of Tn21. *Mol. Gen. Genet.* 213:191–201.

Szpirer, C., E. Top, M. Couturier, and M. Mergeay. 1999. Retrotransfer or gene capture: a feature of conjugative plasmids, with ecological and evolutionary significance. *Microbiol.* 145:3321–3329.

Thomas, C.M. 2000. Paradigms of plasmid organization. *Mol. Microbiol.* 37:485–491.

Top, E.M., Y. Moenne-Loccoz, T. Pembroke, and C.M. Thomas. 2000. Phenotypic traits conferred by plasmids. In Thomas, C.M., ed. *The Horizontal Gene Pool*. Amsterdam: Harwood Academic Publishers, 249–285.

Torres, O.R., R.Z. Korman, S.A. Zahler, and G.M. Dunny. 1991. The conjugative transposon Tn925: enhancement of conjugal transfer by tetracycline in *Enterococcus faecalis* and mobilization of chromosomal genes in *Bacillus subtilis* and E. faecalis. *Mol. Gen. Genet.* 225:395–400.

Towner, K.J. 1995. The genetics of resistance. In D. Greenwood, ed. *Antimicrobial Chemotherapy*. New York: Oxford University Press, 159-167.

Trieu-Cuot, P. 1993. Broad-host range conjugative transfer systems. In R. H. Baltz, G. D. Hegeman and P. L. Skatrud, eds. *Industrial microorganisms: Basic and Applied Molecular Genetics*. Washington, D.C.: American Society for Microbiology, 43–50.

Vaara, M. 1992. Agents that increase the permeability of the outer membrane. *Microbiol. Rev.* 56:395–411.

Valla, S. 1998. Broad-host-range plasmids and their role in gene transfer in nature. *APMIS Suppl.* 84:19–24.

van Elsas, J.D., J. Fry, P. Hirsch, and S. Molin. 2000. Ecology of plasmid transfer and spread. In Thomas, C.M., ed. *The Horizontal Gene Pool*. Amsterdam: Harwood Academic Publishers, 175–206.

Vining, L.C. 1990. Functions of secondary metabolites. *Annu. Rev. Microbiol.* 44:395–427.

Waksman, S.A., E.S. Horning, M. Welsch, and H.B. Woodruff. 1942. Distribution of antagonistic actinomycetes in nature. *Soil Sci.* 54:281–296.

Waksman, S.A., and H.B. Woodruff. 1942. Streptothricin, a new selective bacteriostatic and bactericidal agent, particularly active against gram-negative bacteria. *Proc. Soc. Exper. Biol. Med.* 49:207–210.

Waldor, M.K., and J.J. Mekalanos. 1996. Lysogenic conversion by a filamentous phage encoding cholera toxin. *Science* 172:1910–1914.

Walsh, C. 2000. Molecular mechanisms that confer antibacterial drug resistance. *Nature* 406:775–781.

Watanabe, T., and T. Fukasawa. 1961. Episome-mediated transfer of drug resistance in Enterobacteriaceae. I. Transfer of resistance factors by conjugation. *J. Bacteriol.* 81:669–678.

Webb, V., and J. Davies. 1993. Antibiotic preparations contain DNA: a source of drug resistance genes? *Antimicrob. Agents Chemother.* 37:2379–2384.

Wiedermann, B., J.F. Meyer, and M.T. Zuhlsdorf. 1987. Insertion of resistance genes into Tn21-like transposons. J. *Antimicrob. Chemother.* 18, Suppl. C.:85–92.

World Health Organization. 1997. Monitoring and management of bacterial resistance to antimicrobial agents: a World Health Organization symposium. Geneva, Switzerland, 29 November–2 December. *Clin. Infect. Dis.* 24 Suppl. 1:S1–S176.

Zund, P., and G. Lebek. 1980. Generation time-prolonging R plasmids: correlation between increases in the generation time of *Escherichia coli* caused by R plasmids and their molecular size. *Plasmid* 3:65–69.

7

REGULATORY ACTIVITIES OF THE U.S. FOOD AND DRUG ADMINISTRATION DESIGNED TO CONTROL ANTIMICROBIAL RESISTANCE IN FOODBORNE PATHOGENS

LINDA TOLLEFSON, WILLIAM T. FLYNN, AND MARCIA L. HEADRICK

INTRODUCTION. Development of antimicrobial-resistant bacteria is a hazard associated with antimicrobial drug use in both human and veterinary medicine. The selection of antimicrobial-resistant bacterial populations is a consequence of exposure to antimicrobial drugs and can occur from human, animal, and agricultural uses. The use of antimicrobial drugs in food-producing animals is necessary to maintain their health and welfare. Unfortunately, food-producing animals can become reservoirs of bacteria capable of being transferred on food. Food-carrying resistant bacterial pathogens can cause illness in people consuming the food, which may result in therapeutic failures.

In industrialized countries, *Salmonella* and *Campylobacter* are infrequently transferred from person to person. In these countries, epidemiological data have demonstrated that a significant source of antibiotic-resistant foodborne infections in humans is the acquisition of resistant bacteria originating from animals that are transferred on food (Angulo et al. 1998; Cohen and Tauxe 1986; Edgar and Bibi 1997). This has been demonstrated through several different types of foodborne disease follow-up investigations, including laboratory surveillance, molecular subtyping, outbreak investigations, and studies on infectious dose and carriage rates (Holmberg et al. 1984; Spika et al. 1987; Tacket et al. 1985). Holmberg et al. (1954) were the first in the United States to document an outbreak of salmonellosis in people caused by eating hamburger that contained multiple-drug-resistant *Salmonella* originating from South Dakota beef cattle fed chlortetracycline for growth promotion. In 1986 Cohen and Tauxe published a review of infections caused by drug-resistant *Salmonella* in the United States. Using data from prospective studies on *Salmonella* conducted by the Centers for Disease Control and Prevention (CDC), the authors showed that no correlation existed between antimicrobial drugs used to treat salmonellosis in humans and antimicrobial resistance patterns among *Salmonella* from either human or animal sources. However, a great deal of similarity was found between resistance patterns in *Salmonella* isolates from animals and humans and antimicrobial drugs used in farm animals (Cohen and Tauxe 1986). CDC epidemiologists have recently updated and confirmed these findings with more current data and determined that antimicrobial use in humans contributes little to the emergence and development of antimicrobial-resistant *Salmonella*

in the United States (Angulo et al. 1998). Because human use of antimicrobial drugs is not a significant contributor to development of resistance in *Salmonella*, it is likely that the majority of antimicrobial-resistant *Salmonella* in humans and food-producing animals is the result of antimicrobial use in food-producing animals. This assumption does not require that every food-producing animal carrying resistant *Salmonella* has to have been treated with an antimicrobial drug. Resistant foodborne organisms can be acquired from a contaminated environment due to drug use in a previous herd or flock of animals. Resistant foodborne organisms can also be acquired through contact with other animals during transportation to the slaughter plant and antemortem processing, through cross-contamination at the time of slaughter or harvest or at various post-slaughter or harvest processing points, or through contamination of other foods in the home by raw product.

Bacterial resistance to antimicrobial drugs may develop by mutation of genes that code for antimicrobial drug uptake into cells or the binding sites for antibiotics, or by acquiring foreign DNA mediating resistance. Resistance traits may also be passed to human pathogenic bacteria by mechanisms that allow the exchange of the bacteria's genetic material. Animal enteric bacteria nonpathogenic to humans may then pass resistance traits to pathogenic bacteria. Mobile genetic elements often carry several resistance genes; consequently, the uptake of a single mobile genetic element may confer resistance toward multiple antimicrobials (Edgar and Bibi 1997). Moreover, structurally different classes of antimicrobial drugs may select for resistance to unrelated drug classes that have a common site of action (Leclercq and Courvalin 1991).

CURRENT NEW ANIMAL DRUG APPROVAL PROCESS. Before any animal drug may be legally marketed in the United States, the drug sponsor must have a New Animal Drug Application (NADA) approved by the Food and Drug Administration (FDA), Center for Veterinary Medicine (CVM). To obtain NADA approval, the drug's sponsor must demonstrate that the drug is effective and safe for the animal, safe for the environment, and can be manufactured to uniform standards of purity, strength, and identity. If the animal drug product is intended for use in food-producing animals, the drug's sponsor must also demonstrate that edible products produced from treated animals are safe for consumers (Friedlander et al. 1999). Two major concerns must be addressed by the sponsors of antimicrobial drugs intended for use in food-producing animals: the human food safety of veterinary drug residues and development of antimicrobial resistance caused by the use of the drug in animals.

Human Food Safety of Veterinary Drug Residues. To determine the food safety of residues of an antimicrobial product, the drug sponsor conducts a standard battery of toxicology tests. The battery includes studies that examine the effect of the product on systemic toxicity, genotoxicity/mutagenicity, and reproductive and developmental toxicity. The toxicology studies are designed to show a dose that causes a toxic effect and a dose that causes no observed effect.

After the no-observed-effect level is established for all the toxicity endpoints, the most sensitive effect in the species that is most predictive of effects in humans is identified. The no-observed-effect level is divided by a safety factor to account for uncertainty in extrapolating from animals to humans and for variability, that is, the difference among individuals, to calculate an acceptable daily intake (ADI) for drug residues. The ADI represents the amount of drug residues that can be safely consumed daily for a lifetime. The value of the safety factor used to calculate an ADI from a no-observed-effect level is normally 100. It comprises two factors of ten, the first of which is designed to offset the uncertainty of the no-observed-effect level that arises from the necessarily restricted number of animals used in the toxicological study. It also takes into account the possibility that human beings might be more sensitive to the toxic effects than the most sensitive laboratory animal. If the no-observed-effect level has been determined on the basis of undesirable effects on humans, this second factor of ten is not used. The second factor is designed to account for the genetic variability of consumers, which is much wider than the genetic variability of the laboratory animals used in the toxicological study. The second factor also takes into consideration that some human subpopulations, such as the very young, the very old, pregnant women, and people with illness or with unique metabolic conditions, may be more than tenfold more sensitive than the average healthy adult.

After the ADI is established, the drug's sponsor conducts drug metabolism and depletion studies to determine how the drug is metabolized and excreted. The tissue in which the drug depletes the slowest is established as the target tissue, and the amount of drug that can be measured with a regulatory method is established as the tolerance. When the drug residue in the target tissue depletes below the tolerance, all edible tissues are safe for consumption (Friedlander et al. 1999).

There are special food safety concerns for the residues of antimicrobial drugs (Paige et al. 1997). It is known that therapeutic doses of antimicrobial drugs administered to humans can cause adverse effects on the human intestinal microflora. The FDA has identified the selection of resistant bacteria, the perturbation of the barrier effect, changes in enzymatic activity, and alteration in bacterial counts as potential effects of antimicrobial drug residues present in food on the human intestinal microflora that are of public health concern. A perturbation in the barrier effect is of concern because the gut microflora provides a barrier against the overgrowth and invasion of pathogenic bacteria. When an antibiotic destroys this barrier, overgrowth of pathogenic bacteria may occur.

Consideration of Antimicrobial Resistance. Based upon the increasing evidence that use of antimicrobial drugs in food-producing animals may select for resistant bacteria of public health concern, the FDA published new regulatory guidance in 1999 (U.S. Food and Drug Administration 1999a). The guidance states that FDA believes it is necessary to consider the potential human health impact of the antimicrobial effects associated with all uses of all classes of antimicrobial new animal drugs intended for use in food-producing animals.

Minimizing the emergence of antimicrobial-resistant bacteria in animals and their subsequent spread to humans through the food supply is a complex problem requiring a coordinated, multifaceted approach. The strategy developed by the FDA to address antimicrobial resistance is one component of more broad-reaching strategies being developed at the national level in the form of the Public Health Action Plan to Combat Antimicrobial Resistance (U.S. Government 2001). The Federal Public Health Action Plan describes a coordinated focus on education, surveillance, research, and product development. Implementation of new regulatory requirements for animal drugs is a top priority action item of the Federal plan.

In a 1999 discussion document titled "Proposed Framework For Evaluating and Assuring The Human Safety Of The Microbial Effects Of Antimicrobial New Animal Drugs Intended For Use In Food-Producing Animals" (the Framework Document), the FDA outlined possible strategies for managing the potential risks associated with the use of antimicrobial drugs in food-producing animals (U.S. Food and Drug Administration 1999b). The Framework Document describes both pre-approval and post-approval approaches. The strategies include: (1) categorization of antimicrobial drugs based upon the importance of the drug for human medicine; (2) revision of the pre-approval safety assessment for antimicrobial resistance for new animal drug applications to assess all uses for microbial safety; (3) post-approval monitoring for the development of antimicrobial drug resistance; (4) collection of food animal antimicrobial drug use data; and (5) the establishment of regulatory thresholds. The Framework Document as well as the individual strategies have been the subject of a number of public meetings.

IMPLEMENTATION OF THE FRAMEWORK DOCUMENT.

The FDA is currently drafting guidance for industry that incorporates the concepts previously described in the Framework Document and outlines a general strategy for regulating the use of antimicrobial drugs in food-producing animals. This guidance will incorporate the five basic strategies of the Framework Document discussed previously into a single coordinated and comprehensive plan.

Drug Categorization. A key component of the Framework Document is the concept of categorizing antimicrobial drugs according to their importance for treating disease in humans. The Framework Document discusses three categories, with the most important drugs considered Category I. The categorization process is an integral part of assessing safety in that it provides a mechanism for characterizing the potential human health impact resulting from treatment failure because of resistance. The categorization process also serves to focus the greatest level of attention on those antimicrobial drugs of greatest importance to human medical therapy.

Certain aspects of the drug class as well as the target bacteria need to be considered when developing a ranking of antimicrobials based on their importance in human medicine. The types of infections treated by the drug, the availability of alternative treatments, the uniqueness of the mechanism of action of the drug, the ease with which resistance develops, the range of use by the agent in humans, and the usefulness of the drug in foodborne infections are examples of issues that should be addressed.

In general, drugs that are essential for treatment of a serious or life-threatening disease in humans for which no satisfactory alternative therapy exists would be placed in Category I, the category of highest concern. Drugs that are the only approved therapy or one of only a few approved therapies available to treat certain very serious human infections are particularly important in human medicine. Examples include vancomycin and linezolid in the treatment of certain serious infections and dalfopristin/quinupristin for the treatment of serious bloodstream infections caused by vancomycin-resistant enterococcus species.

Drugs that are not the only therapy for a disease in humans but are still the preferred choice for therapy for a given infection may be considered Category I under certain circumstances. Preferential characteristics of drugs include tolerability, ease of administration, and spectrum of antimicrobial activity. Some of these characteristics are quite important in the management of disease in human medicine. For example, the ability to treat serious infections either wholly or in part with oral therapy allows for easy outpatient therapy in situations when a patient might otherwise require parenteral and therefore possibly hospital therapy. The utility of a drug in treating foodborne infections is another consideration in determining the drug's importance for human medical therapy. For example, fluoro-

quinolones for treating foodborne infections due to multidrug-resistant *Salmonella* or ceftriaxone for treating foodborne infections in children would be classified as Category I drugs.

Drugs that are important for the treatment of potentially serious disease in humans but for which satisfactory alternative therapy exists would be candidates for Category II. Drugs with little or no use in human medicine would likely fall into Category III. However, other factors would need to be considered. For example, if cross-resistance were known to exist between drugs of differing importance ranking, the drug in question would be ranked according to the highest level of importance among the drugs involved.

Pre-Approval Product Review. The evaluation of chemical residues associated with animal-derived foods for their toxic properties uses a battery of well-established animal and laboratory tests. Unfortunately, no such predictive models currently exist to estimate with precision the rate and extent of bacterial resistance that may emerge from the use of antimicrobial drugs in food-producing animals. Despite the lack of such models, certain information can be generated to support a pre-approval antimicrobial resistance risk assessment.

Consistent with the approach outlined in the Framework Document, applications for antimicrobial drugs for food-producing animals would initially undergo an antimicrobial resistance risk assessment. The antimicrobial resistance risk assessment will be qualitative in nature and will combine a broad set of information regarding the drug and the proposed drug use for evaluation. Relevant basic information includes specifics on the drug itself, such as drug class and mechanism of antimicrobial activity, and product-specific facts, such as route of administration, dosing regimen, product formulation, proposed product indication, and intended target animal species. In addition, data on such factors as the drug's spectrum of activity, pharmacokinetics and pharmacodynamics, specific susceptibility data, resistance mechanisms, and the occurrence and rate of transfer of resistance determinants can all be used to support a pre-approval antimicrobial resistance risk assessment.

The qualitative antimicrobial resistance risk assessment will also characterize the drug product as to the potential for resistance to emerge in animals and the potential that such resistance would be transmitted from animals to humans, and will describe the relative importance of the drug, or related drug, to human medical therapy. These factors taken together can be used to characterize the overall level of concern that antimicrobial resistance may emerge in association with the proposed use of the drug in animals and cause an unacceptable human health impact. This approach to characterizing antimicrobial drugs is consistent with the basic concepts outlined in the Framework Document (U.S. Food and Drug Administration 1999b). The regulatory strategy would then be scaled to account for different levels of risk associated with certain drugs and certain drug use conditions.

In addition to the risk assessment, applications for antimicrobial drugs for food-producing animals will undergo a risk management assessment. Based on the conclusions of the antimicrobial resistance risk assessment, possible risk management steps may range from refusing to approve a drug application to approving the application under a spectrum of use conditions imposed to ensure safe use of the product. Because the use of antimicrobial drugs exerts selection pressure on antimicrobial-resistant bacteria, it is relevant to consider drug use conditions when defining steps for managing risks from antimicrobial resistance. A principle consistent with the judicious use concept is that certain drugs, or certain drug use conditions, are more (or less) likely than other drugs, or drug use conditions, to exert pressures favorable to resistance emergence. Use condition restrictions may include limitations regarding product marketing status, extra-label use provisions, or specific dosage and administration instructions.

Post-Approval Monitoring. The surveillance of foodborne pathogens resistant to antimicrobial agents is key to the ability to detect emerging resistance. The identification of emerging resistance and the capability to investigate resistance patterns and trends identified through statistically robust national monitoring systems are essential elements to facilitate timely and appropriate public health response activities. Two types of monitoring systems are important to the control of antimicrobial resistance in foodborne pathogens: monitoring of the prevalence of resistance in enteric bacteria from animal, human, and retail meat sources; and monitoring of amounts of antimicrobials used in food-producing animals.

U.S. NATIONAL ANTIMICROBIAL RESISTANCE MONITORING SYSTEM FOR ENTERIC BACTERIA. A key component of FDA's overall strategy on antimicrobial resistance is a national surveillance program that monitors resistance among enteric pathogens in both animals and humans. In 1996 the FDA, the CDC, and the USDA established the National Antimicrobial Resistance Monitoring System for Enteric Bacteria (NARMS) to monitor prospectively changes in antimicrobial susceptibilities of zoonotic enteric pathogens from human and animal clinical specimens, from carcasses of food-producing animals at slaughter, and from retail food (Tollefson et al. 1998). NARMS is currently monitoring susceptibilities of human and animal isolates of nontyphoid *Salmonella*, *E. coli*, *Campylobacter*, and Enterococcus. Human isolates of *Salmonella* Typhi, *Listeria*, *Vibrio*, and *Shigella* are also tested. Antimicrobials are selected for testing based on their importance in human or veterinary medicine. Minimum inhibitory concentration ranges are chosen to detect incremental changes in resistance based on the previous two years of data.

Animal isolate testing is conducted at the USDA Agricultural Research Service Russell Research Center in Athens, Georgia. Human isolate testing is conducted at the CDC National Center for Infectious Diseases Foodborne Disease Laboratory in Atlanta. All the testing sites follow identical isolation, identification, and susceptibility testing procedures. Retail meat testing was added to NARMS in 2001 at the FDA Center for Veterinary Medicine Laboratory in Laurel, Maryland, to enhance the ability to obtain additional organisms for monitoring from all species of food-producing animals. Also, meat sold at retail represents the point of exposure that is closest to the consumer that we are currently able to sample. Data from retail meat samples combined with data from slaughter plant samples provides a more representative picture of the prevalence of resistant pathogens in products derived from food-producing animals.

Surveillance systems provide an invaluable source of isolates. These isolates can be used for historical analysis of resistance trends and for investigating the molecular mechanisms of resistance. Because of the advancement in molecular techniques, screening large numbers of isolates for identification of resistance determinants is now possible in a timely manner. In addition, surveillance systems provide the opportunity to monitor events over time. Trends can be analyzed to determine changes and may help signal developing resistance. The ability to implement comparable systems worldwide would provide the ability to track emerging resistance and provide some countries the opportunity to implement mitigation strategies prior to the emergence of resistance. Additionally, surveillance systems provide descriptive data that enable the design of more targeted research studies. The complement of hypothesis-testing studies with descriptive surveillance data is a powerful tool for providing answers to complex questions.

The NARMS program is integral to FDA's strategy of enhanced pre-approval assessment, post-approval surveillance, and regulatory controls in order to better characterize and control the development of antimicrobial resistance. Early identification of emerging resistance will facilitate management of the problem as well as development of educational efforts on the appropriate use of antimicrobial agents. Monitoring will also allow assessment of the impact of these efforts.

MONITORING USE OF ANTIMICROBIAL DRUGS IN FOOD-PRODUCING ANIMALS. FDA currently requires the submission of certain drug sales information as part of the annual drug experience report for approved drug products. The Framework Document (U.S. Food and Drug Administration 1999b) identified the need for the pharmaceutical industry to submit more detailed antimicrobial drug sales information as part of its annual report. FDA is in the process of developing new procedures to monitor sales data as part of the regulatory framework for antimicrobial drug products intended for use in food-producing animals. To fully understand the antimicrobial resistance issues, information on drug use is critical to allow more direct correlation between loss of susceptibility or increasing resistance trends and actual use of individual antimicrobial drugs and drug classes.

The goal of drug use monitoring is to obtain objective quantitative information to evaluate usage patterns by antimicrobial agent or class, animal species, route of administration, and type of use (for example, therapeutic or growth promotion) to evaluate antimicrobial exposure. These data are essential for risk analysis and planning, can be helpful in interpreting resistance surveillance data obtained through NARMS, and can assist in the response to emerging antimicrobial resistance in a more precise and targeted way than is currently possible. The data can also be used to evaluate the effectiveness of mitigation strategies implemented in response to trends of increasing resistance and to assess the success of prudent drug use initiatives.

For all approved animal drugs in the United States, the drug sponsor is required to submit information on issues related to the product including promotional activities, adverse experience reports, literature reports, and marketing activities, including quantities of drugs marketed. This information is reported in annual submissions on the anniversary date of approval. Drug sponsors typically provide a quantity for each of the dosage forms marketed for each drug, but the data are not differentiated by animal species or commodity class, actual use conditions, or geographic region. There is also no requirement for the drug sponsor to differentiate between the amount of the product marketed domestically and product exported. Because the reporting year is based on the drug's approval date, the information submitted spans a unique, or nearly unique, reporting period. Therefore it is virtually impossible to aggregate data.

FDA is proposing to amend the current regulations on reporting requirements for certain approved drugs used in food-producing animals through a notice and comment rule-making process. To better evaluate trends in resistance or losses in susceptibilities, expanded information is needed on each labeled species or target animal, routes of administration, labeled claim, and geographic region on a calendar year basis.

Regulatory Thresholds. Development and dissemination of antimicrobial resistance is a dynamic phenomenon, unlike the more traditional human health hazard arising from residues of veterinary drugs in foods. Chemical residues are stable and the safety of the food supply can be ensured through tolerance setting and enforcement. Because resistance to an antimicrobial does not arise until the drug is used, the human health hazard at the time the drug is approved is potential in nature and of unknown magnitude. Resistant pathogens develop only after the drug is marketed and used in animals.

Under the Federal Food, Drug and Cosmetic Act, the FDA cannot approve a new animal drug for use in food-producing animals unless it has been shown to be safe for consumers of the animal products derived from the treated animal. "Safe," in the context of human food safety, means "reasonable certainty of no harm." The definition is derived from language in House Report No. 2284, 85th Cong., 2d. Sess. 4–5, 1958, defining the term "safe" as it appears in section 409 of the Act, 21 U.S.C. 348, which governs food additives. This is a very high standard, and one that Congress has reaffirmed over the years by passing amendments that retain the original consumer-protective and remedial nature of the Food, Drug and Cosmetic Act.

In striving to resolve the issue of antimicrobial resistance, the FDA's goal is to provide for the safe use of antimicrobials in food-producing animals while ensuring that significant human antimicrobial therapies are not compromised or lost. To accomplish this goal, FDA is considering establishing health-based thresholds for the development of resistant bacteria before approving a new animal drug, the goal being to protect public health.

The concept of establishing thresholds was introduced in the Framework Document (U.S. Food and Drug Administration 1999b). The threshold concept was further refined in the December 2000 discussion document entitled "An Approach for Establishing Thresholds in Association with the Use of Antimicrobial Drugs in Food-Producing Animals" (U.S. Food and Drug Administration 2000) and discussed at a public meeting in January 2001. The approach outlined in the December 2000 "Threshold Document" proposed the establishment of two types of thresholds, a human health threshold and a resistance threshold. The human health threshold was to represent the unacceptable level of infections in humans that are treated with the antimicrobial drug of concern, are associated with bacteria resistant to the drug of concern, and for which the resistance is attributable to the use of an antimicrobial drug in animals. The resistance threshold was to represent the maximum allowable level of resistance prevalence in foodborne pathogenic bacteria isolated from food-producing animals that does not pose an unacceptable risk to human health.

The human health threshold represents a predetermined point at which human health is adversely affected. A model, such as that described in the Threshold Document, may be used to relate the predetermined human health threshold to a prevalence of resistance in animals. That model-derived prevalence of resistance in animals is defined as the resistance threshold. The prevalence of resistance would be monitored in animals through the NARMS program to determine if the levels are approaching or have exceeded the threshold. As noted in the Threshold Document, the FDA would initiate procedures to withdraw from the label any animal species that has reached or exceeded its threshold.

The FDA recognizes that establishing health-based thresholds will be difficult and resource intensive, but believes the advantages of thresholds warrants that the idea be given careful consideration. Thresholds would allow the animal drug industry to limit human risk by taking corrective action if loss of susceptibility to a particular drug approaches the unacceptable level of resistance prevalence. This would appear to be the least restrictive approach to regulating antimicrobial drugs in food-producing animals. However, setting a human health threshold would need to be reconciled with the current safety standard in the Federal Food, Drug and Cosmetic Act. Moreover, consumers and international trading partners may not be satisfied that the approach is sufficiently consumer protective, given the long and burdensome process required to remove an animal drug from the market after a human health threat became apparent.

RESEARCH ACTIVITIES. More research is needed on the relationship between antimicrobial use in food animals and the associated human health impact related to antimicrobial resistant bacteria. Such research is important for guiding FDA in further developing its pre-approval and post-approval activities related to the issue. Research provides the knowledge necessary to understand the fundamental processes involved in antimicrobial resistance and the impact on humans, animals, and the environment and to develop appropriate responses. An extensive knowledge base has been developed, but the FDA recognizes that additional research is needed.

CVM supports intramural and collaborative research efforts to investigate factors associated with the development, dissemination, and persistence of bacterial antibiotic resistance in both the animal production environment and the human food supply. Microbiologists from CVM's Office of Research are currently conducting or are participating in projects specifically targeted to gathering data on such issues as the following: (1) the development and persistence of bacterial antibiotic resistance from aquaculture and animal production environments; (2) characterization of mechanisms of resistance dissemination and transfer among pathogenic and commensal bacteria associated with food-producing animals and aquaculture environments; (3) the determination of the roles that animal feeds and feed commodities play in the dissemination of antibiotic resistance and pathogen carriage; and (4) co-selection of antibiotic resistance phenotypes associated with the use of sanitizers and other antimicrobial drugs in animals. In addition, CVM is a contributing laboratory to CDC's PulseNet molecular fingerprinting network involved in the molecular epidemiology of foodborne outbreaks.

In addition to the intramural research, CVM collaborates in extramural research grants and funds extramural research activities through cooperative agreements. This extramural research is designed to complement and augment the intramural research program. Several of the projects currently funded are designed to elucidate the prevalence and risk factors associated with the dissemination of antibiotic resistant *Salmonella*, *E. coli* O157:H7, and enterococci within the animal production environment. Another study seeks to adapt and validate for use in the animal production environment microbial detection methods developed for human food.

CVM is also developing methods for conducting antimicrobial resistance risk assessments to better characterize and quantify the human health impact associated with antimicrobial drug use in food-producing animals. A quantitative risk assessment has been completed with regard to fluoroquinolone use in poultry, and a second risk assessment on the use of virginiamycin is in progress. The fluoroquinolone risk assessment is a model for the direct transfer of resistance, which explains the situation in which the resistant bacteria are transferred from animals to humans as a food contaminant. The virginiamycin risk assessment attempts to model the indirect transfer of resistance, during which resistant determinants are passed among bacteria in the human gut flora.

SUMMARY AND CONCLUSIONS. Increasing resistance to antimicrobial agents is of growing concern to public health officials worldwide. The focus of concern includes infections acquired in hospitals, community infections acquired in outpatient care settings, and resistant foodborne disease associated with drug use in food-producing animals. The FDA's goal in resolving the public health impact arising from use of antimicrobials in food animals is to ensure that significant human antimicrobial therapies are not compromised or lost while providing for the safe use of antimicrobials in food animals. The FDA's approach to the problem is multipronged and innovative as delineated in the Framework Document (U.S. Food and Drug Administration 1999b). The strategy includes revision of the pre-approval safety assessment for new animal drug applications, use of risk assessment to determine the human health impact resulting from the use of antimicrobials in food animals, and robust monitoring for changes in susceptibilities among foodborne pathogens to drugs that are important in both human and veterinary medicine, research, and risk management.

The debate over the impact on human health of antimicrobial drug use in food-producing animals has continued for more than thirty years. The identification and subsequent containment of resistance will help ensure the continued effectiveness of veterinary and human antimicrobial drugs for years to come. The ultimate outcome will be to prolong the efficacy of existing and new antimicrobial drugs that are desperately needed to control both human and animal disease and to minimize the spread of resistant zoonotic pathogens to humans through the food supply.

REFERENCES

Angulo, F.J., R.V. Tauxe, and M.L. Cohen. 1998. The origins and consequences of antimicrobial-resistant nontyphoidal *Salmonella*: Implications for use of fluoroquinolones in food animals. In *Use of quinolones in food animals and potential impact on human health*. Report of a WHO meeting, WHO/EMC/ZDI/98.10, Geneva, Switzerland, 205–219.

Cohen, M.L., and R.V. Tauxe. 1986. Drug-resistant *Salmonella* in the United States: An epidemiologic perspective. *Science* 234:964–969.

Edgar, R., and E. Bibi. 1997. MdfA and *Escherichia coli* multidrug resistance protein with extraordinarily broad spectrum of drug recognition. *J. Bacteriology* 179:2274–2280.

Friedlander, L.G., S.D. Brynes, and A.H. Fernandez. 1999. The human food safety evaluation of *new* animal drugs. In *Chemical Food Borne Hazards and Their Control. Veterinary Clinics of North America: Food Animal Practice* 15(1):1–11.

Holmberg, S.D., M.T. Osterholm, K.A. Senger, and M.L. Cohen. 1984. Drug-resistant *Salmonella* from animals fed antimicrobials. *N. Engl. J. Med.* 311(10):617–622.

Leclercq, R., and P. Courvalin. 1991. Bacterial resistance to the macrolide, lincosamide, and streptogramin antibiotics by target modification. *Antimicrob. Agents Chemother.* 35:1267–1272.

Paige, J.C., L. Tollefson, and M. Miller. 1997. Public health impact of drug residues in animal tissues. *Veterinary and Human Toxicology*. 39(3):162–169.

Spika, J.S., S.H. Waterman, G.W. Soo Hoo, M.E. St. Louis, R.E. Pacer, S.M. James, M.L. Bissett, L.W. Mayer, J.Y. Chiu, B. Hall, K. Greene, M.E. Potter, M.L. Cohen, and P.A. Blake. 1987. Chloramphenicol-resistant *Salmonella* *Newport* traced through hamburger to dairy farms. *N. Engl. J. Med.* 316(10):565–570.

Tacket C.O., L.B. Dominguez, H.J. Fisher, and M.L. Cohen. 1985. An outbreak of multiple-drug-resistant *Salmonella* enteritis from raw milk. *J.A.M.A.* 253(14):2058–2060.

Tollefson L., F.J. Angulo, and P.J. Fedorka-Cray. 1998. National surveillance for antibiotic resistance in zoonotic enteric pathogens. In *Microbial Food Borne Pathogens. Veterinary Clinics of North America: Food Animal Practice* 14(1):141–150.

U.S. Food and Drug Administration. 1999a. Guidance for Industry: Consideration of the human health impact of the microbial effects of antimicrobial new animal drugs intended for use in food-producing animals (GFI #78). *Federal Register* 64:72083. Available at http://www.fda.gov/cvm/guidance/guidad78.html.

U.S. Food and Drug Administration. 1999b. A proposed framework for evaluating and assuring the human safety of the microbial effects of antimicrobial new animal drugs intended for use in food-producing animals. *Federal Register* 64:887. Available at http://www.fda.gov/cvm/index/vmac/antimi18.html.

U.S. Food and Drug Administration. 2000. An approach for establishing thresholds in association with the use of antimicrobial drugs in food-producing animals. *Federal Register* 65:830700. Available at http://www.fda.gov/cvm/antimicrobial/threshold21.doc.

U.S. Government. 2001. Public Health Action Plan to Combat Antimicrobial Resistance. Available at http://www.cdc.gov/drugresistance/actionplan/.

8

PREVENTION AND CONTROL ACTIVITIES TO ADDRESS ANTIMICROBIAL RESISTANCE

LYLE P. VOGEL

INTRODUCTION. The overall relationship between drug use and resistance is well established and based on laboratory studies, ecological studies, cross-sectional and case-control studies, and prospective studies (Bell 2001). But the precise details of the relationship between drug use and the emergence and spread of resistance are complex, are not always understood, and may vary for different drug-pathogen combinations. Transmissibility and virulence are important, as are other factors as yet unknown. For example, antibiotics such as doxycycline and nitrofurantoin have been used in high volume in human medicine for decades without bacteria becoming resistant to them (Cunha 2001). In addition, resistance to one drug in an antibiotic class does not necessarily confer resistance to other drugs in the class. Among the tetracyclines in human medicine, resistance has been described only with tetracycline but not doxycycline or minocycline (Cunha 1998).

Antimicrobial resistance prevention and control activities are based on the premise that the use of antimicrobials will sooner or later result in the development or expression of antimicrobial resistance. The corollary that guides prevention and control activities is that voluntary or regulatory limitations on the use of antimicrobials will lessen the development of antimicrobial resistance, prevent further increases in resistance if resistance is already present, and may result in decreasing levels of resistance.

Animal health and welfare concerns preclude a complete cessation of the use of antimicrobials in animals. The ultimate aim is to prevent the transmission of infections and thereby eliminate the need to use antimicrobials in food animals. Remarkable improvements have been achieved in disease prevention through animal health research, improved nutrition, changes in management practices, genetic selection, and use of vaccines. Yet new diseases emerge and old diseases still occur in food animals that require the use of antimicrobials for prevention, control, and treatment. Therefore the immediate goal is to promote appropriate use of antimicrobials in food animals. The Federal Interagency Task Force on Antimicrobial Resistance defined appropriate use as

use that maximizes therapeutic impact while minimizing toxicity and the development of resistance. In practice, this involves prescribing antimicrobial therapy when and only when it is beneficial to the patient, targeting ther-

apy to the desired pathogens, and using the appropriate drug, dose, and duration. Appropriate antimicrobial drug use should not be interpreted simply as reduced use, because these drugs offer valuable benefits when used appropriately. It is overuse and misuse that must be decreased to reduce the selective pressure favoring the spread of resistance (U.S. Government 2001a).

PREVENTION AND CONTROL STRATEGIES. Released in 2001, *The Public Health Action Plan to Combat Antimicrobial Resistance* provides many recommendations for the prevention and control of antimicrobial resistance in humans and the following for agriculture and veterinary medicine:

- Improve the understanding of the risks and benefits of antimicrobial use and ways to prevent the emergence and spread of resistance;
- Develop and implement the principles for appropriate antimicrobial drug use in the production of food animals and plants;
- Improve animal husbandry and food-production practices to reduce the spread of infection; and
- Provide a regulatory framework to address the need for antimicrobial drug use in agriculture and veterinary medicine while ensuring that such use does not pose a risk to human health (U.S. Government 2001).

The *Public Health Action Plan* is a broad-based consensus by federal agencies on actions needed to address antimicrobial resistance. It also provides a blueprint for specific, coordinated federal actions in the four focus areas of surveillance, prevention and control, research, and product development. Within the focus area of prevention and control, the Public Health Action Plan outlines priority goals and actions. As an example, it addresses ways to:

- Prevent infection transmission through improved infection control methods and use of vaccines.
- Prevent and control emerging antimicrobial resistance problems in agriculture human and veterinary medicine.
- Extend the useful life of antimicrobial drugs through appropriate use policies that discourage overuse and misuse.

To support the goal of promoting infection control through behavioral and educational interventions, the action items specific to foodborne organisms are two-fold: (1) support ongoing public health education campaigns on food safety whose aims are to educate food producers, suppliers, retailers, and consumers to reduce foodborne infections, including antimicrobial resistant infections; and (2) educate the public about the merits and safety of irradiation as one tool to reduce bacterial contamination of food.

To help provide ways for prevention of emerging antimicrobial resistance in agricultural and veterinary settings, several action items are listed. One is to evaluate the nature and magnitude of the impact of using antimicrobials as growth promotants and to use this information in risk-benefit assessments. Others are to conduct additional research to further define the effects of using veterinary drugs on the emergence of resistant bacteria, identify and evaluate new food pasteurization strategies, and assess the risk of antimicrobial resistance emergence and spread caused by environmental contamination by antimicrobial drugs or by resistant bacteria in animal and human waste.

To address appropriate use policies, a top priority action item states, "In consultation with stakeholders, refine and implement the proposed Food and Drug Administration (FDA) Center for Veterinary Medicine (CVM) Framework for approving new antimicrobial drugs for use in food-animal production and, when appropriate, for re-evaluating currently approved veterinary antimicrobial drugs." The FDA Framework effort is further explained in the previous chapter on regulatory activities. Besides the Framework, other actions are mentioned to achieve appropriate antimicrobial use. One is to work with veterinary and agricultural communities to help educate users of antimicrobials about resistance issues and promote the implementation and evaluation of guidelines. Several veterinary and agricultural organizations have previously implemented activities that address this action item. For example, the American Veterinary Medical Association (AVMA) with other veterinary organizations, in cooperation with CVM, developed educational booklets for food-animal veterinarians (Anon. 2000a). The AVMA also wrote a script for an educational video for veterinarians and veterinary students, which the FDA produced and distributed. Producer organizations, such as the National Pork Board, developed educational materials for their members (Anon. 2000b). Another action item is to encourage involvement of veterinarians in decisions regarding the use of systemic antimicrobials in animals, regardless of whether a prescription is required to obtain the drug. An extension of this action item is another recommendation to evaluate the impact of making all systemic antimicrobials for animals available by prescription only.

PRINCIPLES FOR APPROPRIATE ANTIMICROBIAL DRUG USE. A substantial set of clinical

guidelines has been developed for human medicine, including recommendations for healthcare-associated infections (for example, vancomycin resistance), malaria, sexually transmitted diseases, and others. Many of the guidelines can be accessed on the Centers for Disease Control and Prevention Web site at http://www.cdc.gov/drugresistance/technical/clinical.htm. Other professional organizations have developed guidelines as well (Anon. 2001a; Shales et al. 1997).

Veterinary and animal producer organizations in many countries have also developed and implemented judicious or prudent use principles or guidelines. In 1998, the AVMA formed a Steering Committee on Antimicrobial Resistance to address prevention and control measures to minimize the development of resistance in animal and human pathogens potentially resulting from the therapeutic use of antimicrobials in animals. The Steering Committee is advised by public health and human infectious disease experts. The Steering Committee developed general and species-specific judicious use principles and guidelines. The AVMA principles and some examples from other organizations are as follows:

- AVMA: Position Statement and Principles for Judicious Therapeutic Antimicrobial Use by Veterinarians (Anon. 1998a)
- British Veterinary Association: General Guidelines on the Use of Antimicrobials (Anon. 1998b)
- British Veterinary Poultry Association: Antimicrobials Guidelines (Anon. 1998c)
- World Veterinary Association/International Federation of Animal Producers/World Federation of the Animal Health Industry: Prudent Use of Antibiotics: Global Basic Principles (Anon. 1999a)
- Responsible Use of Medicines in Agriculture (RUMA) Alliance Guidelines: Responsible Use of Antimicrobials in Poultry Production (Anon. 1999b)
- RUMA Alliance Guidelines: Responsible Use of Antimicrobials in Pig Production (Anon. 1999c)
- RUMA Alliance Guidelines: Responsible Use of Antimicrobials in Dairy and Beef Cattle Production (Anon. 2000c)
- RUMA Alliance Guidelines: Responsible Use of Antimicrobials in Sheep Production (Anon. 2000d)
- Australian Veterinary Association: Code of Practice for the Use of Antimicrobial Drugs in Veterinary Practice (Anon. 1999d)
- American Association of Swine Veterinarians: Basic Guidelines of Judicious Therapeutic Use of Antimicrobials in Pork Production (Anon. 1999e)
- National Pork Board: Basic Guidelines of Judicious Therapeutic Use of Antimicrobials in Pork Production for Pork Producers (Anon. 2000b)
- American Association of Avian Pathologists: Guidelines to Judicious Therapeutic Use of Antimicrobials in Poultry (Anon. 2000e)
- American Association of Bovine Practitioners: Prudent Drug Usage Guidelines (Anon. 2000f)

- Federation of Veterinarians of Europe: Antibiotic Resistance & Prudent Use of Antibiotics in Veterinary Medicine (Anon., No date)
- National Cattlemen's Beef Association: A Producers Guide for Judicious Use of Antimicrobials in Cattle, National Cattlemen's Beef Association Beef Quality Assurance National Guidelines (Anon. 2001b)

International organizations, such as the World Health Organization (WHO), Office International des Epizooties (OIE), and Codex Alimentarius Commission, also have developed or are developing principles or codes of practice to contain antimicrobial resistance. The WHO published Global Principles for the Containment of Antimicrobial Resistance in Animals Intended for Food (Anon. 2000g). The OIE has written five documents concerning antimicrobial resistance, including "Technical Guidelines for the Responsible and Prudent Use of Antimicrobial Agents in Veterinary Medicine." The other four documents deal with risk analysis methodology, monitoring of use quantities, surveillance programs, and laboratory methodologies (Anon. 2001c). Additionally, several Codex committees are addressing aspects of antimicrobial resistance, including the Codex Committee on Residues of Veterinary Drugs in Foods, Codex Committee on Food Hygiene, and the Ad hoc Intergovernmental Codex Task Force on Animal Feeding. The Codex committees have not yet published guidelines.

The aforementioned documents have substantial differences: some are intended for veterinarians, whereas others are intended for animal producers. Some documents address only therapeutic (treatment, prevention, and control of disease) uses of antimicrobials, whereas others discuss uses for growth promotion. Some present general principles; others are developed for certain species of animals. But the various documents share common tenets. Most have the dual aim of protecting human and animal health. They recognize that any use of antimicrobials, whether for humans, animals, or plants, has the potential to select for antimicrobial resistance. They also recognize that all uses cannot be eliminated or severely constrained. Therefore the intent of the documents is to promote appropriate use of antimicrobials by maximizing efficacy and minimizing resistance development. The documents also recognize the unfortunate fact that little is known about the different conditions of use under which antimicrobials select for resistant bacteria. This lack of knowledge leaves decision makers in the unenviable situation of developing guidelines when the underlying specific causes of antimicrobial resistance are not completely understood. Yet decision makers, including veterinarians and animal producers, cannot wait for the ultimate answer because the existence of antimicrobial resistance is increasingly perceived as a threat to public health. This perception creates a need for immediate action, regardless of whether the action is the most efficient. Implementation of judicious or prudent use guidelines is a reasonable course of action.

AVMA General Principles. As an example of the guidelines, the AVMA general judicious use principles are discussed here. Judicious use principles are a guide for optimal use of antimicrobials but should not be interpreted so restrictively as to replace the professional judgement of veterinarians or to compromise animal health or welfare. In all cases, animals should receive prompt and effective treatment as deemed necessary by the attending veterinarian.

The fifteen principles, with explanatory notes, follow:

1. **Preventive strategies, such as appropriate husbandry and hygiene, routine health monitoring, and immunizations, should be emphasized.**

   Antimicrobial use should not be viewed in isolation from the disciplines of animal management, animal welfare, husbandry, hygiene, nutrition, and immunology. Diseases must be controlled through preventive medicine to reduce the need for antimicrobial use. The objective is to prevent disease to the greatest extent possible so that antimicrobial treatment is not required. In food animals, antimicrobial use is part of, but not a replacement for, integrated disease control programs. These programs are likely to involve hygiene and disinfection procedures, biosecurity measures, management alterations, changes in stocking rates, vaccination, and other measures. Continued antimicrobial use in such control programs should be regularly assessed for effectiveness and whether use can be reduced or stopped. Additional research is needed on economical and effective alternatives (such as probiotics and competitive exclusion products) to the use of antimicrobials and to evaluate the effects of the alternatives on the selection of resistant bacteria.

2. **Other therapeutic options should be considered prior to antimicrobial therapy.**

   Effective therapeutic options other than antimicrobials are viable choices for the treatment and prevention of some diseases. For example, scours may need to be treated only with fluid replacement and not with antimicrobials. Viral-induced disease may be supported through good nutrition and administration of drugs such as nonsteroidal anti-inflammatory drugs with antipyretic properties.

3. **Judicious use of antimicrobials, when under the direction of a veterinarian, should meet all the requirements of a valid veterinarian-client-patient relationship.**

   Use of prescription antimicrobials or the extralabel use of antimicrobials is permitted only under the direction of a veterinarian. This direction may take place only within the context of a valid veterinarian-client-patient relationship (VCPR). A valid VCPR exists when all of the following conditions have been met:
   a) The veterinarian has assumed the responsibility for making clinical judgements regarding the health of the animal(s) and the need for medical

treatment, and the client has agreed to follow the veterinarian's instructions.

b) The veterinarian has sufficient knowledge of the animal(s) to initiate at least a general or preliminary diagnosis of the medical condition of the animal(s). This means that the veterinarian has recently seen and is personally acquainted with the keeping and care of the animal(s) by virtue of an examination of the animal(s) or by medically appropriate and timely visits to the premises where the animal(s) are kept.

c) The veterinarian is readily available for follow-up evaluation, or has arranged for emergency coverage, in the event of adverse reactions or failure of the treatment regimen.

4. **Prescription, Veterinary Feed Directive (VFD), and extra-label use of antimicrobials must meet all the requirements of a valid veterinarian-client-patient relationship (VCPR).**

U.S. Federal regulations mandate a valid VCPR for the use of prescription and VFD drugs, and for extra-label use of drugs. VFD drugs are a special category of drugs approved in the United States that require a veterinarian order to be incorporated in feed.

5. **Extra-label antimicrobial therapy must be prescribed only in accordance with the Animal Medicinal Drug Use Clarification Act amendments to the Food, Drug, and Cosmetic Act and its regulations.**

The first line of choice should be based on the products approved for the species and the indication concerned. When no suitable product is approved for a specific condition or species, or the approved product is considered to be clinically ineffective, the choice of an alternative product should be based, whenever possible, on the results of valid scientific studies and a proven efficacy for the condition and species concerned.

a) For food animals, extra-label drug use (ELDU) is not permitted if a drug exists that is labeled for the food-animal species, is labeled for the indication, is clinically effective, and contains the needed ingredient in the proper dosage form.

b) ELDU is permitted only by or under the supervision of a veterinarian.

c) ELDU is allowed only for FDA-approved animal and human drugs.

d) ELDU is permitted for therapeutic purposes only when an animal's health is suffering or threatened. ELDU is not permitted for production drugs (for example, growth promotion).

e) ELDU in feed is prohibited.

f) ELDU is permitted for preventative purposes when an animal's health is threatened.

g) ELDU is not permitted if it results in a violative food residue or any residue that may present a risk to public health.

h) ELDU requires scientifically based drug withdrawal times to ensure food safety.

i) Record and labeling requirements must be met.

j) The FDA prohibits specific ELDUs. The following drugs are prohibited for extra-label use in food animals: chloramphenicol, clenbuterol, diethylstilbestrol, dimetridazole, ipronidazole, other nitroimidazoles, furazolidone, nitrofurazone, other nitrofurans, sulfonamide drugs in lactating dairy cows (except approved use of sulfadimethoxine, sulfabromomethazine, and sulfaethoxypyridazine), fluoroquinolones, and glycopeptides (for example, vancomycin) (Anon. 2002).

6. **Veterinarians should work with those responsible for the care of animals to use antimicrobials judiciously regardless of the distribution system through which the antimicrobial was obtained.**

Since 1988 the FDA has approved new therapeutic antimicrobials for use in animals as prescription-only products (Anon. 2000h). The prescription-only policy is based on the need to ensure the proper use of antimicrobials through precise diagnosis and correct treatment of disease to minimize animal suffering and to avoid drug residues in food. However many of the older antimicrobials are available for over-the-counter sale to producers. For these drugs, the FDA has determined that the producers can use the antimicrobials safely and effectively as directed on the label. Close veterinary involvement can help assist the producers by providing informed advice and guidance on judicious use. Extra-label use of over-the-counter antimicrobials would require that a veterinarian and the producer follow the constraints of AMDUCA, including the establishment of a valid veterinarian-client-patient relationship.

7. **Regimens for therapeutic antimicrobial use should be optimized using current pharmacological information and principles.**

For labeled use of an antimicrobial, the most accessible source of information is the label, which includes the package insert. For extra-label use, the Food Animal Residue Avoidance Databank (FARAD) can assist with determinations of withdrawal times. FARAD is a computer-based decision support system designed to provide livestock producers, extension specialists, and veterinarians with practical information on how to avoid drug, pesticide, and environmental contaminant residue problems (Anon. 1998d). To assist with decisions about possible alternatives and drug use regimens of antimicrobials, several veterinary and two producer organizations initiated the funding of the Veterinary Antimicrobial Decision Support System (VADS). The FDA is providing funding to continue VADS. The objective of VADS is to provide veterinarians with a source of easily accessible information on the therapy of specific diseases to help them make informed treatment decisions. The new decision support system will allow veterinarians to access current, peer-reviewed information when selecting treatment regimens. The available information will include a full-range of therapeutic options and supporting data for each antimicrobial. The pathogen data will include susceptibility profile information, when available, as well as an interpretation of susceptibility breakpoints as related to clinical efficacy.

The choice of the right antimicrobial needs to take into account pharmacokinetic parameters such as bioavailability, tissue distribution, apparent elimination half-life, and tissue kinetics to ensure the selected therapeutic agent reaches the site of infection. Duration of withdrawal times may be a factor in choosing suitable products. Consideration must also be given to the available pharmaceutical forms and to the route of administration.

8. **Antimicrobials considered important in treating refractory infections in human or veterinary medicine should be used in animals only after careful review and reasonable justification. Other antimicrobials should be considered for initial therapy.**

This principle takes into account the development of resistance or cross-resistance to important antimicrobials. In December 1998, the FDA made available "A Proposed Framework for Evaluating and Assessing the Human Safety of the Microbial Effects of Antimicrobial New Animal Drugs Intended for Use in Food-Producing Animals" (Framework Document) (FDA 1999). A concept introduced by the Framework Document is the categorization of antimicrobials based on their unique or relative importance to human medicine with special consideration given to antimicrobials used to treat human enteric infections. While the criteria for categorization remain under discussion, it is expected that antimicrobials such as the fluoroquinolones and third-generation cephalosporins will be classified in the most important category. The fluoroquinolones are also very important in veterinary medicine for the treatment of colibacillosis in poultry, and the cephalosporins are important for the treatment of respiratory disease in cattle.

9. **Use narrow spectrum antimicrobials whenever appropriate.**

To minimize the likelihood of broad antimicrobial resistance development, an appropriate narrow spectrum agent instead of a broad spectrum should be selected.

10. **Utilize culture and susceptibility results to aid in the selection of antimicrobials when clinically relevant.**

Susceptibility profiles can vary between herds and flocks. Periodic culture and susceptibility testing can provide historical data on which to base future empirical treatment as well as assist in selecting a treatment for refractory infections. Ideally, the susceptibility profile of the causal organism should be determined before therapy is started.

The veterinarian has a responsibility to determine the applicability of the breakpoints used by the laboratory to the specific disease indication being considered. In disease outbreaks involving high mortality or signs of rapid spread of disease, treatment may be started on the basis of a clinical diagnosis and previous susceptibility patterns before current samples are submitted for susceptibility or results are obtained. The susceptibility of the suspected causal organism should, where possible, be determined so that if treatment fails, it can be changed in the light of the results of susceptibility testing. Antimicrobial susceptibility trends should be monitored. These trends can be used to guide clinical judgement on antibiotic usage.

Susceptibility tests are intended to be a guide for the practitioner, not a guarantee for effective antimicrobial therapy. Susceptibility testing can give an indication only of what the clinical activity of the drug will be. The projection of clinical efficacy from an *in vitro* minimum inhibitory concentration (MIC) determination is much more accurate for antimicrobials with validated breakpoints for the specific indication. The effect of the drug *in vivo* depends on its ability to reach the site of infection in a high enough concentration, the nature of the pathological process, and the immune responses of the host.

11. **Therapeutic antimicrobial use should be confined to appropriate clinical indications. Inappropriate uses such as for uncomplicated viral infections should be avoided.**

Veterinarians should use their professional knowledge and clinical judgment to decide whether viral infections may involve or predispose to a superimposed bacterial infection.

12. **Therapeutic exposure to antimicrobials should be minimized by treating only for as long as needed for the desired clinical response.**

Theoretically, infections should be treated with antimicrobials only until the host's defense system is adequate to resolve the infection. Judging the optimal treatment duration is difficult. Limiting the duration of use to only what is required for therapeutic effect will minimize the exposure of the bacterial population to the antimicrobial. The adverse effects on the surviving commensal microflora are minimized and the impact on the remaining zoonotic organisms is reduced. Too long of a treatment duration will increase the chances of antimicrobial resistance. At the same time, a shortened treatment period may be problematic because it can lead to recrudescence of the infection. It is then possible that a higher percentage of the pathogens involved in the recrudescence episode have reduced susceptibility to the antimicrobial.

13. **Limit therapeutic antimicrobial treatment to ill or at-risk animals, treating the fewest animals indicated.**

In some classes of livestock, if a substantial

number of animals in a group have overt signs of disease, both sick and healthy animals may be treated with therapeutic levels of an antimicrobial. This is intended to cure the clinically affected animals, reduce the spread of the disease, and arrest disease development in animals not yet showing clinical signs.

It is recognized that strategic medication of a specific group of animals may be appropriate in certain precisely defined circumstances. However, this type of treatment also exposes more animals to antimicrobials, and the need for such medication should be regularly re-evaluated. The use of antimicrobials in the absence of clinical disease or pathogenic infections should be restricted to situations in which past experience indicates that the risk is high that a group of animals may develop disease if not treated. In addition, long-term administration to prevent disease should not be practiced without a clear medical justification.

14. **Minimize environmental contamination with antimicrobials whenever possible.**

Unused antimicrobials should be properly disposed. Some antimicrobials may be environmentally stable. If the antimicrobials are not bound in an inactive form, environmental exposure could contribute to resistance development. Consideration may need to be given to disposal methods that will not recycle resistant organisms to humans or animals.

15. **Accurate records of treatment and outcome should be used to evaluate therapeutic regimens.**

Outcome records can greatly assist with design of future empiric treatment regimens.

SUMMARY AND CONCLUSIONS. Although the precise details of the relationship between drug use and the development of resistance are not always understood, the overall relationship is well established. Animal health concerns preclude a complete cessation of the use of antimicrobials in food animals, but they must be used judiciously. Several organizations have developed judicious or prudent use principles and guidelines to mitigate the development of antimicrobial resistance as well as education programs to inform veterinarians and food-animal producers of the need to ensure prudent use. Veterinarians and food-animal producers are implementing the principles and guidelines. The implementation of judicious use principles will reduce the development of resistant zoonotic pathogens and commensals in animals and will lessen the risk of a human health impact related to the therapeutic use of antimicrobials in animals. These efforts reinforce the federal agencies' plan to combat antimicrobial resistance associated with the use in animals.

REFERENCES

Anon. 1998a. Position statement and principles of judicious therapeutic antimicrobial use by veterinarians. American Veterinary Medical Association. Available at http://www.avma.org/scienact/jtua/jtua98.asp.

Anon. 1998b. General guidelines on the use of antimicrobials. British Veterinary Association. *Vet. Rec* 143(20)565–566.

Anon. 1998c. Antimicrobial guidelines. British Veterinary Poultry Association. Available at http://www.bvpa.org.uk/medicine/amicguid.htm.

Anon. 1998d. Food Animal Residue Avoidance Databank Website. Available at http://www.farad.org.

Anon. 1999a. Prudent use of antibiotics: Global basic principles. World Veterinary Association/International Federation of Animal Producers/International Federation of Animal Health. Available at http://www.worldvet.org/docs/t-3-2.pru.doc.

Anon. 1999b. RUMA guidelines—responsible use of antimicrobials in poultry production. Responsible Use of Medicines in Agriculture Alliance. Available at http://www.ruma.org.uk.

Anon. 1999c. RUMA guidelines—responsible use of antimicrobials in pig production. Responsible Use of Medicines in Agriculture Alliance. Available at http://www.ruma.org.uk.

Anon. 1999d. Code of practice for the use of antimicrobial drugs in veterinary practice. Australian Veterinary Association. Artarmon, NSW, Australia.

Anon. 1999e. Basic guidelines of judicious therapeutic use of antimicrobials in pork production. American Association of Swine Veterinarians. Available at http://www.avma.org/scienact/jtua/swine/swine99.asp.

Anon. 2000a. Judicious Use of Antimicrobials for Beef Cattle Veterinarians. Judicious Use of Antimicrobials for Dairy Cattle Veterinarians. Judicious Use of Antimicrobials for Poultry Veterinarians. Judicious Use of Antimicrobials for Swine Veterinarians. American Veterinary Medical Association and Food and Drug Administration. Available at http://www.avma.org/scienact/jtua/default.asp.

Anon. 2000b. Basic guidelines of judicious therapeutic use of antimicrobials in pork production for pork producers. National Pork Board. Available at http://www.porkscience.org/documents/Other/psantibicprod.pdf.

Anon. 2000c. RUMA guidelines—responsible use of antimicrobials in dairy and beef cattle production. Responsible Use of Medicines in Agriculture Alliance. Available at http://www.ruma.org.uk.

Anon. 2000d. RUMA guidelines—responsible use of antimicrobials in sheep production. Responsible Use of Medicines in Agriculture Alliance. Available at http://www.ruma.org.uk.

Anon. 2000e. Guidelines to judicious therapeutic use of antimicrobials in poultry. American Association of Avian Pathologists. Available at http://www.avma.org/scienact/jtua/poultry/poultry00.asp.

Anon. 2000f. Prudent drug use guidelines. American Association of Bovine Practitioners. Available at http://www.avma.org/scienact/jtua/cattle/cattle00.asp.

Anon. 2000g. WHO global principles for the containment of antimicrobial resistance in animals intended for food. *Report of a WHO Consultation*, Geneva, Switzerland. World Health Organization. Available at http://www.who.int/emc/diseases/zoo/who_global_principles/index.htm.

Anon. 2000h. Human Use Antibiotics in Livestock Production. *HHS Response to House Report 106-157—Agriculture, Rural Development, Food and Drug Administration, and Related Agencies, Appropriations Bill.* Food and Drug Administration, Center for Veterinary Medicine. Available at http://www.fda.gov/cvm/antimicrobial/HRESP106_157.htm.

Anon. 2001a. Principles and strategies intended to limit the impact of antimicrobial resistance. Infectious Diseases Society of America. Available at http://www.idsociety.

org/pa/ps&p/antimicrobialresistanceprinciplesandstrategies_8-7-01.htm.

Anon. 2001b. Beef quality assurance, national guidelines, a producer's guide for judicious use in cattle. National Cattlemen's Beef Association. Available at http://www.bqa.org.

Anon. 2001c. Office International des Epizooties. *Scientific and Technical Review* 20:3; 797–870.

Anon. 2002. FDA Prohibits Nitrofuran Drug Use In Food-Producing Animals. CVM Update. FDA, Center for Veterinary Medicine. Available at http://www.fda.gov/cvm/index/updates/nitroup.htm.

Anon. No date. Antibiotic resistance and prudent use of antibiotics in veterinary medicine. Federation of Veterinarians of Europe. Available at http://www.fve.org/papers/pdf/vetmed/antbioen.pdf.

Bell, D.M. 2001. Promoting appropriate antimicrobial drug use: Perspective from the Centers for Disease Control and Prevention. *Clin. Infect. Dis.* 33:62–67.

Cunha, B.A. 1998. Antibiotic resistance. Control strategies. *Crit. Care Clin.* 14:309–327.

Cunha, B.A. 2001. Effective antibiotic-resistance control strategies—commentary. *Lancet* 357:1307–1308.

Food and Drug Administration. 1999. A Proposed Framework for Evaluating and Assessing the Human Safety of the Microbial Effects of Antimicrobial New Animal Drugs Intended for Use in Food-Producing Animals. Federal Register 64:887. Available at http://www.fda.gov/cvm/index/vmac/antim18.pdf.

Shales, D.M., D.N. Gerding, J.F. John Jr., et al. 1997. Society for Healthcare Epidemiology of America and Infectious Diseases Society of America Joint Committee on the Prevention of Antimicrobial Resistance: Guidelines for the prevention of antimicrobial resistance in hospitals. *Clin. Infect. Dis.* 25:584–599.

U.S. Government. 2001. A public health action plan to combat antimicrobial resistance. Part I. Domestic issues. Interagency Task Force on Antimicrobial Resistance. Available at http://www.cdc.gov/drugresistance/actionplan/.

NOTES

1. The explanatory notes are largely derived from educational booklets written by the American Veterinary Medical Association with the assistance of the American Association of Swine Veterinarians, American Association of Avian Pathologists, and American Association of Bovine Practitioners. The booklets were published by the Food and Drug Administration and are also available at http://www.avma.org/scienact/jtua/default.asp.

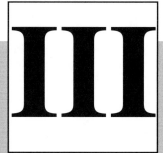

PART

SALMONELLA SPP.

9

THE EPIDEMIOLOGY AND ECOLOGY OF *SALMONELLA* IN MEAT-PRODUCING ANIMALS

CLIFFORD WRAY AND ROBERT H. DAVIES

INTRODUCTION. Salmonellosis is an important zoonotic infection, and human salmonellosis causes widespread morbidity and economic loss. Many countries have had a marked increase in the number of human cases in recent years. Much of this increase has been associated with poultry meat and table eggs (Saeed 1999).The continued importance of *Salmonella* Enteritidis as a human pathogen has overshadowed other *Salmonella* serovars, many of which are capable of causing human disease. Indeed, in the 1990s, a multiple resistant strain of *Salmonella* Typhimurium DT 104 (STM 104) has occurred in many countries. In the United Kingdom, some 17,000 cases of human salmonellosis, predominantly associated with *S*. Enteritidis and STM 104, occurred in 2000. In the United States, some 40,000 cases are reported annually.

The economic losses associated with human salmonellosis are not only associated with investigations, treatment, and prevention of illness but also may affect the whole human chain of food production. Estimated annual costs for salmonellosis have ranged from billions of dollars in the United States to hundreds of million dollars in Canada (Todd 1989). There have been few studies of the cost benefits of preventing *Salmonella* infection, but it has been suggested that for every £1 (about $1.50) spent on investigation and curtailment of an outbreak there is a saving of £5 (about $7.50) (Roberts 1989).

Although primarily intestinal bacteria, *Salmonella* are widespread in the environment and commonly found in farm effluents, human sewage, and in any material subject to fecal contamination. Salmonellosis has been recognized in all countries but appears to be most prevalent in areas of intensive animal husbandry, especially poultry and swine production. Although disease can affect all species of domestic animals, young animals and pregnant animals are the most susceptible. Many animals may also be infected without showing signs of illness.

Approximately 2,500 different *Salmonella* serovars have been described, and the number increases annually as new serovars are recognized. All members of the genus are considered to be potentially pathogenic, although serovars may differ widely in their host range and the pathogenic syndromes that they produce. Some serovars appear to show a degree of host-adaptation and primarily infect one animal species. They also tend to cause more severe illness than the other serovars. For example, *S*. Pullorum and *S*. Gallinarum infect poultry, *S*. Choleraesuis occurs in swine, and *S*. Dublin appears to have a predilection for cattle, although occasional outbreaks of disease caused by this serovar occur in sheep. In contrast, *S.* Typhimurium affects all species of animals and is one of the most common causes of food poisoning in humans. The numerous other serovars are widely distributed, and the predominant serovar for an animal species in a country may vary over the year. Many routes exist through which an animal may be infected. This chapter describes the complex epidemiology of *Salmonella* in meat-producing animals, predominantly in the United States and the United Kingdom. Most of the data on *Salmonella* incidence results from isolations from diseased animals and may not reflect the incidence in healthy animals, although routine monitoring of some species of farm animals is now becoming more widespread in some industries. In recent years efforts have been made to provide better data by including *Salmonella* isolations from abattoirs and studies on sentinel farms, and by commissioning structured surveys of different animal populations. In contrast to human data, which surveys individuals, it is necessary to evaluate populations (such as infected farms) rather than individual animals.

THE INCIDENCE OF *SALMONELLA* INFECTION IN MEAT-PRODUCING ANIMALS. The incidences of *Salmonella* infection in cattle, swine, and poultry in the United States and the United Kingdom are shown in Tables 9.1 and 9.2.

Cattle. The commonest serovars in the United States and the United Kingdom are *S.* Dublin and *S.* Typhimurium (Ferris and Miller 1996; MAFF 2000). Since the 1980s, *S.* Dublin has spread eastward from California to other states and northward into Canada, where it had not been detected previously (Robinson et al. 1984). In 1995–1996 it was detected in twenty-six states (Ferris and Miller 1996).

Many different phage types of *S.* Typhimurium have been isolated from cattle in the United Kingdom, where a particular phage type becomes dominant for a period until it is replaced by another phage type. For example, the multiple-resistant phage type DT204c was common during the 1980s; the penta-resistant DT104 became predominant in the 1990s. Many other different serovars have been isolated from cattle (Wray and Davies 2000). In the United States, 48 percent of the 730 *Salmonella*—other than Dublin and Typhimurium—isolated from cattle were represented by seven serovars (Ferris and Miller 1996). Of current importance is the increasing prevalence of multiple-resistant *S.* Newport, which is also of public health significance.

Swine. Although the prevalence of *S.* Choleraesuis has declined in Western Europe, it still remains a major problem in the United States. The reasons for the differences between the two areas are not known but may be related to husbandry practices. In the United States, the top five serovars recovered during a national prevalence study conducted in 1995 were as follows: *S.*

Table 9.1—The Most Common *Salmonella* Serovars Reported from Food Animals During the Period July 1999 to June 2000 (Ferris et al. 2000)

Chickens	Turkeys	Swine	Cattle
Heidelberg	Heidelberg	Typhimurium	Typhimurium
Kentucky	Senftenberg	Derby	Anatum
Berta	Hadar	Choleraesuis	Dublin
Enteritidis	Typhimurium	Heidelberg	Montevideo
Typhimurium	Muenster	Anatum	Newport
Senftenberg	Reading	Muenchen	Kentucky
Ohio	Agona	Orion	Mbandaka
Agona	Saintpaul	Worthington	Cerro
Infantis	Bredeney	Senftenberg	Agona

Table 9.2—The Most Common *Salmonella* Serovars Reported from Food Animals in the United Kingdom During 2000 (Defra 2000).

Chickens	Turkeys	Swine	Cattle
Senftenberg	Derby	Typhimurium	Dublin
Give	Agona	Derby	Typhimurium
Kedougou	Typhimurium	Kedougou	Agama
Montevideo	Newport	Goldcoast	Goldcoast
Thompson	Fischerkietz	Panama	Newport
Bredeny	Orion	Livingstone	Montevideo
Livingstone	Montevideo	London	Kentucky
Mbandaka	Senftenberg	Heidelberg	Orion
Typhimurium	Kottbus	Agona	Brandenburg

Derby (33.5 percent); *S.* Typhimurium (including var. Copenhagen) (14.7 percent); *S.* Agona (13.0 percent); *S.* Brandenburg (8.0 percent); and *S.* Mbandaka (7.7 percent) (Fedorka-Cray et al. 2000). Their isolation, however, varies by source as well as by geographical location (Davies et al. 1997). Phage typing of veterinary isolates of *S.* Typhimurium commenced in the United States in 1996, and DT104 has been isolated from swine. In 1997 multiple resistance to five or more antimicrobials in Typhimurium was detected in 23.3 percent of diagnostic isolates and 16.8 percent of abattoir isolates (CDC 1998).

Poultry. During the past decade, *S.* Enteritidis (SE) has replaced *S.* Typhimurium as the most common serovar from poultry in many countries. In the United States, the most prevalent serovars in chickens and turkeys during the period July 1999 to June 2000 are listed in Table 9.1. The phage types of SE most commonly associated with human infection are PTs 8, 13a, and 13 (Altekruse et al. 1993), but in 1993, PT 4 was detected in an outbreak associated with table eggs (Boyce et al. 1996) and subsequently its prevalence has increased.

SOURCES OF *SALMONELLA*. As indicated in Figure 9.1, many routes exist through which *Salmonella* may be introduced into a group of animals. These are discussed briefly in the following section. A more complete description can be found in Wray and Wray (2000).

Animal Feed. *Salmonella* can be isolated regularly from animal feeds and their ingredients, which include both animal and vegetable proteins; therefore, feed is a major potential route by which new infections may be introduced into a herd or flock (Davies and Hinton 2000). Indeed, many new serovars have been introduced into countries in imported feed ingredients (Clark et al. 1973). Basic cereal ingredients such as barley or maize are not considered to be particularly susceptible to *Salmonella* contamination but may be contaminated, during storage, by wild animals, especially rodents or birds. Home-grown cereals stored within one kilometer of livestock housing or manure storage areas on cattle and pig farms may be cross-contaminated by wild birds, rodents, and cats that acquire infection from the livestock areas. Other causes of contamination include the storing of cereals in empty livestock buildings that have not been adequately cleaned and disinfected.

Contamination may also occur in the feed mill from either the ingredients or from persistent contamination within a plant. Vegetable proteins are frequently contaminated during cooling after cooking and oil extraction. In fact, some *Salmonella* serovars have been found to persist in individual coolers for years, producing intermittently contaminated batches of feed (Davies and Hinton 2000). Feed, including ingredients, may be contaminated during transport either to the mill or to the farm if the vehicle has not been adequately cleaned and disinfected. The numbers of *Salmonella* in finished feed are usually very low and may not be detected by conventional culture techniques. Although researchers have suggested that these low numbers are unlikely to cause infection in farm animals, it is common to find the same serovar in flocks or herds as in the feed and the feed mill environment.

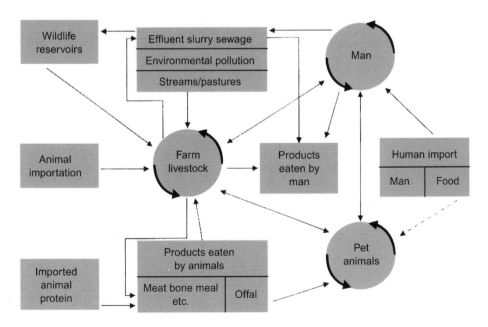

FIG. 9.1—The *Salmonella* cycle: pathways of infection.

Heat treatment of animal feeds has been used to reduce the risk of animal infection, but one concern is post-processing contamination. The prevalence of *Salmonella* in pigs has been shown to be lower in those given semi-solid feed compared with those fed pelleted feed. Thus the animal's diet may have a profound effect on its susceptibility to *Salmonella* infection. In a similar manner, at least in poultry, *Salmonella* colonization can be competitively prevented by the normal adult intestinal microbiota. This has led to the development of competitive exclusion products in the poultry industry (Schneitz and Mead 2000).

The Environment. *Salmonella* are widespread in the environment and can exist in many niches. As a consequence, *Salmonella* perhaps should be thought of as an environmental organism whose dissemination is likely to continue and increase in the future (Murray 2000).

FARM BUILDINGS. Farm buildings may become directly contaminated with *Salmonella* following outbreaks of disease or colonization of animals, or indirectly contaminated from other sources such as contaminated water used for cleaning or from wild animals and birds. Persistent contamination of houses and transport vehicles is an important factor in the maintenance and spread of *Salmonella* in animal populations, and the organism may persist in dry livestock buildings for up to 30 weeks (Davies and Wray 1996). Environmental samples from all types of animal production systems are frequently found to contain *Salmonella* even after cleaning and disinfection. Indeed, inadequate cleansing and disinfection may aggravate the situation, and the use of high-pressure hoses for cleaning may result in the widespread dissemination of the organism. Dust in animal houses originates from the feed, feces, and the animals themselves and may contain large numbers of micro-organisms of which *Salmonella* is a frequent isolate. Contaminated dust may also cause airborne transmission of *Salmonella*.

MANURE. *Salmonella* may survive for long periods in infected feces and slurries where their survival is dependent on a number of factors, especially climatic conditions. In moist, uncomposted feces, *Salmonella* may survive for 3–4 months in temperate climates and for longer periods in hotter climates. In poultry litter, *Salmonella* have been found to survive for periods of 2–20 weeks. However, in properly composted feces, the high temperatures produced rapidly kill *Salmonella* (Poppe 2000). Intensive management systems and the consequent increase in the amount of slurry produced have highlighted the risk of pasture contamination because of disposal problems. *Salmonella* have been found to survive for up to 286 days in slurry, but this is a function of the initial number of organisms, storage temperature, and the serovar. Survival is greatest at temperatures below 10°C and in slurries containing more than 5 percent solids. Sewage sludge is used as a fertilizer, and many samples have been found to be

contaminated with *Salmonella*. To reduce the risk of infection where sludge is used as a fertilizer on pasture, regulations/guidelines have been produced in many countries that stipulate minimum periods before grazing animals are allowed access to sludge-treated pasture.

WATER. Contaminated drinking water may facilitate the rapid spread of *Salmonella* among farm animals, which often defecate in their drinking water. Contamination may occasionally occur in the water storage tanks in a building, from wild animal feces or even from carcasses. Many reports exist of the isolation of *Salmonella* from rivers and streams, and animals may be infected either by drinking such water or when flooding occurs, contaminating the pasture. Many outbreaks of Salmonellosis in cattle have arisen from grazing recently flooded pasture (Williams 1975). Biological treatment of sewage removes most of the suspended and dissolved organic matter but does not necessarily remove pathogenic bacteria; consequently, watercourses may be contaminated from this source. Gay and Hunsaker (1993) described a *Salmonella* outbreak in which contaminated recycled water that had been used to flush out an animal housing area was responsible for the persistence of *Salmonella* on the farm.

OTHER ANIMALS. Although most farms animals are likely to acquire *Salmonella* infection from an animal of the same species, numerous publications have recorded the isolation of *Salmonella* from a wide range of mammals, birds, and arthropods (Murray 2000). The presence of *Salmonella* in wild animals may be of little epidemiological significance and may merely indicate a heavily contaminated environment rather than a primary source of infection unless the ecological balance is disturbed. Rodents, however, by their burrowing, nibbling, defecation, and urination may cause extensive damage and economic loss on poultry farms. Mice have been shown to be of importance in the epidemiology of S. Enteritidis infection in poultry units, where they may defecate in the birds' feed and water and spread *Salmonella* from contaminated to clean houses (Henzler and Opitz 1992). *Salmonella* infections have been detected in many species of wild birds. Seagulls have been found to spread *Salmonella* onto pastures where domestic animals may acquire infection. Conversely, seagulls may also spread infection from contaminated pastures. Similarly, flies and other insects have been shown to be vectors of *Salmonella*.

TRANSMISSION. Historically, transmission of *Salmonella* between hosts is thought to occur via the fecal-oral route of exposure. Because *Salmonella* are often shed in large numbers in the feces, this is likely a major route for transmission of the organism. A large number of studies in farm animals (Wray and Wray 2000) have reproduced experimental infection by the oral route, but high doses generally have had to be used and disease is often difficult to reproduce. Under farm conditions, domestic animals, especially chickens, may become infected by small doses of *Salmonella* during their first few days of life; thereafter, the infective dose in otherwise normal healthy animals becomes progressively higher.

It has long been suggested that aerosol transmission may be of importance, and disease has been reproduced in oesophagectomized calves and piglets by De Jong and Ekdahl (1965) and Fedorka-Cray et al. (1995). Likewise, experimental *Salmonella* infection, by aerosol, of calves and chickens has been reported by Wathes et al. (1988) and Baskerville et al. (1992). It is unclear whether this predilection for the lung is due solely to the pathogen, poor ventilation in large confinement buildings, or some combination of these and other factors. Experimental infection models have not provided good answers because positive lung samples have been regarded as an artifact of intranasal or per os inoculation. However, Gray et al. (1996) demonstrated that the lung becomes colonized in swine that are naturally exposed to other *S.* Choleraesuis-infected pigs; this indicates that lung colonization is not an artifact of experimental inoculation. More recent experiments (Proux, K., personal communication), in which pigs were kept separate except for contact via a common airspace, showed that infection could be transmitted by aerosol. Collectively, these studies indicate that the traditional paradigm of fecal-oral transmission is no longer valid and that other possibilities, such as aerosol or dust-borne transmission, need to be considered where large groups of animals are housed together. In poultry, ovarian transmission may be caused either directly or indirectly by *Salmonella*; this aspect is discussed later.

INTRODUCTION OF INFECTION ONTO THE FARM.
There are many routes by which *Salmonella* infection may be introduced onto a farm, and these are discussed briefly in relation to the different farm animals.

Cattle. Most infection is introduced into *Salmonella*-free herds by the purchase of infected cattle, either as calves for intensive rearing or adult cattle as replacements.

ADULT CATTLE. *Salmonella* are most likely to be introduced onto farms by the purchase of infected animals. Evans and Davies (1996) showed that the introduction of newly purchased cattle to a farm increased the risk of Salmonellosis and that the period of highest risk was within one month of arrival. Purchase through dealers was associated with a fourfold increase in risk as compared with purchasing cattle directly from other farms. Important differences exist, however, in the epidemiological behavior of *S.* Typhimurium and *S.* Dublin in adult cattle. Cattle that recover from *S.* Typhimurium do not tend to become permanent carri-

ers, and the period of *Salmonella* excretion is usually limited to a few weeks or months. In breeding herds, calfhood infections are seldom acquired from adult cattle.

In contrast, some adult cattle that recover from clinical *S.* Dublin infection may become permanently infected and excrete up to 102–104 per g. feces. Others in the herd may be latent carriers that harbor the organism in lymph nodes or tonsils and excrete the organism only in their feces (and milk, genital discharges, and urine, in the case of cows) when stressed, particularly at parturition. These latent carriers are difficult to identify, and it has been suggested that they may produce congenitally infected calves (Richardson 1973; Wray et al. 1989). During an investigation into *S.* Dublin infection, Watson et al. (1971) found that fourteen of fifty-nine cows with negative rectal swabs were infected at post-mortem examination. Smith et al. (1994) found that some cattle with chronic *S.* Dublin infections of the udder would shed the organism in both feces and milk. Such milk may be a source of infection for calves, and House et al. (1993) found that three heifers, which were infected with *S.* Dublin as calves, had infected supra mammary lymph nodes when slaughtered at thirteen months of age.

Ingestion of *Salmonella* does not necessarily lead to infection or disease. Richardson (1975) considered that *S.* Dublin, and presumably other serovars, may be ingested by adult cattle and pass through the alimentary tract with little or no invasion of lymph nodes. Grazing contacts of active carriers often yield positive rectal swabs but when removed from pasture and housed, they cease shedding *Salmonella* in about two weeks. During an outbreak of *S.* Newport infection, Clegg et al. (1983) frequently isolated the organism from a number of animals over a long period, but when the two cows, from which the organism had been detected for the longest period, were housed in clean premises, the organism was isolated neither from rectal swabs nor at post-mortem examination.

A number of studies have shown that *Salmonella* infection may be present on farms in the absence of clinical disease. The patterns of *Salmonella* infection on four farms was studied by Heard et al. (1972), who found that on well-managed farms, little disease was present although the prevalence of *Salmonella* was high. A three-year study on a farm where *S.* Dublin was present detected the organism in cattle and the environment in the absence of clinical disease (Wray et al. 1989). In beef cattle, *Salmonella* was detected on 38 of the 100 feedlots and on 21 of 187 beef cow-calf operations (Fedorka-Cray et al. 1998; Dargatz et al. 2000). Morisse et al. (1992) isolated *Salmonella* from 6.9 percent of 145 cows in 17 herds without a history of the organism. Thus, failure to recognize that the introduction of infection may sometimes precede the development of clinical disease by several months or years may lead investigators to attribute infection to somewhat unlikely sources. A sudden change in herd resistance caused by concurrent disease; for example, fascioliasis, bovine virus diarrhea virus, nutritional stress, or severe weather can result in clinical disease. Brownlie and Grau (1967) found that adult cattle rapidly eliminated *Salmonella*, but reducing the food intake retarded the elimination or permitted *Salmonella* growth in the rumen. Growth of *Salmonella* occurred during starvation, and resumption of feeding after starvation permitted further multiplication. These findings would suggest that adequate nutrition prior to and during transport and within the abattoir could be an important intervention to prevent spread of *Salmonella*.

CALVES. The mixing of young, susceptible calves in markets and dealers' premises and their subsequent transport is an effective means for the rapid dissemination of *Salmonella*, especially *S*. Typhimurium. A number of studies have shown that although few calves may be infected on arrival at a rearing farm, infection may then spread rapidly, often in the absence of clinical disease. Epidemiological studies (Wray et al. 1987) showed that the infection rate for *S*. Typhimurium increases during the first week to reach a peak at 14–21 days, in contrast to the spread of *S*. Dublin, which was slower and peaked at 4–5 weeks. On all the farms studied, little clinical disease was present and the prophylactic use of antibiotics had no effect on the spread of the organism. It was also found that the *Salmonella* infection rate and duration of excretion was higher in single penned calves than in those that were group housed. During an analysis of the spatial and temporal patterns of *Salmonella* excretion by calves that were penned individually, the results showed that noncontagious or indirect routes were more important than direct contagious routes in disease spread (Hardman et al. 1991).

Swine. Surveys have been done in many countries to determine the number of swine farms on which *Salmonella* is present and its prevalence on contaminated farms, but it is often unclear as to how representative the farms are, how they were selected, and how sensitive the detection techniques were.

ON-FARM PREVALENCE. It should not be assumed that all farms are contaminated with *Salmonella*. For example, Stege et al. (2000), by using culture and ELISA, detected 23 of 96 farms that had no positive pen samples and all pigs had negative serological results. In the United Kingdom, Davies and Wray (1997a) detected *Salmonella* on all of the twenty-three farms investigated, whereas Pearce and Revitt (1998) found 14.8 percent (4 of 27) farms positive, with 12.5 percent of the finishers being infected. Data from Denmark show that the herd prevalence in 1993 was 22 percent and in 1998 it was 11.4 percent. Of the different herd types, the percentages positive for breeding/multiplying, farrow to grower, and finisher were 11.7 percent, 16.7 percent, and 11.4 percent, respectively (Christensen et al. 1999). A survey in the United States (Bush et al. 1999) detected *Salmonella* on fifty-eight of 152 farms, where

the most common serovars were *S*. Typhimurium and Derby, the prevalence being higher in the southeastern states (65.5 percent) than in midwestern states (29.9 percent). In the northern states of the United States, ninety-six of 142 (67 percent) herds were positive for *Salmonella* (Damman et al. 1999). In Canada, *Salmonella* were isolated on 16 of 28 herds examined (Quessy et al. 1999).

WITHIN-FARM VARIATION. Even on farms where *Salmonella* has been detected, variation occurs in its distribution, not only with regard to the age of the pigs but also to the buildings and pens. Carlson and Blaha (1999) found a wide variation in different shipments of pigs from the same farm and also in different sites. Van der Wolf (2000) did a longitudinal study on seven high sero-positive and five low sero-positive farms. All of the positive farms remained contaminated, but even on these farms some groups of pigs and pens were negative. The low sero-positive farms remained free from *Salmonella* for long periods but were always at risk of infection from the introduction of infected animals or persistent environmental contamination.

A six-year study by Davies et al. (2001) on a four hundred–sow farrow-to-finish farm initially detected *Salmonella* Typhimurium DT104 in breeding stock and in weaner and finishing pigs as well as three other serovars in the breeding stock. In the second year, a different genotype of DT104 was detected, and by the third year two different phage types of *S*. Typhimurium appeared, as well as a different serovar in the breeding stock. A study by Lo Fo Wong et al. (2001) assessed the stability of an assigned *Salmonella* status of finishing pig herds over a two-year period. Of the finishing herds studied, 62 percent changed from their initial status at least once and steady status periods varied from one month to more than two years. The odds for finishers being positive were ten times higher if growers were positive in the previous round of sampling than if they were seronegative. When *Salmonella* was detected in pen fecal samples in the same sampling round, the herd was four times more likely to be seropositive, as compared with the absence of *Salmonella*.

AGE OF PIGS. Variations have been reported as to the age group of pigs in which *Salmonella* is most prevalent. Van Schie and Overgoor (1987) found a high *Salmonella* prevalence in sows, with the boars' areas being the most contaminated, causing a constant cycle of infection in the non-immune gilts. However, Barber et al. (1999) found *Salmonella* in all age groups, with the highest prevalence in finishers. On the other hand, Weigel et al. (1999) found the prevalence increased with age on the six farrow-to-finish farms that they investigated, with 1.4 percent in suckled pigs and 6.2 percent in sows. Funk et al. (1999) found wide variation in the age of pigs in which the prevalence of *Salmonella* was highest, which suggests that the prevalence may relate to the type of farm and its management. In Canada, Letellier et al. (1999) found that 15.9 percent of the replacement sows and 21.9 percent of the gilts were positive. They suggested that vertical transmission was occurring throughout all the different production stages. In the United Kingdom, Davies and Wray (1997a) found that on breeder farms, gilts and boars were more heavily infected with *S.* Typhimurium than were adult sows, and on a breeder/rearer farm, the prevalence for weaners, growers, and fatteners ranged from 32–44 percent. However, the role of vertical transmission has been questioned because of the success of strategic removal of weaners to clean premises (Dahl et al. 1997). Davies et al. (1998) examined feces obtained from 792 pigs housed on seven farms: one gilt development farm, two breeding farms, one nursery farm, and three finishing farms. *Salmonella* was isolated from all farms and from 12 percent of feces. The prevalence of *Salmonella* ranged from 3.4 percent at the gilt development farm to 18 and 22 percent, respectively, at the breeding farms. The most prevalent serovar on the finishing farms was *S.* Typhimurium, which was not isolated on the breeding or nursery farms. They consequently concluded that vertical transmission was unimportant as a source of *Salmonella* for finishers and that the high prevalence in breeders implicated feed. More recent publications indicate that although breeding herds may be a minor source of infection for finishers, they can play an important role in the maintenance of *Salmonella* on farms and its transmission to other farms (Davies et al. 2000).

TYPE OF PRODUCTION. The *Salmonella* seroprevalence in finishing pigs and sows for regular, free-range, and biologic-dynamic (certified system with extra controls over free range) pig herds in the Netherlands was determined by van der Wolf (2000) (Table 9.3). In the free-range finishers, the *Salmonella* seroprevalence was significantly higher than in the other two systems, although the sows do not differ significantly from each other. The *Salmonella* prevalence was studied on organic, free-range, and conventional pig farms by Wingstrand et al. (1999). They found that the relative risk for free-range as opposed to conventional farms was 1.7 with an apparent increase in risk for organic farms, although the number was too small to make valid conclusions. Wingstrand et al. (1999) suggested that because the number of alternative systems is likely to increase in the future, more studies should focus on the identification of risk factors for these systems.

Sheep. *Salmonella* infection in sheep has been described in most countries. Given the global size of the sheep population, it is perhaps surprising that sheep seldom appear to be associated with human salmonellosis. This may relate to the more extensive range production systems used for sheep, but outbreaks of salmonellosis can occur during lambing and congregating in feedlots prior to transport (Murray 2000).

Poultry. After a *Salmonella* serovar with an affinity for poultry has become established in a primary breeding flock, it can infect poultry in other units via hatcheries by both vertical and lateral transmission. This can have far-ranging and serious effects on the health of both poultry and humans. If a breeding flock is infected with *Salmonella*, a cycle can be established in which the organism passes via the egg to the progeny and even to chicks hatched from eggs laid by these infected progeny. This cycle can occur by true ovarian transmission, as is the case in pullorum disease (*S.* Pullorum) and *S.* Enteritidis and a few other serovars (Poppe 2000). It is, however, much more likely to happen through fecal contamination of the surface of the egg. As the egg passes through the cloaca, *Salmonella* in feces attach to the warm, wet surface of the shell and the organism may be drawn inside as it cools. Penetration of the shell by *Salmonella* will occur more readily if the eggs are stored at above room temperature. When the chicks from infected eggs hatch (86 percent hatchability), there is ample opportunity for lateral spread to contact chicks in hatcheries. Air sampling has shown a marked increase in hatchery contamination at twenty days of incubation (Cason et al. 1994). *Salmonella* are thus commonly introduced into a unit with purchased poultry in which infection may be transmitted by direct contact between chickens or by indirect contact with contaminated environments through ingestion or inhalation. *Salmonella* can also be introduced into a country with imports of live poultry or hatching eggs. A two-year study of the effectiveness of cleaning and disinfecting broiler farms and the persistence of *Salmonella* in two integrated broiler companies was done by Davies et al. (2001). They found that a cleaning and disinfecting regime was highly effective in preventing carry-over of infection in the houses, but both companies had persistent contamination problems in their feed mills. The hatchery-incubators of both companies were also persistently contaminated with *Salmonella*. The results of their study demonstrate the diversity of

Table 9.3—*Salmonella* Seroprevalence for Finishing Pigs and Sows in the Dutch Pig Population (van der Wolf 2000).

	No. of samples	Percentage positive
Regular finishers	1760	11.1
Regular sows	1902	9.9
Free-range finishers	56	25.0
Free-range sows	66	10.6
Biologic-dynamic finishers	16	12.5
Biologic-dynamic sows	42	4.8

sources of *Salmonella* within large poultry organizations and the problems of control.

It is thus of considerable importance to establish how infection can be introduced into a *Salmonella*-free breeding flock (Fig. 9.2). When poultry over four weeks of age are infected with *Salmonella*, the organism may colonize the intestine, but almost all such chickens free themselves of infection within sixty days. However, a small percentage of infected birds will excrete *Salmonella* continuously or intermittently for long periods. Stress, such as coming into lay, may also reactivate infection and excretion. Although the large majority of infected birds carry *Salmonella* for only short periods, the conditions in intensive poultry units allow considerable scope for infection to recycle from bird to bird. The period of *Salmonella* excretion by an infected bird may also be extended by antibiotic treatment. At harvest, it is likely that a few birds will remain infected, and further transmission may take place during transport to the abattoir or during the slaughter process.

SUMMARY AND CONCLUSIONS. *Salmonella* control presents a major challenge to all involved in animal production because of its complex epidemiology and the many routes by which the organism can be transmitted. It is apparent that control must be based on a detailed knowledge of the epidemiology and ecology of the organism, and a detailed control program must be developed for each individual unit. Many strategies have been tried and tested, such as slaughter, immunization, and the use of antibacterials. None was successful on its own, but improved hygiene and disease security, combined with vaccination, has had a major impact in the poultry and swine industries. There is increasing recognition that a farm-to-table approach is necessary to reduce pathogens, and good manufacturing practices (GMPs) using hazard analysis critical control points (HACCP) principles are being used increasingly to effect reduction. Using HACCP systems applied to critical control points on farms and supplies of feed and services will be beneficial in the future.

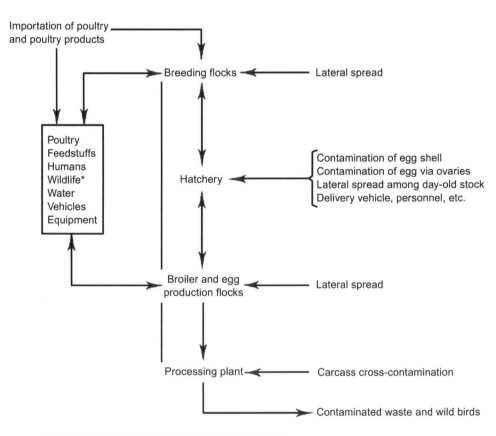

*Wildlife includes vermin, wild birds, flies and insects

FIG. 9.2—Cycles of *Salmonella* infection in poultry.

REFERENCES

Altekuse, S., J. Koehler, F. Hickmann-Brenner, R.V. Tauxe, and K.Ferris. 1993. A comparison of *Salmonella* enteritidis phage types from egg-associated outbreaks and implicated laying flocks. *Epidemiol. and Infect.* 110: 17–22.

Barber, D.A., R.M. Weigel, R.E Isaacson, P.B. Bohnson, and C.J. Jones. 1999. Spatial and temporal patterns of the distribution of *Salmonella* on swine farms in Illinois. *Proceedings of the 3rd International Symposium on the Epidemiology and Control of Salmonella in Pork.* Washington, D.C., Aug 5–7, 1999, 97–100.

Baskerville, A., and C. Dow. 1972. Ultrastructure of phagocytosis of *Salmonella* choleraesuis by pulmonary macrophages *in vivo. Br. J. Exper. Pathol.* 53:641–647.

Boyce, T.G., D. Koo, D.L. Swerdlow, T.M. Gomez, B. Serano, L.N. Nickey, F.W. Hickman-Brenner, G.B. Malcolm, and P.M. Griffin. 1996. Recurrent outbreaks of *Salmonella* Enteritidis infection in a Texas restaurant: phage type 4 arrives in the U.S. *Epidemiol. and Infect.* 117:29–34.

Brownlie, L.E., and F.H. Grau. 1967. Effects of food intake on growth and survival of *Salmonella*s and *Escherichia coli* in the bovine rumen. *J. Gen. Microbiol.* 46:125–134.

Bush, E.J., B. Wagner, and P.J. Fedorka-Cray. 1999. Risk factors associated with shedding of *Salmonella* by U.S. finishing hogs. *Proceedings of the 3rd International Symposium on the Epidemiology and Control of Salmonella in Pork.* Washington, D.C., Aug 5–7, 1999, 106–109.

Carlson, A.R., and T. Blaha. 1999. Investigations into the infection/contamination cycle of zoonotic *Salmonella* on swine farms. Investigations of the simultaneous occurrence of *Salmonella* on 25 selected Minnesota farms. *Proceedings of the 3rd International Symposium on the Epidemiology and Control of Salmonella in Pork.* Washington, D.C., Aug 5–7, 1999, 113–118.

Cason, J.A., N.A. Cox, and J.S. Bailey. 1994. Transmission of *Salmonella* typhimurium during hatching of broiler chicks. *Avian Dis.* 38:583–588.

Center for Disease Control and Prevention. 1998. Annual Report, NARMS National Antimicrobial Resistance Monitoring System: Enteric Bacteria. Atlanta, GA.

Christensen, J., D.L. Baggesen, A.C. Nielsen, B. Nielsen. 1999. *Proceedings of the 3rd International Symposium on the Epidemiology and Control of Salmonella in Pork.* Washington, D.C., Aug 5–7, 1999, 333–335.

Clark, G.M., A.F. Kauffmann, and E.J. Gangerosa. 1973. Epidemiology of an international outbreak of *Salmonella* agona. *Lancet* 2:490–493.

Clegg, F.G., S.N. Chiejina, A.L. Duncan, R.N. Kay, and C. Wray. 1983. Outbreaks of *Salmonella* newport infection in dairy herds and their relationship to management and contamination of the environment. *Vet. Rec.* 112:580–584.

Dahl, J., A. Wingstrand, D.L. Baggesen, and B. Nielsen. 1997. Elimination of S. typhimurium infection by the strategic movement of pigs. *Vet. Rec.* 140:679–681.

Damman, D., P.B. Bahnson, R.M. Weigel, R.E. Isaacson, H.F. Troutt, and J.Y. Kim. 1999. An estimate of *Salmonella* prevalence on Illinois swine farms using mesenteric lymph node cultures. *Proceedings of the 3rd International Symposium on the Epidemiology and Control of Salmonella in Pork.* Washington, D.C., Aug 5–7, 1999, 123–125.

Dargatz, D.A., P.J. Fedorka-Cray, S.R. Ladely, and K.E. Ferris. 2000. Survey of *Salmonella* serotypes shed in feces of beef cows and their antimicrobial susceptibility patterns. *J. Food Prot.* 63:1648–1653.

Davies, P.R., W.E.M. Morrow, F.T. Jones, J. Deen, P.J. Fedorka-Cray, and I.T. Harris. 1997. Prevalence of *Salmonella* serotypes from faeces of pigs raised in different production systems in North Carolina, USA. *Epidemiol. and Infect.* 119:237–244.

Davies, P.R., F.G.E. Bovee, J.A. Funk, W.E.M. Morrow, F.T. Jones, and J. Deen. 1998. Isolation of *Salmonella* serotypes from feces of pigs raised in multiple site production systems. *J.A.V.M.A.* 212:1925–1929.

Davies, P.R., J.A. Funk, and W.E.M. Morrow. 2000. Fecal shedding of *Salmonella* by gilts before and after introduction to a swine breeding farm. *Swine Health and Prod.* 8:25–29.

Davies, R.H., M. Breslin, J.E.L. Corry, W. Hudson, and V.M. Allen. 2001. Observations on the distribution and control of *Salmonella* species in two integrated broiler companies. *Vet. Rec.* 149:227–232.

Davies, R.H., and M.H. Hinton. 2000. *Salmonella* in Animal Feed. In *Salmonella in Domestic Animals.* C. Wray and A. Wray, eds. Wallingford, Oxon, UK: CABI, 285–300.

Davies, R.H., and C. Wray. 1996. Persistence of *Salmonella* Enteritidis in poultry units and poultry food. *Br. Poult. Sci.* 37:589–596.

Davies, R.H., and C. Wray. 1997a. Distribution of *Salmonella* on 23 pig farms in the UK. *Proceedings of the 2nd International Symposium on Epidemiology and Control of Salmonella in Pork.* Copenhagen, Denmark, Aug. 20–22, 1997, 137–141.

Davies, R.H., S. Bedford, and I. M. McLaren. 2001. A six year study of the persistence of S. Typhimurium DT104 on a farrow to finish pig farm. *Proceedings of the 4th International Symposium on Epidemiology and Control of Salmonella and other Food borne Pathogens in Pork.* Leipzig, Germany, 2–5 Sept., 2001, 265–273.

De Jong, H., and M.O. Ekdahl. 1965. Salmonellosis in calves—the effect of dose rate and other factors on transmission. *N. Zeal. Vet. J.* 13:59–64.

Evans, S.J., and R.H. Davies. 1996. Case control of multiple-resistant *Salmonella* typhimurium DT104 Infection of Cattle in Great Britain. *Vet. Rec.* 139:557–558.

Fedorka-Cray, P.J., L.C. Kelley, T.J. Stabel, J.T. Gray, and J.A. Laufer. 1995. Alternate routes of invasion may affect pathogenesis of *Salmonella* typhimurium in swine. *Infect. and Immun.* 63:2658–2664.

Fedorka-Cray, P.J., D.A. Dargatz, L.A. Thomas, and J.T. Gray. 1998. Survey of *Salmonella* serotypes in feedlot cattle. *J. Food Prot.* 61:525–530.

Fedorka-Cray, P.J., J.T. Gray, and C. Wray. 2000. *Salmonella* infection in pigs. In *Salmonella in Domestic Animals.* C. Wray and A. Wray, eds. Wallingford, Oxon, UK: CABI, 191–207.

Ferris, K.E., and D.A. Miller. 1996. *Salmonella* serovars from animals and related sources reported during July 1995–June 1996. *Proceedings of the 104 Annual Meeting of the United States Animal Health Association* 100:505–526.

Ferris, K.E., B.R. Flugrad, J.M. Timm, and A.E. Ticer. 2000. *Salmonella* serovars from animals and related sources reported during July 1999–June 2000. *Proceedings of the 104 Annual Meeting of the United States Animal Health Association,* Birmingham, Alabama, Oct. 20–27, 2000, 521–526.

Funk, J.A., P.R. Davies, and M.A.B. Nichols. 1999. Longitudinal study of *Salmonella* shedding in 2 three-site swine production systems. *Proceedings of the 3rd International Symposium on the Epidemiology and Control of Salmonella in Pork.* Washington, D.C., Aug. 5–7, 1999, 131–136.

Gay, J.M., and M.E. Hunsaker. 1993. Isolation of multiple *Salmonella* serovars from a dairy two years after a clinical salmonellosis outbreak. *J.A.V.M.A.* 203:1314–1320.

Gray, J.T., P.J. Fedorka-Cray, T.J. Stabel, and T.T. Kramer. 1996. Natural transmission of *Salmonella* choleraesuis in swine. *Appl. Environ. Microbiol.* 62:141–146.

Hardman, P.M., C.M. Wathes, and C. Wray. 1991. Transmission of *salmonellae* among calves penned individually. *Vet. Rec.* 129:327–329.

Heard, T.W., N.E. Jennet, and A.H. Linton. 1972. Changing patterns of *Salmonella* excretion in various cattle populations. *Vet. Rec.* 90:359–364.

Henzler, D.J., and H.M. Opitz. 1992. The role of mice in the epizootiology of *Salmonella* enteritidis infection on chicken layer farms. *Avian Dis.* 36:626–631.

House, J.K, B.P. Smith, D.W. Dilling, and L.D. Roden. 1993. Enzyme-linked immunosorbent assay for serological detection of *Salmonella* dublin carriers on a large dairy. *A.J.V.R.* 54:1391–1399.

Letellier, A., S. Meeier, J. Pare, J. Menard, and S. Quessy. 1999. Distribution of *Salmonella* in swine herds in Quebec. *Vet. Microbiol.* 67:299–306.

Lo Fong Wong, D.M.A., J. Dahl, J.S. Andersen, A. Wingstrand, P.J. van der Wolf, A. Von Altrock, and B.M. Thorberg. 2001. A European longitudinal study in *Salmonella* seronegative and seropositive classified finishing herds. *Proceedings of the 4th International Symposium on Epidemiology and Control of Salmonella and other Food borne Pathogens in Pork*. Leipzig, Germany, Sept. 2–5, 262–264.

MAFF. 2000. *Salmonella* in livestock production. Veterinary Laboratories Agency (Weybridge), Addlestone, UK, KT15 3NB.

Morrise, J.P., J.P. Cotte, G. Argente, and L. Daniel. 1992. Approche épidémiologique de l'excrétion de salmonelles dans un réseau de 50 exploitations bovine laitiers avec et sans antécédents cliniques. *Annales de Médicine Vétérinaire* 136:403–409.

Murray, C.J. 2000. Environmental Aspects of *Salmonella* in *Salmonella in Domestic Animals*. C. Wray and A. Wray, eds. Wallingford, Oxon, UK: CABI, 265–284.

Richardson, A. 1973. The transmission of *Salmonella* dublin to calves from adult carrier cows. *Vet. Rec.* 92:112–115.

Richardson, A. 1975. Salmonellosis in cattle (review). *Vet. Rec.* 96:329–321.

Robinson, R.A., B.O. Blackburn, C.D. Murphy, E.V. Morse, and M.E. Potter. 1984. Epidemiology of *Salmonella* dublin in the USA. *In Proceedings of the International Symposium on Salmonella*. New Orleans, 1984. American Association of Avian Pathologists, New Bolton Center, Kennet Square, PA, 182–193.

Pearce, G.D., and D. Revitt. 1998. Prevalence of *Salmonella* excretion in grower finisher pigs in the UK. *Proceedings of the 15th IPVS Congress*. Birmingham, England. July 5–9, 1998, 223.

Poppe, C. 2000. *Salmonella* infections in the domestic fowl. In *Salmonella in Domestic Animals*. C. Wray and A. Wray, eds. Wallingford, Oxon, UK: CABI, 107–132.

Quessy, S., A. Leteelier, and E. Nadeau. 1999. Risk factors associated with the prevalence of *Salmonella* in swine herds in Quebec. *Proceedings of the 3rd International Symposium on the Epidemiology and Control of Salmo-*

nella in Pork. Washington, D.C., Aug 5-7, 1999, 165–168.

Roberts, J.A. 1989. The national health service in the UK: from myths to market. *Health Policy Planning* 4:62–71.

Saeed, A.M. 1999. *Salmonella enterica Serovar Enteritidis in Humans and Animals. Epidemiology, pathogenesis and control*. Ames, IA: Iowa State University Press.

Schneitz, C. and G. Mead. 2000. Competitive Exclusion. In *Salmonella in Domestic Animals*. C. Wray and A. Wray, eds. Wallingford, Oxon, UK: CABI, 301–322.

Smith, B.P., L.D. Roden, M.C. Tharmond, G.W. Dilling, H. Konrad, J.A. Pelton and J.P. Picanso. 1994. Prevalence of *salmonellae* in cattle and in the environment on California dairies. *JAVMA*. 205:467–471.

Stege, H, J. Christensen, J.P. Nielsen D.L. Baggesen, C. Ense, and P. Willeberg. 2000. Prevalence of sub-clinical *Salmonella* enterica infection in Danish finishing herds. *Prev. Vet. Med.* 44:175–188.

Todd, E.C.D. 1989. Preliminary estimates of costs of foodborne disease in the United States. *J. Food Prot.* 595-601.

van der Wolf. 2000. *Salmonella* in the pork production chain: feasibility of *Salmonella* free pig production. Thesis, University of Utrecht, the Netherlands.

van Schie, F.W., and G.H.A. Overgoor. 1987. An analysis of the possible effects of different feed upon the excretion of *Salmonella* bacteria in clinically normal groups of fattening pigs. *Vet. Quart.* 9: 185–188.

Wathes, C.M., W.A.R. Zaiden, G.R. Pearson, M.H. Hinton, and N. Todd. 1988. Aerosol infection of calves and mice with *Salmonella* typhimurium. *Vet. Rec.* 123:590–594.

Watson, W.A., B. Wood, and A. Richardson. 1971. *Salmonella* dublin infection in a beef herd. *Br. Vet. J.* 127:294–298.

Weigel, R.M., D.A. Barber, R.E. Isaacson, P.B. Bahnson, and C.J. Jones. 1999. Reservoirs of *Salmonella* infection on swine farms in Illinois. *Proceedings of the 3rd International Symposium on the Epidemiology and Control of Salmonella in Pork*. Washington, D.C., U.S.A. Aug 5–7, 1999, 180–183.

Williams, B.M. 1975. Environmental considerations in salmonellosis. *Vet. Rec.* 96:318–321.

Wingstrand, A., J. Dahl, and D.M.A. Lo Fo Wong. 1999. *Salmonella* prevalence in Danish organic, free-range conventional and breeding herds. *Proceedings of the 3rd International Symposium on the Epidemiology and Control of Salmonella in Pork*. Washington, D.C., U.S.A. Aug 5–7, 1999, 186–188.

Wray, C., and A. Wray. 2000. *Salmonella in Domestic Animals*. Wallingford, Oxon, UK: CABI.

Wray, C., N. Todd, and M.H. Hinton. 1987. The epidemiology of S typhimurium infection in calves: excretion of S. typhimurium in faeces in different management systems. *Vet. Rec.* 121:293–296.

Wray, C., Q.C. Wadsworth, D.W. Richards, and J.H. Morgan. 1989. A three-year study of *Salmonella* dublin in a closed dairy herd. *Vet. Rec.* 124:532–535.

10

SALMONELLA DETECTION METHODS

CAROL W. MADDOX

INTRODUCTION. Writing a chapter on the detection of *Salmonella* was a somewhat daunting task. The sheer volume of research in the field is awesome. More than 26,900 citations appear following a Google search of the Internet for "*Salmonella* detection." A Medline search yielded 1,761 published references available electronically; the literature, however, goes back over a century, beginning with the naming of *Salmonella* in 1900 after an American bacteriologist, D.A. Salmon. Because *Salmonella* is a leading cause of foodborne illness and a zoonotic agent capable of causing gastroenteritis and septicemia, *Salmonella* detection and identification has become the subject of a great deal of research. Numerous culture protocols have been developed and modified to reliably recover and characterize *Salmonella* spp. from a broad range of sources.

Recent emphasis on food safety issues and the development of guidelines using *Salmonella* as an indicator for enteric contamination of meats and potential pathogen in animal products has led to efforts to standardize detection techniques. Furthermore, the on-farm HACCP approaches to improving food safety and animal health have resulted in attempts to optimize antimortem, post-mortem, and environmental specimen collection, processing, and culture for *Salmonella* detection. Much recent emphasis has been placed upon more rapid detection techniques, streamlined protocols, and automated tests for food safety evaluation and epidemiological surveillance. This has led to a vast array of immunological and nucleic acid–based assays. Thus, the following is by no means a comprehensive review of detection methodology as much as it is an introduction to techniques and some examples of where the latest methods have been applied to solving animal health and food safety issues.

CURRENT CULTURE TECHNIQUES. Selective enrichment and differential media for culture of *Salmonella* has been a subject of debate since the early 1900s, with almost as many methods as investigators. Guth (1916) first exploited the increased susceptibility of coliforms to selenite to formulate a selective media for *Salmonella* as early as 1916. Jungherr, Hall and Pomeroy (1950) demonstrated the superior recovery and differentiation of *Salmonella* plated on bismuth sulfite agar from selenite broth cultures for the detection of *S.* Pullorum and *S.* Gallinarium from poultry organs and intestines. In 1923, Mueller obtained nearly pure cultures of *Salmonella* Typhi and Paratyphi from tetrathionate broth (TTB) despite the presence of a 1,000 to 10,000 fold excess of coliforms. More recent survey results published by Waltman and Mallinson (1995) indicated the great disparity in methods among laboratories performing *Salmonella* culture; none of the seventy-four respondent laboratories reported using the same methods for culture and isolation. Fourteen different media and various incubation conditions were reported, greatly affecting the sensitivity of detection.

Selective Enrichment Broth Media. It has become abundantly clear that the means of enrichment for *Salmonella* can have profound effects on the ability to recover *Salmonella* from animal sources, foods, water, or environmental samples. The method selected will vary with regard to the need for nonselective nutrient broths such as buffered peptone water (BPW) or trypticase soy broth (TSB) for resuscitating damaged cells from heat-stressed, desiccated, or otherwise less than ideal samples. Other specimens containing high numbers of competing bacteria, such as feces or ground meats, may require highly selective media to prevent overgrowth by coliforms that can readily out-compete the *Salmonella*. Incubation time and temperature also will influence the success of the enrichment process based upon the type and source of the specimen cultured. Current AOAC recommendations for raw flesh, highly contaminated foods, and animal feeds include an initial enrichment in Rappaport-Vassiladis (RV) medium followed by secondary selection in TTB rather than selenite broth based upon studies by June et al. (1995). Davies et al. (2001) compared BPW pre-enrichment followed by RV selective enrichment to primary enrichment in Hajna's TTB followed by RV secondary enrichment and found the latter to be more sensitive for detection of *Salmonella* from swine feces. Furthermore, they found that delayed secondary enrichment and incubation of RV at 42°C rather than 37°C increased the ability to detect *Salmonella* from swine. Results of Pennsylvania Egg Quality Assurance Program testing indicated that extended enrichment processing of poultry environmental drag swabs yielded 17 percent increase in the recovery of *Salmonella* while delayed secondary enrichment increased recovery by approximately 23 percent (Maddox, unpublished data). Studies to determine whether the

83

improvement is a result of twice the number of plates being examined (two XLD and BG plates versus one set of plates) have not been performed.

Funk et al. (2000) found a statistically higher proportion of *Salmonella* detected in feces with increasing specimen weight (P < 0.001). At a 10 percent overall prevalence, the predicted apparent prevalence increased from 0.9 percent for rectal swabs to 6.5 percent with 25-gram specimens. There also is a profound effect of specimen size on sample number to reach the 95 percent confidence level. Ten- and 25-gram specimens gave similar results at 25 percent overall prevalence; 30 samples were required to detect one positive animal in a herd of 1,000 hogs. Only six 25-gram samples would be required to detect one animal at a 50 percent overall prevalence in a herd of 1,000 hogs. The sensitivity of detection ranged from 0.57–0.78 over three studies using 25-gram specimens.

Recent reports by Worcman-Barninka et al. (2001) indicated that a modified semisolid Rappapport-Vassiladis media (MSRV) could be used in a more rapid selection process that detected 96.1 percent of *Salmonella*-positive food samples versus 84.6 percent with the current FDA protocol (Andrews et al. 1998). Their MSRV process had 95.5 percent sensitivity and 96.8 percent specificity with *Salmonella*-contaminated food samples. Work by Koivuner et al. (2001) found Preuss TTB superior to previous two-step enrichment methods utilizing BPW and RV for detection of *Salmonella* in wastewater.

Agar Plating Media. The Koivuner study also indicated superior selection and differentiation using xylose lysine deoxycholate agar (XLD) and brilliant green agar with MgCl2 (BM), enabling detection in the range of 3->1100 CFU/100 ml of waste water based on most probable number analysis (Koivuner et al. 2001). Miller and Tate et al. (2001) report further improvement in recovery and correct identification using XLT4 that utilizes tergitol rather than deoxycholate suppression of coliforms. Addition of novobiocin has been used by many investigators to reduce contamination by *Proteus sp.* and *Citrobacter sp.* (Hoban et al. 1973; Poisson 1992; Waltman, Horne, et al. 1995). Komatsu and Restaino (1981), reported that recovery of *Salmonella* increased from 50 percent to 82 percent and false positive colonies decreased from 38 percent to 5 percent with the addition of novobiocin to XLD, XLT4, BG, and HE. One element that cannot be overlooked that consistently affects the sensitivity and specificity of the culture data is the bacteriologist's skill or expertise for recognition of *Salmonella* suspect colony morphologies on differential or selective media with which the bacteriologist may or may not be familiar. This media bias often results in reductions in recovery from media other than that currently used in his or her laboratory.

IMMUNOCAPTURE OF *SALMONELLA* ORGANISMS. Many commercially available methods to

enhance selection of *Salmonella* organisms by means of antibody capture have been increasingly utilized to improve the sensitivity and specificity of antigen detection. Immunomagnetic beads, coated with anti-*Salmonella* LPS or flagellar antibodies, have been used to capture *Salmonella* from enrichment broth suspensions. The beads are then plated by rolling them across the agar surface of differential or selective media (or both) such as XLD or BG. More sophisticated methods of capture are used in the Threshold (Immunoassay system that utilizes solution-based binding of biotin- and fluorescein-labeled antibodies to *Salmonella* followed by filtration capture on a nitrocellulose membrane. Anti-fluorescein-urease conjugate is bound to the immunocomplex and then detected via chip-based light-addressable potentiometric sensor (LAPS) (Dill et al. 1999). Conversion of urea to ammonia and CO2 results in a pH change at the silicon surface reported in μVs-1. Detection levels as low a 119 cfu/ml were reported for chicken carcass washes utilizing the LAPS system. More rapid results have been obtained using a quartz crystal, surface-mounted with anti-*Salmonella* polyclonal antibodies that alter the crystal resonance when they bind *Salmonella* organisms (Pathirana et al. 2000). The investigators report a detection limit of 102 cells per ml with a linear response to 107 cells per ml.

For food safety purposes, especially with processed foods, determining the viability of organisms in specimens is important. A membrane-immunofluorescent–viability staining technique based upon the previous method of Cloak et al. (1999) was investigated for detection of *Salmonella* from fresh and processed meat (Duffy et al. 2000). Following enrichment in BPW, cells were bound to a polycarbonate membrane followed by detection with anti-*Salmonella* monoclonal antibody and the Sytox Green viability stain. Unfortunately, freeze-thaw damage to the bacterial membranes resulted in poor correlations between viable stain numbers and direct plate counts. Likewise, heating disrupts the lipoprotein complexes of the membranes and leads to poor accuracy of the viable stain technique. The correlations of viability staining with culture counts in low pH (3.5) and high salt (4 percent) were also poor (r2 = 0.04 and 0.54, respectively). However, there was much better correlation between the techniques for fresh meats (r2 = 0.83 to 0.92). Application to testing meat juice samples was promising.

The VIDAS (Vitek Immunodiagnostic Assay System) is an automated quantitative fluorescent ELISA that uses antibodies specific for *Salmonella* O and H antigens to capture *Salmonella*. Walker et al. (2001) used the Moore swab conventional culture method (CCM) testing one liter of milk and compared the technique to 25 ml milk samples suggested by VIDAS. Using CCM, they detected an additional seventeen positive samples. When the Moore swab technique was used for both assays, there was good agreement (97.5 percent) for the twenty positive bulk tank milk samples. The authors demonstrated 95.5 percent agreement between CCM and VIDAS for similarly processed in-line milk filters.

SEROPREVALENCE OF *SALMONELLA*. Neilson et al. (1995) prepared an anti-LPS ELISA using a mixture of antigens generated using *S.* Typhimurium and *S.* Livingstone, which produce LPS types O1, O4, O5, O6, O7, and O12. These antibodies detect exposure to more than 90 percent of the *S. enterica* serovars. Using this system, Van der Wolfe et al. (2001) examined the seroprevalence of *Salmonella* among swine from the Netherlands. The prevalence in finishing pigs ranged from 10.4 percent to 23.7 percent, whereas sows averaged 40 percent to 40.5 percent positive. Wiuff et al. (2002) reported moderate seroprevalence of *Salmonella* B and C1 mixed ELISA optical densities (ODs) in slaughter hogs from Sweden, where culture of mesenteric lymph nodes and feces demonstrated only 0.1 percent prevalence. However, using the Swedish ELISA to detect antibodies to *Salmonella* LPS, 45 percent of Swedish pigs had ODs of 10–40 percent (considered a suspect positive) and 5 percent had OD's of less than 10 percent. These results took into account reported cross reactivities against *Salmonella* sergroup E1, *Citrobacter freundii*, *Yersinia enterocolotica* O:3, and *E. coli* O139. Danish reference sera had a much higher avidity than the Swedish sera when evaluated by ELISA. Proux et al. (2000) demonstrated a significant correlation between seroprevalence of *Salmonella* among French swine as compared to bacterial culture of mesenteric lymph nodes (P < 0.01). Using serovars Typhimurium, Anatum, Hadar, Infantis, and Enteritidis in their indirect ELISA, they identified that maternal antibodies persisted for the first seven weeks and seropositive animals were first detected at twenty weeks of age.

Czerny et al. (2001) evaluated the meat juice ELISA for *Salmonella* in Bavaria for 3,048 slaughter hogs from fifty-two farms. The indirect ELISA using diaphragm pillar extract demonstrated *Salmonella*-specific antibodies in forty-eight carcasses from twelve farms, of which one farm represented 68.8 percent of the positive samples. Ninety–eight percent (fifty-one farms) belonged to category I with less than 20 percent prevalence for the samples.

Galland et al. (2000) tested 2,731 beef steers from Kansas feedlots by both culture and for evidence of *Salmonella* infection as measured by serum ELISA, using the methods of Smith et al. (1989). The positive culture rate varied from 40 percent immediately after placement of the steers into pens, to none detected when the steers were shipped for slaughter. The number of animals with positive titers for group B antibodies increased to greater than 50 percent on day sixty, and then decreased to half that number by slaughter. There were no significant changes in C1, C3, D1, or E1 titers; they remained low throughout. This finding was consistent with the fact that most clinical infections were predominantly caused by *S.* Typhimurium.

House et al. (2001) examined serological responses to *S.* Dublin in dairy cattle and found a much higher ratio of IgG2:IgG1 LPS titers in infected versus vaccinated cattle. Vaccination resulted in rapidly rising IgG1 titers to both LPS and porins. Wedderkopp et al. (2001) found the ELISA useful for predicting *S.* Dublin prevalence in Danish herds where the conversion from test-negative to test-positive status corresponded to the incidence of new *S.* Dublin outbreaks (r2 = 0.48, n = 19, P < 0.025).

Gast et al. (1997) and many other investigators have successfully used ELISA tests based upon egg-yolk antibodies to *Salmonella* flagella to predict *S.* Enteritidis (SE) infection rates in layers. Although the predictive value on a per-bird basis does not correlate well with culture results because of the extremely low incidence of SE shedding in eggs, the assay is a very good predictor of flock infectious status, as good or better than fecal culture.

NUCLEIC ACID DETECTION. DNA probes prepared from the virA gene, characterized by Gulig and Curtiss (1988) have been used to detect *Salmonella* via colony blot hybridization techniques, dot blots, and others (Maddox and Fales 1991). The genus-specific probe was highly specific but the sensitivity of the assay was only 107 cfu/ml. Using a commercially available colorimetric DNA/rRNA sandwich hybridization kit, Namimatsu et al. (2000) demonstrated probe sensitivity of 105 cfu/ml and 99.5 percent specificity for TTB enriched poultry samples. The ninety-six well format allowed for high throughput but the sensitivity was still less than can be achieved with newer polymerase chain reaction (PCR) techniques.

A number of direct-detection protocols have been developed for identifying animals exposed to *Salmonella*, potential carriers or shedders. Screening of horses was one of the first applications of polymerase chain reaction technology to detect carrier animals at stud farms and veterinary hospitals. Amavist et al. (2001) recently reported on the use of an ompC based primer set. Using spin columns to purify DNA from feces, the authors were able to detect 100 organisms in a 50 µl sample obtained by suspending 1 g of feces in 10 ml of H_2O. Similar sensitivity was obtained when one gram of feces was inoculated into 10 ml of selenite broth and incubated 16–24 hours. Inhibition of the polymerase reaction enzymes by fecal components poses a significant issue. Commercially available purification columns or kits now contain substances that inactivate or capture most of the natural inhibitors, substantially increasing the success rate for PCR assays. The increased sensitivity of PCR over direct culture was considerable: 40.6 percent were positive by PCR after selenite enrichment; 38.5 percent were positive by direct PCR of feces; whereas only 2 percent were positive by culture. Incorporation of an internal control to determine whether inhibition of the PCR reaction has occurred is suggested to avoid reporting false negative results.

Use of selective enrichment culture does enhance the ability to detect low-level shedders by means of PCR as exhibited by the results of Maciorowski et al. (2000). These investigators also reported reduction of several days in the time required for *Salmonella* identification using the BAX system as opposed to culture alone. Cooke et al. (2001) reported a proportional agreement of 98.9 percent for BAX PCR versus an invA PCR protocol modified from Chiu and Ou (1996).

A nested PCR assay by Waage et al. (1999) was evaluated for use in water, sewage, and foods. The assay correctly identified 128 of 129 strains of *Salmonella* subspecies I, II, IIIb, and IV, and identified none of the 31 non-*Salmonella* strains with a sensitivity of 2 cfu (10cfu/100ml of water) and <10cfu/g of food with enrichment broths (1:10 overnight growth in trypticase soy broth) or with no enrichment in 1 percent peptone water.

Determining the viability of *Salmonella* in animal products is critical, because processing may not effectively destroy many newly emerging heat and chemical resistant strains. The use of reverse transcriptase to prepare cDNA from mRNA of specific target genes has provided a means of distinguishing between viable *Salmonella* (newly transcribed mRNA) and non-viable *Salmonella* (static amplification). The choice of gene and resultant mRNA may affect the validity of this approach. For example, Szabo and Mackey (1999) found that *sefK* transcription is affected by variations in temperature and pH, thus altering the sensitivity of the RT-PCR assay. Their modified protocol was able to detect ten viable cells of *Salmonella* Enteritidis PT4 from minced beef or whole eggs following sixteen-hour incubation in buffered peptone water, pH 7.2, 37°C. However, other housekeeping genes are transcribed at a more constant rate, particularly genes involved in DNA replication or transcription. This approach has been exploited by Simpkins et al. (2000) using a technique called nucleic acid sequence-based amplification (NASBA) for the detection of viable *Salmonella* enterica. Using NASBA, RNA is amplified at temperatures below the denaturing temperature of dsDNA (<80°C). Although rRNA is often the target amplification of rRNA, the stability of rRNA is such that products may be made from non-viable cells. In contrast, mRNA has a very short half-life and is rapidly degraded by nucleases (Rnases) and, as a result, is a better target for evaluating viable versus non-viable bacterial populations. With primers derived from the *Salmonella* Typhimurium *dnaK*, the investigators used the NucliSens kit (Organon Teknika, Cambridge, UK) to perform NASBA on silica-guanidine thiocyanate-extracted mRNA in 8M urea. Amplicons were detected using a NucliSens ECL reader at 620nm that detects consumption of tripropylamine caused by oxidation-reduction of rutheniumion.

Comparison of Fecal Culture to PCR Detection.
Though results varied widely based upon the enrichment technique used (73.3 percent for MC, 68.8 per-

cent for TSB, and 24 percent for RV), Feder et al. (2001) demonstrated 76 percent agreement between pre-enriched culture results and PCR; however, their PCR assay failed to detect 28 percent of the *Salmonella* culture positive samples. A more recent study by Cooke et al. (2001) compared the use of an invA-based PCR reaction, a commercial BAX PCR detection method, and culture using a primary enrichment with TTB and a secondary enrichment with RV of bovine fecal samples. The authors reported 98.9 percent agreement between the PCR assays, 95.9 percent agreement for the *invA* assay versus culture, and 94.9 percent agreement for the BAX PCR versus culture.

SUMMARY AND CONCLUSIONS. It is important to realize the significance of the selective and enrichment processes, not only for the successful culture of *Salmonella* but also concerning the capability of subsequent detection methods now in practice. Most methods, such as immunoassays and nucleic acid–based detection systems, utilize an initial selective enrichment broth step. Although this technique represents an overnight delay as compared to assays performed directly on specimens, the selective enrichment step has several advantages. First, it provides for natural amplification by growth of viable organisms. Next, it provides a source of organism for further testing such as antimicrobial susceptibilities, serotyping, and phenotypic or genotypic analysis. The selective enrichment culture usually represents at least a 1:10 dilution of the specimen; often, a subsequent 1:100 dilution is made, which is sufficient to dilute competing coliforms in the immunoassays and dilution of many inhibitors in the PCR assays. Immunoassays impart a specificity that facilitates detection of *Salmonella* amidst mixtures of other organisms, simplifying the identification process and increasing the sensitivity.

In addition to time savings, many of the newer assays represent significant improvements in sensitivity. Nucleic acid probes resulted in detection limits from 10^5 to 10^7 cfu/ml, whereas PCR detection limits approach 10^2 to 10^3 cfu/ml. Nucleic acid assays also represent a substantial time and cost savings because only probe or PCR positive enrichment broth cultures need to be plated. The reliability of PCR results (>95 percent correlation with culture) indicate that it is suitable means of screening large numbers of isolates for animal health surveillance, HACCP or food safety testing, or epidemiological studies.

REFERENCES
Amavist, P., G.F. Browning, D. Lightfoot, S. Church, G.A. Anderson, K.G. Whitehear, and P.F. Markham. 2001. Rapid PCR detection of *Salmonella* in horse faecal samples. *Vet. Microbiol.* 79:63–74.
Andrews, W.H., T.S. Hammack, and R.M. Amaguana. 1998. Chapter 5: *Salmonella*. In *U.S. FDA Bacteriological Analytical Manual*, 8th Edition. Arlington, VA: AOAC International.

Chiu, C.H., and J.T. Ou. 1996. Rapid identification of *Salmonella* serovars in feces by specific detection of virulence genes invA and spvC, by an enrichment broth culture-multiplex PCR combination assay. *J. Clin. Microbol.* 34:2629–2622.

Cooke, C.L., R.S. Singer, C.W. Maddox, and R.E. Isaacson. 2001. A PCR screen for *Salmonella* in pooled enrichment broth cultures of bovine feces. *82nd Annual Meeting of Conf. Res. Workers in Animal Dis.* Nov.11, 2001, St. Louis, MO.

Cloak O.M., G. Duffy, J.J, Sheridan, D.A. McDowell, and I.S. Blair. 1999. Development of a surface adhesion immuno-fluorescent technique for the rapid detection of *Salmonella* spp. from meat and poultry. J. *Appl. Microbiol.* 86:583–590.

Czerny, C.P., K. Osterkorn, G. Wittkowski, and M. Huber. 2001. Meat juice ELISA for determination of *Salmonella* incidence in slaughter pig herds in *Bavaria. Berl Munch Tieraztl Wochenschr* 114:35–39.

Davies, P.R., P.K. Turkson, J.A. Funk, M.A. Nichols, S.R. Ladely, and P.J. Fedorka-Cray. 2001. Comparison of methods for isolating *Salmonella* bacteria from faeces of naturally infected pigs. *J. Appl. Microbiol.* 89:169–177.

Dill, K., L. Stanker, and C.R. Young. 1999. Detection of *Salmonella* in poultry using a silicon chip-based biosensor. *J. Biochem. Biophys. Meth.* 41:61–67.

Duffy, G., B. Kilbride, J.J. Sheridan, I.S. Blair, and D.A. McDowell. 2000. A membrane-immunofluorescent-viability staining technique for the detection of *Salmonella* spp. from fresh and processed meat samples. *J. Appl. Microbiol.* 89:587–594.

Feder, I., J.C. Nietfeld, J. Galland, J. Yearley, J.M. Sargeant, R. Oberst, M. Tamplin, and J.B. Luchanansky. 2001. Comparison of cultivation and PCR-hybridization for detection of *Salmonella* on porcine fecal and water samples. *J. Clin. Microbiol.* 39(7):2477–2484.

Feld, N.C., L. Ekeroth, K.O. Gradel, S. Kabell, and M. Madsen. 2000. Evaluation of a serological *Salmonella* mix-ELISA for poultry used in a national surveillance programme. *Epidemiol. Infect.* 125(2):263–268.

Funk, J.A., P.R. Davies, and M.A. Nichols. 2000. The effect of sample weight on detection of *Salmonella* enterica in swine feces. *J. Vet. Diag. Invest.* 12:412–418.

Galland, J.C., J.K. House, D.R. Hyatt, L.L. Hawkins, N.V. Anderson, C.K. Irwin, and B.P. Smith. 2000. Prevalence of *Salmonella* in beef feeder steers as determined by bacterial culture and ELISA serology. *Vet. Microbiol.* 76:143–151.

Gast, R.T., R.E. Porter Jr., and P.S. Holt. 1997. Applying tests for specific yolk antibodies to predict contamination by *Salmonella* enteritidis in eggs from experimentally infected laying hens. *Avian Dis.* 41:195–202.

Gulig, P.A., and R. Curtiss III. 1988. Cloned and transposon insertion mutagenesis of virulence genes of the 100-kilo-based plasmid of *Salmonella* typhimurium. *Infect. Immun.* 56:3262–3271.

Guth, F. 1916. Selennährbörden für die elective Züchtung von Typhusbacillen. *Centr. Bakt. I Abt. Orig.* 77:487.

Hoben, D.A., D.H. Ashton, and A.C. Peterson. 1973. Some observations on the incorporation of novobiocin into Hektoen enteric agar for improved *Salmonella* isolation. *Appl. Microbiol.* 26:126–137.

House, J.K., B.P. Smith, K. O'Connell, and D.C. VanMetre. 2001. Isotype-specific antibody responses of cattle to *Salmonella* Dublin lipopolysaccharide and porin following *Salmonella* Dublin vaccination and acute and chronic infection. *J. Vet. Diag. Invest.* 13(3):213–218.

June, G.A., P.S. Sherrod, T.S. Hammack, R.M. Amaguana, and W.H. Andrews. 1995. Relative effectiveness of selenite cystine broth, tetrathionate broth, and Rappa-

port-Vassiliadis medium for the recovery of *Salmonella* from raw flesh and other highly contaminated foods: precollaborative study. *J. AOAC Int.* 78:375–380.

Jungherr, E., W.J. Hall, and B.S. Pomeroy. 1951. Techniques for the bacteriologic examination of reactors to pullorum disease antigen. *Proc. 87th Ann. Meet. Amer. Vet. Med. Assoc.*, 1950, 260–263.

Koivunen, J., E. Lanki, R.L. Rajala, A. Siltonen, and H. Heinonen-Tanski. 2001. Determination of *Salmonella* from municipal waste waters. *Water Sci. and Technol.* 43(12):221–224.

Komatsu, K.K., and L. Restaino. 1981. Determination of the effectiveness of novobiocin added to two agar plating media for the isolation of *Salmonella* from fresh meat products. *J. Food Saf.* 3:183–192.

Kristensen, M., V. Lester, and A. Jurgens. 1925. Use of trypsinized casein, brom-thymol blue, brom cresol purple, phenol red and brilliant green for bacteriological nutrient media. *Brit. J. Exp. Path.* 6:291–299.

Maciorowski, K.G., S.D. Pillai, and S.C. Ricke. 2000. Efficacy of a commercial polymerase chain reaction-based assay for detection of *Salmonella* spp. in animal feeds. *J. Appl. Microb.* 89:710–718.

Maddox, C. and W. Fales. 1991. Use of a *Salmonella* typhimurium–derived virulence probe in the detection of *Salmonella* sp. and in the characterization of *Salmonella* cholerae-suis virulence plasmids. *J. Vet. Diag. Invest.* 3:218–222.

Miller, R.G., C.R. Tate, E.T. Mallinson, and J.A. Scherrer. 1991. Xylose lysine tergitol 4: an improved selective agar medium for the isolation of *Salmonella*. *Poultry Sci.* 70:2429–2432.

Mueller, L. 1923. Un nouveau milieu d'enrichissement pour la recherché du bacilli typhique et des paratyphiques. *Compt. Rend. Soc. Boil.* (Paris) 89:434–437.

Namimatsu, T., M. Tsuna, Y. Imai, S. Futo, S. Mitsuse, T. Sakano, and S. Sato. 2000. Detection of *Salmonella* by using the colorimetric DNA/rRNA sandwich hybridization in microtiter wells. *J. Vet. Med. Sci.* 62(6):615–619.

Nielsen, B., D.L. Baggesen, F. Bager, J. Haugegaard, and P. Lind. 1995. The serological response to *Salmonella* serovars typhimurium and infantis in experimentally infected pigs. The time course followed with an indirect anti-LPS ELISA and bacterioiological examinations. *Vet. Microbiol.* 47:205–218.

Pathirana, S.T., J. Barbaree, B.A. Chin, M.G. Hartell, W.C. Neely, and V. Vodyanoy. 2000. Rapid and sensitive biosensor for *Salmonella. Biosensors and Bioelectronics* 15:135–141.

Poisson, D.M. 1992. Novobiocin, brilliant green, glycerol, lactose agar: a new medium for the isolation of *Salmonella* strains. *Res. Microbiol.* 143:211–216.

Proux, K., C. Houndayer, F. Humbert, R. Cariolet, V. Rose, E. Eveno, and F. Madee. 2000. Development of a complete ELISA using *Salmonella* lipopolysaccharides of various serogroups allowing to detect all infected pigs. *Vet. Res.* 31(5):481–490.

Simpkins, S.A., A.B. Chan, J. Hays, B. Popping, and N. Cook. 2000. An RNA transcription-based amplification technique (NASBA) for the detection of viable *Salmonella* enterica. *Lett. Appl. Microbiol.* 30:75–79.

Smith, B.P., D.G. Oliver, P. Singh, G. Dilling, P.A. Martin, B.P. Ram, L.S. Jang, N. Sarkov, and J.S. Orsborn. 1989. Detection of *Salmonella* dublin mammary gland infection in carrier cows, using an enzyme-linked immunosorbent assay for antibody in milk or serum. *Am. J. Vet. Res.* 50:1352–1360.

Szabo, E.A., and B.M. Mackey. 1999. Detection of *Salmonella* enteritidis by reverse transcription-polymerase chain reaction (RT-PCR). *Internat. J. Food Microbiol.* 51:113–122.

Van der Wolfe, P.J., A.R. Elbers, H.M.J.F. van der Heijden, F.W. van Schie, W.A. Hunneman, and M.J.M. Tielen. 2001. *Salmonella* seroprevalence at the population and herd level in pigs in The Netherlands. *Vet. Microbiol.* 80:171–184.

Waage, A.S., T. Vardund, V. Lund, and G. Kapperud. 1999. Detection of low numbers of *Salmonella* in environmental water, sewage and food samples by a nested polymerase chain reaction assay. *J. Appl. Microbiol.* 87:418–428.

Walker, R.L., H. Kinde, R.J. Anderson, A.E. Brown. 2001. Comparison of VIDAS enzyme-linked fluorescent immunoassay using Moore swab sampling and conventional culture method for *Salmonella* detection on bulk tank milk and in-line milk filters in California dairies. *Internat. J. Food Microbiol.* 67:123–129.

Waltman, W.D. 1999. Methods for isolating *Salmonella*e from poultry and the poultry environment. In *Salmonella enterica Serovar Enteritidis in Humans and Animals: Epidemiology, Pathogenesis, and Control.* A.M. Saeed, ed. Ames, IA: Iowa State University Press, 419–432.

Waltman, W.D., A.M. Horne, and C. Pirkle. 1995. Comparative analysis of media and methods for isolating *Salmonella* from poultry and environmental samples. In *Pro-*

ceedings of the symposium on the diagnosis of Salmonella infections. United States Animal Health Association and American Association of Laboratory Veterinary Diagnosticians, 1–14.

Waltman, W.D., and E.T. Mallinson. 1995. Isolation of *Salmonella* from poultry tissue and environmental samples: a nationwide survey. *Avian Dis.* 39:45–54.

Wedderkopp, A., U. Stroger, V. Bitsch, and P. Lind. 2001. Testing of bulk tank milk for *Salmonella* Dublin infections in Danish dairy herds. *Can. J. Vet. Res.* 65(1):15–21.

Wells, J.G., G.K. Morris, P.S. Brackman. 1971. New method for isolating *Salmonella* from milk. *Appl. Microbiol.* 21:235–239.

Wilson, W.J. 1923. Reduction of sulphites by certain bacteria in media containing a fermentable carbohydrate and metallic salts. *J. Hyg.* 21:392–398.

Wiuff, C., B.M. Thorberg, A. Engvall, and P. Lind. 2002. Immunochemical analyses of serum antibodies from pig herds in a *Salmonella* non-endemic region. *Vet. Microbiol.* 85:69–82.

Worcman-Barninka, D., M.T. Destro, S.A. Fernandes, and M. Landgraf. 2001. Evaluation of motility enrichment on modified semi-solid Rappaport-Vassiladis medium (MSRV) for the detection of *Salmonella* in foods. *Internat. J. Food. Microbiol.* 64:387–393.

11

GENETICS AND PATHOGENESIS OF *SALMONELLA*

SHEILA PATTERSON AND RICHARD E. ISAACSON

INTRODUCTION. *Salmonella* is one of the most studied bacteria. It is known for its genetic pliability and for that reason has been used extensively to study general pathways of metabolism and physiology. It also is well known as a pathogen and has been used to identify numerous virulence determinants and strategies used by pathogenic bacteria. *Salmonellae* are Gram-negative, facultatively anaerobic, rod-shaped, motile bacteria that belong to the family *Enterobacteriaceae*.

Salmonellae can cause a number of diseases ranging from a self-limiting diarrhea to enteric fever (Darwin and Miller 1999). Typhoid fever is generally caused by *Salmonella* Typhi or *Salmonella* Paratyphi.

Gastroenteritis is the most common disease caused by *Salmonella*. Symptoms of gastroenteritis generally appear six to twenty-four hours after ingestion of a contaminated food. Symptoms resolve in twenty-four to forty-eight hours but can last as long as one week. Fever, abdominal pain, diarrhea, vomiting, headache, and dehydration are typical symptoms and severity ranges from mild pain and little diarrhea to extreme pain and bloody, severe diarrhea. Secondary disease syndromes can be chronic and include endocarditis, meningitis, pneumonia and arthritis (Buzby et al. 1996). *Salmonella* can be shed in the feces after resolution of symptoms. In humans, children younger than five years old have shed bacteria for up to twenty weeks. Adults have been shown to shed the organism for eight weeks (Gomez and Cleary 1998). Animals also become carriers. Swine have been shown to shed the organism sporadically for up to twenty-eight weeks (Wood et al. 1989).

Most cases of gastroenteritis are caused by ingestion of contaminated food of animal origin such as poultry, eggs, pork, beef, and dairy products (Bean et al. 1996). However, fresh fruits and vegetables, water, contact with reptiles, marijuana and person-to-person contact also have been documented (Buzby et al. 1996; Olsen et al. 2001). Most *Salmonella* infections occur as sporadic infections rather than outbreaks (Darwin and Miller 1999; Olsen et al. 2001).

PATHOGENESIS OF DISEASE. There is a huge amount known about the pathogenesis of salmonellosis. The following is a brief summary of the salient features of pathogenesis.

Acid Tolerance Response. *Salmonella* infections usually begin through a fecal-oral route. After ingestion in contaminated food or water, *Salmonella* must pass through several hostile environments before colonization and invasion. Coming from a natural environment, which may have had low osmolarity with fluctuations in pH and nutrient availability, *Salmonella* pass through the low acid of the stomach (Foster and Spector 1995). *S.* Typhimurium has several strategies for surviving exposure to low acid conditions (Foster and Spector 1995; Slauch et al. 1997). Acid stress responses are regulated by the alternative sigma factor σ encoded by *rpoS*, the two-component sensor regulator PhoP/Q and the ferric uptake regulator Fur (Gahan and Hill 1999). These regulators also play important roles in the regulation of virulence genes of *Salmonella* (Gahan and Hilll 1999 Slauch et al. 1997).

Adhesion. After it moves beyond the stomach, *Salmonella* can colonize the small intestine, cecum, and colon. The first step in colonization is adhesion. A number of different adhesions have been identified in *Salmonella*. The adhesions are fimbriae, which are proteinaceous, filamentous structures located on the bacterial surface. They usually are arranged peritrichously but can be expressed in a polar orientation (Edwards and Puente 1998). A number of different fimbriae have been identified in *S.* Typhimurium (Baumler et al. 1996; Edwards et al. 2000; Emmerth et al. 1999; Morrow et al. 1999) (see Table 11.1). Several fimbriae have specific cell tropisms for a particular cell type or cells from certain animal species, which may contribute to the ability of *Salmonella* to infect a number of different hosts (Baumler et al. 1997; Baumler et al. 1996a; Darwin and Miller 1999; Edwards et al. 2000; Althouse and Isaacson, submitted). Evolutionarily, the acquisition of different fimbrial operons may be one mechanism by which *Salmonella* serotypes have been able to expand their host range to many different domesticated animal species (Baumler et al. 1997).

Many fimbriae of *Salmonella* undergo phase variation between a fimbriated and nonfimbriated state (Duguid et al. 1966). This mechanism allows cells to produce adhesins when necessary and shuts off expression when the adhesins are not needed. Several known mechanisms of phase variation are known including invertible promoter elements (Bloomfield 2001), dif-

Table 11.1—Some Fimbriae of *Salmonella*

Fimbriae	Tropism	Reference
Type 1 fimbriae	Enterocyte	Thankavel et al. 1999; Althouse, submitted
Long polar fimbriae	M-cells	Baumler and Heffron 1995; Baumler et al. 1997
Plasmid encoded fimbriae	Mouse intestine	Friedrich et al. 1993
Thin aggregative fimbriae	Biofilms	Doran et al. 1993; Romling et al. 1998
SEF14	Macrophages	Clouthier et al. 1993; Edwards et al. 2000
Salmonella Typhimurium fimbriae (Stf)	Putative	Emmerth et al. 1999
Salmonella atypical fimbriae (Sap)	Putative	Folkesson et al. 1999
Asp	Putative	Folkesson et al. 1999
Bovine colonization factor (Bcf)	Bovine	Tsolis et al. 1999a

ferential methylation of GATC sequences by Dam methylase (Nicholson and Low 2000), and DNA strand slippage (Dehio et al. 1998; Isaacson and Patterson 1994).

Mucosal Invasion. Following attachment, *Salmonella* are able to cross the intestinal barrier through invasion or uptake by M-cells. Several studies have shown that *Salmonella* preferentially invade M-cells of the Peyer's patches (Jones et al. 1994; Penheiter et al. 1997). *Salmonella* also invade non-phagocytic enterocytes (Takeuchi 1967). *Salmonella* are able to invade through the expression of a type III secretion system (TTSS). Type III secretion systems are composed of twenty or more proteins that span the inner and outer membranes of Gram-negative bacteria and are involved in the targeted delivery of virulence factors into host cells (Darwin and Miller 1999; Galan 1996; Galan and Collmer 1999; Hueck 1998; Lee 1997; Lucas and Lee 2001). *Salmonella* species contain at least three type III secretion systems. The invasion associated TTSS is encoded on the 40 Kb *Salmonella* pathogenicity island 1 (SPI1) at centisome 63. A second TTSS is encoded by SPI2 and is located at centisome 31 (Hensel 2000). The third TTSS in *Salmonella* consists of the fifty-plus genes involved in flagellar gene expression and assembly (Aizawa 2001; Chilcott and Hughers 2000).

The invasion genes of SPI1 were identified for their roles in invasion whereby bacteria induce their own uptake into the normally non-phagocytic epithelial cells of the distal lumen (Galan and Curtiss 1989; Penheiter et al. 1997). Membrane ruffles and actin cytoskeletal rearrangements by eucaryotic host cells (M-cells and enterocytes) allow the bacterium to be engulfed in a manner termed bacteria-mediated endocytosis (Finlay et al. 1991; Francis et al. 1993; Slauch et al. 1997; Takeuchi 1967). Secreted proteins from the needle complex of the TTSS help to stimulate this uptake (Cornelis and Gijsegem 2000; Galan and Collmer 1999; Hueck 1998). The invasion genes also contribute to virulence by affecting a number of other processes including destruction of M-cells in Peyer's patches (Jensen et al. 1998), activation of cytokine expression in epithelial cells (Hobbie et al. 1997), induction of neutrophil migration across the epithelium (Gewirtz et al. 1999; McCormick et al. 1995), and the stimulation of apoptosis in macrophages (Chen et al.

1996; Hersh et al. 1999; Monack et al. 1996). Invasion into epithelial cells is under complex regulation and this regulation of invasion is thought to be modulated by control of invasion gene expression (Lucas et al. 2000). A brief description of the apparatus and effector genes of SPI1 and their roles in invasion follows.

REGULATION OF TTSS BY HILA. The TTSS of SPI1 is under complex regulation by a number of different systems. Optimal expression of the TTSS of SPI1 has been found to occur under conditions of high osmolarity, low aeration, and a slightly basic pH (Bajaj et al. 1996; Daefler 1999; Lostroh and Lee 2001; Lucas and Lee 2000). Other conditions for maximal expression include 37°C and late log phase of growth (Darwin and Miller 1999; Lee and Falkow 1990; Lucas and Lee 2000). The following is a brief description of the major regulators and how they are thought to affect type III secretion.

hilA is a member of the OmpR/ToxR family of transcriptional activators and is considered to be the major regulator of the TTSS of SPI1 (Darwin and Miller 1999; Lostroh and Lee 2001a, 2001b; Lostroh and Lee 2001). *hilA* is at the top of the hierarchal system of regulation in the type III secretion system. HilA directly activates the expression of the *inv* and *prg* genes, which encode the apparatus genes of the needle complex (Bajaj and Lee 1995; Lostroh and Lee 2001a). There also is some read through into the *sicAsipBCDA* operon (Lostroh and Lee 2001b). HilA directly activates transcription of *invF*, another transcriptional activator that then activates the expression of the TTSS effector genes in conjunction with SicA (Darwin and Miller 1999; Darwin and Miller 2001; Eichelberg and Galan 1999). *hilA* expression is regulated by the same conditions that effect the TTSS, and several regulators of *hilA* have been identified (Darwin and Miller 1999; Lostroh and Lee 2001; Lucas and Lee 2000). The conditions under which *hilA* is maximally induced are high osmolarity, low oxygen tension, near-neutral pH, and late log phase of growth. Changes in any of these conditions repress *hilA* expression (Bajaj et al. 1996; Lostroh and Lee 2001a). Bile also has been reported to repress *hilA* expression (Lostroh and Lee 2001a; Prouty and Gunn 2000).

HilA is the best characterized regulator of TTSS. The promoter region has been mapped. A region upstream of the promoter, called the URS (upstream repressing sequence) is required for repression by all the environmental conditions and other regulatory factors identified thus far (Schechter et al. 1999; Schechter and Lee 2001). Mutations in *hilA* greatly reduced invasion into epithelial cells and invasion gene expression in laboratory conditions (Bajaj et al. 1995; Eichelberg and Galan 1999; Johnson et al. 1996; Schechter and Lee 2001). Over-expression of *hilA* resulted in a hyperinvasive phenotype and counteracted the effects of invasion-repressing signals (Bajaj et al. 1995, 1996; Fahlen et al. 2000; Lee et al. 1992). The probable DNA binding site of HilA was identified as a HilA box consisting of two hexamers separated by five nucleotides such that the hexamers are direct repeats. This consensus was identified in the promoter regions of *invF* and *prgH* and artificial constructs containing this sequence were activated by HilA (Lostroh et al. 2000).

hilA itself is regulated by a number of different regulatory systems, and this regulation leads to regulation of the TTSS through changes in the levels of HilA (Lostroh and Lee 2001a). Mutations in several genes decrease the expression of *hilA*. These include *ompR*, *envZ*, *pstS*, and *fadD*. These mutations were found to act independently to reduce *hilA* expression, but the reduction was fairly mild (2–7-fold) (Lucas et al. 2000). The influence of such diverse pathways such as fatty acid biosynthesis (*fadR*), a high affinity inorganic phosphate uptake system (*pstS*), and the osmolarity sensing EnvZ/OmpR two component system on *hilA* expression shows how complex the expression of *hilA* is. Because the effect of these mutations is mild, the suggestion is that there may be intermediate levels of HilA when different regulatory signals are present (Lucas et al. 2000).

Other genes that alter *hilA* expression include *fliZ* (Ikebe et al. 1998; Kutsukake et al. 1999), *csrA/B* system (Altier et al. 2000a, 2000b), *phoP* (Bajaj et al. 1996; Groisman 2001), SirA/BarA (Ahmer et al. 1999), and *fis*, *hha*, and *hupA/hupB* (Lostroh and Lee 2001a). Fis is a site-specific DNA binding protein that induces sharp bends (Perez-Martin and de Lorenzo 1997).

SYSTEMIC INFECTION. After *Salmonella* invade the mucosa, they are capable of entering the reticuloendothelial system and can spread systemically to various organs including the spleen, liver and bone marrow (Slauch et al. 1997). When *Salmonella* encounter activated macrophages, they induce apoptosis (Chen et al. 1996; Lindgren et al. 1996; Monack et al. 1996). When they are within the macrophage, *Salmonella* are able to survive and replicate. *Salmonella* are able to use the macrophage to disseminate via the host lymphoid system. After the bacteria accumulate and replicate, organ failure and sepsis lead to death (Hueck 1998). A number of genes have been identified that contribute to survival within the macrophage (Cotter and DiRita 2000; Galan 2001). A MarA-type transcriptional activator, *slyA*, is involved in resistance to reactive oxygen species (Buchmeier 1997; Daniels et al. 1996; Libby et al. 1994). An inducible magnesium transport system, *mgtCB*, is important in intra-macrophage survival (Blanc-Potard and Groisman 1997; Cotter and DiRita 2000). Another TTSS encoded by *Salmonella* pathogenicity island 2 (SPI2), located at 30 minutes, also is required for survival in the macrophage (Cirillo et al. 1998; Hensel 2000; Ochman et al. 1996; Shea et al. 1996; Valdivia and Falkow 1997). The SPI2-TTSS is induced within host cells and is completely dependent on SsrA-SsrB, a two-component regulatory system encoded within SPI2 (Cirillo et al. 1998; Deiwick and Hensel 1999; Valdivia and Falkow 1997). SifA is an SPI2 effector that is required for maintaining the integrity of the *Salmonella*-containing vacuole (Beuzon et al. 2001; Unsworth and Holden 2000). An operon located on the *Salmonella* virulence plasmid, *spvRABCD*, is necessary for resisting clearance by macrophages in a murine model (Gulig et al. 1998). Recently it was shown that *spvB* and *spvC* could replace the entire virulence plasmid in a mouse model of infection (Matsui et al. 2001). The PhoP/Q regulatory system positively regulates a number of the genes involved in macrophage survival (Cotter and DiRita 2000). The OmpR/EnvZ system also is important for *Salmonella* survival and replication within macrophages (Beuzon et al. 2001; Lee et al. 2000).

ENTERIC SALMONELLOSIS. The mouse model of infection has been used to study *Salmonella* pathogenesis for years. Mice carrying mutations in *nramp* are highly susceptible to killing by *S.* Typhimurium. However, these mice do not exhibit the usual symptoms of diarrhea. Therefore it is difficult to study the virulence determinants required for induction of enteric disease in the mouse model. Cattle infected with *S.* Typhimurium exhibit many of the symptoms associated with human infections (Tsolis et al. 1999). The symptoms of enteritis include diarrhea, vomiting, dehydration, and fever (Santos et al. 2001a). In cattle, *S.* Typhimurium infections remain localized in the intestine and mesenteric lymph nodes. *S.* Typhimurium colonizes the spleen in only 50 percent of infected animals (Tsolis et al. 2000). Studies with cattle have identified several differences in the requirements for enteric disease and the induction of diarrhea versus systemic disease (Santos et al. 2001a; 2001b; Tsolis et al. 1999a, 1999b, 1999c; Wallis and Gaylov 2000). Some of these differences are discussed next.

Recent studies on induction of diarrhea in cattle showed that *sopB* (*sigD*) mutants elicit significantly less fluid accumulation and polymorphonuclear leukocyte (PMN) influx in an ileal ligated loop model (Santos et al. 2001; Tsolis et al. 1999). SopB is an inositol phosphate phosphatase and is encoded on another pathogenicity island located at centisome 20 termed SPI5 (Norris et al. 1998; Wallis et al. 1999). SPI5 encodes *pipA*, *pipB*, and *pipC* as well, and these were implicated in enteropathogenicity but it is not known how (Wood et al. 1998). SopD has no homologies to any proteins in the database and its function is unknown (Jones et al. 1998). A *sopD* mutant had reduced fluid accumulation and inflammation in a bovine ligated ileal loop assay (Jones et al. 1998). SopA is another protein identified as a secreted protein from analysis of a *sip* mutant of *S.* Dublin (Wood et al. 1996). The function of SopA is unknown but it appears to be involved in fluid accumulation and orchestrating the movement of PMNs across epithelial monolayers (Wood et al. 2000). SopB, SopD, and SopA appear to play a role in the induction of diarrhea in cattle in *S.* Dublin; however, their role in *S.* Typhimurium mediated diarrhea is controversial because *S.* Dublin causes a slightly different disease in cattle than *S.* Typhimurium (Santos et al. 2001). The work in *S.* Dublin led to a model of diarrhea induction in which the *Salmonella* infection results in elevated levels of inositol 1,4,5,6-tetrakisphosphate (Ins1,4,5,6-P_4), which can antagonize the closure of chloride channels. This leads to fluid secretion (Norris et al. 1998). *SopB* mutants in *S.* Typhimurium were still able to cause diarrhea, intestinal lesions, and death in orally inoculated calves, although the infective dose was high (Tsolis et al. 1999; Wallis and Galyov 2000).

PERSISTENCE. Persistence of the same *Salmonella* strain/phage type over a period of time has been shown to occur in poultry, swine, and cattle (Baloda et al. 2001; Davies and Wray 1996; McLaren and Wray 1991; Sandvang et al. 2000; Twiddy et al. 1988). Persistent colonization of swine by *S.* Typhimurium frequently occurs, resulting in an asymptomatic carrier state that can lead to contamination of food products and, hence, foodborne disease. It is believed that carrier animals are a major source of infection for both animals and humans (Fedorka-Cray 1994; Gray et al. 1996; Letellier et al. 1999). Swine have been shown to be colonized early in life and to then be persistently colonized by *Salmonella* (Wood et al. 1989). Even though pigs show no sign of infection or disease and may not continually shed the organism, stress may allow a resumption of shedding at the time of slaughter (Isaacson et al. 1999). This allows swine to act as a reservoir for the spread of *Salmonella* throughout the herd, within the packing plant, and during processing to the finished product. Thus the establishment of persistent infections is an important step in which *Salmonella* is introduced into the food chain. Although the proportion of *Salmonella* infections due to contaminated pork and pork products in the United States is unknown (Blaha 1997), studies from Denmark (Nielsen and Wegener 1997) have estimated that 10–15 percent of infections were associated with pork. The level of contamination of pork in the United States is estimated to be much higher than in Denmark. In 1995–1996, nearly 9 percent of swine from meat packing plants was found to be contaminated with *Salmo-*

nella (1996), whereas the study by Nielsen and Wegener (1997), showed only 0.8 percent of pork to be contaminated in Denmark. The CDC and the Center for Science in the Public Interest (CSPI) have estimated that 6–9 percent of foodborne *Salmonella* infections in the United States are associated with pork (Frenzen et al. 1999).

The detection of asymptomatically infected swine is problematic. The isolation of *Salmonella* in feces and other samples is laborious and not very sensitive (Eyigor et al. 2002; Hoorfar and Baggesen 1998; Hoorfar and Mortensen 2000; Wallace et al. 1999). Time-consuming enrichment techniques take five to eleven days to perform (Feder et al. 2001; Hoorfar and Baggesen 1998). Bacteriological culture remains the gold standard for identification of *Salmonella*, but PCR techniques are beginning to be evaluated for reliability and accuracy (Feder et al. 2001; Wallace et al. 1999).

In swine, *S.* Typhimurium can cause diarrhea, but it usually is not as severe as that seen in humans. After an initial bout of diarrhea, pigs recover and appear quite healthy. Doses required for severe disease are high ($\sim 10^{10}$ CFU) (Fedorka-Cray et al. 1994). The animals may shed the organism in feces for several months after infection (Wood et al. 1989). However, *S.* Typhimurium may also undergo a quiescent state in which no organisms are shed. The site of infection during persistence is unknown. It has been shown that stress from withdrawal of feed and transport can promote resumption of shedding, leading to re-infection and contamination of previously uninfected animals (Isaacson 1996; Isaacson et al. 1999). *Salmonella* can be isolated for up to twenty-eight weeks after exposure (Wood et al. 1989), although at lower challenge doses, it usually can be isolated for a shorter period of time. Several studies show that persistent infections are easily established (Baggesen et al. 1999; Kampelmacher et al. 1969; Wood et al. 1989). Rampant shedding at slaughter or fecal contamination during processing can easily lead to contamination of food products.

In humans, chronic, asymptomatic infections with *S.* Typhi have been associated with colonization of the gall bladder. *S.* Dublin also localizes in the gall bladder and spleen of cattle, and carriers may shed the organism for years (Wigley et al. 2000). Clinical and persistent, asymptomatic infections of swine by *S.* Typhimurium differ from those seen in humans, including the lack of gall bladder involvement in persistent infections.

Most studies on persistence focus on the prevalence of *Salmonella* in the herd, the environment, or both. Where and how *Salmonella* persist has not been well researched. Because persistence is becoming more important, it is obvious that studies on the mechanisms of persistence will be necessary to determine how *Salmonella* are able to persist for such long periods of time.

REFERENCES

Anonymous. 1996. USDA, Food Safety and Inspection Service. Nationwide pork microbiological baseline data collection program: market hogs. April 1995–March 1996. USDA Food Safety and Inspection Service.

Ahmer, B.M., J. van Reeuwijk, P.R. Watson, T.S. Wallis, and F. Heffron. 1999. *Salmonella* SirA is a global regulator of genes mediating enteropathogenesis. *Mol. Microbiol.* 31:971–982.

Aizawa, S.I. 2001. Bacterial flagella and type III secretion systems. *FEMS Microbiol. Lett.* 202:157–164.

Altier, C., M. Suyemoto, and S.D. Lawhon. 2000a. Regulation of *Salmonella enterica* serovar typhimurium invasion genes by *csrA. Infect. Immun.* 68:6790–6797.

Altier, C., M. Suyemoto, A.I. Ruiz, K.D. Burnham, and R. Maurer. 2000b. Characterization of two novel regulatory genes affecting *Salmonella* invasion gene expression. *Mol. Microbiol.* 35:635–646.

Baggesen, D.L., J. Christensen, A.C. Nielsen, B. Svenmark, and B. Nielsen. 1999. Presented at the ISECSP, Washington, D.C.

Bajaj, V., C. Hwang, and C.A. Lee. 1995. *hilA* is a novel *ompR/toxR* family member that activates the expression of *Salmonella* typhimurium invasion genes. *Mol. Microbiol.* 18:715–727.

Bajaj, V., R.L. Lucas, C. Hwang, and C.A. Lee. 1996. Coordinate regulation of *Salmonella* typhimurium invasion genes by environmental and regulatory factors is mediated by control of *hilA* expression. *Mol. Microbiol.* 22:703–714.

Baloda, S. B., L. Christensen, and S. Trajcevska. 2001. Persistence of a *Salmonella enterica* serovar typhimurium DT12 clone in a piggery and in agricultural soil amended with *Salmonella*-contaminated slurry. *Appl. Environ. Microbiol.* 67:2859–2862.

Baumler, A. J., and F. Heffron. 1995. Identification and sequence analysis of *lpfABCDE*, a putative fimbrial operon of *Salmonella* typhimurium. *J. Bacteriol.* 177:2087–2097.

Baumler, A.J., A.J. Gilde, R.M. Tsolis, A.W. van der Velden, B.M. Ahmer, and F. Heffron. 1997. Contribution of horizontal gene transfer and deletion events to development of distinctive patterns of fimbrial operons during evolution of *Salmonella* serotypes. *J. Bacteriol.* 179:317–322.

Baumler, A.J., R.M. Tsolis, and F. Heffron. 1997. Fimbrial adhesins of *Salmonella* typhimurium. Role in bacterial interactions with epithelial cells. *Adv. Exp. Med. Biol.* 412:149–158.

Baumler, A.J., R.M. Tsolis, F.A. Bowe, J.G. Kusters, S. Hoffmann, and F. Heffron. 1996. The *pef* fimbrial operon of *Salmonella* typhimurium mediates adhesion to murine small intestine and is necessary for fluid accumulation in the infant mouse. *Infect. Immun.* 64:61–68.

Bean, N.H., J.S. Goulding, C. Lao, and F. J. Angulo. 1996. Sureveillance for foodbourne-disease outbreaks: United States, 1988–1992. *Mor. Mortal. Wkly. Rep. CDC Surveill. Summ.* 45:1–66.

Beuzon, C.R., K.E. Unsworth, and D.W. Holden. 2001. In vivo genetic analysis indicates that PhoP-PhoQ and the *Salmonella* pathogenicity island 2 type III secretion system contribute independently to *Salmonella enterica* serovar Typhimurium virulence. *Infect. Immun.* 69:7254–7261.

Blaha, T. 1997. Public health and pork: pre-harvest food safety and slaughter perspectives. *Rev. Sci. Tech.* 16:489–495.

Blanc-Potard, A.B., and E.A. Groisman. 1997. The *Salmonella selC* locus contains a pathogenicity island mediating intramacrophage survival. *EMBO J.* 16:5376–5385.

Blomfield, I. C. 2001. The regulation of *pap* and type 1 fimbriation in *Escherichia coli. Adv. Microb. Physiol.* 45:1–49.

Buchmeier, N., S. Bossie, C.Y. Chen, F.C. Fang, D.G. Guiney, and S.J. Libby. 1997. SlyA, a transcriptional regulator of

Salmonella typhimurium, is required for resistance to oxidative stress and is expressed in the intracellular environment of macrophages. *Infect. Immun.* 65:3725–3730.

Buzby, J.C., T. Roberts, C.T. Jordan-Lin, and J.M. Macdonald. 1996. Bacterial foodborne disease medical costs and productivity losses. 741. USDA, Economic Research Service report number 741.

Chen, L.M., K. Kaniga, and J.E. Galan. 1996. *Salmonella* spp. are cytotoxic for cultured macrophages. *Mol. Microbiol.* 21:1101–1115.

Chilcott, G.S., and K.T. Hughes. 2000. Coupling of flagellar gene expression to flagellar assembly in *Salmonella enterica* serovar typhimurium and *Escherichia coli. Microbiol. Mol. Biol. Rev.* 64:694–708.

Cirillo, D.M., R.H. Valdivia, D.M. Monack, and S. Falkow. 1998. Macrophage-dependent induction of the *Salmonella* pathogenicity island 2 type III secretion system and its role in intracellular survival. *Mol. Microbiol.* 30:175–188.

Clouthier, S.C., K.H. Muller, J.L. Doran, S.K. Collinson, and W.W. Kay. 1993. Characterization of three fimbrial genes, *sefABC*, of *Salmonella enteritidis. J. Bacteriol.* 175:2523–2533.

Cornelis, G.R., and F. Van Gijsegem. 2000. Assembly and function of type III secretory systems. *Annu. Rev. Microbiol.* 54:735–774.

Cotter, P.A., and V.J. DiRita. 2000. Bacterial virulence gene regulation: an evolutionary perspective. *Annu. Rev. Microbiol.* 54:519–565.

Daefler, S. 1999. Type III secretion by *Salmonella* typhimurium does not require contact with a eukaryotic host. *Mol. Microbiol.* 31:45–51.

Daniels, J.J., I.B. Autenrieth, A. Ludwig, and W. Goebel. 1996. The gene *slyA* of *Salmonella* typhimurium is required for destruction of M cells and intracellular survival but not for invasion or colonization of the murine small intestine. *Infect. Immun.* 64:5075–5084.

Darwin, K.H., and V.L. Miller. 1999. Molecular basis of the interaction of *Salmonella* with the intestinal mucosa. *Clin. Microbiol. Rev.* 12:405–428.

Darwin, K.H., and V.L. Miller. 2001. Type III secretion chaperone-dependent regulation: activation of virulence genes by SicA and InvF in *Salmonella* typhimurium. *EMBO J.* 20:1850–1862.

Davies, R.H., and C. Wray. 1996. Persistence of *Salmonella enteritidis* in poultry units and poultry food. *Br. Poult. Sci.* 37:589–596.

Dehio, C., S.D. Gray-Owen, and T.F. Meyer. 1998. The role of *neisserial* Opa proteins in interactions with host cells. *Trends Microbiol.* 6:489–495.

Deiwick, J., and M. Hensel. 1999. Regulation of virulence genes by environmental signals in *Salmonella typhimurium. Electrophoresis* 20:813–817.

Doran, J.L., S.K. Collinson, J. Burian, G. Sarlos, E.C. Todd, C.K. Munro, C.M. Kay, P.A. Banser, P.I. Peterkin, and W.W. Kay. 1993. DNA-based diagnostic tests for *Salmonella* species targeting *agfA*, the structural gene for thin, aggregative fimbriae. *J. Clin. Microbiol.* 31:2263–2273.

Duguid, J.P., E.S. Anderson, and I. Campbell. 1966. Fimbriae and adhesive properties in *Salmonellae. J. Pathol. Bacteriol.* 92:107–138.

Edwards, R.A., and J.L. Puente. 1998. Fimbrial expression in enteric bacteria: a critical step in intestinal pathogenesis. *Trends Microbiol.* 6:282–287.

Edwards, R.A., D.M. Schifferli, and S.R. Maloy. 2000. A role for *Salmonella* fimbriae in intraperitoneal infections. *Proc. Natl. Acad. Sci. USA* 97:1258–1262.

Eichelberg, K., and J.E. Galan. 1999. Differential regulation of *Salmonella* typhimurium type III secreted proteins by pathogenicity island 1 (SPI-1)-encoded transcriptional activators InvF and HilA. *Infect. Immun.* 67:4099–4105.

Emmerth, M., W. Goebel, S.I. Miller, and C.J. Hueck. 1999. Genomic subtraction identifies *Salmonella* typhimurium prophages, F-related plasmid sequences, and a novel fimbrial operon, *stf*, which are absent in *Salmonella* typhi. *J. Bacteriol.* 181:5652–5661.

Eyigor, A., K.T. Carli, and C.B. Unal. 2002. Implementation of real-time PCR to tetrathionate broth enrichment step of *Salmonella* detection in poultry. *Lett. Appl. Microbiol.* 34:37–41.

Fahlen, T.F., N. Mathur, and B.D. Jones. 2000. Identification and characterization of mutants with increased expression of *hilA*, the invasion gene transcriptional activator of *Salmonella* typhimurium. *FEMS Immunol. Med. Microbiol.* 28:25–35.

Feder, I., J.C. Nietfeld, J. Galland, T. Yeary, J.M. Sargeant, R. Oberst, M.L. Tamplin, and J. B. Luchansky. 2001. Comparison of cultivation and PCR-hybridization for detection of *Salmonella* in porcine fecal and water samples. *J. Clin. Microbiol.* 39:2477–2484.

Fedorka-Cray, P.J., S.C. Whipp, R.E. Isaacson, N. Nord, and K. Lager. 1994. Transmission of *Salmonella* typhimurium to swine. *Vet. Microbiol.* 41:333–344.

Finlay, B.B., S. Ruschkowski, and S. Dedhar. 1991. Cytoskeletal rearrangements accompanying *Salmonella* entry into epithelial cells. *J. Cell Sci.* 99:283–296.

Folkesson, A., A. Advani, S. Sukupolvi, J.D. Pfeifer, S. Normark, and S. Lofdahl. 1999. Multiple insertions of fimbrial operons correlate with the evolution of *Salmonella* serovars responsible for human disease. *Mol. Microbiol.* 33:612–622.

Foster, J.W., and M.P. Spector. 1995. How *Salmonella* survive against the odds. *Annu. Rev. Microbiol.* 49:145–174.

Francis, C.L., T.A. Ryan, B.D. Jones, S.J. Smith, and S. Falkow. 1993. Ruffles induced by *Salmonella* and other stimuli direct macropinocytosis of bacteria. *Nature* 364:639–642.

Frenzen, P.D., T.L. Riggs, J.C. Buzby, T. Bruer, T. Roberts, D. Voetsch, S. Reddy, and F.W. Group. 1999. *Salmonella* cost estimate updated using FoodNet data. *FoodReview* 22:10–15.

Friedrich, M.J., N.E. Kinsey, J. Vila, and R.J. Kadner. 1993. Nucleotide sequence of a 13.9 kb segment of the 90 kb virulence plasmid of *Salmonella* typhimurium: the presence of fimbrial biosynthetic genes. *Mol. Microbiol.* 8:543–558.

Gahan, C.G., and C. Hill. 1999. The relationship between acid stress responses and virulence in *Salmonella* typhimurium and *Listeria monocytogenes*. *Int. J. Food Microbiol.* 50:93–100.

Galan, J.E. 1996. Molecular genetic bases of *Salmonella* entry into host cells. *Mol. Microbiol.* 20:263–271.

Galan, J.E. 2001. *Salmonella* interactions with host cells: type III secretion at work. *Annu. Rev. Cell. Dev. Biol.* 17:53–86.

Galan, J.E., and A. Collmer. 1999. Type III secretion machines: bacterial devices for protein delivery into host cells. *Science* 284:1322–1328.

Galan, J.E., and R. Curtiss, 3rd. 1989. Cloning and molecular characterization of genes whose products allow *Salmonella* typhimurium to penetrate tissue culture cells. *Proc. Natl. Acad. Sci. USA* 86:6383–6387.

Gewirtz, A.T., A. M. Siber, J.L. Madara, and B.A. McCormick. 1999. Orchestration of neutrophil movement by intestinal epithelial cells in response to *Salmonella* typhimurium can be uncoupled from bacterial internalization. *Infect. Immun.* 67:608–617.

Gomez, H.F., and G.G. Cleary. 1998. *Salmonella*, 4 ed., vol. 1. Philadelphia, PA: W.B. Saunders Co.

Gray, J.T., T.J. Stabel, and P.J. Fedorka-Cray. 1996. Effect of dose on the immune response and persistence of *Salmonella choleraesuis* infection in swine. *Am. J. Vet. Res.* 57:313–319.

Groisman, E.A. 2001. The pleiotropic two-component regulatory system PhoP-PhoQ. *J. Bacteriol.* 183:1835–1842.

Gulig, P.A., T.J. Doyle, J.A. Hughes, and H. Matsui. 1998. Analysis of host cells associated with the Spv-mediated increased intracellular growth rate of *Salmonella* typhimurium in mice. *Infect. Immun.* 66:2471–2485.

Hensel, M. 2000. *Salmonella* pathogenicity island 2. *Mol. Microbiol.* 36:1015–1023.

Hersh, D., D.M. Monack, M.R. Smith, N. Ghori, S. Falkow, and A. Zychlinsky. 1999. The *Salmonella* invasin SipB induces macrophage apoptosis by binding to caspase-1. *Proc. Natl. Acad. Sci. USA* 96:2396–2401.

Hobbie, S., L.M. Chen, R.J. Davis, and J.E. Galan. 1997. Involvement of mitogen-activated protein kinase pathways in the nuclear responses and cytokine production induced by *Salmonella* typhimurium in cultured intestinal epithelial cells. *J. Immunol.* 159:5550–5559.

Hoorfar, J., and D.L. Baggesen. 1998. Importance of pre-enrichment media for isolation of *Salmonella* spp. from swine and poultry. *FEMS Microbiol. Lett.* 169:125–130.

Hoorfar, J., and A.V. Mortensen. 2000. Improved culture methods for isolation of *Salmonella* organisms from swine feces. *Am. J. Vet. Res.* 61:1426–1429.

Hueck, C.J. 1998. Type III protein secretion systems in bacterial pathogens of animals and plants. *Microbiol. Mol. Biol. Rev.* 62:379–433.

Ikebe, T., S. Iyoda, and K. Kutsukake. 1999. Structure and expression of the *fliA* operon of *Salmonella* typhimurium. *Microbiology* 145:1389–1396.

Isaacson, R.E. 1996. Pathogenesis of enteric bacterial infections. *Advances in Swine in Biomedical Research*, 365–384.

Isaacson, R.E., L.D. Firkins, R.M. Weigel, F.A. Zuckermann, and J.A. DiPietro. 1999. Effect of transportation and feed withdrawal on shedding of *Salmonella typhimurium* among experimentally infected pigs. *Am. J. Vet. Res.* 60:1155–1158.

Isaacson, R.E., and S. Patterson. 1994. Analysis of a naturally occurring K99+ enterotoxigenic *Escherichia coli* strain that fails to produce K99. *Infect. Immun.* 62:4686–4689.

Jensen, V.B., J.T. Harty, and B.D. Jones. 1998. Interactions of the invasive pathogens *Salmonella* typhimurium, *Listeria monocytogenes*, and *Shigella flexneri* with M cells and murine Peyer's patches. *Infect. Immun.* 66:3758–3766.

Johnston, C., D.A. Pegues, C.J. Hueck, A. Lee, and S.I. Miller. 1996. Transcriptional activation of *Salmonella* typhimurium invasion genes by a member of the phosphorylated response-regulator superfamily. *Mol. Microbiol.* 22:715–727.

Jones, B.D., N. Ghori, and S. Falkow. 1994. *Salmonella typhimurium* initiates murine infection by penetrating and destroying the specialized epithelial M cells of the Peyer's patches. *J. Exp. Med.* 180:15–23.

Jones, M.A., M.W. Wood, P.B. Mullan, P.R. Watson, T.S. Wallis, and E.E. Galyov. 1998. Secreted effector proteins of *Salmonella dublin* act in concert to induce enteritis. *Infect. Immun.* 66:5799–5804.

Kampelmacher, E.H., W. Edel, P.A. Guinee, and L.M. van Noorle Jansen. 1969. Experimental *Salmonella* infections in pigs. *Zentralbl Veterinarmed [B]* 16:717–724.

Kutsukake, K., T. Ikebe, and S. Yamamoto. 1999. Two novel regulatory genes, *fliT* and *fliZ*, in the flagellar regulon of *Salmonella*. *Genes Genet. Syst.* 74:287–292.

Lee, A.K., C.S. Detweiler, and S. Falkow. 2000. OmpR regulates the two-component system *ssrA-ssrB* in *Salmonella* pathogenicity island 2. *J. Bacteriol.* 182:771–781.

Lee, C.A. 1997. Type III secretion systems: machines to deliver bacterial proteins into eukaryotic cells? *Trends Microbiol.* 5:148–156.

Lee, C.A., and S. Falkow. 1990. The ability of *Salmonella* to enter mammalian cells is affected by bacterial growth state. *Proc. Natl. Acad. Sci. USA* 87:4304–4308.

Lee, C.A., B.D. Jones, and S. Falkow. 1992. Identification of a *Salmonella* typhimurium invasion locus by selection for hyperinvasive mutants. *Proc. Natl. Acad. Sci. USA* 89:1847–1851.

Libby, S.J., W. Goebel, A. Ludwig, N. Buchmeier, F. Bowe, F.C. Fang, D.G. Guiney, J.G. Songer, and F. Heffron. 1994. A cytolysin encoded by *Salmonella* is required for survival within macrophages. *Proc. Natl. Acad. Sci. USA* 91:489–493.

Lindgren, S.W., I. Stojiljkovic, and F. Heffron. 1996. Macrophage killing is an essential virulence mechanism of *Salmonella* typhimurium. *Proc. Natl. Acad. Sci. USA* 93:4197–4201.

Lostroh, C.P., V. Bajaj, and C.A. Lee. 2000. The *cis* requirements for transcriptional activation by HilA, a virulence determinant encoded on SPI-1. *Mol. Microbiol.* 37:300–315.

Lostroh, C.P., and C.A. Lee. 2001a. The HilA box and sequences outside it determine the magnitude of HilA-dependent activation of *prgH* from *Salmonella* pathogenicity island 1. *J. Bacteriol.* 183:4876–4885.

Lostroh, C. P., and C.A. Lee. 2001b. The *Salmonella* pathogenicity island-1 type III secretion system. *Microbes Infect.* 3:1281–1291.

Lucas, R.L., and C.A. Lee. 2000. Unravelling the mysteries of virulence gene regulation in *Salmonella* typhimurium. *Mol. Microbiol.* 36:1024–1033.

Lucas, R.L., C.P. Lostroh, C.C. DiRusso, M.P. Spector, B.L. Wanner, and C.A. Lee. 2000. Multiple factors independently regulate *hilA* and invasion gene expression in *Salmonella enterica* serovar typhimurium. *J. Bacteriol.* 182:1872–1882.

Matsui, H., C.M. Bacot, W.A. Garlington, T.J. Doyle, S. Roberts, and P.A. Gulig. 2001. Virulence plasmid-borne *spvB* and *spvC* genes can replace the 90-kilobase plasmid in conferring virulence to *Salmonella enterica* serovar Typhimurium in subcutaneously inoculated mice. *J. Bacteriol.* 183:4652–4658.

McCormick, B.A., P.M. Hofman, J. Kim, D.K. Carnes, S.I. Miller, and J.L. Madara. 1995. Surface attachment of *Salmonella* typhimurium to intestinal epithelia imprints the subepithelial matrix with gradients chemotactic for neutrophils. *J. Cell Biol.* 131:1599–1608.

McLaren, I.M., and C. Wray. 1991. Epidemiology of *Salmonella* typhimurium infection in calves: persistence of *salmonellae* on calf units. *Vet. Rec.* 129:461–462.

Monack, D.M., B. Raupach, A.E. Hromockyj, and S. Falkow. 1996. *Salmonella* typhimurium invasion induces apoptosis in infected macrophages. *Proc. Natl. Acad. Sci. USA* 93:9833–9838.

Morrow, B.J., J.E. Graham, and R. Curtiss, 3rd. 1999. Genomic subtractive hybridization and selective capture of transcribed sequences identify a novel *Salmonella* typhimurium fimbrial operon and putative transcriptional regulator that are absent from the *Salmonella typhi* genome. *Infect. Immun.* 67:5106–5116.

Nicholson, B., and D. Low. 2000. DNA methylation-dependent regulation of *pef* expression in *Salmonella* typhimurium. *Mol. Microbiol.* 35:728–742.

Nielsen, B., and H.C. Wegener. 1997. Public health and pork and pork products: regional perspectives of Denmark. *Rev. Sci. Tech.* 16:513–524.

Norris, F.A., M.P. Wilson, T.S. Wallis, E.E. Galyov, and P.W. Majerus. 1998. SopB, a protein required for virulence of *Salmonella dublin*, is an inositol phosphate phosphatase. *Proc. Natl. Acad. Sci. USA* 95:14057–14059.

Norris, T.L., and A.J. Baumler. 1999. Phase variation of the *lpf* operon is a mechanism to evade cross-immunity between *Salmonella* serotypes. *Proc. Natl. Acad. Sci. USA* 96:13393–13398.

Ochman, H., F.C. Soncini, F. Solomon, and E.A. Groisman. 1996. Identification of a pathogenicity island required for *Salmonella* survival in host cells. *Proc. Natl. Acad. Sci. USA* 93:7800–7804.

Olsen, S.J., R. Bishop, F.W. Brenner, T.H. Roels, N. Bean, R.V. Tauxe, and L. Slutsker. 2001. The changing epidemiology of *salmonella*: trends in serotypes isolated from humans in the United States, 1987–1997. *J. Infect. Dis.* 183:753–761.

Olsen, S.J., L.C. MacKinnon, J.S. Goulding, N.H. Bean, and L. Slutsker. 2000. Surveillance for foodborne-disease outbreaks—United States, 1993–1997. *Mor. Mortal. Wkly. Rep. CDC Surveill. Summ.* 49:1–62.

Penheiter, K.L., N. Mathur, D. Giles, T. Fahlen, and B.D. Jones. 1997. Non-invasive *Salmonella* typhimurium mutants are avirulent because of an inability to enter and destroy M cells of ileal Peyer's patches. *Mol. Microbiol.* 24:697–709.

Perez-Martin, J., and V. de Lorenzo. 1997. Clues and consequences of DNA bending in transcription. *Annu. Rev. Microbiol.* 51:593–628.

Prouty, A.M., and J.S. Gunn. 2000. *Salmonella enterica* serovar typhimurium invasion is repressed in the presence of bile. *Infect. Immun.* 68:6763–6769.

Romling, U., W.D. Sierralta, K. Eriksson, and S. Normark. 1998. Multicellular and aggregative behaviour of *Salmonella* typhimurium strains is controlled by mutations in the *agfD* promoter. *Mol. Microbiol.* 28:249–264.

Salyers, A., and D. Witt. 1994. *Salmonella* infections. In *Bacterial Pathogenesis A Molecular Approach*. Washington, D.C.: ASM Press, 229–243.

Sanderson, K.E., A. Hessel, and K.E. Rudd. 1995. Genetic map of *Salmonella* typhimurium, edition VIII. *Microbiol. Rev.* 59:241–303.

Sandvang, D., L.B. Jensen, D.L. Baggesen, and S.B. Baloda. 2000. Persistence of a *Salmonella enterica* serotype typhimurium clone in Danish pig production units and farmhouse environment studied by pulsed field gel electrophoresis (PFGE). *FEMS Microbiol. Lett.* 187:21–25.

Santos, R.L., R.M. Tsolis, S. Zhang, T.A. Ficht, A.J. Baumler, and L.G. Adams. 2001a. *Salmonella*-induced cell death is not required for enteritis in calves. *Infect. Immun.* 69:4610–4617.

Santos, R.L., S. Zhang, R.M. Tsolis, R.A. Kingsley, L. Garry Adams, and A.J. Baumler. 2001b. Animal models of *Salmonella* infections: enteritis versus typhoid fever. *Microbes Infect.* 3:1335–1344.

Schechter, L.M., S.M. Damrauer, and C.A. Lee. 1999. Two AraC/XylS family members can independently counteract the effect of repressing sequences upstream of the *hilA* promoter. *Mol. Microbiol.* 32:629–642.

Schechter, L.M., and C.A. Lee. 2001. AraC/XylS family members, HilC and HilD, directly bind and derepress the *Salmonella* typhimurium *hilA* promoter. *Mol. Microbiol.* 40:1289–1299.

Shea, J.E., M. Hensel, C. Gleeson, and D.W. Holden. 1996. Identification of a virulence locus encoding a second type III secretion system in *Salmonella* typhimurium. *Proc. Natl. Acad. Sci. USA* 93:2593–2597.

Slauch, J., R. Taylor, and S. Maloy. 1997. Survival in a cruel world: how *Vibrio cholerae* and *Salmonella* respond to an unwilling host. *Genes Dev.* 11:1761–1774.

Takeuchi, A. 1967. Electron microscope studies of experimental *Salmonella* infection. I. Penetration into the intestinal epithelium by *Salmonella* typhimurium. *Am. J. Pathol.* 50:109–136.

Thankavel, K., A.H. Shah, M.S. Cohen, T. Ikeda, R.G. Lorenz, R. Curtiss, 3rd, and S.N. Abraham. 1999. Molecular basis for the enterocyte tropism exhibited by *Salmonella* typhimurium type 1 fimbriae. *J. Biol. Chem.* 274:5797–5809.

Tsolis, R.M., L.G. Adams, T.A. Ficht, and A.J. Baumler. 1999a. Contribution of *Salmonella* typhimurium virulence factors to diarrheal disease in calves. *Infect. Immun.* 67:4879–4885.

Tsolis, R.M., R.A. Kingsley, S.M. Townsend, T.A. Ficht, L.G. Adams, and A.J. Baumler. 1999b. Of mice, calves, and men. Comparison of the mouse typhoid model with other *Salmonella* infections. *Adv. Exp. Med. Biol.* 473:261–274.

Tsolis, R.M., S.M. Townsend, E.A. Miao, S.I. Miller, T.A. Ficht, L.G. Adams, and A. J. Baumler. 1999A. Identification of a putative *Salmonella enterica* serotype typhimurium host range factor with homology to IpaH and YopM by signature-tagged mutagenesis. *Infect. Immun.* 67:6385–6393.

Twiddy, N., D.W. Hopper, C. Wray, and I. McLaren. 1988. Persistence of *S.* typhimurium in calf rearing premises. *Vet. Rec.* 122:399.

Unsworth, K.E., and D.W. Holden. 2000. Identification and analysis of bacterial virulence genes in vivo. *Philos. Trans. R. Soc. Lond. B. Biol. Sci.* 355:613–622.

Valdivia, R.H., and S. Falkow. 1997. Fluorescence-based isolation of bacterial genes expressed within host cells. *Science* 277:2007–2011.

Wallace, H.A., G. June, P. Sherrod, T.S. Hammack, and R.M. Amaguana. 1999. *Salmonella.* In L. A. Tomlison, ed., *Food and Drug Administration Bacteriological Analytical Manual.*

Wallis, T.S., and E.E. Galyov. 2000. Molecular basis of *Salmonella*-induced enteritis. *Mol. Microbiol.* 36:997–1005.

Wallis, T.S., M. Wood, P. Watson, S. Paulin, M. Jones, and E. Galyov. 1999. Sips, Sops, and SPIs but not stn influence *Salmonella* enteropathogenesis. *Adv. Exp. Med. Biol.* 473:275–280.

Wigley, P., A. Berchieri, Jr., K.L. Page, A.L. Smith, and P.A. Barrow. 2001. *Salmonella enterica* serovar Pullorum persists in splenic macrophages and in the reproductive tract during persistent, disease-free carriage in chickens. *Infect. Immun.* 69:7873–7879.

Wood, M.W., M.A. Jones, P.R. Watson, S. Hedges, T.S. Wallis, and E.E. Galyov. 1998. Identification of a pathogenicity island required for *Salmonella* enteropathogenicity. *Mol. Microbiol.* 29:883–891.

Wood, M.W., M.A. Jones, P.R. Watson, A.M. Siber, B.A. McCormick, S. Hedges, R. Rosqvist, T.S. Wallis, and E.E. Galyov. 2000. The secreted effector protein of *Salmonella dublin*, SopA, is translocated into eukaryotic cells and influences the induction of enteritis. *Cell. Microbiol.* 2:293–303.

Wood, M.W., R. Rosqvist, P.B. Mullan, M.H. Edwards, and E.E. Galyov. 1996. SopE, a secreted protein of *Salmonella dublin*, is translocated into the target eukaryotic cell via a *sip*-dependent mechanism and promotes bacterial entry. *Mol. Microbiol.* 22:327–338.

Wood, R.L., A. Pospischil, and R. Rose. 1989. Distribution of persistent *Salmonella* typhimurium infection in internal organs of swine. *Am. J. Vet. Res.* 50:1015–1021.

12

FOODBORNE *SALMONELLA* INFECTIONS

WOLFGANG RABSCH, CRAIG ALTIER, HELMUT TSCHÄPE, AND ANDREAS J. BÄUMLER

HUMAN DISEASE SYNDROMES CAUSED BY *SALMONELLA* SEROTYPES. There are currently 2,449 known *Salmonella* serotypes (Brenner et al. 2000). From the standpoint of human disease, these pathogens can be divided into three groups causing distinct clinical syndromes: typhoid fever, bacteremia, and enterocolitis (for a recent review see Santos et al. 2001). Typhoid fever (caused by the strictly human-adapted *Salmonella enterica* serotypes Typhi and Paratyphi A, B and C) and bacteremia (caused by the bovine-adapted *S. enterica* serotype Dublin and the porcine-adapted *S. enterica* serotype Choleraesuis) are currently rare in the United States and Europe (Saphra and Wassermann 1954; Werner et al. 1979; Fang and Fierer 1991; Mead et al. 1999). Enterocolitis is by far the most common clinical syndrome, representing the second most frequent cause of bacterial foodborne disease of known etiology in the United States with an estimated 1,412,498 illnesses per year (Mead et al. 1999). Furthermore, *Salmonella*-induced enterocolitis is the single most common cause of death from foodborne illnesses associated with viruses, parasites or bacteria in the United States (Mead et al. 1999). Between 1987 and 1997, 61 percent of human cases reported to the Centers of Disease Control and Prevention (CDC) were caused by five serotypes, including *S. enterica* serotype Typhimurium (23 percent), *S. enterica* serotype Enteritidis (21 percent), *S. enterica* serotype Heidelberg (8 percent), *S. enterica* serotype Newport (5 percent) and *S. enterica* serotype Hadar (4 percent) (Olsen et al. 2001).

SOURCES OF HUMAN INFECTION. *S. enterica* serotypes causing typhoid fever (typhoidal serotypes) do not have an animal reservoir and persist in the human population by person-to-person transmission. In contrast, *S. enterica* serotypes causing bacteremia or enterocolitis (nontyphoidal serotypes) are associated with animal reservoirs and outbreak investigations reveal that infections in humans commonly result from animal-to-human transmission (St. Louis et al. 1988; Mishu et al. 1994). In outbreaks in the United States in which the sources were identified, the food vehicles most commonly implicated were chicken, beef, turkey, and eggs (Tauxe 1991). Meat, meat products, eggs, or egg products may contain *Salmonella* serotypes either because animals are infected or because fecal contamination occurs during processing (Galbraith 1961). These observations show that the persistence of *Salmonella* serotypes in livestock and domestic fowl is directly responsible for their subsequent introduction into the derived food products. The pre-harvest occurrence of *Salmonella* serotypes in food animals is hence of central importance for food safety and is the focus of this chapter. Vehicles of transmission that are less important for the spread of nontyphoidal *Salmonella* serotypes and are not discussed in this article include produce (Mahon et al. 1997), water (Angulo et al. 1997) and reptiles (D'Aoust et al. 1990; Woodward et al. 1997; Olsen et al. 2001).

***SALMONELLA* SEROTYPES IN CATTLE.** The occurrence of *Salmonella* serotypes in cattle is a problem for both food safety and animal health. The following is a discussion of *Salmonella* in cattle.

Cattle As a Source of Human Infections with *Salmonella* Serotypes. Cattle are an important source of human infections with *Salmonella* serotypes. Because of the widespread pasteurization of milk and dairy products, the risk to public health from these food items is of minor importance in the United States (Tauxe 1991). In contrast, beef is a major source of human infections with *Salmonella* serotypes. According to a survey performed by CDC between 1983 and 1987, beef is the second most common source of human infections (Tauxe 1991). *Salmonella* serotypes are associated with 48 percent of all beef-related outbreaks of foodborne illness in the United States (Anon. 1993). *S.* Typhimurium, the serotype most frequently associated with human cases of disease in the United States, is often cattle associated (Tauxe 1991). Humans become primarily infected from this source because they consume raw or undercooked beef containing fecal contamination. According to the nationwide beef microbiological baseline data collection program conducted by the U.S. Department of Agriculture (USDA), Food Safety and Inspection Service (FSIS) between 1992 and 1994, the prevalence of *Salmonella* serotypes in carcasses is 3 percent for cows and bulls and 1 percent for steers and heifers (USDA-FSIS website). The Nationwide Federal Plant Raw Ground Beef Microbiological Survey performed by FSIS between 1993 and 1994 shows that *Salmonella* serotypes can be cultured from 8 percent of raw ground beef samples in the United States (USDA-FSIS website).

The recent emergence of the multiresistant *S.* Typhimurium definitive phage type (DT) 104 in the United States has intensified the threat that *Salmonella* serotypes pose for food safety (Angulo 1996). *S.* Typhimurium strain DT104 was first isolated in the United Kingdom in 1984 from cattle and has attracted attention because it has chromosomally encoded resistance to five antibiotics, which complicates antimicrobial therapy of systemic infections (Threlfall et al. 1994). The recent increase in the incidence of ciprofloxacin resistance in DT104 isolates is of concern because fluoroquinolones and third-generation cephalosporins are the drugs of choice for invasive *Salmonella* infections in humans (Angulo et al. 2000). Human infections with DT104 have been traced back to consumption of contaminated food and direct contact with farm animals, in particular cattle (Fone and Barker 1994; Threlfall et al. 1994; Wall et al. 1994; Davies et al. 1996; Low et al. 1996; Penny et al. 1996). It should be mentioned in this context that the prevalence of DT104 has recently declined in England (Threlfall et al. 2000). For a more detailed discussion on multidrug resistance in *Salmonella* serotypes, the reader is referred to recent review articles (Cohen and Tauxe 1986; Angulo et al. 2000; Rabsch et al. 2001).

Salmonella Serotypes Associated with Disease in Cattle. *Salmonella* serotypes are frequently isolated from ill cattle. A recent evaluation of the relative importance of health issues affecting dairy cattle ranked salmonellosis as the second most important dairy cattle disease after mastitis (Wells et al. 1998). A survey performed in Britain revealed that *Salmonella* serotypes are associated with 12 percent of diarrhea outbreaks among calves (Reynolds et al. 1986). The mortality rate among calves with salmonellosis is estimated to be between 19 percent and 24 percent (Rothenbacher 1965). The average age of calves that die from salmonellosis is fourteen days (Rothenbacher 1965). The serotypes responsible for the vast majority of cases of disease in cattle in Europe and the United States are *S. enterica* serotype Dublin and *S.* Typhimurium (Gibson 1961; Rothenbacher 1965; Sojka and Field 1970; Hughes et al. 1971; Sojka and Wray 1975; Smith et al. 1994; Wells et al. 1998). *S.* Typhimurium is the serotype most commonly isolated from ill cattle in the United States (Rothenbacher 1965; Ferris and Thomas 1995; Wells et al. 1998). Between 1994 and 1995, *S.* Typhimurium was associated with 48 percent of salmonellosis cases in cattle in the United States, followed by *S.* Dublin (10 percent of cases) and *S. enterica* serotype Kentucky (6 percent of cases) (Ferris and Thomas 1995).

The disease syndrome associated with *S.* Typhimurium infection in cattle is enterocolitis. *S.* Typhimurium can be transmitted either vertically, through milk (Giles et al. 1989) or horizontally, by the fecal-oral route (that is, fecal contamination of the environment) (Wray et al. 1987; McLaren and Wray 1991; Wray et al. 1991). Upon oral infection with *S.* Typhimurium, calves develop clinical signs of disease, including diarrhea, anorexia, fever, dehydration, and prostration within twelve to forty-eight hours (Wray and Sojka 1978; Smith et al. 1979; Tsolis et al. 1999). Morbidity and mortality are inversely proportional to age (Smith et al. 1979) and approximately 75 percent of natural *S.* Typhimurium infections occur in calves fewer than two months of age, before the animals are weaned (Sojka and Field 1970). Signs of disease subside within ten days after onset. Animals may shed *S.* Typhimurium with their feces for a few months after clinical recovery (Sojka and Field 1970). The inability to detect *S.* Typhimurium in the feces may not be a reliable indicator that the infection has been cleared, because it has been reported that dairy cows may excrete this pathogen intermittently from their udder rather than in feces for 2.5 years (Giles et al. 1989).

Prevalence of Salmonella Serotypes in Cattle and Beef. In addition to their association with disease in cattle, *Salmonella* serotypes are frequently isolated from the feces of apparently healthy animals in the United States. The fractions of animals shedding *Salmonella* serotypes with their feces are 1 percent for beef cows (Dargatz et al. 2000), 6 percent for feedlot cattle (Fedorka-Cray et al. 1998) and 5 percent for milk cows (Wells et al. 2001). Although a study performed by the National Animal Health Monitoring System (NAHMS, USDA) between 1991 and 1992 shows that *S.* Typhimurium is frequently isolated from dairy calves (27 percent of total *Salmonella* isolates), it comprises less than 1 percent of the total number of *Salmonella* serotypes in fecal samples collected from milk cows on the farm (Wells et al. 2001). Similarly, *S.* Typhimurium is not among the most frequently isolated *Salmonella* serotypes in beef cows, or feedlot cattle (Fedorka-Cray et al. 1998; Dargatz et al. 2000), despite its frequent association with disease in calves. The serotypes most commonly cultured from fecal samples are *S. enterica* serotype Oranienburg (22 percent of isolations), *S. enterica* serotype Cerro (22 percent), and *S. enterica* serotype Anatum (10 percent) for beef cows, (Dargatz et al. 2000); *S.* Anatum (28 percent of isolations), *S. enterica* serotype Montevideo (13 percent), and *S. enterica* serotype Muenster (12 percent) for feedlot cattle (Fedorka-Cray et al. 1998); and *S.* Montevideo (21 percent of isolations), *S. enterica* serotype Kentucky (12 percent), and *S. enterica* serotype Menhaden (12 percent) for dairy cows (Wells et al. 2001). However, the prevalence of *S.* Typhimurium in cattle increases before slaughter. Although *S.* Typhimurium comprises less than 1 percent of isolates from milk cows on the farm, it comprises 2 percent of *Salmonella* serotypes isolated from fecal samples of milk cows expected to be culled within seven days, and 4 percent of isolates from fecal samples of culled dairy cows at markets (Wells et al. 2001). Concomitantly, the fraction of animals shedding

Salmonella serotypes with their feces increases from 5 percent for milk cows on the farm to 15 percent for culled dairy cows at markets (Wells et al. 2001). Finally, after slaughter, *S.* Typhimurium accounts for 14 percent of *Salmonella* isolations from cattle carcasses in the United States (Schlosser et al. 2000), thereby making it the serotype most frequently isolated from cattle carcasses in the United States. Other serotypes frequently isolated from cattle carcasses include *S.* Montevideo (10 percent of isolations) and *S.* Muenster (8 percent of isolations) (Schlosser et al. 2000). The results of several surveys suggest that the increased prevalence of *Salmonella* serotypes in the intestine of cattle before slaughter is caused by transport and long periods with intermittent feeding (Brownlie and Grau 1967; Grau et al. 1968; Samuel et al. 1979; Moo et al. 1980; Samuel et al. 1981). Nonetheless, it is not clear why the prevalence of *S.* Typhimurium rises more dramatically before slaughter than that of other *Salmonella* serotypes (including *S.* Oranienburg, *S.* Anatum, *S.* Montevideo, *S.* Cerro, *S.* Muenster, *S.* Kentucky, and *S.* Menhaden). It is conceivable that the relatively high virulence of *S.* Typhimurium (indicated by its frequent association with disease in cattle) may allow this organism to more effectively take advantage of the increased susceptibility of animals caused by transport stress and intermittent feeding. The result may be a reactivation of existing *S.* Typhimurium infections associated with extensive growth in the intestine, thereby promoting further spread of the infection to other animals by the fecal oral route.

Control Measures. Pre-harvest control strategies should be directed toward the *Salmonella* serotypes of greatest concern for animal and human health. *S.* Typhimurium causes the highest morbidity and mortality among cattle, has acquired multiple antibiotic resistance, which complicates treatment, and is isolated most frequently from cattle carcasses and from patients with salmonellosis in the United States. Thus the pre-harvest occurrence of *S.* Typhimurium may be considered a major food-safety hazard in the production of beef (Wells et al. 2001). Preventive measures to reduce the pre-harvest occurrence of *S.* Typhimurium can be aimed at reducing the number of animals excreting this organism and blocking vertical transmission on the farm through milk or horizontal transmission by fecal contamination of the environment. To this end, a number of critical control points could be targeted with established preventive measures. For instance, adequate cleaning and disinfection routines can minimize the spread of infection by the fecal oral route (McLaren and Wray 1991). Excreters spreading the infection through milk or by the fecal oral route can be identified and isolated or culled (Giles et al. 1989). Vaccination can be used to reduce the number of fecal excreters and to protect the herd against new infections (Jones et al. 1991; Weber et al. 1993). Analysis of fecal and milk samples prior to procurement could reduce the risk of purchasing infected animals that may reintroduce the infection on the farm. The Pathogen Reduction, Hazard Analysis and Critical Control Point (HACCP) System implemented by FSIS to improve food safety of meat and poultry products does not target all these critical control points. Nonetheless, preliminary data from the CDC's Emerging Infections Program Foodborne Diseases Active Surveillance Network (FoodNet) show that implementation of HACCP between 1998 and 2000 in the United States was accompanied by a reduction in the incidence of human salmonellosis from 13.6 per 100,000 population in 1997 to an estimated 12 per 100,000 population in 2000 (Anon. 2001).

***SALMONELLA* SEROTYPES IN SWINE.** The genus *Salmonella* obtained its name from the American veterinarian Daniel E. Solmon, who first isolated *S. enterica* serotype Choleraesuis from pigs in 1885. The following discusses *Salmonella* isolated from swine.

Swine As a Source of Human Infections with *Salmonella* Serotypes. Swine represent an important source of *Salmonella* serotypes causing disease in humans, although less so than do other production animals. Pork products were implicated in 2.9 percent of all *Salmonella* outbreaks during the years 1983 to 1987, compared to 18.2 percent of those originating from poultry and eggs, and 11.2 percent from beef (Tauxe 1991). Between 1988 and 1992, 18 percent of the outbreaks caused by the consumption of contaminated meat reported to the CDC were associated with ham and pork (Bean et al. 1996). Swine also shed antimicrobial-resistant *Salmonella* serotypes that pose a threat to food safety. With increasing frequency, *Salmonella* isolates obtained from pigs are resistant to one or multiple antimicrobials. Recent studies have shown that at least half, and in some cases more than 90 percent, of *Salmonella* isolates obtained from commercially raised swine in the United States are multiresistant (Gebreyes et al. 2000; Farrington et al. 2001). A common resistance is to tetracyclines, an antibiotic class used routinely in swine feed to improve feed efficiency. In one study, more than 85 percent of *Salmonella* isolates from swine were resistant to tetracycline (Gebreyes et al. 2000); in another, greater than 95 percent were resistant to chlortetracycline (Farrington et al. 2001). Resistance to several other antibiotics has commonly been found. A national survey of swine slaughter samples in 1999 showed resistance to sulfamethoxazole (30.7 percent), streptomycin (29.3 percent), ampicillin (10.8 percent), chloramphenicol (8.0 percent), and kanamycin (6.7 percent) (National Antibiotic Resistance Monitoring System, USDA). Decreased susceptibility to the quinolone class of antibiotics is rare among *Salmonella* isolates obtained from swine in the United States, but has commonly been reported among isolates from Europe (Malorny et al. 1999). This decrease in susceptibility has coincided with the use of quinolones in animal agriculture in Europe, although no member of this drug class has been approved for use in swine in the United States.

The pattern of resistance observed in *Salmonella* from swine is related to the phage type of the isolate. Phage type DT104, a cause of disease outbreaks throughout the world, is usually resistant to ampicillin, chloramphenicol, tetracycline, streptomycin, and sulfonamide (resistance type ACTSSu). It has been shown that *Salmonella* strains from swine that are resistant to chloramphenicol are usually of this phage type (Gebreyes et al. 2000). DT104 also encodes all of its five resistance genes as a small chromosomal element (Briggs and Fratamico 1999). This close physical linkage among resistance genes may explain why resistance to chloramphenicol, an antibiotic not used in food animals for more than a decade in the United States, persists in the population. A second phage type commonly found among swine isolates and presenting a concern to public health is DT193. It has been reported to have various resistance patterns including resistance to tetracycline alone; ampicillin, tetracycline, streptomycin, and sulfonamide; or, most recently, ampicillin, kanamycin, tetracycline, streptomycin, and sulfonamide (AKTSSu) (Hampton et al. 1995). Isolates from swine with the AKTSSu penta-resistance pattern, and those more generally with resistance to kanamycin, are often of the DT 193 phage type (Gebreyes et al. 2000).

Salmonella Serotypes Associated with Disease in Swine.

Two distinct clinical syndromes in swine are caused by *Salmonella*. Bacteremia is caused by *S. enterica* serotype Choleraesuis, often the Kunzendorf variant (Wilcock et al. 1976). Clinical signs are severe, including fever, anorexia, and lethargy, and may be potentiated by concomitant infection with porcine reproductive and respiratory syndrome virus (PRRSV) (Schwartz 1999; Wills et al. 2000). The disease can lead to ulcerative colitis, hepatic necrosis, or pneumonia, and produces a high rate of mortality (Turk et al. 1992; Schwartz 1999). The serotype is host-adapted to swine but can also cause bacteremia in humans (Saphra and Wassermann 1954). Infection of humans by *S.* Choleraesuis is rare but important because it produces a severe disease syndrome. Swine can also suffer clinical disease in the form of enterocolitis caused by one of a number of serotypes, *S.* Typhimurium being the most notable. Enterocolitis is most commonly seen in young pigs, or in those suffering from other illnesses (Schwartz 1999).

Prevalence of *Salmonella* Serotypes in Swine and Pork.

Although *Salmonella* can cause clinical disease in swine, the vast majority of infections are subclinical. These infections can be caused by a large number of serotypes; more than thirty have been isolated from pigs on farms, and more than fifty have been identified from swine carcasses at slaughter (Fedorka-Cray et al. 2000; Gebreyes et al. 2000). Surveys have shown that up to six serotypes can be isolated from clinically normal pigs on a single farm (Davies et al. 1998; Fedorka-Cray et al. 2000). A national survey for fecal *Salmonella* shedding by pigs most frequently produced *S. enterica* serotypes Derby, Agona, Typhimurium, Brandenburg, Mbendaka, and Heidelberg (NAHMS). Among this group, *S.* Typhimurium, *S.* Heidelberg, and *S.* Agona are also known to be common isolates from humans. *Salmonella* is also found among pork products both in processing plants and retail stores. Pork and pork sausage samples taken from plants were positive for *Salmonella* at a rate of 5.8 percent (Duffy et al. 2001). Between 3.3 percent and 8.3 percent of whole-muscle pork from stores was positive for *Salmonella* serotypes (Duffy et al. 2001; Zhao et al. 2001), while pork sausage was contaminated to a greater degree, from 7.3 percent to 12.5 percent (Duffy et al. 2001). The serotypes most frequently cultured from swine carcasses are *S.* Derby (28 percent of isolations), *S. enterica* serotype Johannesburg (10 percent of isolations) and *S.* Typhimurium (10 percent of isolations) (Schlosser et al. 2000). The serotypes most frequently cultured from raw ground pork are *S.* Typhimurium (18 percent of isolations), *S.* Derby (17 percent of isolations), and *S.* Anatum (7 percent of isolations) (Schlosser et al. 2000).

Control Measures.

A number of studies have attempted to identify the environmental and managerial factors that affect the shedding of *Salmonella* by pigs in commercial swine herds. On the farm, general cleanliness has been cited as the most important factor in reducing *Salmonella* shedding (Berends et al. 1996; Dahl et al. 1997; Funk et al. 2001). Such factors as farm hygiene, including cleaning and disinfection, housing, and feed and water sources affect *Salmonella* shedding by pigs (Berends et al. 1996). Biosecurity, including the number of people and other domestic species at the site, also appears to play an important role (Berends et al. 1996; Funk et al. 2001). The use of barns with slotted floors has been shown to decrease *Salmonella* prevalence in swine finishing barns (Davies et al. 1997). *Salmonella* serotypes have been isolated from feed on swine farms, but often in association with other risk factors such as lack of pest control or mixing of feeds on the farm, suggesting that contamination of feed occurs at the farm and that prudent measures to improve farm sanitation might be an effective deterrent to infection (Harris et al. 1997). The use of all in/all out management, a method used to successfully decrease the incidence of a number of other infectious diseases of swine, does not seem to offer a significant advantage over its alternative, continuous flow, in decreasing the prevalence of *Salmonella* serotypes (Davies et al. 1997).

After animals leave the farm, transport to the slaughterhouse presents an additional risk to the contamination of pork products with *Salmonella* serotypes. Serotypes not found in fecal samples from pigs while on farms have been isolated from their intestinal contents at slaughter (Hurd et al. 2001), suggesting infection during transport or at the slaughterhouse. Swine can be rapidly infected experimentally with *Salmonella*, harboring viable organisms less than one hour after infection (Hurd et al. 2001). The organism can be cultured from the intestinal contents of a larger proportion of animals after transport than on the farm (Isaacson et al. 1999). The effect of transport may, however, be offset by withholding feed from animals for twenty-four hours prior to slaughter, suggesting that feed withdrawal is a means of decreasing the number of *Salmonella* serotypes in the intestinal tracts of transported animals (Isaacson et al. 1999). Contamination of animals in the slaughterhouse itself can be important. Samples obtained from pens used for lairage, the holding of animals prior to slaughter, show them often to be contaminated with *Salmonella* serotypes, thus better cleaning and reduced lairage time might reduce bacterial contamination of meat (Swanenburg et al. 2001a, 2001b).

SALMONELLA SEROTYPES IN CHICKENS.

While *Salmonella* serotypes associated with disease in chickens have been unallicited in the U.S., chicken eggs have emerged as an importnt source for human outbreaks.

Chickens and Chicken Eggs as Sources of Human Infections with *Salmonella* Serotypes. Food products derived from chickens (that is, chicken meat and eggs) are the most important sources of human infection with *Salmonella* serotypes in the United States. According to a survey performed by CDC, chicken was the most common source of human infections between 1983 and 1987, while eggs were the third most common source (Tauxe 1991). Chicken carcasses may contain *Salmonella* serotypes either because animals are infected or because fecal contamination occurs during processing. According to the Nationwide Broiler Chicken Microbiological Baseline Data Collection Program conducted by the Food Safety and Inspection Service between 1994 and 1995 (FSIS, USDA), *Salmonella* serotypes can be cultured from 20 percent of broiler carcasses in the United States (USDA-FSIS website). *Salmonella* serotypes are also highly prevalent in ground chicken meat, with 45 percent of samples in the United States testing positive in 1995 (Nationwide Raw Ground Chicken Microbiological Baseline Data Collection Program, FSIS, USDA) (USDA-FSIS website).

Chicken eggs can become contaminated through cracks in the shell after contact with chicken feces, or by trans-ovarian contamination with *Salmonella* serotypes (Snoeyenbos et al. 1969). National U.S. surveys of nonpasteurized liquid egg show that *Salmonella* serotypes were present in 53 percent of samples collected from twenty plants between 1991 and 1992 (Ebel et al. 1993) and in 48 percent of samples collected between 1994 and 1995 (Hogue et al. 1997). Considering the fact that liquid egg samples were drawn from bulk tanks containing on average the contents of approximately 7,500 eggs, these data suggest that the prevalence of *Salmonella* serotypes in eggs is much lower than that in animal carcasses. Nonetheless, eggs are implicated in the majority of infections with *S.* Enteritidis, the second most common isolate from humans (21 percent of cases reported to CDC between 1987 and 1997) (Coyle et al. 1988; St. Louis et al. 1988; Cowden et al. 1989b, 1989b; Henzler et al. 1994; Olsen et al. 2001).

***Salmonella* Serotypes Associated with Disease in Chickens.** The avian-adapted *S. enterica* serotype Gallinarum includes two biotypes: biotype Pullorum is the causative agent of pullorum disease, and biotype Gallinarum causes fowl typhoid. At the beginning of the twentieth century, *S.* Gallinarum posed a serious economic problem for the poultry industry (Bullis 1977). A macroscopic tube agglutination and a whole-blood test for agglutination of stained-antigen were introduced in 1913 and 1931, respectively, for the detection of birds infected with *S.* Gallinarum (Jones 1913; Schaffer and MacDonald 1931). A National Poultry Improvement Plan, designed to reduce the level of *S.* Gallinarum infection in the United States, was adopted in 1935 and included large-scale, voluntary testing of flocks (Bullis 1977). By this detection-and-slaughter method of disease control, the percentage of chickens testing positive for *S.* Gallinarum was reduced from 3.6 percent in 1935–1936 to 0.000006 percent in 1974. In 1975, no evidence for *S.* Gallinarum infection was detected in commercial flocks in the United States (Bullis 1977).

Since the eradication of *S.* Gallinarum in the United States, avian paratyphoid is the only disease syndrome caused by *Salmonella* serotypes in chickens. The disease can be caused by a large number of *Salmonella* serotypes, but it is not associated with high morbidity and mortality rates. Since the eradication of *S.* Gallinarum in England, the *Salmonella* serotypes most frequently associated with avian paratyphoid in chickens are *S.* Enteritidis and *S.* Typhimurium (Sojka and Wray 1975).

Prevalence of *Salmonella* Serotypes in Chickens, Chicken Carcasses, and Eggs. *Salmonella* serotypes are frequently isolated from apparently healthy broiler chickens in the United States. *Salmonella* serotypes are cultured from the crops of 2 percent of broiler chickens, and the prevalence increases to 10 percent by feed withdrawal before slaughter (Corrier et al. 1999). A recent survey suggests that *S.* Heidelberg (30 percent of isolations) and *S. enterica* serotype Kentucky (20 percent of isolations) are the serotypes most commonly isolated from commercial broiler hatcheries by a conventional tray liner culture method (Byrd et al. 1999). Similarly, the serotypes isolated most frequently by the drag swab assembly culture method from broiler farms include *S.* Heidelberg (36 percent of isolations) and *S.* Kentucky (23 percent of isolations) (Byrd et al. 1999).

S. Heidelberg (26 percent of isolations), *S.* Kentucky (20 percent of isolations), and *S.* Typhimurium (10 percent of isolations) are the serotypes most frequently cultured from chicken carcasses (Schlosser et al. 2000). Finally, *S.* Heidelberg (30 percent of isolations) and *S.* Kentucky (14 percent) are also the serotypes most frequently cultured from ground raw chicken (Schlosser et al. 2000). Thus, it appears that organisms that already become introduced into flocks at the time of hatching, specifically *S.* Heidelberg and *S.* Kentucky, are responsible for the contamination of chicken meat in the United States.

Salmonella serotypes are also frequently isolated from apparently healthy laying hens. The National Spent Hen Survey performed by USDA in 1991 showed that *Salmonella* serotypes were recovered from 24 percent of pooled cecal samples (five ceca per pool) collected from United States commercial egg-production flocks, thereby yielding an estimated prevalence of 5 percent for these pathogens in laying hens (Ebel et al. 1992). However, the prevalence of *S.* Enteritidis in hens was estimated by this study to be only 0.5 percent (Ebel et al. 1992). A separate study performed in California showed that the majority of *Salmonella* serotypes isolated by drag swabbing manure piles from sixty egg-producing ranches in 1996 were *S.* Heidelberg, *S.* Cerro, and *S.* Kentucky, whereas *S.* Enteritidis was isolated infrequently (Riemann et al. 1998). The low prevalence of *S.* Enteritidis in the feces of laying hens compared to that of other *Salmonella* serotypes is interesting, especially since the above data were collected just one year after the peak of the *S.* Enteritidis epidemic in the United States. Surveys on the prevalence of *Salmonella* serotypes in ovaries of laying hens show that *S.* Heidelberg is the most common isolate from this source in the United States (Barnhardt et al. 1991; Barnhart et al. 1993). That is, *S.* Heidelberg represented 57 percent of total isolates from ovary samples, followed by *S. enterica* serotype Agona (13 percent of isolates) and *S.* Oranienburg (6 percent of isolates). *S.* Enteritidis represented only 1 percent of total isolates from ovary samples (Barnhardt et al. 1991). Importantly, *S.* Enteritidis was present in 13 percent of nonpasteurized liquid egg samples collected from U.S. breaking plants between 1991 and 1992 (Ebel et al. 1993) and in 19 percent of unpasteurized liquid egg samples collected between 1994 and 1995 (Hogue et al. 1997). Laying hens naturally infected with *S.* Enteritidis produce between 0.1 percent and 1 percent of eggs that contain this pathogen (Humphries 1999). Although 45 percent of chicken laying flocks were found to be *S.* Enteritidis positive in the 1995 National Spent Hen Survey (Hogue et al. 1997), only a small fraction of hens in each flock carry the organism. The overall prevalence of *S.* Enteritidis in eggs produced in the United States is thus estimated by FSIS to be only one in every twenty thousand (Ebel and Schlosser 2000). Nonetheless, it is widely accepted that the dramatic increase in human *S.* Enteritidis infections reported in the 1980s and 1990s is related to the consumption of raw or undercooked chicken eggs (Coyle et al. 1988; St. Louis et al. 1988; Cowden et al. 1989a, 1989b; Henzler et al. 1994). These data therefore suggest that even a very low prevalence of *S.* Enteritidis in eggs must be considered a significant food-safety problem.

Control Measures. Human infections with *S.* Enteritidis steadily increased in frequency since reporting began in 1963, and between 1990 and 1996 it was the serotype most frequently isolated in the United States (Aserkoff et al. 1970; Mishu et al. 1994; Olsen et al. 2001). Given the dramatic increase in egg-associated human *S.* Enteritidis infections in the United States in the 1980s and early 1990s, the pre-harvest occurrence of this pathogen in laying flocks and eggs is considered to be a major food-safety hazard for the egg industry. Importantly, egg-associated *S.* Enteritidis illnesses are a prominent food-safety problem despite the infrequent (relative to other *Salmonella* serotypes such as *S.* Heidelberg) isolation of this organism from chickens in the United States and its low prevalence in chicken eggs (only one in every twenty thousand eggs produced is estimated to contain *S.* Enteritidis). These considerations suggest that for pre-harvest control measures to be effective, a complete elimination of *S.* Enteritidis from laying flocks would be needed. The USDA "trace back" regulation from 1990 and intensified efforts to educate food handlers and to enforce safe food-handling practices were initially unsuccessful in reducing the incidence of human illness caused by *S.* Enteritidis (Mason 1994). That is, the number of annual *S.* Enteritidis isolations from patients reported by CDC increased from 8,734 in 1990 to 10,200 in 1995 (Olsen et al. 2001). However, these programs and the implementation of quality assurance-programs by the egg-industry (Hogue et al. 1997) may be responsible for the subsequent decline in the number of annual *S.* Enteritidis isolations to 7,924 cases reported by CDC in 1997 (Olsen et al. 2001). More recently, the HACCP System was implemented by FSIS between 1998 and 2000 (Hogue et al. 1998) and the President's Council on Food Safety developed the Egg Safety Action Plan in 1999 (http://www.foodsafety.gov/~fsg/ceggs.html). Pre-harvest prevention strategies supported by these programs include: environmental testing to identify *S.* Enteritidis–contaminated houses (eggs produced by these flocks should be diverted to the pasteurized egg products market); the use of *S.* Enteritidis free feed; the use of *S.* Enteritidis free breeders; adequate cleaning and disinfection of houses and equipment; improved rodent control; and training of the agency inspection force. During implementation of these strategies, the incidence of human *S.* Enteritidis infections in the United States has declined from 1.9 cases per 100,000 population in 1998 (http://www.foodsafety.gov/~fsg/ceggs.html) to an estimated 1.8 cases per 100,000 population in 2000 (Anon. 2001). It remains to be seen whether these strategies will succeed in achieving the goal of the Egg Safety Action Plan, namely a 50 percent reduction of egg-associated *S.* Enteritidis illnesses by 2005 and an elimination of egg-associated *S.* Enteritidis illnesses in the United States by 2010.

An alternate strategy to eliminate egg-associated *S.* Enteritidis illnesses would be to target factors that are responsible for the recent emergence of *S.* Enteritidis as an egg-associated pathogen. This alternative is appealing because it would not merely treat the symptoms of the *S.* Enteritidis epidemic by improving surveillance and sanitation, but rather eliminate its cause. The emergence of *S.* Enteritidis as an egg-associated pathogen, which was first noted in the early 1970s (Rabsch et al. 2001), has recently been connected to the eradication from the United States and European countries of *S.* Gallinarum in the mid 1970s (Bäumler et al. 2000; Rabsch et al. 2000). Theoretical models suggest that the high prevalence of *S.* Gallinarum (serogroup D1) in chicken flocks during the first half of the twentieth century generated flock immunity against the O9-antigen, thereby preventing other serogroup D1 serotypes, such as *S.* Enteritidis, from infecting chicken flocks (Kingsley and Bäumler 2000; Rabsch et al. 2000). Eradication of *S.* Gallinarum by the test-and-slaughter method of disease control resulted in loss of flock immunity against the O9-antigen, thereby enabling *S.* Enteritidis to become established in chicken flocks (Bäumler et al. 2000). Thus, reestablishment of flock immunity by vaccination with *S.* Gallinarum is a critical control point, which has the potential to eliminate a factor responsible for the emergence of *S.* Enteritidis as an egg-associated pathogen. In recent field trials in the Netherlands, vaccination of layer flocks with *S.* Gallinarum has proven to be safe and more effective in reducing the flock level occurrence of *S.* Enteritidis than vaccination with *S.* Enteritidis (Feberwee et al. 2000, 2001). In light of the above considerations, it appears beneficial to include vaccination as a critical control point during HACCP implementation.

SALMONELLA SEROTYPES IN TURKEYS.
According to a survey performed by CDC between 1983 and 1987, turkey is the third most common source of human infections (Tauxe 1991). The nationwide young turkey microbiological baseline data collection program conducted by the Food Safety and Inspection Service (FSIS, USDA) between 1996 and 1997 shows that 19 percent of turkey carcasses in the United States are contaminated with *Salmonella* serotypes (USDA-FSIS website). The prevalence of *Salmonella* serotypes in ground turkey meat is higher, with 50 percent of samples in the United States testing positive in 1995 (Nationwide Raw Ground Turkey Microbiological Baseline Data Collection Program, FSIS, USDA, 1995) (USDA-FSIS website).

Clinical disease associated with *Salmonella* serotypes in turkeys is uncommon (Borland 1975) and their prevalence in apparently healthy birds in the United States is not well documented. In one survey performed in California between 1984 and 1989, the serotypes isolated

most frequently from turkeys and their environment were *S.* Kentucky, *S.* Anatum, and *S.* Heidelberg (Hird et al. 1993). The serotypes currently most frequently cultured from turkey carcasses are *S.* Hadar (15 percent of isolations), *S.* Heidelberg (14 percent), and *S.* Agona (9 percent) (Schlosser et al. 2000). The serotypes most frequently cultured from ground raw turkey are *S.* Hadar (24 percent of isolations) and *S.* Agona (9 percent) (Schlosser, et al. 2000). *S.* Hadar, *S.* Heidelberg, and *S.* Agona are also among the six serotypes isolated most frequently from patients in the United States between 1987 and 1997 (Olsen et al. 2001).

COMPARISON OF SURVEILLANCE DATA FROM GERMANY AND THE UNITED STATES.
The most striking difference between surveillance data collected in Germany and the United States is the relative importance of *S.* Enteritidis. In Germany, *S.* Enteritidis replaced *S.* Typhimurium in 1986 as the serotype most commonly isolated from humans, and the highest incidence of human cases was reported in 1992 (Tschäpe et al. 1999). In this year, *S.* Enteritidis caused 74 percent of human cases reported by the Robert Koch Institute (compared to only 25 percent of cases reported by the CDC at the peak of the *S.* Enteritidis epidemic in the United States). Despite a subsequent decline in the incidence of *S.* Enteritidis infections in Germany, this pathogen continued to be the *Salmonella* serotype most frequently isolated from humans, accounting for 58 percent of cases in 1999, followed by *S.* Typhimurium with 28 percent of cases (Hartung 2000). The high incidence of human *S.* Enteritidis infections in Germany relative to the United States is accompanied by a relatively high prevalence of this pathogen in the egg industry but not in the meat industry. For instance, German surveillance data from 1999 show that *S.* Enteritidis is the *Salmonella* serotype most frequently isolated from breeding fowl during egg production (86 percent of *Salmonella* isolations), from commercial egg-production flocks (66 percent of isolations), and from eggs (78 percent of isolations) (Hartung 2000). In contrast, *S.* Heidelberg is the most common isolate from laying hens in the United States (Barnhardt et al. 1991; Barnhart et al. 1993; Riemann et al. 1998). In Germany, *S.* Enteritidis can be cultured from approximately one in every 380 egg shells and from one in every five thousand egg yolks (Hartung 2000), whereas it is estimated to be present in only one in every twenty thousand eggs produced in the United States (Ebel and Schlosser 2000). Comparison of surveillance data from Germany and the United States therefore underscores the importance of eggs in the spread of human *S.* Enteritidis infections.

It is noteworthy that laying hens in Germany need to be vaccinated either with a *S.* Typhimurium or a *S.* Enteritidis vaccine since 2000. It is currently a matter of debate of how effective vaccination with *S.* Typhimurium (serogroup B) will be in reducing the prevalence of *S.* Enteritidis (serogroup D1). Studies using the mouse model show that vaccination with *S.* Typhimurium elicits a short-lived, nonspecific response (conferring protection against *S.* Typhimurium and *S.* Enteritidis) and a long-lived, serogroup-specific response (conferring protection against *S.* Typhimurium but not against *S.* Enteritidis) (Hormaeche et al. 1991). That is, vaccination with *S.* Typhimurium can cross-protect mice against *S.* Enteritidis, if the challenge is performed within five weeks post-immunization (Hormaeche et al. 1991; Heithoff et al. 2001). Protection against a heterologous challenge observed within five weeks post-immunization is not uncommon, because persistence of a live attenuated vaccine causes activation of nonspecific defense mechanisms, which confer cross-protection between organisms as divergent as *S.* Typhimurium and mouse hepatitis virus (Fallon et al. 1989, 1991). After ten weeks post-immunization of mice with *S.* Typhimurium, when nonspecific defense mechanisms have returned to their normal value, cross-protection against a lethal *S.* Enteritidis challenge is no longer observed, but mice are still protected against lethal challenge with *S.* Typhimurium (Hormaeche et al. 1991, 1996; Norris and Bäumler 1999). In contrast, it has been reported that chickens immunized with *S.* Typhimurium are protected against organ colonization with *S.* Enteritidis for up to eleven months post-vaccination (Hassan and Curtiss 1997). It remains to be seen whether immunization with a *S.* Typhimurium vaccine (serogroup B) will prove effective in reducing the prevalence of *S.* Enteritidis (serogroup D1) in German laying hens.

SUMMARY AND CONCLUSIONS. Review of the available data shows that animal-derived food products continue to be the most important source for human cases of salmonellosis in the United States. The prevalence of some *Salmonella* serotypes in animal-derived food sources correlates well with their frequency of isolation from patients in the United States. For instance, *S.* Typhimurium, the serotype most frequently isolated from human cases of disease in the United States (23 percent of cases reported between 1987 and 1997) (Olsen et al. 2001), is also the serotype most frequently isolated from beef carcasses (14 percent of isolations) and raw ground pork (18 percent of isolations). Furthermore, *S.* Typhimurium is the third most common isolate from chicken carcasses (10 percent of isolations), raw ground beef (15 percent of isolations), and swine carcasses (10 percent of isolations) in the United States (Schlosser et al. 2000). However, some *Salmonella* serotypes frequently isolated from patients are infrequently isolated from food sources. *S.* Enteritidis, the second most frequently isolated serotype from patients (21 percent of cases reported between 1987 and 1997) (Olsen et al. 2001), is not among the serotypes frequently isolated from meat or meat products (Schlosser et al. 2000). Furthermore, *S.* Enteritidis is estimated to be present in only one of every twenty thousand chicken eggs produced in the United States

(Ebel and Schlosser 2000). The fact that the majority of human *S.* Enteritidis infections can be traced back to chicken eggs (Coyle et al. 1988; St. Louis et al. 1988; Cowden et al. 1989a, 1989b; Henzler et al. 1994) suggests that even a low prevalence of this pathogen in the egg industry represents a significant human health hazard. The recent implementation of nationwide control programs was accompanied by a reduction in the incidence of human salmonellosis between 1996 and 2000 (Anon. 2001). Nonetheless, *Salmonella* serotypes continue to be the second most common source of human bacterial food-borne infections in the United States. Additional pre-harvest control measures, such as vaccination, should be considered to further reduce the prevalence of these pathogens in our food supply.

REFERENCES

Angulo, F. 1996. *Salmonella* infection in people. *Research on salmonellosis in the food safety consortium.* United States Animal Health Association, Little Rock, Arkansas.

Angulo, F.J., K.R. Johnson, R.V. Tauxe, and M.L. Cohen. 2000. Origins and consequences of antimicrobial-resistant nontyphoidal *Salmonella*: implications for the use of fluoroquinolones in food animals. *Microb. Drug Resist.* 6:77–83.

Angulo, F.J., S. Tippen, D.J. Sharp, B.J. Payne, C. Collier, J.E. Hill, T.J. Barrett, R.M. Clark, E.E. Geldreich, H.D. Donnell, Jr. and D.L. Swerdlow. 1997. A community waterborne outbreak of salmonellosis and the effectiveness of a boil water order. *Am. J. Public Health* 87:580–584.

Anon. 1993. National advisory committee on microbiological criteria for foods: Generic HACCP for raw beef. *Food Microbiol.* 10.

Anon. 2001. Preliminary FoodNet data on the incidence of foodborne illnesses—selected sites, United States, 2000. *MMWR* 50:241–246.

Aserkoff, B., S.A. Schroeder, and P.S. Brachman. 1970. Salmonellosis in the United States—a five-year review. *Am. J. Epidemiol.* 92:13–24.

Barnhardt, H.M., D.W. Dreesen, R. Basyien, and O.C. Pancorbo. 1991. Prevalence of *Salmonella enteritidis* and other serovars in ovaries of layer hens at the time of slaughter. *J. Food Prot.* 54:488–491.

Barnhart, H.M., D.W. Dreesen, and J.L. Burke. 1993. Isolation of *Salmonella* from ovaries and oviducts from whole carcasses of spent hens. *Avian Dis.* 37:977–980.

Bäumler, A.J., B.M. Hargis, and R.M. Tsolis. 2000. Tracing the origins of *Salmonella* outbreaks. *Science* 287:50–52.

Bean, N.H., J.S. Goulding, C. Lao, and F.J. Angulo. 1996. Surveillance for foodborne-disease outbreaks—United States, 1988–1992. *MMWR.* 45:1–66.

Berends, B.R., H.A. Urlings, J.M. Snijders, and F. Van Knapen. 1996. Identification and quantification of risk factors in animal management and transport regarding *Salmonella* spp. in pigs. *Int. J. Food Microbiol.* 30:37–53.

Borland, E.D. 1975. *Salmonella* infection in poultry. *Vet. Rec.* 97:406–408.

Brenner, F.W., R.G. Villar, F.J. Angulo, R. Tauxe, and B. Swaminathan. 2000. *Salmonella* nomenclature. *J. Clin. Microbiol.* 38:2465–2467.

Briggs, C.E., and P.M. Fratamico. 1999. Molecular characterization of an antibiotic resistance gene cluster of *Salmonella typhimurium* DT104. *Antimicrob. Agents. Chemother.* 43:846–849.

Brownlie, L.E., and F.H. Grau. 1967. Effect of food intake on growth and survival of *Salmonellas* and *Escherichia coli* in the bovine rumen. *J. Gen. Microbiol.* 46:125–134.

Bullis, K.L. 1977. The history of avian medicine in the U.S. II. Pullorum disease and fowl typhoid. *Avian Dis.* 21:422–429.

Byrd, J.A., J.R. DeLoach, D.E. Corrier, D.J. Nisbet, and L.H. Stanker. 1999. Evaluation of *Salmonella* serotype distributions from commercial broiler hatcheries and grower houses. *Avian. Dis.* 43:39–47.

Cohen, M.L., and R.V. Tauxe. 1986. Drug-resistant *Salmonella* in the United States: an epidemiologic perspective. *Science* 234:964–969.

Corrier, D.E., J.A. Byrd, B.M. Hargis, M.E. Hume, R.H. Bailey, and L.H. Stanker. 1999. Presence of *Salmonella* in the crop and ceca of broiler chickens before and after preslaughter feed withdrawal. *Poult. Sci.* 78:45–49.

Cowden, J.M., D. Chisholm, M. O'Mahony, D. Lynch, S.L. Mawer, G.E. Spain, L. Ward, and B. Rowe. 1989a. Two outbreaks of *Salmonella enteritidis* phage type 4 infection associated with the consumption of fresh shell-egg products. *Epidemiol. Infect.* 103:47–52.

Cowden, J.M., D. Lynch, C.A. Joseph, M. O'Mahony, S.L. Mawer, B. Rowe, and C.L. Bartlett. 1989b. Case-control study of infections with *Salmonella enteritidis* phage type 4 in England. *BMJ* 299:771–773.

Coyle, E.F., S.R. Palmer, C.D. Ribeiro, H.I. Jones, A.J. Howard, L. Ward, and B. Rowe. 1988. *Salmonella enteritidis* phage type 4 infection: association with hen's eggs. *Lancet* 2:1295–1297.

D'Aoust, J.Y., E. Daley, M. Crozier, and A.M. Sewell. 1990. Pet turtles: a continuing international threat to public health. *Am. J. Epidemiol.* 132:233–238.

Dahl, J., A. Wingstrand, B. Nielsen, and D.L. Baggesen. 1997. Elimination of *Salmonella typhimurium* infection by the strategic movement of pigs. *Vet. Rec.* 140:679–681.

Dargatz, D.A., P.J. Fedorka-Cray, S.R. Ladely, and K.E. Ferris. 2000. Survey of *Salmonella* serotypes shed in feces of beef cows and their antimicrobial susceptibility patterns. *J. Food Prot.* 63:1648–1653.

Davies, A., P. O'Neill, L. Towers, and M. Cooke. 1996. An outbreak of *Salmonella typhimurium* DT104 food poisoning associated with eating beef. Communicable Disease Report. *Cdr. Rev.* 6.

Davies, P.R., F.G. Bovee, J.A. Funk, W.E. Morrow, F.T. Jones, and J. Deen. 1998. Isolation of *Salmonella* serotypes from feces of pigs raised in a multiple-site production system. *J.A.V.M.A.* 212:1925–1929.

Davies, P.R., W.E. Morrow, F.T. Jones, J. Deen, P.J. Fedorka-Cray, and J.T. Gray. 1997. Risk of shedding *Salmonella* organisms by market-age hogs in a barn with open-flush gutters. *J.A.V.M.A.* 210:386–389.

Davies, P.R., W.E. Morrow, F.T. Jones, J. Deen, P.J. Fedorka-Cray, and I.T. Harris. 1997. Prevalence of *Salmonella* in finishing swine raised in different production systems in North Carolina, USA. *Epidemiol. Infect.* 119:237–244.

Duffy, E.A., K.E. Belk, J.N. Sofos, G.R. Bellinger, A. Pape, and G.C. Smith. 2001. Extent of microbial contamination in United States pork retail products. *J. Food Prot.* 64:172–178.

Ebel, E., and W. Schlosser. 2000. Estimating the annual fraction of eggs contaminated with *Salmonella enteritidis* in the United States. *Int. J. Food Microbiol.* 61:51–62.

Ebel, E.D., M.J. David, and J. Mason. 1992. Occurrence of *Salmonella enteritidis* in the U.S. commercial egg industry: report on a national spent hen survey. *Avian Dis.* 36:646–654.

Ebel, E.D., J. Mason, L.A. Thomas, K.E. Ferris, M.G. Beckman, D.R. Cummins, L. Scroeder-Tucker, W.D. Sutherlin, R.L. Glasshoff, and N.M. Smithhisler. 1993. Occur-rence of *Salmonella enteritidis* in unpasteurized liquid egg in the United States. *Avian Dis.* 37:135–142.

Fallon, M.T., W.H. Benjamin, Jr., T.R. Schoeb, and D.E. Briles. 1991. Mouse hepatitis virus strain UAB infection enhances resistance to *Salmonella typhimurium* in mice by inducing suppression of bacterial growth. *Infect. Immun.* 59:852–856.

Fallon, M.T., T.R. Schoeb, W.H. Benjamin, Jr., J.R. Lindsey and D.E. Briles. 1989. Modulation of resistance to *Salmonella typhimurium* infection in mice by mouse hepatitis virus (MHV). *Microb. Pathog.* 6:81–91.

Fang, F.C., and J. Fierer. 1991. Human infection with *Salmonella dublin*. *Medicine (Baltimore)* 70:198–207.

Farrington, L.A., R.B. Harvey, S.A. Buckley, R.E. Droleskey, D.J. Nisbet, and P.D. Inskip. 2001. Prevalence of antimicrobial resistance in *Salmonellae* isolated from market-age swine. *J. Food. Prot.* 64:1496–1502.

Feberwee, A., T.S. de Vries, A.R. Elbers, and W.A. de Jong. 2000. Results of a *Salmonella enteritidis* vaccination field trial in broiler-breeder flocks in The Netherlands. *Avian Dis.* 44:249–255.

Feberwee, A., T.S. de Vries, E.G. Hartman, J.J. de Wit, A.R. Elbers, and W.A. de Jong. 2001. Vaccination against *Salmonella enteritidis* in Dutch commercial layer flocks with a vaccine based on a live *Salmonella gallinarum* 9R strain: evaluation of efficacy, safety, and performance of serologic *Salmonella* tests. *Avian Dis.* 45:83–91.

Fedorka-Cray, P.J., D.A. Dargatz, L.A. Thomas, and J.T. Gray. 1998. Survey *Salmonella* serotypes in feedlot cattle. *J. Food Prot.* 61:525–530.

Fedorka-Cray, P.J., J.T. Gray, and C. Wray. 2000. *Salmonella* infections in pigs. *Salmonella in Domestic Animals.* C. Wray and A. Wray, eds. New York: CABI Publishing, 191–207.

Ferris, K. and L.A. Thomas. 1995. *Salmonella* serotypes from animals and related sources reported during July 1994 and June 1995. *99th Annual Meeting of the U.S. Animal Health Organization*, Reno, Nevada.

Fone, D.L., and R.M. Barker. 1994. Associations between human and farm animal infections with *Salmonella typhimurium* DT104 in Herefordshire. Communicable Disease Report. *Cdr. Rev.* 4.

Funk, J.A., P.R. Davies, and W. Gebreyes. 2001. Risk factors associated with *Salmonella* enterica prevalence in three-site swine production systems in North Carolina, USA. *Berl Munch Tierarztl Wochenschr* 114:335–338.

Galbraith, N.S. 1961. Studies of human salmonellosis in relation to infection in animals. *Vet. Rec.* 73:1296–1303.

Gebreyes, W.A., P.R. Davies, W.E. Morrow, J.A. Funk, and C. Altier. 2000. Antimicrobial resistance of *Salmonella* isolates from swine. *J. Clin. Microbiol.* 38:4633–4636.

Gibson, E.A. 1961. Salmonellosis in calves. *Vet. Rec.* 73:1284–1296.

Giles, N., S.A. Hopper, and C. Wray. 1989. Persistence of *S. typhimurium* in a large dairy herd. *Epidemiol. Infect.* 103:235–241.

Grau, F.H., L.E. Brownlie, and E.A. Roberts. 1968. Effect of some preslaughter treatments on the *Salmonella* population in the bovine rumen and faeces. *J. Appl. Bacteriol.* 31:157–163.

Hampton, M.D., E.J. Threlfall, J.A. Frost, L.R. Ward, and B. Rowe. 1995. *Salmonella typhimurium* DT 193: differentiation of an epidemic phage type by antibiogram, plasmid profile, plasmid fingerprint and *Salmonella* plasmid virulence (spv) gene probe. *J. Appl. Bacteriol.* 78:402–408.

Harris, I.T., P.J. Fedorka-Cray, J.T. Gray, L.A. Thomas, and K. Ferris. 1997. Prevalence of *Salmonella* organisms in swine feed. *J. Am. Vet. Med. Assoc.* 210:382–385.

Hartung, M. 2000. Bericht Über die epidemiologische Situation der Zoonosen in Deutschland für 1999. Berlin, *Bundesinstitut für gesundheitlichen Verbraucherschutz und Veterin. rmendizin* (BgVV):34–102.

Hassan, J.O., and R. Curtiss, 3rd. 1997. Efficacy of a live avirulent *Salmonella typhimurium* vaccine in preventing colonization and invasion of laying hens by *Salmonella typhimurium* and *Salmonella enteritidis*. *Avian Dis.* 41:783–791.

Heithoff, D.M., E.Y. Enioutina, R.A. Daynes, R.L. Sinsheimer, D.A. Low, and M.J. Mahan. 2001. *Salmonella* DNA adenine methylase mutants confer cross-protective immunity. *Infect. Immun.* 69:6725–6730.

Henzler, D.J., E. Ebel, J. Sanders, D. Kradel, and J. Mason. 1994. *Salmonella enteritidis* in eggs from commercial chicken layer flocks implicated in human outbreaks. *Avian Dis.* 38:37–43.

Hird, D.W., H. Kinde, J. T. Case, B.R. Charlton, R.P. Chin, and R.L. Walker. 1993. Serotypes of *Salmonella* isolated from California turkey flocks and their environment in 1984–89 and comparison with human isolates. *Avian Dis.* 37:715–719.

Hogue, A., P. White, J. Guard-Petter, W. Schlosser, R. Gast, E. Ebel, J. Farrar, T. Gomez, J. Madden, M. Madison, A.M. McNamara, R. Morales, D. Parham, P. Sparling, W. Sutherlin, and D. Swerdlow. 1997. Epidemiology and control of egg-associated *Salmonella enteritidis* in the United States of America. *Rev. Sci. Tech.* 16:542–553.

Hogue, A.T., E.D. Ebel, L.A. Thomas, W. Schlosser, N. Bufano, and K. Ferris. 1997. Surveys of *Salmonella enteritidis* in unpasteurized liquid egg and spent hens at slaughter. *J. Food Prot.* 60:1194–1200.

Hogue, A.T., P.L. White, and J.A. Heminover. 1998. Pathogen Reduction and Hazard Analysis and Critical Control Point (HACCP) systems for meat and poultry. USDA. *Vet. Clin. North Am. Food. Anim. Pract.* 14:151–164.

Hormaeche, C.E., H.S. Joysey, L. Desilva, M. Izhar, and B.A. Stocker. 1991. Immunity conferred by *Aro-Salmonella* live vaccines. *Microb. Pathog.* 10:149–158.

Hormaeche, C.E., P. Mastroeni, J.A. Harrison, R. Demarco de Hormaeche, S. Svenson, and B.A. Stocker. 1996. Protection against oral challenge three months after i.v. immunization of BALB/c mice with live Aro *Salmonella typhimurium* and *Salmonella enteritidis* vaccines is serotype (species)-dependent and only partially determined by the main LPS O antigen. *Vaccine* 14:251–259.

Hughes, L.E., E.A. Gibson, H.E. Roberts, E.T. Davies, G. Davies, and W.J. Sojka. 1971. Bovine salmonellosis in England and Wales. *Br. Vet. J.* 127:225–238.

Humphries, T.J. 1999. Contamination of eggs and poultry meat with *Salmonella* enterica serovar Enteritidis. *Salmonella* enterica serovar Enteritidis in humans and animals. A.M. Saeed, R.K. Gast, M.E. Potter, and P.G. Wall, eds. Ames, IA: Iowa State University Press, 183–192.

Hurd, H.S., J.K. Gailey, J.D. McKean, and M.H. Rostagno. 2001. Experimental rapid infection in market swine following exposure to a *Salmonella* contaminated environment. *Berl Munch Tierarztl. Wochenschr.* 114:382–384.

Hurd, H.S., J.K. Gailey, J.D. McKean, and M.H. Rostagno. 2001. Rapid infection in market-weight swine following exposure to a *Salmonella typhimurium*-contaminated environment. *Am. J. Vet. Res.* 62:1194–1197.

Isaacson, R.E., L.D. Firkins, R.M. Weigel, F.A. Zuckermann, and J.A. DiPietro. 1999. Effect of transportation and feed withdrawal on shedding of *Salmonella typhimurium* among experimentally infected pigs. *Am. J. Vet. Res.* 60:1155–1158.

Jones, F.S. 1913. The value of the macroscopic agglutination test in detecting fowls that are harboring *Bact. pullorum*. *J. Med. Res.* 27:481–495.

Jones, P.W., G. Dougan, C. Hayward, N. Mackensie, P. Collins, and S.N. Chatfield. 1991. Oral vaccination of calves against experimental salmonellosis using a double aro mutant of *Salmonella typhimurium*. *Vaccine* 9:29–34.

Kingsley, R.A., and A.J. Bäumler. 2000. Host adaptation and the emergence of infectious disease: the salmonella paradigm. *Mol. Microbiol.* 36:1006–1014.

Low, J.C., G. Hopkins, T. King, and D. Munro. 1996. Antibiotic resistant *Salmonella typhimurium* DT104 in cattle. *Vet. Rec.* 138:650–651.

Mahon, B.E., A. Ponka, W. N. Hall, K. Komatsu, S.E. Dietrich, A. Siitonen, G. Cage, P.S. Hayes, M.A. Lambert-Fair, N.H. Bean, P.M. Griffin, and L. Slutsker. 1997. An international outbreak of *Salmonella* infections caused by alfalfa sprouts grown from contaminated seeds. *J. Infect. Dis.* 175:876–882.

Malorny, B., A. Schroeter, and R. Helmuth. 1999. Incidence of quinolone resistance over the period 1986 to 1998 in veterinary *Salmonella* isolates from Germany. *Antimicrob. Agents Chemother.* 43:2278–2282.

Mason, J. 1994. *Salmonella enteritidis* control programs in the United States. *Int. J. Food Microbiol.* 21:155–169.

McLaren, I.M., and C. Wray. 1991. Epidemiology of *Salmonella typhimurium* infection in calves: persistence of salmonellae on calf units. *Vet. Rec.* 129:461–462.

Mead, P.S., L. Slutsker, V. Dietz, L.F. McCaig, J.S. Bresee, C. Shapiro, P.M. Griffin, and R. V. Tauxe. 1999. Food-related illness and death in the United States. *Emerg. Infect. Dis.* 5:607–625.

Mishu, B., J. Koehler, L.A. Lee, D. Rodrigue, F.H. Brenner, P. Blake, and R.V. Tauxe. 1994. Outbreaks of *Salmonella enteritidis* infections in the United States, 1985–1991. *J. Infect. Dis.* 169:547–552.

Moo, D., D. O'Boyle, W. Mathers, and A.J. Frost. 1980. The isolation of *Salmonella* from jejunal and caecal lymph nodes of slaughtered animals. *Austral. Vet. J.* 56:1813.

Norris, T.L., and A.J. Bäumler. 1999. Phase variation of the lpf fimbrial operon is a mechanism to evade cross immunity between *Salmonella* serotypes. *Proc. Natl. Acad. Sci. USA.* 96:13393–13398.

Olsen, S.J., R. Bishop, F.W. Brenner, T.H. Roels, N. Bean, R.V. Tauxe, and L. Slutsker. 2001. The changing epidemiology of salmonella: trends in serotypes isolated from humans in the United States, 1987–1997. *J. Infect. Dis.* 183:753–761.

Penny, C.D., J.C. Low, P.F. Nettleton, P.R. Scott, N.D. Sargison, W. D. Strachan, and P. C. Honeyman. 1996. Concurrent bovine viral diarrhoea virus and *Salmonella typhimurium* DT104 infection in a group of pregnant dairy heifers. *Vet. Rec.* 138:485–489.

Rabsch, W., B.M. Hargis, R.M. Tsolis, R.A. Kingsley, K.H. Hinz, H. Tschäpe, and A.J. Bäumler. 2000. Competitive exclusion of *Salmonella enteritidis* by *Salmonella gallinarum* from poultry. *Emerg. Infect. Dis.* 6. In press.

Rabsch, W., H. Tschäpe, and A.J. Bäumler. 2001. Nontyphoidal salmonellosis: emerging problems. *Microbes Infect.* 3:237–247.

Reynolds, D.J., J.H. Morgan, N. Chanter, P.W. Jones, J.C. Bridger, T.G. Debney, and K.J. Bunch. 1986. Microbiology of calf diarrhoea in southern Britain. *Vet. Rec.* 119:34–39.

Riemann, H., S. Himathongkham, D. Willoughby, R. Tarbell, and R. Breitmeyer. 1998. A survey for *Salmonella* by drag swabbing manure piles in California egg ranches. *Avian Dis.* 42:67–71.

Rothenbacher, H. 1965. Mortality and morbidity in calves with salmonellosis. *J.A.V.M.A.* 147:1211–1214.

Samuel, J.L., J.A. Eccles, and J. Francis. 1981. *Salmonella* in the intestinal tract and associated lymph nodes of sheep and cattle. *J. Hyg.* 87: 225–232.

Samuel, J.L., D.A. O'Boyle, W.J. Mathers, and A.J. Frost. 1979. Isolation of *Salmonella* from mesenteric lymph nodes of healthy cattle at slaughter. *Res. Vet. Sci.* 28:238–241.

Santos, R.L., S. Zhang, R.M. Tsolis, R.A. Kingsley, L.G. Adams, and A.J. Bäumler. 2001. Animal models of Salmonella infections: enteritis vs. typhoid fever. *Mircrob. Infect.* 3:1335–1344.

Saphra, I., and M. Wassermann. 1954. *Salmonella Cholerae suis.* A clinical and epidemiological evaluation of 329 infections identified between 1940 and 1954 in the New York *Salmonella* Center. *Am. J. Med. Sci.* 228:525–533.

Schaffer, J.M., and A.D. MacDonald. 1931. A stained antigen for the rapid whole blood test for pullorum disease. *J.A.V.M.A.* 32:236–240.

Schlosser, W., A. Hogue, E. Ebel, B. Rose, R. Umholtz, K. Ferris, and W. James. 2000. Analysis of *Salmonella* serotypes from selected carcasses and raw ground products sampled prior to implementation of the Pathogen Reduction; Hazard Analysis and Critical Control Point Final Rule in the U.S. *Int. J. Food Microbiol.* 58:107–111.

Schwartz, K.J. 1999. Salmonellosis. *Diseases of Swine.* B.E. Shaw, ed. Ames, IA: Iowa State University Press, 535–551.

Smith, B.P., L. DaRoden, M.C. Thurmond, G.W. Dilling, H. Konrad, J.A. Pelton, and J.P. Picanso. 1994. Prevalence of salmonellae in cattle and in the environment of California dairies. *J.A.V.M.A.* 205:467–471.

Smith, B.P., F. Habasha, M. Reina-Guerra, and A.J. Hardy. 1979. Bovine salmonellosis: experimental production and characterization of the disease in calves, using oral challenge with *Salmonella typhimurium. Am. J. Vet. Res.* 40:1510–1513.

Snoeyenbos, G.H., C.F. Smyser, and H. Van Roekel. 1969. *Salmonella* infections of the ovary and peritoneum of chickens. *Avian Dis.* 13:668–670.

Sojka, W.J., and H.I. Field. 1970. Salmonellosis in England and Wales 1958–1967. *Vet. Bull.* 40:515–531.

Sojka, W.J., and C. Wray. 1975. Incidence of *Salmonella* infection in animals in England amd Wales, 1968–73. *Vet. Rec.* 96:280–284.

St. Louis, M.E., D.L. Morse, M.E. Potter, T.M. DeMelfi, J.J. Guzewich, R.V. Tauxe, and P.A. Blake. 1988. The emergence of grade A eggs as a major source of *Salmonella enteritidis* infections. New implications for the control of salmonellosis [see comments]. *JAMA* 259:2103–2107.

Swanenburg, M., B.R. Berends, H.A. Urlings, J.M. Snijders, and F. van Knapen. 2001. Epidemiological investigations into the sources of *Salmonella* contamination of pork. Berl Munch *Tierarztl. Wochenschr.* 114:356–359.

Swanenburg, M., H.A. Urlings, D.A. Keuzenkamp, and J.M. Snijders. 2001. Salmonella in the lairage of pig slaughterhouses. *J. Food Prot.* 64:12–16.

Tauxe, V.T. 1991. *Salmonella*: A postmodern pathogen. *J. Food Prot.* 54:563–568.

Threlfall, E.J., J.A. Frost, L.R. Ward, and B. Rowe. 1994. Epidemic in cattle and humans of *Salmonella* typhimurium DT104 with chromosomally integrated multiple drug resistance. *Vet. Rec.*134:577.

Threlfall, E.J., L.R. Ward, J.A. Skinner, and A. Graham. 2000. Antimicrobial drug resistance in non-typhoidal salmonellas from humans in England and Wales in 1999: decrease in multiple resistance in *Salmonella enterica* serotypes Typhimurium, Virchow, and Hadar. *Microbiol. Drug Resist.* 6:319–325.

Tschäpe, H., A. Liesegang, B. Gericke, R. Prager, W. Rabsch, and R. Helmuth. 1999. Ups and downs of *Salmonella enterica* serovar Enteritidis in Germany. *Salmonella enterica serovar Enteritidis in humans and animals.* A.M. Saeed, R.K. Gast, M.E. Potter, and P.G. Wall, eds. Ames, IA: Iowa State University Press, 51–61.

Tsolis, R.M., L.G. Adams, T.A. Ficht, and A.J. Bäumler. 1999. Contribution of *Salmonella typhimurium* virulence factors to diarrheal disease in calves. *Infect. Immun.* 67(9):4879–4885.

Turk, J.R., W.H. Fales, C. Maddox, M. Miller, L. Pace, J. Fischer, J. Kreeger, G. Johnson, S. Turnquist, J.A. Ramos, et al. 1992. Pneumonia associated with *Salmonella choleraesuis* infection in swine: 99 cases (1987–1990). *J.A.V.M.A.* 201:1615–1616.

USDA Food Safety and Inspection. Service. The Baseline Data Collection Program. http://www.fsis.usda.gov/OPHS/baseline/contents.htm)

Wall, P.G., D. Morgan, K. Lamden, M. Ryan, M. Griffin, E.J. Threlfall, L.R. Ward, and B. Rowe. 1994. A case control study of infection with an epidemic strain of multiresistant *Salmonella typhimurium* DT104 in England and Wales. Communicable Disease Report. *Cdr. Rev.* 4(11).

Weber, A., C. Bernt, K. Bauer, and A. Mayr. 1993. The control of bovine salmonellosis under field conditions using herd-specific vaccines. *Tierarztl Prax* 21:511–516.

Wells, S.J., P.J. Fedorka-Cray, D.A. Dargatz, K. Ferris, and A. Green. 2001. Fecal shedding of *Salmonella* spp. by dairy cows on farm and at cull cow markets. *J. Food Prot.* 64:3–11.

Wells, S.J., S.L. Ott, and A.H. Seitzinger. 1998. Key health issues for dairy cattle—new and old. *J. Dairy Sci.* 81:3029–3035.

Werner, S.B., G.L. Humphrey, and I. Kamei. 1979. Association between raw milk and human *Salmonella dublin* infection. *Br. Med. J.* 2:238–241.

Wilcock, B.P., C.H. Armstrong, and H.J. Olander. 1976. The significance of the serotype in the clinical and pathological features of naturally occurring porcine salmonellosis. *Can. J. Comp. Med.* 40:80–88.

Wills, R.W., J.T. Gray, P.J. Fedorka-Cray, K.J. Yoon, S. Ladely, and J.J. Zimmerman. 2000. Synergism between porcine reproductive and respiratory syndrome virus (PRRSV) and *Salmonella choleraesuis* in swine. *Vet. Microbiol.* 71:177–192.

Woodward, D.L., R. Khakhria, and W.M. Johnson. 1997. Human salmonellosis associated with exotic pets. *J. Clin. Microbiol.* 35:2786–2790.

Wray, C., and W.J. Sojka. 1978. Experimental *Salmonella typhimurium* infection in calves. *Res. Vet. Sci.* 25:139–143.

Wray, C., J.N. Todd, and M. Hinton. 1987. Epidemiology of *Salmonella typhimurium* infection in calves: excretion of *S. typhimurium* in the faeces of calves in different management systems. *Vet. Rec.* 121:293–296.

Wray, C., N. Todd, I.M. McLaren, and Y.E. Beedell. 1991. The epidemiology of *Salmonella* in calves: the role of markets and vehicles. *Epidemiol. Infect.* 107:521–525.

Zhao, C.,B. Ge, J. De Villena, R. Sudler, E. Yeh, S. Zhao, D. G. White, D. Wagner, and J. Meng. 2001. Prevalence of *Campylobacter* spp., *Escherichia coli*, and *Salmonella* Serovars in Retail Chicken, Turkey, Pork, and Beef from the Greater Washington, D.C., Area. *Appl. Environ. Microbiol.* 67:5431–5436.

13

MOLECULAR PATHOBIOLOGY AND EPIDEMIOLOGY OF EGG-CONTAMINATING *SALMONELLA ENTERICA* SEROVAR ENTERITIDIS

JEAN GUARD-PETTER, ERNESTO LIEBANA, TOM J. HUMPHREY, AND FRIEDA JORGENSEN

INTRODUCTION. *Salmonella enterica* serovar Enteritidis (*S.* Enteritidis) is the only one of more than 2,000 *Salmonellae* serovars to propagate human illness through a unique association with eggs that follows silent infection in hens (Guard-Petter 2001). Its emergence was first detected in the early 1980s (Anon. 1988; Baumler et al. 2000; PHLS 1989; St Louis et al. 1988) and case incidence peaked in the mid-1990s (Anon. 2001; Fisher 2001; Olsen et al. 2001). Continued outbreaks indicate that the threat to public health is still present, and currently, *S.* Enteritidis is still the leading cause of foodborne salmonellosis in the world, whereas it is second in the United States. In England and Wales, another eight-fold reduction in human cases is needed before the pandemic is reduced to pre-emergent levels (Anon. 2001). In the United States, outbreaks of *S.* Enteritidis have remained at approximately forty-five outbreaks per year during the years 1997 to 2000, and new strains continue to emerge. The persistence of this unusual foodborne pathogen in the face of heightened control measures provokes an examination of the most current topics in microbiology. *S.* Enteritidis is particularly useful as a model for foodborne disease because the pandemic occurred when modern methods for analysis were available. The emergence of *Salmonella enterica* serovar Typhimurium (*S.* Typhimurium) during the 1960s was the last time a *Salmonella* serovar became a persistent cause of foodborne salmonellosis that lasted for more than a decade (Gunn and Markakis 1978; Kilic et al. 1987; Kohler et al. 1979; Laszlo et al. 1985; Ling et al. 1987; McDonough et al. 1986; Metz and Lieb 1980; Milch et al. 1985; Przybylska 1990; Wachsmuth 1986). Central to understanding the factors that contribute to the emergence of *S.* Enteritidis are those that contribute to strain heterogeneity and the ability of *S.* Enteritidis to infect multiple hosts and to colonize a variety of ecological niches.

SALMONELLA ENTERITIDIS AROUND THE WORLD.

Although it must be recognized that reporting, surveillance, and egg consumption patterns differ among countries, the more central European countries of Germany, the Czech Republic, Austria, Hungary and Poland seem to have been the worst afflicted (greater than one hundred reported cases per 100,000 capita). Great Britain had an intermediate problem (forty cases per 100,000 capita) (Buchrieser et al. 1997; Glosnicka and Kunikowska 1994; Hasenson et al. 1992; Schulte 1994; Sramova et al. 1991; Ullmann and Scholtze 1989), as did Belgium, the Netherlands, Spain and Italy (Fantasia and Filetici 1994; Perales et al. 1989; Scuderi et al. 2000; Van Loock et al. 2000). France was mildly affected (seven cases per 100,000) (Watier et al. 1993) and Sweden, Norway, Finland, and Denmark even less so (Johansson et al. 1996; Kapperud et al. 1998; Lester et al. 1991; Wierup et al. 1995). Greece, Turkey, and Ireland experienced outbreaks of *S.* Enteritidis, but not to the degree that suggested that an overwhelming pandemic had occurred (Aysev et al. 2001; Badi et al. 1992; Wilson et al. 1996). Eastern European countries were possibly as severely afflicted as were those in central Europe region, but data from this region are limited (Hasenson et al. 1995; Hasenson et al. 1992; Solodovnikov et al. 1996).

In the United States, the northeast and midwest states were the most severely affected during the first fifteen years of the pandemic (Anon. 1999; Hogue et al. 1997; St Louis et al. 1988). One report suggested that flock incidence of egg contamination in Pennsylvania sometimes occurred at levels observed in European countries (Henzler et al. 1998). In 1994, California reported a marked increase in *S.* Enteritidis in poultry that was followed soon thereafter by an increase in human illness (Kinde et al. 1996; Passaro et al. 1996). Texas, New Mexico, Utah, and Hawaii also experienced a surge in isolation of *S.* Enteritidis during this time (Betit et al. 1997; Anon. 1986; Anon. 1993; Boyce et al. 1996; Hogue et al. 1997; Kremer 1999). Disturbingly, this surge was associated with isolation of phage type (PT) 4 *S.* Enteritidis from hens. Canada reported incidence levels lower than that in the United States, and isolation of PT 4 in this country appears to be associated with travel abroad (Khakhria et al. 1997; Poppe 1994).

In Mexico and South America, the reported incidence appeared to be close to that in the United States, although regional outbreaks were often large and *S.* Enteritidis appeared to be displacing *S.* Typhimurium (Fica et al. 1997; Gutierrez-Cogco et al. 2000; Kaku et al. 1995). In India, illness from *Salmonella enterica* serovar Typhi (*S.* Typhi) remained prevalent, and this country did not report the same experience as detected elsewhere in the world. Indonesia, Saudi Arabia, and Australia reported some illness caused by *S.* Enteritidis, but not, apparently, at European levels of incidence (al-Nakhli et al. 1999; Kudoh 1997; Narimatsu et al. 1997; Ueda et al. 1999). Japan reports a number of outbreaks of *S*. Enteritidis and increased illness in people that appears similar to incidence levels in Europe and North America (Kudoh 1997).

COMPETITION BETWEEN S. ENTERITIDIS AND OTHER *SALMONELLA* SEROVARS. An intriguing hypothesis used by some to explain the pandemic of egg-associated salmonellosis is that *S*. Enteritidis has filled an ecological niche vacated by *Salmonella enterica* serovar Pullorum (*S*. Pullorum) (Baumler et al. 2000), which is an avian-adapted serovar historically associated with egg contamination and illness in chickens, but not with human salmonellosis (Shivaprasad 2000). There is little doubt that some sort of cycling between *Salmonella* serovars occurs on a periodic basis that influences relative prevalence. However, for every experiment that claims no cross-immunity exists between serovars with different immunodominant O-chains (region 3 of the lipopolysaccharide molecule) (Kingsley and Baumler 2000; Martin et al. 1996), another describes finding evidence for it (Gherardi et al. 2000; Heithoff et al. 2001). Conflicting evidence about cross-immunity suggests that detection of the phenomenon is perhaps greatly influenced by the strain, the antigenic composition of its outer membrane, and the experimental design (Hormaeche et al. 1990, 1996). There is evidence that a *Salmonella enterica* serovar Gallinarum (*S*. Gallinarum)–modified live vaccine licensed for use in Europe is efficacious for reducing colonization of hens by *S*. Enteritidis (Feberwee et al. 2001). Because this strain lacks lipopolysaccharide (LPS) that has an attached O-chain, cross-protective immunity is more likely elicited by other antigens or by LPS core antigens shared among the Salmonellae. Evidence suggests that *S*. Typhimurium is better at providing cross-protection against *S*. Enteritidis than the other way around (Hassan and Curtiss 1997; Martin et al. 1996).

There is epidemiological evidence that *S*. Pullorum and *S*. Enteritidis are both commonly encountered within countries that do not have a stringent Pullorum control program in place (Hinz et al. 1989; Hoop 1997; Hoop and Albicker-Rippinger 1997; Stanley et al. 1992), which is a fact that contradicts the hypothesis that *S*. Enteritidis is filling a niche vacated by *S*. Pullorum. It is possible that endemic *S*. Typhi interferes with outbreaks of *S*. Enteritidis in humans (Gugnani 1999), but this relationship between serovar group D1 Salmonellae has not been assessed except in India, where it is clear that *S*. Typhi predominates and that *S*. Enteritidis is a minor problem. The current focus on LPS O-chain serovar as a predictor of environmental prevalence does not take into account new information about LPS O-chain structural micro-heterogeneity that does not alter immunodominance imparted by dideoxyhexose sugars in the O-chain region of lipopolysaccharide (that is,

tyvelose, which is O-antigen factor 9, the group determinant of group D serotype) (Guard-Petter et al. 1999; Parker et al. 2001b; Rahman et al. 1997). In regards to new information about LPS O-chain structure, *S*. Enteritidis is sometimes able to produce the capsular-like LPS hydrophilic structure historically attributed to *S*. Typhi, especially if isolated from eggs (Hellerqvist et al. 1969; Parker et al. 2001b; Rahman et al. 1997). Thus, the absence of *S*. Typhi in developed countries may increase susceptibility of humans to *S*. Enteritidis found in eggs. In summary, no clear evidence exists that the absence of *S*. Pullorum in birds is a risk factor for emergence of *S*. Enteritidis.

THE MOUSE IN THE HEN HOUSE. It is not known whether the practice of using *S*. Enteritidis as a rodenticide in Eastern European countries began a cycle of infection in poultry that could not be stopped (Friedman et al. 1996; Rabsch et al. 2001; Rybin 1977). What is known about rodents is that they are important intermediate hosts for the spread of *Salmonella* in the on-farm environment in general (Savage and Read 1913). Modern methods used to assess the role of the house mouse *Mus musculus* in the *S*. Enteritidis pandemic include molecular analysis, epidemiological monitoring, and DNA fingerprinting (Davies and Wray 1995b, 1996b; Guard-Petter et al. 1997; Healing 1991; Henzler and Opitz 1992; Parker et al. 2001b). These approaches, coupled with epidemiological monitoring, identified that more than 20 percent of mice caught in Pennsylvania hen houses harbored organ-invasive strains of *S*. Enteritidis (Guard-Petter et al. 1997; Parker et al. 2001b). DNA fingerprinting using fine map analysis demonstrated a close relationship between orally invasive phenotypes that resulted in egg contamination and isolates obtained from naturally infected mice (Parker et al. 2001b). In some cases, failure to rid hen houses of mice has been correlated with rapid colonization of clean replacement birds (Davies and Wray 1995b, 1996b), and statistical analyses consistently identify mice as a risk factor for egg contamination (USDA 1998). The role attributed to mice in the infection route leading to human illness is that they constantly re-introduce orally invasive phenotypes into the environment of birds (Parker et al. 2001b). Instability of orally invasive phenotypes appears to be the result of microbial competition, where outgrowth of new phenotypes that are clonally related occurs in response to adaptation of bacteria to a variety of environmental stresses and growth conditions (Allen-Vercoe et al. 1998; Guard-Petter et al. 1996; Jorgensen et al. 2000; Zambrano et al. 1993). Identification of the mouse as an intermediate host of orally invasive phenotypes does not preclude other sources of infection (Davies and Wray 1995a; Davies and Wray 1996a; Hubalek et al. 1995; Kohler 1993; Olsen and Hammack 2000). Indeed, the existence of multiple avenues of *S*. Enteritidis into the hen house indicates why control measures must be stringently followed.

THE CONVOLUTED PHENOTYPE AND ORAL INVASION OF *S.* ENTERITIDIS.

It is possible to visualize a progression of naturally occurring colony morphologies for strains of *S.* Enteritidis that differ in their ability to cause oral infection in hens and subsequent egg contamination. The most commonly recognized colony morphology for *Salmonella* has a smooth surface, is tan in color, and is typically no more than 5mm in diameter after sixteen hours' growth at 37°C on rich agar media (Fig. 13.1a). The colony morphology of a convoluted phenotype has a pitted surface caused by the production of a simple biofilm (Fig. 13.1b). The lacy phenotype is a variant of the convoluted phenotype because it produces a more complex biofilm that results in an elaborately filigreed colony surface and larger size (Fig. 13.1c). Strains can produce morphologic variants concomitantly in culture, which is observed as either a lacy colony morphology that retains a smooth center while producing a filigreed edge (Fig. 13.1d), or as a mixture of individual colonies that are either convoluted or smooth, but not both (Parker et al. 2001a). A weak convoluted phenotype is also frequently encountered (picture not shown). Thus, final arbitration of the pathogenic potential of any one strain in a host animal may involve the representation and proportion of a mixture of phenotypes in the inoculating dose (Guard-Petter 2001) (Fig. 13.1e).

The composition of the cell-surface molecules that contribute to the convoluted morphology gives impor-

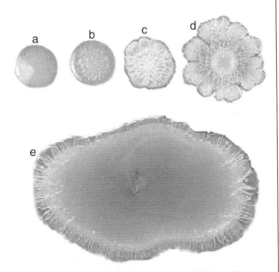

FIG. 13.1—Variant colony morphologies of *Salmonella* serovar Enteritidis. At top, from left to right is a) a smooth tan colony typical of *S.* Enteritidis and usually referred to as wildtype morphology, b) the convoluted colony morphology, c) the lacy variant of the convoluted colony morphology, and d) a colony showing a mixture of wildtype (center) and lacy (rest of colony). Below the small colonies is a large swarm colony of *S.* Enteritidis that forms a narrow edge of biofilm (e). Photographs have been digitally manipulated for purposes of presentation, but details of colony morphology have been preserved. The smallest colony is 5mm in diameter.

tant information about features that contribute to environmental resistance, surface adherence, and oral invasiveness. The *Salmonella* convoluted colony morphology at its most basic originates from production of a temperature- and media-dependent organic matrix, or biofilm, that has as essential components thin aggregative fimbriae and cellulose (Allen-Vercoe et al. 1997; Austin et al. 1998; Collinson et al. 1993; Romling et al. 2000; Zogaj et al. 2001) (Fig. 13.1b). In animal trials, the convoluted phenotype did not result in high-incidence egg contamination or overwhelming systemic infection, but it adhered well to surfaces (Allen-Vercoe et al. 1999; Allen-Vercoe and Woodward 1999; Woodward et al. 2000). The convoluted phenotype can also entrap flagella and glucosylated lipopolysaccharide within biofilm, which is a phenomenon that contributes to the filigreed colony morphology called lacy (Allen-Vercoe et al. 1997; Austin et al. 1998; Guard-Petter et al. 1996; Romling and Rohde 1999; Zogaj et al. 2001) (Fig. 13.1c). The lacy variant of the convoluted phenotype had enhanced oral invasiveness in poultry in experimental infections, but it did not contaminate eggs (Guard-Petter et al. 1996; Parker et al. 2001a; Woodward et al. 2000). The *Salmonella* convoluted phenotype is resistant to a number of environmental stressors and is invariably linked with production of sigma factor RpoS and the SEF17 fimbriae (Jorgensen et al. 2000; Romling et al. 1998). Nonconvoluted strains are sometimes encountered that lack SEF17 expression due to spontaneous mutation in *rpoS* (Ivanova et al. 1992; Jorgensen et al. 2000; Visick and Clarke 1997; Waterman and Small 1996; Zambrano et al. 1993). Not all flagellated phenotypes of *S.* Enteritidis make the convoluted phenotype (Guard-Petter et al. 1997; Parker and Guard-Petter 2001; Allen-Vercoe et al. 1999; Allen-Vercoe and Woodward 1999; Parker and Guard-Petter 2001). Hydrophobic glucosylated low-molecular-mass (LMM) LPS O-chain rather than the hydrophilic glucosylated high-molecular-mass (HMM) structure is recovered from strains producing the convoluted colony morphology (Humphrey and Guard-Petter, personal communication) (Parker et al. 2001b). In addition, loss of LPS O-chain correlates with loss of biofilm; thus, glucosylated LMM O-chain may be a third essential component of the convoluted phenotype (Guard-Petter et al. 1996). Analysis of biofilm from the lacy phenotype with monoclonal antibodies detected a physical association between flagella entrapped in biofilm and glucosylated lipopolysaccharide (Guard-Petter et al. 1996). Considering that glucose-rich cellulose is also a component of biofilm, it is possible that the high concentration of glucose surrounding proteins in biofilm may drive covalent adduction between molecules and be a final stage in colony development, imparting a cohesiveness that makes it difficult to solubilize biofilm and that may contribute to environmental persistence (Creighton et al. 1980; Kasai et al. 1983; Monnier and Cerami 1983; Stevens et al. 1978).

What is now most important to determine about the undoubtedly complex convoluted phenotype of *S.*

Enteritidis is its prevalence and stability between different continents. Preliminary discussions suggest that recovery of the convoluted phenotype is so common in Western Europe that it is considered wildtype (Allen-Vercoe et al. 1998; Humphrey et al. 1998; Romling et al. 1998). In North America, the convoluted phenotype has not been the predominant phenotype and thus is not considered to be wildtype (Guard-Petter et al. 1996, 1997). If the emergence or stability of this distinctive phenotype in the hen house environment differs greatly between continents, then oral invasion of hens could be a stage in the infectious process that is rate-limiting for incidence of egg contamination and that contributes to regional differences. However, a different phenotype is the final arbiter of egg contamination, and the stages of infection that occur after mucosal invasion with systemically adapted strains are equally important to the entire process. There is evidence that mixtures of orally invasive and parenterally adapted strains produce high-incidence egg contamination, even when exposure is by aerosolization of bacteria (Gast et al. 2002). Oral infection of chickens with a strain of S. Enteritidis that naturally produced a mixture of phenotypes in culture resulted in high-incidence egg contamination in vaccinated birds (Parker et al. 2001a).

HOW *S.* ENTERITIDIS DIFFERS FROM OTHER PATHOGENIC *SALMONELLA*.

Although the role of the convoluted phenotype is important to the entire infectious process, it does not distinguish S. Enteritidis from other *Salmonellae* or even from *Escherichia coli* (Allen-Vercoe et al. 1997; Collinson et al. 1991, 1993; Romling et al. 1998, 2000; Woodward et al. 2000). In many other ways, S. Enteritidis resembles S. Typhimurium with regard to known virulence mechanisms central to mammalian cell invasion and survival and growth in the host. Both pathogens have the highly conserved pathogenicity islands, Type III secretion mechanisms, virulence effecter proteins, two-component regulatory systems; and both harbor a large virulence plasmid, are motile, and produce a galactose-rhamnose-mannose repeat unit of the LPS O-chain backbone decorated with a dideoxyhexose that determines serovar (Blanc-Potard et al. 1999; Galan and Curtiss 1991; Jones and Falkow 1996; Marcus et al. 2000; Ochman and Groisman 1996; Reeves 1993). It has therefore not been obvious how S. Enteritidis is especially able to follow the human infection route and above all, at a critical step, how it so successfully contaminates and grows in egg contents. Following are strain characteristics of S. Enteritidis that correlate with the final stages in the process of egg contamination associated with systemic infection of the hen and that may be central to the impact of this pathogen as a public health problem.

HMM LPS and Parenteral Adaptation.

Variants of egg-contaminating S. Enteritidis often produce glucosylated high-molecular-mass (GHMM) LPS, which is one of four LPS structural variants obtained from S. Enteritidis (Guard-Petter et al. 1995; Parker et al. 2001b) (Fig. 13.2). This capsule-like LPS structure is similar to that produced by S. Typhi, the host-adapted agent of Typhoid fever in people (Hellerqvist et al. 1969; Rahman et al. 1997). S. Typhi has a much longer average O-chain than does S. Typhimurium, which can be accurately differentiated only from what is referred to as "long" or "smooth" O-chain by chemical methods (Parker et al. 2001b). As the chain length of LPS increases, so does hydrophilicity of the outer cell membrane (Guard-Petter et al. 1999). Variants producing GHMM LPS have been isolated most often from the egg and less often from the spleen of the naturally infected mouse (Parker et al. 2001b). Current estimates are that nearly 40 percent of variants from eggs and 10 percent from the mouse spleen produce GHMM LPS. In addition to the egg being a selective environment for O-chain micro-heterogeneity, there is also evidence that other types of selection pressures alter LPS structure, as evidenced by generation of resistance to naturally occurring antimicrobial peptides found in soil (Gunn 2001; Gunn et al. 2000).

A role for HMM LPS in the infection route from bird to human is that HMM LPS appears to cause unusual pathology of the avian reproductive tract, because injected S. Enteritidis culture has a striking ability to induce regression of the uterus (Parker et al. 2002). This effect was not seen with a regulator of length:chain length determinant (*rol:cld*) mutant, which can only make low molecular-mass (LMM) LPS. However, both the wildtype and mutant retained the ability to cause ovarian lesions in these studies. Detection of oviduct regression by one strain and causation of ovarian pathology by both suggests that different graphic locations may be populated by strains that vary in their ability to colonize different anatomical regions of the avian reproductive tract. Evidence from European and North American investigators support this concept, because differences in how often S. Enteritidis is recovered from ovum (yolk) and albumen (egg white, which is produced in the oviduct) have been described (Gast and Holt 2000; Humphrey 1994). Results from hens infected by contact or with lower oral dosages suggested that a primary role of HMM LPS was to mitigate signs of illness in hens (Parker et al. 2002). The concept that S. Enteritidis has a mechanism to reduce illness in infected birds suggests that HMM LPS production contributes greatly to the current problem, because farmers have difficulty detecting flocks that are actively producing contaiminated eggs. Altered physiology of the avian reproductive tract in response to infection alters shell quality, and thus, a marginally defective shell could provide a third route (besides yolk and albumen) for S. Enteritidis and possibly other bacteria to enter eggs (Parker et al. 2002). Computerized analysis of shell quality has been suggested as a high throughput method for detection of clusters of contaminated eggs within the larger supply coming from farms, but it is not yet known whether this

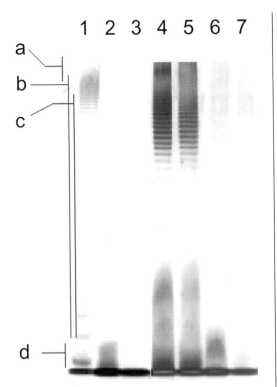

FIG. 13.2—Frequently encountered lipopolysaccharide O-chain structural variants of *Salmonella*. Shown is a SDS-polyacrylamide gel that has been silver stained to detect LPS. LPS samples are from: **lane 1**, *S.* Typhimurium that has a multiple repeat unit O-chain linked to lipid A-core (callout b); **lane 2**, O-chain polymerase mutant (*wzy*) that links only one O-chain repeat unit to core (callout d); **lane 3**, core mutant (*wza*) of *S.* Typhimurium that produces no O-chain at all; **lane 4**, *S.* Enteritidis producing a substantial amount of HMM LPS (callout a); **lane 5**, *S.* Enteritidis producing primarily LMM LPS (callout c); **lane 6**, *wzy* mutant of *S.* Enteritidis; **lane 7**, core mutant of *S.* Enteritidis. LPS O-chain has a ladder-like appearance caused by variation in the number of repeat units per LPS molecule. Glucosylation of LPS O-chain and quantification analysis of O-chain micro-heterogeneity is further analyzed by gas chromatographic analysis of alditol acetate derivatives and sometimes by immunoblot analysis. Detailed reviews of lipopolysaccharide structure are available (Heinrichs et al. 1998; Raetz 1996).

approach is feasible, because many factors contribute to shell quality (Guard-Petter 2001). A fourth route for bacteria to enter eggs is by venereal infection, which involves infection of the rooster as well as the hen (Cox et al., submitted; Okamura et al. 2000). However, breeding between infected birds does not contribute directly to the pandemic associated with *S.* Enteritidis, because eggs for consumption are not fertilized.

LPS O-chain Glucosylation. Lipopolysaccharide (LPS) obtained from *S.* Enteritidis differs greatly between strains in the degree of glucosylation that occurs in the O-chain region (Fig. 13.2). This ranges from a lack of glucosylation in stored isolates to an exceptionally high degree in variants freshly cultured from the spleens of naturally infected mice and from eggs (Parker et al. 2001b). *S.* Typhimurium is on average well glucosylated, even when strains have been stored for years or examined as low passage isolates (Parker et al. 2001b). In contrast, glucosylation by *S.* Enteritidis is a cell-surface characteristic that is quickly lost in response to environmental conditions.

Motility. The avian-adapted, egg-contaminating serovars *S.* Pullorum and *S.* Gallinarum are the only *Salmonellae* that lack flagella and associated motility. Mutation of *flhD* (part of the *flhDC* master operon regulator of flagellar biosynthesis) to eliminate flagellation in *S.* Enteritidis unexpectedly enhanced oral invasiveness in poultry, whereas mutation of *fliC* (the flagellin structural gene) did not (Parker and Guard-Petter 2001). In other Enterobacteriaceae, it has been shown that alterations in gene expression from the flagellar master operon has effects on other cell functions, such as cell division and possibly virulence (Claret and Hughes 2000a, 2000b; Pruss et al. 1997). Because mutation of *fliC* did not enhance oral invasiveness, flagella and motility may not be important for achieving oral invasion in chickens. Instead, the ability of some *S.* Enteritidis isolates to greatly alter cellular division is considered the bacterial characteristic that is more likely to contribute to enhanced oral invasiveness (Pruss et al. 1997). Some *Salmonellae* classified as *S.* Pullorum have been further identified as *S.* Enteritidis that has temporarily suppressed flagellation, which further supports the view that extreme variation in the regulation of flagellar biosynthesis could be occurring in field isolates (Chaubal and Holt 1999; Guard-Petter 1997). The peculiar behavior of *flhD* mutants toward enhanced oral invasiveness in hens probably involves altered host cell interactions that have been described for the same mutant in *S.* Typhimurium (Schmitt et al. 2001).

Swarm Cell Differentiation. Swarm cell differentiation is a developmental pathway characterized by cell elongation and hyperflagellation, and in the Enterobacteriaceae, it is governed largely by upregulation of the flagellar master operon *flhDC* (Dufour et al. 1998; Fraser et al. 1999). Swarm cell differentiation results in enhanced virulence and migration of multicellular aggregates across surfaces (Allison et al. 1992; Darzins 1994; Gardel and Mekalanos 1996; Liaw et al. 2000; MacFarlane et al. 2001; Szymanski et al. 1995) (refer to Fig. 13.1e). Conditions for inducing swarm cell differentiation of *S.* Enteritidis across normally inhibitory solid agar have been identified (Guard-Petter 1997; Guard-Petter et al. 1996). The structure of LPS is important for swarm cell migration in soft agar by *Salmonellae* (Guard-Petter 1997; Toguchi et al. 2000). Because *S.* Enteritidis produces a broad range of LPS structures (Parker et al. 2001b), it is possible that it migrates more aggressively within the environment of the hen house or the bird than do serovars such as *S.* Typhimurium, which produces a relatively homogeneous LPS structure. Spontaneous swarm cell migration on inhibitory hard agar was detected in a group of mouse spleen variants of *S.* Enteritidis that yielded HMM LPS (Guard-Petter et al. 1997). Interestingly, wild type strains of *S.* Enteritidis that can undergo spontaneous swarm cell differentiation are virulent in chicks when injected, but not when given orally, which is an inverse result to that seen in chicks infected with the nonflagellated *flhD* mutant (Parker and Guard-Petter 2001). These results further suggest that the ability of *S.* Enteritidis to alternate between extremes of flagellation enhances its pathogenicity overall.

High Cell Density Growth. Quorum-sensing is a broadly conserved virulence mechanism among many Gram-negative pathogens that infect both plants and animals (Withers et al. 2001). In quorum-sensing, basal levels of freely diffusible acylated homoserine lactones reach a critical level that alter expression of *luxR*-regulated genes in a coordinated manner throughout the affected population of bacterial cells (Fuqua and Greenberg 1998; Hastings and Greenberg 1999). It has been surprising that no evidence of quorum-sensing has been detected in pathogenic *Salmonellae*, especially because mathematical models argue that it is a principle common to bacterial populations (Ward et al. 2001). High cell density growth of *S.* Enteritidis results in autoinduction of luciferase from a *luxR*-containing broad spectrum sensor plasmid (Guard-Petter 1998), but assay of cell free concentrated supernatant has not detected an acylated homoserine lactone, or any other diffusible molecule that is capable of inducing *luxR*-regulated transcription (P. Greenberg, personal communication) (Holden et al. 1999; Surette et al. 1999). Thus, at this time, luciferase production by *S.* Enteritidis appears to occur in response to intracellular factors that are not diffusible. It is possible that a high concentration of homoserine lactone produced as a byproduct of methionine metabolism has the potential to initiate LuxR transcription despite a lack of acylation (P. Greenberg and J. Guard-Petter, personal communication). This signal is not unimportant, because its production correlates with egg contamination in hens and virulence potential in chicks (Guard-Petter 1998). The Salmonellae maintain an AHL sensor, SdiA, which suggests that cell communication is at least important for co-existence of *Salmonella* with other intestinal and environmental bacteria (Michael et al. 2001). *Salmonella* has a second quorum-sensing mechanism, AI-2, that may be important for communication between gram-negative bacteria (Schauder et al. 2001); however, a direct role in virulence is not evident (Day and Maurelli 2001). Methods for interrupting bacterial cell communication are important to pursue, because they may result in new ways to control *S.* Enteritidis in the hen house by manipulation of the environment (Holden et al. 1999). Future research may yet reveal that the

Salmonellae have alternative autoinducers and quorum-sensing capabilities.

EPIDEMIOLOGICAL MONITORING OF *S.* ENTERITIDIS. A frequently asked question is what genetic changes in this pathogen contribute to its remarkable strain heterogeneity. *S.* Enteritidis produces some mutants in the environment of hens that have spontaneously lost LPS O-chain, fimbriae, and resistance to oxidation (Allen-Vercoe et al. 1998; Guard-Petter et al. 1999; Humphrey et al. 1998). Although these obvious mutations may be a general indication that nonselective adaptation is occurring (Finkel and Kolter 1999; Zambrano et al. 1993), they result in attenuation and do not explain phenotypic divergence that specifically contributes to egg contamination. Thus, the question of which genetic method is the best for subtyping smooth strains of *S.* Enteritidis that vary in their ability to contaminate eggs is important to answer.

Currently, no coordination exists between laboratories that do *S.* Enteritidis genetic typing (E. Ribot, CDC, personal communication). However, most methods detect that *S.* Enteritidis bacteriophage types (PT) 4, 7, 6, and 5 represent one lineage that predominates in Eurasia, whereas bacteriophage types 2, 8, 13a, and 13 predominate in North America (Desai et al. 2001; Gruner et al. 1994; Hickman-Brenner et al. 1991; Humphrey et al. 1989; Liebana et al. 2001a; Martinetti and Altwegg 1990; Usera et al. 1994; Wachsmuth 1986). Comparison of antibiotic resistance patterns (Ling et al. 1998), plasmid profiles (Stubbs et al. 1994; Wachsmuth et al. 1991), *IS*200 RFLP, ribotype (Landeras et al. 1996, 1998; Landeras and Mendoza 1998; Liebana et al. 2001a, 2001b), and pulsed field gel electrophoresis patterns (Lukinmaa et al. 1999; Ridley et al. 1998; Thong et al. 1995, 1998) have been used for intra-PT differentiation of *S.* Enteritidis strains with varying degrees of success. Furthermore, several PCR-based methods such as RAPD (random amplification of polymorphic DNA) may lack sensitivity and reproducibility for routine use (Landeras and Mendoza 1998; Tyler et al. 1997). There is considerable debate about the reliability of amplified fragment length polymorphisms (AFLP) (Desai et al. 2001). A recent interlaboratory study aimed at comparing similar methodologies in two European laboratories, namely VLA in Weybridge, United Kingdom, and ID Lelystad, the Netherlands, did not find that the combination *Eco*RI(0)-*Mse*I(C) approach provided an adequate level of sensitivity to separate closely related isolates (E. Liebana, unpublished data).

Inter-continental collaboration suggests that ribotyping is a sensitive method of subtyping that aids communication between countries (Brosious et al. 1981; Liebana et al. 2001a; Martinetti and Altwegg 1990). This method has been used successfully to identify single nucleotide polymorphisms (SNPs) that correlate with LPS O-chain micro-heterogeneity, phage type, and the source of the isolate (Liebana et al. 2001a, 2001b; Parker et al. 2001b; Sokurenko et al. 2001). rRNA operons are highly conserved and are present in several copies on the bacterial chromosome. The stable regions of the gene can act as molecular chronometers of the phylogenetic relationship of organisms, with variable rRNA gene and flanking regions allowing discrimination between strains (Woese 1987). The number and location of rRNA operon copies and restriction sites within the genes and in their flanking regions differ between bacterial clones. The value of ribotyping depends on the restriction enzyme used for digestion and the gene probe (Sokurenko et al. 2001; Weide-Botjes et al. 1998). It is especially important that genetic typing methods distinguish reliably between PT4 and non-PT4 lineages and between strains that vary in their ability to contaminate eggs and to be orally invasive.

VACCINATION TO CONTROL EGG CONTAMINATION. No control program substitutes for practicing stringent rodent control and aggressive biocontainment, or for stocking the hen house with pathogen-free birds. However, vaccination is also used to improve control (Barbezange et al. 2000; Davison et al. 1999; Feberwee et al. 2001; Hassan and Curtiss 1994, 1997; Heithoff et al. 2001; Hormaeche et al. 1990; Martin et al. 1996). Current estimates are that 14 percent of flocks are vaccinated yearly in the United States (USDA 1998). Vaccination of laying hens is required in Great Britain and Germany, and its general use throughout Europe is credited in part for the steady decline in human illness during the late 1990s. Currently, no modified live vaccines are licensed for use in laying hens in the United States, although there are products licensed for use in young, growing birds. In the United States, producers generally use killed bacterins, although off-label immunization with modified live products is reported.

It has been difficult to directly test vaccines for their ability to reduce or eliminate internal egg contamination, because animal models have been unreliable for producing the high incidence of contamination needed to obtain label claims from federal regulatory agencies. Currently, claims for efficacy against egg contamination have been made based upon culture of the entire egg, including the shell, which confuses the issue of gut colonization with egg contamination. Most vaccine trials base efficacy on the incidence of recovery of bacteria from reproductive tract tissue (Hassan and Curtiss 1994, 1997). Oral infection of hens with a strain that produced two phenotypes concomitantly in culture was used to test the ability of a modified live vaccine undergoing testing to prevent internal egg contamination, and results indicated that protection could be overridden (Parker et al. 2001a). Further refinements to the egg contamination animal model include co-infection of hens with defined stable strains (Gast et al. 2002). This approach appears promising for testing vaccines

that are intended to directly reduce or eliminate *Salmonella* in the internal contents of eggs. Attenuated modified live vaccines have a small risk of reversion to virulence after being released into the environment of birds; therefore, licensure has required the introduction of double mutations and additional data about persistence of the vaccine in the environment and ancillary hosts after birds are vaccinated (Feberwee et al. 2001).

Competitive exclusion is also used to reduce *S.* Enteritidis in chickens and their environment. It is based on the concept that colonization of the intestinal tract by beneficial bacteria reduces colonization by pathogens (Hirn et al. 1992; Nurmi et al. 1992; Wierup et al. 1988). It has been used most in hatchlings as a method to aid the rapid acquisition of a protective gut flora (Methner et al. 1997; Nuotio et al. 1992; Schneitz et al. 1992). Competitive exclusion does not work as well in mature birds, but it may be useful to protect birds that are under stresses which alter gut flora (Seo et al. 2000). Modified live vaccines have a competitive exclusion effect in addition to eliciting a protective immune response, which suggests that any number of attenuated mutant strains obtained from homologous and heterologous *Salmonella* serotypes are likely to reduce gut colonization (Barrow 2000; Dueger et al. 2001; Hassan and Curtiss 1997; Hollister et al. 1999; Methner et al. 1997; Rabsch et al. 2000).

SUMMARY AND CONCLUSIONS. *S.* Enteritidis generates a striking degree of strain heterogeneity and there is a complex network of characteristics underlying its diverse behavior. It is evident that agricultural practices apply selection pressures on the pathogen that increase the risk of egg contamination (Himathongkham et al. 1999; Mattick et al. 2000; Henzler et al. 1998; Holt et al. 1998; Smith et al. 2000; USDA 1998). *S.* Enteritidis as a causative agent of pandemic disease is thus a modern-day example of how a pathogenic bacteria interacts with the on-farm environment. Further reductions of *S.* Enteritidis in the human food chain, and application of new knowledge to the prevention of new types of foodborne outbreaks, will continue to require a coordinated approach between scientists and producers as the demand for food increases worldwide.

REFERENCES
Allen-Vercoe, E., R. Collighan, and M.J. Woodward. 1998. The variant rpoS allele of *S. enteritidis* strain 27655R does not affect virulence in a chick model nor constitutive curliation but does generate a cold-sensitive phenotype. *FEMS Microbiol. Lett.* 167:245–253.
Allen-Vercoe, E., M. Dibb-Fuller, C.J. Thorns, and M.J. Woodward. 1997. SEF17 fimbriae are essential for the convoluted colonial morphology of *Salmonella* enteritidis. *FEMS Microbiol. Lett.* 153:33–42.
Allen-Vercoe, E., A.R. Sayers, and M.J. Woodward. 1999. Virulence of *Salmonella enterica* serotype Enteritidis aflagellate and afimbriate mutants in a day-old chick model. *Epidemiol. Infect.* 122:395–402.

Allen-Vercoe, E., and M.J. Woodward. 1999. Colonisation of the chicken caecum by afimbriate and aflagellate derivatives of *Salmonella enterica* serotype Enteritidis. *Vet. Microbiol.* 69:265–275.

Allison, C., H.C. Lai, and C. Hughes. 1992. Co-ordinate expression of virulence genes during swarm-cell differentiation and population migration of *Proteus mirabilis. Mol. Microbiol.* 6:1583–1591.

al-Nakhli, H.M., Z.H. al-Ogaily, and T.J. Nassar. 1999. Representative Salmonella serovars isolated from poultry and poultry environments in Saudi Arabia. *Rev. Sci. Tech.* 18:700–709.

Anon. 1986. *Salmonella* heidelberg outbreak at a convention—New Mexico. *MMWR* 35:91.

Anon. 1988. *Salmonella* enteritidis phage type 4: chicken and egg. *Lancet* 2:720–722.

Anon. 1993. Outbreaks of *Salmonella* enteritidis gastroenteritis—California, 1993. *MMWR* 42:793–797.

Anon. 1999. Update on *Salmonella* serotype Enteritidis infections. Centers for Disease Control and Prevention, Atlanta, GA.

Anon. 2001. *Salmonella* in humans: England and Wales, 1981–1999. Public Health Laboratory Service (www.phls.co.uk).

Austin, J.W., G. Sanders, W.W. Kay, and S.K. Collinson. 1998. Thin aggregative fimbriae enhance Salmonella enteritidis biofilm formation. *FEMS Microbiol. Lett.* 162:295–301.

Aysev, A.D., H. Guriz, and B. Erdem. 2001. Drug resistance of Salmonella strains isolated from community infections in Ankara, Turkey, 1993–1999. *Scand. J. Infect. Dis.* 33:420–422.

Badi, M.A., N. Iliadis, K. Sarris, and E. Artopios. 1992. [Salmonella infection sources in poultry flocks in northern Greece]. *Berl Munch Tierarztl Wochenschr.* 105:236–239.

Barbezange, C., G. Ermel, C. Ragimbeau, F. Humbert, and G. Salvat. 2000. Some safety aspects of Salmonella vaccines for poultry: in vivo study of the genetic stability of three *Salmonella* typhimurium live vaccines. *FEMS Microbiol. Lett.* 192:101–106.

Barrow, P.A. 2000. The paratyphoid salmonellae. *Rev. Sci. Tech.* 19:351–375.

Baumler, A.J., B.M. Hargis, and R.M. Tsolis. 2000. Tracing the origins of Salmonella outbreaks. *Science* 287:50–52.

Betit, R., C. Brokoop, and C.R. Nichols. 1997. 1996: The year in review. Utah Department of Health, Bureau of Epidemiology. 1–5.

Blanc-Potard, A.B., F. Solomon, J. Kayser, and E.A. Groisman. 1999. The SPI-3 pathogenicity island of *Salmonella enterica. J. Bacteriol.* 181:998–1004.

Boyce, T.G., D. Koo, D.L. Swerdlow, T.M. Gomez, B. Serrano, L.N. Nickey, F.W. Hickman-Brenner, G.B. Malcolm, and P.M. Griffin. 1996. Recurrent outbreaks of *Salmonella Enteritidis* infections in a Texas restaurant: phage type 4 arrives in the United States. *Epidemiol. Infect.* 117:29–34.

Brosious, J., A. Ullrich, M.A. Raker, A. Gray, T.J. Dull, R.R. Gutell, and H.F. Noller. 1981. Construction and fine mapping of recombinant plasmids containing the rnB ribosomal operon of *E. coli. Plasmid.* 6:112–118.

Buchrieser, C., R. Brosch, O. Buchrieser, A. Kristl, J.B. Luchansky, and C.W. Kaspar. 1997. Genomic analyses of Salmonella enteritidis phage type 4 strains from Austria and phage type 8 strains from the United States. *Zentralbl. Bakteriol.* 285:379–388.

Chaubal, L.H., and P.S. Holt. 1999. Characterization of swimming motility and identification of flagellar proteins in *Salmonella* pullorum isolates. *Am. J. Vet. Res.* 60:1322–1327.

Claret, L., and C. Hughes. 2000a. Functions of the subunits in the FlhD(2)C(2) transcriptional master regulator of bacterial flagellum biogenesis and swarming. *J. Mol. Biol.* 303:467–478.

Claret, L., and C. Hughes. 2000b. Rapid turnover of FlhD and FlhC, the flagellar regulon transcriptional activator proteins, during Proteus swarming. *J. Bacteriol.* 182:833–836.

Collinson, S.K., P.C. Doig, J.L. Doran, S. Clouthier, T.J. Trust, and W.W. Kay. 1993. Thin, aggregative fimbriae mediate binding of *Salmonella* enteritidis to fibronectin. *J. Bacteriol.* 175:12–18.

Collinson, S.K., L. Emody, K.H. Muller, T.J. Trust, and W.W. Kay. 1991. Purification and characterization of thin, aggregative fimbriae from *Salmonella* enteritidis. *J. Bacteriol.* 173:4773–4781.

Creighton, M.O., P.J. Stewart-DeHaan, W.M. Ross, M. Sanwal, and J.R. Trevithick. 1980. Modelling cortical cataractogenesis: 1. In vitro effects of glucose, sorbitol and fructose on intact rat lenses in medium 199. *Can. J. Ophthalmol.* 15:183–188.

Darzins, A. 1994. Characterization of a *Pseudomonas aeruginosa* gene cluster involved in pilus biosynthesis and twitching motility: sequence similarity to the chemotaxis proteins of enterics and the gliding bacterium *Myxococcus xanthus. Mol. Microbiol.* 11:137–153.

Davies, R.H., and C. Wray. 1995a. Contribution of the lesser mealworm beetle (*Alphitobius diaperinus*) to carriage of *Salmonella* enteritidis in poultry. *Vet. Rec.* 137:407–408.

Davies, R.H., and C. Wray. 1995b. Mice as carriers of *Salmonella enteritidis* on persistently infected poultry units. *Vet. Rec.* 137:337–341.

Davies, R.H., and C. Wray. 1996a. Persistence of *Salmonella* enteritidis in poultry units and poultry food. *Br. Poult. Sci.* 37:589–596.

Davies, R.H., and C. Wray. 1996b. Studies of contamination of three broiler breeder houses with *Salmonella* enteritidis before and after cleansing and disinfection. *Avian Dis.* 40:626–633.

Davison, S., C.E. Benson, D.J. Henzler, and R.J. Eckroade. 1999. Field observations with *Salmonella* enteritidis bacterins. *Avian Dis.* 43:664–669.

Day, W.A., Jr., and A.T. Maurelli. 2001. Shigella flexneri LuxS quorum-sensing system modulates virB expression but is not essential for virulence. *Infect. Immun.* 69:15–23.

Desai, M., E.J. Threlfall, and J. Stanley. 2001. Fluorescent amplified–fragment length polymorphism subtyping of the *Salmonella enterica* serovar enteritidis phage type 4 clone complex. *J. Clin. Microbiol.* 39:201–206.

Dueger, E.L., J.K. House, D.M. Heithoff, and M.J. Mahan. 2001. *Salmonella* DNA adenine methylase mutants elicit protective immune responses to homologous and heterologous serovars in chickens. *Infect. Immun.* 69:7950–7954.

Dufour, A., R.B. Furness, and C. Hughes. 1998. Novel genes that upregulate the *Proteus mirabilis* flhDC master operon controlling flagellar biogenesis and swarming. *Mol. Microbiol.* 29:741–751.

Fantasia, M., and E. Fileteci. 1994. Salmonella enteritidis in Italy. *Int. J. Food Microbiol.* 21:7–13.

Feberwee, A., T.S. de Vries, E.G. Hartman, J.J. de Wit, A.R. Elbers, and W.A. de Jong. 2001. Vaccination against *Salmonella* enteritidis in Dutch commercial layer flocks with a vaccine based on a live Salmonella gallinarum 9R strain: evaluation of efficacy, safety, and performance of serologic Salmonella tests. *Avian Dis.* 45:83–91.

Fica, A., A. Fernandez, S. Prat, O. Figueroa, R. Gamboa, I. Tsunekawa, and I. Heitmann. 1997. [Salmonella enteritidis, an emergent pathogen in Chile]. *Rev. Med. Chil.* 125:544–551.

Finkel, S.E., and R. Kolter. 1999. Evolution of microbial diversity during prolonged starvation. *Proc. Natl. Acad. Sci. USA.* 96:4023–4027.

Fisher, I. 2001. Enter-net Quarterly Salmonella Report. Public Health Laboratory Service. www2.phls.co.uk/reports/latest.html.

Fraser, G.M., R.B. Furness, and C. Hughes. 1999. Swarming migration by *Proteus* and related bacteria. In Prokaryotic Development. L. Shimketts and Y. Brun, eds. American Society for Microbiology, Washington. 381–401.

Friedman, C.R., G. Malcolm, J.G. Rigau-Perez, P. Arambulo, and R.V. Tauxe. 1996. Public health risk from Salmonella-based rodenticides. *Lancet.* 347:1705–1706.

Fuqua, C., and E.P. Greenberg. 1998. Cell-to-cell communication in Escherichia coli and Salmonella typhimurium: they may be talking, but who's listening? *Proc. Natl. Acad. Sci. U S A.* 95:6571–6572.

Galan, J.E., and R. Curtiss. 1991. Distribution of the invA, -B, -C, and -D genes of Salmonella typhimurium among other Salmonella serovars: invA mutants of Salmonella typhi are deficient for entry into mammalian cells. *Infect. Immun.* 59:2901–2908.

Gardel, C.L., and J.J. Mekalanos. 1996. Alterations in Vibrio cholerae motility phenotypes correlate with changes in virulence factor expression. *Infect. Immun.* 64:2246–2255.

Gast, R., J. Guard-Petter, and P.S. Holt. 2002. Characteristics of Salmonella Enteritidis contamination in eggs after oral, aerosol, and intravenous inoculation of laying hens. *Avian Dis.* 92:196–209.

Gast, R.K., and P.S. Holt. 2000. Deposition of phage type 4 and 13a Salmonella enteritidis strains in the yolk and albumen of eggs laid by experimentally infected hens. *Avian Dis.* 44:706–710.

Gherardi, M.M., M.I. Gomez, V.E. Garcia, D.O. Sordelli, and M.C. Cerquetti. 2000. Salmonella enteritidis temperature-sensitive mutants protect mice against challenge with virulent Salmonella strains of different serotypes. *FEMS Immunol. Med. Microbiol.* 29:81–88.

Glosnicka, R., and D. Kunikowska. 1994. The epidemiological situation of Salmonella enteritidis in Poland. *Int. J. Food Microbiol.* 21:21–30.

Gruner, E., G. Martinetti Lucchini, R.K. Hoop, and M. Altwegg. 1994. Molecular epidemiology of Salmonella enteritidis. *Eur. J. Epidemiol.* 10:85–89.

Guard-Petter, J. 1997. Induction of flagellation and a novel agar-penetrating flagellar structure in *Salmonella enterica* grown on solid media: possible consequences for serological identification. *FEMS Microbiol Lett.* 149:173–180.

Guard-Petter, J. 1998. Variants of smooth *Salmonella enterica* serovar Enteritidis that grow to higher cell density than the wild type are more virulent. *Appl. Environ. Microbiol.* 64:2166–2172.

Guard-Petter, J. 2001. The chicken, the egg, and *Salmonella* Enteritidis. *Environ. Microbiol.* 3:421–430.

Guard-Petter, J., D.J. Henzler, M.M. Rahman, and R.W. Carlson. 1997. On-farm monitoring of mouse-invasive *Salmonella enterica* serovar enteritidis and a model for its association with the production of contaminated eggs. *Appl. Environ. Microbiol.* 63:1588–1593.

Guard-Petter, J., L.H. Keller, M.M. Rahman, R.W. Carlson, and S. Silvers. 1996. A novel relationship between O-antigen variation, matrix formation, and invasiveness of *Salmonella enteritidis*. *Epidemiol. Infect.* 117:219–231.

Guard-Petter, J., B. Lakshmi, R. Carlson, and K. Ingram. 1995. Characterization of lipopolysaccharide heterogeneity in Salmonella enteritidis by an improved gel electrophoresis method. *Appl. Environ. Microbiol.* 61:2845–2851.

Guard-Petter, J., C.T. Parker, K. Asokan, and R.W. Carlson. 1999. Clinical and veterinary isolates of Salmonella enterica serovar enteritidis defective in lipopolysaccharide O-chain polymerization. *Appl. Environ. Microbiol.* 65:2195–2201.

Gugnani, H.C. 1999. Some emerging food and water borne pathogens. *J. Commun. Dis.* 31:65–72.

Gunn, J.S. 2001. Bacterial modification of LPS and resistance to antimicrobial peptides. *J. Endotoxin. Res.* 7:57–62.

Gunn, J.S., S.S. Ryan, J.C. Van Velkinburgh, R.K. Ernst, and S.I. Miller. 2000. Genetic and functional analysis of a PmrA-PmrB-regulated locus necessary for lipopolysaccharide modification, antimicrobial peptide resistance, and oral virulence of Salmonella enterica serovar typhimurium. *Infect. Immun.* 68:6139–6146.

Gunn, R.A., and G. Markakis. 1978. Salmonellosis associated with homemade ice cream. An outbreak report and summary of outbreaks in the United States in 1966 to 1976. *JAMA* 240:1885–1886.

Gutierrez-Cogco, L., E. Montiel-Vazquez, P. Aguilera-Perez, and M.C. Gonzalez-Andrade. 2000. [Salmonella serotypes identified in Mexican health services]. *Salud. Publica. Mex.* 42:490–495.

Hasenson, L., B. Gericke, A. Liesegang, H. Claus, J. Poplawskaja, N. Tscherkess, and W. Rabsch. 1995. [Epidemiological and microbiological studies on salmonellosis in Russia]. *Zentralbl Hyg Umweltmed.* 198:97–116.

Hasenson, L.B., L. Kaftyreva, V.G. Laszlo, E. Woitenkova, and M. Nesterova. 1992. Epidemiological and microbiological data on Salmonella enteritidis. *Acta. Microbiol. Hung.* 39:31–39.

Hassan, J.O., and R. Curtiss. 1994. Development and evaluation of an experimental vaccination program using a live avirulent Salmonella typhimurium strain to protect immunized chickens against challenge with homologous and heterologous Salmonella serotypes. *Infect. Immun.* 62:5519–5527.

Hassan, J.O., and R. Curtiss. 1997. Efficacy of a live avirulent Salmonella typhimurium vaccine in preventing colonization and invasion of laying hens by Salmonella typhimurium and Salmonella enteritidis. *Avian Dis.* 41:783–791.

Hastings, J.W., and E.P. Greenberg. 1999. Quorum sensing: the explanation of a curious phenomenon reveals a common characteristic of bacteria. *J. Bacteriol.* 181:2667–2668.

Healing, T.D. 1991. Salmonella in rodents: a risk to man? *CDR (Lond Engl Rev).* 1:R114–116.

Heinrichs, D.E., J.A. Yethon, and C. Whitfield. 1998. Molecular basis for structural diversity in the core regions of the lipopolysaccharides of Escherichia coli and Salmonella enterica. *Mol. Microbiol.* 30:221–232.

Heithoff, D.M., E.Y. Enioutina, R.A. Daynes, R.L. Sinsheimer, D.A. Low, and M.J. Mahan. 2001. Salmonella dna adenine methylase mutants confer cross-protective immunity. *Infect. Immun.* 69:6725–6730.

Hellerqvist, C.G., B. Lindberg, S. Svensson, T. Holme, and A.A. Lindberg. 1969. Structural studies on the O-specific side chains of the cell wall lipopolysaccharides from *Salmonella typhi* and *S. enteritidis*. *Acta. Chem. Scand.* 23:1588–1596.

Henzler, D.J., D.C. Kradel, and W.M. Sischo. 1998. Management and environmental risk factors for Salmonella enteritidis contamination of eggs. *Am. J. Vet. Res.* 59:824–829.

Henzler, D.J., and H.M. Opitz. 1992. The role of mice in the epizootiology of Salmonella enteritidis infection on chicken layer farms. *Avian Dis.* 36:625–631.

Hickman-Brenner, F.W., A.D. Stubbs, and J.J. Farmer, 3rd. 1991. Phage typing of Salmonella enteritidis in the United States. *J. Clin. Microbiol.* 29:2817–2823.

Himathongkham, S., S. Nuanualsuwan, and H. Riemann. 1999. Survival of Salmonella enteritidis and Salmonella typhimurium in chicken manure at different levels of water activity. *FEMS Microbiol. Lett.* 172:159–163.

Hinz, K.H., G. Glunder, S. Rottmann, and M. Friederichs. 1989. [Salmonella gallinarum field isolates of the biovars pullorum and gallinarum]. *Berl. Munch. Tierarztl. Wochenschr.* 102:205–208.

Hirn, J., E. Nurmi, T. Johansson, and L. Nuotio. 1992. Long-term experience with competitive exclusion and salmonellas in Finland. *Int. J. Food Microbiol.* 15:281–285.

Hogue, A., P. White, J. Guard-Petter, W. Schlosser, R. Gast, E. Ebel, J. Farrar, T. Gomez, J. Madden, M. Madison, A.M. McNamara, R. Morales, D. Parham, P. Sparling, W. Sutherlin, and D. Swerdlow. 1997. Epidemiology and control of egg-associated Salmonella enteritidis in the United States of America. *Rev. Sci. Tech.* 16:542–553.

Holden, M.T., S. Ram Chhabra, R. de Nys, P. Stead, N.J. Bainton, P.J. Hill, M. Manefield, N. Kumar, M. Labatte, D. England, S. Rice, M. Givskov, G.P. Salmond, G.S. Stewart, B.W. Bycroft, S. Kjelleberg, and P. Williams. 1999. Quorumsensing cross talk: isolation and chemical characterization of cyclic dipeptides from Pseudomonas aeruginosa and other gram-negative bacteria. *Mol. Microbiol.* 33:1254–1266.

Hollister, A.G., D.E. Corrier, D.J. Nisbet, and J.R. DeLoach. 1999. Effects of chicken-derived cecal microorganisms maintained in continuous culture on cecal colonization by Salmonella typhimurium in turkey poults. *Poult. Sci.* 78:546–549.

Holt, P.S., B.W. Mitchell, and R.K. Gast. 1998. Airborne horizontal transmission of Salmonella enteritidis in molted laying chickens. *Avian Dis.* 42:45–52.

Hoop, R.K. 1997. The Swiss control programme for Salmonella enteritidis in laying hens: experiences and problems. *Rev. Sci. Tech.* 16:885–890.

Hoop, R.K., and P. Albicker-Rippinger. 1997. [Salmonella gallinarum-pullorum infection of poultry: experiences in Switzerland]. *Schweiz. Arch. Tierheilkd.* 139:485–489.

Hormaeche, C.E., H.S. Joysey, L. Desilva, M. Izhar, and B.A. Stocker. 1990. Immunity induced by live attenuated Salmonella vaccines. *Res. Microbiol.* 141:757–764.

Hormaeche, C.E., P. Mastroeni, J.A. Harrison, R. Demarco de Hormaeche, S. Svenson, and B.A. Stocker. 1996. Protection against oral challenge three months after i.v. immunization of BALB/c mice with live Aro Salmonella typhimurium and Salmonella enteritidis vaccines is serotype (species)-dependent and only partially determined by the main LPS O antigen. *Vaccine* 14:251–259.

Hubalek, Z., W. Sixl, M. Mikulaskova, B. Sixl-Voigt, W. Thiel, J. Halouzka, Z. Juricova, B. Rosicky, L. Matlova, M. Honza, V. Hajek, and J. Sitko. 1995. Salmonellae in gulls and other free-living birds in the Czech Republic. *Cent. Eur. J. Public Health.* 3:21–24.

Humphrey, T.J. 1994. Contamination of egg shell and contents with Salmonella enteritidis: a review. *Int. J. Food Microbiol.* 21:31–40.

Humphrey, T.J., A. Baskerville, S. Mawer, B. Rowe, and S. Hopper. 1989. Salmonella enteritidis phage type 4 from the contents of intact eggs: a study involving naturally infected hens. *Epidemiol. Infect.* 103:415–423.

Humphrey, T.J., A. Williams, K. McAlpine, F. Jorgensen, and C. O'Byrne. 1998. Pathogenicity in isolates of Salmonella enterica serotype Enteritidis PT4 which differ in RpoS expression: effects of growth phase and low temperature. *Epidemiol. Infect.* 121:295–301.

Ivanova, A., M. Renshaw, R.V. Guntaka, and A. Eisenstark. 1992. DNA base sequence variability in *katF* (putative sigma factor) gene of *Escherichia coli*. *Nucleic Acids Res.* 20:5479–5480.

Johansson, T.M., R. Schildt, S. Ali-Yrkko, A. Siitonen, and R.L. Maijala. 1996. The first Salmonella enteritidis phage type 1 infection of a commercial layer flock in Finland. *Acta. Vet. Scand.* 37:471–479.

Jones, B.D., and S. Falkow. 1996. Salmonellosis: host immune responses and bacterial virulence determinants. *Annu. Rev. Immunol.* 14:533–561.

Jorgensen, F., S. Leach, S.J. Wilde, A. Davies, G.S. Stewart, and T. Humphrey. 2000. Invasiveness in chickens, stress resistance and RpoS status of wild-type Salmonella enterica subsp. enterica serovar typhimurium definitive type 104 and serovar enteritidis phage type 4 strains. *Microbiology* 146 Pt 12:3227–3235.

Kaku, M., J.T. Peresi, A.T. Tavechio, S.A. Fernandes, A.B. Batista, I.A. Castanheira, G.M. Garcia, K. Irino, and D.S. Gelli. 1995. [Food poisoning outbreak caused by Salmonella Enteritidis in the northwest of Sao Paulo State, Brazil]. *Rev. Saude. Publica.* 29:127–131.

Kapperud, G., J. Lassen, and V. Hasseltvedt. 1998. Salmonella infections in Norway: descriptive epidemiology and a case-control study. *Epidemiol. Infect.* 121:569–577.

Kasai, K., T. Nakamura, N. Kase, T. Hiraoka, R. Suzuki, F. Kogure, and S.I. Shimoda. 1983. Increased glycosylation of proteins from cataractous lenses in diabetes. *Diabetologia.* 25:36–38.

Khakhria, R., D. Woodward, W.M. Johnson, and C. Poppe. 1997. Salmonella isolated from humans, animals and other sources in Canada, 1983–1992. *Epidemiol. Infect.* 119:15–23.

Kilic, Z., C. Kalyoncu, S. Durmus, M.A. Aksit, Y. Akgun, and F. Aksit. 1987. Distribution and typing in salmonellosis in children 1978–1983. *Mikrobiyol. Bul.* 21:103–109.

Kinde, H., D.H. Read, A. Ardans, R.E. Breitmeyer, D. Willoughby, H.E. Little, D. Kerr, R. Gireesh, and K.V. Nagaraja. 1996. Sewage effluent: likely source of Salmonella enteritidis, phage type 4 infection in a commercial chicken layer flock in southern California. *Avian Dis.* 40:672–676.

Kingsley, R.A., and A.J. Baumler. 2000. Host adaptation and the emergence of infectious disease: the Salmonella paradigm. *Mol. Microbiol.* 36:1006–1014.

Kohler, B. 1993. [Example of the concentration of salmonellae in the environment]. *Dtsch. Tierarztl. Wochenschr.* 100:264–274.

Kohler, B., K. Vogel, H. Kuhn, W. Rabsch, H.J. Rummler, L. Schulze, and W. Scholl. 1979. Epizootiology of Salmonella typhimurium infection in chickens. *Arch. Exp. Veterinarmed.* 33:281–298.

Kremer, E.T. 1999. The emergence of Salmonella Enteritidis in Hawai'i. Maui District health Office.

Kudoh, Y. 1997. [Current status of gastrointestinal infection in Japan—with special reference to enterohaemorrhagic Escherichia coli infection]. *Rinsho. Byori.* 45:242–248.

Landeras, E., M.A. Gonzalez-Hevia, R. Alzugaray, and M.C. Mendoza. 1996. Epidemiological differentiation of pathogenic strains of Salmonella enteritidis by ribotyping. *J. Clin. Microbiol.* 34:2294–2296.

Landeras, E., M.A. Gonzalez-Hevia, and M.C. Mendoza. 1998. Molecular epidemiology of Salmonella serotype Enteritidis. Relationship between food, water and pathogenic strains. *Int. J. Food Microbiol.* 43:81–90.

Landeras, E., and M.C. Mendoza. 1998. Evaluation of PCR-based methods and ribotyping performed with a mixture of PstI and SphI to differentiate strains of Salmonella serotype Enteritidis. *J. Med. Microbiol.* 47:427–434.

Laszlo, V.G., E.S. Csorian, and J. Paszti. 1985. Phage types and epidemiological significance of Salmonella enteritidis strains in Hungary between 1976 and 1983. *Acta. Microbiol. Hung.* 32:321–340.

Lester, A., N.H. Eriksen, H. Nielsen, P.B. Nielsen, A. Friis-Moller, B. Bruun, J. Scheibel, K. Gaarslev, and H.J. Kolmos. 1991. Non-typhoid Salmonella bacteraemia in Greater Copenhagen 1984 to 1988. *Eur. J. Clin. Microbiol. Infect. Dis.* 10:486–490.

Liaw, S.J., H.C. Lai, S.W. Ho, K.T. Luh, and W.B. Wang. 2000. Inhibition of virulence factor expression and swarming differentiation in Proteus mirabilis by p-nitrophenylglycerol. *J. Med. Microbiol.* 49:725–731.

Liebana, E., L. Garcia-Migura, J. Guard-Petter, S.W.J. McDowell, S. Rankin, H.M. Opitz, F.A. Clifton-Hadley, and R.H. Davies. 2001a. *Salmonella enterica* sero-type Enteritidis phage types 4, 7, 6, 8, 13a, 29, and 34 : A comparative analysis of genomic fingerprints from geographically distant isolates. *J. App. Microbiol.*. In press.

Liebana, E., D. Guns, L. Garcia-Migura, M.J. Woodward, F.A. Clifton-Hadley, and R.H. Davies. 2001b. Molecular typing of salmonella serotypes prevalent in animals in england: assessment of methodology. *J. Clin. Microbiol.* 39:3609–3616.

Ling, J., P.Y. Chau, and B. Rowe. 1987. Salmonella serotypes and incidence of mulitiply-resistant Salmonellae isolated from diarrhoel patients in Hong Kong from 1973–1982. *Epidemiol. Infect.* 99:295–306.

Ling, J.M., I.C. Koo, K.M. Kam, and A.F. Cheng. 1998. Antimicrobial susceptibilities and molecular epidemiology of Salmonella enterica serotype enteritidis strains isolated in Hong Kong from 1986 to 1996. *J. Clin. Microbiol.* 36:1693–1699.

Lukinmaa, S., R. Schildt, T. Rinttila, and A. Siitonen. 1999. Salmonella enteritidis phage types 1 and 4: pheno- and genotypic epidemiology of recent outbreaks in Finland. *J. Clin. Microbiol.* 37:2176–2182.

MacFarlane, S., M.J. Hopkins, and G.T. MacFarlane. 2001. Toxin Synthesis and Mucin Breakdown Are Related to Swarming Phenomenon in Clostridium septicum. *Infect. Immun.* 69:1120–1126.

Marcus, S.L., J.H. Brumell, C.G. Pfeifer, and B.B. Finlay. 2000. Salmonella pathogenicity islands: big virulence in small packages. *Microbes Infect.* 2:145–156.

Martin, G., U. Methner, G. Steinbach, and H. Meyer. 1996. [Immunization with potential Salmonella enteritidis mutants—2. Investigations on the attenuation and immunogenicity for mice and young hens]. *Berl. Munch. Tierarztl. Wochenschr.* 109:369–374.

Martinetti, G., and M. Altwegg. 1990. rRNA gene restriction patterns and plasmid analysis as a tool for typing Salmonella enteritidis. *Res. Microbiol.* 141:1151–1162.

Mattick, K.L., F. Jorgensen, J.D. Legan, M.B. Cole, J. Porter, H.M. Lappin-Scott, and T.J. Humphrey. 2000. Survival and filamentation of Salmonella enterica serovar enteritidis PT4 and Salmonella enterica serovar typhimurium DT104 at low water activity. *Appl. Environ. Microbiol.* 66:1274–129.

McDonough, P.L., S.J. Shin, and J.F. Timoney. 1986. Salmonella serotypes from animals in New York State, 1978–1983. *Cornell Vet.* 76:30–37.

Methner, U., P.A. Barrow, G. Martin, and H. Meyer. 1997. Comparative study of the protective effect against Salmonella colonisation in newly hatched SPF chickens using live, attenuated Salmonella vaccine strains, wild-type Salmonella strains or a competitive exclusion product. *Int. J. Food Microbiol.* 35:223–230.

Metz, H., and U. Lieb. 1980. Enteritis salmonellae in man and animal from 1953 to 1975 in Southern Bavaria. *Zen-

tralbl. Bakteriol. Mikrobiol. Hyg.* 171:231–255.

Michael, B., J.N. Smith, S. Swift, F. Heffron, and B.M. Ahmer. 2001. SdiA of Salmonella enterica is a LuxR homolog that detects mixed microbial communities. *J. Bacteriol.* 183:5733–5742.

Milch, H., V.G. Laszlo, and E.S. Csorian. 1985. Epidemiological analysis of Salmonella typhimurium infections on the basis of laboratory methods. I. Distribution of phage types and biotypes of Salmonella typhimurium isolated in Hungary in the period 1960–1981. *Acta. Microbiol. Hung.* 32:75–86.

Monnier, V.M., and A. Cerami. 1983. Detection of nonenzymatic browning products in the human lens. *Biochim. Biophys. Acta.* 760:97–103.

Narimatsu, H., K. Ogata, Y. Fuchi, and K. Hoashi. 1997. [Bacteriological studies on sporadic diarrhea diseases in Oita District, 1985–1996]. *Kansenshogaku. Zasshi.* 71:644–651.

Nuotio, L., C. Schneitz, U. Halonen, and E. Nurmi. 1992. Use of competitive exclusion to protect newly-hatched chicks against intestinal colonisation and invasion by Salmonella enteritidis PT4. *Br. Poult. Sci.* 33:775–779.

Nurmi, E., L. Nuotio, and C. Schneitz. 1992. The competitive exclusion concept: development and future. *Int. J. Food Microbiol.* 15:237–240.

Ochman, H., and E.A. Groisman. 1996. Distribution of pathogenicity islands in Salmonella spp. *Infect. Immun.* 64:5410–5412.

Okamura, M., Y. Kamijima, T. Miyamoto, H. Tani, K. Sasai, and E. Baba. 2001. Differences among six Salmonella serovars in abilities to colonize reproductive organs and to contaminate eggs in laying hens. *Avian Dis.* 45:61–69.

Olsen, A.R., and T.S. Hammack. 2000. Isolation of Salmonella spp. from the housefly, Musca domestica L., and the dump fly, Hydrotaea aenescens (Wiedemann) (Diptera: Muscidae), at caged-layer houses. *J. Food Prot.* 63:958–960.

Olsen, S.J., R. Bishop, F.W. Brenner, T.H. Roels, N. Bean, R.V. Tauxe, and L. Slutsker. 2001. The Changing Epidemiology of Salmonella: Trends in Serotypes Isolated from Humans in the United States, 1987–1997. *J. Infect. Dis.* 183:753–761.

Parker, C.T., K. Asokan, and J. Guard-Petter. 2001a. Egg contamination by *Salmonella* serovar Enteritidis following vaccination of chickens with delta-*aroA Salmonella* serovar Typhimurium. *FEMS Microbiol. Lett.* 195:73–78.

Parker, C.T., and J. Guard-Petter. 2001. Contribution of flagella and *Salmonella* invasion proteins to pathogenesis of salmonella enterica serovar enteritidis in chicks. *FEMS Microbiol. Lett.* 204:287–291.

Parker, C.T., E. Liebana, D.J. Henzler, and J. Guard-Petter. 2001b. Lipopolysaccharide O-antigen micro-heterogeneity of *Salmonella* serotypes Enteritidis and Typhimurium. *Environ. Microbiol.* Accepted.

Parker, C.T., B. Harmon, and J. Guard-Petter. 2002. Mitigation of avian reproductive tract function by *Salmonella entertitidis* producing high-molecular-mass lipopolysaccharide. *Environ. Microbiol.* 4:538–545.

Passaro, D.J., R. Reporter, L. Mascola, L. Kilman, G.B. Malcolm, H. Rolka, S.B. Werner, and D.J. Vugia. 1996. Epidemic Salmonella enteritidis infection in Los Angeles County, California. The predominance of phage type 4. *West. J. Med.* 165:126–130.

Perales, I., J. Muniz, C. Zigorraga, M. Dorronsoro, F. Martin, and M.A. Garcia-Calabuig. 1989. [Infectious food poisoning in the Autonomous Basque Community (1984–1986)]. *Enferm. Infecc. Microbiol. Clin.* 7:525–529.

PHLS. 1989. First report on Salmonella in eggs. PHLS evidence. Agricultural Committee of the House of Commons, London. 14–53.

Poppe, C. 1994. Salmonella enteritidis in Canada. *Int. J. Food Microbiol.* 21:1–5.

Pruss, B.M., D. Markovic, and P. Matsumura. 1997. The *Escherichia coli* flagellar transcriptional activator flhD regulates cell division through induction of the acid response gene cadA. *J. Bacteriol.* 179:3818–3821.

Przybylska, A. 1990. Foci of outbreaks of food poisoning and intestinal infections in Poland 1945–1989. *Przegl. Epidemiol.* 44:309–316.

Rabsch, W., B.M. Hargis, R.M. Tsolis, R.A. Kingsley, K.H. Hinz, H. Tschape, and A.J. Baumler. 2000. Competitive exclusion of Salmonella enteritidis by Salmonella gallinarum in poultry. *Emerg. Infect. Dis.* 6:443–448.

Rabsch, W., H. Tschape, and A.J. Baumler. 2001. Nontyphoidal salmonellosis: emerging problems. *Microbes Infect.* 3:237–247.

Raetz, C.R.H. 1996. Bacterial lipopolysaccharides: a remarkable family of bioactive macroamphiphiles. *In Escherichia coli* and *Salmonella.* F.C. Neidhardt, editor. ASM Press, Washington. 1035–1063.

Rahman, M.M., J. Guard-Petter, and R.W. Carlson. 1997. A virulent isolate of *Salmonella enteritidis* produces a *Salmonella typhi*-like lipopolysaccharide. *J. Bacteriol.* 179:2126–2131.

Reeves, P. 1993. Evolution of *Salmonella* O antigen variation involved in interspecies gene transfer on a large scale. *Trends Genet.* 9:17–22.

Ridley, A.M., E.J. Threlfall, and B. Rowe. 1998. Genotypic characterization of Salmonella enteritidis phage types by plasmid analysis, ribotyping, and pulsed-field gel electrophoresis. *J. Clin. Microbiol.* 36:2314–2321.

Romling, U., Z. Bian, M. Hammar, W.D. Sierralta, and S. Normark. 1998. Curli fibers are highly conserved between Salmonella typhimurium and Escherichia coli with respect to operon structure and regulation. *J. Bacteriol.* 180:722–731.

Romling, U., and M. Rohde. 1999. Flagella modulate the multicellular behavior of Salmonella typhimurium on the community level. *FEMS Microbiol. Lett.* 180:91–102.

Romling, U., M. Rohde, A. Olsen, S. Normark, and J. Reinkoster. 2000. AgfD, the checkpoint of multicellular and aggregative behaviour in Salmonella typhimurium regulates at least two independent pathways. *Mol. Microbiol.* 36:10–23.

Rybin, A.P. 1977. [Use of bactorodencide in rodent control]. *Veterinariia* 52–53.

Savage, W.G., and W.J. Read. 1913. Gaertner group bacilli in rats and mice. *J. Hyg.* 13.

Schauder, S., K. Shokat, M.G. Surette, and B.L. Bassler. 2001. The LuxS family of bacterial autoinducers: biosynthesis of a novel quorum-sensing signal molecule. *Mol. Microbiol.* 41:463–476.

Schmitt, C.K., J.S. Ikeda, S.C. Darnell, P.R. Watson, J. Bispham, T.S. Wallis, D.L. Weinstein, E.S. Metcalf, and A.D. O'Brien. 2001. Absence of all components of the flagellar export and synthesis machinery differentially alters virulence of Salmonella enterica serovar Typhimurium in models of typhoid fever, survival in macrophages, tissue culture invasiveness, and calf enterocolitis. *Infect. Immun.* 69:5619–5625.

Schneitz, C., L. Nuotio, G. Mead, and E. Nurmi. 1992. Competitive exclusion in the young bird: challenge models, administration and reciprocal protection. *Int. J. Food Microbiol.* 15:241–244.

Schulte, T. 1994. [The 1990 salmonella epidemic in the commercial city of Lubeck—an epidemiologic study as a contribution to determining the etiology of the salmonella outbreak]. *Gesundheitswesen* 56:606–610.

Scuderi, G., M. Fantasia, and T. Niglio. 2000. Results from the first computerized Italian surveillance of human Salmonella isolates. The Italian SALM-NET Working Group. *New Microbiology* 23:367–382.

Seo, K.H., P.S. Holt, R.K. Gast, and C.L. Hofacre. 2000. Combined effect of antibiotic and competitive exclusion treatment on Salmonella enteritidis fecal shedding in molted laying hens. *J. Food Prot.* 63:545–548.

Shivaprasad, H.L. 2000. Fowl typhoid and pullorum disease. *Rev. Sci. Tech.* 19:405–424.

Smith, A., S.P. Rose, R.G. Wells, and V. Pirgozliev. 2000. The effect of changing the excreta moisture of caged laying hens on the excreta and microbial contamination of their egg shells. *Br. Poult. Sci.* 41:168–173.

Sokurenko, E.V., V. Tchesnokova, A.T. Yeung, C.A. Oleykowski, E. Trintchina, K.T. Hughes, R.A. Rashid, J.M. Brint, S.L. Moseley, and S. Lory. 2001. Detection of simple mutations and polymorphisms in large genomic regions. *Nucleic Acids Res.* 29:E111.

Solodovnikov, I.P., I.N. Lytkina, N.N. Filatov, G.G. Chistiakova, B.E. Zaitsev, and S.V. Serzhenko. 1996. [Salmonellosis in Moscow: its epidemiological characteristics and the prevention tasks]. *Zh. Mikrobiol. Epidemiol. Immunobiol.* 46–49.

Sramova, H., D. Dedicova, P. Petras, and C. Benes. 1991. [Epidemic occurrence of alimentary bacterial infections in the Czech Republic 1979–1989]. *Cesk. Epidemiol. Mikrobiol. Imunol.* 40:74–84.

St Louis, M.E., D.L. Morse, M.E. Potter, T.M. DeMelfi, J.J. Guzewich, R.V. Tauxe, and P.A. Blake. 1988. The emergence of grade A eggs as a major source of Salmonella enteritidis infections. New implications for the control of salmonellosis. *JAMA* 259:2103–2107.

Stanley, J., A.P. Burnens, E.J. Threlfall, N. Chowdry, and M. Goldsworthy. 1992. Genetic relationships among strains of Salmonella enteritidis in a national epidemic in Switzerland. *Epidemiol. Infect.* 108:213–220.

Stanley, J., C.S. Jones, and E.J. Threlfall. 1991. Evolutionary lines among Salmonella enteritidis phage types are identified by insertion sequence IS200 distribution. *FEMS Microbiol. Lett.* 66:83–89.

Stevens, V.J., C.A. Rouzer, V.M. Monnier, and A. Cerami. 1978. Diabetic cataract formation: potential role of glycosylation of lens crystallins. *Proc. Natl. Acad. Sci. U S A.* 75:2918–2922.

Stubbs, A.D., F.W. Hickman-Brenner, D.N. Cameron, and J.J. Farmer, 3rd. 1994. Differentiation of Salmonella enteritidis phage type 8 strains: evaluation of three additional phage typing systems, plasmid profiles, antibiotic susceptibility patterns, and biotyping. *J. Clin. Microbiol.* 32:199–201.

Surette, M.G., M.B. Miller, and B.L. Bassler. 1999. Quorum sensing in Escherichia coli, Salmonella typhimurium, and Vibrio harveyi: a new family of genes responsible for autoinducer production. *Proc. Natl. Acad. Sci U S A.* 96:1639–1644.

Szymanski, C.M., M. King, M. Haardt, and G.D. Armstrong. 1995. Campylobacter jejuni motility and invasion of Caco-2 cells. *Infect. Immun.* 63:4295–4300.

Thong, K.L., Y.F. Ngeow, M. Altwegg, P. Navaratnam, and T. Pang. 1995. Molecular analysis of Salmonella enteritidis by pulsed-field gel electrophoresis and ribotyping. *J. Clin. Microbiol.* 33:1070–1074.

Thong, K.L., S. Puthucheary, and T. Pang. 1998. Outbreak of Salmonella enteritidis gastroenteritis: investigation by pulsed-field gel electrophoresis. *Int. J. Infect. Dis.* 2:159–163.

Toguchi, A., M. Siano, M. Burkart, and R.M. Harshey. 2000. Genetics of swarming motility in Salmonella enterica serovar typhimurium: critical role for lipopolysaccharide. *J. Bacteriol.* 182:6308–6321.

Tyler, K.D., G. Wang, S.D. Tyler, and W.M. Johnson. 1997. Factors affecting reliability and reproducibility of amplification- based DNA fingerprinting of representative bacterial pathogens. *J. Clin. Microbiol.* 35:339–346.

Ueda, Y., N. Suzuki, T. Furukawa, Y. Takegaki, N. Takahashi, K. Miyagi, K. Noda, H. Hirose, S. Hashimoto, H. Miyamoto, S. Yano, Y. Miyata, M. Taguchi, M. Ishibashi, and T. Honda. 1999. [Bacteriological studies of traveller's diarrhoea (6). Analysis of enteropathogenic bacteria at Kansai Airport Quarantine Station from September 4th, 1994 through December 1996]. *Kansenshogaku Zasshi* 73:110–121.

Ullmann, R., and H.K. Scholtze. 1989. [Salmonella occurrence in the Erfurt district]. *Z Gesamte Hyg.* 35:676–679.

USDA. 1998. *Salmonella* Enteritidis Risk Assessment: Shell Eggs and Egg Products. U. S. Dept. of Agriculture, Washington, D.C.

Usera, M.A., T. Popovic, C.A. Bopp, and N.A. Strockbine. 1994. Molecular subtyping of Salmonella enteritidis phage type 8 strains from the United States. *J. Clin. Microbiol.* 32:194–198.

Van Loock, F., G. Ducoffre, J.M. Dumont, M.L. Libotte-Chasseur, H. Imberechts, M. Gouffaux, J. Houins-Roulet, G. Lamsens, K. De Schrijver, N. Bin, A. Moreau, L. De Zutter, and G. Daube. 2000. Analysis of foodborne disease in Belgium in 1997. *Acta. Clin. Belg.* 55:300–306.

Visick, J.E., and S. Clarke. 1997. RpoS- and OxyR-independent induction of HP1 catalase at stationary phase in *Escherichia coli* and identification of *rpoS* mutations in common laboratory strains. *J. Bacteriol.* 179:4158–4163.

Wachsmuth, I.K., J.A. Kiehlbauch, C.A. Bopp, D.N. Cameron, N.A. Strockbine, J.G. Wells, and P.A. Blake. 1991. The use of plasmid profiles and nucleic acid probes in epidemiologic investigations of foodborne, diarrheal diseases. *Int. J. Food Microbiol.* 12:77–89.

Wachsmuth, K. 1986. Molecular epidemiology of bacterial infections: examples of methodology and of investigations of outbreaks. *Rev. Infect. Dis.* 8:682–692.

Ward, J.P., J.R. King, A.J. Koerber, P. Williams, J.M. Croft, and R.E. Sockett. 2001. Mathematical modelling of quorum sensing in bacteria. *IMA J. Math. Appl. Med. Biol.* 18:263–292.

Waterman, S.R., and P.L.C. Small. 1996. Characterization of the acid resistance phenotype and *rpoS* alleles of shiga-like toxin producing *Escherichia coli. Infect. Immun.* 64:2808–2811.

Watier, L., S. Richardson, and B. Hubert. 1993. Salmonella enteritidis infections in France and the United States: characterization by a deterministic model. *Am. J. Public Health.* 83:1694–1700.

Weide-Botjes, M., B. Kobe, and S. Schwarz. 1998. Inter- and intra-phage type differentiation of Salmonella enterica subsp. enterica serovar enteritidis isolates using molecular typing methods. *Zentralbl. Bakteriol.* 288:181–193.

Wierup, M., B. Engstrom, A. Engvall, and H. Wahlstrom. 1995. Control of Salmonella enteritidis in Sweden. *Int. J. Food Microbiol.* 25:219–226.

Wierup, M., M. Wold-Troell, E. Nurmi, and M. Hakkinen. 1988. Epidemiological evaluation of the Salmonella-controlling effect of a nationwide use of a competitive exclusion culture in poultry. *Acta. Vet. Scand. Suppl.* 84:309–311.

Wilson, I.G., T.S. Wilson, and S.T. Weatherup. 1996. Salmonella in retail poultry in Northern Ireland. *Commun. Dis. Rep CDR Rev.* 6:R64–66.

Withers, H., S. Swift, and P. Williams. 2001. Quorum sensing as an integral component of gene regulatory networks in Gram-negative bacteria. *Curr Opin Microbiol.* 4:186–193.

Woese, C.R. 1987. Bacterial evolution. *Microbiol. Rev.* 51:221–271.

Woodward, M.J., M. Sojka, K.A. Sprigings, and T.J. Humphrey. 2000. The role of SEF14 and SEF17 fimbriae in the adherence of Salmonella enterica serotype Enteritidis to inanimate surfaces. *J. Med. Microbiol.* 49:481–487.

Zambrano, M.M., D.A. Siegele, M. Almiron, A. Tormo, and R. Kolter. 1993. Microbial competition: Escherichia coli mutants that take over stationary phase cultures. *Science* 259:1757–1760.

Zogaj, X., M. Nimtz, M. Rohde, W. Bokranz, and U. Romling. 2001. The multicellular morphotypes of Salmonella typhimurium and Escherichia coli produce cellulose as the second component of the extracellular matrix. *Mol. Microbiol.* 39:1452–1463.

14

MULTIPLE ANTIBIOTIC RESISTANCE AND VIRULENCE OF *SALMONELLA ENTERICA* SEROTYPE TYPHIMURIUM PHAGE TYPE DT104

STEVE A. CARLSON, MAX T. WU, AND TIMOTHY S. FRANA

INTRODUCTION. *Salmonella* infections are an important health concern in both industrialized and developing countries (Slutsker 1998). Treatment of this microorganism is compromised by its ability to acquire resistance to multiple antibiotics. Monitoring of *Salmonella* isolates in the United States and abroad has shown that increasing proportions are resistant to several antimicrobial agents (Low 1997; Glynn 1998). Of particular concern is a unique strain of *Salmonella enterica* serotype Typhimurium, characterized as definitive type 104 (DT104), that is commonly resistant to five antibiotics (ampicillin, chloramphenicol, streptomycin, sulfonamides, and tetracycline; ACSSuT antibiogram). First reported in the United Kingdom in 1984, DT104 is now the second most prevalent *Salmonella* isolated from humans in England and Wales (Ridley and Threlfall 1998). *Salmonella* infections caused by multiresistant DT104 have also been reported in cattle (Imberechts 1998), swine (Poppe 1998), marine wildlife (Foster 1998), cats (Low 1996), and a variety of other species (Besser 1997). Additionally, DT104 and related multiresistant *Salmonella* were recently detected in approximately 50 percent of retail ground meats (White 2001).

The virulence of DT104 is also of concern because, relative to an infection with a non-DT104 *Salmonella*, an infection with DT104 is more likely to lead to hospitalization for a human (Wall 1994) or death for a calf (Evans and Davies 1996). These phenomena suggest that DT104 may be more virulent due to alterations in inherent pathogenic characteristics, acquisition of new virulence characteristics, or treatment failures resulting from the use of inappropriate antibiotics. Therefore, DT104 represents a major obstacle for preserving food safety.

CHARACTERISTICS OF DT104. The current questions about DT104 are the following: When did DT104 arise? How did DT104 become antibiotic resistant? How do we detect DT104? Is DT104 more virulent than antibiotic-sensitive cohorts? The sections that follow address these questions.

Origin and Emergence of DT104. Multiple-antibiotic-resistant DT104 appears to have originated when a strain of antibiotic-sensitive DT104 acquired a segment of chromosomal DNA in a bacteriophage-mediated transduction event (Schmieger and Schicklmaier 1999). Although the time and place of this event is unknown, it appears to have occurred prior to 1984. It is possible that DT104 is a variant of one of the multiple-antibiotic-resistant DT29 or DT36 that were identified in the 1960s (Anderson 1968). Regardless of the date when the gene transfer occurred, it seems likely that years of exposure to antibiotics led to an amplification of the original clone by selective pressures. However, evidence suggests that antibiotic-independent forces might have a role. Davis et al. (1999) have pointed out that ACSSuT *S. enterica* serotype Typhimurium (predominantly DT104) increased preferentially over ASSuT *S. enterica* serotype Typhimurium in cattle during a period when chloramphenicol or its relatives were banned for use in food animals. Also, DT104 has disseminated into wildlife (Besser 1997), and fluoroquinolone-resistant DT104 have been isolated from domestic animal reservoirs not receiving fluoroquinolones (Molbak 1999). Other researchers cite examples of broad dissemination of *S. enterica* serotype Typhimurium clones susceptible to antibiotics in common use (Khakhria 1983; Passaro 1996). This research suggests that the ecology of *S. enterica* serotype Typhimurium be such that certain phagetypes become dominant in the environment, replacing previously dominant phagetypes, by means not yet known. The previous dominant phagetype may have been DT204 that was prevalent in the 1980s (Threlfall 1985) but has since mostly disappeared.

Antibiotic Resistance of DT104. Acquisition of exogenous genes appears to be the molecular basis for multiple antibiotic resistance in DT104 (Table 14.1). The acquired genes are often part of plasmids, transposons, and mobile cassettes of DNA called *integrons* (Hall and Stokes 1993; Recchia and Hall 1997). For DT104, the ACSSuT antibiotic resistance pattern is conferred by a chromosomal arrangement of genes (Ridley and Threlfall 1998; Sandvang 1998; Briggs and Fratamico 1999). This genetic arrangement is composed largely of two integrons containing the genes *pse-1*, a *cmlA* homologue, *aadA2*, *sul1*, and *tetA* that respectively encode resistance to ampicillin, chloramphenicol, streptomycin/spectinomycin, sulfonamides, and tetracycline. This integron structure has similarities to genes found in multiple-antibiotic-resistant *Pasteurella piscicida* (Kim and Aoki 1996a, 1996b), and the transfer of the integrons appears to be linked to P22-like bacteriophage transduction (Schmieger and Schicklmaier 1999). Insertion of the integrons into DT104 is dependent upon the presence of a retron, a segment of DNA introduced as a result of an infection with an RNA bacteriophage (Lampson 1990) that is present in *S. enterica* serotype Typhimurium but absent in non-Typhimurium serotypes (Boyd 2000) (Fig. 14.1).

The acquisition of ACSSuT-independent, or *extra-integron*, genes is determined by conjugative plasmids (Low 1997). Plasmids bearing aminoglycoside/ aminocyclitol resistance genes, that is, genes encoding resistance to apramycin, amikacin, gentamicin, or kanamycin, have been identified in DT104. Interestingly, single conserved genes appear to confer these resistances (refer to Table 14.1). Other plasmids have been identified for cephalosporin resistance genes (Bauernfeind 1992; Fey 2000; Zhao 2001). These plasmids can be transferred from nonpathogenic *E. coli* to *Salmonella* (Winokur 2001). One of these plasmids was implicated in a putative animal-to-human transfer of antibiotic resistance in DT104 (Fey 2000). This study implicated the use of ceftiofur, a ceftriaxone-related cephalosporin used in veterinary medicine, in cattle as the cause of the ceftriaxone resistance in a DT104 isolate that infected a child. The importance of this study is underscored by the fact that ceftriaxone is the current drug of choice against DT104 in children. Ongoing studies are aimed at developing rapid diagnostic tests for identifying DT104 that harbor the ceftriaxone/ceftiofur resistance plasmid.

Fluoroquinolone resistance is uncommon in DT104 although this may change given the current widespread use of ciprofloxacin as a prophylactic against anthrax. Although ciprofloxacin is one of a number of antibiotics effective against *Bacillus anthracis*, ciprofloxacin and related fluoroquinolones are considered to be the current last line of defense against DT104 in adults. Fluoroquinolone resistance is conferred by silent point mutations in genes encoding DNA gyrase, an enzyme that assists in unwinding and uncoiling DNA (Bryan 1989). A number of mutations confer fluoroquinolone resistance in DT104 (Table 14.1). The polymerase chain reaction (PCR) can be used for detecting these types of mutations in DT104 (Walker 2001). However, gyrase mutation-independent fluoroquinolone resistance can occur as described in the ensuing paragraph.

A minor way for DT104 to become antibiotic resistant is via activation of innate antibiotic efflux systems. The predominant efflux system involves the multiple antibiotic resistance (*mar*) operon. The *mar* operon regulates an efflux system thought to be involved in *enteroadaptation*, that is, the *mar* operon may have originated when bacteria sought refuge in a hostile environment consisting of bile acids (Ma 1996). This efflux system endows the ability to expel bile acids, along with other foreign substances, from the bacterium. The hallmark of an active *mar* operon is resistance to ampicillin, chloramphenicol, ciprofloxacin, rifampin, and tetracycline (Cohen 1989). The *mar* operon is activated by physiologic stressors, such as sublethal antibiotics (George and Levy 1983; Carlson and Ferris 2000) and high salinity (Ma 1996), and by alterations in a protein that represses the operon (Cohen 1993; Martin 1995; Sulavik 1997). Activation of the *mar* system typically will lead to low-level antibiotic resistance (Giraud 2000; Randall and Woodward 2001), although high-level antibiotic resistance has been observed in *S. enterica* serotype Typhi with a *mar*-like phenotype (Toro 1990). A very small number of DT104 isolates exhibit an unusual *mar*-like response in which penicillin derivatives, when used inappropriately, can activate the system (Carlson and Ferris 2000). Thus, prudent use of antibiotics is especially important for treating potential DT104 infections. Another facet of the *mar* response is the recent finding that the *mar* operon can be activated in response to insults from other bacteria (Carlson 2001a). This finding opens the possibility that environmental signals, independent of therapeutic or subtherapeutic antibiotics, can spur on antibiotic resistance in *Salmonella*.

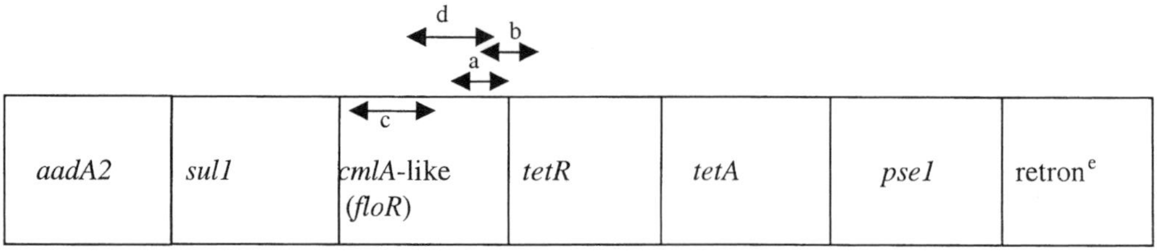

a: *floR* amplicon (Bolton 1999)
b: *cmlA*-like/*tetR* amplicon (Carlson 1999)
c: section of *cmlA*-like gene missing in DT104 isolates exhibiting the ASSuT antibiogram (Briggs and Fratamico 1999)
d: *floR* amplicon (Khan 2000)
e: integron insertion site that is found only in *Salmonella enterica* serotype Typhimurium (Boyd 2000)

FIG. 14.1—Integron, retron and integron-specific amplicon relationships in *Salmonella enterica* serotype Typhimurium DT104. Arrangement is presented 5' to 3' but is not to scale.

Table 14.1—Antibiotic Resistances and Genetic Sources of Resistance in *Salmonella enterica* serotype Typhimurium phagetype DT104, Including the Relative Prevalence of These Genetic Determinants from a Sample Population

Resistance	Integron-associated	Gene acquired	Genetic mutation	Relative prevalence[a]
amikacin	No	*aacA4* (Tosini 1998)	—	0%[b]
ampicillin/ amoxicillin/ ticarcillin	Yes	*pse-1* (Briggs and Fratamico 1999)	—	97%[c]
apramycin	No	*aacC4* (Frana 2001)	—	6%
chloramphenicol/ florfenicol	Yes	*cmlA*-like *(floR)* (Briggs and Fratamico 1999)	—	97%[c]
gentamicin	No	*aadB* (Frana 2001)	—	8%
kanamycin	No	*aphA1-Iab* (Frana 2001)	—	29%
ceftriaxone	No	*cmy*-2 (Fey 2000); *cmy*-4 (White 2002)	—	ND[d]
fluoroquinolone	No	—	GyrA S83→F/Y, GyrA D87→N /G/Y (Walker 2001); GyrA G81→C (Giraud 2000); GyrA A119→E (Griggs 1996); GyrA A67→P plus G81→S (Reyna 1995); GyrB S463→Y (Gensberg 1995)	0%[b]
streptomycin/ spectinomycin	Yes	*aadA2* (Briggs and Fratamico 1999)	—	100%[e]
sulfamethoxazole	Yes	*sul1* (Briggs and Fratamico 1999)	—	ND
tetracycline	Yes	*tetR* (Briggs and Fratamico 1999)	—	ND

[a] Relative prevalence of the acquired gene with a sample population of 500 isolates of multiresistant *Salmonella enterica* serotype Typhimurium. This sample pool contained DT104, U302, DT120, DT193, and DT208 phage types from clinical (livestock and companion animals) and nonclinical sources. Some of the isolates are described in previous studies (Carlson 2000a; Carlson 2000b; Frana 2001), whereas other isolates have been added recently.
[b] No amikacin- or fluoroquinolone-resistant isolates were part of this pool.
[c] Although DT208 displays the ACSSuT phenotype, these isolates (n=12) do not possess the same integron structure as DT104 (Frana 2001).
[d] The genetic basis for the ceftriaxone was not determined in the 44 isolates (9%) that were ceftriaxone resistant.
[e] The DT208 isolates possess the *aadA2* gene (Frana 2001).

Detection of DT104. Antibiotic resistance genes are often used as markers for detecting DT104. Integron sequences have been targeted for detecting DT104 in various samples. One of these sequences is the *cmlA*-like gene [a.k.a. *floR* (White 2000) or *flo*$_{st}$ (Bolton 1999)]. The *cmlA*-like gene, when used in a multiplex PCR with a *Salmonella* virulence gene (*invA*), can be used for detecting DT104 and other definitive types such as U302, DT193 and DT120 (Bolton 1999; Khan 2000). Unfortunately, recent studies have shown that the *cmlA*-like gene of DT104 has a homologue, designated as *floR*, in *E. coli* K99 isolated from neonatal calves presenting with diarrhea (White 2000). Therefore, the *floR/invA* PCR would have false positives in samples containing both *E. coli* K99/*floR* and a non-DT104 *Salmonella* possessing *invA*. Another problem with detection of the *cmlA*-like amplicon is that some DT104 are ASSuT resulting from an incomplete *cmlA*-like gene (Briggs and Fratamico 1999; Carlson 1999) (see Fig. 14.1). Thus,

detection schemes targeting the 5' end of the *cmlA*-like gene (Khan 2000) will result in false negatives for the ASSuT phenotype. Another integron sequence used for PCR-based detection of DT104 is the junction between the *cmlA*-like gene and *tetR*, the gene that represses the tetracycline resistance gene (Fig. 14.1). The *cmlA-tetR* segment, along with the junctional segment between two virulence genes (*sipB* and *sipC*), can be used for detecting DT104 and the other definitive types mentioned previously (Carlson 1999). However, false positives, like those postulated for the *cmlA*-like/*invA* multiplex PCR, can occur if a non-*Salmonella* microbe acquires the specific integron structure that currently resides only in DT104 and related phagetypes. An extra-integron sequence has been used for detecting DT104 and U302. This detection method utilizes PCR primers specific to a unique 16S-23S spacer region in DT104 (Pritchett 2000). The drawback to this assay is that antibiotic-sensitive DT104 would be detected.

Virulence and the Assessment of Hypervirulence in DT104. The status of DT104 as a hypervirulent superpathogen remains a controversy. Although clinical outcomes of DT104 infections clearly can be more problematic, it is unclear whether this phenomenon is related to a true alteration in virulence. Some of the reports of hypervirulence could be attributed to antibiotic resistance and inappropriate antibiotic use. However, it remains possible that DT104 has evolved such that it has virulence characteristics not observed in other *Salmonella*.

The basis for true hypervirulence in DT104 has several explanations. First, DT104 may have acquired exogenous virulence genes when it acquired the pentaresistant gene cluster. This explanation does not seem likely because no virulence genes have been identified in the integrons. Second, the pentaresistant gene cluster may have been inserted in a virulence repressor gene, although the integron/retron complex appears to lie between two virulence-independent genes, *thdf* and *yidY* (Boyd 2000). Third, integron gene products may be regulating the expression of cryptic virulence genes. Finally, individual clones of DT104 may have acquired exogenous virulence genes.

The study of DT104 virulence has been performed mostly *in vitro* or *ex vivo*. Tissue culture invasion studies indicate that most DT104 isolates exhibit invasive characteristics that are similar to standards set for *Salmonella* (Carlson 2000a, 2000b; Allen 2001). From a pool of approximately five hundred DT104 and DT104-related isolates, twelve isolates exhibited diminished invasiveness (Carlson 2000a) while four isolates displayed enhanced invasiveness (Fig. 14.2). Notably, however, the hyperinvasive isolates do not have altered murine LD_{50} values (Table 14.2). Therefore, the altered invasiveness observed occasionally in DT104 does not appear to be related to alterations in overt pathogenicity. However, these studies do not necessarily extrapolate to human infections. Tissue culture assays can predict *in vivo* virulence, but host factors may play an undetectable role *in vitro/ex vivo*. For example, DT104 may have a tropism for adhering to human intestinal cells, evading human immunocytes, or perturbing human microflora. Because *in vivo* studies in this area are not possible, the generalizations described herein must suffice.

Intramacrophage survival is another virulence property that has been assessed in DT104. One study indicates that increased macrophage survival is not con-

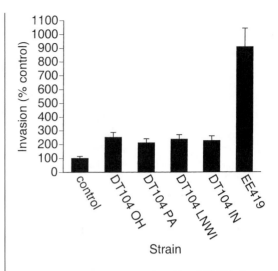

FIG. 14.2.—Invasive characteristics of four DT104 isolates found to be hyperinvasive. Invasion was determined in HEp-2 cells, as previously described (Carlson 2000a), and compared to the invasion of S. enterica serotype Typhimurium SL1344 (Wray and Sojka 1978). Invasion was also compared to S. enterica serotype Typhimurium EE419 (Lee 1992), a genetically-engineered hyperinvasive strain.

tributing to hypervirulence of DT104 (Allen 2001). However, this study is limited by the small sample size (n=1) and the availability of species-specific macrophage cell lines. Because murine macrophage cell lines were evaluated in this study, and because intramacrophage survival may be host specific (Schwan 2000), studies with porcine and bovine macrophages will be needed to complete this line of study.

The area of DT104 pathogenesis that remains open is cytotoxicity. Recent studies from our laboratory indicate that some DT104 isolates are capable of secreting a cytotoxin (Carlson 2001b; Carlson 2002). DT104 was recovered from calves displaying signs that are not consistent with salmonellosis, that is, peritonitis (independent of intestinal perforation), abomasitis, and polyserositis. Most of the affected calves died despite therapeutic interventions. Abomasitis and occasional peritonitis were observed in calves experimentally infected with one of these DT104 isolates (Carlson 2002). Further studies revealed that these strains were

Table 14.2—Murine LD_{50} values and invasiveness of *Salmonella enterica* serotype Typhimurium phagetype DT104 isolates found to be hypoinvasive or hyperinvasive *ex vivo*.

Strain	Relative invasiveness[a]	Relative murine LD_{50}[a]	Reference
DT104 strain 12781	10%	50%	(Carlson 2000a)
DT104 strain LNWI (Carlson 2001b)	225%	100%	(unpublished observations)

[a]Compared to *S. enterica* sero

capable of secreting a cytotoxin (Carlson 2001b). Recent studies have putatively identified the cytotoxin as a collagenase (Wu 2002) that appears to be constitutively repressed until derepressed by host factors (unpublished observations). The breadth and scope of this cytotoxic phenotype is under investigation.

An area of DT104 virulence that has not been studied is the induction of diarrhea. *Salmonella*-mediated diarrhea is related to expression of an enterotoxin (Chary 1993). Although numerous assays exist for evaluating diarrheal responses, some of these assays are not quantitative, and some are not suitable for assessing large numbers of isolates. Interpreting these assays would also be potentially challenging because the hypervirulence of DT104 could be caused by host factor–induced expression of the classic *Salmonella* enterotoxin or expression of an uncharacterized enterotoxin. Nonetheless, this area of study is the most open with respect to DT104 virulence.

SUMMARY AND CONCLUSIONS. Clearly, DT104 represents the multiple-antibiotic-resistant *Salmonella* strain that most threatens food safety. This microbe has two properties, antibiotic resistance and hypervirulence, which separate it from other *Salmonella*. DT104 is the reigning multiresistant *Salmonella*, although other phage types, such as DT208, may be the *Salmonella* that occupies the DT104 niche in the future. Alternatively, another *S. enterica* serotype, such as Agona (Boyd 2001), may be elevated to the status of DT104. Nonetheless, the problems associated with DT104 are significant and awaiting more study. Two major concerns exist regarding the control of this pathogen in the future. First, the dissemination of ceftriaxone and fluoroquinolone resistances will be detrimental for treating DT104 infections. The transfer of the ceftriaxone resistance plasmid will drive the former part of the problem, thus the monitoring of ceftriaxone/ceftiofur resistance will be of great importance. Monitoring of fluoroquinolone resistance will also be important, especially given the widespread use of ciprofloxacin versus anthrax. Second, the dissemination of cytotoxic DT104 will be a problem for clinical diagnosis of DT104 because the signs of these strains are atypical for salmonellosis. Recently we developed an RT-PCR assay for cytotoxin expression in DT104 (Wu 2002) that may become an important part of DT104 diagnosis in the future. The search for other contributors to hypervirulence will obviously augment our future approaches to controlling DT104 or its emerging relative. Continued vigilance will be important for assuaging DT104.

REFERENCES

Allen, C., P. Fedorka-Cray, A. Vazquez-Torres, M. Suyemoto, C. Altier, L.R. Ryder, F.C. Fang, and S.J. Libby. 2001. *In vitro* and *in vivo* assessment of *Salmonella enterica* serovar Typhimurium DT104 virulence. *Infect. and Immun.* 69(7):4673–4677.

Anderson, E. 1968. The ecology of transferable drug resistance in the Enterobacteria. *Ann. Rev. Microbiol.* 22:131–180.

Bauernfeind, A., J. Casellas, M. Goldberg, M. Holley, R. Jungwirth, P. Mangold, T. Rohnisch, S. Schweighart, and R. Wilhelm. 1992. A new plasmidic cefotaximase from patients infected with *Salmonella typhimurium*. *Infection* 20(3):158–163.

Besser, T.E., C.C. Gay, J.M. Gay, D.D. Hancock, D. Rice, L.C. Pritchett, and E.D. Erickson. 1997. Salmonellosis associated with *S. typhimurium* in the USA. *Vet. Rec.* 140(3):75.

Bolton, L., L. Kelley, M. Lee, P. Fedorka-Cray, and J. Maurer. 1999. Detection of multidrug-resistant *Salmonella enterica* serotype *typhimurium* DT104 based on a gene which confers cross-resistance to florfenicol and chloramphenicol. *J. Clin. Microbiol.* 37(5):1348–1351.

Boyd, D., G. Peters, L. Ng, and M. Mulvey. 2000. Partial characterization of a genomic island associated with the multidrug resistance region of *Salmonella enterica* Typhimurium DT104. *FEMS Microbiol. Lett.* 189(2):285–291.

Boyd, D., G. Peters, A. Cloeckaert, K. Boumedine, E. Chaslus-Dancla, H. Imberechts, and M.R. Mulvey. 2001. Complete nucleotide sequence of a 43-kilobase genomic island associated with the multidrug resistance region of *Salmonella enterica* serovar Typhimurium DT104 and its identification in phage type DT120 and serovar Agona. *J. Bacteriol.* 183(19):5725–5732.

Briggs, C.E., and P.M. Fratamico. 1999. Molecular characterization of an antibiotic resistance gene cluster of *Salmonella typhimurium* DT104. *Antimicrob. Ag. Chemother.* 43(4):846–849.

Bryan, L., J. Bedard, S. Wong, and S. Chamberland. 1989. Quinolone antimicrobial agents: mechanism of action and resistance development. *Clin. Investig. Medicine* 12(1):14–19.

Carlson, S.A., L.F. Bolton, C.E. Briggs, H.S. Hurd, V.K. Sharma, P. Fedorka-Cray, and B.D. Jones. 1999. Detection of *Salmonella typhimurium* DT104 using multiplex and fluorogenic PCR. *Mol. Cell. Probes* 13(3):213–222.

Carlson, S.A., and K.E. Ferris. 2000. Augmentation of antibiotic resistance in multiple antibiotic resistant *Salmonella typhimurium* DT104 following exposure to penicillin derivatives. *Vet. Microbiol.* 73(1):25–35.

Carlson, S., M. Browning, K. Ferris, and B. Jones. 2000a. Identification of diminished tissue culture invasiveness among multiple antibiotic resistant *Salmonella typhimurium* DT104. *Microb. Pathogen.* 28(1):37–44.

Carlson, S., R. Willson, A. Crane, and K. Ferris. 2000b. Evaluation of invasion-conferring genotypes and antibiotic-induced hyperinvasive phenotypes in multiple antibiotic resistant *Salmonella typhimurium* DT104. *Microb. Pathogen.* 28(6):373–378.

Carlson, S., T. Frana, and R. Griffith. 2001a. Antibiotic resistance in *Salmonella enterica* serovar Typhimurium exposed to microcin-producing *Escherichia coli*. *Appl. Environ. Microbiol.* 67(8):3763–3766.

Carlson, S., D. Meyerholz, T. Stabel, and T. Jones. 2001b. Secretion of a putative cytotoxin in multiple antibiotic resistant *Salmonella enterica* serotype Typhimurium phagetype DT104. *Microb. Pathogen.* 31(4):201–204.

Carlson, S., W. Stoffregen, and S. Bolin. 2002. Abomasitis due to multiple antibiotic resistant *Salmonella enterica* serotype Typhimurium phagetype DT104. *Vet. Microbiol.* 85(3):233–240.

Chary, P., R. Prasad, A. Chopra, and J.W. Peterson. 1993. Location of the enterotoxin gene from *Salmonella typhimurium* and characterization of the gene products. *FEMS Microbiol. Lett.* 111(1):87–92.

Cohen, S.P., L.M. McMurry, D.C. Hooper, J.S. Wolfson, and S.B. Levy. 1989. Cross-resistance to fluoroquinolones in multiple-antibiotic-resistant (Mar) *Escherichia coli* selected by tetracycline or chloramphenicol: decreased drug accumulation associated with membrane changes in addition to OmpF reduction. *Antimicrob. Ag. Chemother.* 33(8):1318–1325.

Cohen, S.P., H. Hachler, and S.B. Levy. 1993. Genetic and functional analysis of the multiple antibiotic resistance (*mar*) locus in *Escherichia coli. J. Bacteriol.* 175(5):1484–1492.

Davis, M., D. Hancock, T. Besser, D. Rice, J. Gay, C. Gay, L. Gearhart, and R. DiGiacomo. 1999. Changes in antimicrobial resistance among *Salmonella enterica* Serovar typhimurium isolates from humans and cattle in the Northwestern United States, 1982–1997. *Emerg. Infect. Dis.* 5(6):802–806.

Evans, S., and R. Davies. 1996. Case control study of multiple-resistant *Salmonella typhimurium* DT104 infection of cattle in Great Britain. *Vet. Rec.* 139(23):557–558.

Fey, P., T.J. Safranek, M.E. Rupp, E.F. Dunne,E. Ribot, P.C. Iwen, P.A. Bradford, F.J. Angulo, and S.H. Hinrichs. 2000. Ceftriaxone-resistant salmonella infection acquired by a child from cattle. *N. Engl. J. Med.* 342(17):1242–1249.

Foster, G., H.M. Ross, I.A. Patterson, R.J. Reid, and D.S. Munro. 1998. *Salmonella typhimurium* DT104 in a grey seal. *Vet. Rec.* 142(22):615.

Frana, T., S. Carlson, and R. Griffith. 2001. Relative distribution and conservation of genes encoding aminoglycoside-modifying enzymes in *Salmonella enterica* serotype Typhimurium phage type DT104. *Appl. Environ. Microbiol.* 67(1):445–448.

Gensberg, K., Y. Jin, and L.J. Piddock. 1995. A novel *gyrB* mutation in a fluoroquinolone-resistant clinical isolate of *Salmonella typhimurium. FEMS Microbiol. Lett.* 132(1–2):57–60.

George, A.M., and S.B. Levy. 1983. Amplifiable resistance to tetracycline, chloramphenicol, and other antibiotics in *Escherichia coli*: involvement of a non-plasmid-determined efflux of tetracycline. *J. Bacteriol.* 155(2):531–540.

Giraud, E., A. Cloeckaert, D. Kerboeuf, and E. Chaslus-Dancla. 2000. Evidence for active efflux as the primary mechanism of resistance to ciprofloxacin in *Salmonella enterica* serovar typhimurium. *Antimicrob. Ag. Chemother.* 44(5):1223–1228.

Glynn, M.K., C. Bopp, W. Dewitt, P. Dabney, M. Molktar, and F.J. Angulo. 1998. Emergence of multidrug-resistant *Salmonella enterica* serotype *typhimurium* DT104 infections in the United States. *N. Engl. J. Med.* 338(19):1333–1338.

Griggs, D., K. Gensberg, and L.J. Piddock. 1996. Mutations in *gyrA* gene of quinolone-resistant *Salmonella* serotypes isolated from humans and animals. *Antimicrob. Ag. Chemother.* 40(4):1009–1013.

Hall, R.M., and H.W. Stokes. 1993. Integrons: novel DNA elements which capture genes by site-specific recombination. *Genetica* 90(2–3):115–132.

Imberechts, H., M.D. Filette, C. Wray, Y. Jones, C. Godard, and P. Pohl. 1998. *Salmonella typhimurium* phage type DT104 in Belgian livestock. *Vet. Rec.* 143(15):424–425.

Khakhria, R., G. Bezanson, D. Duck, and H. Lior. 1983. The epidemic spread of *Salmonella typhimurium* phage type 10 in Canada (1970–1979). *Canadian J. Microbiol.* 29(11):1583–1588.

Khan, A., M. Nawaz, S. Khan, and C. Cerniglia. 2000. Detection of multidrug-resistant *Salmonella typhimurium* DT104 by multiplex polymerase chain reaction. *FEMS Microbiol. Lett.* 182(2):355–360.

Kim, E., and T. Aoki. 1996a. Sulfonamide resistance gene in a transferable R plasmid of *Pasteurella piscicida. Microbiol. and Immunology* 40(5):397–399.

Kim, E., and T. Aoki. 1996b. Sequence analysis of the florfenicol resistance gene encoded in the transferable R-plasmid of a fish pathogen, *Pasteurella piscicida. Microbiol. and Immunology* 40(9):665–669.

Lampson, B., M. Viswanathan, and M. Inouye. 1990. Reverse transcriptase from *Escherichia coli* exists as a complex with msDNA and is able to synthesize double-stranded DNA. *J. Biol. Chem.* 265(15):8490–8496.

Lee, C.A., B.D. Jones, and S. Falkow, S. 1992. Identification of a *Salmonella typhimurium* invasion locus by selection for hyperinvasive mutants. *Proc. Natl. Acad. Sci. USA* 89(5):1847–1851.

Low, J.C., B. Tennant, and D. Munro. 1996. Multiple-resistant *Salmonella typhimurium* DT104 in cats. *Lancet* 348(9038):1391.

Low, J.C., M. Angus, G. Hopkins, D. Munro, and S.C. Rankin. 1997. Antimicrobial resistance of *Salmonella enterica* Typhimurium DT104 isolates and investigation of strains with transferable apramycin resistance. *Epidemiol. Infect.* 118(2):97–103.

Ma, D., M. Alberti, C. Lynch, H. Nikaido, and J.E. Hearst. 1996. The local repressor AcrR plays a modulating role in the regulation of *acrAB* genes of *Escherichia coli* by global stress signals. *Mol. Microbiol.* 19(1):101–112.

Martin, R.G., and J.L. Rosner. 1995. Binding of purified multiple antibiotic-resistance repressor protein (MarR) to mar operator sequences. *Proc. Natl. Acad. Sci. USA* 92(12):5456–5460.

Molbak, K., D. Baggesen, F. Aarestrup, J. Ebbesen, J. Engberg, K. Frydendahl, P. Gerener-Smidt, A.M. Petersen, and H.C. Wegener. 1999. An outbreak of multidrug-resistant, quinolone-resistant *Salmonella enterica* serotype typhimurium DT104. *N. Engl. J. Med.* 341(19):1420–1425.

Passaro, D., R. Reporter, L. Mascola, L. Kilman, G. Malcolm, H. Rolka, S.B. Werner, and D.J. Vugia. 1996. Epidemic *Salmonella enteritidis* infection in Los Angeles County, California. The predominance of phage type 4. *West. J. Med.* 165(3):126–130.

Poppe, C., N. Smart, R. Khakhria, W. Johnson, J. Spika, and J. Prescott. 1998. *Salmonella typhimurium* DT104: a virulent and drug-resistant pathogen. *Canadian Vet. J.* 39(9):559–565.

Pritchett, L., M. Konkel, J. Gay, and T.E. Besser. 2000. Identification of DT104 and U302 phage types among *Salmonella enterica* serotype typhimurium isolates by PCR. *J. Clin. Microbiol.* 38(9):3484–3489.

Randall, L., and M. Woodward. 2001. Multiple antibiotic resistance (*mar*) locus in *Salmonella enterica* serovar Typhimurium DT104. *Appl. Environ. Microbiol.* 67(3):1190–1197.

Recchia, G.D., and R.M. Hall. 1997. Origins of the mobile gene cassettes found in integrons. *Trends Microbiol.* 5(10):389–394.

Reyna, F., M. Huesca, V. Gonzalez, and L.Y. Fuchs. 1995. *Salmonella typhimurium gyrA* mutations associated with fluoroquinolone resistance. *Antimicrob. Ag. Chemother.* 39(7):1621–1623.

Ridley, A., and E.J. Threlfall. 1998. Molecular epidemiology of antibiotic resistance genes in multiresistant epidemic *Salmonella typhimurium* DT 104. *Microb. Drug Resist.* 4(2):113–118.

Sandvang, D., F.M. Aarestrup, and L.B. Jensen. 1998. Characterisation of integrons and antibiotic resistance genes in Danish multiresistant *Salmonella enterica* Typhimurium DT104. *FEMS Microbiol. Lett.* 160(1):37–41.

Schmieger, H., and P. Schicklmaier. 1999. Transduction of multiple drug resistance of *Salmonella enterica* serovar *typhimurium* DT104. *FEMS Microbiol. Lett.* 170(1):251–256.

Schwan, W., X. Huang, L. Hu, and D.J. Kopecko. 2000. Differential bacterial survival, replication, and apoptosis-inducing ability of *Salmonella* serovars within human and murine macrophages. *Infect. Immun.* 68(3):1005–1013.

Slutsker, L., S.F. Altekruse, and D.L. Swerdlow. 1998. Foodborne diseases. Emerging pathogens and trends. *Infect. Dis. Clin. N. Amer.* 12(1):199–216.

Sulavik, M.C., M. Dazer, and P.F. Miller. 1997. The *Salmonella typhimurium* mar locus: molecular and genetic analyses and assessment of its role in virulence. *J. Bacteriol.* 179(6):1857–1866.

Threlfall, E.J., B. Rowe, J.L. Ferguson, and L.R. Ward. 1985. Increasing incidence of resistance to gentamicin and related aminoglycosides in *Salmonella typhimurium* phage type 204c in England, Wales and Scotland. *Vet. Rec.* 117(14):355–357.

Toro, C., S. Lobos, I. Calderon, M. Rodriguez, and G. Mora. 1990. Clinical isolate of a porinless *Salmonella typhi* resistant to high levels of chloramphenicol. *Antimicrob. Ag. Chemother.* 34(9):1715–1719.

Tosini, F., P. Visca, I. Luzzi, A. Dionisi, C. Pezzella, A. Petrucca, and A. Carattoli. 1998. Class 1 integron-borne multiple-antibiotic resistance carried by IncFI and IncL/M plasmids in *Salmonella enterica* serotype typhimurium. *Antimicrob. Ag. Chemother.* 42(12):3053–3058.

Walker, R., N. Saunders, A. Lawson, E. Lindsay, M. Dassama, L. Ward, M.J. Woodward, R.H. Davies, E. Liebana, and E.J. Threlfall. 2001. Use of a LightCycler *gyrA* mutation assay for rapid identification of mutations conferring decreased susceptibility to ciprofloxacin in multiresistant *Salmonella enterica* serotype Typhimurium DT104 iso-

lates. *J. Clin. Microbiol.* 39(4):1443–1448.

Wall, P., D. Morgan, K. Lamden, M. Ryan, M. Griffin, E. Threlfall, L.R. Ward, and B. Rowe. 1994. A case control study of infection with an epidemic strain of multiresistant *Salmonella typhimurium* DT104 in England and Wales. *Commun. Dis. Rep. CDR Rev.* 4(11):R130–R135.

White, D., C. Hudson, J. Maurer, S. Ayers, S. Zhao, M. Lee, L. Bolton, T. Foley, and J. Sherwood. 2000. Characterization of chloramphenicol and florfenicol resistance in *Escherichia coli* associated with bovine diarrhea. *J. Clin. Microbiol.* 38(12):4593–4598.

White, D., S. Zhao, R. Sudler, S. Ayers, S. Friedman, S. Chen, P.F. McDermott, S. McDermott, D.D. Wagner, and J. Meng. 2001. The isolation of antibiotic-resistant *Salmonella* from retail ground meats. *N. Engl. J. Med.* 345(16):1147–1154.

Winokur, P., D. Vonstein, L. Hoffman, E. Uhlenhopp, and G.B. Doern. 2001. Evidence for transfer of CMY-2 AmpC beta-lactamase plasmids between *Escherichia coli* and *Salmonella* isolates from food animals and humans. *Antimicrob. Ag. Chemother.* 45(10):2716–2722.

Wray, C., and W.J. Sojka. 1978. Experimental *Salmonella typhimurium* in calves. *Res. Vet. Sci.* 25(2):139–143.

Wu, M.T., S.A. Carlson, and D.K. Meyerholz. 2002. Identification and characterization of a cytotoxin in multiple antibiotic resistant *Salmonella enterica* serotype Typhimurium phagetype DT104. *Microb. Pathogen.*, in press.

Zhao, S., D.G. White, P.F. McDermott, S. Friedman, L. English, S. Ayers, J. Meng, J.J. Maurer, R. Holland, and R.D. Walker. 2001. Identification and expression of cephamycinase bla(CMY) genes in *Escherichia coli* and *Salmonella* isolates from food animals and ground meat. *Antimicrob. Ag. Chemother.* 45(12):3647–50.

PART IV

ESCHERICHIA COLI O157:H7

15

THE EPIDEMIOLOGY OF *ESCHERICHIA COLI* O157:H7

JAN M. SARGEANT AND DAVID R. SMITH

The enterohemorrhagic *Escherichia coli* (EHEC) are organisms that cause various forms of human illness including mild or self-limiting diarrhea, hemorrhagic colitis, hemolytic uremic syndrome (HUS), and thrombotic thrombocytopenic purpura (TTP) (Boyce et al. 1995). The young and elderly are especially susceptible to severe disease or death following infection. Humans may become infected with EHEC following direct contact with infected humans or animals, or indirectly after consuming contaminated food or water (Feng 1995). Animals, cattle in particular, appear to provide an important reservoir for human exposure (Armstrong et al. 1996).

THE ENTEROHEMORRHAGIC *E. COLI* AND HUMAN DISEASE.
It is important to consider the epidemiology of human EHEC disease to understand the causal role of animal agriculture.

Clinical Presentation of EHEC Infection. Assuming that 95 percent of cases are unreported, each year as many as 110,220 people in the United States may be sickened and 91 people may die as the result of EHEC infection (Mead et al. 1999). Most illness caused by

EHEC infection presents as diarrhea or hemorrhagic colitis (Boyce et al. 1995). Illness associated with human EHEC infection is more likely to be severe in the very young (CDC 1997a; Paton and Paton 1998). In the United States, approximately two-thirds of human EHEC infections are caused by *E. coli* O157:H7 (Mead et al. 1999), although a survey of diarrheal stool samples from Nebraska found non-O157 serogroups as prevalent as O157 serogroups (Fey et al. 2000).

Approximately 5–10 percent of individuals with hemorrhagic colitis progress to HUS, especially for children less than five years of age (CDC 1997a; Paton and Paton 1998). Clinically, HUS is characterized by hemolytic anemia, thrombocytopenia, oliguria-anuria, and, rarely, seizures (Tan et al. 2001). In the United States, *E. coli* O157:H7 infection is a leading cause of HUS and the most common cause of acute kidney failure in children (Altekruse et al. 1997). Three to 5 percent of HUS patients die, and 12 percent suffer severe sequella including end-stage renal disease, hypertension, and neurologic injury (Altekruse et al. 1997). Infected adults may develop TTP; the clinical signs of TTP resemble HUS but also include neurologic abnormalities (Tan et al. 2001).

Virulence of the EHEC. The factors that provide for the virulence of the EHEC include mechanisms for attachment to gut epithelium and the production of cellular toxins (Levine 1987; Muhldorfer and Hacker 1994; Paton and Paton 1998). The genetic characteristics of EHEC are the presence of a locus for enterocyte effacement (LEE), a large plasmid that encodes for a hemolysin, and genes coding for cytotoxins known as verotoxins or Shiga toxins (Stx1 and Stx2) (Levine 1987).

As with the enteropathogenic *E. coli* (EPEC), the EHEC cause intestinal attaching and effacing (A/E) lesions (Levine 1987). LEE is a large cluster of genes responsible for intestinal cell attachment and effacing lesions. Intimin, an outer membrane protein encoded by the eaeA gene, serves as the mediator of attachment and is necessary for A/E lesions (Goosney et al. 1999). To facilitate attachment, the EPEC and EHEC insert a receptor protein, Tir, into the mammalian host cells, allowing intimin to attach to the cell surface and then trigger other host signaling events (Park et al. 1999).

The EHEC primarily differ from EPEC because they produce Shiga toxin (Levine 1987; Muhldorfer and Hacker 1994; Paton and Paton 1998). Shiga-toxin-producing *E. coli* are associated with severe human disease (Park et al. 1999) including HUS and TTP (Boyce et al. 1995; Paton and Paton 1998). Intestinal EHEC infection results in inflammatory damage to the large intestinal epithelium and allows Stx to bind to intestinal endothelial cells, inducing vascular thrombosis, ischemia, and worsening dysentery (O'Loughlin 1997; Paton and Paton 1998). Damage to tissue in distant organs such as the brain, kidney, pancreas, or skin occurs if sufficient Stx is absorbed into the systemic circulation (O'Loughlin 1997).

Escherichia coli O157:H7 is the predominant EHEC serotype causing human disease in North America and Europe. Other EHEC serotypes, including O111:H- (Park et al. 1999) and O26:H11 (Paton and Paton 1998) are also responsible for human illness (Fey et al. 2000). Some researchers synonymously classify EHEC as either verotoxigenic *E. coli* (VTEC) or Shiga-toxin-producing *E. coli* (STEC) without regard to somatic (O) and flagellar (H) antigen serotyping. Further, the methods used to define *E. coli* O157:H7 vary between studies; some authors report the presence of Shiga toxin genes, and some authors exclude testing for the H7 antigen.

Temporal Trends of Human EHEC Disease. It appears, based on increased reporting of *E. coli* O157–related illnesses, that the organism is an emerging disease pathogen in humans (Armstrong et al. 1996). Less clear, however, is whether the increased reporting of *E. coli* O157 illness is an artifact of increased awareness and better diagnostic methods or is due to a true increase in infections—perhaps because of an increasingly susceptible population, recent introduction of the agent to human or animal populations, or greater opportunities for transmission (Armstrong et al. 1996).

The incidence of HUS has been used to study trends in disease occurrence. This outcome measure is particularly well suited for examining the emergence of EHEC-related diseases because the clinical presentation of HUS is fairly distinctive and perhaps less susceptible to misdiagnosis than other related illnesses. The annual incidence of HUS in King County, Washington, averaged over five-year periods, increased significantly from 0.69 cases per 100,000 children under age fifteen (1971–1975) to 1.77 cases per 100,000 (1976–1980) and then was unchanged at 1.74 cases per 100,000 from 1981 to 1986 (Tarr et al. 1989). A study of Minnesota children under eighteen revealed a significant increase in incidence of HUS from a low of 0.5 cases per 100,000 in 1979 to a high of 2.0 cases per 100,000 in 1988. Most of the increase in HUS incidence was observed in children less than five years of age (Martin et al. 1990). However, a study of Utah children under eighteen found no temporal trend in the incidence of HUS over a twenty-year period from 1971 to 1990 (Siegler et al. 1994).

Surveillance of laboratory-confirmed *E. coli* O157–related illnesses in the selected regions of the United States from 1996–2000 show a relatively stable annual incidence of *E. coli* O157 infection, ranging from 2.1 to 2.9 diagnosed infections per 100,000 population (CDC 2001a). However, a seasonal pattern is evident with greater rates of disease occurring during the summer months (Martin et al. 1990; CDC 1999a).

Sources of EHEC Transmission. Humans may become infected with EHEC following direct contact with infected humans or animals, or indirectly after consuming contaminated food or water (Feng 1995). A summary of 139 outbreak investigations conducted between 1982 and 1996 revealed twenty-eight outbreaks associated with person-to-person contact, fifty-three outbreaks attributed to consumption of beef or unpasteurized milk, nineteen associated with the consumption of produce, ten with swimming in pools or ponds, and three with consumption of contaminated water (Sparling 1998).

Direct human-to-human transmission may occur secondary to cases of foodborne illness; however, the cause of the index case is often unknown. Attendance at daycare centers (Spika et al. 1986; Belongia et al. 1993) and swimming in lakes and pools (Keene et al. 1994; Hildebrand and Maguire 1996; CDC 1996a; Friedman et al. 1999) have been linked to human illness, presumably due to direct transmission. The importance of direct transmission between humans has been demonstrated in case-control studies that found that the greatest risk factors for the development of childhood HUS was having previous exposure to individuals with diarrhea (Rowe et al. 1993) or attending large daycare centers (Martin et al. 1990). Children recovering from *E. coli* O157:H7 infection have shed the organism in their feces for many weeks and therefore may serve as an important source of infection to others (Belongia et al. 1993).

Direct exposure to animals may present an important risk for human *E. coli* O157 infection. Infection with VTEC, including *E. coli* O157:H7, has been demonstrated among dairy farm families and their cattle, although the infections were not associated with illness (Wilson et al. 1998). Farm visitors are assumed to be at greater risk for disease because they might lack protective immunity from prior exposure. *Escherichia coli* O157 infection and illness, especially in children, have been documented following visits to farms (Prichard et al. 2000; CDC 2001b) or rural areas (Licence et al. 2001). Human infection with *E. coli* O157 has been associated with exposure to species other than cattle, including horses (Chalmers et al. 1997) and deer (Keene et al. 1997a).

The EHEC are most widely recognized as a cause of foodborne illnesses. The first time that *E. coli* O157:H7 was recognized as a significant human pathogen was in 1982 after outbreaks of gastrointestinal illness related to the consumption of fast-food hamburgers (Riley et al. 1983). Since that time, consumption of ground beef has continued to be an important risk factor for sporadic (Slutsker et al. 1998) and outbreak-associated EHEC illnesses (Ryan et al. 1986; Bell et al. 1994; Cieslak et al. 1997; CDC 1997b; Tuttle et al. 1999).

Not all EHEC-related foodborne illnesses are related to beef consumption. In a single large outbreak affecting more than 8,300 school children and others in Japan, the consumption of white radish sprouts originating from a single farm was epidemiologically associated with *E. coli* O157:H7 illness (Michino et al. 1999). Foodborne illness has been associated with a number of EHEC-contaminated raw foods including fresh vegetables (Ackers et al. 1998), unpasteurized cider (Besser et al. 1993; CDC 1996b), and unpasteurized milk from cattle (Keene et al. 1997b; Clark et al. 2000) and goats (Bielaszewska et al. 1997). An outbreak of human illness was associated with jerky prepared from dehydrated, hunter-killed venison (Keene et al. 1997a).

Acidity is an important hurdle to bacterial growth for many foods (Genigeorgis 1981). However, foodborne illness has occurred after consuming low pH foods such as cider (Besser et al. 1993; CDC 1996b) and dry-cured salami (CDC 1995; Tilden, Jr. et al. 1996; Williams et al. 2000) that were contaminated with *E. coli* O157:H7. The organism is relatively acid-tolerant and may survive for weeks at pH 4.2, thereby enhancing the organism's ability to survive in food and in the environment (Glass et al. 1992; Attenborough and Matthews 2000).

Consumption of drinking water contaminated with *E. coli* O157 has resulted in illness and death. Two notable outbreaks occurred in Cabool, Missouri (Swerdlow et al. 1992), and Walkerton, Ontario (Kondro 2000). In each outbreak, several deaths and hundreds of illnesses were attributed to *E. coli* O157–contaminated municipal water supplies. The outbreak in Cabool was attributed to sewage contamination of the water lines, and in Walkerton, faulty municipal water

sanitation practices apparently allowed manure in surface run-off from nearby farms to enter the ground water. Leakage from a dormitory septic system may have been the cause for contamination of unchlorinated well water at a fairgrounds in Washington County, New York, resulting in an outbreak of 921 illnesses and two deaths among fairgoers who consumed drinks prepared from the water. Stool cultures recovered *E. coli* O157:H7 from 116 persons, and *Campylobacter jejuni* may have been responsible for some illness (CDC 1999b). Visitors to rural areas have become ill after consuming *E. coli* O157–contaminated water obtained from a spring (Licence et al. 2001). A sixteen-month-old child from an Ontario farm family developed HC apparently after drinking water from a contaminated well. Genetically related isolates of *E. coli* O157:H7 were recovered from the child, cattle on the farm, and water from the well (Jackson et al. 1998).

Humans may be exposed to EHECs from a variety of sources (Chart 1998). However, it is clear that animals, and cattle in particular, play an important role in the transmission of *E. coli* O157:H7 to humans, either directly or by fecal contamination of food or the environment (Altekruse et al. 1998). Further, the percentage of beef carcasses contaminated with *E. coli* O157:H7 has been correlated to the percentage of cattle with the organism present in feces or on their hide (Elder et al. 2000). This relationship suggests that actions taken on the farm to reduce the prevalence of cattle shedding the organism might help to reduce contamination of meat. It is, therefore, important to examine the ecology of EHEC in animal populations.

THE ENTEROHEMORRHAGIC *E. COLI* IN ANIMALS. Animals and their environment may serve as a reservoir for EHEC exposure of humans. Understanding the epidemiology of the organism in animal populations may help prevent human illness.

Prevalence of *E. coli* O157 in Cattle and Cattle Operations. The identification of bovine sources in human cases of *E. coli* O157 illness has led to extensive investigations into the occurrence of this pathogen in cattle operations. Because of differences in sampling methods, laboratory techniques, and laboratory criteria to define an EHEC, comparing prevalence values between studies is difficult. However, point-in-time surveys in North America, with or without the inclusion of the H7 antigen, initially reported a prevalence of fecal shedding in individual animals of less than 6 percent in dairy cattle (Wells et al. 1991; Hancock et al. 1994; Zhao et al. 1995), range beef cattle (Hancock et al. 1994; Sargeant et al. 2000), and feedlot beef cattle (Hancock et al. 1994; Hancock et al. 1997a; Galland et al. 2001). Similar prevalence levels in cattle were reported in surveys in European cattle herds (Montenegro et al. 1990; Blanco et al. 1996; Vold et al. 1998; Lahti et al. 2001). *Escherichia coli* O157 have been recovered from the oral cavity of cattle (Gray et al.

2000) and rumen contents (Rasmusssen et al. 1993). Continued improvements have been made in detection methods for *E. coli* O157 (Meyer-Broseta et al. 2001; Rasmussen and Casey 2001). Recent surveys have found a much higher prevalence of fecal shedding of *E. coli* O157 in individual cattle. A study of weaned beef calves reported an overall prevalence of 7.4 percent, with up to 20 percent of calves positive in some herds (Laegreid et al. 1999). In a study of fed cattle at slaughter, a 28 percent prevalence was reported (Elder et al. 2000). Recently reported prevalence levels in feedlot cattle range from 10-23 percent (National Animal Health Monitoring System 2001; Sargeant et al. 2001; Smith et al. 2001). Serological evidence suggests that most cattle have been exposed to *E. coli* O157 (Laegreid et al. 1999). It is not clear whether the apparent increases in reported prevalence are caused entirely by improvements in diagnostic techniques or because the prevalence in cattle populations has actually increased.

At the herd level, *E. coli* O157 is ubiquitous in cattle operations. Longitudinal sampling has resulted in the identification of *E. coli* O157 in the feces of cattle in most, if not all, dairy and beef cattle operations (Hancock et al. 1997b; Sargeant et al. 2000). Even studies that sampled herds at a single point in time have observed herd-level prevalence values of 87 to 100 percent (Laegreid et al. 1999; National Animal Health Monitoring System 2001; Sargeant et al. 2001; Smith et al. 2001). In spite of the ubiquity of the organism in feedyards the pen-level prevalence of fecal shedding varies widely even within a season (Smith et al. 2001).

Epidemiology of *E. coli* O157 in the Bovine. Although the epidemiology of *E. coli* O157 in the bovine is not completely understood, several consistent observations have emerged. The prevalence of fecal shedding of *E. coli* O157 appears to vary by age, with higher prevalence values generally reported in younger animals in field studies (Hancock et al. 1994; Wells et al. 1991; VanDonkersgoed et al. 1999). Increased magnitude and duration of fecal shedding in calves versus adults also has been described in experimental challenge studies (Cray and Moon 1995; Wray et al. 2000). However, several field studies have reported a higher prevalence in weaned dairy calves compared to pre-weaned calves (Hancock et al. 1994; Zhao et al. 1995; Garber et al. 1995; Hancock et al. 1997b). The reasons for age-related differences are not known, but may include age-related differences in ruminal development, the level or duration of fecal shedding, or management differences such as dietary factors (Meyer-Broseta et al. 2001).

Longitudinal studies have shown that individual animals shed *E. coli* O157 transiently (Zhao et al. 1995; Besser et al. 1997; Rahn et al. 1997; Shere et al. 1998). Transient shedding has also been observed in calves and adult cattle experimentally inoculated with *E. coli* O157 (Brown et al. 1997; Wray et al. 2000). Calves inoculated with the same strain of *E. coli* O157 on more than one occasion can shed the bacteria after each inoculation, suggesting that previous exposure to the pathogen is not protective (Cray and Moon 1995; Johnson et al. 1996; Sanderson et al. 1999; Wray et al. 2000).

Fecal shedding of *E. coli* O157 in cattle has not been associated with disease in field studies (reviewed in Armstrong et al. 1996). In one experimental challenge study, calves less than thirty-six hours old developed enterocolitis within eighteen hours of inoculation with *E. coli* O157:H7 and had attaching and effacing lesions in the large and small intestine at necropsy (Dean-Nystrom et al. 1997). However, other challenge studies have not reported illness in pre-weaned calves inoculated with *E. coli* O157 (Cray and Moon 1995; Sanderson et al. 1999; Woodward et al. 1999).

Studies in cattle have documented a seasonal pattern to fecal shedding, with higher prevalence values in warmer months (Hancock et al. 1994; Chapman et al. 1997; Hancock et al. 1997b; VanDonkersgoed et al. 1999). A seasonal variation in disease incidence in humans also is apparent, with higher rates observed in the summer months (Wallace et al. 2000). It is not known whether a direct relationship exists between the seasonal patterns in cattle and humans, or whether the increase in human cases observed during the warmer months is caused by behavioral factors in humans. However, a warm season peak in contamination rates with generic *E. coli* and *E. coli* O157 in the feces of cattle and sheep at slaughter, in carcasses, and in meat samples has recently been reported (Chapman et al. 2001).

Molecular typing techniques, such as pulse-field gel-electrophoresis, have been used to investigate the epidemiology of *E. coli* O157 in cattle. A diverse population of genetic subtypes is found in dairy (Faith et al. 1996; Rice et al. 1999) and beef herds (Rice et al. 1999; Galland et al. 2001; Renter et al. 2001a). Multiple subtypes may be present within a farm at a single point in time, and some subtypes appear to be unique to a specific herd (Rice et al. 1999). The subtypes found in a herd may change over time (Faith et al. 1996; Rice et al. 1999). This diversity may be caused by mutation events or the introduction of subtypes that briefly exist in a cattle operation but fail to maintain within the cattle or the environment. When multiple isolates from the same fecal sample are genetically typed, different subtypes may be observed (Faith et al. 1996; Galland et al. 2001). Within herds, subtypes with indistinguishable genetic patterns may persist for months (Faith et al. 1996) or for up to two years (Shere et al. 1998; Rice et al. 1999). A single subtype may occur at a relatively high frequency within a herd (Faith et al. 1996; Shere et al. 1998; Rice et al. 1999). Isolates with indistinguishable subtypes have been found in herds separated geographically by long distances (Faith et al. 1996; Rice et al. 1999). Although it is biologically plausible that this finding is caused by movement of cattle, Rice et al. (1999) did not find a difference in the number of subtypes in dairy farms with cattle introduced to the farm compared to dairy farms with no new animal introductions. Other possible explanations for the finding of identical sub-types over wide geographic regions include movement of wildlife species or humans, or a common source of contamination such as purchased feed.

Non-Bovine Livestock: Swine, Poultry, Sheep. Livestock other than cattle may shed *E. coli* O157 in their feces. *Escherichia coli* O157, both naturally occurring (Kudva et al. 1996; Chapman et al. 1997; Kudva et al. 1997; Heuvelink et al. 1998; Chapman et al. 2001) and following experimental challenge (Cornick et al. 2000), have been isolated from the feces of healthy sheep. The epidemiology in sheep is apparently similar to cattle: fecal shedding is transient, with a higher prevalence observed in the summer; sheep may shed more than one strain of *E. coli* O157 simultaneously; and the strains shed by individual animals change over time (Kudva et al. 1996; Kudva et al. 1997). *Escherichia coli* O157 and other EHECs have been isolated, but not consistently, in studies of other non-ruminant domestic animals such as poultry and swine (Doyle and Schoeni 1987; Chapman et al. 1997; Heuvelink et al. 1999; Pilipcinec et al. 1999). Interestingly, little evidence exists for a role for non-ruminant domestic species in human disease.

Environmental Sources of *E. coli* O157. Cattle are a well-documented source of *E. coli* O157 in human outbreaks of disease. However, debate continues over whether the bovine species is the reservoir for these bacteria. Reservoir species are defined as those without which the agent cannot perpetuate (Martin et al. 1987). The transient nature of fecal shedding in individual animals, the persistence of specific genetic strains of *E. coli* O157 within farms, and the widespread distribution of genetic strains across geographic distances suggests that other reservoirs or niches exist to sustain these bacteria within agricultural environments. Given that existing diagnostic tests for *E. coli* O157 do not have perfect sensitivity, cattle might maintain the bacteria below detection thresholds. However, it is also essential to investigate and understand other potential sources of *E. coli* O157 in the agricultural environment that may be important to the maintenance of these bacteria in cattle populations. To this end, considerable research has occurred in non-bovine sources of *E. coli* O157 such as water, nondomestic species, feed, and soil.

AGRICULTURAL LAND. Possible routes of entry of *E. coli* O157 into the environment include the direct deposition of feces by pastured cattle, spreading of human or animal wastes onto agricultural lands, leeching of bacteria from fertilized pastures into water, and direct contamination of garden produce caused by fertilization with infected feces.

Escherichia coli O157 has been shown to replicate in bovine feces at 22°C and 37°C and survive for up to eight weeks; at 5°C, the bacteria did not grow, but were capable of surviving for up to ten to eleven weeks

(Wang et al. 1996). In addition to *E. coli* O157, other Shiga-toxin-producing serotypes can survive for long periods in bovine feces (Fukushima et al. 1999). Kudva et al. (1998) sequentially isolated *E. coli* O157 from an ovine manure pile over a twenty-one–month period. However, *E. coli* O157 was not detected after seventy-two hours of composting at 45°C (Lung et al. 2001).

Slurry refers to a mixture of manure, urine, feed debris, and water (Kudva et al. 1998). Slurry may be applied directly to land or following treatment by aeration or retention for a period of time. Even though *E. coli* O157 can survive for short periods (< 9 days) in cattle slurry under experimental conditions, survival is reduced in cattle slurry compared to manure (Maule 1997; Kudva et al. 1998). Aeration of slurry significantly reduces pathogen survival (Kudva et al. 1998).

Laboratory models have shown that *E. coli* O157 may persist for at least 130 days in soil cores containing rooted grass at 18°C, although the organism survived for shorter periods in sieved, grass-free soil (Maule 1999). Other experimental models have documented survival in soil for periods longer than two weeks, with survival times influenced by temperature, moisture content, soil type, and method of pathogen delivery (Gagliardi and Karns 2000; Odgen et al. 2001).

WATER. *Escherichia coli* O157 have been identified in various types of cattle water sources, including troughs, ponds, tanks, and free-flowing sources such as streams and creeks (Faith et al. 1996; Hancock et al. 1998; Sargeant et al. 2000; LeJeune et al. 2001a). LeJeune et al. (2001b) used an experimental microcosm to simulate cattle water troughs and found that *E. coli* O157 could survive, persist, and remain infectious to calves for months in these microcosms. Using pulse-field gel electrophoresis to determine genetic strain, identical strains of *E. coli* O157 have been identified in water troughs and cattle on the same farm (Faith et al. 1996; Hancock et al. 1998; Shere et al. 1998).

The role played by water in the survival or transmission of *E. coli* O157 is not entirely known. Cattle water sources could conceivably be contaminated by cattle feces, cud, saliva, or by mechanical vectors such as feed and bedding or other animals with access to the water source. The extent to which water serves as a transmission point between cattle, and the extent to which water serves as a bacterial multiplication point, are unknown. However, water may be an important component of on-farm control programs. Additionally, free-flowing water sources may encompass multiple farm operations, and watersheds may encompass both livestock and human habitats.

FEED. The possibility that livestock feed could serve as a vehicle for *E. coli* O157 has been proposed as one explanation for the widespread geographic distribution of identical genetic strains of the bacteria (Lee et al. 1996; Rice et al. 1999). Enteric bacteria, such as *Escherichia coli* and *Salmonella,* have been detected in livestock feeds, both on-farm and at the feed mill, suggesting that fecal contamination of livestock feed occurs (Veldman et al. 1995; Harris et al. 1997; Krytenburg et al. 1998; Lynn et al. 1998). Under experimental conditions, *E. coli* O157 have been demonstrated to survive and replicate in a variety of feeds (Lynn et al. 1998), including poorly fermented silage (Fenlon and Wilson 2000). Several studies have failed to identify *E. coli* O157 in cattle feed samples (Faith et al. 1996; Hancock et al. 1998; Lynn et al. 1998), or failed to show that recovery from feed was associated with the prevalence of cattle shedding the organism (Smith et al. 2001). However, a recent study identified *E. coli* O157 in 16.6 percent of 446 feed samples collected from cattle feed bunks in commercial feedlots (Dodd et al. 2001). In this study, it was not known whether the feed was contaminated in the bunk or prior to feeding. Possible routes of fecal contamination of feed include contamination of the feedstuffs due to manure applications in the field, contamination at the feedmill, contamination during storage perhaps by bird or rodent feces, contamination of equipment used to administer feed, or contamination by cattle feces in the bunk. Identifying when contamination of feeds occurs, and implementing means to control this occurrence, may aid in the on-farm control of *E. coli* O157.

INSECTS. *Escherichia coli* O157 have been isolated from flies on cattle operations (Rahn et al. 1997; Hancock et al. 1998; Iwasa et al. 1999) although the detection methods did not allow a determination as to whether the isolates were from the body surface or from the gut of the flies. Experimental studies have shown that *E. coli* O157 ingested from pure cultures by houseflies were viable in the excreta of the flies for at least three days after feeding (Kobayashi et al. 1999), suggesting that flies may be more than simple mechanical vectors. It is not known whether the amount of bacteria excreted by flies could be sufficient to infect cattle or could lead to replication in feed or water contaminated by flies. Therefore, further work is necessary to determine the extent to which flies or other insects may be involved in the on-farm ecology of *E. coli* O157.

WILDLIFE. In contrast to many other livestock species, cattle on most operations spend some or all of their time outside. This provides the potential for considerable contact with nondomestic animal species. More than half of feedlot producers participating in a national survey considered canines, felines, rodents, small mammals, and birds to be a problem on their operation (National Animal Health Monitoring System 2000). In a survey of dairy cattle producers, more than half reported that deer, dogs, and cats had the opportunity for physical contact with female dairy cattle and/or their feed, mineral, or water supply (National Animal Health Monitoring System 1996). Almost 70 percent of cow-calf producers reported seeing deer within one mile of pastured cattle more than four times per month (National Animal Health Monitoring System 1998). In a national survey, 20.4 percent of livestock or poultry producers reported wildlife-caused losses, with regional differences apparent (Wywialowske 1994).

Escherichia coli O157 have been identified in a number of nondomestic species in livestock environments, including rats (Cizek et al. 1999), birds (Hancock et al. 1998), horses (Hancock et al. 1998), dogs (Hancock et al. 1998), possum (Renter et al. 2000), raccoon (Shere et al. 1998) and deer (Rice et al. 1995; Keene et al. 1997; Sargeant et al. 1999; Fischer et al. 2001; Renter et al. 2001b). *Escherichia coli* O157 also have been isolated from a small percentage of wild birds (primarily gulls) from an urban landfill site and intertidal sediments (Wallace et al. 1997). Many of the reports of *E. coli* O157 in nondomestic species come from small numbers of samples collected in studies investigating the pathogen in cattle populations and environments. Thus the number of samples examined in any one study precludes the determination of prevalence or incidence. However, surveys have been conducted to determine the prevalence of *E. coli* O157 in deer, with reported prevalence values ranging from less than 1 percent to 2.4 percent (Sargeant et al. 1999; Fischer et al. 2001; Renter et al. 2001b). Oral inoculation of young deer with *E. coli* O157 resulted in transient shedding patterns similar to those seen in experimentally infected calves (Fisher et al. 2001). Experimental infection of rats and domestic pigeons illustrated that both species can shed *E. coli* O157 following oral inoculation, with pigeons shedding the bacteria for longer periods than rats (Cizek et al. 2000).

Nondomestic species shedding *E. coli* O157 in their feces may provide a direct source of contamination to humans or may be indirectly involved in the on-farm ecology of these bacteria because of shared water and feed or contamination of cattle environments. Wildlife species are not confined to an area, pen, or even farm. Many of these animals may travel between farms as part of their normal range or may have access to areas where cattle feeds are stored. The finding of identical genetic strains of *E. coli* O157 in isolates from a human case of illness, deer jerky consumed by the patient, and the source carcass for the jerky illustrate that wildlife may be a direct source of infection (Keene et al. 1997). The potential for environmental contamination was illustrated by a study where the feces of experimentally infected rats remained positive for *E. coli* O157 for months when stored under a variety of environmental conditions (Cizek et al. 2000).

Despite the evidence that nondomestic species may be involved in the on-farm ecology of *E. coli* O157, further work is required to clarify the importance of these species in the on-farm survival and dissemination of this pathogen.

SUMMARY AND CONCLUSIONS. Humans have many sources of exposure to *E. coli* O157 including direct contact with infected humans or animals, and consumption of contaminated food or water. Cattle are a major source of contamination for both food and the environment. Within cattle production systems are a myriad of routes of transmission between cattle and their environments. These complex interactions are not well understood.

REFERENCES

Ackers, M.L., B.E. Mahon, E. Leahy, B. Goode, T. Damrow, P.S. Hayes, W.R. Bibb, D.H. Rice, T.J. Barrett, L. Hutwagner, P.M. Griffin, and L. Slutsker. 1998. An outbreak of *Escherichia coli* O157:H7 infections associated with leaf lettuce consumption. *J. Infect. Dis.* 177(6):1588–1593.

Altekruse, S.F., M. L. Cohen, and D.L. Swerdlow. 1997. Emerging foodborne diseases. *Emerg. Infect. Dis.* 3(3):285–293.

Altekruse, S.F., D.L. Swerdlow, and S.J. Wells. 1998. Factors in the emergence of food borne diseases. *Vet. Clinics of N. America: Food Animal Practice* 14(1):1–15.

Armstrong, G.L., J. Hollingsworth, and J.G. Morris.1996. Emerging foodborne pathogens: *Escherichia coli* O157:H7 as a model of entry of a new pathogen into the food supply of the developed world. *Epidemiol. Rev.* 18:29–51.

Attenborough, M., and K.R. Matthews. 2000. Food safety through the meat supply chain. *J. Appl. Microbiol. Symposium Supplement* 88:144S–148S.

Bell, B.P., M. Goldoft, P.M. Griffin, M.A. Davis, D.C. Gordon, P.I. Tarr, C.A. Bartleson, J.H. Lewis, T.J. Barret, and J.G. Wells. 1994. A multistate outbreak of *Escherichia coli* O157:H7-associated bloody diarrhea and hemolytic uremic syndrome from hamburgers. The Washington experience. *J.A.M.A.* 272(17):1349–1353.

Belongia, E.A., M.T. Osterholm, J.T. Soler, D.A. Ammend, J.E. Braun, and K.L. MacDonald. 1993. Transmission of *Escherichia coli* O157:H7 infection in Minnesota child day-care facilities. *J. A.M.A.* 269(7):883–888.

Besser, R.E., S.M. Lett, J.T. Weber, M.P. Doyle, T.J. Barrett, J.G. Wells, and P.M. Griffin. 1993. An outbreak of diarrhea and hemolytic uremic syndrome from *Escherichia coli* O157:H7 in Fresh-Pressed Apple Cider. *J.A.M.A.* 269(17):2217–2220.

Besser, T.E., D.D. Hancock, L.C. Pritchett, E.M. McRae, D.H. Rice, and P.I. Tarr. 1997. Duration and detection of fecal excretion of *Escherichia coli* O157:H7 in cattle. *J. Infect. Dis.* 175:726–729.

Bielaszewska, M., J. Janda, K. Blahova, H. Minarikova, E. Jikova, M.A. Karmali, J. Laubova, J. Sikulova, M.A. Preston, R. Khakhria, H. Karch, H. Klazarova, and O. Nyc. 1997. Human *Escherichia coli* O157:H7 infection associated with the consumption of unpasteurized goat's milk. *Epidemiol. and Infect.* 119:299–305.

Blanco, M., J.E. Blanco, J. Blanco, E.A. Gonzalez, A. Mora, C. Prado, L. Fernandez, M. Rio, J. Ramos, and M.P. Alonso. 1996. Prevalence and characteristics of *Escherichia coli* serotype O157:H7 and other verotoxin-producing *E. coli* in healthy cattle. *Epidemiol. and Infect.* 117:251–257.

Boyce, T.G., D.L. Swerdlow, and P.M. Griffin. 1995. *Escherichia coli* O157:H7 and the Hemolytic-Uremic Syndrome. *N. Engl. J. Med.* 333(6):364–368.

Brown, C.A., B.G. Harmon, T. Zhao, and M.P. Doyle. 1997. Experimental *Escherichia coli* O157:H7 carriage in calves. *Appl. Environ. Microbiol.* 63:27–32.

Centers for Disease Control and Prevention. 1995. *Escherichia coli* O157:H7 outbreak linked to commercially distributed dry-cured salami—Washington and California, 1994. *M.M.W.R.* 44(9):157–160.

Centers for Disease Control and Prevention. 1996a. Lake-associated outbreak of *Escherichia coli* O157:H7—Illinois, 1995. *M.M.W.R.* 45(21):437–439.

Centers for Disease Control and Prevention. 1996b. Outbreak of *Escherichia coli* O157:H7 infections associated with drinking unpasteurized commercial apple juice-British Columbia, California, Colorado, and Washington, October 1996. *M.M.W.R.* 45(44):975.

Centers for Disease Control and Prevention. 1997a. *Escherichia coli* O157:H7 infections associated with eating a nationally distributed commercial brand of frozen ground beef patties and burgers—Colorado, 1997. *M.M.W.R.* 46(33):777–778.

Centers for Disease Control and Prevention. 1997b. Isolation of *E. coli* O157:H7 from sporadic cases of hemorrhagic colitis—United States, 1982. *M.M.W.R.* 46(30):700–704.

Centers for Disease Control and Prevention. 1999a. Incidence of foodborne illnesses: preliminary data from foodborne diseases active surveillance network—United States, 1998. *M.M.W.R.* 48(9):189–194.

Centers for Disease Control and Prevention. 1999b. Outbreak of *Escherichia coli* O157:H7 and *Campylobacter* among attendees of the Washington County Fair—New York, 1999. *M.M.W.R.* 48(36):803.

Centers for Disease Control and Prevention. 2001a. Outbreaks of *Escherichia coli* O157:H7 infections among children associated with farm visits—Pennsylvania and Washington, 2000. *M.M.W.R.* 50(15):293–297.

Centers for Disease Control and Prevention. 2001b. Preliminary Foodnet data on the incidence of foodborne illnesses—selected sites, United States, 2000. *M.M.W.R.* 50(13):241–246.

Chalmers, R.M., R.L. Salmon, G.A. Willshaw, T. Cheasty, N. Looker, I. Davies, and C. Wray. 1997. Vero-cytotoxin-producing *Escherichia coli* O157 in a farmer handling horses. *Lancet* 349(9068):1816.

Chapman, P.A., C.A. Siddons, A.T. Cerdan Malo, and M.A. Harkin. 1997. A 1-year study of *Escherichia coli* O157:H7 in cattle, sheep, pigs, and poultry. *Epidemiol. and Infect.* 119:245–250.

Chapman, P.A., A.T. Cerdan Malo, M. Ellin, R. Ashtom, and M.A. Harkin. 2001. *Escherichia coli* O157 in cattle and sheep at slaughter, on beef and lamb carcasses and in raw beef and lamb products in South Yorkshire, UK. *Intl. J. Food Microbiol.* 64:39–150.

Chart, H. 1998. Are all infections with *Escherichia coli* O157 associated with cattle? *Lancet* 352(9133):1005.

Cieslak, P.R., S.J. Noble, D.J. Maxson, L.C. Empey, O. Ravenholt, G. Legarza, J. Tuttle, M.P. Doyle, T.J. Barret, J.G. Wells, A.M. McNamara, and P.M. Griffin. 1997. Hamburger-associated *Escherichia coli* O157:H7 infection in Las Vegas: a hidden epidemic. *Am. J. Publ. Health* 87(2):176–80.

Cizek, A., P. Alexa, I. Literak, J. Hamrik, P. Novak, and J. Smola. 1999. Shiga toxin-producing *Escherichia coli* O157 in feedlot cattle and Norwegian rats from a large-scale farm. *Letters in Appl. Microbiol.* 28:435–439.

Cizek, A., I. Literak, and P. Scheer. 2000. Survival of *Escherichia coli* O157 in faeces of experimentally infected rats and domestic pigeons. *Letters in Appl. Microbiol.* 31:349–352.

Clark, A., S. Morton, and P. Wright. 1997. A community outbreak of Vero cytotoxin producing *Escherichia coli* O157 infection linked to a small farm dairy. *Commun. Dis. Rept. CDR Review* 7(13):R206–R211.

Cornick, N.A., S.L. Booher, T.A. Casey, and H.W. Moon. 2000. Persistent colonization of sheep by *Escherichia coli* O157 and other *E. coli* pathotypes. *Appl. Environ. Microbiol.* 66:4926–4934.

Cray, W.C., and H.W. Moon. 1995. Experimental infection of calves and adult cattle with *Escherichia coli* O157:H7. *Appl. Environ. Microbiol.* 61:1586–1590.

Dean-Nystrom, E.A., B.T. Bosworth, W.C. Cray, and H.W. Moon. 1997. Pathogenicity of *Escherichia coli* O157:H7 in the intestines of neonatal calves. *Infect. and Immun.* 65:1842–1848.

Dodd, C.C., M.W. Sanderson, J.M. Sargeant, T.G. Nagaraja, and R.P. Church. 2001. Prevalence of *Escherichia coli* O157 in cattle feed. *82nd Annual meeting of the Conference of Research Workers in Animal Diseases.* November 11–13, 2001. St. Louis, MO.

Doyle, M.P., and J.L. Schoeni. 1987. Isolation of *Escherichia coli* O157:H7 from retail fresh meats and poultry. *Appl. Environ. Microbiol.* 53:2394–2396.

Elder, R.O., J.E. Keen, G.R. Siragusa, G.A. Barkocy-Gallagher, M. Koohmaraie, and W.W. Laegreid. 2000. Correlation of enterohemorrhagic *Escherichia coli* O157 prevalence in feces, hides, and carcasses of beef cattle during processing. *Proc. Natl. Acad. Sci., U.S.A.* 97:2999–3003.

Faith, N.G., J.A. Shere, R. Brosch, K.W. Arnold, S.E. Ansay, M.S. Lee, J.B. Luchansky, and C.W. Kaspar. 1996. Prevalence and clonal nature of *Escherichia coli* O157:H7 on dairy farms in Wisconsin. *Appl. Environ. Microbiol.* 62:1519–1525.

Feng P. 1995. *Escherichia coli* serotype O157:H7: novel vehicles of infection and emergence of phenotypic variants. *Emerg. Infect. Dis.* 1(2):47–52.

Fenlon, D.R., and J. Wilson. 2000. Growth of *Escherichia coli* O157 in poorly fermented laboratory silage: A possible environmental dimension in the epidemiology of *E. coli* O157. *Letters in Appl. Microbiol.* 30:118–121.

Fey, P.D., R.S. Wickert, M.E. Rupp, T.J. Safranek, and S.H. Hinrichs. 2000. Prevalence of Non-O157:H7 Shiga Toxin-Producing *Escherichia coli* in Diarrheal Stool Samples from Nebraska. *Emerg. Infect. Dis.* 6(5):530–533.

Fischer, J.R., T. Zhao, M.P. Doyle, M.R. Goldberg, C.A. Brown, C.T. Sewell, D.M. Kavanaugh, and C.D. Bauman. 2001. Experimental and field studies of *Escherichia coli* O157:H7 in while-tailed deer. *Appl. Environ. Microbiol.* 67:1218–1224.

Friedman, M.S., T. Roels, J.E. Koehler, L. Feldman, W.F. Bibb, and P. Blake. 1999. *Escherichia coli* O157:H7 outbreak associated with an improperly chlorinated swimming pool. *Clin. Infect. Dis.* 29:298–303.

Fukushima, H., K. Hoshina, and M.Gomyoda. 1999. Long-term survival of Shiga toxin-producing *Escherichia coli* 026, 0111, and 0157 in bovine feces. *Appl. Environ. Microbiol.* 65:5177–5181.

Gagliardi, J.V., and J.S. Karns. 2000. Leaching of *Escherichia coli* O157:H7 in diverse soils under various agricultural management practices. *Appl. Environ. Microbiol.* 66:877–883.

Galland, J.C., D.R. Hyatt, S.S. Crupper, and D.W. Acheson. 2001. Prevalence, antibiotic susceptibility, and diversity of *Escherichia coli* O157:H7 isolates from a longitudinal study of beef cattle feedlots. *Appl. Environ. Microbiol.* 67:1619–1627.

Garber, L.P., S.J. Wells, D.D. Hancock, M.P. Doyle, J. Tuttle, J.A. Shere, and T. Zhao. 1995. Risk factors for fecal shedding of *Escherichia coli* O157:H7 in dairy calves. *J.A.V.M.A.* 207:46–49.

Genigeorgis, C.A. 1981. Factors affecting the probability of growth of pathogenic microorganisms in foods. *J.A.V.M.A.* 179:1410–1416.

Glass, K.A., J.M. Loeffelholz, J.P. Ford, and M.P. Doyle. 1992. Fate of *Escherichia coli* O157:H7 as affected by pH or sodium chloride and in fermented, dry sausage. *Appl. Environ. Microbiol.* 58(8):2513–6.

Goosney, D.L., D.G. Knoechel, and B.B. Finlay. 1999. Enteropathogenic *E.coli, Salmonella,* and *Shigella*: masters of host cell cytoskeletal exploitation. *Emerg. Infect. Dis.* 5(2):216–223.

Gray, J.T., D.R. Smith, R. Moxley, C. Rolfes, L. Hungerford, S. Younts, M. Blackford, D. Bailey, R. Milton, and T. Klopfenstein. 2000. Comparison of oral and rectal sampling to classify the *Escherichia coli* O157:H7 status of pens of feedlot cattle. *A.S.M. 2000 General Meeting.* May 21–25, 2000. Los Angeles, CA.

Hancock, D.D., T.E. Besser, M.L. Kinsel, P.I. Tarr, D.H. Rice, and M.G. Paros. 1994. The prevalence of *Escherichia coli* O157:H7 in dairy and beef cattle in Washington State. *Epidemiol. and Infect.* 133:199–207.

Hancock, D.D., D.H. Rice, L.A. Thomas, D.A. Dargatz, and T.E. Besser. 1997. Epidemiology of *Escherichia coli* O157 in feedlot cattle. *J. Food Prot.* 60:462–465.

Hancock, D.D., T.E. Besser, D.H. Rice, D.E. Herriott, and P.I. Tarr. 1997. A longitudinal study of *Escherichia coli* O157 in fourteen cattle herds. *Epidemiol. and Infect.* 118:193–195.

Hancock, D.D., T.E. Besser, D.H. Rice, E.D. Ebel, D.E. Herriott, and L.V. Carpenter. 1998. Multiple sources of *Escherichia coli* O157 in feedlots and dairy farms in the Northwestern USA. *Prev. Vet. Med.* 35:11–19.

Harris, I.T., P.J. Fedorka-Cray, J.T. Gray, L.A. Thomas, and K. Ferris. 1997. Prevalence of *Salmonella* organisms in swine feed. *J.A.V.M.A.* 210:382–385.

Heuvelink, A.E., F.L. van den Biggelaar, E. de Boer, R.G. Herbes, W.J. Melchers, J.H. Huis in 't Veld, and L.A. Monnens. 1998. Isolation and characterization of verocytotoxin-producing *Escherichia coli* O157 strains from Dutch cattle and sheep. *J. Clin. Microbiol.* 36:878–882.

Heuvelink, A.E., J.T. Zwartkruis-Nahuis, F.L. van den Biggelaar, W.J. van Leeuwen, and E. de Boer. 1999. Isolation and characterization of verocytotoxin-producing *Escherichia coli* O157 from slaughter pigs and poultry. *Intl. J. Food Microbiol.* 52:67–75.

Hildebrand, J.M., and H.C. Maguire. 1996. An outbreak of *Escherichia coli* O157 infection linked to paddling pools. *Commun. Dis. Rpt. CDR review* 6(2):R33–R36.

Iwasa, M., S. Makino, H. Asakura, H. Kobori, and Y. Morimoto. 1999. Detection of *Escherichia coli* O157:H7 from *Musca domestica* (Diiptera: Muscidae) at a cattle farm in Japan. *J. Med. Entomol.* 36:108–112.

Jackson, S.G., R.B. Goodbrand, and R.P. Johnson. 1998. *Escherichia coli* O157:H7 diarrhoea associated with well water and infected cattle on an Ontario farm. *Epidemiol. and Infect.* 120:17–20.

Johnson, R.P., W.C. Cray, and S.T. Johnson. 1996. Serum antibody responses of cattle following experimental infection with *Escherichia coli* O157:H7. *Infect. and Immun.* 64:1879–1883.

Keene, W.E., K. Hedberg, D.E. Herriot, D.D. Hancock, R.W. McKay, T.J. Barret, and D.W. Fleming. 1997. A prolonged outbreak of *Escherichia coli* O157:H7 infections caused by commercially distributed raw milk. *J. Infect. Dis.* 176:815–818.

Keene, W.E., J.M. McAnulty, F.C. Hoesly, L.P. Williams, K. Hedberg, G.L. Oxman, T.J. Barrett, M.A. Pfaller, and D.W. Fleming. 1994. A swimming-associated outbreak of hemorrhagic colitis caused by *Escherichia coli*

O157:H7 and *Shigella sonnei*. *N. Engl. J. Med.* 331(9):579–584.

Keene, W.E., E. Sazie, J. Kok, D.H. Rice, D.D. Hancock, V.K. Balan, T. Zhao, and M.P. Doyle. 1997. An outbreak of *Escherichia coli* O157:H7 infections traced to jerky made from deer meat. *J.A.M.A.* 277:1229–1231.

Kobayashi, M., T. Sasake, N. Saito, K. Tamura, K. Suzuki, H. Watanabe, and N. Agui. 1999. Houseflies: Not simple mechanical vectors of enterohemorrhagic *Escherichia coli* O157:H7. *Am. J. of Trop. Med. and Hygiene* 61:625–629.

Kondro, W. 2000. *E. coli* outbreak deaths spark judicial inquiry in Canada. *Lancet* 355:2058.

Krytenburg, D.S., D.D. Hancock, and D.H. Rice. 1998. A pilot survey of *Salmonella enterica* contamination of cattle feeds in the Pacific northeastern USA. *Animal Feed Sci. and Tech.* 75:75–79.

Kudva, I.T., P.G. Hatfield, and C.J. Hovde. 1996. *Escherichia coli* O157:H7 in microbial flora of sheep. *J. Clin. Microbiol.* 34:431–433.

Kudva, I.T., P.G. Hatfield, C.J. Hovde. 1997. Characterization of *Escherichia coli* O157:H7 and other Shiga toxin-producing *E. coli* serotypes isolated from sheep. *J. Clin. Microbiol.* 35:892–899.

Kudva, I.T., K. Balnch, and C.J. Hovde. 1998. Analysis of *Escherichia coli* O157:H7 survival in ovine or bovine manure and manure slurry. *Appl. Environ. Microbiol.* 64:3166–3174.

Laegreid, W.W., R.O. Elder, and J.E. Keen. 1999. Prevalence of *Escherichia coli* O157:H7 in range beef calves at weaning. *Epidemiol. and Infect.* 123:291–298.

Lahti, E., M. Keskimaki, L. Rantala, P. Hyvonen, A. Siitonen, and T. Honkanen-Buzalski. 2001. Occurrence of *Escherichia coli* O157 in Finnish cattle. *Vet. Microbiol.* 79:239–251.

Lee, M.S., C.W. Kaspar, R. Brosch, J. Shere, and J.B. Luchansky. 1996. Genomic analysis using pulse-field gel electrophoresis of *Escherichia coli* O157:H7 isolated from dairy calves during the United States National Dairy Heifer Evaluation Project (1991–1992). *Vet. Microbiol.* 48:223–230.

LeJeune, J.T., T.E. Besser, N.L. Merrill, D.H. Rice, and D.D. Hancock. 2001. Livestock drinking water microbiology and the factors influencing the quality of drinking water offered to cattle. *J. Dairy Sci.* 84:1856–1862.

LeJeune, J.T., T.E. Besser, and D.D. Hancock. 2001. Cattle water troughs as reservoirs of *Escherichia coli* O157. *Appl. Environ. Microbiol.* 67:3053–3057.

Levine, M.M. 1987. *Escherichia coli* that cause diarrhea: enterotoxigenic, enteropathogenic, enteroinvasive, enterohemorrhagic, and enteroadherent. *J. Infect. Dis.* 155(3):377–389.

Licence, K., K.R. Oates, B.A. Synge, and T.M.S. Reid. 2001. An outbreak of *E. coli* O157:H7 infection with evidence of spread from animals to man through contamination of a private water supply. *Epidemiol. and Infect.* 126:135–138.

Lung, A.J., C.M. Lin, J.M. Kim, M.R. Marshall, R. Nordstedt, N.P. Thompson, and C.I. Wei. 2001. Destruction of *Escherichia coli* O157:H7 and *Salmonella* enteritidis in cow manure composting. *J. Food Prot.* 64:1309–1314.

Lynn, T.V., D.D. Hancock, T.E. Besser, J.H. Harrison, D.H. Rice, N.T. Stewart, and L.L. Rowan. 1998. The occurrence and replication of *Escherichia coli* in cattle feeds. *J. Dairy Sci.* 81:1102–1108.

Martin, D.L., K.L. MacDonald, K.E. White, J.T. Soler, and M.T. Osterholm. 1990. The epidemiology and clinical aspects of the hemolytic uremic syndrome in Minnesota. *N.Engl.J. Med.* 323(17):1161–1167.

Martin, S.W., A.H. Meek, and P. Willegerg. 1987. In *Veterinary Epidemiology: Principles and Methods*. Ames, IA: Iowa State Press, 343.

Maule, A. 1997. Survival of the verotoxigenic strain *E. coli* O157 in laboratory-scale microcosms. In *Coliforms and E. coli: Problem or Solution?* Eds. D. Kay, and C. Fricker. London: Royal Society of Chemistry, 61–65.

Maule, A. 1999. Environmental aspects of *E. coli* O157. *Intl. Food Hygiene* 9:21–23.

Mead, P.S., L. Slutsker, V. Dietz, L.F. McCaig, J.S. Bresee, C. Shapiro, P.M. Griffin, and R.V. Tauxe. 1999. Food-related illness and death in the United States. *Emerg. Infect. Dis.* 5(5):607–625.

Meyer-Broseta, S., S.N. Bastian, P.D. Arne, O. Cerf, and M. Sanaa. 2001. Review of epidemiological surveys on the prevalence of contamination of health cattle with *Escherichia coli* serogroup O157:H7. *Intl. J. of Hygiene and Environ. Health* 203:347–361.

Michino, H., K. Araki, S. Minami, S. Takaya, N. Sakai, M. Miyazaki, A. Ono, and H. Yanagawa. 1999. Massive outbreak of *Escherichia coli* O157:H7 infection in school children in Sakai City, Japan, associated with consumption of white radish sprouts. *Am. J. Epidemiol.* 150(8):787–796.

Montenegro, M.A., M. Bulte, T. Trumpf, S.A. Aleksic, G. Reuter, E. Bulling, and R. Helmuth. 1990. Detection and characterization of fecal verotoxin-producing *Escherichia coli* from healthy cattle. *J. Clin. Microbiol.* 28:1417–1421.

Muhldorfer, I., and J. Hacker. 1994. Genetic aspects of *Escherichia coli* virulence. *Microb. Pathogenesis* 16:171–81.

National Animal Health Monitoring System. 1996. In *Dairy '97. Part I: Reference of 1996 dairy management practices*. p. 35.

National Animal Health Monitoring System. 1998. In *Beef '97. Part III: Reference of 1997 beef cow-calf production management and disease control*. p. 25.

National Animal Health Monitoring System. 2000. In *Feedlot '99. Part III: Health management and biosecurity in U.S. feedlots, 1999*. Section F. Biosecurity.

National Animal Health Monitoring System. 2001. *Escherichia coli* O157 in United States feedlots. USDA:APHIS InfoSheet #N234.2001.

Odgen, I.D., D.R. Fenlon, A.J.A. Vinten, and D. Lewis. 2001. The fate of *Escherichia coli* O157 in soil and its potential to contaminate drinking water. *Intl. J. Food Microbiol.* 66:111–117.

O'Loughin, E. 1997. *Escherichia coli* O157:H7. *Lancet* 349:1553.

Park, S., R.W. Worobom, and R.A. Durstm. 1999. *Escherichia coli* O157:H7 as an emerging foodborne pathogen: a literature review. *Crit. Rev. in Food Sci. and Nutr.* 39(6):481–502.

Paton, J.C., and A.W. Paton. 1998. Pathogenesis and diagnosis of Shiga toxin-producing *Escherichia coli* infections. *Clin. Microbiol. Rev.* 11(3):450–479.

Pilipcinec, E., L. Tkacikova, H.T. Naas, R. Cabadaj, and I. Mikula. 1999. Isolation of verotoxigenic *Escherichia coli* O157 from poultry. *Folia Microbiologica* 44:455–456.

Prichard, G.C., G.A. Willshaw, J.R. Bailey, T. Carson, and T. Cheasty. 2000. Verocytotoxin-producing *Escherichia coli* O157 on a farm open to the public: outbreak investigation and longitudinal bacteriological study. *Vet. Rec.* 147:259–64.

Rahn, K., S.A. Renwick, R.P. Johnson, J.B. Wilson, R.C. Clarke, D. Alves, S. McEwen, H. Lior, and J. Spika. 1997. Persistence of *Escherichia coli* O157:H7 in dairy cattle and the dairy farm environment. *Epidemiol. and Infect.* 119:251–259.

Rasmussen, M.A., W.C. Cray, T.A. Casey, and S.C. Whipp. 1993. Rumen contents as a reservoir of enterohemorrhagic *Escherichia coli*. *FEMS Microbiology Letters* 114:79–84.

Rasmussen, M.A., and T.A. Casey. 2001. Environmental and food safety aspects of *Escherichia coli* O157:H7 infections in cattle. *Crit. Rev. Microbiol.* 27:57–73.

Renter, D.G., J.M. Sargeant, L.L. Hungerford, and S.E. Hygnstrom. 2000. Distribution of *E. coli* O157 in agricultural range environments: preliminary results. *9th Symposium of the International Society for Veterinary Epidemiology and Economics*. Breckenridge, CO.

Renter, D.G., J.M. Sargeant, and R.D. Oberst. 2001. The distribution and diversity of *E. coli* O157 subtypes in agricultural rangelands. *82nd Annual meeting of the Conference of Research Workers in Animal Diseases*. November 11–13, 2001. St. Louis, MO.

Renter, D.G., J.M. Sargeant, S.E. Hygnstrom, J.E. Hoffman, and J.R. Gillespie. 2001. *Escherichia coli* O157:H7 in free-ranging deer. *J. Wildlife Dis.* 37:755–760.

Rice, D.H., D.D. Hancock, and T.E. Besser. 1995. Verotoxigenic *E. coli* O157 colonisation of wild deer and range cattle. *Vet. Rec.* (Letter) 137:524.

Rice, D.H., K.M. McMenamin, L.C. Pritchett, D.D. Hancock, and T.E. Besser. 1999. Genetic subtyping of *Escherichia coli* O157 isolates from 41 Pacific Northwest USA cattle farms. *Epidemiol. and Infect.* 122:479–484.

Riley, L.W., R.S. Remis, S.D. Helgerson, H.B. McGee, J.G. Wells, B.R. Davis, R.J. Hebert, E.S. Olcott, L.M. Johnson, N.T. Hargrett, P.A. Blake, and M.L. Cohen. 1983. Hemorrhagic colitis associated with a rare *Escherichia coli* serotype. *N. Engl. J. Med.* 308(12):681–685.

Rowe, P.C., E. Orrbine, H. Lior, G.A. Wells, and P.N. McLaine. 1993. Diarrhoea in close contacts as a risk factor for childhood haemolytic uraemic syndrome. *Epidemiol. and Infect.* 110:9–16.

Ryan, C.A., R.V. Tauxe, G.W. Hosek, J.G. Wells, P.A. Stoesz, H.W. McFadden, Jr., P.W. Smith, G.F. Wright, and P.A. Blake. 1986. *Escherichia coli* O157:H7 diarrhea in a nursing home: clinical, epidemiological, and pathological findings. *J. Infect. Dis.* 154(4):631–638.

Sanderson, M.W., T.E. Besser, J.M. Gay, C.C. Gay, and D.D. Hancock. 1999. Fecal *Escherichia coli* O157:H7 shedding patterns of orally inoculated calves. *Vet. Microbiol.* 69:199–205.

Sargeant, J.M., D.J. Hafer, J.R. Gillespie, R.D. Oberst, and S.J.A. Flood. 1999. Prevalence of *Escherichia coli* O157:H7 in white-tailed deer sharing rangeland with cattle. *J.A.V.M.A.* 215:792–794.

Sargeant, J.M., J.R. Gillespie, R.D. Oberst, R.K. Phebus, D.R. Hyatt, L.K. Bohra, and J.C. Galland. 2000. Results of a longitudinal study of the prevalence of *Escherichia coli* O157 on cow-calf farms. *Am. J. Vet. Res.* 61:1375–1379.

Sargeant, J.M, D.D. Griffin, R.A. Smith, and M.W. Sanderson. 2001. Associations between management, climate and *Escherichia coli* O157 in feedlot cattle. *82nd Annual meeting of the Conference of Research Workers in Animal Diseases*. November 11–13, 2001. St. Louis, MO.

Shere, J.A., K.J. Bartlett, and C.W. Kaspar. 1998. Longitudinal study of *Escherichia coli* O157:H7 dissemination on four dairy farms in Wisconsin. *Appl. Environ. Microbiol.* 64:1390–1399.

Siegler, R.L., A.T. Pavia, R.D. Christofferson, and M.K. Milligan. 1994. A 20-year population-based study of postdiarrheal hemolytic uremic syndrome in Utah. *Pediatrics* 94(1):35–40.

Slutsker, L., A.A. Ries, K. Maloney, J.G. Wells, K.D. Greene, and P.M. Griffin. 1998. A nationwide case-control study

of *Escherichia coli* O157:H7 infection in the United States. *J. Infect. Dis.* 177:962–966.

Smith, D.R., M. Blackford, S. Younts, R. Moxley, J. Gray, L. Hungerford, T. Milton, and T. Klopfenstein. 2001. Ecological relationships between the prevalence of cattle shedding *Escherichia coli* O157:H7 and characteristics of the cattle or conditions of the feedlot pen. *J. Food Prot.* 64:1899–1903.

Sparling, P.H. 1998. *Escherichia coli* O157:H7 outbreaks in the United States, 1982–1996. *J.A.V.M.A.* 213(12):1733.

Spika, J.S., J.E. Parsons, D. Nordenberg, J.G. Wells, R.A. Gunn, and P.A. Blake. 1986. Hemolytic uremic syndrome and diarrhea associated with *Escherichia coli* O157:H7 in a day care center. *J. Pediatrics* 109(2):287–291.

Swerdlow, D.L., B.A. Woodruff, R.C. Brady, P.M. Griffin, S. Tippen, H.D. Donnell, Jr., E. Geldreich, B.J. Payne, A. Meyer, Jr., and J.G. Wells. 1992. A waterborne outbreak in Missouri of *Escherichia coli* O157:H7 associated with bloody diarrhea and death. *Annals Intern. Med.* 117(10):812–819.

Tan, L.J., J. Lyznicki, P.M. Adcock, E. Dunne, J. Smith, E. Parish, A. Miller, H. Seltzer, and R. Etzel. 2001. Diagnosis and management of food borne illnesses: A primer for physicians. *M.M.W.R.* 50(RR-2):1–69.

Tarr, P.I., M.A. Neill, J. Allen, C.J. Siccardi, S.L. Watkins, and R.O. Hickman. 1989. The increasing incidence of the hemolytic-uremic syndrome in King County, Washington: lack of evidence for ascertainment bias. *Am. J. Epidemiol.* 129(3):582–586.

Tilden, J., Jr., W. Young, A.M. McNamara, C. Custer, B. Boesel, M.A. Lambert-Fair, J. Majkowski, D. Vugia, S.B. Werner, J. Hollingsworth, and J.G. Morris. 1996. A new route of transmission for *Escherichia coli*: infection from dry fermented salami. *Am. J. Publ. Health* 86(8):1142–1145.

Tuttle, J., T. Gomez, M.P. Doyle, J.G. Wells, T. Zhao, R.V. Tauxe, and P.M. Griffin. 1999. Lessons from a large outbreak of *Escherichia coli* O157:H7 infections: insights into the infectious dose and method of widespread contamination of hamburger patties. *Epidemiol. and Infect.* 1999(122):185–192.

VanDonkersgoed, J., T. Graham, and V. Gannon. 1999. The prevalence of verotoxins, *Escherichia coli* O157:H7, and *Salmonella* in the feces and rumen of cattle at processing. *Can. Vet. J.* 40:332–338.

Veldman, A., H.A. Vahl, G.J. Borggreve, and D.C. Fuller. 1995. A survey of the incidence of *Salmonella* species and enterobacteriaceae in poultry feeds and feed components. *Vet. Rec.* 136:169–172.

Vold, L., B. Klungseth Johansen, H. Kruse, E. Skjerve, and Y. Wasteson. 1998. Occurrence of shigatoxinogenic *Escherichia coli* O157 in Norwegian cattle herds. *Epidemiol. and Infect.* 120:21–28.

Wallace, J.S., T. Cheasty, and K. Jones. 1997. Isolation of Vero cytotoxin-producing *Escherichia coli* O157 from wild birds. *J. Appl. Microbiol.* 82:399–404.

Wallace, D.J., T. VanGilder, S. Shallow, T. Fiorentino, S.D. Segler, K.E. Smith, B. Shiferaw, R. Etzel, W.E. Garthright, F.J. Angulo, and Foodnet Working Group. 2000. Incidence of foodborne illnesses reported by the Foodborne Diseases Active Surveillance Network (Foodnet)—1997. *J. Food Prot.* 63:807–809.

Wang, G., T. Zhao, and M.P. Doyle. 1996. Fate of Enterohemorrhagic *Escherichia coli* O157:H7 in bovine feces. *Appl. Environ. Microbiol.* 62:2567–2570.

Wells, J.G., L.D. Shipman, K.D. Greene, E.G. Sowers, J.H. Green, D.N. Cameron, F.P. Downes, M.L. Martin, P.M. Griffin, S.M. Ostroff, M.E. Potter, R.V. Tauxe, and I.K. Wachsmuth. 1991. Isolation of *Escherichia coli* serotype

O157:H7 and other shiga-like-toxin-producing *E. coli* from dairy cattle. *J. Clin. Microbiol.* 29:985–989.

Williams, R.C., S. Isaacs, M.L. Decou, E.A. Richardson, M.C. Buffett, R.W. Slinger, M.H. Brodsky, B.W. Ciebin, A. Ellis, and J. Hockin. *E.coli* O157:H7 Working Group. 2000. Illness outbreak associated with Escherichia coli O157:H7 in Genoa salami. *Can. Med. Assoc. J.* 162(10):1409–1413.

Wilson, J., J. Spika, R. Clarke, S. McEwen, R. Johnson, K. Rahn, S. Renwick, M. Karmali, H. Lior, D. Alves, C. Gyles, and K. Sandhu. 1998. Verocytotoxigenic *Escherichia coli* infection in dairy farm families. *Can. Commun. Dis. Rpt.* 24(3):17–20.

Wray, C., I.M. McLaren, L.P. Randall, and G.R. Pearson. 2000. Natural and experimental infection of normal cattle with *Escherichia coli* O157. *Vet. Rec.* 147:65–68.

Woodward, M.J., D. Gavier-Widen, I.M. McLaren, C. Wray, M. Sozmen, and G.R. Pearson. 1999. Infection of gnotobiotic calves with *Escherichia coli* O157:H7 strain A84. *Vet. Rec.* 144:466–470.

Wywialowski, A.P. 1994. Agricultural producers' perceptions of wildlife-caused losses. *Wildlife Soc. Bull.* 22:370–382.

Zhao, T., M.P. Doyle, J. Shere, and L. Garber. 1995. Prevalence of enterohemorrhagic *Escherichia coli* O157:H7 in a survey of dairy herds. *Appl. Environ. Microbiol.* 61:1290–1293.

16

DETECTION AND DIAGNOSIS OF *ESCHERICHIA COLI* O157:H7 IN FOOD-PRODUCING ANIMALS

RODNEY A. MOXLEY

INTRODUCTION. *Escherichia coli* O157:H7 is an increasingly recognized cause of diarrhea in humans and the bacterial pathogen most commonly isolated from visibly bloody stool specimens (Slutsker et al. 1997). *E. coli* O157:H7 infections are most often food-borne and may cause not only nonbloody diarrhea or bloody diarrhea (hemorrhagic colitis) but also life-threatening post-diarrheal sequelae, such as hemolytic uremic syndrome (HUS) and thrombotic thrombocytopenic purpura (TTP; Armstrong et al. 1996; Karmali et al. 1989; Levine 1987; Nataro and Kaper 1998). *E. coli* O157:H7 is the prototype of a class of diarrheagenic *E. coli* known as enterohemorrhagic (EHEC; Levine 1987; Nataro and Kaper 1998). EHEC strains are thought to cause disease by their ability to attach intimately to the intestinal mucosa and cause *attaching-effacing* lesions, and also by their ability to cause vascular damage through the production of potent toxins known as Shiga toxins (Stx, formerly Shiga-like toxin, SLT; Karmali 1989; Levine 1987; Moon et al. 1983; Nataro and Kaper 1998; O'Brien and Holmes 1987; Paton and Paton 1998). Consequently, these organisms also have been classified as attaching-effacing *E. coli* (AEEC) and Stx-producing *E. coli* (STEC; Levine 1987; Moon et al. 1983; Nataro and Kaper 1998; Paton and Paton 1998). Because Stxs cause toxic effects on Vero cells, these toxins are also commonly known as verotoxins or verocytotoxins (VT) and *E. coli* strains that produce them as VTEC (Karmali 1989). Cattle are an important reservoir of *E. coli* O157:H7 infection for humans, because the organism is highly prevalent in cattle and shed in the feces of these animals (Armstrong et al. 1996; Chapman et al. 1997a; Elder et al. 2000; Smith et al. 2001). Most *E. coli* O157:H7 infections are caused by consumption of contaminated food and water; contact with contaminated environments is also a risk factor, and bovine feces is the most common source of the organism in these cases, both as sporadic cases and epidemics (Armstrong et al. 1996; Borczyk et al. 1987; Chapman et al. 1993; Locking et al. 2001; Michel et al. 1999; Wells et al. 1991).

The first reports of isolation of *E. coli* O157:H7 from cattle occurred while investigating sources of human infection (Borczyk et al. 1987; Chapman, Wright and Norman 1989; Chapman et al. 1993a; Chapman, Wright and Higgins 1993b; Martin et al. 1986; Ostroff et al. 1990; Wells et al. 1983). Increased interest in cattle as reservoir hosts led to further epidemiological studies with resultant improved methods to detect the organism in this species (Chapman et al. 1997b; Elder et al. 2000; Smith et al. 2001). Similar studies have been done on other food-producing animals, including sheep (Chapman et al. 1996, 1997a; Fegan and Desmarchelier 1999; Heuvelink et al. 1998; Kudva et al. 1995; Kudva et al. 1997; Licence et al. 2001), goats (Bielaszewska et al. 1997), deer (Chapman and Ackroyd 1997; Fischer et al. 2001), swine (Heuvelink et al. 1999; Nakazawa and Akiba 1999), turkeys (Heuvelink et al. 1999), and geese (Allison et al. 1997). Epidemiologic studies have included chickens, but the organism was not detected in this species (Chapman et al. 1997a; Heuvelink et al. 1999).

GENERAL CULTURE PROCEDURES AND DETECTION OF *E. COLI* O157:H7 IN BOVINE FECES. Methods for the detection and diagnosis of *E. coli* O157:H7 in food-producing animals are largely modifications of those first used to detect the organism in food and human patients. Several excellent reviews have been published that include methods for detection of the organism in food and human patients (Armstrong et al. 1996; Meng and Doyle 1998; Padhye and Doyle 1992; Paton and Paton 1998; Riley 1987; Strockbine et al. 1998; Su et al. 1995). Described herein is an overview of detection and diagnostic methods for *E. coli* O157:H7 in cattle and other food-animal species.

Sorbitol-MacConkey Agar. In the initial report in which *E. coli* O157:H7 was associated with food-borne outbreaks of hemorrhagic colitis, standard methods for isolation of *E. coli* (for example, direct plating of fecal material onto MacConkey medium) from human patients were used (Riley et al. 1983; Wells et al. 1983). In that study, *E. coli* O157:H7 was observed to not ferment D-sorbitol within seven days of incubation, whereas previous reports had indicated that 93 percent of all *E. coli* strains of human origin did so (Wells et al. 1983). Other workers later noted that *E. coli* O157:H7 strains failed to ferment sorbitol after overnight incubation, whereas approximately 95 percent of other *E. coli* were sorbitol-positive (Farmer and Davis 1985; Harris et al. 1985; March and Ratnum 1986). Farmer and Davis (1985) exploited this feature to differentiate

143

E. coli O157:H7 from other *E. coli* in the direct plating step. They reported that MacConkey base plus 1 percent D-sorbitol (sorbitol-MacConkey agar, SMAC) could be utilized as an effective primary isolation medium for *E. coli* O157:H7.

Non-sorbitol-fermenting organisms, which primarily include non-Stx-producing *E. coli* strains, *Proteus* spp., and *Aeromonas* spp., are normally found in human fecal specimens (Strockbine et al. 1998). To distinguish *E. coli* O157:H7 from these other non-sorbitol-fermenters on SMAC plates, rhamnose was added at a concentration of 0.5 percent (w/v; Chapman et al. 1991). *E. coli* O157:H7 isolates do not usually ferment rhamnose rapidly, in contrast to about 60 percent of other non-sorbitol-fermenting *E. coli* strains (Chapman et al. 1991). To inhibit *Proteus* spp., cefixime was added at a concentration of 0.05 µg/ml, to form cefixime-rhamnose-SMAC (CR-SMAC) (Chapman et al. 1991). To inhibit *Providencia* spp. and *Aeromonas* spp., which are prevalent in the feces of humans and cattle, potassium tellurite was added at a concentration of 2.5 µg/ml, to create CT-SMAC (Zadik et al. 1993). Chapman et al. (1994) found that *E. coli* O157:H7 was easier to obtain in pure culture from CT-SMAC than CR-SMAC in artifically inoculated bovine feces. The mechanism of tellurite resistance was later discovered when it was shown that *E. coli* O157:H7 strains carry a pathogenicity island termed TAI (tellurite resistance- and adherence-conferring island) that contains genes encoding both a novel adherence-conferring protein and tellurite resistance (Tarr et al. 2000).

β-Glucuronidase Tests. Approximately 96 percent of all *E. coli* strains produce functional β-glucuronidase (GUD), in contrast to *E. coli* O157:H7 strains, which produce a nonfunctional form of the enzyme (Doyle and Schoeni 1984; Feng and Lampel 1994; Thompson et al. 1990; Trepeta and Edberg 1984). *E. coli* strains that produce functional GUD can be positively detected by a reaction in which GUD cleaves the substrate 4-methylumbelliferyl-β-D-glucuronide (MUG) to form a fluorescent product (4-methylumbelliferone) that is detectable with long-wave (366-nm) UV light (MUG-positive reaction; Thompson et al. 1990). Feng and Hartman (1982) developed a rapid assay using MUG as an indicator for *E. coli*. Doyle and Schoeni (1984) found that all *E. coli* O157:H7 strains they tested were MUG-negative. Trepeta and Edberg (1984) demonstrated that MUG could be incorporated directly into a modified MacConkey agar, which would allow for the simultaneous testing for lactose fermentation and GUD activity. Krishnan et al. (1987) added MUG to SMAC and found that all *E. coli* O157:H7 isolates from a nursing home outbreak of hemorrhagic colitis were both sorbitol- and MUG-negative. Thompson et al. (1990) found that suspect *E. coli* O157:H7 isolates could be rapidly tested for GUD activity by inoculation of tubes containing MUG medium, incubating for twenty minutes, and checking for fluorescence under long-wave UV light. Okrend et al. (1990) reported that screening

of sorbitol-negative colonies for functional β-glucuronidase reduced the number of false-positive identifications by 36 percent in cultures of ground beef. Feng and Lampel (1994) found that *E. coli* O157:H7 expresses GUD, but the enzyme is nonfunctional. Sequencing of the GUD structural gene (*uidA*) revealed nucleotide base substitutions that were thought to account for the absence of GUD activity. Absence of GUD activity is a consistent trait of *E. coli* O157:H7, as there appears to be only one report of isolation of a sorbitol-negative, MUG-positive *E. coli* O157:H7 strain in the United States in a patient with bloody diarrhea (Hayes et al. 1995).

Sorbitol-Fermenting *E. coli* O157:NM. Although the use of CT-SMAC significantly increased the sensitivity of detection of *E. coli* O157:H7 in fecal specimens, researchers in Germany recognized that EHEC O157:NM isolates from some cases of HUS could not grow on this media, but did grow and rapidly ferment sorbitol on SMAC agar (Karch et al. 1996). These strains were found to lack tellurite resistance and, in addition to the ability to ferment sorbitol within forty-eight hours, produced functional GUD (Karch and Bielaszewska 2001). Although non-motile, these strains have been shown to carry the *fliC* (H7 structural) gene (Bielaszewska et al. 2000). Sorbitol-fermenting EHEC O157:NM strains have mainly been found in Europe but represent the most frequent serotype of STEC isolated from HUS patients in Germany (Karch and Bielaszewska 2001). Sorbitol-fermenting EHEC O157:NM were isolated from a dairy cow in Germany (Bielaszewska et al. 2000). The cow in this case was epidemiologically associated with two HUS patients, and the same strains were isolated from both the cow and the patients. These findings raised the hypothesis that cattle may be a reservoir of this organism (Bielaszewska et al. 2000). Sorbitol-fermenting EHEC O157:NM have not been frequently found in humans or cattle in North America (Strockbine et al. 1998); however, because many investigators have selected only sorbitol-negative colonies in their studies, the prevalence of sorbitol-fermenting EHEC O157:NM in North America may be underestimated.

Enrichment and Sample Size. Sanderson et al. (1995) found that enrichment of bovine feces in tryptic soy broth (TSB) containing cefixime (0.05 µg/ml) and vancomycin (40 µg/ml; TSBcv) or TSB containing the same antibiotics and tellurite (2.5 µg/ml; TSBcvt) prior to plating on CT-SMAC significantly increased the sensitivity of *E. coli* O157:H7 detection in bovine feces. However, no differences between the broth enrichment media (that is, TSBcv and TSBcvt) were found. In directly inoculated fecal samples, enrichment decreased the 50 percent detection limit (CFU/g) from 251 (direct plating) to 13 (TSBcv) or 16 (TSBcvt).

Brown et al. (1997) experimentally inoculated eight-week-old calves with *E. coli* O157:H7 and found that enrichment was necessary to detect low fecal shedding levels that occurred after twenty days post-inoculation. The calves were inoculated with a nalidixic acid-resistant strain of *E. coli* O157:H7, and fecal shedding of this strain was monitored by direct plating of feces on SMAC containing nalidixic acid (NA-SMAC), or by enrichment in TSB containing nalidixic acid (NA-TSB) followed by plating on NA-SMAC. Ten gram fecal samples were obtained and initially mixed with 15 ml of Cary-Blair medium (that is, a 40 percent w/v mixture). Fecal shedding was quantified by direct plating on NA-SMAC, with 0.1 ml of each dilution plated in duplicate. In addition, 0.1 ml of undiluted 40 percent feces-Cary-Blair mixture was plated in quadruplicate on NA-SMAC, yielding a detection limit of 6.25 CFU/g. Enrichment was used to determine whether shedding occurred at levels below that detectable by direct plating. Enrichment used a 10 g fecal sample inoculated into 100 ml NA-TSB. After twenty days post-inoculation, shedding was detectable only by enrichment, reflecting *E. coli* O157:H7 levels ≤5 CFU/g. Similar results were found previously by Cray and Moon (1995).

Sanderson et al. (1995) found that incubation of 10 g fecal samples in TSBcv and subsequent plating on CT-SMAC was more sensitive for detecting *E. coli* O157:H7 in samples from naturally colonized cattle than incubation of swab fecal samples (containing approximately 0.1 g feces) in TSBcv with subsequent plating on SMAC containing cefixime (C-SMAC). More positive samples were detected by using 10 g fecal samples than swab samples for plating on CT-SMAC; however, the difference was not statistically significant.

Immunomagnetic Separation (IMS). Because *E. coli* O157:H7 is often shed in low numbers in the feces of cattle, Chapman et al. (1994) noted that more sensitive enrichment culture methods were needed. In an attempt to improve sensitivity, they tested the effect of IMS following incubation in selective enrichment broth on detection of the organism following CT-SMAC plating (Chapman et al. 1994). Fecal swabs or 100 µl of inoculated fecal suspensions were placed in 5 ml of enrichment culture medium consisting of buffered peptone water supplemented with vancomycin (8 µg/ml), cefixime (0.05 µg/ml), and cefsulodin (10 µg/ml; BPW-VCC) to inhibit the growth of Gram-positive organisms, aeromonads, and *Proteus* spp., respectively. After vortex mixing, broths were incubated at 37°C for 6 h and 1 ml of the culture was added to 20 µl of superparamagnetic polystyrene beads coated with anti-*E. coli* O157 antibodies (Dynabeads, Dynal A. S., Oslo) in a 1.5 ml microcentrifuge tube. Following repeated mixing, magnetic separation, and washing steps, the beads were resuspended in approximately 25 µl of PBS, inoculated onto CT-SMAC medium, and incubated overnight at 37°C. BPW-VCC-IMS followed by CT-SMAC plating (BPW-VCC-IMS-CT-SMAC) was approximately 100-fold more sensitive for detection of *E. coli* O157:H7 than direct culture on either CT-SMAC or CR-SMAC.

During monitoring of a dairy herd for *E. coli* O157:H7, 72.6 percent of all the strains isolated were detected only by the BPW-VCC/IMS/CT-SMAC procedure, compared to only 27.4 percent by direct culture on CT-SMAC, all of which were also detected by BPW-VCC-IMS-CT-SMAC (Chapman et al. 1994). Chapman et al. (1997a) applied the BPW-VCC-IMS-CT-SMAC to an epidemiological study in cattle and demonstrated an overall prevalence of 15.7 percent, a figure higher than had been shown in previous studies, highlighting the importance of sensitive culture methods for more accurate estimation of prevalence.

Sanderson et al. (1995) studied the effect of IMS on detection of a NA-resistant *E. coli* O157:H7 strain in the feces of neonatal dairy calves following inoculation, using TSBcv as the enrichment medium. Swab samples were incubated for 6 h in 3 ml TSBcv; culture medium was then serially diluted in TSB to 10^{-6} and plated on C-SMAC, or to 10^{-3} or 10^{-4} and plated on CT-SMAC. For comparison, 6 h TSBcv cultures were subjected to IMS and plated on CT-SMAC, or 10 g fecal samples were incubated overnight in TSBcv, diluted to 10^{-4}, and plated on CT-SMAC. The 6-h IMS method detected the most positive samples but was not significantly more sensitive than plating dilutions of broth enrichments of swab fecal samples on CT-SMAC.

Chapman et al. (1997a) compared the following procedures for detection of *E. coli* O157:H7 in bovine fecal samples obtained from rectal swabs: direct plating on CT-SMAC, BPW-VCC-CT-SMAC, BPW-VCC-IMS-CT-SMAC, and a commercial enzyme immunoassay (EIA). The EIA used enrichment culture in modified *E. coli* (EC) broth supplemented with novobiocin (20 µg/ml; mECn) prior to heat treatment and detection of *E. coli* O157 antigen by a standard antibody-based EIA. For confirmation of the EIA results, the IMS procedure and CT-SMAC culture were conducted on stored mECn broths used in the EIA. Both the EIA and IMS methods gave a 10- to 100-fold increase in sensitivity compared to direct culture on CT-SMAC or BPW-VCC-CT-SMAC enrichment (without IMS) followed by subculture.

Dynal Biotech, Inc. (Oslo, Norway), the manufacturer of Dynabeads, recommends the use of GN broth for enrichment culture of animal and human fecal samples (10 ml broth per swab) prior to IMS with an incubation time of 5 h at 37°C followed by subculturing onto CT-SMAC plates (Dynabeads product insert). Studies testing the effect of GN broth for enrichment with IMS were initially conducted by Cudjoe (1995). Karch et al. (1996) used GN broth enrichment with IMS to detect *E. coli* O157:H7 in stool samples of human HUS patients. Stool samples were diluted 1:2 in PBS, and 1 ml of the suspension was inoculated into 10 ml GN broth and incubated at 37°C for 4 h; following magnetic separation and washing, the beads were plated on CT-SMAC. This procedure allowed the detection of *E. coli* O157 strains at 10^2 CFU/g of stools in the presence of 10^7 coliform background flora organisms, and detected more positive samples than did cytotoxicity assays for the detection of Stx, or direct PCR for marker genes conducted on DNA extracted from stools. Using this method, *E. coli* O157 strains were isolated from 90 percent of HUS patients with a positive O157 serology, whereas in previous studies, this pathogen was isolated from only 20–26 percent of HUS patients (Karch et al. 1996).

Laegreid et al. (1999) made modifications to the GN-IMS culture methods by increasing the sample size and adding antibiotics previously shown by Chapman et al. (1994) to enhance detection in bovine feces. Laegreid et al. (1999) inoculated 90 ml GN broth supplemented with vancomycin (8 μg/ml), cefixime (0.05 μg/ml), and cefsulodin (10 μg/ml; GN-VCC) with 10 g fecal samples, incubated for 6 h at 37°C, and subcultured onto CT-SMAC plates (GN-VCC-IMS-CT-SMAC). Using this procedure, they found that 87 percent of all range beef herds tested had at least one fecal culture positive isolation of *E. coli* O157:H7, and within positive herds, 7.4 percent of calves shed the organism in the feces at weaning. Also using the GN-VCC-IMS-CT-SMAC procedure, Elder et al. (2000) detected a 28 percent prevalence of *E. coli* O157:H7 in the feces of cattle at slaughter at four midwestern beef processing facilities. Their data revealed the prevalence of *E. coli* O157:H7 in cattle and on carcasses to be much higher than previously estimated (Gansheroff and O'Brien 2000). Similarly, Smith et al. (2001), using the GN-VCC-IMS-CT-SMAC method, found the prevalence of *E. coli* O157:H7 in the feces of cattle in five midwestern U.S. feedlots to be 23 percent.

Heuvelink et al. (1998) used the following culture procedures to isolate *E. coli* O157:H7 from cattle feces. Twenty g of feces was added to 180 ml of modified *E. coli* broth (mEC) containing novobiocin (20 μg/ml) (mECn). Enrichment cultures were incubated 18 to 20 h at 37°C with aeration, serially diluted, and then plated onto CT-SMAC. Alternatively, 6 h enrichment cultures were made, subjected to IMS, and then plated on CT-SMAC. CT-SMAC plates were incubated at 42°C for 20 h, and sorbitol-non-fermenting colonies were subcultured onto SMAC-MUG and Levine's eosin methylene blue agar (L-EMB). MUG-negative colonies that also had a metallic sheen on L-EMB were tested further by latex agglutination, API 20E, serotyping, and confirmation by molecular methods. The protocol that included IMS was 7-fold more sensitive, detecting the organism in 10.0 and 11.1 percent of adult Dutch cattle in 1995 and 1996, respectively.

Other investigators used minor variations of the aforementioned procedures to isolate *E. coli* O157:H7 from bovine feces (Besser et al. 1997; Hancock et al. 1994; Martin et al. 1986; McDonough et al. 2000; Rahn et al. 1997; Shere et al. 1998; Wells et al. 1991).

Detection of *E. coli* O157 and H7 Antigens. In the initial investigation in which *E. coli* O157:H7 was associated with hemorrhagic colitis, isolates were serotyped by standard procedures (Wells et al. 1983). Soon thereafter, more expeditious means of detecting

the *E. coli* O157 and H7 antigens were sought. Farmer and Davis (1985) developed an H7 antiserum-sorbitol fermentation medium that allowed colonies on primary isolation plates to be screened for H7 expression upon subculture. In the late 1980s, latex agglutination reagents for detection of *E. coli* O157 and H7 became available commercially and were shown to be a rapid and economical alternative to tube agglutination (Chapman 1989). Chan et al. (1998) compared four commercial latex kits for detection of *E. coli* O157:H7 and found all to be 99–100 percent sensitive and specific; however, precautions must be taken with their use. Borczyk et al. (1990) noted that false-positive identification of *E. coli* O157 could occur if the latex controls were not routinely used. The factor responsible for the nonspecific agglutination of untagged latex particles (control) was heat labile (100°C, 5–10 min); hence, if agglutination was found to occur with control particles, it was recommended that cells be boiled and retested with control and O157-tagged particles because following boiling, the only intact antigens remaining should be O antigens. False-negative H7 antigen determinations often occur because isolates may require multiple passages in motility medium before flagellae are expressed. Detection of the H7 flagellar gene, *fliC*, by polymerase chain reaction (PCR) is now often used to avoid false-negative results associated with lack of expression in primary cultures.

Detection of *E. coli* O157:H7 in Fecal Specimens by Direct Immunofluorescence and Enzyme-Linked Immunosorbent Assay (ELISA). Methods for the immunological detection of *E. coli* O157 antigen directly in the feces have been developed. The main incentive for development of these tests was to provide a more rapid diagnosis than can be made by culture (Dylla et al. 1995). Park et al. (1994) developed and evaluated a direct immunofluorescence (DIF) assay for the detection of *E. coli* O157 in human feces. The DIF was tested on fecal smears that were either treated or not with bleach. The fecal smears were then stained with commercial fluorescein-conjugated anti-*E. coli* O157 polyclonal antibodies and examined microscopically. The DIF assay detected all (336) isolates of *E. coli* O157 that were recovered by culture. No false-negative results were obtained with bleach-pretreated specimens. The turnaround time for the DIF assay was < 2 h.

Dylla et al. (1995) and Park et al. (1994) compared an ELISA (LMD Laboratories, Inc., Carlsbad, CA) with culture for the detection of *E. coli* O157:H7 in human fecal specimens. This assay utilized plastic microwell test strips coated with anti-*E. coli* O157 polyclonal antibodies and detected *E. coli* O157 antigen. The ELISA was found to be an acceptably sensitive, specific, and rapid method for directly screening stool samples for *E. coli* O157:H7 and required only about 1 h to complete. However, the ELISA was unable to distinguish toxigenic from nontoxigenic strains, and culture isolates of *Citrobacter freundii*, *E. hermannii*, and *Salmonella urbana* produced positive ELISA results.

Chapman et al. (1997a) compared a commercial enzyme immunoassay (EIA; *E. coli* O157 Visual Immunoassay; Tecra Diagnostics) performed on enrichment cultures in mECn with IMS performed on enrichment cultures in BPW-VCC for the detection of *E. coli* O157 in bovine fecal samples. Tests on fecal suspensions inoculated with each of twelve different strains of *E. coli* O157 showed that both the EIA and IMS methods were 10- to 100-fold more sensitive than direct culture or enrichment culture methods for detection of the organism. EIA and IMS were also compared for the detection of *E. coli* O157 in bovine rectal swabs. Both methods were found to be sensitive, with the numbers of positives detected by the two assays not significantly different. One problem was that of positive EIA results that could not be confirmed by culture. Eight samples gave positive EIA results that could not be confirmed by either the IMS or another confirmation procedure. However, when fifty randomly selected colonies of sorbitol-fermenting *E. coli* per sample were screened by a latex agglutination test, three of these samples were found to contain sorbitol-fermenting *E. coli* O157. These strains were nontoxigenic and *eaeA* negative; this study confirmed previous observations that assays based strictly on detection of the O157 antigen are prone to false-positive results.

Biochemical Identification of Isolates. The O antigens of several bacterial species are known to cross-react with *E. coli* O157, including *Salmonella* O30$_1$ (group N), *Yersinia enterocolitica* O9, *C. freundii*, and *E. hermannii* (Caroff et al. 1984; Perry and Bundle 1990; Perry et al. 1986a, 1986b). Hence, presumptive O157 isolates should be identified as *E. coli*. Because *E. hermannii* cross-reacts serologically with O157, is biochemically very similar to *E. coli*, and is sorbitol- and MUG-negative, special precautions should be taken to differentiate the two organisms. *E. hermannii* can presumptively be differentiated on the basis of the appearance of its colonies, which are gold colored. Definitive differentiation of *E. coli* and *E. hermannii* can be made by using one or both of two tests: growth in the presence of potassium cyanide (KCN) and fermentation of cellobiose. *E. coli* cannot grow in the presence of KCN and does not ferment cellobiose, whereas *E. hermannii* is positive in both these tests.

Commercial automated systems and test kits may be used for biochemical identification of *E. coli* strains, and these methods have also shown differences between classical strains and *E. coli* O157:H7. Using the API 20E system (bioMérieux Vitek, Hazelwood, MO), Haldane et al. (1986) found that 91.9 percent (thirty-four of thirty-seven) of the *E. coli* O157:H7 isolates tested had a numerical taxonomic profile of 5144172 ("low discrimination; choice 1: *E. coli*, choice 2: *Kluyvera*"). Abbott et al. (1994), using the MicroScan conventional Gram-negative identification panel (Baxter Diagnostics, Inc. West Sacramento, CA),

tested seventy-three confirmed *E. coli* O157:H7 strains; sixty-two (85 percent) generated a biochemical profile number (6711501-0) that was not observed among any of the other non-O157:H7 strains tested. This profile consisted of positive reactions for D-glucose, sucrose, raffinose, rhamnose, L-arabinose, melibiose, indole, lysine decarboxylase, ornithine decarboxylase, and *o*-nitrophenyl-β-D-galactopyranoside (ONPG); negative reactions for D-sorbitol, *m*-inositol, adonitol, urea, H$_2$S, arginine dihydrolase, tryptophan deaminase, esculin, Voges-Proskauer, citrate, malonate, and oxidase; and the absence of growth in the presence of colistin (4 µg/ml) and cephalothin (8 µg/ml).

Leclercq et al. (2001), using a recent upgrade of the API 20E system (ID 32E, bioMérieux Vitek, Hazelwood, MO), found that *E. coli* O157:H7 or O157:NM strains do not display a unique profile, but instead display several particular biochemical profiles. *E. coli* O157:H7 showed significant divergent biochemical activities from classical *E. coli* for ornithine decarboxylase, arginine dihydrolase, urease, 5-ketoglutarate, β-glucuronidase, sorbitol, and, to a lesser extent, rhamnose, adonitol, D-arabitol, trehalose, and inositol. Although no single biochemical profile number could account for all *E. coli* O157:H7 isolates, two-thirds of these isolates were distributed into only five profiles, with one (that is, 54465743000) representing 25 percent of the tested strains.

Enterohemolysin (Ehly) and EHEC-hemolysin (Ehx). *E. coli* O157:H7 and many other STEC strains express weak hemolytic activity that was originally referred to as being caused by enterohemolysin (Ehly). Enterohemolysin (Ehly) was first described by Beutin et al. (1988) as being produced by enteropathogenic *E. coli* strains (EPEC) of the O26 and O111 serogroups. Two different phage-encoded hemolysins were later shown to be responsible for Ehly phenotype of these strains, and the genes encoding these toxins, *ehly1* (Stroeher et al. 1993) and *ehly2* (Beutin et al. 1993b), respectively, were both found to be carried on the chromosome. In contrast to β-hemolysin, these Ehlys were not secreted from the cell and could be detected only on blood plates containing washed erythrocytes (Beutin et al. 1988). Hemolysis on plates was turbid and seen only beneath the colonies. This hemolytic phenotype was later noted to be associated with VTEC (STEC) strains of different serotypes including O157:H7 (Beutin et al. 1989). Beutin et al. (1993a) used the Ehly phenotype on tryptose blood agar plates containing washed sheep erythrocytes to screen fecal samples from seven different animal species for VTEC (STEC). Isolates were then tested for Stx activity by the Vero cell assay and also subjected to colony blot DNA hybridization for *ehly1* (Stroeher et al. 1993). A close association between Stx production and the Ehly phenotype was found in STEC serotype O5:NM, O146:H21, O128:H2, O77:H4, L119:H25, and O123:H10 strains (Beutin et al. 1993a). However, only 12.4 percent of these strains hybridized with the *ehly1* probe. Soon thereafter, a second phage-encoded enterohemolysin (Ehly2) was discovered (Beutin et al. 1993b). Ehx of *E. coli* O157:H7 was later found to be distinct from Ehly1 and Ehly2, to be located on the 90-kb plasmid instead of the chromosome, and also to be a member of the RTX family of toxins (Bauer and Welch 1996; Beutin et al. 1993; Schmidt et al. 1995; Stroeher et al. 1993). Other STEC strains were found also to carry the *ehx* gene on a 90-kb plasmid, which was responsible for their enterohemolytic phenotype (Fratamico et al. 1995). Because of the commonality of the enterohemolytic phenotype and distribution of *ehx* among STEC strains, the usefulness of this phenotype or respective gene as a tool for identification of *E. coli* O157:H7 is limited.

DNA Probes for Virulence Genes. Montenegro et al. (1990), Samadpour et al. (1990), and Beutin et al. (1993a) detected VTEC (STEC) in cattle feces by colony blot DNA hybridization for *stx1* and *stx2*. Although this procedure specifically detected STEC, other tests would still be required for confirmation of isolates as *E. coli* O157:H7. In addition, polymerase chain reaction (PCR) would prove to be a more advantageous molecular diagnostic method because it is more rapid, less labor intensive, and can be modified into a multiplex format to detect multiple genes in one reaction.

Polymerase Chain Reaction (PCR). PCR has been used both to confirm the identity of *E. coli* O157:H7 isolates and to directly detect the organism in feces or enrichment broth cultures. Because of problems caused by PCR inhibitors in feces, most PCR assays have been used to confirm isolates as *E. coli* O157:H7; however, even for this purpose, most methods have had pitfalls. Fratamico et al. (1995) developed a multiplex PCR to detect *eaeA*, *stx*, and *ehx*. *E. coli* O157:H7 isolates were shown to carry all three genes, whereas other STEC isolates were usually negative for at least one. This procedure, especially when conducted on biochemically characterized isolates, increased the level of confirmation as *E. coli* O157:H7; however, none of the amplified targets was actually specific for this organism. Cebula et al. (1995) developed a multiplex PCR using mismatch primers specific for the unique base substitution in the *uidA* gene of *E. coli* O157:H7 (Feng and Lampel 1994) in combination with primers for *stx1* and *stx2*. This assay was specific for *E. coli* O157:H7, but did not allow for detection of variants lacking the specific *uidA* mutation. Li et al. (1997) developed a PCR assay for *E. coli* O157:H7 based on a unique 16-bp deletion in *fimA* (the structural gene for type 1 fimbriae). This assay, too, was reportedly specific for *E. coli* O157:H7 but did not confirm the presence of EHEC virulence genes (that is, *stx* or *eaeA*) or the *E. coli* O157 antigen gene cluster. Gannon et al. (1997) developed a multiplex PCR for *E. coli* O157:H7 in which H7-specific primers were coupled with primers that targeted *stx* and a region of *eaeA* gene relatively specific for *E. coli* O157 (*eaeA*$_{O157}$). This assay, although based on the virulence genes of *E. coli* O157:H7, did not include primers to confirm the presence of the *E. coli* O157 antigen gene cluster (Wang and Reeves 1998); in addition, the *eaeA*$_{O157}$ primers detected other *E. coli* serotypes; for example, O55:H7 and O145:NM. Bilge et al. (1996) identified the *rfbE* gene, located on the *E. coli* O157 antigen gene cluster (Wang and Reeves 1998). Based on the sequence of this gene, Desmarchelier et al. (1998) designed a PCR assay for the detection of *E. coli* O157:H7. Although this assay detected serogroup O157 *E. coli*, it did not detect either *stx* or *eaeA*, and failed to specifically identify isolates as EHEC. Nagano et al. (1998) developed a multiplex PCR using primers to amplify the *rfbE*, *stx*, and *fliC* genes, respectively. This assay specifically identified isolates as EHEC O157:H7 as long as *stx* was present, but did not identify which *stx* genes were present, nor did it confirm the presence of *eaeA*. Detection of both *stx* and *eaeA* are important because isolates frequently lose the *stx* gene upon subcultivation (Karch et al. 1992). Feng and Monday (2000) developed a multiplex PCR using primers to amplify *uidA*, β-intimin *eaeA* gene, *stx1*, *stx2*, and *ehxA* genes. The β-intimin *eaeA* gene is carried both by O157:H7 and O55:H7 serotypes of *E. coli*; this assay also did not specifically detect the O157 antigen gene cluster. Fratamico et al. (2000) developed a multiplex PCR for detection of *E. coli* O157:H7 in food and bovine fecal samples. The assay utilized primers to amplify the *ehx*, *fliC*, *stx1*, *stx2*, and *eaeA* genes from enrichment cultures. This assay could detect *E. coli* O157:H7 in bovine feces at a concentration of 1 CFU/g based on seeding experiments; however, the assay was not designed to detect the O157 antigen gene cluster. Hu et al. (1999) developed a multiplex PCR using primers to amplify the *eaeA*$_{O157}$, *stx1*, *stx2*, *fliC*, and *rfbE* genes. The collection of primers selected by these authors provided confirmation of *E. coli* O157:H7 when performed on isolated colonies.

Hu et al. (1999) tested their multiplex PCR directly on bovine feces and on TSB enrichment cultures of these specimens. Previous studies had shown that feces contain inhibitors of the PCR reaction, and assays conducted directly on feces were often unsuccessful or relatively insensitive (Brian et al. 1992). Other problems have included a potentially low number of target organisms and high levels of competing microflora; hence, enrichment is usually needed, as well as techniques to reduce PCR-inhibiting substances (Jinneman et al. 1995). However, these techniques must not decrease the amount of DNA template below a level that allows amplification (Jinneman et al. 1995). Hu et al (1999) were unsuccessful in their attempts to detect the organism directly from feces, a procedure that utilized the QIAamp test kit (Qiagen) to extract template DNA. However, they found that their multiplex PCR could detect 1 CFU/g of feces following TSB enrichment and QIAamp treatment based on seeding experiments. Stewart et al. (1998) tested several different methods for preparing template DNA for detection of the *stx1* gene of *E. coli* O157:H7 in bovine feces using PCR. These included boiling, enzyme treatment, enzyme treatment plus phenol-chloroform extraction, and enzyme treatment plus phenol-chloroform extraction plus Geneclean (Bio 101, Vista, CA) purification. The boiling method provided the best results for template preparation from both enrichment cultures and direct detection in feces (without enrichment). However, the procedure was insensitive without enrichment; the detection limits were approximately 3 CFU/g of feces based on seeding experiments with enrichment and 10^5 CFU/g without enrichment (direct detection method). The boiling method was also combined with IMS to detect *E. coli* O157:H7 in less than eight hours, but with a sensitivity of only 10^3 CFU/g feces. Oberst et al. (1998) used an automated *eaeA*-based *E. coli* O157:H7 5′ nuclease detection assay (TaqMan; PE Applied Biosystems, Foster City, CA) to detect *E. coli* O157:H7 in cultures of experimentally seeded ground beef. They found that the TaqMan assay, when integrated with an effective DNA recovery process, was capable of rapid, semiautomated, presumptive detection of *E. coli* O157:H7 if 10^3 CFU/ml was present in pure culture in mTSB or mEC, or if 10^4 CFU/ml was present in a ground beef-mTSB mixture. Collectively, these studies suggest that PCR or multiplex PCR can potentially be applied to enrichment cultures for rapid, sensitive, and specific detection of *E. coli* O157:H7 in bovine feces; however, more improvements are needed to increase the sensitivity of the assay for this purpose. Multiplex PCR has been shown to be very useful for confirmation of isolates as *E. coli* O157:H7.

Detection of Stx. The association of idiopathic HUS (Karmali et al. 1985) and TTP (Anonymous 1986; Morrison et al. 1986) with infection by *E. coli* O157:H7 and other VTEC (STEC) serotypes made apparent the need to test for the presence of STEC in patients with these specific clinical presentations. Karmali et al. (1985) determined that detection of free fecal VT (Stx) in the stools of HUS patients was a sensitive method for detecting enteric infection with VTEC (STEC) strains. However, these methods, which utilized Vero or HeLa cell assays, are labor intensive and many clinical laboratories are not equipped to conduct these tests. In addition, studies showed that cattle commonly carry a variety of STEC serotypes. Beutin et al. (1993a) identified forty-one different O:H serotypes among VTEC (STEC) isolates from different animal species, including cattle, sheep, pigs, goats, dogs and cats. VTEC from cattle were most heterogeneous with fifteen different O groups and eighteen O-untypeable strains among thirty-three isolates (Beutin et al. 1993a). Hence, assays for fecal Stx are of little value for diagnosis of *E. coli* O157:H7 infection in cattle for all practical purposes. Molecular detection methods, such as PCR to detect *stx* (described previously), soon replaced cell culture methods as the means to detect STEC.

Serology. Tests to detect serum antibodies directed against *E. coli* O157 antigen have been used in cattle to detect exposure to *E. coli* O157:H7 (Johnson et al. 1996; Laegreid et al 1998). Because the *E. coli* O157 antigen shares structural elements with the LPS of *B. abortus*, *Y. enterocolitica* O9, and some other bacterial species, exposure of cattle to these other bacterial species gave false-positive test results for *E. coli* O157 antibodies (Perry and Bundle 1990). Laegreid et al. (1998), using monoclonal antibodies specific for an epitope of the O157 antigen not shared by *B. abortus* or *Y. enterocolitica* (Westerman et al. 1997), developed a blocking ELISA with enhanced specificity for detection of serum anti-*E. coli* O157 antibodies. These authors proposed use of this assay as a diagnostic method for detection of *E. coli* O157:H7 infection in cattle. However, one potential drawback is that cattle are frequently infected with non-EHEC *E. coli* O157 strains (Chapman et al. 1997b; Whittam et al. 1988), which would be expected to cause seroconversion to *E. coli* O157 antigen and false-positive test results. In addition, these tests do not actually detect presence of *E. coli* O157:H7 in the animal.

An indirect ELISA was developed to detect serum antibody responses against *E. coli* O157:H7 secreted proteins in cattle following experimental infection with *E. coli* O157:H7 (Berberov et al. 2001). The procedures for this assay were modifications of an indirect ELISA used to detect antibodies to *E. coli* O157:H7 secreted proteins in HUS patients (Li et al. 2000). Serum antibodies detected against these proteins would be expected to cross-react with the secreted proteins of other EHEC bacteria; hence, these tests also would not be diagnostically useful for routine detection of *E. coli* O157:H7 infection in cattle.

DETECTION AND DIAGNOSIS IN OTHER FOOD-ANIMAL SPECIES.

In different epidemiologic studies, *E. coli* O157:H7 has been isolated from the feces of other food-animal species, including sheep, pigs, goats, deer, and turkeys. In all cases, the methods used were similar to those used to detect the organism in humans and cattle.

Sheep. Kudva et al. (1995; 1996; 1997) inoculated 50 ml TSB enrichment broth supplemented with cefixime (0.05 µg/ml), potassium tellurite (2.5 µg/ml), and vancomycin (40 µg/ml) (TSB-CTV) with 1 g fecal samples and incubated 18 h at 37°C with aeration. Following incubation, TSB-CTV cultures were serially diluted and plated onto CT-SMAC plates containing MUG (100 µg/ml) (CTM-SMAC). Chapman et al. (1997a) used the BPW-VCC-IMS-CT-SMAC culture method as described previously for cattle. Heuvelink et al. (1998) used the same methods as those this group used to isolate the organism from cattle.

Fegan and Desmarchelier (1999) detected STEC in the feces of sheep (adults) and lambs using the following procedure. One gram fecal samples were diluted

1:50 and cultured in mEC broth with novobiocin without bile salts at 37°C for 18–20 h. Crude cell lysates were prepared from enrichment broths by pelleting the cells, resuspending in water, boiling, and recentrifuging. The supernatant (crude cell lysate) was used as a template for PCR with *stx* primers. Enrichment broths positive for *stx* were streaked onto Eosin Methylene Blue agar (Oxoid) and incubated for 18–20 h. At least twenty colonies showing typical *E. coli* characteristics per plate were inoculated into 150 µl of nutrient broth in microtiter tray wells. After overnight incubation, broth cultures were replicated onto nylon membranes with a 96-pin replicator (Nunc, Inc., Naperville, IL) and probed for *stx1* and *stx2* by DNA hybridization. Isolates positive for *stx1* and *stx2* were further tested by PCR for *eae* and *ehx*, and biochemically identified using the Microbact 12E system (Oxoid). *E. coli* isolates positive for *stx* and *eae* were tested for O157 antigen by latex agglutination and tested for flagella by successive passaging through motility media. One hundred twenty-seven of 144 (88 percent) sheep and 53 of 72 (74 percent) lamb fecal samples were positive for *stx*. All *stx*-positive isolates from sheep were identified as *E. coli*, but none of these isolates contained *eae*. The *eae* gene was present in three (8 percent) of thirty-six STEC isolated from lambs, and all three isolates were identified as *E. coli* O157:NM. In this study, the authors noted that the procedures used to isolate STEC from PCR-positive samples were not very sensitive. PCR was used as an initial screening procedure because they were searching for all STEC, as opposed to a single serotype.

Sidjaba-Tambunan and Bensink (1997) detected VTEC in the feces of sheep by the following procedure. Fecal samples collected at an abattoir were frozen at −20°C until further processed. Samples were later thawed, and about 500 mg was inoculated into 10 ml Luria-Bertani (LB) broth and incubated at 37°C for 4 h, after which 10 ml was transferred into 10 ml of fresh LB broth and incubated at 37°C overnight. Cells in 1 ml of broth culture were pelleted by centrifugation; bacterial pellets were then resuspended in sterile distilled water, boiled for 10 min, and aliquots were used as DNA templates in PCR assays using *stx* primers. All PCR-positive broth samples were diluted and plated on MacConkey agar. After overnight incubation, colonies were subjected to colony DNA hybridization using *stx* probes. No apparent attempts were made to serotype the isolates. VTEC were detected in 68 percent of sheep samples, 18 percent of calf samples, and 1.5 percent of pig samples.

Swine. Nakazawa and Akiba (1999) detected *E. coli* O157:H7 in 1.4 percent of 221 pigs cultured by the following procedures. Rectal swabs were placed in Cary-Blair transport media and kept refrigerated until processing (usually about 48 h). Swabs were then incubated overnight at 42°C in 10 ml of mEC broth containing 20 µg/ml novobiocin. One loopful of enrichment culture was spread onto SMAC plates, and

after overnight incubation at 37°C, sorbitol-negative colonies were tested by latex agglutination for O157 antigen, and strains that agglutinated were confirmed as *E. coli* using the API 20E system. *E. coli* O157 test positive colonies were then subcultured onto motility test medium and tested for H7 with anti-H7 antiserum. Isolates were confirmed as *E. coli* O157:H7 by PCR using primers to detect *stx1*, *stx2*, *eae*, and the pO157 plasmid.

Heuvelink et al. (1999) used the same methods to culture *E. coli* O157:H7 from swine as those used previously by this group to isolate the organism from cattle (Heuvelink et al. 1998). *E. coli* O157:H7 was isolated from 1 of 145 (0.7 percent) of normal slaughter pigs. The isolate was biochemically typical, produced Stx, and contained the *eae* and *ehx* genes.

Goats. Bielaszewska et al. (1997) screened fecal samples for *E. coli* O157 antigen with a commercial ELISA (LMD Laboratories, Carlsbad, CA) and by the Vero cell assay for free fecal Stx (Karmali et al. 1985). Fecal samples containing detectable *E. coli* O157 antigen or Stx were then cultured on SMAC for *E. coli* O157:H7. Stx2-producing *E. coli* O157:H7 was isolated from one farm goat; the same strain (same pulsed field gel electrophoresis pattern and phage type) was isolated from children with HUS who had consumed milk originating from the goat.

Deer. Chapman and Ackroyd (1997) isolated *E. coli* O157 from the feces of three of ten farm-raised deer using the BPW-VCC-IMS-CT-SMAC culture methods that had been developed previously for bovine fecal culture (Chapman et al. 1994). All three isolates were biochemically typical of *E. coli* O157; two were positive for both *stx1* and *stx2* by DNA hybridization; the other strain was nontoxigenic and *eaeA* negative. Fischer et al. (2001) isolated *E. coli* O157:H7 from wild deer by using Cary-Blair as a transport medium and enrichment in TSB-novobiocin. Broth cultures were screened with an *E. coli* O157 ELISA, and antigen-positive broth cultures were subjected to IMS and plating on SMAC-MUG.

Turkeys. Heuvelink et al. (1999) used the same methods to culture *E. coli* O157:H7 from turkeys as those used previously by this group to isolate the organism from cattle. In this study, *E. coli* O157:H7 was detected in 1 of 459 (0.2 percent) pooled fecal samples from turkey flocks. The isolate was biochemically typical, produced Stx, and contained the *eae* and *ehx* genes.

SUMMARY AND CONCLUSIONS. Methods for the detection and diagnosis of *E. coli* O157:H7 in food-producing animals are largely modifications of those first used to detect the organism in food and human patients. Fecal shedding of *E. coli* O157:H7 in cattle often occurs at levels low enough that selective enrichment and immunomagnetic separation (IMS) are required for detection. GN broth or buffered peptone water containing cefixime, vancomycin, and tellurite followed by IMS and plating is a relatively sensitive method that is most often used to detect the organism in the feces of food-producing animals. Sorbitol-MacConkey agar containing cefixime and tellurite is the standard plating method. PCR as a primary detection method has required prior enrichment culture to give acceptable sensitivity. Multiplex or collective individual PCR reactions targeting the O157 antigen gene cluster (such as *rfbE*), H7 antigen (*fliC*), Stxs (*stx1* and *stx2*), and intimin (*eaeA*) on suspect isolates are recommended for confirmation as *E. coli* O157:H7.

REFERENCES

Abbott, S.L., D.F. Hanson, T.D. Felland, S. Connell, A.H. Shum, and J.M. Janda. 1994. *Escherichia coli* O157:H7 generates a unique biochemical profile on MicroScan conventional gram-negative identification panels. *J. Clin. Microbiol.* 32:823–824.

Allison, L.J., P.E. Carter, and F.M. Thomson-Carter. 1997. Evidence of a direct transmission of *E. coli* O157 infection between animals and humans. In *Enterohemorrhagic Escherichia coli. Fortschritte der Medizin* 115 Monographie 84:45.

Anonymous. 1986. Thrombotic thrombocytopenic purpura associated with *Escherichia coli* O157:H7—Washington. *MMWR* 35:549–551.

Armstrong, G.L., J. Hollingsworth, and J.G. Morris. 1996. Emerging foodborne pathogens: *Escherichia coli* O157:H7 as a model of entry of a new pathogen into the food supply of the developed world. *Epidemiol. Rev.* 18:29–51.

Bauer, M.E., and R.A. Welch. 1996. Characterization of an RTX toxin from enterohemorrhagic *Escherichia coli* O157:H7. *Infect. Immun.* 64:167–175.

Berberov, E.M., D.L. Bailey, B.B. Finlay, and R.A. Moxley. 2001. Serum antibody responses of weanling calves to *Escherichia coli* attaching-effacing proteins. *Pro. 82nd Annual Meeting of the Conf. of Res. Workers in Animal Dis.*, St. Louis MO, Nov. 11–13, Abstract No. 42P.

Besser, T.E., D.D. Hancock, L.C. Pritchett, E.M. McRae, D.H.Rice, and P.I. Tarr. 1997. Duration of detection of fecal excretion of *Escherichia coli* O157:H7 in cattle. *J. Infect. Dis.* 175:726–729.

Beutin, L., D. Geier, H. Steinruck, S. Zimmermann, and F. Scheutz. 1993a. Prevalence and some properties of verotoxin (Shiga-like toxin)-producing *Escherichia coli* in seven different species of healthy domestic animals. *J. Clin. Microbiol.* 31:2493-2488.

Beutin, L., M.A. Montenegro, I. Orskov, F. Orskov, J. Prada, S. Zimmermann, and R. Stephan. 1989. Close association of verotoxin (Shiga-like toxin) production with enterohemolysin production in strains of *Escherichia coli*. *J. Clin. Microbiol.* 27:2559–2564.

Beutin, L., J. Prada, S. Zimmermann, R. Stephan, I. Orskov, and F. Orskov. 1988. Enterohemolysin, a new type of hemolysin produced by some strains of enteropathogenic *E. coli* (EPEC). *Zentralb. für Bakteriol., Mikrobiol. und Hyg., 1. Abt. Orig. A, Med. Mikrobiol., Infekt. und Parasitol.* 267:576–588.

Beutin, L., U.H. Stroeher, and P.A. Manning. 1993b. Isolation of enterohemolysin (Ehly2)-associated sequences encoded on temperate phages of *Escherichia coli*. *Gene* 132:95–99.

Bielaszewska, M., J. Janda, K. Blahova, H. Minarikova, E. Jikova, M.A. Karmali, J. Laubova, J. Sikulova, M.A. Pre-

ston, R. Khakhria, H. Karch, H. Klazarova, and O. Nyc. 1997. Human *Escherichia coli* O157:H7 infection associated with the consumption of unpasteurized goat's milk. *Epidemiol. Infect.* 119:299–305.

Bielaszewska, M., H. Schmidt, A. Liesegang, R. Prager, W. Rabsch, H. Tschape, A. Cizek, J. Janda, K. Blahova, and H. Karch. 2000. Cattle can be a reservoir of sorbitol-fermenting Shiga toxin-producing *Escherichia coli* O157:H⁻ strains and a source of human diseases. *J. Clin. Microbiol.* 38:3470–3473.

Bilge, S. S.J. C. Vary, Jr., S.F. Dowell, and P.I. Tarr. 1996. Role of the *Escherichia coli* O157:H7 O side chain in adherence and analysis of an *rfb* locus. *Infect. Immun.* 64:4795–4801.

Borczyk, A.A., M.A. Karmali, H. Lior, and L.M.C. Duncan. 1987. Bovine reservoir for verotoxin-producing *Escherichia coli* O157:H7. *Lancet* 1:98.

Borczyk, A.A., N. Harnett, M. Lombos, and H. Lior. 1990. False-positive identification of *Escherichia coli* O157 by commercial latex agglutination tests. *Lancet* 336:946–947.

Brian, M.J., M. Frosolono, B.E. Murray, A. Miranda, E.L. Lopez, H.F. Gomez, and T.G. Cleary. 1992. Polymerase chain reaction for diagnosis of enterohemorrhagic *Escherichia coli* infection and hemolytic-uremic syndrome. *J. Clin. Microbiol.* 30:1801–1806.

Brown, C.A., B.G. Harmon, T. Zhao, and M.P. Doyle. 1997. Experimental *Escherichia coli* O157:H7 carriage in calves. *Appl. Environ. Microbiol.* 63:27–32.

Caroff, M., D.R. Bundle, and M.B. Perry. 1984. Structure of the O-chain of the phenol-phase soluble cellular lipopolysaccharide of *Yersinia enterocolitica* serotype O9. *Eur. J. Biochem* 139:195–200.

Cebula, T.A., W.L. Payne, and P. Feng. 1995. Simultaneous identification of strains of *Escherichia coli* serotype O157:H7 and their Shiga-like toxin type by mismatch amplification mutation assay-multiplex PCR. *J. Clin. Microbiol.* 33:248–250.

Chan, E., L. Ball, and G.B. Horsman. 1998. Comparison of four latex kits for detection of *E. coli* O157. *Clin. Lab. Sci.* 11:266–268.

Chapman, P. A. 1989. Evaluation of commercial latex slide test for identifying *Escherichia coli* O157. *J. Clin. Pathol.* 42:1109–1110.

Chapman, P.A., and H.J. Ackroyd. 1997. Farmed deer as a potential source of verocytotoxin-producing *Escherichia coli* O157:H7. *Vet. Rec.* 141:314–315.

Chapman, P.A., A.T. Cerdan Malo, C. A. Siddons, and M. Harkin. 1997a. Use of commercial enzyme immunoassays and immunomagnetic separation systems for detecting *Escherichia coli* O157:H7 in bovine fecal samples. *Appl. Environ. Microbiol.* 63:2549–2553.

Chapman, P.A., C.A. Siddons, and M.A. Harkin. 1996. Sheep as a potential source of verocytotoxin-producing *Escherichia coli* O157. *Vet. Rec.* 138:23–24.

Chapman, P.A., C.A. Siddons, A.T. Cerdan Malo, and M.A. Harkin. 1997b. A 1-year study of *Escherichia coli* O157 in cattle, sheep, pigs and poultry. *Epidemiol. Infect.* 119:245–250.

Chapman, P.A., C.A. Siddons, D.J. Wright, P. Norman, J. Fox, and E. Crick. 1993. Cattle as a possible source of verocytotoxin-producing *Escherichia coli* O157 infections in man. *Epidemiol. Infect.* 111:439–447.

Chapman, P.A., C.A. Siddons, P.M. Zadik, and L. Jewes. 1991. An improved selective medium for the isolation of *Escherichia coli* O157. *J. Med. Microbiol.* 35:107–110.

Chapman, P.A., D.J. Wright, and R. Higgins. 1993b. Untreated milk as a source of verotoxigenic *E. coli* O157. *Vet. Rec.* 133:171–172.

Chapman, P.A., D.J. Wright, and P. Norman. 1989. Verotoxin-producing *Escherichia coli* infections in Sheffield: cattle as a possible source. *Epidemiol. Infect.* 102:439–442.

Chapman, P.A., D.J. Wright, and C.A. Siddons. 1994. A comparison of immunomagnetic separation and direct culture for the isolation of verocytotoxin-producing *Escherichia coli* O157 from bovine faeces. *J. Med. Microbiol.* 40:424–427.

Cray, W.C., and H.W. Moon. 1995. Experimental infection of calves and adult cattle with *Escherichia coli* O157:H7. *Appl. Environ. Microbiol.* 61:1586–1590.

Cudjoe, K. 1995. Protocol for isolation of *E. coli* O157 in stool samples using Dynabeads® anti-*E. coli* O157: a clinical study protocol prepared for the U.S. market. Dynal A.S., Oslo, Norway.

Desmarchelier, P.M., S.S. Bilge, N. Fegan, L. Mills, J.C. Vary, Jr., and P.I. Tarr. 1998. A PCR specific for *Escherichia coli* O157 based on the *rfb* locus encoding O157 lipopolysaccharide. *J. Clin. Microbiol.* 36:1801–1804.

Doyle, M.P., and J.L. Schoeni. 1984. Survival and growth characteristics of *Escherichia coli* associated with hemorrhagic colitis. *Appl. Environ. Microbiol.* 48:855–856.

Dylla, B.L., E.A. Vetter, J.G. Hughes, and F.R. Cockerill, III. 1995. Evaluation of an immunoassay for direct detection of *Escherichia coli* O157 in stool specimens. *J. Clin. Microbiol.* 33:222–224.

Elder, R.O., J.E. Keen, G.R. Siragusa, G.A. Barkocy-Gallagher, M. Koohmaraie, and W.W. Laegreid. 2000. Correlation of enterohemorrhagic *Escherichia coli* O157 prevalence in feces, hides, and carcasses of beef cattle during processing. *Proc. Natl. Acad. Sci. USA* 97:2999–3003.

Farmer, J.J., and B.R. Davis. 1985. H7 antiserum-sorbitol fermentation medium: a single tube screening medium for detecting *Escherichia coli* O157:H7 associated with hemorrhagic colitis. *J. Clin. Microbiol.* 22:620–625.

Fegan, N., and P. Desmarchelier. 1999. Shiga toxin-producing *Escherichia coli* in sheep and pre-slaughter lambs in eastern Australia. *Lett. Appl. Microbiol.* 28:335–339.

Feng, P.C., and P.A. Hartman. 1982. Fluorogenic assays for immediate confirmation of *Escherichia coli. Appl. Environ. Microbiol.* 43:1320–1329.

Feng, P., and K.A. Lampel. 1994. Genetic analysis of *uidA* gene expression in enterohemorrhagic *Escherichia coli* serotype O157:H7. *Microbiol.* 140:2101–2107.

Feng, P., and S.R. Monday. 2000. Multiplex PCR for detection of trait and virulence factors in enterohemorrhagic *Escherichia coli* serotype. *Molec. and Cell. Probes* 14:333–337.

Fischer, J., T. Zhao, M.P. Doyle, M.R. Goldberg, C.A. Brown, C.T. Sewell, D.M. Kavanaugh, and C.D. Bauman. 2001. Experimental and field studies of *Escherichia coli* O157:H7 in white-tailed deer. *Appl. Environ. Microbiol.* 67:1218–1224.

Fratamico, P.M., S.K. Sackitey, M. Wiedmann, and M.Y. Deng. 1995. Detection of *Escherichia coli* O157:H7 by multiplex PCR. *J. Clin. Microbiol.* 33:2188–2191.

Fratamico, P.M., L.K. Bagi, and T. Pepe. 2000. A multiplex polymerase chain reaction assay for rapid detection and identification of *Escherichia coli* O157:H7 in foods and bovine feces. *J. Food Prot.* 63:1032–1037.

Gannon, V.P., S. D'Souza, T. Graham, R.K. King, K. Rahn, and S. Read. 1997. Use of the flagellar H7 gene as a target in multiplex PCR assays and improved specificity in identification of enterohemorrhagic *Escherichia coli* strains. *J. Clin. Microbiol.* 35:656–662.

Gansheroff, L.J., and A.D. O'Brien. 2000. *Escherichia coli* O157:H7 in beef cattle presented for slaughter in the U.S.: higher prevalence rates than previously estimated. *Proc. Natl. Acad. Sci. USA* 97:2959–2961.

Haldane, D.J. M., M.A. S. Damm, and J.D. Anderson. 1986. Improved biochemical screening procedure for small clinical laboratories for Vero (Shiga-like)-toxin-producing strains of *Escherichia coli* O157:H7. *J. Clin. Microbiol.* 24:652–653.

Hancock, D.D., T.E. Besser, M.L. Kinsel, P.I. Tarr, D.H. Rice, and M.G. Paros. 1994. The prevalence of *Escherichia coli* O157:H7 in dairy and beef cattle in Washington State. *Epidemiol. Infect.* 113:199–207.

Harris, A.A., R.L. Kaplan, L.J. Goodman, M. Doyle, W. Landau, J. Segreti, K. Mayer, and S. Levin. 1985. Results of a screening method used in a 12-month stool survey for *Escherichia coli* O157:H7. *J. Infect. Dis.* 152:775–777.

Hayes, P.S., K. Blom, P. Feng, J. Lewis, N.A. Strockbine, and B. Swaminathan. 1995. Isolation and characterization of a (-D-glucuronidase-producing strain of *Escherichia coli* O157:H7 in the United States. *J. Clin. Microbiol.* 33:3347–3348.

Heuvelink, A.E., F.L. A.M. van den Biggelaar, E. de Boer, R.G. Herbes, W.J.G. Melchers, J.H.J. Huis In'T Veld, and L.A.H. Monnens. 1998. Isolation and characterization of verocytotoxin-producing *Escherichia coli* O157:H7 strains from Dutch cattle and sheep. *J. Clin. Microbiol.* 36:878–882.

Heuvelink, A.E., J.T. Zwartkruis-Nahuis, F.L. van den Biggelaar, W.J. van Leeuwen, and E. de Boer. 1999. Isolation and characterization of verocytotoxin-producing *Escherichia coli* O157 from slaughter pigs and poultry. *Int. J. Food Microbiol.* 52:67–75.

Hu, Y., Q. Zhang, and J.C. Meitzler. 1999. Rapid and sensitive detection of *Escherichia coli* O157:H7 in bovine faeces by a multiplex PCR. *J. Appl. Microbiol.* 87:867–876.

Jinneman, K.C., P.A. Trost, W.E. Hill, S.D. Weagant, J.L. Bryant, C.A. Kaysner, and M.M. Wekell. 1995. Comparison of template preparation methods from foods for amplification of *Escherichia coli* O157 Shiga-like toxins type I and II DNA by multiplex polymerase chain reaction. *J. Food Prot.* 58:722–726.

Johnson, R.P., W.C. Cray, and S.T. Johnson. 1996. Serum antibody responses of cattle following experimental infection with *Escherichia coli* O157:H7. *Infect. Immun.* 64:1879–1883.

Karch, H., and M. Bielaszewska. 2001. Sorbitol-fermenting Shiga toxin-producing *Escherichia coli* O157:H⁻ strains: epidemiology, phenotypic and molecular characteristics, and microbiological diagnosis. *J. Clin. Microbiol.* 39:2043–2049.

Karch, H., C. Janetzki-Mittmann, S. Aleksic, and M. Datz. 1996. Isolation of enterohemorrhagic *Escherichia coli* O157 strains from patients with hemolytic-uremic syndrome by using immunomagnetic separation, DNA-based methods, and direct culture. *J. Clin. Microbiol.* 34:516–519.

Karch, H., T. Meyer, H. Russmann, and J. Heesemann. 1992. Frequent loss of Shiga-like toxin genes in clinical isolates of *Escherichia coli* upon subcultivation. *Infect. Immun.* 60:3464–3467.

Karmali, M.A. 1989. Infections by verocytotoxin-producing *Escherichia coli. Clin. Microbiol. Rev.* 2:15–38.

Karmali, M.A., M. Petric, C. Lim, P.C. Fleming, G.S. Arbus, and H. Lior. 1985. The association between hemolytic uremic syndrome and infection by Verotoxin-producing *Escherichia coli. J. Infect. Dis.* 151:775–782.

Krishnan, C., V.A. Fitzgerald, S.J. Dakin, and R.J. Behme. 1987. Laboratory investigation of outbreak of hemorrhagic colitis caused by *Escherichia coli* O157:H7. *J. Clin. Microbiol.* 25:1043–1047.

Kudva, I.T., P.G. Hatfield, and C.J. Hovde. 1995. Effect of diet on the shedding of *Escherichia coli* O157:H7 in a sheep model. *Appl. Environ. Microbiol.* 61:1363–1370.

Kudva, I.T., P.G. Hatfield, and C.J. Hovde. 1996. *Escherichia coli* O157:H7 in microbial flora of sheep. *J. Clin. Microbiol.* 34:431–433.

Kudva, I.T., P.G. Hatfield, and C.J. Hovde. 1997. Characterization of *Escherichia coli* O157:H7 and other Shiga toxin-producing *E. coli* serotypes isolated from sheep. *J. Clin. Microbiol.* 35:892–899.

Laegreid, W.W., R.O. Elder, and J.E. Keen. 1999. Prevalence of *Escherichia coli* O157:H7 in range beef calves at weaning. *Epidemiol. Infect.* 123:291–298.

Laegreid, W.W., M. Hoffman, J. Keen, R. Elder, and J. Kwang. 1998. Development of a blocking enzyme-linked immunosorbent assay for detection of serum antibodies to O157 antigen of *Escherichia coli. Clin. Diagn. Lab. Immunol.* 5:242–246.

Leclercq, A.,B. Lambert, D. Pierard, and J. Mahillon. 2001. Particular biochemical profiles for enterohemorrhagic *Escherichia coli* O157:H7 isolates on the ID 32E system. *J. Clin. Microbiol.* 39:1161–1164.

Levine, M.M. 1987. *Escherichia coli* that cause diarrhea: enterotoxigenic, enteropathogenic, enteroinvasive, enterohemorrhagic, and enteroadherent. *J. Infect. Dis.* 155:377–389.

Li, B., W.H. Koch, and T.A. Cebula. 1997. Detection and characterization of the *fimA* gene of *Escherichia coli* O157:H7. *Molec. Cell. Probes* 11:397–406.

Li, Y., E. Frey, A.M. R. MacKenzie, and B.B. Finlay. 2000. Human response to *Escherichia coli* O157:H7 infection: antibodies to secreted virulence factors. *Infect. Immun.* 68:5090–5095.

Licence, K., K.R. Oates, B.A. Synge, and T.M.S. Reid. 2001. An outbreak of *E. coli* O157 infection with evidence of spread from animals to man through contamination of a private water supply. *Epidemiol. Infect.* 126:135–138.

Locking, M.E., S.J. O'Brien, W.J. Reilly, E.M. Wright, D.M. Campbell, J.E. Coia, L.M. Browning, and C.N. Ramsay. 2001. Risk factors for sporadic cases of *Escherichia coli* O157 infection: the importance of contact with animal excreta. *Epidemiol. Infect.*127:215– 220.

March, S.B., and S. Ratnum. 1986. Sorbitol-MacConkey medium for detection of *Escherichia coli* O157:H7 associated with hemorrhagic colitis. *J. Clin. Microbiol* 23:869–872.

Martin, M.L., L.D. Shipman, J.G. Wells, M.E. Potter, K. Hedberg, I.K. Wachsmuth, R.V. Tauxe, J.P. Arnoldi, and J. Tilleli. 1986. Isolation of *Escherichia coli* O157:H7 from dairy cattle associated with two cases of haemolytic uraemic syndrome. *Lancet* 2:1043.

McDonough, P.L., C.A. Rossiter, R.B. Rebhun, S.M. Stehman, D.H. Lein, and S.J. Shin. 2000. Prevalence of *Escherichia coli* O157:H7 from cull dairy cows in New York State and comparison of culture methods used during preharvest food safety investigations. *J. Clin. Microbiol.* 38:318–322.

Meng, L, and M.P. Doyle. 1998. Microbiology of Shiga toxin-producing *Escherichia coli* in foods. In *Escherichia coli and Other Shiga Toxin-Producing E. coli Strains,* Ed. J. B. Kaper and A. D. O'Brien, Washington, D.C., American Society for Microbiology, 92–108.

Michel, P., J.B. Wilson, S.W. Martin, R.C. Clarke, S.A. McEwen, and C.L. Gyles. 1999. Temporal and geographical distributions of reported cases of *Escherichia coli* O157:H7 infection in Ontario. *Epidemiol. Infect.* 122:193–200.

Moon, H.W., S.C. Whipp, R.A. Argenzio, M.M. Levine, and R.A. Gianella. 1983. Attaching and effacing *Escherichia coli* in pig and rabbit intestines. *Infect. Immun.* 41:1340–1351.

Montenegro, M.A., M. Bulte, T. Trumpf, S. Aleksic, G. Reuter, E. Bulling, and R. Helmuth. 1990. Detection and characterization of fecal verotoxin producing

Escherichia coli from healthy cattle. *J. Clin. Microbiol.* 28:1417–1421.

Morrison, D.C., D.L. J. Tyrrell, and L.D. Jewell. 1986. Colonic biopsy in verotoxin-induced hemorrhagic colitis and thrombotic thrombocytopenic purpura (TTP). *Am. J. Clin. Pathol.* 86:108–112.

Nagano, I., M. Kunishima, Y. Itoh, Z. Wu, and Y. Takahashi. 1998. Detection of verotoxin-producing *Escherichia coli* O157:H7 by multiplex polymerase chain reaction. *Microbiol. Immunol.* 42:371–376.

Nakazawa, M. and M. Akiba. 1999. Swine as a potential reservoir of Shiga toxin-producing *Escherichia coli* O157:H7 in Japan. *Emerging Infectious Diseases* 5:833–834.

Nataro, J.P., and J.B. Kaper. 1998. Diarrheagenic *Escherichia coli. Clin. Microbiol. Rev.* 11:142–201.

Oberst, R.D., M.P. Hays, L.K. Bohra, R.K. Phebus, C.T. Yamashiro, C. Paszko-Kolva, S.J.A. Flood, J.M. Sargeant, and J.R. Gillespie. 1998. PCR-based DNA amplification and presumptive detection of *Escherichia coli* O157:H7 with an internal fluorogenic probe and the 5' nuclease (TaqMan) assay. *Appl. Environ. Microbiol.* 64:3389–3396.

O'Brien, A.D., and R.K. Holmes. 1987. Shiga and Shiga-like toxins. *Microbiol. Rev.* 51:206–220.

Okrend, A.J.G., B.E. Rose, and C.P. Lattuada. 1990. Use of 5-bromo-4-chloro-3-indoxyl-(-D-glucuronide in MacConkey sorbitol agar to aid in the isolation of *Escherichia coli* O157:H7 in ground beef. *J. Food Prot.* 53:941–943.

Ostroff, S.M., P.M. Griffin, R.V. Tauxe, L.D. Shipman, K.D. Greene, J.G. Wells, J.H. Lewis, P.A. Blake, and J.M. Kobayashi. 1990. A statewide outbreak of *Escherichia coli* O157:H7 infections in Washington state. *Am. J. Epidemiol.* 132:239–247.

Padhye, N.V., and M.P. Doyle. 1992. *Escherichia coli* O157:H7: epidemiology, pathogenesis and methods for detection in food. *J. Food Prot.* 55:555–565.

Park, C.H., D.L. Hixon, W.L. Morrison, and C.B. Cook. 1994. Rapid diagnosis of enterohemorrhagic *Escherichia coli* O157:H7 directly from fecal specimens using immunofluorescence stain. *Am. J. Clin. Pathol.* 101:91–94.

Park, C.H., N.M. Vandel, and D.L. Hixon. 1996. Rapid immunoassay for detection of *Escherichia coli* O157 directly from stool specimens. *J. Clin. Microbiol.* 34:988–990.

Paton, J.C., and A.W. Paton. 1998. Pathogenesis and diagnosis of Shiga toxin-producing *Escherichia coli* infections. *Clin. Microbiol. Rev.* 11:450–479.

Perry, M.B., and D.R. Bundle. 1990. Antigenic relationships of the lipopolysaccharides of *Escherichia hermannii* strains with those of *Escherichia coli* O157:H7, *Brucella melitensis,* and *Brucella abortus. Infect. Immun.* 58:1391–1395.

Perry, M.B., D.R. Bundle, L. Machean, J.A. Perry, and D.W. Griffith. 1986a. The structure of the antigenic lipopolysaccharide O-chains produced by *Salmonella urbana* and *Salmonella godesburg. Carb. Res.* 156:107–122.

Perry, M.B., L. Machean, and D.W. Griffith. 1986b. Structure of the O-chain polysaccharide of the phenol-phase soluble lipopolysaccharide of *Escherichia coli* O157:H7. *Biochem Cell. Biol.* 64:21–28.

Rahn, K., S.A. Renwick, R.P. Johnson, J.B. Wilson, R.C. Clarke, D. Alves, S. McEwen, H. Lior, and J. Spika. 1997. Persistence of *Escherichia coli* O157:H7 in dairy cattle and the dairy farm environment. *Epidemiol. Infect.* 119:251–259.

Riley, L.W. 1987. The epidemiologic, clinical, and microbiological features of hemorrhagic colitis. *Ann. Rev. Microbiol.* 41:383–407.

Riley, L.W., R.S. Remis, S.D. Helgerson, H.B. McGee, J.G. Wells, B.R. Davis, R.J. Hebert, E.S. Olcott, L.M. Johnson, N.T. Hargrett, P.A. Blake, and M.L. Cohen. 1983. Hemorrhagic colitis associated with a rare *Escherichia coli* serotype. *N. Engl. J. Med.* 308:681–685.

Samadpour, M., J. Liston, J.E. Ongerth, and P.I. Tarr. 1990. Evaluation of DNA probes for the detection of Shiga-like toxin producing *Escherichia coli* in food and calf fecal samples. *Appl. Environ. Microbiol.* 56:1212–1215.

Sanderson, M.W., J.M. Gay, D.D. Hancock, C.C. Gay, L.K. Fox, and T.E. Besser. 1995. Sensitivity of bacteriologic culture for detection of *Escherichia coli* O157:H7 in bovine feces. *J. Clin. Microbiol.* 33:2616–2619.

Schmidt, H., L. Beutin, and H. Karch. 1995. Molecular analysis of the plasmid-encoded hemolysin of *Escherichia coli* O157:H7 strain EDL 933. *Infect. Immun.* 63:1055–1061.

Shere, J.A., K.J. Bartlett, and C.W. Kaspar. 1998. Longitudinal study of *Escherichia coli* O157:H7 dissemination on four dairy farms in Wisconsin. *Appl. Environ. Microbiol.* 64:1390–1399.

Sidjabat-Tambunan, H., and J.C. Bensink. 1997. Verotoxin-producing *Escherichia coli* from the faeces of sheep, calves and pigs. *Austrail. Vet. J.* 75:292–293.

Slutsker, L., A.A. Ries, K.D. Greene, J.G. Wells, L. Hutwagner, and P.M. Griffin. 1997. *Escherichia coli* O157:H7 diarrhea in the United States: clinical and epidemiologic features. *Ann Internal Med.* 126:505–513.

Smith, D., M. Blackford, S. Younts, R. Moxley, J. Gray, L. Hungerford, T. Milton, and T. Klopfenstein. 2001. Ecological relationships between the prevalence of cattle shedding *Escherichia coli* O157:H7 and characteristics of the cattle or conditions of the feedlot pen. *J. Food Prot.* 64:1899–1903.

Stewart, D.S., M.L. Tortorello, and S.M. Gendel. 1998. Evaluation of DNA preparation techniques for detection of the SLT-1 gene of *Escherichia coli* O157:H7 in bovine faeces using the polymerase chain reaction. *Lett. Appl. Microbiol.* 26:93–97.

Stroeher, U.H., L. Bode, L. Beutin, and P.A. Manning. 1993. Characterization and sequence of a 33-kDa enterohemolysin (Ehly)-associated protein in *Escherichia coli*. *Gene* 132:89–94.

Strockbine, N.A., J.G. Wells, C.A. Bopp, and T.J. Barrett. 1998. Overview of detection and subtyping methods. In *Escherichia coli and Other Shiga Toxin-Producing E. coli Strains*, eds. J.B. Kaper and A.D. O'Brien, Washington, D.C.: American Society for Microbiology, 331–356.

Su, C., and L. J. Brandt. 1995. *Escherichia coli* O157:H7 infection in humans. *Ann. Internal Med.* 123:698–714.

Tarr, P.I., S.S. Bilge, J.C. Vary, S. Jelacic, R.L Habeeb, T.R. Ward, M.R. Baylor, and T.E. Besser. 2000. Iha: a novel *Escherichia coli* O157:H7 adherence-conferring molecule encoded on a recently acquire chromosomal island of conserved structure. *Infect. Immun.* 68:1400–1407.

Thompson, J.S., D.S. Hodge, and A.A. Borczyk. 1990. Rapid biochemical test to identify verocytotoxin-positive strains of *Escherichia coli* O157:H7. *J. Clin. Microbiol* 28:2165–2168.

Trepeta, R.W., and S.C. Edberg. 1984. Methylumbelliferyl-β-D-glucuronide based medium for rapid isolation and identification of *Escherichia coli*. *J. Clin. Microbiol.* 19:172–174.

Wang, L., and P.R. Reeves. 1998. Organization of *Escherichia coli* O157 O antigen gene cluster and identification of its specific genes. *Infect. Immun.* 66:3545–3551.

Wells, J.G., B.R. Davis, I.K. Wachsmuth, L.W. Riley, R.S. Remis, R. Sokolow, and G.K. Morris. 1983. Laboratory investigation of hemorrhagic colitis outbreaks associated with a rare *Escherichia coli* serotype. *J. Clin. Microbiol.* 18:512–520.

Wells, J.G., L.D. Shipman, K.D. Greene, E.G. Sowers, J.H. Green, D.N. Cameron, F.P. Downes, M.L. Martin, P.M. Griffin, S.M. Ostroff, M.E. Potter, R.V. Tauxe, and I.K. Wachsmuth. 1991. Isolation of *Escherichia coli* O157:H7 and other shiga-like toxin-producing *E. coli* from dairy cattle. *J. Clin. Microbiol.* 29:985–989.

Westerman, R.B., Y. He, J.E. Keen, E.T. Littledike, and J. Kwang. 1997. Production and characterization of monoclonal antibodies specific for the lipopolysaccharide of *Escherichia coli* O157:H7. *J. Clin. Microbiol.* 35:679–684.

Whittam, T.S., and R.A. Wilson. 1988. Genetic relationships among pathogenic *Escherichia coli* of serogroup O157. *Infect. Immun.* 56:2467–2473.

Zadik, P.M., P.A. Chapman, and C.A. Siddons. 1993. Use of tellurite for the selection of verocytotoxigenic *Escherichia coli* O157. *J. Med. Microbiol.* 39:155–158.

17

MOLECULAR AND POPULATION GENETICS OF VIRULENCE TRAITS IN *ESCHERICHIA COLI* O157:H7

ANDREW BENSON

INTRODUCTION. *E. coli* O157:H7 is a relatively robust foodborne pathogen, capable of entering the food supply from food animals and a variety of ecological niches within or linked to food animal production areas. Although its membership in the genus *Escherichia* portends adaptation to intestinal environments, the virulence machinery that has been acquired by this organism offers an alternative means for occupying intestinal niches in human hosts. The focus of this chapter is on the organization of its unique machinery, the characteristics it confers in human and ruminant hosts, and the evolutionary pathways in which the machinery converged into the O157:H7 genome.

E. coli O157:H7 is a member of the enterohemorrhagic *E. coli* (EHEC), which are well known for their ability to cause diarrhea, hemorrhagic colitis, and serious secondary complications such as hemolytic uremic syndrome (HUS) and thrombotic thrombocytopenic purpura (TTP) in children and adults, respectively. EHEC strains further comprise a subgroup of the diverse Shiga toxin-producing *E. coli* (STEC), all of which produce at least one of the two known Shiga toxin types. The EHEC are distinguished from most STEC by their unique attachment phenotype known as attachment and effacement (Levine 1987). Phylogenetically, the EHEC pathotype appears to have emerged on at least two independent occasions (Whittam 1998), apparently through parallel pathways in which common sets of virulence genes were successively acquired by two distinct lineages (Reid et al. 2000). O157:H7 strains, and the related enterohemorrhagic O157:H-derivatives, comprise one lineage of highly related genotypes referred to as the EHEC 1, whereas O26:H11, O111:H8, and O111:H- comprise the genetically distinct EHEC 2 lineage (Whittam 1998). The EHEC lineages share the characteristic of attachment and effacement with enteropathogenic *E. coli* (EPEC) strains that are known to cause infantile diarrhea, but EHEC and EPEC differ in many of the details of their pathogenesis. Whereas EPEC colonize the small intestine and cause primarily diarrhea, EHEC preferentially colonize the colon and cause diarrhea, bloody diarrhea, and serious secondary complications such as HUS and TTP. Nonetheless, molecular phylogenetic analysis indicates that the two EHEC lineages share common ancestry with two EPEC lineages; the EHEC 1 lineage shows relatively close genetic relationship to the EPEC 1 lineage whereas EHEC 2 and EPEC 2 appear to share recent common ancestry. This finding has led to the view that the EHEC 1/EPEC 1 and EHEC 2/EPEC 2 lineages descended from two progenitors that independently acquired the locus of enterocyte effacement. In this model, the contemporary populations in the two lineages are products of independent, but somewhat parallel, courses of virulence gene acquisition and physiological diversification (see Donnenberg and Whittam 2001, for recent review).

OVERVIEW OF PATHOGENESIS. EHEC pathogenesis is a consequence of specialized attachment systems and the synthesis of exotoxins, known as Shiga toxins (also called Shiga-like toxins or Verotoxins). Following ingestion, the histopathological sequence of events begins by formation of microcolonies at sites in the large intestine, followed by subsequent formation of attaching and effacing (A/E) lesions, a hallmark of both EHEC and EPEC pathogenesis (Moon et al. 1993; Knutton et al. 1989). Within these lesions, bacteria become tightly bound to host cells via specialized host-derived pedestal structures. Microvilli are displaced from regions adjacent to the bound bacteria, and local cytoskeletal rearrangements lead to formation of pedestals at the site of bacterial attachment (Frankel et al. 1998). Although many details of the A/E process have been defined (see the next section), the mechanism of diarrhea remains elusive for both EPEC and EHEC. Loss of absortive microvilli in the small intestine (Natarro and Kaper 1998), changes in ion secretion (Collington et al. 1998), breeches of integrity of tight junctions between enterocytes (Philpott et al. 1996), and changes in signal transduction (Philpott et al. 1998) have all been hypothesized to play a role in the genesis of diarrhea in EPEC. The mechanism in EHEC remains unknown.

Though EPEC and EHEC share the A/E characteristic, vascular damage and secondary complications of hemolytic uremic syndrome (HUS) and thrombotic thrombocytopenic purpura (TTP) have been associated only with EHEC and specifically with the ability of EHEC populations to synthesize Shiga toxins (Karmali et al. 1983). These toxins, which are secreted by the bacteria colonizing the colon, pass through the intestinal wall via transcytosis (Acheson et al. 1996) and subsequently spread hemotogenously to target tissues. Ultimately, the toxins preferentially attack renal tissue because of their high affinity for receptors that are found predominantly in this tissue. After internalization by receptor-mediated endocytosis, the active subunit of the toxin catalytically inactivates ribosomes by cleaving a residue on the 28 S ribosomal RNA (Sandvig and van Deurs 1996). Cell death within the glomeruli ultimately results in tissue damage that can be further complicated by the ensuing inflammatory response (Inward et al. 1997).

Attachment. The process of A/E lesion formation in EPEC appears to comprise two stages: an initial phase of diffuse adherence followed by a secondary phase of high-affinity binding, formation of organized pedestal structures, and displacement of microvilli. Diffuse adherence in EPEC in the small intestine is mediated by the bundle-forming pilus, which facilitates formation of highly structured microcolonies on the host cell surface (Frankel et al. 1998; Donnenberg et al. 1997). The *bfp* genes, which collectively encode the bundle forming pilli, are present on the large virulence plasmid found in EPEC strains.

In EHEC and other A/E populations of *E. coli* the *bfp* genes are absent, suggesting that multiple mechanisms of diffuse adherence have emerged among different populations of A/E pathogens (Natarro and Kaper 1998). The precise mechanism of diffuse adherence in O157:H7 and other EHEC remains unknown, although several candidate loci have been identified. At least one of the candidate loci is encoded on pO157, the large virulence plasmid carried by O157:H7 and other EHEC strains, suggesting the phenotype may be plasmid mediated, as in EPEC, but nonetheless by distinct machinery. Initial reports described a fimbrial mechanism in O157:H7 that was presumably encoded by pO157 since the phenotype was lost in plasmid-cured strains (Karch et al. 1987). DNA sequence analysis of pO157 did not reveal any fimbrial genes (Makino et al. 1998; Burland et al. 1998), although it remains formally possible that a plasmid-encoded protein could mediate expression or function of chromosomally encoded fimbrial genes. Indeed, fimbrial gene clusters are present on the O157:H7 chromosome (Perna et al. 2001; Hayashi et al. 2001); their function, however, remains to be determined. Other studies did not detect fimbriae in O157:H7 strains bearing pO157, but nonetheless showed dependence of adherence phenotypes on the pO157 plasmid (Toth et al. 1990). Conflicting data from Fratamico et al (1993) indicated that pO157 was not required for attachment, implying that the diffuse adherence phenotype could be mediated by multiple systems in a strain-dependent and environment-dependent fashion.

Recent studies by Tatsuno et al. (2001) provide compelling evidence that the *toxB* locus of pO157 is necessary for diffuse adherence, because attachment of pO157-cured derivatives was shown to be defective and the phenotype was restored by introduction of a mini-pO157 derivative consisting only of the pO157 origin and the *toxB* gene. ToxB shares significant similarity with the chromosomal *eaf-1* gene from O111 EHEC

and Tn-phoA insertions into *eaf-1* reduce adherence in CHO cells (Nicholls et al. 2000). In addition to a function in adherence, ToxB may also have cytotoxicity activity, because it shares substantial similarity to the lymphotoxin LifA from O126 EPEC (Klapproth et al. 2000). ToxB and Eaf-1 also share N-terminal homology with the *toxA* and *toxB* genes of *Clostridium dificile* (Burland et al. 1998; Makino et al. 1998, Nicholls et al. 2000), further supporting a possible cytotoxicity activity.

The *toxB*⁻ phenotype with respect to attachment may be mediated by its effect on expression of EspA, rather than direct function of ToxB as a receptor. EspA expression is substantially decreased in the *toxB*⁻ background (Tatsuno et al. 2001). This finding is also congruent with studies of attachment in O26 EHEC showing that EspA was necessary for early stages in attachment, prior to pedestal formation (Ebel et al. 1998). EspA is better known for its function in formation of the type III secretion tube required to mediate the high-affinity binding phase of A/E lesion formation (see later in this section). Thus, EspA could play distinct roles by mediating early stages of diffuse adherence and subsequently serving as a conduit for secretion of effector proteins required for high affinity binding.

In addition to plasmid-encoded genes, a chromosomal gene from O157:H7 has been shown to confer diffuse adherence characteristics when cloned in K-12 strains of *E. coli*. The gene responsible for conferring the phenotype shares significant similarity to the iron-regulated protein IrgA of *Vibrio cholerae* and was subsequently named *iha* (Iron regulated protein Homologue Adhesin) (Tarr et al. 2000). The *iha* gene is part of an island, termed the tellurite adherence island (TAI), which encodes tellurite resistance in addition to the adhesin (Tarr et al. 2000). Genome sequence analysis of the U.S. O157:H7 strain EDL 933 (Perna et al. 2001) and the Japanese O157:H7 Sakai strain (Hayashi et al. 2001) revealed that the TAI island is actually part of a larger 88Kb island comprising three functional segments encoding a urease system, tellurite resistance, and the *iha* adhesin. The tellurite resistance region is similar to genes encoded by *Serratia marcessans* (Tarr et al. 2000), whereas the urease cluster is similar to a urease gene cluster of *Klebsiella aerogenes* (Heimer et al. 2002). The pattern of homology suggests that the region is a mosaic of functional segments acquired on independent occasions. Indeed, each of the functional regions of the island is separated by stretches of DNA encoding multiple insertion sequences and putative transposases. Whether the functional mosaic was amassed before entry into the O157:H7 lineage or built up in a stepwise fashion within O157:H7 remains to be determined.

Comparison of the EDL933 and Sakai genomes shows that the entire 88 Kb island is duplicated in EDL933. In both genomes, a copy of the island is inserted in the *serW* tRNA gene, but EDL933W also contains an additional identical copy inserted 370 Kb downstream in the *serX* tRNA locus. The simplest model is that the ancestral island was duplicated and subsequently transposed in EDL933 and that the TAI found at *serW* reflects the ancestral state.

The TAI island is found among O157:H7 strains that are non-sorbitol fermenting and β-glucuronidase negative. Populations of sorbitol-fermenting, β-glucuronidase positive O157:H- EHEC (SFO157 EHEC) strains that are common to Germany and central Europe, however, lack this island (Tarr et al. 2000) and are believed to have diverged early during the ontogeny of the O157:H7 lineage (Feng et al. 1998) (see "Phylogeny of the O157:H7 Lineage," later in this chapter). In addition to their unique sorbitol and β-glucuronidase phenotypes, the virulence plasmid in SFO157EHEC is also distinct from the pO157 (referred to as pSFO157) and encodes a unique fimbrial system that facilitates mannose-sensitive hemeagglutination (Brunder et al. 2001). Although the contribution of this fimbrial system and other systems described previously to the diffuse adherence phenotype remains to be fully elucidated, it seems plausible that independent mechanisms for diffuse adherence have been acquired on different occasions during radiation of divergent subpopulations within the O157:H7 EHEC 1 lineage.

After an initial phase of diffuse adherence, the characteristic A/E lesion begins to appear, marked by the emergence of pedestals located immediately below the bacterium on the host cell surface and the associated displacement of neighboring microvilli. The known genes required for A/E formation are encoded within a pathogenicity island known as the locus of enterocyte effacement (LEE). LEE has been found in various pathogenic types of *E. coli* as well as *Citrobacter rodentum*, (Fig. 17.1), and DNA sequence analyses of the island

from these organisms demonstrates that it is a highly conserved element (McDaniel et al. 1995; Perna et al. 1998; Zhu et al. 2001; Deng et al. 2001). Genes encoding components of the type III secretion system (*esc* genes) comprise a large proportion of the island. Several of the secreted effector proteins (*espA, espB, espD,* and *espF*) are encoded at one end of the island and the adhesin intimin, along with its Translocated Intimin Receptor (Tir), are encoded in the central portion by the *eaeA* and *tir* genes, respectively.

The type III secretion system plays a critical role in the intimate attachment process by providing a specialized conduit for secretion of the effector proteins from the bacterium into host cells (for a recent review, see Vallance and Finlay 2000). The secreted effector proteins set the stage for translocation and function of a receptor protein, Tir, which serves as a receptor for bacterial binding and as an organizational center for remodeling the host cell cytoskeleton at the point of attachment. After its arrival in the host cell cytoplasm, Tir localizes to the plasma membrane of the host cell, where it serves as the receptor for attachment of the bacterium (Kenney et al. 1997; Ebel et al. 1998; DeVinney et al. 1999). The bacterium attaches to Tir via the intimin protein (EaeA), which is expressed on the outer membrane of the bacterium (Rosenshine et al. 1996). The intimin-Tir interaction is crucial to pathogneesis of EPEC and EHEC, and strains lacking either gene are avirulent in animal models (Marches et al. 2000; Dean-Nystrom et al. 1998; McKee et al. 1995; Donnenberg et al. 1993a, 1993b).

The effector proteins appear to promote the attachment process by facilitating delivery and trafficking of

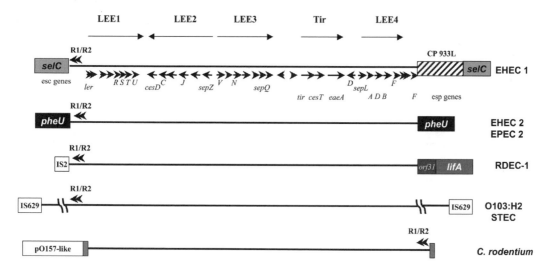

FIG. 17.1—Map and alignment of LEE pathogenicity islands from different lineages of *E. coli* and *C. rodentium*. Arrows at the top of the map indicate orientation and composition of the five major transcription units (LEE 1–4 and Tir) within the LEE island. Small arrows indicate the relative position of different genes in the LEE island (not to scale). Genes with characterized functions are indicated below the arrows. Genes and elements adjacent to the island are indicated in the various strains that have been examined. The R1/R2 region is also indicated in the EHEC 1 lineage and the *C. rodentium* lineage, suggestive of the mechanism proposed by Deng et al. 2001.

Tir to its appropriate place in the membrane. EspA appears to play a role in formation of the "barrel" of the type III secretion system (Knutton et al. 1998). EspB and EspD, which have hemolytic activity, may play roles as pore-forming molecules that provide a portal for the EspA secretion tube to breech the host cell membrane (Warawa et al. 1999). CesT, on the other hand, likely serves as a chaperone to traffic Tir into the host cell membrane (Abe et al. 1999; Elliott et al. 1999a).

In addition to serving as a receptor for intimin, plasma membrane-bound Tir also coordinates remodeling of the host cell cytoskeleton. This is accomplished by binding of Tir to the host cell protein Nck and subsequent recruitment of N-WASP and the Arp 2/3 complex (Gruenheid et al. 2001; Kalman et al. 1999). Together the N-WASP and Arp2/3 complexes serve to reshape local actin filaments, leading to localized complexes of actin polymerization crosslinked with F-actin, β-actin, talin, ezrin, and villin. Tir itself appears to be anchored to the actin complexes via β-actinin (Goosney et al. 2000).

The LEE island genes are organized into four primary transcription units, LEE1-LEE4 (Elliott et al. 1999b; Mellies et al. 1999). Transcription of these operons is coupled to at least two different regulatory systems. In O157:H7 and in EPEC, the Ler protein—encoded by the LEE1 operon—functions as a positive regulator of LEE2, LEE3, and *tir* operon transcription. The Per locus, encoded by the EAF plasmid in EPEC, also serves to modulate Ler activity; however, these genes are not found in EHEC (Mellies et al. 1999). A quorum-sensing molecule, whose synthesis is encoded by the *luxS* gene, also appears to exert some degree of regulation on these operons. In both EPEC and O157:H7 EHEC, LEE1 and LEE2 are directly activated by the quorum sensing system, whereas LEE3 and *tir* are indirectly activated via the enhanced levels of Ler produced from LEE1 transcripts (Sperandio et al 1999). The recently discovered *luxS*-dependent quorum sensing system may be a widespread, highly conserved pathway for synthesis of a common autoinducer molecule that can be detected by a wide variety of Gram-positive and –negative bacteria (Schrauder et al. 2001). Thus, instead of being a system dedicated to sensing its own density, it may provide *E. coli* with information regarding density of bacteria in its environment, and thus a way to distinguish sparsely populated extraintestinal niches from cell-dense intestinal environments. In addition to the effects of LuxS on LEE gene expression, analysis of global gene expression patterns in *luxS*+ and *luxS*- backgrounds yielded several categories of differentially regulated genes, including positive regulation of flagellar, chemotaxis genes, and Shiga toxin genes and negative regulation of cell division and ribosomal protein genes (Sperandio et al. 2001). This suggests that the role of quorum sensing in general is to downregulate growth and upregulate virulence genes (in lineages such as EHEC) as the bacteria enter a potential host environment and encounter competition from other enteric species.

Although sequence analysis of LEE shows that the islands are highly homologous in the different A/E *E. coli* populations, the islands are not necessarily equivalent. One of the first indications of functional diversity came from studies demonstrating that cloned copies of LEE from EPEC could confer A/E characteristics to K-12 strains of *E. coli*, whereas LEE from O157:H7 EHEC could not (Karaolis et al. 1997; McDaniel et al. 1997; Elliott et al. 1999b). Subsequent analysis of Tir function between EPEC and EHEC demonstrated significant differences in their phosphorylation patterns and further showed that the two are not functionally interchangeable (DeVinney et al. 1999; DeVinney et al. 2001). Co-infection studies provided evidence that additional effector proteins are necessary for function of EHEC Tir (DeViney et al. 2001).

Further support for diversification and specialization of LEE in different lineages comes from comparative sequence analysis of LEE-encoded proteins. In general, greater diversity is observed among genes coding for host-interacting proteins (Esps, Tir, and intimin) versus the type III secretion machinery, and the diversity is typically greater than that observed between the various genome backbones of the lineages in which they reside (Perna et al. 1998; Zhu et al. 2001). Nonetheless, sequence divergence seems to follow a pattern more consistent with the bacterial lineage in which LEE resides rather than a pattern reflecting adaptation to a particular host species (Zhu et al. 2001). This would imply that adaptation of LEE likely reflects that context of the genome backbone into which it has been acquired, including the niche specialization characteristics of the lineage and any additional virulence characteristics the lineage might already possess.

The LEE element constitutes a pathogenicity island because of its alien nucleotide composition and codon usage patterns relative to the *E. coli* genome (McDaniel et al. 1995; Perna et al. 1998; Zhu et al. 2001), and its presence at various positions in the *E. coli* genome indicate that it has been acquired on several independent occasions. In both O127:H6 and O157:H7, LEE is inserted at the *selC* gene; however, in the EPEC 2/EHEC 2 lineage, it is inserted at *pheU* (Wieler et al. 1997; Sperandio et al. 1998). In the more distantly related rabbit diarrhegenic *E. coli* RDEC-1, it remains unknown precisely where LEE is inserted, but it is not present at *selC* or *pheU* (Zhu et al. 2001). LEE of O157:H7 contains an additional thirteen genes downstream of the LEE4 operon that encode components of the cryptic prophage 933L (Perna et al. 1998). At this same position, LEE of the EPEC 1 strain, E2348/69, which is also inserted at *selC*, contains a copy of IS 600 (McDaniel et al. 1995; Sperandio et al. 1998). The corresponding position of LEE present in the RDEC-1 representative strain contains the *lifA* gene and an unknown orf (Zhu et al. 2001), leading these authors to speculate that the region adjacent to the LEE4 operon is subject to frequent recombination.

Insights as to a mechanism by which LEE is acquired were recently obtained from DNA sequence analysis of the LEE island from *C. rodentium*, where regions of substantial similarity to the large virulence plasmid pO157 and the R100 plasmid of *Shigella* lie adjacent to LEE, in the position normally occupied by the R1/R2 region (Deng et al. 2001). R1/R2 in *C. rodentium* is positioned at the opposite end of LEE, downstream of the LEE4 operon. The adjacent plasmid sequences suggest a model in which LEE is transferred laterally on conjugative plasmids. Moreover, the alternative location of R1/R2 in *C. rodentium* versus EPEC and EHEC implies that the ancestral LEE-containing plasmid carried the R1/R2 segment downstream of LEE4; a single recombination event between LEE4 and R1/R2, followed by loss of plasmid sequences, would lead to the orientation of LEE found in EPEC and EHEC, whereas a single recombination event downstream of R1/R2 would lead to the architecture observed in *C. rodentium* (Deng et al. 2001). This simple but elegant model for mobility of LEE is attractive because it relies upon known mechanisms for plasmid integration (that is, Campbell-type integration) and leaves only site selection and loss of plasmid sequences as unexplained components of the contemporary LEE architectures. However, studies from Jores et al. (2001) show that there may be multiple mechanisms for mobility of LEE, because LEE is integrated in O103:H2 strains within a larger complex island and is flanked by inverted copies of IS629, suggestive of a composite transposon.

It is clear from the detailed studies of the A/E process in EPEC and EHEC that multiple populations of *E. coli* and other enteric species have emerged around the central characteristic of A/E lesion formation. Specialization of the A/E systems in different populations of *E. coli*, coupled with a variety of additional attachment factors found in these populations, is likely to be a major factor in defining the particular tissue that is affected as well as the degree of virulence. Other virulence factors, such as toxins, serve as accessory factors in EHEC and mediate further aspects of pathogenesis, namely secondary complications in the host.

Toxin Synthesis and Secretion. Synthesis and secretion of the Shiga toxins is one of the major characteristics that discriminates EHEC from other A/E-forming bacteria, and this difference is generally accepted as the explanation for why EHEC are capable of causing extraintestinal complications such as HUS and TTP. The toxin name is derived from the prototype Shiga toxin from *Shigella dysenteriae*. Two types of toxins, Stx1 and Stx2, are produced by STEC. Stx1 and Shiga toxin from *S. dysenteriae* are nearly identical, whereas Stx2 shares only 55–60 percent amino acid identity with Shiga toxin and Stx1 (Paton and Paton 1998). Structurally, the Shiga toxins are members of a larger group of A_1B_5 toxins, which share a similar three-dimensional structure of a pentameric binding (B) component, comprised of five identical StxB subunits, and a monomeric catalytic (A) subunit derived from the StxA protein. The binding pentamers of these toxins all bind to glycolipid receptors and further serve to deliver the catalytic subunit to the appropriate cellular compartment. The A chain of Stx has two domains that can be separated proteolytically with trypsin: an A1 domain that cleaves the N-glycosidic bond of a specific adenine on the 28S rRNA, and an A2 domain that serves a structural role in tethering the A1 chain to the B pentamer (Endo et al. 1988; Saxena et al. 1989). Structural analyses of the Shiga holotoxin and Stx1 B-pentamer show striking structure relationships with B-pentamers of the Cholera toxin and *E. coli* Labile toxin, despite the fact that no observable sequence identity exists between them (Fraser et al. 1994; Sixma et al. 1993; Stein et al. 1992). Cholera Toxin, Labile Toxin, and Stx also share a common substructure, the oligonucleotide/oligosaccharide binding (OB) fold with Staphylococcal nuclease and tRNA synthetases (Muzrin 1993). Because these proteins adopt the OB fold structure in the absence of recognizable primary sequence similarity, it is not clear whether they share deep but common ancestry or whether they represent convergent evolution.

Because EHEC are not known to be highly invasive, toxin synthesis likely occurs exclusively during colonization of the colon. After they are synthesized and secreted, the toxins are transported by transcytosis through the lumen and spread by hemotogenous routes to their primary target organs, the kidney and central nervous system tissue (Tzipori et al. 1988; Wadolkowski et al. 1990). The toxins bind glycolipid receptors comprised of globotriaosylceramide (Gb3) (Jacewicz et al. 1986; Lindberg et al. 1987; Waddell et al. 1988). Indeed, this characteristic of Stx spears to define the target cell range, because cells lacking Gb3 are resistant to the toxins (Weinstein et al. 1989). After receptor binding, the toxin is transported via retrograde transport to the golgi and ER, a process that displays some cell-type specificity and likely provides additional points of divergence in susceptibility of certain cell types (Sandvig et al. 1992; Sandvig and van Deurs 1996). Catalytic inactivation of ribosomes ultimately results in death of the cell by apoptosis (Yoshida et al. 1999). In addition to their function as cytotoxins, recent studies also point to roles of the toxins in mediating tissue necrosis in the intestine by stimulating production of proinflamitory mediators (Thorpe et al. 1999, 2001). On top of tissue damage caused by cell death, localized inflammatory responses could presumably further serve to mediate necrosis of colonic tissue and facilitate adsorption of toxin into the bloodstream.

Stx1 genes that have been sequenced show only limited polymorphism. Of the three initially identified, all were identical to one another and differed from the *S. dysenteriae stx* by three nucleotides (DeGrandis et al. 1987; Jackson et al. 1987; Kozlov et al. 1988). Only one variant, *stx1c*, has been identified in STEC; it has been observed in strains isolated from humans and sheep but is not observed among O157:H7 strains that have been tested (Zhang et al. 2002). In contrast, several variants of *stx2* have been found, including *stx2c*, *stx2d*, *stx2e*, and *stx2f*, which, relative to *stx2*, range from 99–63 percent identity in the A subunits and 95–75 percent identity in the B subunits (Schmitt et al. 1991; Pierard et al. 1998; Weinstein et al. 1988; Schmidt et al. 2000). Given the allelic diversity observed within *stx2* and the lack thereof in *stx1*, it is believed that *stx1* has only recently been introduced into STEC.

Although both Stx1 and Stx2 share similarity in structure and mechanism of action, epidemiological studies and *in vitro* studies suggest that Stx2 is more toxic than Stx1 and has a higher propensity to cause HUS than does Stx1 (Boerlin et al. 1999; Ostroff et al. 1989; Tesh et al. 1993; Wadolkowski et al. 1990). Even among the Stx2 variants, there appear to be significant differences in the likelihood of causing HUS (Friedrich et al. 2002). The most significant risk seems to be posed by strains carrying Stx2, regardless of the EHEC lineage. Stx2 is by far the most common among clinical strains, both in the O157:H7 and the non-O157:H7 EHEC lineages. However, there is a skewed distribution of the variant alleles; *stx2c* is the only variant allele found among O157:H7 strains, whereas *stx2d* is more common among non-O157:H7 EHEC 2 lineage strains (Friedrich et al. 2002).

Wide dissemination of *stx* genes among *E. coli* populations is a consequence of the fact that they are phage encoded. The ancestral *stxAB* genes encoding Stx in *S. dysenteriae* are chromosomal but are flanked by remnants of prophage-like sequences (Unkmeir and Schmidt 2000). In every instance that has been examined, *stx* genes in *E. coli* are found encoded on lambdoid prophage genomes or are associated with phage-like genes (Unkmeir and Schmidt 2000; Karch et al. 1999; Makino et al. 1999; Plunkett et al. 1998; Neely and Friedman 1998; Huang et al. 1987; Newland et al. 1985; O'Brien et al. 1984). Of the *stx* genes whose positions have been mapped, all lie in the same region of the prophage, immediately downstream of the Q antiterminator of the prophage genome (Plunkett III et al. 1999; Neely and Friedman 1998; Makino et al. 1999; Karch et al. 1999; Unkmeir and Schmidt 2000). The region of the prophage genome between the Q gene and the R and S holin genes is known to be highly polymorphic among lambdoid phages (Campbell 1994, 1998). Congruently, the corresponding segment immediately downstream of the *stx* genes is indeed quite diverse (Unkmeir and Schmidt 2000; Karch et al. 1999). In the instance of stx2 prophages, three tRNA genes are also found preceding the *stxA* coding region. These segments putatively encode an isoleucine and two arginine tRNAs (Plunkett et al. 1999). The anticodon loops of the latter are found among other bacteriophage encoding tRNAs but apparently are not observed among known bacterial tRNAs (Plunkett et al. 1999). Codon usage analysis of the 933W prophage genome and the K-12 genome indicates that the putative codon recognized by the phage-borne isoleucine tRNA is very low-frequency in the K-12 genome but significantly inflated in genes encoded by the 933 prophage, particularly the *stxA* gene (Plunkett et al.

1999). Whether these unusual tRNAs actually increase expression of stxA and other phage genes, however, remains to be determined.

Given the position of the *stx* genes in the prophage genome, it is generally held that *stx* expression is driven from the PR′ promoter late in the lytic cycle and it depends on antitermination from PR′ by the Q gene (Plunkett III et al. 1999; Neely and Friedman 1998). This view was supported by studies showing that expression of a *stx2A::phoA* translational fusion, containing the *Q-stxA* region, was activated in trans by one or more genes present on 933W or H19B phages (Muhldorfer et al. 1996). Although transcripts originating immediately upstream of *stx1A* and *stx2A* can be detected during lysogeny (Calderwood et al. 1987 and Sung et al. 1990), the most significant toxin gene expression seems to derive from late gene transcription because deletion of the PR′ promoter nearly abolishes toxin transcription *in vitro* and leads to substantially decreased detectable toxin in the stools of mice infected with these mutant strains (Wagner et al. 2001). Because high-level toxin expression seems to depend on prophage induction, several studies have examined the effect of antibiotics on prophage induction and toxin expression (Kohler et al. 2000; Matshuiro et al. 1999; Grif et al. 1998), demonstrating that several classes of antibiotics do, in fact, stimulate prophage induction and subsequent toxin expression. This finding, coupled with ineffectiveness and potential risk of some antibiotics in clinical settings, has led some to contraindicate antibiotics as therapeutic agents in treating EHEC infections (Wong et al. 2000).

Genome analysis and recent genome sequencing has demonstrated that O157:H7 strains typically carry multiple lambdoid prohages and cryptic prophages, in addition to Stx encoding phage. Genome sequences of the EDL933 and Sakai O157:H7 strains reveal substantial evidence of strain level diversity caused by prophage content; eighteen different propohage or cryptic prophage in the EDL933 genome and twenty two in the Sakai genome (Perna et al. 2001; Hayashi et al. 2001). Even within the Stx2-encoding prophages in these genomes, which are found at identical positions in the genomes, six different regions of non-homology can be observed between the 933W and VT2Sakai prophages (Makino et al. 1999). As is typical of phage genomes (Campbell 1994, 1998), these regions of non-homology are flanked by regions of near identity. Prophage diversity has also been observed among O157:H7 strains by hybridization with lambda or lambdoid phage DNA (Samadpour et al. 1993; Kim et al. 1999), hybridization with lambda O and P replication genes (Datz et al. 1996), OBGS (Kim et al.1999), and by identification of regions giving rise to distinct PFGE profiles (Kudva et al. 2002). Though the diversity in prophage content can produce different patterns of superinfection immunity—a characteristic that could be important upon entry into a new host—the more relevant question is whether any of the genes encoded by non-Stx prophage or cryptic prophage can effect toxin expression or entry of the Stx prophage into the lytic cycle. If Stx prophage induction is subject to modulation by other prophage, then the tremendous diversity observed with regard to prophage content of O157:H7 and other EHEC could lead to quantitative differences in toxin production, and ultimately lead to differences in outcome of human infection.

PHYLOGENY OF THE O157:H7 LINEAGE. *E. coli* O157:H7 is believed to be an emerging pathogen on the basis of its recent association with clinical disease (Riley 1987). A framework for the ontogeny of the O157:H7 lineage was first provided by Multi-Locus Enzyme Electrophoresis (MLEE) studies of O157:H7 and other EHEC and EPEC strains (Whittam et al. 1993). This study demonstrated that the EHEC phenotype was distributed among two phylogenetically distinct lineages. One lineage comprised a cluster of related sorbitol nonfermenting β-glucuronidase–negative O157:H7 strains and the SFO157 EHEC strains (EHEC 1), whereas the second comprised O111:H8, O111:H-, and O26:H11 strains (EHEC 2). The EHEC 1 lineage was similar to a cluster of EPEC strains, typified by O127:H6, O5:H6, and O142:H6 (EPEC 1); however, EHEC 1 was most similar to an unusual O55:H7 EPEC strain, leading the authors to speculate that the two shared recent common ancestry (Whittam et al. 1993). The EHEC 2 lineage, on the other hand, shared similarity with a second lineage of EPEC comprised of O111:H2, O128:H2, O111:H21, and O128:H21.

In addition to genetic relationship predicted by MLEE, other lines of study also support the hypothesis that EHEC 1 / EPEC 1 and EHEC 2 / EPEC 2 lineages share common ancestry. First, EPEC 1, the unusual O55:H7 EPEC, and EHEC 1 all carry the LEE island inserted at the *selC* locus, whereas EPEC 2 and EHEC 2 strains carry LEE in the *pheU* locus (Wieler et al. 1997). In the instance of the EHEC 1 / EPEC 1 lineage, comparison of single nucleotide polymorphisms in the *uidA* gene of O55:H7, O157:H7, and SFO157:H- EHEC strains show a progressive pattern of accumulation, as would be predicted from stepwise buildup of mutations among populations sharing recent common ancestry (Feng et al. 1998; Monday et al. 2001), and multilocus sequence comparison of other loci also give rise to a similar inferred phylogeny (Reid et al. 2000). Based on these studies, a model (Fig. 17.2) for the stepwise evolution of the contemporary sorbitol nonfermenting, β-glucuronidase–negative O157:H7 populations has been proposed (Feng et al. 1998; Whittam et al. 1998; Reid et al. 2000). In this model, the EHEC 1 and EPEC 1 lineages descended from a common ancestor that acquired LEE at the *selC* locus. From this ancestor, an O55:H7 EPEC population emerged, which subsequently acquired Stx2, presumably by phage conversion. The O157:H7 population arose by transfer of the *rfb* encoding the O157 antigen into a LEE+, stx2+ O55:H7 EPEC strain. The transfer event has occurred on multiple occasions since a common O157 antigen-encoding *rfb* region has been found among EHEC and

unrelated non-EHEC O157 strains (Tarr et al. 2001). Transfer of this region into an Stx2+ O55:H7-like population therefore led to a LEE+, Stx2+, sorbitol-fermenting, β-glucuronidase–positive ancestor. This ancestor subsequently was lysogenized by an Stx1 prophage and acquired the pO157 plasmid and the TAI locus, leading to the immediate ancestor of contemporary sorbitol nonfermenting, β-glucuronidase–negative O157:H7 strains that are found worldwide. Prior to acquisition of TAI and loss of β-glucuronidase and sorbitol fermentation characteristics, the SFO157:H7 clone diverged and subsequently lost motility. Substantial differences between pSFO157 and pO157 have been observed, making it unclear whether pSF1057 plasmid is a derived state from the pO157 or whether it was independently acquired (Brunder et al. 1999, Brunder et al. 2001). Despite the differences between virulence factors of O157:H7 and the SFO157 EHEC, they are nonetheless capable of causing disease. If the inferred phylogeny is correct and if the recently described fimbrial system encoded on pSFO157 mediates diffuse adherence, then independent mechanisms for diffuse adherence are likely to have evolved even within the EHEC 1 lineage.

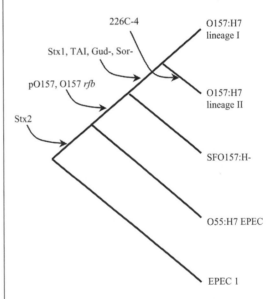

FIG. 17.2—Proposed steps marking descent of contemporary *E. coli* O157:H7 populations. In this model, the pathway proceeds from ancestral (left) to contemporary (right) and is based on models proposed by Feng et al. 1998 and Kim et al. 2001. Major steps include the acquisition of the Stx2-converting prophage in a variant O55:H7 EPEC population, acquisition of the O157 *rfb* region and the pO157-like plasmid, prior to divergence of the SFO157:H- population, acquisition of the Stx1 converting phage and the TAI island, and loss of the β-glucuronidase production (Gud-) and Sorbitol fermentation (Sor-) characteristics. The last step is marked by a 9-base duplication in the *folD-sfmA* intergenic region in lineage II.

Despite the fact that sorbitol nonfermenting, β-glucuronidase–negative O157:H7 strains are found worldwide, studies using PFGE to track these populations in epidemiological studies consistently reveal significant genome diversity among different strains (for example, see Faith et al. 1996; Shere et al. 1998; Gouveia et al. 1998). The diversity implies that the multiple subpopulations have recently radiated from this clone. Indeed, high-resolution genome comparison of strains isolated from human clinical samples and bovine samples subsequently showed that the common sorbitol nonfermenting, β-glucuronidase–negative O157:H7 clone comprises at least two major subpopulations, termed lineage I and lineage II (Kim et al. 1999). Probing of strain sets with lambdoid phage genomes, and phylogenetic analyses of the resulting hybridization patterns, also produced a concordant inferred phylogeny showing two lineages, supporting the conclusion that phage-mediated events were associated with, or even could have caused divergence of, the lineage I and lineage II subpopulations. Comparison of O157:H7 and O157:H-EHEC strains from the United States and Australia provided insight into the ancestry of these lineages, showing that populations on both continents were descendants of lineage I and lineage II strains (Kim et al. 2001). This finding is consistent with a model in which a unique population of sorbitol nonfermenting, β-glucuronidase–negative O157:H7 diverged at a single point in time and space (for example, it was subjected to an evolutionary bottleneck) and the descendants of the event have subsequently spread globally. Several markers of the divergence were detected in these studies, including the 226C-4 marker, which corresponds to an 8-base duplication in the intergenic region of the divergent *folD* and *sfmA* genes and is found in all lineage II strains examined to date (Kim et al. 2001).

Perhaps the most surprising finding from these high-resolution population studies was that human clinical isolates and bovine isolates were nonrandomly distributed between the two lineages; most lineage I strains were human derived, whereas most lineage II strains were derived from cattle (Kim et al. 1999). Biased distribution of clinical and bovine strains suggests a model in which the two lineages are not equally capable of causing disease in humans or are not equally capable of being transmitted to humans from contaminated beef. When strains from Australia were examined, however, several human isolates were detected in lineage II, suggesting that the relatively small sample sizes of strains used in the studies are biased or that the Australian lineage II strains are distinct from those found in the United States (Kim et al. 2001).

Insights into the differential virulence hypothesis have been provided also by phenotypic studies. In support of the differential virulence hypothesis, studies reported by Baker et al. (1997) detected quantitative differences between human clinical and bovine O157:H7 strains in their ability to induce brain lesions in the gnotobiotic pig model. McNally et al. (2001) recently compared the amounts of secreted LEE island proteins from bovine and human clinical isolates of O157:H7 and found statistically significant differences in secretion of EspD and Tir secretion in the two types of strains. Although most human strains secreted EspD and Tir, these proteins were nearly undetectable in the supernatants of most bovine strains. The two groups of strains also showed significant differences in their ability to adhere to HeLa cells. It remains to be determined, however, whether the bovine and human strains in this study correspond to the lineage I and II populations detected by Kim et al. (1999).

SUMMARY AND CONCLUSIONS. The combination of molecular and population genetic analysis of O157:H7 and other EHEC strains has provided a glimpse into the vast array of mechanisms that have evolved in enteric bacteria to facilitate colonization of the intestinal tract. Based on the current understanding, several key questions remain. First, there are still many unknown aspects of virulence in these organisms, such as the basis for diffuse adherence and diarrheagenesis in O157:H7. Also, many unresolved issues remain regarding lateral gene transfer. Though lateral transfer clearly has played a large role in shaping the genomes of contemporary *E. coli* populations, little is known regarding the mechanisms of transfer events, the frequency at which such events occur, and whether common patterns or sequences of transfer events reflect chance or selective pressures. Of course, one of the holy grails for microbiologists, epidemiologists, and evolutionary biologists is to understand whether the apparent emergence of O157:H7 as a human pathogen was the consequence of manmade events, such as changes in agricultural practices, or an event entirely beyond our control. Though answers to the latter are likely to be complex, the combined approaches offered by epidemiology, molecular biology, population genetics, and genomics provide a powerful arsenal to aim at this question.

REFERENCES

Abe, A., M. de Grado, R.A. Pfuetzner, C. Sanchez-Sanmartin, R. Devinney, R., J.L. Puente, N.C. Strynadka, and B.B. Finlay. 1999. Enteropathogenic *Escherichia coli* translocated intimin receptor, Tir, requires a specific chaperone for stable secretion. *Mol. Microbiol.* 33:1162–1175.

Acheson, D.W.K., R. Moore, and S. DeBreuler. 1996. Translocation of Shiga-like toxins across polarized intestinal cells in tissue culture. *Infect. Immun.* 64:3294–3300.

Baker, D.R., R.A. Moxley, and D.H. Francis. 1997. Variation in virulence in the gnotobiotic pig model of O157:H7 *Escherichia coli* strains of bovine and human origin. *Adv. Exp. Med. Biol.* 412:53–58.

Boerlin, P., S.A. McEwen, F. Boerlin-Petzold, J.B. Wilson, R.P. Johnson, and C.L. Gyles. 1999. Associations between virulence factors of Shiga toxin-producing *Escherichia coli* and disease in humans. *J. Clin. Microbiol.* 37:497–503.

Brunder, W., A. S. Kahn, J. Hacker, and H. Karch. 2001. Novel type of fimbriae encoded by the large plasmid of sorbitol-fermenting enterohemorrhagic *Escherichia coli* O157:H-. *Infect. Immun.* 69:4447–4457.

Brunder, W., H. Schmidt, M. Frosch, and H. Karch. 1999. The large plasmids of Shiga-toxin-producing *Escherichia coli* (STEC) are highly variable genetic elements. *Microbiology* 145:1005–1014.

Burland, V., Y. Shao, N.T. Perna, G. Plunkett, J.J. Sofia, and F.R. Blattner. 1998. The complete DNA sequence and analysis of the large plasmid of Escherichia coli O157:H7. *Nucleic Acids Res.* 26:4196–4204.

Calderwood, S.B., F. Auclair, A. Donahue-Rolfe, G.T. Kneusch, and J.J. Meklanos. 1987. Nucleotide sequence of the Shiga-like toxin genes of *Escherichia coli*. *Proc. Natl. Acad. Sci. USA* 84:4364–4368.

Campbell, A.M. 1994. Comparative molecular biology of lambdoid phages. *Annu. Rev. Microbiol.* 48:193–222.

Campbell, A.M. 1988. Phage evolution and speciation. In *The Bacteriophages*. R. Calendar, ed. New York, NY: Plenum Press, 1–14.

Collington, G.K., I.W. Booth, M.S. Donnenberg, J.B. Kaper, and S. Knutton. 1998. Enteropathogenic *Escherichia coli* virulence genes encoding secreted signalling proteins are essential for modulation of Caco-2 cell electrolyte transport. *Infect. Immun.* 66:6049–6053.

Datz, M., C. Janetzki-Milkman, S. Franke, F. Gunzer, H. Schmidt, and H. Karch. 1996. Analysis of the enterohemorrhagic *Escherichia coli* O157 DNA region containing lambdoid phage gene p and Shiga-like toxin structural genes. *Appl. Environ. Microbiol.* 62:791–797.

Dean-Nystrom, E.A., B.T. Bosworth, H.W. Moon, and A.D. O'Brien. 1998. *Escherichia coli* O157:H7 requires intimin for enteropathogenicity in calves. *Infect. Immun.* 66:4560–4563.

DeGrandis, S., J. Ginsberg, M. Toone, S. Climie, J. Friesen, and J. Brunton. 1987. Nucleotide sequence and promoter mapping of the *Escherichia coli* Shiga-like toxin operon of bacteriophage H19B. *J. Bacteriol.* 169:4313–4319.

Deng, W., Y. Li, B.A. Vallance, and B.B. Finlay. 2001. Locus of enterocyte effacement from Citrobacter rodentium: Sequence analysis and evidence for horizontal transfer among attaching and effacing pathogens. *Infect. Immun.* 69:6323–6335.

DeVinney, R., M. Stein, D. Reinscheid, A. Abe, S. Ruschkowski, and B.B. Finlay. 1999. Enterohemorrhagic Escherichia coli O157:H7 produces Tir, which is translocated to the host cell membrane but is not tyrosine phosphorylated. *Infect. Immun.* 67:2389–2398.

DeVinney, R., J.L. Puente, A. Gauthier, D. Goosney, and B.B. Finlay. 2001. Enterohaemorrhagic and enteropathogenic *Escherichia coli* use a different Tir-based mechanism for pedestal formation. *Mol. Microbiol.* 41:1445–58.

Donnenberg, M., C. Tacket, S. James, G. Losonsky, J. Nataro, S. Wasserman, J. Kaper, and M. Levine. 1993a. Role of the *eaeA* gene in experimental enteropathogenic *Escherichia coli* infection. *J. Clin. Investig.* 92:1412–1417.

Donnenberg, M., S. Tzipori, M. Mckee, A.D. O'Brien, J. Alroy, and J. Kaper. 1993. The role of the *eaeA* gene of enterohemorrhagic *Escherichia coli* in intimate attachment in vitro and in a porcine model. *J. Clin. Investig.* 92:1418–1424.

Donnenberg, M.S., J.B. Kaper, and B.B. Finlay. 1997. Interactions between enteropathogenic Escherichia coli and host epithelial cells. *Trends Microbiol.* 5:109–114.

Donnenberg, M.S. and T.S. Whittam. 2001. Pathogenesis and evolution of virulence in enteropathogenic and enterohemorrhagic *Escherichia coli*. *J. Clin. Invest.* 107:539–548.

Ebel, F., T. Podsadel, M. Rohde, A.U. Kresse, S. Kramer, C. Diebel, C.A. Guzman, and T. Chakraborty. 1998. Initial binding of Shiga toxin-producing *Escherichia coli* to host cells and subsequent induction of actin rearrangement depend on filamentous EspA-containing surface appendages. *Mol. Microbiol.* 30:147–161.

Elliott, S.J., S.W. Hutcheson, M.S. Dubois, J.L. Mellies, L.A. Wainwright, M. Batchelor, G. Frankel, S. Knutton, and J.B. Kaper. 1999a. Identification of CesT, a chaperone for the type III secretion of Tir in enteropathogenic Escherichia coli. *Mol. Microbiol.* 33:1176–1189.

Elliott, S.J., J. Yu, and J.B. Kaper. 1999b. The cloned locus of enterocyte effacement from enterohemorrhagic *Escherichia coli* O157:H7 is unable to confer the attaching and effacing phenotype upon *E. coli* K-12. *Infect. Immun.* 67:4260–4263.

Endo, Y., K. Tsurugi, T. Yutsudo, Y. Takeda, Y. Ogasawara, and K. Igarashi. 1988. Site of action of a Verotoxin (VT2) from *Escherichia coli* O157:H7 and of Shiga toxin on eukaryotic ribosomes. *Eur. J. Biochem.* 171:45–50.

Faith, N.G., J.A. Shere, R. Brosch, K.W. Arnold, S.E Ansay, M.S. Lee, J.B. Luchansky, and C.W. Kaspar 1996. Prevalence and clonal nature of *Escherichia coli* O157:H7 on dairy farms in Wisconsin. *Appl. Environ. Microbiol.* 62:1519–1525.

Feng, P., K. A. Lampel, H. Karch, and T.S. Whittam. 1998. Genotypic and phenotypic changes in the emergence of *Escherichia coli* O157:H7. *J. Infect. Dis.* 177:1750–1753.

Frankel G., A.D. Phillips, I. Rosenshine, G. Dougan, J.B. Kaper, and S. Knutton. 1998. Enteropathogenic and enterohaemorrhagic Escherichia coli: more subversive elements. *Mol. Microbiol.* 30:911–21.

Fraser, M.E., M.M. Chernaia, Y.V. Kozlov, and M.N. James. 1994. Crystal structure of the holotoxin from Shigella dysenteriae at 2.5 A resolution. *Nat. Struct. Biol.* 1:59–64.

Fratamico, P.M., S. Bhaduri, and R.L. Buchanan. 1993. Studies on Escherichai coli serotype O157:H7 strains containing a 60-MDa plasmid and on 60-MDa plasmid-cured derivatives. *J. Med. Microbiol.* 39:371–381.

Friedrich, A. W., M. Bielaszewska, W.L. Zhang, M. Pulz, T. Kuczius, A. Ammon, and H. Karch. 2002. *Escherichia coli* harboring Shiga toxin 2 gene variants: frequency and association with clinical symptoms. *J. Infect. Dis.* 185:74–84.

Goosney D.L., R. DeVinney, R.A. Pfuetzner, E.A. Frey, N.C. Strynadka, and B.B. Finlay. 2000. Enteropathogenic *E. coli* translocated intimin receptor, Tir, interacts directly with alpha-actinin. *Curr. Biol.* 10:735–738.

Gouveia, S., M.E. Proctor, M.S. Lee, J.B. Luchansky, and C.W. Kaspar. 1998. Genomic comparisons and Shiga toxin production among Escherichia coli O157:H7 isolates from a day care center outbreak and sporadic cases in southeastern Wisconsin. *J. Clin. Microbiol.* 36:727–733.

Grif, K., M.P. Dierich, H. Karch, and F. Allerberger. 1998. Strain-specific differences in the amount of Shiga toxin released from enterohemorrhagic *Escherichia coli* O157 following exposure to subinhibitory concentrations of antimicrobial agents. *Eur. J. Clin. Microbiol. Infect. Dis.* 17:761–766.

Gruenheid, S., R. DeVinney, F. Bladt, D. Goosney, S. Gelkop, G.D. Gish, T. Pawson, and B.B. Finlay. 2001. Enteropathogenic E. coli Tir binds Nck to initiate actin pedestal formation in host cells. *Nature Cell Biol.* 3:856–859.

Hayashi, T. K. Makino, M. Onishi, K. Kurokawa, K. Ishi, K. Yokoyama, C.G. Han, E. Ohtsubo, K. Nakayama, T. Murata, M. Tanaka, T. Tobe, T. Iida, H. Takami, T.

Honda, C. Saskawa, N. Ogasawara, T. Yasunaga, S. Kuhara, T. Shiba, M. Hattori, and H. Shinagawa. 2001. Complete genome sequence of enterohemorrhagic *Escherichia coli* O157:H7 and genomic comparison with a laboratory strain. *DNA Research* 8:11–22.

Heimer, S.R., R.A. Welch, N.T. Perna, G. Pfosai, P.S. Evans, J.B. Kaper, F.R. Blattner, and H.T. Mobley. 2002. Urease of enterohemorrhagic *Escherichia coli*: Evidence for regulation by Fur and a *trans*-acting factor. *Infect. Immun.* 70:1027–1031.

Huang, A., J. Friesen, and J.L. Brunton. 1987. Characterization of a bacteriophage that carries the genes for production of Shiga-like toxin 1 in *Escherichia coli*. *J. Bacteriol.* 169:4308–4312.

Inward, C.D., A.J. Howie, M.M. Fitzpatrick, F. Rafaat, D.V. Milford, and C.M. Taylor. 1997. Renal histopathology in fatal cases of diarrhea-associated haemolytic uraemic syndrome. *Pediatr. Nephrol.* 11:556–559.

Jacewicz, M., H. Clausen, E. Nudelman, A. Donohue-Rolfe, and G.T. Keusch. 1986. Pathogenesis of Shigella diarrhea. XI. Isolation of a Shigella toxin-binding glycolipid from rabbit jejunum and HeLa cells and its identification as globtriaosylceramide. *J. Exp. Med.* 163:1391–1404.

Jackson, M.P., R.J. Neil, A.D. O'Brien, R.K. Holmes, and J.W. Newland. 1987. Nucleotide sequence analysis of the structural genes for Shiga-like toxin I encoded by bacteriophage 933J from *Escherichia coli*. *Microb. Pathog.* 2:147–153.

Jores, J., L. Rumer, S. Kiessling, J.B. Kaper, and L.H. Wieler. 2001. A novel locus of enterocyte effacement (LEE) pathogenicity island inserted at *pheV* in bovine Shiga toxin-producing *Escherichia coli* strain O103:H2. *FEMS Microbiol. Lett.* 204:75–79.

Kalman D., O.D. Weiner, D.L. Goosney, J.W. Sedat, B.B. Finlay, A. Abo, and J.M. Bishop. 1999. Enteropathogenic *E. coli* acts through WASP and Arp2/3 complex to form actin pedestals. *Nature Cell Biol.* 1:389–391.

Karaolis, D.K., T.K. McDaniel, J.B. Kaper, and E.C. Boedeker. 1997. Cloning of the RDEC-1 locus of enterocyte effacement (LEE) and functional analysis of the phenotype on HEp-2 cells. *Adv Exp Med Biol.* 412:241–245.

Karch, H., J. Heesemann, R. Laufs, A.D. O'Brien, C.O. Tackett, and M.M. Levine. 1987. A plasmid of enterohemorrhagic *Escherichia coli* O157:H7 is required for expression of a new fimbrial antigen and for adhesion to epithielal cells. *Infect. Immun.* 55:455–461.

Karch, H., H. Schmidt, C. Janetzki-Mittman, J. Scheef, and M. Kroger. 1999. Shiga toxins even when different are encoded at identical positions in the genomes of related temperate bacteriophages. *Mol. Gen. Genet.* 262:600–607.

Karmali, M.A., B.T. Steele, M. Petric, and C. Lim. 1983. Sporadic cases of hemolytic-uremic syndrome associated with fecal verotoxin and cytotoxin producing *Escherichia coli* in stools. *Lancet* 1:619–620.

Kenney, B., R. DeVinney, M. Stein, D.J. Reinscheid, E.A. Frey, and B.B. Finlay. 1997. Enteropathogenic *E. coli* (EPEC) transfers its receptor for intimate adherence into mammalian cells. *Cell* 91:511–520.

Kim, J., J. Nietfeldt, and A. K. Benson. 1999. Octamer based genome scanning distinguishes a unique subpopulation of *Escherichia coli* O157:H7 strains in cattle. *Proc. Natl. Acad. Sci. USA* 96:13288–13293.

Kim, J., J. Nietfeldt, J. Ju, J. Wise, N. Fegan, P. Desmarchelier, and A.K. Benson. 2001. Ancestral divergence, genome diversification, and phylogeographic variation in subpopulations of sorbitol-negative, β-glucuronidase-negative enterohemorrhagic *Escherichia coli* O157. *J. Bacteriol.* 183:6885–6897.

Klapproth, J.M. A., I.C.Scaletsky, B.P. McNamara, L.C. Lai, C. Malstrom, S.P. James, and M.S. Donnenberg. 2000. A large toxin from pathogenic *Escherichia coli* strains that inhibits lymphocyte activation. *Infect. Immun.* 68:2148–2155.

Knutton, S., T. Baldwin, P.H. Williams, and A.S. McNeish. 1989. Actin accumulation at sites of bacterial adhesion to tissue culture cells: basis of a new diagnostic test for enteropathogenic and enterohemorrhagic Escherichia coli. *Infect. Immun.* 57:1290–1298.

Knutton, S., I. Rosenshine, M.J. Pallen, I. Nisan, B.C. Neves, C. Bain, C. Wolff, G. Dougan, and G. Frankel. 1998. A novel EspA-associated surface organelle of enteropathogenic *Escherichia coli* involved in protein translocation into epithelia cells. *EMBO J.* 17:2166–2176.

Kohler, B., J. Karch, and H. Schmidt. 2000. Antibacterials that are used as growth promoters in animal husbandry can affect the release of Shiga-toxin-2-converting bacteriophages and Shiga toxin 2 from *Escherichia coli* strains. *Microbiol.* 146:1085–1090.

Kozlov, Y.V., A.A. Kabishev, E.V. Lukyanov, and A.A. Bayev. 1988. The primary structure of the operons encoding for *Shigella dysenteriae* toxin and temperate phage H30 Shiga-like toxin. *Gene.* 67:213–221.

Kudva, I.T., P.S. Evans, N.T. Perna, T.J. Barrett, F.M. Ausbel, F.R. Blattner, and S.C. Calderwood. 2002. Strains of *Escherichia coli* O157:H7 differ primarily by insertions or deletions, not single-nucleotide polymorphisms. *J. Bacteriol.* 184:1873–1879.

Levine, M.M. 1987. *Escherichia coli* that cause diarrhea: enterotoxigenic, enterophathogenic, enteroinvasive, enterohemorrhagic, and enteroadherent. *J. Infect. Dis.* 155:377–389.

Lindberg, A.A., J.E. Brown, N. Stromberg, M. Westling-Ryd, J.E. Schultz, and K.A. Karlson. 1987. Identification of the carbohydrate receptor for Shiga toxin produced by *Shigella dysenteriae* type I. *J. Biol. Chem.* 262:1779–1785.

Makino, K., K. K. Ishii, T. Yasunaga, M. Hattori, K. Yokoyama, C.H. Yutsudo, Y. Kubota, Y. Yamaichi, T. Iida, K. Yamamoto, T. Honda, C.G. Han, E. Ohtsubo, M. Kasamatsu, T. Hayashi, S. Kuhara, and H. Shinagawa. 1998. Complete nucleotide sequences of 93-Kb and 3.3-Kb plasmids of an enterohemorrhagic *Escherichia coli* O157:H7 derived from Sakai outbreak. *DNA Res.* 28:1–9.

Makino, K., K. Yokoyama, Y. Kubota, C.H. Yutsudo, S. Kimura, K. Kurokawa, K. Ishii, M. Hattori, I. Tatsuno, H. Abe, T. Iida, K. Yamamoto, M. Onishi, T. Hayashi, T. Yasunaga, T. Honda, C. Sasakawa, and H. Shinagawa. 1999. Complete nucleotide sequence of the prophage VT2-Sakai carrying the verotoxin 2 genes of the enterohemorrhagic *Escherichia coli* O157:H7 derived from the Sakai outbreak. *Genes Genet. Syst.* 74:227–239.

Marches, O., J.P. Nougayrede, S. Boullier, J. Mainil, G. Charlier, I. Raymond, P. Pohl, M. Boury, J. De Rycke, A. Milon, and E. Oswald. 2000. Role of tir and intimin in the virulence of rabbit enteropathogenic *Escherichia coli* serotype O103:H2. *Infect. Immun.* 68:2171–2182.

Matsushiro, A., K. Sato, H. Miyamoto, T. Yamamura, and T. Honda. 1999. Induction of prophages of enterohemorrhagic *Escherichia coli* O157:H7 with norfloxacin. *J Bacteriol.* 181:2257–2260.

McDaniel, T.K., K. G. Jarvis, M.S. Donnenberg, and J.B. Kaper. 1995. A genetic locus of enterocyte effacement conserved among diverse enterobacterial pathogens. *Proc. Natl. Acad. Sci. USA* 92:1664–1668.

McDaniel, T.K. and J.B. Kaper. 1997. A cloned pathogenicity island from enteropathogenic *Escherichia coli* confers the attaching and effacing phenotype on *E. coli* K12. *Mol. Microbiol.* 23:399–407.

McKee, M.L. A.R. Melton-Celsa, R.A. Moxley, D.H. Francis, and A.D. O'Brien. 1995. Enterohemorrhagic *Escherichia coli* O157:H7 requires intimin to colonize the gnotobiotic pig intestine and to adhere to HEp-2 cells. *Infect. Immun.* 63:3739–3744.

McNally, A., A.J. Roe, S. Simpson, F.M. Thomson-Carter, D.E. Hoey, C. Currie, T. Chakraborty, D.G. Smith, and D.L. Gally. 2001. Differences in levels of secreted locus of enterocyte effacement proteins between human disease-associated and bovine *Escherichia coli* O157. *Infect. Immun.* 69:5107–14.

Mellies, J. L., S.J. Elliott, V. Sperandio, M.S. Donnenberg, and J.B. Kaper. 1999. The Per regulon of enteropathogenic *Escherichia coli*: identification of a regulatory cascade and a novel transcriptional activator, the locus of enterocyte effacement (LEE)-encoded regulator (Ler)) *Mol. Microbiol.* 33:296–306.

Miyamoto H, W. Nakai, N. Yajima, A. Fujibayashi, T. Higuchi, K. Sato, and A. Matsushiro. 1999. Sequence analysis of Stx2-converting phage VT2-Sa shows a great divergence in early regulation and replication regions. *DNA Res.* 6:235–240.

Monday, S.R., T.S. Whittam, and P.C. Feng. 2001. Genetic and evolutionary analysis of mutations in the *gusA* gene that cause the absence of beta-glucuronidase activity in *Escherichia coli* O157:H7. *J. Infect. Dis.* 184:918–921.

Moon, H.W., S.C. Whipp, R.A. Argenzio, M.M. Levine, and R.A. Gianella. 1993. Attaching and effacing activities of rabbit and human enteropathogenic *Escherichia coli* in pig and rabbit intestines. *Infect. Immun.* 41:1340–1351.

Muhldorfer, I., J. Hacker, G.T. Keusch, D.W. Acheson, H. Tschape, A.V. Kane, A. Ritter, T. Olschlager, and T. Donahue-Rolfe. 1996. Regulation of the Shiga-like toxin II operon in *Escherichia coli*. *Infect. Immun.* 64:495–502.

Murzin, A.G. 1993. OB(oligonucleotide/oligosaccharide binding)-fold: common structural and functional solution for non-homologous sequences. *EMBO J.* 12:861–867.

Nataro, J. M. and J. B. Kaper. 1998. Diarrheagenic *Escherichia coli*. *Clin. Microbiol. Rev.* 1998. 11:142–201.

Neely, M.N. and D.I. Friedman. 1998. Functional and genetic analysis of regulatory regions of coliphage H-19B: location of shiga-like toxin and lysis genes suggest a role for phage functions in toxin release. *Mol. Microbiol.* 28:1255–1267.

Newland, J.W., N.A. Stockbrine, F.F. Miller, A.D. O'Brien, and R.K. Holmes. 1985. Cloning of shiga-like toxin genes from a toxin converting phage of *Escherichia coli*. *Science* 230:179–181.

Nicholls, L., T.H. Grant, and R.M. Robins-Browne. 2000. Identification of a novel genetic locus that is required for in vitro adhesion of a clinical isolate of enterohemorrhagic *Escherichia coli* to adhere to epithelial cells. *Mol. Microbiol.* 35:275–288.

O'Brien, A.D., J.W. Newland, S.F. Miller, R.K. Holmes, H.W. Smith, and S.B. Formal. 1984. Shiga-like toxin converting phages from *Escherichia coli* that causes hemorrhagic colitis or infantile diarrhea. *Science* 236:694–696.

Ostroff, S.M., P.I. Tarr, M.A. Neill, J.H. Lewis, N. Hargrett-Bean, and J.M. Kobayashi. 1989. Toxin genotypes and plasmid profiles as determinants of systemic sequelae in *Escherichia coli* O157:H7 infections. *J. Infect. Dis.* 160:994–998.

Paton, J.C. and A.W. Paton. 1998. Pathogenesis and diagnosis of Shiga toxin-producing *Escherichia coli* infections. *Clin. Microbiol. Rev.* 11:450–479.

Perna, N.T., G.F. Mayhew, G. Posfai, S. Elliott, M.S. Donnenberg, J.B Kaper, and F.R. Blattner. 1998. Molecular evolution of a pathogenicity island from enterohemorrhagic *Escherichia coli* O157:H7. *Infect. Immun.* 66:3810–7.

Perna, N.T., G. Plunkett III, V. Burland, B. Mau, J. D. Glasner, D. J. Rose, G. F. Mayhew, P. S. Evans, J. Gregor, H. A. Kirkpatrick, G. Posfai, J. Hackett, S. Klink, A. Boutin, Y. Shao, L. Miller, E. J. Grotbeck, N. W. Davis, A. Lim, E. T. Dimalanta, K. D. Potamousis, J. Apodaca, T. S. Anantharaman, J. Lin, G. Yen, D. C. Schwartz, R. A. Welch, and F. R. Blattner. 2001. Genome sequence of enterohaemorrhagic *Escherichia coli* O157:H7. *Nature* 409:529–33

Philpott, D.J., D.M. McKay, P.M. Sherman, and M.H. Perdue. 1996. Infection of T84 cells with enteropathogenic *Escherichia coli* alters barrier and transport functions. *Am. J. Physiol.* 270:G634–G635.

Philpott, D.J., D.M. McKay, W. Mak, M.H.Perdue, and P.M. Sherman. 1998. Signal transduction pathways involved in enterohemorrhagic *Escherichia coli*–induced alterations in T84 epithelial permeability. *Infect. Immun.* 66:1680–7.

Pierard, D., G. Muyldermans, L Moriau, D. Stevens, and S. Lauwers. 1998. Identification of new verocytotoxin type 2 variant B-subunit genes in human and animal *Escherichia coli* isolates. *J. Clin. Microbiol.* 36:3317–3322.

Plunkett, G. III, D.J. Rose, T.J. Durfee, and F.R. Blattner. 1999. Sequence of shiga toxin 2 phage 933W from *Escherichia coli* O157:H7: shiga toxin as a phage late-gene product. *J. Bacteriol.* 181:1767–1778.

Reid, S.D., C.J. Herbelin, A.C. Brumbaugh, R. K. Selander, and T. S. Whittam. 2000. Parallel evolution of virulence in pathogenic *Escherichia coli*. *Nature* 406:64–67.

Riley, L.W. 1987. The epidemiologic, clinical, and microbiologic features of hemorrhagic colitis. *Annu. Rev. Microbiol.* 41:383–407.

Rosenshine, I., S. Ruschkowski, M. Stein, D.J.Reinscheid, S.D. Mills, and B.B. Finlay. 1996. A pathogenic bacterium triggers epithelial signals to form a functional bacterial receptor that mediates actin pseudopod formation. *EMBO J.* 15:2613–2624.

Samadpour, M., Grimm, L.M., Desai, B., Alfi, D., Ongerth, J.E., and P.I. Tarr. 1993. Molecular epidemiology of *Escherichia coli* O157:H7 strains by bacteriophage lambda restriction fragment length polymorphism analysis: application to a multistate foodborne outbreak and a day-care center cluster. *J. Clin. Microbiol.* 31:3179–3183.

Sandvig, K., O. Garred, K. Prydz, J. Koslov, S. Hansen, and B. van Deurs. 1992. Retrograde transport of endocytosed Shiga toxin: a glycolipid-binding protein from *Shigella dysenteriae*. *Nature* 358:510–512.

Sandvig, K. and B. van Deurs. 1996. Endocytosis, intracellular transport, and cytotoxic action of Shiga toxin and ricin. *Physiol. Rev.* 76:949–96.

Saxena, S.S., A.D. O'Brien, and E.J. Ackerman. 1989. Shiga toxin, Shiga-like toxin II variant, and ricin are all single-site RNA N-glycosidases of 28 S RNA when microinjected into *Xenopus* oocytes. *J. Biol. Chem.* 264:596–601.

Schauder, S., K. Shokat, M.G. Surette, and B.L.Bassler. 2001. The LuxS family of bacterial autoinducers: biosynthesis of a novel quorum-sensing signal molecule. *Mol. Microbiol.* 41:463–476.

Schmitt, C.K., M.L. McKee, and A.D. O'Brien. 1991. Two copies of shiga-like toxin II-related genes common in enterohemorrhagic *Escherichia coli* strains are responsible for the antigenic heterogeneity of the O157:H- strain E32511. *Infect. Immun.* 59:1065–1073.

Schmidt, H., J. Sheef, S. Morabito, A. Caprioli, L.H. Wieler, and H. Karch. 2000. A new Shiga toxin variant (Stx2f) from *Escherichia coli* isolated from pigeons. *Appl. Environ. Microbiol.* 66:1205–1208.

Shere, J.A., K.J. Bartlett, and C.W. Kaspar. 1998. Longitudinal study of Escherichia coli O157:H7 dissemination on four dairy farms in Wisconsin. *Appl. Environ. Microbiol.* 64:1390–1399.

Sixma, T.K., P.E. Stein, W.G. Hol, and R.J. Read. 1993. Comparison of the B-pentamers of heat-labile enterotoxin and verotoxin-1: two structures with remarkable similarity and dissimilarity. *Biochemistry* 32:191–198.

Sperandio, V, J.B. Kaper, M.R. Bortolini, B.C. Neves, R. Keller, and L.R. Trabulsi. 1998. Characterization of the locus of enterocyte effacement (LEE) in different enteropathogenic *Escherichia coli* (EPEC) and Shiga-toxin producing *Escherichia coli* (STEC) serotypes. *FEMS Microbiol. Lett.* 164:133–139.

Sperandio, V., J.L. Mellies, W. Nguyen, S. Shin, and J.B. Kaper. 1999. Quorum sensing controls expression of the type III secretion gene transcription and protein secretion in enterohemorrhagic and enteropathogenic *Escherichia coli*. *Proc. Natl. Acad. Sci. USA* 96**:**15196–15201.

Sperandio, V. A.G. Torres, J.A. Giron, and J.B. Kaper. 2001. Quorum sensing is a global regulatory mechanism in enterohemorrhagic *Escherichia coli* O157:H7. *J. Bacteriol.* 183:5187–5197.

Stein, P.E., A. Boodhoo, G.J. Tyrrell, J.L. Brunton, and R.J. Read. 1992. Crystal structure of the cell-binding B oligomer of verotoxin-1 from *E. coli*. *Nature.* 355:748–750.

Sung, M., M.P. Jackson, A.D. O'Brien, and R.K. Holmes. 1990. Transcription of the shiga-like toxin type II and shiga-like toxin type II variant operons of *Escherichia coli*. *J. Bacteriol.* 172:6386–6395.

Tarr, P.I., S.S. Bilge, J.C. Vary Jr., S. Jelacic, R.C. Habeeb, T.R. Ward, M.R. Baylor, and T.E. Besser. 2000. Iha: a novel *Escherichia coli* O157:H7 adherence-conferring molecule encoded on a recently acquired chromosomal island of conserved structure. *Infect. Immun.* 68:1400–1407.

Tarr, P.I., L.M. Schoening, Y.L. Yea, T.R. Ward, S. Jelacic, and T. S. Whittam. 2000. Acquisition of the *rfb-gnd* cluster in evolution of *Escherichia coli* O55 and O157. *J. Bacteriol.* 182:6183–6191.

Tatsuno, I., M. Horie, H. Abe, T. Miki, K. Makino, H. Shinigawa, H. Taguchi, S. Kamiya, T. Hayashi, and C. Sasakawa. 2001. *toxB* gene on pO157 of enterohemorrhagic *Escherichia coli* O157:H7 is required for full epithielal cell adherence phenotype. *Infect. Immun.* 69:6660–6669.

Tesh, V.L., J.A. Burris, J.W. Owens, V.W. Gordon, E.A. Wadolkowski, A.D. O'Brien, and J.E. Samuel. 1993. Comparison of the relative toxicities of Shiga-like toxins type I and type II for mice. *Infect. Immun.* 61:3392–3402.

Thorpe, C.M., B.P. Hurley, L.L. Lincicome, M.S. Jacewicz, G.T. Keusch, and D.W.K. Acheson. 1999. Shiga toxins stimulate secretion of interleukin-8 from intestinal epithelial cells. *Infect. Immun.* 67:5985–5993.

Thorpe, C.M., W.E. Smith, B.P. Hurley, and D.W. Acheson. 2001. Shiga toxins induce, superinduce, and stabilize a variety of C-X-C chemokine mRNAs in intestinal epithelial cells, resulting in increased chemokine expression. *Infect Immun.* 69:6140–6147.

Toth, I., M.L. Cohen, H. Rumshlag, L.W. Riley, E.H. White, J.H. Carr, W.W. Bond, and I.K. Wachsmuth. 1990. Influence of the 60-megadalton plasmid on adherence of *Escherichia coli* O157:H7 and genetic derivatives. *Infect. Immun.* 58:1223–1231.

Tzipori, S., C.W. Chow, and H.R. Powell. 1988. Cerebral infection with Escherichia coli O157:H7 in humans and gnotobiotic piglets. *J. Clin. Pathol.* 41:1099–1103.

Unkmeir, A., and H. Schmidt. 2000. Structural analysis of phage-borne *stx* genes and their flanking sequences in Shiga toxin-producing *Escherichia coli* and *Shigella dysernteriae* type 1 strains. *Infect. Immun.* 68:4856–4864.

Vallance, B.A. and B.B. Finlay. 2000. Exploitation of host cells by enteropathogenic *Escherichia coli*. *Proc. Natl. Acad. Sci. USA* 97:8799–8806.

Waddell, T., S. Head, M. Petric, A. Cohen, and C. Lingwood. 1988. Globtriosyl ceramide is specifically recognized by the *Escherichia coli* Verocytotoxin 2. *Biochem. Biophys. Res. Commun.* 152:674–679.

Wadolkowski, E A., L. Sung, J.A. Burris, J.E. Samuel, and A.D. O'Brien. 1990. Acute renal tubular necrosis and death of mice orally infected with *Escherichia coli* strains that produce Shiga-like toxins II. *Infect. Immun.* 58:3959–3965.

Wadolkowski, E.A., J.A. Burris, and A.D. O'Brien. 1990. Acute renal tubular necrosis and death of mice orally infected with *Escherichia coli* strains that produce Shiga-like toxin type II. *Infect. Immun.* 58. 3959–3965.

Wagner, P.L., M.N. Neely, X. Zhang, D.W.K. Acheson, M.K. Waldor, and D.I. Friedman. 2001. Role for a phage promoter in shiga toxin 2 expression from a pathogenic *Escherichia coli* strain. *J. Bacteriol.* 183:2081–2085.

Warawa, J., B.B. Finlay, and B. Kenney. 1999. A pathogenic bacterium triggers epithelial signals to form a functional bacterial receptor that mediates actin pseudopod formation. *Infect. Immun.* 67:5538–5540.

Weinstein, D.L., M.P. Jackson, J.E. Samuel, R.K. Holmes, and A.D. O'Brien. 1988. Cloning and sequencing of a Shiga-like toxin type II variant from an *Escherichia coli* strain responsible for edema disease of swine. *J. Bacteriol.* 170:4223–4230.

Weinstein, D.L., M. P. Jackson, L.P. Perera, R.K. Holmes, and A.D. O'Brien.1989. In vivo formation of hybrid toxins comprising Shiga toxin and the Shiga-like toxins and role of the B subunit in localization and cytotoxic activity. *Infect Immun.* 57:3743–3750.

Wieler, L.H., T.K. McDaniel, T.S. Whittam, and J.B. Kaper. 1997. Insertion site of the locus of enterocyte effacement in enteropathogenic and enterohemorrhagic *Escherichia coli* differs in relation to the clonal phylogeny of the strains. *FEMS Microbiol Lett.* 156:49–53.

Whittam, T.S., Wolfe, M.L., Wachsmuth, I.K., Orskov, F., Orskov, I., and R.A. Wilson. 1993. Clonal relationships among *Escherichia coli* strains that cause hemorrhagic colitis and infantile diarrhea. *Infect. Immun.* 61:1619–1929.

Whittam, T.S. 1998. Evolution of STEC strains. In *Escherichia Coli O157:H7 and Other Shiga Toxin-Producing E. Coli Strains*. J.B. Kaper and A.D. O'Brien, eds. Washington, D.C.: American Society for Microbiology, 195–209.

Wong, C.S., S. Jelacic, R.L. Habeeb, S.L. Watkins, and P.I. Tarr. 200. The risk of the hemolytic-uremic syndrome after antibiotic treatment of Escherichia coli O157:H7 infections. *New Engl. J. Med.* 342:1930–19366.

Yoshida, T., M. Fukada, N. Koide, H. Ikeda, T. Sugiyama, Y. Kato, N. Ishikawa, and T. Yokochi. 1999. Primary cultures of human endothelial cells are susceptible to low doses of Shiga toxins and undergo apoptosis. *J. Infect. Dis.* 180:2048–2052.

Zhang, W., M. Bielaszewska, T. Kuczius, and H. Karch. 2002. Identification, characterization, and distribution of shiga toxin I gene variant (stx_{1c}) in *Escherichia coli* strains isolated from humans. *J. Clin. Microbiol.* 40:1441–1446.

Zhu C., T.S. Agin, S.J. Elliott, L.A. Johnson, T.E. Thate, J.B. Kaper, and E. C. Boedeker. 2001. Complete nucleotide sequence and analysis of the locus of enterocyte effacement from rabbit diarrheagenic *Escherichia coli* RDEC-1. *Infect. Immun.* 69:2107–2115.

18

PREVENTION AND CONTROL OF *ESCHERICHIA COLI* O157:H7

THOMAS E. BESSER, JEFF T. LEJEUNE, DAN RICE, AND DALE D. HANCOCK

INTRODUCTION. Nearly twenty years after the first identified foodborne outbreak of disease caused by *E. coli* O157:H7 and ten years into intensive research efforts into pre-harvest control measures, there are still no well-documented management interventions shown to affect the prevalence of cattle infection with this agent. However, there are grounds for optimism that this situation will soon change because of the improved understanding of the reservoirs and vehicles of bovine infection and the dynamics of gastrointestinal carriage of *E. coli* O157:H7 by cattle. This chapter reviews the epidemiology of *E. coli* O157:H7 in cattle populations, with particular attention to some of the promising potential control measures that have been identified so far.

POTENTIAL OF PRE-HARVEST INTERVENTIONS TO REDUCE HUMAN EXPOSURE TO *E. COLI* O157:H7. Both theoretical models and empirical data suggest that successful efforts to reduce the prevalence of cattle infection with *E. coli* O157:H7 would be beneficial. The theoretical models suggest that reduction of prevalence of *E. coli* O157:H7 in cattle would reduce both the frequency of contamination of beef (Jordan 1999) and the resultant human disease (USDA:FSIS 2000). These conclusions of these modeling studies are supported by the important observation that the pre-slaughter presence of *E. coli* O157:H7 in individual cattle and in specific pens of cattle prior to slaughter correlate with the frequency of *E. coli* O157:H7 contamination detected in the slaughter plant (Chapman et al. 1993; Elder et al. 2000).

Reducing the pre-harvest prevalence of *E. coli* O157:H7 is expected to reduce two important routes of human exposure. In addition to reducing the frequency of carcass contamination and the consequent food poisoning, it is also expected to reduce the frequency of human infection resulting from direct contact with infected cattle and their environments. These sources of human infection are increasingly recognized as substantial and important (Coia et al. 1998; MacDonald et al. 1996; Michel et al. 1999; Trevena et al. 1999).

KEY FEATURES OF THE EPIDEMIOLOGY OF *E. COLI* O157:H7 IN CATTLE. A growing number of well-established epidemiologic features of *E. coli* O157:H7 infection of cattle and other animals have

been documented, as is recently summarized in Hancock et al. (1998). Some of these features are briefly described in the following sections.

Characterization of the Bovine Infection. The fecal shedding of *E. coli* O157:H7 by cattle, whether naturally infected or following experimental oral challenge, is typically transient (ranging from a few days to a few months, but most frequently limited to a few weeks) and variable in degree but rarely exceeding 10^6 CFU per gram of feces (Besser et al. 2001; Mechie et al. 1997). The nature of the interaction of the agent with the bovine intestinal epithelium is undefined, except for the occurrence of attaching and effacing lesions, which seem to be limited to very young calves (Dean-Nystrom et al. 1997). The molecular mechanisms that may underlie the interaction between *E. coli* O157:H7 and the intestinal epithelium of weaned cattle, if any, are currently unknown (Stevens et al. 2002).

The observed prevalence of fecal *E. coli* O157:H7 shedding in cattle depends strongly upon the sensitivity of the detection method used, and many fecal positive cattle detected with highly sensitive methods may have feces with very low concentrations of *E. coli* O157:H7 (Besser et al. 2001). Most well-established features of fecal *E. coli* O157:H7 shedding by cattle, such as the higher prevalence in weaned immature calves, the transient duration of shedding, and the summertime seasonal peaks of prevalence, were established with poorly sensitive methods that probably only consistently detected high levels ($>10^4$ cfu/g) of *E. coli* O157:H7. These observations are consistent with the possibility that many or most of the cattle detected with low levels of *E. coli* O157:H7 in their feces are simply passively excreting ingested *E. coli* O157:H7 cells or may be truly infected but with restricted proliferation of the agent. While animals excreting low concentration of E. coli O157 in their feces are still of food safety significance, they may have relatively little epidemiological significance.

Cattle with feces culture positive for *E. coli* O157:H7 typically do not remain culture positive for long periods of time, and most evidence favors durations of cattle infection typically limited to weeks or months. Nevertheless, it is difficult or impossible to rule out persistently infected cattle absolutely, and some evidence suggests that some individual cattle may show prolonged consistent shedding (Magnuson et al. 2000). The best evidence supporting the generally transient nature of bovine infection includes concordance between the lack of persistence of shedding of marked strains in experimental infection (where the sensitivity of detection is likely very much higher than for nonmarked strains) and lack of persistence of detection from cattle on farms. More important, persistently infected cattle are clearly not the principal reservoir of maintenance of *E. coli* O157:H7 in feedlots, where the entire population changes several times a year, yet individual persistent strain types may remain as predominant isolates of *E. coli* O157:H7 over several years, at least (Hancock and Besser, unpublished data).

Seasonal and Regional Variation in Distribution of *E. coli* O157:H7. The prevalence of fecal shedding of *E. coli* O157:H7 by cattle in some geographical locations varies strongly by season, with peak prevalence occurring during the summertime or early fall. Originally identified in North America (Hancock at al. 1994), very similar seasonal variation of *E. coli* O157:H7 prevalence has now been observed in many locations (Bonardi et al. 1999; Heuvelink et al. 1998). Logically, the similar patterns of summertime increase in human disease caused by *E. coli* O157:H7 (Wallace et al. 1997; Michel et al. 1999; PHLS 1999) is probably influenced by seasonal variation in exposure to contaminated food products, environmental reservoirs, and direct contact with shedding cattle rather than to food preparation practices alone. Consistent with this explanation, a summertime increase in the frequency of contamination of retail foods has recently been observed in England (Chapman et al. 2001). However, this correlation is not always seen, as peak prevalences of bovine shedding of *E. coli* O157:H7 in England and Scotland tended to occur in spring and fall rather than summer (Chapman et al 1997; Synge, personal communication), and human disease with *E. coli* O157:H7 in Scotland (the country with the highest documented rate of *E. coli* O157:H7 human infection in the world) peaked seasonally during summer months (Douglas and Kurien 1997).

Within several well-studied areas, including most areas of North America as well as the United Kingdom and other central and southern European countries, cattle shedding *E. coli* O157:H7 in feces are common enough to suggest that the organism is ubiquitously present on cattle farms. Moreover, we have observed apparently identical strain types (as determined by PFGE analysis) on farms hundreds of kilometers apart with no traceable animal movements to explain their transfer (Rice et al. 1999).

Despite the ubiquitous distribution of *E. coli* O157:H7 in some regions, other regions, such as Scandinavia, report an apparently lower prevalence of this organism in cattle populations (Lahti et al. 2001; Johnson et al. 2001; Vold et al. 1998). This may be due to climatic factors or farm management practices less conducive to cattle colonization with *E. coli* O157:H7, or perhaps because *E. coli* O157:H7 has not yet been introduced into these areas. These differences in cattle prevalence do not appear to be caused entirely by differences in detection

methodology; highly sensitive methods have been utilized in some of these low-prevalence regions.

Supporting the possibility of true regional differences regarding the frequency of occurrence of *E. coli* O157:H7 is the varying occurrence of the haemolytic uraemic syndrome (HUS). This sequelae to enteric *E. coli* O157:H7 infection is sufficiently severe and unique to make its reported occurrence unlikely to be affected by differences in reporting artifacts. Some regions, notably northern Scandinavian countries, report both very low HUS and other VTEC (verotoxigenic *E. coli*)-associated disease incidence compared to other world locations (Decludt et al. 2000; Huber et al. 2001; Gianviti et al. 1994; Keskimaki et al. 1998; Siegler et al. 1994; Waters et al. 1994) and very low *E. coli* O157:H7 fecal prevalence in cattle (Vold et al. 1998).

A "northern tier" effect has been observed in the United States, where a tendency exists for northern states to report a higher incidence of HUS and other VTEC-related human disease than states more to the south (Wallace et al. 2000). This situation apparently is not simply a reflection of exposure, because feedlot cattle did not show any tendency toward prevalence differences by latitude in a national study utilizing identical sampling and testing methods (Hancock et al. 1997). The relatively low rate of HUS prevalence in Italy compared to many other European countries, and the relatively high rate of *E. coli* O157:H7–related infection in Scotland compared to England and Wales, suggest that a similar northern tier effect may be present on the European continent (Gianviti et al. 1994).

Therefore, at least two sources of regional variation may exist: variation in the prevalence of the agent in cattle populations, which results in reduced exposure and infection of human populations as is apparently the case in northern Scandinavian countries; and the northern tier effect, where an apparent increased susceptibility of human populations exists in some northern regions to *E. coli* O157:H7 infection, despite the lack of evidence of increased exposure in these regions. In the case of the northern Scandinavian countries, an effect similar to the first source of regional varation also occurs with human salmonellosis, apparently at least in part because of domestic livestock having very low prevalence of *salmonellae* (Kapperud et al. 1998).

Persistence of Environmental *E. coli* O157:H7 Contamination. Despite the fact that most fecal shedding by cattle is not persistent beyond a period of several weeks or months, evidence exists for persistence of specific *E. coli* O157:H7 strain types in specific environments. On individual dairy farms and cattle feedlots, identical or very closely related *E. coli* O157:H7 strain types may be found over periods of several years (Chapman et al. 1997; Hancock and Besser, unpublished data). In the case of feedlots, this persistence occurs in the face of several complete turnovers of the animal population several times per year. One logical possibility is that this persistence results from survival of the organism in environmental reservoirs such as manure (Fukushima et al. 1999; Bolton et al. 1999;

Kudva et al. 1998) or water troughs (Lejeune et al. 2001). Clearly, the multiple new strain types of *E. coli* O157:H7 that must surely be introduced by the movement of thousands or tens of thousands of animals onto a feedlot in the course of the year do not result in complete turnover of the strain types on the premises. On the other hand, multiple pulsed field gel electropherotypes (PFGE) subtypes can usually be detected on single farms, and turnover in the predominant PFGE subtypes usually occurs with periodic additions of new strain types, even on farms closed to the addition of new animals from outside the farm.

Alternative Hosts. Many animals in addition to cattle shed *E. coli* O157:H7, and in most cases do so in the absence of any signs of enteric disease. Domestic animals identified with fecal shedding of *E. coli* O157:H7 include cattle, sheep, goats, horses, dogs, and reindeer (Hancock 1998). In addition, a variety of wildlife species have been found to shed *E. coli* O157:H7 in feces (deer, birds, and rabbits) or carry it on their mouthparts (flies) (Hancock 1998; Pritchard et al. 2001). The epidemiologic significance of wild or domestic alternative hosts of *E. coli* O157:H7 to cattle remains largely undefined. Venison has been associated with human foodborne infection (Keene et al. 1997), and deer share *E. coli* O157:H7 strain types with cattle on at least some occasions (Sargeant et al. 1999; Rice et al. 1995). Whether deer or other wildlife reservoirs are subject to similar regional and temporal variation in prevalence as occur in cattle is currently unknown. Similarly, most demonstrations of *E. coli* O157:H7 in bird feces or flies result from studies associated with domestic animal production facilities, and whether these are representative of populations less closely associated with domestic animals is unknown.

Shedding Levels and Infectious Dose. Experimentally infected animals vary widely in the concentrations of *E. coli* O157:H7 shed, both from animal-to-animal and temporally within individual animals. In experimental studies of natural transmission, we typically observe that a large majority (>90 percent) of the total *E. coli* O157:H7 shed in feces from the group is caused by shedding from the single, highest-shedding-level animal. It seems likely that a minority of animals in a group shedding the highest concentrations in their feces plays a disproportionately large role in the epidemiology of *E. coli* O157:H7 on cattle farms. For example, all the best established features of natural infection of cattle with *E. coli* O157:H7 (seasonal peaks, higher prevalence in younger animals) were originally observed with very insensitive crude fecal isolation methods. It is also possible that passive shedding, that is, excretion of ingested bacteria without infection, could explain many of the animals testing positive through the use of very sensitive culture techniques; as described later, contamination of drinking water and feedbunk feeds must frequently result in ingestion of the agent by large numbers of animals. If so, this would argue that development of tests for fecal

E. coli O157:H7 to reliably distinguish high-level fecal shedders would be useful for research into on-farm interventions to reduce fecal shedding.

Most experimental infection studies with *E. coli* O157:H7 have used oral doses that are relatively large compared to the exposures likely to occur from contact with feces of naturally infected animals (Cray et al. 1995; Dean-Nystrom et al. 1997). However, studies have shown the potential for at least some animals to become infected following challenge with low (<10^3) doses of *E. coli* O157:H7 (Besser et al. 2001; LeJeune et al. 2001). These resulted in shedding patterns similar to those of the higher-challenge-dose studies, suggesting that the high-dose challenge models may be relevant to natural infection. More important, the low-dose infections demonstrate the potential epidemiologic significance of environmental contamination with relatively low levels of *E. coli* O157:H7.

ERADICATION OF *E. COLI* O157:H7 IS UNLIKELY. The epidemiologic features described in the previous section represent constraints on some approaches to pre-harvest control of *E. coli* O157:H7, as discussed elsewhere (Hancock et al. 1998). The widespread distribution of the agent, its stable persistence in many environmental reservoirs, the lack of resistance of cattle to re-infection, and the wide host range of the agent (including many wildlife species) would seem to make eradication of the agent an implausible prospect at best. The transient nature of bovine infection, together with the lack of apparent resistance of animals to re-infection and the environmental persistence of the agent, make slaughter, removal, or segregation of test-positive animals of little likely benefit. Testing of animals to divert positive animals from slaughter until infections are cleared is problematic for several reasons, including the frequent occurrence of hide contamination in the absence of fecal contamination and the likelihood that a reasonable percentage of test-negative animals will become test positive between testing and slaughter.

For these reasons, the most realistic likely outcome for pre-harvest control of *E. coli* O157:H7 is a significant reduction in prevalence, rather than regional or total elimination of the agent from the cattle populations. In this regard, the pre-harvest level is no different from interventions that take place at or subsequent to slaughter (that is, processing, retail food handling, and consumer education), and incremental progress at each of these levels will probably be needed to contribute to achieve best control of this agent.

PRE-HARVEST INTERVENTIONS TO PREVENT AND CONTROL *E. COLI* O157:H7. In our laboratory, efforts to reduce the prevalence of *E. coli* O157:H7 shedding have focused on identification and removal of on-farm reservoirs where *E. coli* O157:H7 persists and identification and removal of the vehicles that result in cattle infection are possible; interventions that further reduce the duration of bovine infection with *E. coli* O157:H7; and interventions that reduce the magnitude of cattle fecal shedding of *E. coli* O157:H7 (and other sources of environmental contamination with the agent).

Identification and Removal of Environmental Reservoirs. The persistence over periods of years of *E. coli* O157:H7 strain types on farms, including feedlots with multiple complete changes in animal populations each year, the lack of prolonged shedding of *E. coli* O157:H7 in individual cattle, and the occurrence of summertime seasonal peaks of prevalence of fecal shedding of *E. coli* O157:H7 by cattle after long periods of low or no detectable shedding, are all consistent with a major reservoir for *E. coli* O157:H7 in the farm environment besides cattle. Clearly, this persistence could occur in many habitats in the farm environment; however, it seems logical to pay particular attention to those potential reservoirs that also would be the most likely vehicles of transmission of *E. coli* O157:H7 to cattle.

Cattle are exposed to *E. coli* from consumption of their feeds, ingestion of water, grooming activities of themselves and of other cattle, and from their environment. However, as described later, feeds and water are likely to be the largest sources of oral *E. coli* exposure to cattle. Feed and water sources of *E. coli* are of potential significance for additional reasons: Feeds represent a potential source by which *E. coli* O157:H7 strains are introduced from outside the farm, and water represents a potential location for *E. coli* O157:H7 proliferation and amplification.

WATER TROUGHS. Water sources for cattle are frequently contaminated with relatively high numbers of *E. coli*, and both the numbers of contaminating total *E. coli* and the volume of water consumed by cattle demonstrably increase in the warm summer months coincident with the increased prevalence of *E. coli* O157:H7 infections in cattle. Observed water trough concentrations of total *E. coli* in dairy farm water troughs in our area range from <1 cfu/ml to 100 cfu per ml (LeJeune et al. 2001b). In heavily stocked feedlot pens the concentrations can reach significantly higher values, and the typical range for summertime feedlot water troughs is from 50 to 50,000 cfu per ml. Based on water consumption of 40 liters or more per day in hot summer weather, cattle may ingest with drinking water oral doses of *E. coli* exceeding 10^8 cfu, the *E. coli* equivalent of hundreds of grams of feces. Frequent, even daily cleaning of feedlot water troughs is insufficient to reduce this *E. coli* population significantly during the heat of summer. In fact, frequent manual cleaning may result in increased *E. coli* concentrations in some cases, because high concentrations of protozoa and metazoa will develop in uncleaned troughs, and these organisms are significant predators of *E. coli* (LeJeune et al. 2001b). The cause(s) of the summertime increases in *E. coli* concentrations in feedlot and dairy water troughs are not clearly defined, but some plausible contributing factors include more frequent use (and contamination) by cattle as their water consumption increases (including, in some cases, cattle physically stepping into the water troughs in the hottest weather) as well as proliferation of *E. coli* in the sediments and possibly in the water column in water troughs warmed by summer weather.

In an ongoing study, we are examining the correlation between water contamination, feed contamination, and fecal prevalence of *E. coli* O157:H7 in thirty dairy herds, longitudinally through one-year periods. To date, *E. coli* O157:H7 has been isolated from more than 3 percent of water trough samples in this study, and the *E. coli* O157:H7 prevalence in cattle feces and in water trough water are positively correlated. These observations (evidence for summertime proliferation in the water as described in the previous paragraph, and the positive correlation between water and cattle fecal *E. coli* O157:H7) suggest that water could be both a reservoir and a vehicle of cattle infection. This possibility has been clearly demonstrated experimentally; when water troughs were intentionally contaminated with feces from infected cattle, the agent persisted for more than six months and remained infectious to calves at the end of this period (LeJeune et al. 2001a). However, it is also probable that some of the correlation between water and cattle fecal *E. coli* O157:H7 prevalence is caused by fecal contamination of the water that increases when more infected cattle are using the trough. Clear dissection of the relative importance of water troughs as reservoirs and vehicles of *E. coli* O157:H7 infection must await development of effective methods of prevention and control of *E. coli* contamination of cattle water sources.

FEEDS. Feedstuffs also represent large sources of oral exposure to *E. coli*. In most situations, ingestion of *E. coli* in feeds will greatly exceed that in water consumption in all except the most heavily contaminated water troughs. Therefore, feeds may logically be suspected as important in the transmission of *E. coli* O157:H7 on farms. Total mixed rations fed to dairy cattle contain from <1 to >10^4 cfu *E. coli* per gram, with little seasonal variation (Lynn et al. 1998). Given the large feed intake typical of high-producing dairy cattle, total daily *E. coli* ingested in the feed likely frequently exceed 10^8 cfu per day throughout the year, or similar to the highest summertime expected *E. coli* intakes caused by contaminated water sources.

In contrast to an earlier study by our group (Lynn et al. 1998), we have not been able to subsequently document the occurrence of significant replication of *E. coli* in cattle feedbunks either in observations of farm feedlots or experimentally. Unlike *E. coli* counts in water troughs, concentrations in on-farm feedbunks do not vary seasonally, and summertime counts are similar to other seasons' concentrations. Silage, a major component of many cattle feeds, significantly inhibits growth of *E. coli* when the feed contains as little as 20 percent silage. The observed concentrations of *E. coli* in feedbunk feeds are lower than expected from the weighted average of the feed components, presumably because of inhibition by the silage component of the bunk feeds.

Similarly, in laboratory experiments, silage consistently inhibits growth (and concentrations) of *E. coli,* including *E. coli* O157:H7 and *Salmonella* spp. (Hancock, unpublished data). Therefore, the high numbers of *E. coli* found in feedbunk mixed rations seems to be the result of heavy contamination of one or more ration component feeds prior to mixing. In our studies, the feed components most likely to contain very high *E. coli* concentrations are forages such as alfalfa hay or green chopped corn. Therefore, although feeds are frequently contaminated at the source with *E. coli* (rarely including *E. coli* O157:H7), and although feedbunks are frequently contaminated with bovine fecal matter (often including *E. coli* O157:H7, particularly in the summertime), the contaminating bacteria are inhibited from proliferation by feed components and are unlikely to replicate to reach high concentrations in the feedbunk environment.

Another potential role for feed contamination is dissemination of *E. coli* O157:H7 on a regional or even international scale. Isolates indistinguishable on PFGE after restriction by XbaI and NotI have been identified on farms hundreds of kilometers apart and lacking any known shared animal sources or transfers (Rice et al. 1999). Similarly, a widely disseminated clonal subtype has been reported in Scotland that is both frequently isolated from cattle populations and that accounts for a significant percentage of human cases, and which also raises the question of the mechanism of regional dissemination (Allison et al. 2000). Livestock feedstuffs would seem to be a plausible mechanism for regional dissemination of *E. coli* O157:H7. Feeds frequently are contaminated to varying degrees with *E. coli* (Lynn et al. 1998). In a current study, we have observed occasional (<1 percent) contamination of purchased feedstuffs with *E. coli* O157:H7, when the feeds were sampled at the farm bulk storage facility where the feeds were delivered. Significantly, this contamination, when observed, was associated with increased feedbunk (>6-fold), water trough (4.9-fold), and bovine fecal *E. coli* O157:H7 prevalence (>2-fold in adults, >3-fold for heifers), suggesting that amplification occurred following introduction via the feeds. These results are consistent with an important role of contaminated, purchased feed components in the dissemination of *E. coli* O157:H7.

MANURE AND BEDDING. Manure and bedding are of course certain to become contaminated with *E. coli* O157:H7 in the course of cattle infection and fecal excretion. The process of grooming and play with objects in the pen undoubtedly results in cattle ingesting a certain amount of fecal bacteria, in many cases in more concentrated form than is likely in contaminated water and feed sources. Furthermore, it is known that *E. coli* O157:H7 can persist both on skin and hair (which is probably one of the more common methods of carcass contamination in slaughter plants) and in manure piles (Fukushima et al. 1999; Bolton et al. 1999; McGee et al. 2001). However, it is doubtful that exposure to contaminated manure and bedding is the most important source for cattle infection. First of all, manure and bedding contamination does not account

very well for the marked seasonal variation of bovine infection, at least without speculating about proliferation in response to summertime warmth in these substrates. The temporal clustering suggestive of point source exposure typical of *E. coli* O157:H7 fecal shedding in cattle herds is also difficult to explain by general environmental contamination. Given a typical adult cow fecal *E. coli* concentration of less than 10^6 cfu/g, hundreds of grams or even several kilograms of feces would need to be ingested daily to match the oral *E. coli* exposure accounted for in the normal daily water and feed intake.

Increasing Resistance to Infection. There are several means by which cattle might be rendered more resistant to infection from *E. coli* O157:H7, including microbes and microbial products that kill *E. coli* O157:H7, microbes or feed components that compete for or eliminate the niche for *E. coli* O157:H7 in the bovine intestine, and vaccination to produce an immune response that protects cattle from *E. coli* O157:H7 infection.

PHAGE. Lytic bacteriophage have been identified that are capable of specifically lysing *E. coli* O157:H7. Such phages could theoretically reduce the effective fecal excretion rates of *E. coli* O157:H7 by cattle, increase the resistance of cattle to *E. coli* O157:H7 colonization, or even reduce the numbers of infectious *E. coli* O157:H7 cells in foods of bovine origin to reduce the risk of human infection (Kudva et al. 1999). Currently, however, no published data exists to date demonstrating the use of *E. coli* O157:H7 lytic bacteriophage to affect the cattle or foodborne prevalence of *E. coli* O157:H7.

COMPETITIVE MICROFLORA. Other microflora may be able to reduce the likelihood of cattle colonization with *E. coli* O157:H7, either by direct inhibitory or bacteriocin-mediated effects, or indirectly by efficient utilization of the habitat that otherwise might be occupied by *E. coli* O157:H7 in the gastrointestinal tract of cattle (Jordi et al. 2001; Ohya et al. 2000; Zhao et al. 1998). The use of competitive microflora to reduce colonization with zoonotic enteropathogens is best established in poultry, where application of the competitive flora strongly reduces the ability of *Salmonella* spp. to colonize chickens. Use of competitive microflora in cattle is theoretically more difficult because of the animals' longer pre-harvest life spans, the weaning process, the typically multiple major changes in the composition of the animals' feeds during their productive lives, and other factors. Nevertheless, research groups are actively pursuing competitive inhibition (Zhao et al. 1998) and have produced initially very promising results. At present, no published data demonstrates the use of competitive microflora to affect the cattle or foodborne prevalence of *E. coli* O157:H7.

VACCINATION. Cattle are demonstrated to respond to *E. coli* O157:H7 colonization by producing antibodies to LPS and to other surface antigens of *E. coli* O157:H7 (Johnson et al. 1996). Typically, these responses have not been sufficient to result in sterile immunity to challenge. Nevertheless, there is reason to think that vaccination of cattle to reduce the likelihood of colonization has the potential to affect prevalence of this agent. For example, even though natural exposure seems to be limited in its effect on resistance of cattle to re-colonization, tailoring of the antigens present in the putative vaccine can potentially result in responses with much more potent effects on susceptibility to colonization, especially if adherence factors used by the bacterium to efficiently colonize cattle can be identified. To date, there are no published data demonstrating the use of vaccination to affect the cattle prevalence of *E. coli* O157:H7.

FEEDS THAT REDUCE SUSCEPTIBILITY. An alternative approach to identifying competitive microflora or *E. coli* O157:H7 lytic phage or flora would be to identify cattle feeds that affect either the ability of *E. coli* O157:H7 to colonize the bovine gastrointestinal tract or the virulence of those isolates that do succeed in bovine colonization. In risk factor identification studies, a number of feed practices have been identified that are associated with *E. coli* O157:H7 prevalence in the cattle population. Among these specific ration ingredients, whole cottonseed (associated with decreased *E. coli* O157:H7 prevalence, Garber et al. 1995) and rations containing corn silage or barley (associated with increased *E. coli* O157:H7 prevalence, Herriott et al. 1998; Buchko et al. 2000) have each been associated with *E. coli* O157:H7 prevalence in multiple-risk-factor studies. Hovde et al. (1999) showed prolonged excretion of *E. coli* O157:H7 by experimentally challenged cattle fed all hay diets compared to those on high-grain-content rations. It's important to point out that dietary risk factors such as these identified in surveillance studies require confirmation by prospective studies.

It has also been proposed that transition to an all-hay diet prior to slaughter will reduce the degree to which colonic *E. coli* O157:H7 bacteria are conditioned to acid, and thereby reduce the risk of survival of the *E. coli* O157:H7 through the gastric acid barrier of the human stomach (Diez-Gonzalez et al. 1998). However, the principal data on which this proposal was based used generic *E. coli* rather than *E. coli* O157:H7, and *E. coli* O157:H7 is known to exhibit differences in acid resistance when compared to most *E. coli*. In what would seem to be a test of this hypothesis, Hovde et al. (1999) compared cattle on hay and grain-based rations, and failed to identify any difference in the acid resistance of *E. coli* O157:H7 excreted by these groups. Furthermore, as mentioned previously, the cattle on the all-hay diet were found to excrete *E. coli* O157:H7 for a longer period. More work is required to demonstrate an unequivocal effect of acidifying grain-based diets on either prevalence or virulence of *E. coli* O157:H7.

Reducing *E. coli* O157:H7 Carried on the Hide and Hair Coat. Most bacteria on the surface of the carcass at slaughter result from contamination from the soiled hides of the incoming cattle, rather than from direct fecal or intestinal content contamination (Grau 1987). Therefore, reduction in the degree to which the hides and hair of cattle entering slaughter plants are contaminated with *E. coli* O157:H7 would be expected to have directly beneficial results for food safety. Toward this end, some slaughter plants discriminate against animals that are visibly heavily soiled. However, modeling studies have estimated only limited beneficial effects of reduction of visible hide soiling ("tags") (Jordan et al. 1999). Furthermore, apparently little correlation exists between visible soiling and actual bacterial counts on the hides (Van Donkersgoed et al. 1997). Very vigorous (three minutes' power wash) physical washes are required to reduce *E. coli* O157:H7 counts on the hide (Byrne et al. 2000). Therefore, although reduction of *E. coli* O157:H7 surface contamination on cattle is theoretically beneficial, it is likely that, as with other potential pre-harvest interventions, the benefits obtained will be on the margin.

SUMMARY AND CONCLUSIONS. At present, no specific interventions have been unequivocally shown to affect the prevalence or virulence of *E. coli* O157:H7 shed by cattle in their feces or carried by cattle on their skin and hair. Nevertheless, sound theoretical and experimental reasons exist to pursue development of such interventions, and the emerging understanding of the epidemiology and ecology of the agent in cattle populations suggest that such interventions may be identified in the future.

REFERENCES

Allison, L.J., P.E. Carter, and F.M. Thomson-Carter. 2000. Characterization of a recurrent clonal type of *Escherichia coli* O157:H7 causing major outbreaks of infection in Scotland. *J. Clin. Microbiol.* 38:1632–1635.

Benjamin, M.M., and A.R. Datta. 1995. Acid tolerance of enterohemorrhagic *Escherichia coli*. *Appl. Environ. Microbiol.* 61:1669–1672.

Besser, T.E., B.L. Richards, D.H. Rice, and D.D. Hancock. 2001. *Escherichia coli* O157:H7 infection of calves: infectious dose and direct contact transmission. *Epidemiol. Infect.* 127:555–560.

Bettelheim, K.A., E.M. Cooke, S. O'Farrell, and R.A. Shooter. 1979. The effect of diet on intestinal *Escherichia coli*. *J. Hyg. London* 79:43–45.

Bolton, D.J., C.M. Byrne, J.J. Sheridan, D.A. McDowell, and I.S. Blair. 1999. The survival characteristics of a non-toxigenic strain of *Escherichia coli* O157:H7. *J. Appl. Microbiol.* 86:407–411.

Bonardi, S., E. Maggi, A. Bottarelli, M.L. Pacciarini, A. Ansuini, G. Vellini, S. Morabito, and A. Caprioli. 1999. Isolation of verocytotoxin-producing *Escherichia coli* O157:H7 from cattle at slaughter in Italy. *Vet. Microbiol.* 67:203–211.

Buchko, S.J., R.A. Holley, W.O. Olson, V.P. Gannon, and D.M. Veira. 2000. The effect of different grain diets on fecal shedding of *Escherichia coli* O157:H7 by steers. *J. Food Protect.* 63:1467–1474.

Byrne, C.M., D.J. Bolton, J.J. Sheridan, D.A. McDowell, and I.S. Blair. 2000. The effects of preslaughter washing on the reduction of *Escherichia coli* O157:H7 transfer from cattle hides to carcasses during slaughter. *Lett. Appl. Microbiol.* 30:142–145.

Chapman, P.A., A.T. Cerdan Malo, M. Ellin, R. Ashton, and M.A. Harkin. 2001. *Escherichia coli* O157 in cattle and sheep at slaughter, on beef and lamb carcasses and in raw beef and lamb products in South Yorkshire, UK. *Int. J. Food Microbiol.* 64:139–150.

Chapman, P.A., C.A. Siddons, D.J. Wright, P. Norman, J. Fox, and E. Crick. 1993. Cattle as a possible source of verocytotoxin-producing *Escherichia coli* O157 infections in man. *Epidemiol. Infect.* 111:439–447.

Chapman, P.A., C.A. Siddons, A.T. Cerdan Malo, and M.A. Harkin. 2000. A one-year study of *Escherichia coli* O157 in raw beef and lamb products. *Epidemiol. Infect.* 124:207–213.

Coia, J.E., J.C.M. Sharp, D.M. Campbell, J. Curnow, and C.N. Ramsay. 1998. Environmental risk factors for sporadic *Escherichia coli* O157 infection in Scotland: results of a descriptive epidemiology study. *J. Infect.* 36:317–321.

Cray, W.C., Jr., and H.W. Moon. 1995. Experimental infection of calves and adult cattle with *Escherichia coli* O157:H7. *Appl. Environ. Microbiol.* 61:1586–1590.

Dean-Nystrom, E.A., B.T. Bosworth, W.C. W.C. Cray, Jr., and H.W. Moon. 1997. Pathogenicity of *Escherichia coli* O157:H7 in the intestines of neonatal calves. *Infect. Immun.* 65:1842–1848.

Decludt, B., P. Bouvet, P. Mariani-Kurkdjian, F. Grimont, P.A. Grimont, B. Hubert, B., and C. Loirat. 2000. Haemolytic uraemic syndrome and Shiga toxin-producing *Escherichia coli* infection in children in France. The Société de Néphrologie Pédiatrique. *Epidemiol. Infect.* 124 :215–220.

Diez-Gonzalez, F., T.R. Callaway, M.G. Kizoulis, and J.B. Russell. 1998. Grain feeding and the dissemination of acid-resistant *Escherichia coli* from cattle. *Science* 281:1666–1668.

Douglas, A.S., and A. Kurien. 1997. Seasonality and other epidemiological features of haemolytic uraemic syndrome and *E. coli* O157 isolates in Scotland. *Scott. Med. J.* 42:166–171.

Elder, R.O., J.E. Keen, G.R. Siragusa, G.A. Barkocy-Gallagher, M. Koohmaraie, and W.W. Laegreid. 2000. Correlation of enterohemorrhagic *Escherichia coli* O157 prevalence in feces, hides, and carcasses of beef cattle during processing. *Proc. Natl. Acad. Sci. USA* 97:2999–3003.

Freter, R., H. Brickner, J. Fakete, M.M. Vickerman, and K.E. Carey. 1983. Survival and implantation of *Escherichia coli* in the intestinal tract. *Infect. Immun.* 39:686–703.

Fukushima, H., K. Hoshina, and M. Gomyoda. 1999. Longterm survival of shiga toxin-producing *Escherichia coli* O26, O111, and O157 in bovine feces. *Appl. Environ. Microbiol.* 65:5177–5181.

Garber, L.P., S.J. Wells, D.D. Hancock, M.P. Doyle, J. Tuttle, J.A. Shere, and T. Zhao. 1995. Risk factors for fecal shedding of *Escherichia coli* O157:H7 in dairy calves. *J.A.V.M.A.* 207:46–49.

Gianviti, A., F. Rosmini, A. Caprioli, R. Corona, M.C. Matteucci, F. Principato, I. Luzzi, and G. Rizzoni. 1994. Haemolytic–uraemic syndrome in childhood: surveillance and case-control studies in Italy. Italian HUS Study Group. *Pediatr. Nephrol.* 8:705–709.

Gill, C.O., and C. McGinnis. 1993. Changes in the microflora on commercial beef trimmings during their collection, distribution and preparation for retail sale as ground beef. *Int. J. Food Microbiol.* 18:321–332.

Grau, F.H. 1987. Prevention of microbial contamination in the export beef abattoir. In *Elimination of Pathogenic Organisms from Meat and Poultry.* F.J.M. Smulders, ed. Amsterdam: Elsevier, 221–233.

Hancock, D.D., T.E. Besser, M.L. Kinsel, P.I. Tarr, D.H. Rice, and M.G. Paros. 1994. The prevalence of *Escherichia coli* O157.H7 in dairy and beef cattle in Washington State. *Epidemiol. Infect.* 113:199–207.

Hancock, D.D., T.E. Besser, and D.H. Rice. 1998. Ecology of *E. coli* O157:H7 in cattle and impact of management practices. In *Escherichia coli O157:H7 and Other Shiga Toxin Producing E. coli strains.* J.B. Kaper, and A.D. O'Brien, A.D., eds. Washington, D.C.: American Society for Microbiology Press, 85–91.

Hancock, D.D., D.H. Rice, L.A. Thomas, D.A. Dargatz, and T.E. Besser. 1997. Epidemiology of *Escherichia coli* O157 in feedlot cattle. *J. Food Protect.* 60:462–465.

Herriott, D.E., D.D. Hancock, E.D. Ebel, L.V. Carpenter, D.H. Rice, and T.E. Besser. 1998. Association of herd management factors with colonization of dairy cattle by Shiga toxin-positive *Escherichia coli* O157. *J. Food Protect.* 61:802–807.

Heuvelink, A.E., F.L. van den Biggelaar, J. Zwartkruis-Nahuis, R.G. Herbes, R. Huyben, N. Nagelkerke, W.J. Melchers, L.A. Monnens, and E. de Boer. 1998. Occurrence of verocytotoxin-producing *Escherichia coli* O157 on Dutch dairy farms. *J. Clin. Microbiol.* 36:3480–3487.

Hovde, C.J., P.R. Austin, K.A. Cloud, C.J. Williams, and C.W. Hunt. 1999. Effect of cattle diet on *Escherichia coli* O157:H7 acid resistance. *Appl. Environ. Microbiol.* 65:3233–3235.

Huber, H.C., R. Kugler, and B. Liebl. 1998. Infections with enterohemorrhagic *Escherichia coli* EHEC—results of an epidemiologic survey in Bavaria for the April 1996 to May 1997 time frame. *Gesundheitswesen* 60:159–165.

Jordi, B.J., K. Boutaga, C.M. van Heeswijk, F. van Knapen, and L.J. Lipman. 2001. Sensitivity of Shiga toxin-producing *Escherichia coli* (STEC) strains for colicins under different experimental conditions. *FEMS Microbiol. Lett.* 204:329–334.

Johnsen, G., Y. Wasteson, E. Heir, O.I. Berget, and H. Herikstad. 2001. *Escherichia coli* O157:H7 in faeces from cattle, sheep and pigs in the southwest part of Norway during 1998 and 1999. *Int. J. Food Microbiol.* 65:193–200.

Johnson, R.P., W.C. Cray, Jr., and S.T. Johnson. 1996. Serum antibody responses of cattle following experimental infection with *Escherichia coli* O157:H7. *Infect. Immun.* 64:1879–1883.

Jordan, D., S.A. McEwen, A.M. Lammerding, W.B. McNab, and J.B. Wilson. 1999. Pre-slaughter control of *Escherichia coli* O157 in beef cattle: a simulation study. *Prev. Vet. Med.* 41:55–74.

Kapperud, G., J. Lassen, and V. Hasseltvedt. 1998. Salmonella infections in Norway: descriptive epidemiology and a case-control study. *Epidemiol. Infect.* 121:569–577.

Keene, W.E., E. Sazie, J. Kok, D.H. Rice, D.D. Hancock, V.K. Balan, T. Zhao, and M.P. Doyle. 1997. An outbreak of *Escherichia coli* O157:H7 infections traced to jerky made from deer meat. *J.A.M.A.* 277:1229–1231.

Keskimaki, M., M. Saari, T. Heiskanen, and A. Siitonen. 1998. Shiga toxin-producing *Escherichia coli* in Finland from 1990 through 1997: prevalence and characteristics of isolates. *J. Clin. Microbiol.* 36:3641–3646.

Krytenburg, D., D.D. Hancock, D.H. Rice, T.E. Besser, C.C. Gay, and J.M. Gay. 1998. *Salmonella enterica* in cattle feeds from the Pacific Northwest. *Anim. Feed Sci. Technol.* 75:75–79.

Kudva, I.T., K. Blanch, and C.J. Hovde. 1998. Analysis of *Escherichia coli* O157:H7 survival in ovine or bovine manure and manure slurry. *Appl. Environ. Microbiol.* 64:3166–3174.

Kudva, I.T., S. Jelacic, P.I. Tarr, P. Youderian, and C.J. Hovde. 1999. Biocontrol of *Escherichia coli* O157 with O157-specific bacteriophages. *Appl. Environ. Microbiol.* 65:3767–3773.

Lahti, E., M. Keskimaki, L. Rantala, P. Hyvonen, A. Siitonen, and T. Honkanen-Buzalski. 2001. Occurrence of *Escherichia coli* O157 in Finnish cattle. *Vet. Microbiol.* 79:239–251.

LeJeune, J.T., T.E. Besser, and D.D. Hancock. 2001a. Cattle water troughs as reservoirs of *Escherichia coli* O157. *Appl. Environ. Microbiol.* 67:3053–3057.

LeJeune, J.T., T.E. Besser, N.L. Merrill, D.H. Rice, and D.D. Hancock. 2001b. Livestock drinking water microbiology and the factors influencing the quality of drinking water offered to cattle. *J. Dairy Sci.* 84:1856–1862.

Lynn, T.V., D.D. Hancock, T.E. Besser, J.H. Harrison, D.H. Rice, N.T. Stewart, and L.L. Rowan. 1998. The occurrence and replication of *Escherichia coli* in cattle feeds. *J. Dairy Sci.* 81:1102–1108.

MacDonald, I.A., I.M. Gould, and J. Curnow. 1996. Epidemiology of infection due to *Escherichia coli* O157: a 3-year prospective study. *Epidemiol. Infect.* 116:279–284.

Magnuson, B.A., M. Davis, S. Hubele, P.R. Austin, I.T. Kudva, C.J. Williams, C.W. Hunt, and C.J. Hovde. 2000. Ruminant gastrointestinal cell proliferation and clearance of *Escherichia coli* O157:H7. *Infect. Immun.* 68:3808–3814.

Manning, J.G., B.M. Hargis, A. Hinton, Jr., D.E. Corrier, J.R. De-Loach, and C.R. Creger. 1994. Effect of selected antibiotics and anticoccidials on *Salmonella enteritidis* cecal colonization and organ invasion in Leghorn chicks. *Avian Dis.* 38:256–261.

McGee, P., D.J. Bolton, J.J. Sheridan, B. Earley, and N. Leonard. 2001. The survival of Escherichia coli O157:H7 in slurry from cattle fed different diets. *Lett Appl Microbiol.* 32:152–5.

McHan, F., and R.B. Shotts. 1993. Effect of short-chain fatty acids on the growth of *Salmonella typhimurium* in an in vitro system. *Avian Dis.* 37:396–398.

Mechie, S.C., P.A. Chapman, and C.A. Siddons. 1997. A fifteen month study of *Escherichia coli* O157:H7 in a dairy herd. *Epidemiol. Infect.* 118:17–25.

Michel, P., J.B. Wilson, S.W. Martin, R.C. Clarke, S.A. McEwen, and C.L. Gyles. 1999. Temporal and geographical distributions of reported cases of *Escherichia coli* O157: H7 infection in Ontario. *Epidemiol. Infect.* 122:193–200.

Nisbet, D.J. 1998. Use of competitive exclusion in food animals. *J.A.V.M.A.* 213:1744–1746.

Ohya, T., T. Marubashi, and H. Ito. 2000. Significance of fecal volatile fatty acids in shedding of *Escherichia coli* O157 from calves: experimental infection and preliminary use of a probiotic product. *J. Vet. Med. Sci.* 62:1151–1155.

Public Health Laboratory Service of England and Wales PHLS. 1999. Disease facts: *Escherichia coli* O157. http://www.phls.co.uk/facts/Gastro/ecoliQua.htm.

Que, J.U., S.W. Casey, and D.J. Hentges. 1986. Factors responsible for increased susceptibility of mice to intestinal colonization after treatment with streptomycin. *Infect. Immun.* 53:116–123.

Pritchard, G.C., S. Williamson, T. Carson, J.R. Bailey, L. Warner, G. Willshaw, and T. Cheasty. 2001. Wild rabbits—a novel vector for verocytotoxigenic *Escherichia coli* O157.*Vet. Rec.* 149:567.

Radostits, O.M., D.C. Blood, and C. Gay. 1994. *Veterinary Medicine.* London: Balliere Tindall, 733.

Rice, D.H., D.D. Hancock, and T.E. Besser. 1995. Verotoxigenic E. coli O157 colonisation of wild deer and range cattle. *Vet. Rec.* 137:524.

Rice, D.H., K.M. McMenamin, L.C. Pritchett, D.D. Hancock, and T.E. Besser. 1999. Genetic subtyping of *Escherichia coli* O157 isolates from 41 Pacific Northwest USA cattle farms. *Epidemiol. Infect.* 122:479–484.

Russell, J.B., F. Diez-Gonzalez, and G.N. Jarvis. 2000. Invited review: effects of diet shifts on *Escherichia coli* in cattle. *J. Dairy Sci.* 83:863–873.

Sargeant, J.M., D.J. Hafer, J.R. Gillespie, R.D. Oberst, and S.J. Flood. 1999. Prevalence of *Escherichia coli* O157:H7 in white-tailed deer sharing rangeland with cattle. *J.A.V.M.A.* 215:792–794.

Siciliano-Jones, J., and M.R. Murphy. 1989. Production of volatile fatty acids in the rumen and cecum-colon of steers as affected by forage:concentrate and forage physical form. *J. Dairy Sci.* 72:485–492.

Siegler, R.L., A.T. Pavia, R.D. Christofferson, and M.K. Milligan. 1994. A 20-year population-based study of postdiarrheal hemolytic uremic syndrome in Utah. *Pediatrics* 94:35–40.

Stevens, M.P., O. Marches, J. Campbell, V. Huter, G. Frankel, A.D. Phillips, E. Oswald, and T.S. Wallis. 2002. Intimin, tir, and shiga toxin 1 do not influence enteropathogenic responses to shiga toxin-producing *Escherichia coli* in bovine ligated intestinal loops. *Infect. Immun.* 70:945–952.

Trevena, W.B., G.A. Willshaw, T. Cheasty, G. Domingue, and C. Wray. 1999. Transmission of Vero cytotoxin producing *Escherichia coli* O157 infection from farm animals to humans in Cornwall and West Devon. *Commun. Dis. Public Health* 2:263–268.

Uljas, H.E., and S.C. Ingham. 1998. Survival of *Escherichia coli* O157:H7 in synthetic gastric fluid after cold and acid habituation in apple juice or trypticase soy broth acidified with hydrochloric acid or organic acids. *J. Food Protect.* 61:939–947.

United States Department of Agriculture USDA: Food Safety Inspection Service FSIS, 2000. Risk assessment for *E. coli* O157:H7 in ground beef. http://www.fsis.usda.gov/OPHS/ecolrisk/.

Van Donkersgoed, J., K.W.F. Jericho, H. Grogan, and B. Thorlakson. 1997. Preslaughter hide status of cattle and the microbiology of carcasses. *J. Food Protect.* 60:1502–1508.

Van Donkersgoed, J., T. Graham, and V. Gannon. 1999. The prevalence of verotoxins, *Escherichia coli* O157:H7, and Salmonella in the feces and rumen of cattle at processing. *Can. Vet. J.* 40:332–338.

Vold, L., B. Klungseth Johansen, H. Kruse, E. Skjerve, and Y. Wasteson. 1998. Occurrence of shigatoxinogenic *Escherichia coli* O157 in Norwegian cattle herds. *Epidemiol. Infect.* 120:21–28.

Wallace, D.J., T. Van Gilder, S. Shallow, T. Fiorentino, S.D. Segler, K.E. Smith, B. Shiferaw, R. Etzel, W.E. Garthright, and F.J. Angulo. 2000. Incidence of foodborne illnesses reported by the foodborne diseases active surveillance network Food-Net—1997. *J. Food Protect.* 63:807–809.

Waterman, S.R., and P.L. Small. 1998. Acid-sensitive enteric pathogens are protected from killing under extremely acidic conditions of pH 2.5 when they are inoculated onto certain solid food sources. *Appl. Environ. Microbiol.* 64:3882–3886.

Waters, J.R., J.C. Sharp, and V.J. Dev. 1994. Infection caused by *Escherichia coli* O157:H7 in Alberta, Canada, and in Scotland: a five-year review, 1987–1991. *Clin. Infect. Dis.* 19:834–843.

Zhao, T., M.P. Doyle, B.G. Harmon, C.A. Brown, P.O. Mueller, and A.H. Parks. 1998. Reduction of carriage of enterohemorrhagic *Escherichia coli* O157:H7 in cattle by inoculation with probiotic bacteria. *J. Clin. Microbiol.* 36:641–647.

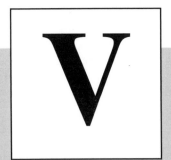
CAMPYLOBACTER SPP.

19

EPIDEMIOLOGY OF *CAMPYLOBACTER* SPP. IN ANIMALS

JOHN. B. KANEENE AND RACHEL CHURCH POTTER

INTRODUCTION. *Campylobacter jejuni* subsp. *jejuni* and *C. coli* are the organisms responsible for approximately 95 percent of *Campylobacter*-attributed disease in humans (Lastovica and Skirrow 2000). *Campylobacter upsaliensis, C. hyointestinalis,* and *C. lari* have also been linked, although rarely, to human enteritis (Lawson et al. 1999). *Campylobacter jejuni* is the most common cause of human bacterial gastroenteritis in the United States today. The yearly incidence rate was 15.7 / 100,000 person-years in 2000, according to the CDC FoodNet active surveillance system (CDC 2001). Illness occurs three to five days after an ingestion of a dose as low as 500 organisms (Robinson 1981). Clinical signs and symptoms include fever, profuse and often bloody diarrhea, abdominal pain, and nausea. Clinical signs may remain severe for one or two days before resolving over a period of one week or longer. Rarely, infection occurs at other sites, causing cholecystitis, pancreatitis, cystitis, or septic abortion. Fatal bacteremia may occur in immunocompromised hosts (Manfredi et al. 1999). Campylobacteriosis is a significant cause of morbidity and lost productivity, not only because of the severe gastroenteritis it produces but also because of sequelae such as reactive arthritis and Guillain-Barré Syndrome (Smith

1995). Because *C. jejuni* and the other *Campylobacter* spp. cannot multiply or survive long in the environment, warm-blooded animals are considered the reservoir for the organism and the ultimate source of human infection.

Animals themselves do not typically exhibit clinical signs related to carriage of the organism, and they shed the organism intermittently. Understanding the epidemiology of campylobacters in animals is critical because of the risk to humans from animals and animal products.

POULTRY. *Campylobacter jejuni* has commonly been isolated from poultry carcasses and live poultry. Retail meat isolation rates have been measured at 23.1 percent, 40 percent, 45.9 percent, and 59 percent (Harris et al. 1986; Varma et al. 2000; Atanassova and Ring 1999; Osano and Arimi 1999). Poultry also carry *Campylobacter coli*, but at a much lower rate. Isolation rates from flocks are between 1 percent and 2.7 percent (Atanassova and Ring 1999; Refregier-Petton et al. 2001). At slaughter, the isolation rate can be 15.8 percent (Pokamunski et al. 1986). Poultry are not major transmitters of *C. upsaliensis* (Hald and Madsen 1997).

Vertical Transmission. The importance of vertical transmission is still uncertain, but horizontal transmission from environmental sources appears to occur readily, and the organism can spread rapidly throughout a flock. Early studies indicated a lack of vertical transmission. Several studies included the failure of investigators to find *Campylobacter* spp. in 650 fertile eggs and 230 twenty-four-hour-old chicks (Pearson et al. 1993), chicks younger than one week (Pokamunski et al. 1986), ten days (Genigeorgis et al. 1986), and thirteen days (Jacobs-Reitsma et al. 1995). A prospective study of broilers followed seven flocks through a broiler house and found that six were *C. jejuni*–free. Because at least four different mother flocks were involved in supplying this producer, and because eggs are commonly mixed at hatcheries (that is, one broiler flock could descend from more than one mother flock), the authors concluded that vertical transmission was unlikely given the high prevalence of *C. jejuni* infection in chickens (van de Giessen et al. 1992). Only one of two flocks originating from the same parents and hatch but residing on different farms was *C. jejuni* positive, whereas flocks of different origin but residing on the same farm were infected with the same *C. jejuni* serotypes (Jacobs-Reitsma 1995).

More recent studies, however, indicate the possibility of vertical transmission. It has been shown, under experimental conditions, that it is possible for *C. jejuni* to penetrate the eggshell and contaminate its contents (Shane et al. 1986; Shanker et al. 1986). A study of eggs from a highly infected (74 percent) breeder flock found that 1 percent of eggs tested were infected with *C. jejuni* (Shanker et al. 1986). In another study, using the very sensitive method of Southern blot hybridization (SBH), *C. jejuni* was found in two of fifty-one eighteen-day-old chicken embryos and three of forty newly hatched chickens. It was not possible for the investigators to culture and isolate the organism, however, so it is not clear whether dead organisms, viable but not culturable organisms, or DNA alone was responsible for the positive test (Chuma et al. 1996). Therefore, the findings may or may not represent a potential source of flock infection. Circumstantial evidence for vertical transmission has also been found. Investigators prospectively studied the incidence of *C. jejuni* in broiler flocks for five years and found that new serotypes dominated successive flocks, suggesting a common source of introduction to the farm. The source of infection for these flocks could have been either the hatchery that supplied the one-day old chicks or the layers that supplied eggs to the hatchery (Pearson et al. 1996). More convincing evidence has been reported. Identical isolates were obtained from a breeder flock and its six-week-old progeny that were more than twenty miles away (Stern et al. 2000).

Horizontal Transmission. Experimental evidence suggests that chicks are very susceptible to infection and colonization by *C. jejuni*. As few as ninety bacteria are capable of inducing diarrhea in 90 percent of three-day-old chicks. These chicks can start to shed the organism after only twenty-four hours, and shedding can persist until market age (Ruiz-Palacios et al. 1981; Achen et al. 1998). Because of this susceptibility and high rates of shedding, *Campylobacter* spp. can spread rapidly through the flock. A large study of four ranches with a total of twenty-four houses of birds found that the infection rate increased from 2.3 percent of ten-day-old birds to 81.8 percent of slaughter-age birds (Genigeorgis et al. 1986). Another study found that 40 percent of birds were infected by four weeks of age and more than 90 percent by seven weeks (Evans and Sayers 2000). Similarly, turkeys rapidly acquire infection. Colonization of turkey chicks began seven days after arrival on one farm and reached 100 percent within three weeks (Wallace et al. 1997). *Campylobacter* spp. spread throughout flocks by means of coprophagia, fomites, and vectors. More than one serotype may be present in contaminated flocks, which may indicate that campylobacters from multiple sources can enter a flock (Jacobs-Reitsma et al. 1995). Many risk factors for infection of flocks caused by horizontal transmission have been elucidated, but contaminated water is of the most concern.

In a multivariable analysis, feeding undisinfected water, caring for other poultry before entering the broiler house, and caring for pigs before entering the broiler house all increased the odds for colonization. Of these, disinfection of the drinking water would have the greatest impact and would reduce the incidence of disease by 53 percent (Kapperud et al. 1993). Persistently contaminated water on a broiler farm can infect generations of flocks and, in one case, was linked to an outbreak and multiple sporadic cases in the population served by the farm for more than eighteen months (Pearson et al. 1993). A longitudinal study of one hundred broiler flocks in England confirmed that adequate disinfection of the water header tank reduces the risk of infection. This study also found protective effects with the use of a boot dip that is changed weekly and used after house disinfection (Evans and Sayers 2000). Groundwater can also become contaminated with *C. jejuni* from environmental sources or other farm animals and provide a source of contamination for flocks using this water source (Stanley et al. 1998). Other hygienic practices found to increase contamination by *Campylobacter* spp. include static ventilation in the broiler house, more than two chick houses on a farm, more than two persons involved in taking care of the house, the presence of litter-beetles in the change room, and acidification of drinking water (Refregier-Petton et al. 2001). Acidification of water is an unexpected risk factor for infection, given that its use has been suggested to prevent *Salmonella* infection. In this case, the authors suggested that acidification may be an indicator that unsanitary surface water, a source of *Campylobacter* spp., was given to the birds.

Insects, such as litter beetles or houseflies, can serve as vectors, physically moving the organism from one part of the house or farm to another. Wild birds can also harbor and transmit *Campylobacter* spp. (Khalil et al. 1994; Riordan et al. 1993). Investigators in Georgia collected fecal droppings and intestinal and cloacal samples of wild birds within a 200-foot radius of broiler chicken houses. Ten percent of these total samples were positive for *C. jejuni*, indicating that they carry the organism and could transmit it (Craven et al. 2000). Litter and feed are not thought to be sources of infection to flocks because the material is too dry for the organism to survive (Pokamunski et al. 1986). Season also plays a role in the risk of contamination for broiler flocks. In one study, birds reared in summer or autumn had six times the odds of contamination compared to birds reared in winter; another study showed that birds reared in autumn had more than three times the odds of infection compared with other seasons (Refregier-Petton et al. 2001; Kapperud et al. 1993). This seasonal variation correlates positively with minimum temperature and sun light hours (Wallace et al. 1997). Management strategies to control the environmental temperature within broiler houses should remove the influence of season. However, the risk of contamination of a flock does not end at the farm but continues on into processing.

Processing. Transport to the slaughterhouse and equipment in the slaughterhouse can increase carcass contamination. Defecation and crowding during transport leads to significant increases in contamination of poultry carcasses compared to levels before transport (Stern et al. 1995). Additionally, contaminated flocks can contaminate slaughterhouse equipment so that even *C. jejuni*–free flocks processed later will become contaminated (Genigeorgis et al. 1986). The percentage of positive carcasses increased throughout the week in one slaughterhouse when only hot water was used to clean the equipment after each day (Harris et al. 1986). The isolation rate from carcasses after processing can be higher than the isolation rate from live birds (Atanassova and Ring 1999). Molecular characterization of *C. jejuni* isolates reveals that this contamination of carcasses is caused by leakage of intestinal contents and cross-contamination of carcasses (Rivoal et al. 1999). Good management practices at the farm level will need to be coupled with good hygiene at the slaughterhouse to decrease carcass contamination.

CATTLE. One of the major causes of *C. jejuni* enteritis outbreaks in humans is consumption of raw milk; this is associated with sporadic cases as well (Evans et al. 1996; Kornblatt et al. 1985; Kalman et al. 2000; Studahl and Andersson 2000; Schmid et al. 1987; Hopkins et al. 1984; Eberhart-Philips et al. 1997). Direct contact with cattle and the consumption of undercooked beef do not seem to be major risk factors to humans, although sporadic cases attributable to these causes have occurred (Eberhart-Philips 1997). *Campylobacter jejuni* can be isolated from the feces of both beef and dairy cattle. A cross-sectional study found

37.7 percent of individual dairy cattle fecal samples and 80.6 percent of farm operations were positive (Wesley et al. 2000). Another study found that ten of twelve dairy herds carried the organism (Humphrey and Beckett 1986). *C. coli*, *C. fetus*, and *C. hyointestinalis* have also been isolated from cattle, but at lower rates (Grau 1988). Calves carry the organism more frequently than adult cattle. Samples obtained from four-week-old calves at slaughter isolated the organism from the feces of 54 percent of the calves but only 12.5 percent of the adult cattle. Calves may be more susceptible to infection because the rumen is not yet fully developed and their milk consumption may allow *Campylobacter* spp. to survive and pass through the rumen to colonize the upper gatrointestinal tract (Grau 1988). Isolations from the rumen indicate that cattle are probably ingesting the organism regularly.

Horizontal Transmission. Cattle become carriers through horizontal transmission of the organism. Risk factors for infection at the herd level include the use of a broadcast spreader to dispose of manure, feeding alfalfa or cotton seed or hulls, and accessibility of feed to birds. At the individual level, a large herd size (more than one hundred cows) and feeding brewers' by products to lactating cows increased odds for positivity (Wesley et al. 2000). Herd size was also found to increase risk in a study of beef cattle in California (Hoar et al. 2001). Diet may influence the ability of *Campylobacter* spp. to survive in the enteric flora. Increased contact with birds that are carriers and shedders of *C. jejuni* is also a risk factor. Large herd size or increased animal-to-animal contact may be required to maintain the infection in a herd. Lot feeding compared to pasture feeding increases the odds for carriage of both *C. jejuni* and *C. hyointestinalis* (Grau 1988) but whether this increase is caused by diet, crowding, exposure to a large number of animals, or all three remains unclear. Cattle can also acquire the infection from environmental sources. A study in the United Kingdom found that positive herds obtained water from contaminated rivers and streams, whereas negative herds were supplied with municipal water (Humphrey et al. 1986). Dairy herds show a seasonal variation in shedding, with peaks in the spring and the fall of each year that coincide with calving and changes in housing and diet (Stanley et al. 1998b). These peaks correspond with the seasonality of human *C. jejuni* enteritis outbreaks.

Raw Milk. Raw milk may become contaminated with *C. jejuni* in one of two ways. It may be excreted directly by a mastitic udder, or feces may contaminate the product. *Campylobacter jejuni* mastitis has been induced experimentally, and *C. jejuni* has been re-isolated from the milk (Lander and Gill 1979). Contamination of milk from *C. jejuni* mastitis is thought to be rare, but direct milk excretion leading to human cases has been reported (Orr et al. 1995). A survey of milk samples from 1,501 cows with mastitis was negative for *C. jejuni* (Waterman et al. 1984). *C. jejuni* has been isolated from the feces, but not the milk, of cows linked to raw milk–associated outbreaks (Kalman et al. 2000; Blaser et al. 1987; Kornblatt et al. 1985). Surveys of raw milk from bulk tanks have revealed variable rates of infection. Only 1.4 percent of raw milk samples from twenty-seven farms was positive for *C. jejuni*, but these farms were chosen specifically because they were known to produce high-quality milk (Desmasures et al. 1997). Another survey of 256 raw milk samples in Manitoba found that only 1.2 percent of the samples were contaminated, whereas 12.3 percent of the milk shipped to an east Tennessee processing plant tested positive (Davidson et al. 1989; Rohrbach et al. 1992). Only five hundred organisms in milk can induce enteritis in humans (Robinson 1981). Campylobacters enter the milk through improper hygiene when milking or when instruments come in contact with the milking parlor floor (Dilworth et al. 1988; Humphrey and Beckett 1986). Although *Campylobacter* cannot multiply in milk, it can survive for up to twenty-one days at refrigerator temperatures (Doyle and Roman 1982).

SWINE. Pigs have long been known as carriers of *Campylobacter coli*. Isolation rates vary from 66 percent to 97 percent (Manser and Dalziel 1985; Nielsen 1997; Munroe et al. 1983; Cabrita et al. 1992). They are now known to commonly carry *C. jejuni* as well. In the Netherlands, 79 percent of pigs were intestinal carriers of *C.jejuni* (Oosterom et al. 1985), and the cecal contents of 31 percent of Texas pigs tested positive (Harvey et al. 1999). Only 4 percent of fecal samples were positive for *C. jejuni* in a study in Denmark. Carriage is higher in pigs with enteric disease than in healthy pigs (Manser and Dalziel 1985).

Transmission. Pigs are not born with *Campylobacter* spp., but horizontal transmission from dam to offspring occurs at an early age. A study in the Netherlands followed sows from one week before delivery to eight weeks after delivery and their piglets from one week after delivery to eight weeks after delivery. All sows were infected with *Campylobacter* spp., and parturition was shown to enhance excretion. Consequently, 90 percent of piglets were infected by eight weeks of age (Weijtens et al. 1997). In the United States, investigators compared the colonization rates of gilts, pregnant sows and their newborn piglets, and weaned piglets. They found that the average prevalence of *Campylobacter* spp. in gilts was 76 percent and of these, 76.3 percent were *C. jejuni*, 21 percent *C. coli*, and 2.6 percent *C. lari*. One hundred percent of pregnant sows were positive. *Campylobacter jejuni* was isolated from 89 percent and *C. coli* from 11 percent. Newborn piglets had an average incidence of *Campylobacter* of 57.8 percent within twenty-four hours of birth. Among these newborns, the incidence of *C. jejuni* was lower (31.7 percent) than *C. coli* (68.3 percent). The inci-

dence increased to 85 percent by weaning of the piglets, with *C. jejuni* at 82 percent and *C. coli* at 18 percent (Young et al. 2000). An experimental study compared the prevalence of colonization in pigs reared on- and off-sow. Twenty-four hours after birth, piglets were removed from the sow and placed into wire-floored nurseries. Although there was no difference at the first day, after twenty days the prevalence in piglets reared off-sow was significantly lower than those reared with the sow. Repeated or constant exposure to contaminated fecal material from the dam or other pigs may be required for *Campylobacter* spp. to colonize the intestinal tract. These and other studies suggest that starting out with *Campylobacter*-free pigs would involve intervention strategies at birth or in the nursery and must be combined with good hygiene practices. Wild birds, rats, and mice can all carry and shed *Campylobacter* spp. (Kapperud and Rosef 1983; Meanger and Marshall 1989). Good hygiene would involve adequate control of these animals, but may be impossible in swine reared outdoors.

As with poultry, swine carcasses may become contaminated in the slaughterhouse. More than 32 percent of swabs from slaughterhouse surfaces and equipment were positive for *C. jejuni* when sampled during slaughter, but only 2.5 percent were positive after cleaning and drying overnight. In the same study, 9 percent of the carcasses were positive at evisceration, but none were positive after cooling (Oosterom et al. 1985). Cooling is accompanied by drying, which probably has the direct effect of killing the organism. Pig offal can provide a source for human infection, because 6 of samples were contaminated in one study (Bolton et al. 1985). Although people usually consume pork offal that is cooked, infection may occur through cross-contamination or by feeding raw offal to pets, who would then become carriers and shedders of the organism. However, *Campylobacter* spp. is not commonly isolated from pork for retail sale (Oosterom et al. 1985; Duffy et al. 2001).

OVINES. The average isolation rate of thermophilic campylobacters from 360 lambs at slaughter was 91.7 percent, but the isolation rate from the fecal samples of sheep at pasture was only 29.3 percent. Of these isolates, *Campylobacter jejuni* accounted for 88.2 percent of isolates, followed by *C. coli* (10.8 percent), *C. hyointestinalis* (0.8 percent) and *C. lari* (0.2 percent) (Stanley et al. 1998c). Similarly, another study found that shedding of *Campylobacter* spp. at pasture was between one-third and one-half of the carriage rate (92 percent) of the intestinal contents of sheep at slaughter. *Campylobacter jejuni* was the species found in 90 percent of *campylobacter*-positive grazing sheep, followed by *C. coli* (8 percent) and *C. lari* (2 percent) (Jones et al. 1999). Another study had an isolation rate of 13.6 percent from the feces of healthy sheep and 4.3 percent from diarrheic sheep (Prescott and Bruin-Mosch 1981).

Isolation rates in sheep are affected by type of pasture, season, and lambing. Poor grazing and digestibility may be associated with higher rates of shedding. Isolation rates are highest in the spring, which corresponds with change of pasture, the spring rise in parasites, lambing, and weaning. Shedding is often low during gestation and increases dramatically after lambing; this allows for the horizontal transfer of *Campylobacter* spp. from ewes to their lambs (Jones et al. 1999). Risk to humans is primarily through carcass contamination. *Campylobacter* spp. were not detected in a study of unpasteurized goats' and ewes' milk in the United Kingdom (Little and de Louvois 1999).

GOATS. Fecal isolation rates vary from 0 percent (Cabrita et al. 1992; Manser and Dalziel 1985) to 2.7 percent of healthy goats and 3.7 percent of diarrheic goats (Prescott and Bruin-Mosch 1981). *Campylobacter jejuni* enteritis has been associated with consumption of raw goat's milk in the United States, but was not recovered from any raw milk samples in a U.K. study (Harris et al. 1987; Little and de Louvois 1999).

DOMESTIC PETS. Dogs and cats are also known to carry and shed thermophilic *Campylobacters* (*C. jejuni* and *C. coli*). Because they are often members of a household, they are in close contact with humans and potentially represent a large source of infection. They may also facilitate transmission of the organism from animal to animal on the farm. Epidemiologic studies have associated cats and diarrheic animals, including dogs and puppies, with sporadic cases of human enteritis (Hopkins et al. 1984; Deming et al. 1987; Saeed et al. 1993).

SUMMARY AND CONCLUSIONS. Of all the food animals, poultry may be the most responsible for the foodborne disease in humans caused by *Campylobacter* spp. Infection may be foodborne or caused by direct contact with the animals (Studahl and Andersson 2000). The incidence of human enteritis decreased by 40 percent when Belgian poultry was removed from the market during the dioxin scare of 1999 (Vellinga and VanLoock 2002). Dairy cattle, though not traditionally thought of, are emerging as important carriers of the organism. They may cause human disease through direct contact, fecal-contaminated raw milk, or consumption of undercooked meats. Swine were long thought to be carriers of only *C. coli* and therefore not contributing to human disease. It is now known, however, that they also carry *C. jejuni* at high rates.

The control of *Campylobacter* spp. will be difficult because the epidemiology is complex. In addition, the organism has been difficult to isolate with traditional methods. Future efforts of control will require education and interventions at many levels. Strategies on the farm should be aimed at further identification of risk factors and the practice of good hygiene and pest control to limit the introduction of *Campylobacter* spp. Consumers will need to be educated about proper handling and cooking of meat, avoiding the consumption of raw milk, and practicing good hygiene when in contact with animals.

REFERENCES

Achen, M., T.Y. Morishita, E.C. Ley. 1998. Shedding and colonization of *Campylobacter jejuni* in broilers from day-of-hatch to slaughter age. *Avian Dis.* 42:732–737.

Atanassova, V., and C. Ring. 1999. Prevalence of *Campylobacter* spp. in poultry and poultry meat in Germany. *Intl. J. Food Microbiol.* 51:187–190.

Blaser, M.J., E. Sazie, and P. Williams. 1987. The influence of immunity on raw milk-associated *Campylobacterial* infection. *J.A.M.A.* 257:43–46.

Bolton, F.J., H.C. Dawkins, and D.N. Hutchinson. 1985. Biotypes and serotypes of thermophilic campylobacters isolated from cattle, sheep, and pig offal and other red meats. *J. Hygiene* 95:1–6.

Burnens, A.P., B. Angeloz-Wick, and J. Nicolet. 1992. Comparison of *Campylobacter* carriage rates in diarrheic and healthy pet animals. *J.A.V.M.A. Series B.* 39:175–180.

Cabrita, J., J. Rodrigues, F. Braganca, C. Morgado, I. Pires, and A.P. Goncalves. 1992. Prevalence, biotypes, plasmid profile and antimicrobial resistance of *Campylobacter* isolated from wild and domestic animals from Northeast Portugal. *J. Appl. Bacteriol.* 73:279–285.

CDC. 2001. Preliminary FoodNet data on the incidence of foodborne illnesses—selected sites, United States, 2000. *M.M.W.R.* 50(13):241–246.

Chuma, T., K. Yano, O. Hayato, K. Okamoto, and H. Yugi. 1997. Direct detection of *Campylobacter jejuni* in chicken cecal contents by PCR. *J.A.V.M.A.* 59(1):85–87.

Craven, S.E., N.J. Stern, E. Line, J.S. Bailey, N.A. Cox, and P. Fedorka-Cray. 2000. Determination of the incidence of *Salmonella spp.*, *Campylobacter jejuni*, and *Clostridium perfringens* in wild birds near broiler chicken houses by sampling intestinal droppings. *Avian Dis.* 44:715–720.

Davidson, R.J., D.W. Sprung, C.E. Park, and M.K. Rayman. 1989. Occurrence of *Listeria monocytogenes*, *Campylobacter* spp. and *Yersinia enterocolitica* in Manitoba raw milk. *Can. Inst. Food Sci. and Techn. J.* 22:70–74.

Deming, M.S., R.V. Tauxe, P.A. Blake, S.E. Dixon, B.S. Flowler, T.S. Jones, E.A. Lockamy, C.M. Patton, and R.O. Sikes. 1987. *Campylobacter* enteritis at a university: transmission from eating chicken and from cats. *Am. J. Epidemiol.* 126(3):526–534.

Desmasures, N., F. Bazin, and M. Gueguen. 1997. Microbiological composition of raw milk from selected farms in the Camembert region of Normandy. *J. Appl. Microbiol.* 83:53–58.

Dilworth, C.R., H. Lior, and M.A. Belliveau. 1988. *Campylobacter Enteritis* acquired from cattle. *Can. J. Pub. Health* 79:60–62.

Doyle, M.P., and D.J. Roman. 1982. Prevalence and survival of *Campylobacter jejuni* in unpasteruized milk. *Appl. Environ. Microbiol.* 44:1154–1158.

Duffy, E.A., K.E. Belk, J.N. Sofos, G.R. Bellinger, A. Pape, and G.C. Smith. 2001. Extent of microbial contamination in United States pork retail products. *J. Food Prot.* 64:172–178.

Eberhart-Philips J., N. Walker, N. Garrett, D. Bell, D. Sinclair, W. Rainger, and M. Bates. 1997. Campylobacteriosis in New Zealand: results of a case-control study. *J. Epidemiol. and Comm. Health* 51:686–691.

Evans, S.J., and A.R. Sayers. 2000. A longitudinal study of campylobacter infection of broiler flocks in Great Britain. *Prev. Vet. Med.* 46:209–223.

Genigeorgis, C., M. Hassuneh, and P. Collins. 1986. *Campylobacter jejuni* infection of poultry farms and its effect on poultry meat contamination during slaughtering. *J. Food Prot.* 49:895–903.

Gondrosen, B., T. Knaevelsrud, and K. Dommarsnes. 1985. Isolation of thermophilic Campylobacters from Norwegian dogs and cats. *Acta Veterinaria Scandinavica* 26:81–90.

Grau, F.H. 1988. *Campylobacter jejuni* and *Campylobacter hyointestinalis* in the intestinal tract and on the carcasses of calves and cattle. *J. Food Prot.* 51:857–861.

Hald, B., and M. Madsen. 1997. Healthy puppies and kittens as carriers of *Campylobacter* spp., with special reference to *Campylobacter upsaliensis*. *J. Clin. Microbiol.* 35:3351–3352.

Harris, N.V., D. Thompson, D.C. Martin, and C.M. Nolan. 1986. A survey of *Campylobacter* and other bacterial contaminants of pre-market chicken and retail poultry and meats, King County, Washington. *Am. J. Pub. Health* 76:401–406.

Harris, N.V., T.J. Kimball, P. Bennett, Y. Johnson, D. Wakely, and C.M. Nolan. 1987. *Campylobacter jejuni* enteritis associated with raw goat's milk. *Am. J. Epidemiol.* 126:179–186.

Harvey, R.B., C.R. Young, R.L. Ziprin, M.E. Hume, K.J. Genovese, R.C. Anderson, R.E. Droleskey, L.H. Stanker, D.J. Nisbet. 1999. Prevalence of *Campylobacter* spp. isolated from the intestinal tract of pigs raised in an integrated swine production system. *J.A.V.M.A.* 215:1601–1603.

Hoar, B.R., E.R. Atwill, C. Elmi, and T. B. Farver. 2001. An examination of risk factors associated with beef cattle shedding pathogens of potential zoonotic concern. *Epidemiol. and Infect.* 127:147–155.

Hopkins, R.S., R. Olmsted, and G.R. Istre. 1984. Endemic *Campylobacter jejuni* infection in Colorado: identified risk factors. *Am. J. Pub. Health* 74:249–250.

Humphrey, T.J., and P. Beckett. 1987. *Campylobacter jejuni* in dairy cows and raw milk. *Epidemiol. and Infect.* 98:263–269.

Jacobs-Reitsma, W.F. Campylobacter bacteria in breeder flocks. 1995. *Avian Dis.* 39:355–359.

Jacobs-Reitsma, W.F., A.W. van de Giessen, N.M. Bolder, and R.W.A.W. Mulder. 1995. Epidemiology of *Campylobacter* spp. at two Dutch broiler farms. *Epidemiol. and Infect.* 114:413–421.

Jones, K., S. Howard, and J.S. Wallace. 1999. Intermittent shedding of thermophlic campylobacters by sheep at pasture. *J. Appl. Microbiol.* 86:531–536.

Kalman, M, E. Szollosi, B. Czermann, M. Zimanyi, S Sekeres, and M. Kalman. 2000. Milkborne *Campylobacter* infection in Hungary. *J. Food Prot.* 63:1426–1429.

Kapperud, G., and O. Rosef. 1983. Avian wildlife reservoir of *Campylobacter fetus* subsp. *jejuni*, *Yersinia* spp., and *Salmonella* spp. in Norway. *Appl. Environ. Microbiol.* 45:375–380.

Kapperud, G., E. Skjerve, L. Bik, K. Hauge, A. Lysaker, I. Aalmen, S.M. Ostroff, and M. Potter. 1993. Epidemiological investigation of risk factors for campylobater colonization in Norwegian broiler flocks. *Epidemiol. and Infect.* 111:245–255.

Khalil, K., G.-B. Lindblom, K. Mazhar, and B. Kaijser. 1994. Flies and water as reservoirs for bacterial enteropathogens in urban and rural areas in and around Lahore, Pakistan. *Epidemiol. and Infect.* 113:435–444.

Kornblatt, A.N., T. Barrett, G.K. Morris, and F.E. Tosh. 1985. Epidemiologic and laboratory investigation of an outbreak of *Campylobacter* enteritis associated with raw milk. *Am. J. Epidemiol.* 122:884–889.

Lander, K.P., and K.P.W. Gill. 1980. Experimental infection of the bovine udder with *Campylobacter coli / jejuni. J. Hygiene* 84:421–428.

Lastovica, A.J., and M.B. Skirrow. 2000. Clinical significance of *Campylobacter* and related species other than *Campylobacter jejuni* and *C.coli*. In *Campylobacter*, 2nd ed. Irving Nachamkin and Martin J. Blaser, eds. Washington, D.C: ASM Press, 89.

Lawson, A.J., M.J. Logan, G.L. O'Neill, M. Desai, and J. Stanley. 1999. Large-scale survey of *Campylobacter* species in human gastroenteritis by PCR and PCR-enzyme-linked immunosorbent assay. *J. Clin. Microbiol.* 37(12):3860–3864.

Little, C.L., and J. de Louvois. 1999. Health risks associated with unpasteurized goats' and ewes' milk on retail sale in England and Wales. A PHLS Dairy Products Working Group Study. *Epidemiol. and Infect.* 122:403–408.

Manfredi, R., A. Nanetti, M. Ferri, and F. Chiodo. 1999. Fatal *Campylobacter jejuni* bacteraemia in patients with AIDS. *J. Med. Microbiol.* 48:601–603.

Manser, P.A., and R.W. Dalziel. 1985. A survey of campylobacter in animals. *J. Hygiene* 95:15–21.

Meanger, J.D., and R.B. Marshall. 1989. *Campylobacter jejuni* infection within a laboratory animal production unit. *Laboratory Animals* 23:126–132.

Munroe, D.L., J.F. Prescott, and J.L. Penner. 1983. *Campylobacter jejuni* and *Campylobacter coli* serotypes isolated from chickens, cattle, and pigs. *J. Clin. Microbiol.* 18:877–881.

Nielsen, E.M., J. Engberg, and M. Madsen. 1997. Distribution of serotypes of *Campylobacter jejuni* and *C. coli* from Danish patients, poultry, cattle and swine. *FEMS Immunol. and Med. Microbiol.* 19:47–56.

Oosterom, J., R. Dekker, G.J.A. deWilde, F. vanKempen-deTroye, and G.B. Engels. 1985. Prevalence of *Campylobacter jejuni* and *Salmonella* during pig slaughtering. *The Vet. Quarterly* 7:31–34.

Orr, K.E., N.F. Lightfoot, P.R. Sisson, B.A. Harkis, J.L. Tweddle, P. Boyd, A. Carroll, C.J. Jackson, D.R.A. Wareing, and R. Freeman. 1995. Direct milk excretion of *Campylobacter jejuni* in a dairy cow causing cases of human enteritis. *Epidemiol. and Infect.* 114:15–24.

Osano, O. and S.M. Arimi. 1999. Retail poultry and beef as sources of *Campylobacter jejuni*. *East African Med. J.* 76:141–143.

Pearson, A.D., M. Greenwood, T.D. Healing, D. Rollins, M. Shahamat, J. Donaldson, and R.R. Colwell. 1993. Colonization of broiler chickens by waterborne *Campylobacter jejuni*. *Appl. Environ. Microbiol.* 59:987–996.

Pearson, A.D., M.H. Greenwood, R.K.A. Feltham, T.D. Healing, J. Donaldson, D.M. Jones, and R.R. Colwell. 1996. Microbial ecology of *Campylobacter jejuni* in a United Kingdom chicken supply chain: intermittent common source, vertical transmission, and amplification by flock propagation. *Appl. Environ. Microbiol.* 62(12):4614–4620.

Pearson, A.D., M.H. Greenwood, J. Donaldson, T.D. Healing, D.M. Jones, M. Shahamat, R.K.A. Feltham, and R.R. Colwell. 2000. Continuous source outbreak of Campylobacteriosis traced to chicken. *J. Food Prot.* 63:309–314.

Pokamunski, S., N. Kass, E. Borochovich, B. Marantz, and M. Rogol. 1986. Incidence of *Campylobacter* spp. in broiler flocks monitored from hatching to slaughter. *Avian Pathol.* 15:83–92.

Prescott, J.F., and C.W. Bruin-Mosch. 1981. Carriage of *Campylobacter jejuni* in healthy and diarrheic animals. *Am. J. Vet. Res.* 42:164–165.

Refregier-Petton, J., N. Rose, M. Denis, and G. Salvat. 2001. Risk factors for *Campylobacter* spp. contamination in French broiler-chicken flocks at the end of the rearing period. *Prev. Vet. Med.* 50:89–100.

Riordan, T., T.J. Humphrey, and A. Fowles. 1993. A point source outbreak of campylobacter infection related to bird-pecked milk. *Epidemiol. and Infect.* 110:261–265.

Rivoal, K., M. Denis, G. Salvat, P. Colin, and G. Ermel. 1999. Molecular characterization of the diversity of *Campylobacter* spp. isolates collected from a poultry slaughterhouse: analysis of cross-contamination. *Letters in Appl. Microbiol.* 29:370–374.

Robinson, D.A. 1981. Infective dose of *Campylobacter jejuni* in milk. *Br. Med. J.* 282:1584.

Rohrbach, B.W., F.A. Draughon, P.M. Davidson, and S.P. Oliver. 1992. Prevalence of *Listeria monocytogenes*, *Campylobacter jejuni*, *Yersinia enterocolitica*, and *Salmonella* in bulk milk: risk factors and risk of human exposure. *J. of Food Prot.* 55:93–97.

Ruiz-Palacios, G.M., E. Escamilla, N. Torres. 1981. Experimental *Campylobacter* diarrhea in chickens. *Infect. and Immun.* 34:250–255.

Saeed, A.M., N.V. Harris, R.F. DiGiacomo. 1993. The role of exposure to animals in the etiology of *Campylobacter jejuni / coli* enteritis. *Am. J. Epidemiol.* 137:108–114.

Schmid, G.P., R.E. Schaefer, B.D. Plikaytis, J.R. Schaefer, J.H. Bryner, L.A. Wintermeyer, and A.F. Kauffman. A one-year study of endemic Campylobacteriosis in a midwestern city: association with consumption of raw milk. *J. Infectious Dis.* 156(1):218–222.

Shane, S.M., D.H. Gifford, K. Yogasundram. 1986. *Campylobacter jejuni* contamination of eggs. *Vet. Res. Communications* 10:487–492.

Shanker, S., A. Lee, T.C. Sorrell. 1986. *Campylobacter jejuni* in broilers: the role of vertical transmission. *J. Hygiene* 96:153–159.

Smith, J.L. 1995. Arthritis, Guillain-Barre Syndrome, and other sequelae of *Campylobacter jejuni* enteritis. *J. Food Prot.* 58(10):1153–1170.

Stanley, K., R. Cunningham, and K. Jones. 1998. Isolation of *Campylobacter jejuni* from groundwater. *J. Appl. Microbiol.* 85:187–191.

Stanley, K.N., J.S. Wallace, J.E. Currie, P.J. Diggle, and K. Jones. 1998b. The seasonal variation of thermophilic campylobacters in beef cattle, dairy cattle and calves. *J. Appl. Microbiol.* 85:472–480.

Stanley, K.N., J.S. Wallace, J.E. Currie, P.J. Diggle, and K. Jones. 1998c. Seasonal variation of thermophilic campylobacters in lambs at slaughter. *J. Appl. Microbiol.* 84:1111–1116.

Stern, N.J., M.R.S. Clavero, J.S. Bailey, N.A. Cox, and M.C. Robach. 1995. *Campylobacter* spp. in broilers on the farm and after transport. *Poultry Sci.* 74:937–941.

Stern, N.J., K.L. Hiett, N.A. Cox, G.A. Alfredsson, K.G. Kristinsson, and J.E. Line. 2000. Recent developments pertaining to *Campylobacter*. *Irish J. Agricultural and Food Res.* 39:183–187.

Studahl, A. and Y. Andersson. 2000. Risk factors for indigenous campylobacter infection: Swedish case-control study. *Epidemiol. and Infect.* 125:269–275.

Van de Giessen, A., S. Mazurier, W. Jacobs-Reitsma, W. Jansen, P. Berkers, W. Ritmeester, and K. Wernars. 1992. Study on the epidemiology and control of *Campylobacter jejuni* in poultry broiler flocks. *Appl. Environ. Microbiol.* 58:1913–1917.

Varma, K.S., N. Jagadeesh, H.K. Mukhopadhyay, and N. Dorairajan. 2000. Incidence of *Campylobacter jejuni* in poultry and their carcasses. *J. Food Sci. Techn.* 37:639–641.

Vellinga, A., and F. Van Loock. 2002. The dioxin crisis as experiment to determine poultry-related *Campylobacter* enteritis. *Emerg. Infect. Dis.* 8(1):19–21.

Wallace, J.S., K.N. Stanley, J.E. Currie, P.J. Diggle, and K. Jones. 1997. Seasonality of thermophilic *Campylobacter* populations in chickens. *J. Appl. Microbiol.* 82:219–224.

Wallace, J.S., K.N. Stanley, and K. Jones. 1998. The colonization of turkeys by thermophilic campylobacters. *J. Appl. Microbiol.* 85:224–230.

Waterman, S.C., R.W.A. Park, and A.J. Bramley. 1984. A search for the source of *Campylobacter jejuni* in milk. *J. Hygiene* 92:333–337.

Weijtens, M.J.B.M., J. van der Plas, P.G.H. Bijker, H.A.P. Urlings, D. Koster, J.G. van Logtestijn, and J.H.J. Huis In't Veld. 1997. The transmission of campylobacter in piggeries; an epidemiological study. *J. Appl. Microbiol.* 83:693–698.

Wesley, I.V., S.J. Wells, K.M. Harmon, A. Green, L. Schroeder-Tucker, M. Glover, and I. Siddique. 2000. Fecal shedding of *Campylobacter* and *Arcobacter* spp. in dairy cattle. *Appl. Environ. Microbiol.* 66:1994–2000.

Young, C.R., R. Harvey, R. Anderson, D. Nisbet, H. Stanker. 2000. Enteric colonisation following natural exposure to *Campylobacter* in pigs. *Res. Vet. Sci.* 68:75–78.

20

DETECTION OF *CAMPYLOBACTER*

ORHAN SAHIN, QIJING ZHANG, AND TERESA Y. MORISHITA

INTRODUCTION. The genus *Campylobacter* comprises a number of phenotypically and genotypically diverse species and subspecies. Within the genus, thermophilic species (*C. jejuni*, *C. coli*, and *C. lari*) are of the primary importance to food safety, with *C. jejuni* being responsible for the majority of human campylobacteriosis, followed by *C. coli*, and rarely by other species such as *C. lari*. Members of this genus are Gram-negative curved or spiral rods (0.2 to 0.9 μm wide and 0.5 to 5 μm long) with a single polar flagellum; however, they may transform to coccoid forms as the culture ages or in response to environmental stress and starvation. *Campylobacter* organisms are widespread in food animals, including both livestock and poultry, which serve as reservoirs for human campylobacteriosis. Therefore, successful control of *Campylobacter* infections in farm animals at the pre-harvest stage would have a significant impact on the prevention of human foodborne illnesses. To achieve this goal, reliable detection methods are required to understand the ecology and epidemiology of *Campylobacter* in food animals and their production environments. Despite the fact that *Campylobacter* organisms are more difficult to isolate than other foodborne pathogens, a great variety of culture media and protocols have been successfully used for detecting *Campylobacter*. In this chapter we review various methods used for the detection and identification of thermophilic campylobacters, with special attention given to the newly developed ones that can be used to detect *Campylobacter* in animal reservoirs. Detection of *Campylobacter* spp. from processed meat and food samples has been recently reviewed elsewhere (Corry et al.1995; Jacobs-Reitsma 2000) and is not a main topic of this chapter.

CULTURE-BASED DETECTION METHODS. Thermophilic campylobacters are fastidious and slow-growing, and require a microaerophilic atmosphere (containing 5 percent O_2, 10 percent CO_2, 85 percent N_2) and elevated temperature (42°C) for optimal growth under laboratory conditions (Corry et al. 1995; Doyle and Roman 1981; Hazeleger et al. 1998; Nachamkin et al. 2000; Skirrow 1977). Therefore, culturing *Campylobacter* spp. from fecal material with a high level of background flora is cumbersome, time consuming (Bolton et al. 1983; Madden et al. 2000;

Musgrove et al. 2001; Solomon and Hoover 1999), and requires the use of special culture media and conditions. Usually, a selective medium is necessary to recover *Campylobacter* from field samples. The first selective medium for culturing *C. jejuni* and *C. coli* was developed in 1977 by Skirrow (1977). Since then, approximately forty solid and liquid selective media for culturing of *Campylobacter* from clinical and food samples have been reported. Therefore, methods have been reviewed by Corry et al. (1995). Some of the most commonly used ones are Skirrow, Preston, Karmali, modified charcoal cefoperazone deoxycholate agar (mCCDA), cefoperazone amphotericin teicoplanin (CAT) agar, Campy-CVA (cefoperazone vancomycin amphotericin), and Campy-Cefex medium. Depending on the type and origin of the samples, these media may give varying recovery rates of thermophilic *Campylobacter* spp. (Bolton et al. 1983; Gun-Munro et al. 1987). Both solid agar plates and liquid enrichment media contain a variety of different combinations of antibiotics to which thermophilic campylobacters are intrinsically resistant, such as polymyxin, vancomycin, trimethoprim, rifampicin, cefoperazone, cephalothin, colistin, cycloheximide, and nystatin (Corry et al. 1995). These antibiotics inhibit the growth of many background microbial flora present in fecal specimens, thereby facilitating the isolation of slow-growing *Campylobacter* spp.

Because *Campylobacter* spp. are sensitive to oxygen levels above 5 percent, all selective media for these organisms contain various oxygen-quenching agents in order to neutralize the toxic effect of oxygen radicals on campylobacters (Corry et al. 1995; Solomon and Hoover 1999). The commonly used oxygen-quenching agents include several types of blood, such as lysed horse blood and defibrinated blood from various animals, which are used in Skirrow and Campy-CVA media. Blood is also thought to counteract the inhibitory effects of thymidine in formulations containing trimethoprim as one of the selective agents (Corry et al. 1995). Other commonly used oxygen-quenching agents include a combination of ferrous sulfate, sodium metabisulfite and sodium pyruvate (as in Campy-Cefex agar), charcoal (as in mCCDA agar), and hematin (as in Karmali agar). Several of these selective media may contain multiple oxygen-quenching agents (Corry et al. 1995). Another method for reducing the effect of oxygen is the addition of

Oxyrase, an oxygen scavenger enzyme, into culture media to avoid the need for a microaerophilic atmosphere (Tran 1995; Wonglumsom et al. 2001). As reported by Wonglumsom et al. (2001), enrichment broth with added Oxyrase incubated under aerobic conditions was as efficient as microaerophilic incubation for recovery of *C. jejuni* from artificially inoculated food samples. However, the performance of Oxyrase varied among different types of samples tested and among different media used (Tran 1995), which raises concerns about the consistency of Oxyrase when used to replace the microaerophilic culture procedures.

Depending on the type of specimen, selective media can be used either for direct plating or for an enrichment step followed by plating for *Campylobacter* colonies. In processed foods in which bacteria are in low numbers, are in an "injured" state, or both, an enrichment step in liquid medium followed by plating on solid agar plates is usually superior for isolation compared to only direct plating. (Corry et al. 1995; Jacobs-Reitsma 2000). The procedure may not always perform better than direct plating when testing fecal samples. Feces from chickens usually contain high enough numbers (up to 10^8 CFU/g feces) of *Campylobacter* organisms to be detected by direct plating (Kaino et al. 1988; Musgrove et al. 2001; Stern 1992). When isolating *Campylobacter* spp. from animal feces, one should consider the species of animal tested and the sites of specimen collection. Even though fecal samples from cattle and sheep can be enriched before plating for better recovery rates (Bolton et al. 1983; Jones et al. 1999; Stanley et al. 1998b; Stanley et al. 1998c), enrichment may greatly reduce the recovery rate of *Campylobacter* spp. from different sites of the intestinal tract of poultry and pigs (Madden et al. 2000; Musgrove et al. 2001). Recently, Musgrove et al. (2001) compared detection of *Campylobacter* spp. in ceca and crops employing enrichment and direct plating. Direct plating of cecal samples on a selective agar resulted in a significantly higher recovery rate than the enrichment method; however, enrichment was slightly better than direct plating for the recovery of *Campylobacter* spp. from crop samples. Madden et al. (2000) showed how the choice of selective medium (Preston vs. mCCD media) and isolation procedure (enrichment vs. direct plating) can greatly affect the recovery of *Campylobacter* spp. from different parts of the intestinal tract of pigs. For the anal swabs, direct plating was superior to enrichment (twenty-four hours) because the positive rate with direct plating (100 percent) was higher than that with the enrichment method (88 percent). However, enrichment (twenty-four hours) gave better recovery than direct plating for isolation of campylobacters from ileal contents (86 percent recovery rate with enrichment versus 56 percent with direct plating). In all cases, the isolation rate of *Campylobacter* spp. from fecal specimens was markedly higher with mCCD medium than with Preston medium. For optimal recovery, enrichment should be controlled for about twenty-four hours, because a prolonged enrichment to seventy-two hours decreased the isolation rate to a level even below that of direct plating (Madden et al. 2000). Cecal and anal swabs are different from the crop and ileal contents and contain much higher numbers of fast-growing background flora. These flora may overgrow *Campylobacter* spp. during the enrichment procedure, leading to reduced recovery of the target organism.

To isolate thermophilic *campylobacters* from environmental water, two methods can be used to increase the detection sensitivity. A large volume of water can be filtered through a single membrane with a pore size of 0.2 µm, which concentrates the organism onto the membrane. Subsequently, the membrane can either be placed directly on a selective agar plate or first cultured in an enrichment broth followed by selective plating (Oyofo and Rollins 1993; Pearson et al. 1993; Stanley et al. 1998a). If large particles are present in water, prefiltering with a membrane of a larger pore size may be required prior to the final filtering with a membrane of 0.2 µm. Alternatively, water samples can be concentrated by high-speed centrifugation from which the supernatant is discarded and the pellet is cultured by direct plating or enriched in broth followed by plating.

Usually, typical *Campylobacter* colonies are visible on solid media after forty-eight hours' incubation (Bolton et al. 1983; Corry et al. 1995; Tran 1995), but it may take seventy-six to ninety-six hours to observe visible colonies of some slow-growing strains (Nachamkin et al. 2000; Sahin et al. 2001b). Depending on the media used, colonies of *Campylobacter* spp. may appear differently. If the agar is moist, typical colonies look grey, flat, irregular, and thinly spreading. Colonies may form atypical shapes (round, convex, glistening), especially when plates are dry (Corry et al. 1995; Nachamkin et al. 2000). After *Campylobacter*-like colonies appear on plates, identification at the genus or species level should be performed. Doing so requires the use of multiple tests. Presumptive identification of thermophilic *Campylobacter* spp. can be done according to colony morphology, typical cellular shapes (spiral or curved rods), and characteristic rapid darting motility as observed under phase-contrast microscope. The most commonly used phenotypic tests for identification of *Campylobacter* to genus or species level include biochemical tests (catalase, oxidase, nitrate reduction, hippurate hydrolysis, indoxyl acetate hydrolysis), antibiotic susceptibility patterns (nalidixic acid, cephalothin), and growth characteristics at different temperatures (25°C, 37°C, and 42°C). Some of the commonly used tests are summarized in Table 20.1. It is important to note that thermophilic *Campylobacter* spp. as a genus and hippurate-positive *C. jejuni* as a species can be easily identified with these phenotypic tests. However, hippurate-negative *C. jejuni* isolates, which have been reported, require further testing by other methods to confirm the identity at the species level (Steinbrueckner et al. 1999). Overall, these phenotypic identification methods are simple, rapid, inexpensive, and can be performed with commercial kits (such as API Campy), but they may misidentify hippurate-negative *C. jejuni* isolates as *C. coli*. Differentiation between *C. jejuni* and *C. coli* is of importance for food safety, because *C. jejuni* is the primary etiologic agent of human campylobacteriosis. Interestingly, *C. jejuni* is the predominant *Campylobacter* species in poultry and is also frequently isolated from cattle and sheep (Corry and Atabay 2001; Hoorfar et al. 1999; Jones et al. 1999; Stanley et al. 1998b; Stern 1992), whereas swine *Campylobacter* isolates are predominantly *C. coli* (Madden et al. 2000; Nielsen et al. 1997; Stern 1992). Nonthermophilic *Campylobacter* spp., such as *Campylobacter hyointestinalis*, are often isolated from cattle using the isolation conditions for thermophilic *Campylobacter* (Hoorfar et al. 1999; Nielsen et al. 1997).

IMMUNOLOGY-BASED DETECTION/IDENTIFICATION METHODS.

Various immunological assays, based on antigen-antibody interactions, have been used for detection and identification of *Campylobacter* species. The most commonly used ones are discussed in the next section.

Agglutination Assays. Agglutination tests that are often used for culture confirmation and identification purposes. Examples of the commercially available agglutination assays include Campyslide (BLL Microbiology Systems), Meritec-Campy (Meridian Diagnostics), Microscreen (Mercia Diagnostics), and ID Campy (Integrated Diagnostics). A detailed discussion on the usefulness of these commercial tests has been published elsewhere (On 1996). In all these tests, latex particles are coated with different polyclonal anti-*Campylobacter* spp. antibodies to react with putative *Campylobacter* colonies. A positive reaction results in the agglutination of the latex particles, which can be observed visually. Because of their low sensitivity and specificity, latex agglutination tests are not designed for the direct detection of campylobacters from field samples and are used mainly for culture confirmation purposes. False-positive results can occur from nonspecific agglutination of the latex particles (On 1996). Therefore, results from these tests need to be inter-

preted in conjunction with additional data obtained from other methods, such as biochemical or DNA-based tests.

Enzyme Immunoassays. The second category of immunology-based detection methods include mainly enzyme immunoassays (EIA), which can be used for direct detection of *Campylobacter* spp. in animal feces or processed food. These assays are commercially available in a very similar format to sandwich-ELISA assays, which use two different antibodies, to detect *Campylobacter* spp. directly in field samples (Endtz et al. 2000; Hindiyeh et al. 2000) or after a selective enrichment step (Hoorfar et al. 1999; Lilja and Hanninen 2001). Examples of these commercial kits include VIDAS Campylobacter (bioMerieux), EIA-Foss Campylobacter (Foss Electric), and ProSpecT (Alexon-Trend). VIDAS Campylobacter and EIA-Foss assays have been compared with the traditional culture method for detecting *Campylobacter* in bovine and swine fecal samples following overnight enrichment (Hoorfar et al. 1999). Compared to the culture method, VIDAS Campylobacter gave adequate sensitivity (95 percent with cattle samples and 88 percent with swine samples) and specificity (71 percent with cattle and 76 percent with swine samples), whereas EIA-Foss test yielded substantially lower sensitivity and specificity with both types of samples (Hoorfar et al. 1999). Another sensitive and specific EIA was developed using a monoclonal antibody specific for *C. jejuni* and *C. coli* (Chaiyaroj et al. 1995). In this assay, 10 percent fecal suspensions were used directly to coat microtiter plates, and the presence of *C. jejuni* and *C. coli* was detected following sequential addition of the monoclonal antibody and enzyme-labeled secondary antibodies. This assay's detection limit was shown to be >10^3 CFU/ml as determined by artificial seeding of stool specimens. Although EIAs are more sensitive and more specific than latex agglutination tests, they are not as sensitive as culture methods for detecting *Campylobacter* spp. and are not suitable for testing samples in which *Campylobacter* spp. are suspected to be in low numbers. They are more rapid than traditional culture methods and can be automated for easy handling (Endtz et al. 2000; Hindiyeh et al. 2000; Hoorfar et al. 1999). Thus, the EIA assays may be good choices

Table 20.1—Phenotypic Characteristics of Thermophilic *Campylobacter* Species[a]

Test	*C. jejuni*	*C. coli*	*C. lari*
Catalase	+	+	+
Oxidase	+	+	+
Nitrate reduction	+	+	+
Glucose utilization	−	−	−
Hippurate hydrolysis	+[b]	−	−
Indoxyl acetate hyd.	+	+	−
Growth at 42°C	+	+	+

[a]Adapted from references On 1996 and Steinbrueckner et al. 1999.
[b]Hippurate negative isolates of *C. jejuni* have been reported.

for large-scale surveillance programs that deal with a large number of samples and require high throughput screening methods.

Colony Blotting. A third category of immunoassays for detecting *Campylobacter* are developed in the formats of colony blotting. In an earlier study by Taylor and Chang (1987), a polyclonal antibody against the major outer membrane protein of *C. jejuni* was used to distinguish thermophilic *Campylobacter* spp. (*C. jejuni*, *C. coli*, and *C. lari*) from the other genus members using immunoblotting, ELISA, and colony blotting techniques. Rice et al. (1996) developed a colony lift immunoassay (CLI) for specific identification and enumeration of *Campylobacter* colonies grown on selective agar or filter membranes using a commercial goat anti-*Campylobacter* antibody. With CLI, colonies of thermophilic campylobacters could be differentiated from background bacterial colonies and quantified in a short period of time (eighteen to twenty-eight hours) following plating. A critical requirement for achieving reliable colony blotting data is the use of highly pure and specific antibodies. The antibodies must be thoroughly tested with the same type of *Campylobacter*-free samples prior to use. Otherwise, cross-reaction may occur with background colonies in field samples, resulting in misleading results.

Immunomagnetic Separation. A recent development in detecting *Campylobacter* is the application of immunomagnetic separation (IMS). For IMS, magnetic beads are first coated with pathogen-specific polyclonal or monoclonal antibodies and then mixed with the sample following a short enrichment period in a selective medium. As a result, bacterial cells are captured and concentrated onto the magnetic beads. The beads allow the easy separation of the bacteria from other contaminants by use of magnetic fields. This separation step increases the sensitivity of subsequent culture or PCR procedures. Recently, Yu et al. (2001) used tosylactivated Dynabeads and streptavidin-coated beads conjugated with polyclonal anti-*Campylobacter* antibodies to detect *Campylobacter* spp. in artificially inoculated ground poultry meat and culture suspensions. Plating of the bead-captured cells onto a selective medium indicated that the IMS could detect 10^4 CFU/g of ground meat without enrichment; however, no comparison was made with the regular culture procedure without IMS. Other studies employed IMS to capture *Campylobacter* cells from enriched food samples followed by microtiter hybridization (Lamoureux et al. 1997) or PCR (Docherty et al. 1996). In both cases, the IMS method was sensitive enough to recover an initial number of <10 CFU/g of food after one to two days of selective enrichment. Despite the great potential of IMS, it has not been commonly used for detection of *Campylobacter* spp. in food animal production systems, probably because great antigenic diversities exist in *Campylobacter* spp., making it difficult to detect every strain of this organism using a single antibody.

Antibody-Based Detection. Immunoassays can also be used to detect *Campylobacter*-specific antibodies in animal hosts. Although a positive antibody reaction does not necessarily mean the presence of *Campylobacter* in the tested animals, it would certainly indicate that they have been exposed to this organism. Thus, serological data can be very useful for estimating the prevalence of *Campylobacter* infections at the herd or flock level. We recently used ELISA and immunoblotting to measure the prevalence of anti-*Campylobacter* antibodies in broiler production systems, demonstrating the widespread presence of the antibodies in parent flocks, eggs, and broiler chickens (Sahin et al. 2001b, 2001c). This finding is consistent with the fact that most broiler chickens on poultry farms are colonized by *C. jejuni*. ELISA and immunoblotting were also used for detecting anti-*Campylobacter* antibodies in humans (Martin et al. 1989). For the ELISA assays, purified antigens (outer membrane proteins, glycine-extracts, or flagellin) were used to coat microtiter plates followed by the sequential addition of the serum samples and enzyme-conjugated secondary antibodies. Compared with the lengthy isolation procedures for *Campylobacter*, measuring serum antibodies can be done much more quickly and with lower cost. This approach can be combined with isolation methods for understanding the ecology and epidemiology of *Campylobacter* in animal reservoirs.

NUCLEIC ACID-BASED DETECTION METHODS.
DNA-based methods have been widely used for detection and identification of *Campylobacter* spp. These methods are generally designed in the formats of hybridization assays or PCR tests, which can be used for culture confirmation or for direct detection of *Campylobacter* from field samples.

Hybridization Assays. The majority of DNA probes used for hybridization assays are designed from the 16S rRNA gene of *Campylobacter spp.*, although probes constructed from other gene sequences also exist (Lamoureux et al. 1997; On 1996; Taylor and Hiratsuka 1990). Several probes directed to the 16S rRNA gene of thermophilic *Campylobacter* spp. have been commercialized, such as SNAP (Syngene) and AccuProbe (Gen-Probe). These systems use nonradioactive probes in either membrane or liquid hybridization format and have been tested on both pure cultures and clinical (or food) samples. For the direct detection, DNA is usually extracted from fecal or food samples and used in membrane or liquid hybridization assays. The performance of these commercial kits and other noncommercial probes in direct detection was generally less sensitive than traditional culture methods (Olsen et al. 1995; On 1996; Taylor and Hiratsuka 1990). Consequently, DNA probes are often combined with culture methods to increase detection efficiency. For example, DNA probes can be used for colony hybridization, in which bacterial colonies grown on selective agar plates are transferred to nylon membranes and hybridized with a *Campylobacter*-probe. This approach could detect low numbers of *C. jejuni* in chicken feces (Chuma et al. 1993) and increase the detection efficiency of the conventional culture method (Chuma et al. 1994). Colony hybridization is especially useful when typical *Campylobacter* colonies cannot be identified easily with conventional methods because of the overgrowth by background flora or other reasons. DNA probe-based methods can also be in the format of "dot blot hybridization" and can be very useful for differentiation of *Campylobacter* spp. For example, two types of *Campylobacter* colonies were isolated from a single human patient and identified as *C. jejuni* and *C. coli* by conventional phenotypic tests. However, a dot blot assay, in which purified genomic DNA of the two isolates were hybridized with species-specific DNA probes, identified both of the isolates as *C. jejuni* (Quentin et al. 1993).

Hybridization reactions with specific probes can also be performed in microtiter plates, facilitating the testing of multiple samples concurrently. Lamaureux et al. (1997) utilized this system recently, through a process in which rDNA or RNA probes specific for thermophilic *Campylobacter* spp. and *C. jejuni* were immobilized onto the wells of microtiter plates. To increase sensitivity of the assay, immunomagnetic beads coated with a monoclonal antibody specific for thermophilic *Campylobacter* spp. were used to capture the bacterial cells in contaminated chicken meat and milk following forty-eight hours of enrichment in a selective medium. Both DNA and RNA were extracted directly from bacteria bound to the beads and added to the wells containing *Campylobacter*-specific probes, followed by detection of RNA-DNA hybrids with a monoclonal anti–RNA-DNA antibody. The assay detected as few as three cells of *C. jejuni* in 10 g of meat within twelve hours following a two-day enrichment step (Lamoureux et al. 1997).

Polymerase Chain Reaction. PCR-based methods are much more commonly used than nucleic acid probes for the detection and identification of foodborne pathogens including *Campylobacter* spp. PCR allows exponential amplification of the targeted sequences within a short period of time, thereby permitting the rapid detection of low numbers of organisms. Depending on the purposes, oligonucleotide primers can be designed from either variable or conserved gene sequences of *Campylobacter*. PCR primers directed to conserved regions are usually used for general detection, whereas primers designed from variable regions can be used for differentiation of species or strains. A variety of PCR assays targeting genus- or species-specific sequences have been developed to detect and identify *Campylobacter* spp. in food (Allmann et al. 1995; Docherty et al. 1996; Grennan et al. 2001; Konkel et al. 1999; Manzano et al. 1995; O'Sullivan et al. 2000; Waller and Ogata 2000; Winters and Slavik 2000), feces (Chuma et al. 1997; Collins et al. 2001; Houng et al. 2001; Itoh et al. 1995; Lawson et al. 1998, 1999; Linton et al. 1997; Metherell et al. 1999; Rasmussen et al. 1996; Waegel and Nachamkin 1996), and environmental samples (Studer et al. 1999; Waage et al. 1999). Table 20.2 lists some representative PCR assays used for detection of *Campylobacter*.

PCR can be combined with other procedures, such as Southern blotting (al Rashid et al. 2000; Chuma et al. 1997; Konkel et al. 1999; Lawson et al. 1998; O'Sullivan et al. 2000), reverse blot hybridization (Collins et al. 2001; O'Sullivan et al. 2000; van Doorn et al. 1999), and ELISA (Grennan et al. 2001; Hald et al. 2001; Metherell et al. 1999; Rasmussen et al. 1996; Waller and Ogata 2000), to enhance detection efficiency. Reverse blot hybridization and PCR-ELISA employ specific capture probes immobilized onto membranes or microtiter plates, which are then hybridized with the PCR products amplified with labeled primers. Subsequent detection of the label is achieved using colorimetric systems. The procedures for reverse blot

Table 20.2—Representative PCR Methods for Detection and Identification of Thermophilic Campylobacter Species

Specific for	Target gene	Remarks	References
Campylobacter spp.	16S-23S rRNA intergenic region	PCR-reverse hybridization	Collins et al. 2001; O'Sullivan et al. 2000
Campylobacter spp.	16S rRNA	PCR-ELISA	Grennan et al. 2001; Waller and Ogata 2000; Metherell et al. 1999
Campylobacter spp.	*flaA*	Direct PCR-RFLP typing	Waegal and Nachamkin 1996
Campylobacter spp.	16S rRNA	Real-time PCR	Logan et al. 2001
Campylobacter spp.	23S rRNA	PCR-RFLP	Fermer and Engvall 1999
Campylobacter spp.	*glyA*	PCR-Southern Blot	al Rashid et al. 2000
Campylobacter spp.	not named	RT-PCR	Sails et al. 1998
C. jejuni/coli	*ceuE*	Multiplex-PCR	Houng et al. 2001
C. jejuni/coli	*flaA-flaB* intergenic region	Seminested-PCR	Waage et al. 1999
C. jejuni/coli	*cadF*	PCR-Southern Blot	Konkel et al. 1999
C. jejuni	*hip*	Regular PCR	Linton et al. 1997
C. jejuni	*flaA*	Regular PCR	Itoh et al. 1995
C. jejuni	*mapA*	Regular PCR	Stucki et al. 1995

hybridization and PCR-ELISA are illustrated in Figure 20.1. Both systems have the capacity of working with large numbers of samples and can successfully detect, identify, and differentiate *Campylobacter* species, depending on the design of the capture probes used. Because multiple probes specific for different species of *Campylobacter* can be immobilized onto a single membrane strip or in a microtiter plate, detection and differentiation of mixed populations can be simultaneously identified with a single test (Collins et al. 2001; Grennan et al. 2001; O'Sullivan et al. 2000).

Real-time PCR has also been used to detect and identify *Campylobacter* spp. (Logan et al. 2001; Nogva et al. 2000). Logan et al. (2001) recently developed a real-time PCR assay using the biprobes detection system to detect and identify *Campylobacter* spp. in stool samples from gastroenteritis patients. In the first phase, the genus-specific 16S rRNA gene for *Campylobacter* spp. was amplified using regular PCR. In the second phase, a LightCycler was used to identify of the genus-specific amplicon through the analysis of its melting peak profiles with three specific biprobes. The test was very specific for *Campylobacter* spp. and did not cross-react with any other closely related species, but it was unable to differentiate between *C. jejuni* and *C. coli*. The assay had a 100 percent correlation with the con-

ventional culture method and PCR-ELISA tests. One of the main advantages of real-time PCR is that it can accurately measure the quantity of template DNA and consequently provide an estimation on the number of organisms present in a sample. However, this test requires special equipment and reagents and is more expensive than the conventional PCR assays.

Many PCR tests are designed for general detection and cannot discriminate between *C. jejuni* and *C. coli* isolates (Konkel et al. 1999; Lawson et al. 1998; Logan et al. 2001; Rasmussen et al. 1996; Studer et al. 1999; Waage et al. 1999; Waegel and Nachamkin 1996). Differentiation between these two very closely related species can be achieved using species-specific primer sets (Day, Jr. et al. 1997; Eyers et al. 1993; Gonzalez et al. 1997; Linton et al. 1997; Stucki et al. 1995), or using conserved primers for the amplification process followed by hybridization to species-specific probes (al Rashid et al. 2000; Metherell et al. 1999; van Doorn et al. 1999). Alternatively, digestion of PCR amplicons derived from the polymorphic region of the 23S rRNA gene by restriction enzymes can be used to distinguish between closely related *Campylobacter* spp. (Fermer and Engvall 1999). Additionally, PCR can be designed in a multiplex format, in which multiple sets of broad or species-specific primers, or both, are included in one

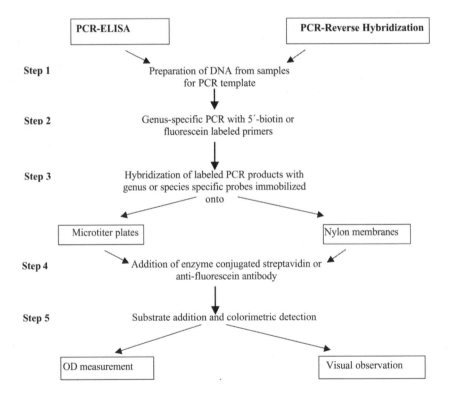

FIG. 20.1—Stepwise illustration of PCR-ELISA and PCR-Reverse Hybridization assays for detection and differentiation of *Campylobacter* spp.

reaction so that detection and differentiation can be simultaneously accomplished in a single test. Several multiplex PCRs have been published by different laboratories (Chuma et al. 2000; Denis et al. 1999; Gonzalez et al. 1997; Harmon et al. 1997; Houng et al. 2001; Korolik et al. 2001; Winters and Slavik 2000); some of these are listed in Table 20.2, shown previously.

The main advantages of PCR are its sensitivity, specificity, and rapidness. However, PCR usually works best with pure bacterial cultures, and its performance on direct testing of field samples may be reduced dramatically because of the presence of PCR-inhibitors in animal feces and other types of samples. Purification of DNA from fecal, food, and environmental samples using a reliable method is essential for achieving satisfactory results with PCR. A number of techniques have been reported for the extraction of DNA from fecal or food materials prior to PCR. The most commonly used ones include phenol-chloroform extraction followed by ethanol precipitation (Itoh et al. 1995; Konkel et al. 1999), commercial DNA purification kits (Collins et al. 2001; Denis et al. 2001), immunomagnetic separation (Docherty et al. 1996; Lamoureux et al. 1997; Nogva et al. 2000; Waller and Ogata 2000), silicon dioxide extraction (Houng et al. 2001), column chromatography (Waegel and Nachamkin 1996), and differential centrifugation (Allmann et al. 1995; Chuma et al. 1997; Manzano et al. 1995; Rasmussen et al. 1996; Studer et al. 1999; Waage et al. 1999; Wang et al. 1999). Although commercial kits tend to provide rapid and reproducible preparations of DNA for PCR reactions, many of them do not provide good purification of DNA from animal feces or other samples of great complexity. In our laboratory, we have successfully used buoyant density centrifugation with Percoll (Pharmacia Biotech) gradient medium for very sensitive detection of *C. jejuni* in artificially spiked egg contents with PCR (Sahin et al. 2001a). Despite the use of various purification methods, PCR inhibitors may not be completely removed from the reaction, decreasing the performance of PCR for detecting samples with low numbers of *Campylobacter*. An enrichment step is generally required prior to PCR (Denis et al. 2001; Grennan et al. 2001; Manzano et al. 1995; O'Sullivan et al. 2000), which not only increases the number of *Campylobacter* in the samples but also dilutes the concentration of PCR inhibitors. Fecal samples from *Campylobacter*-shedding animals and humans may contain high enough numbers of the organisms to be detected directly without the need for an enrichment step (Chuma et al. 1997; Houng et al. 2001; Itoh et al. 1995; Linton et al. 1997; Waegel and Nachamkin 1996).

Another drawback of PCR-based detection methods is that they are unable to discriminate between dead and live cells, which may cause problems in interpreting a positive PCR result under certain circumstances. Some studies have attempted to differentiate between viable and dead cells using RNA molecules as the PCR targets because RNA molecules, especially mRNA, are much less stable than DNA and are regarded as markers for cell viability (Sails et al. 1998; Uyttendaele et al. 1997). Uyttendaele et al. (1997) employed a nucleic acid sequence-based amplification (NASBA) system designed for selective amplification of the 16S rRNA sequences isothermally to assess cell viability. Although this method was able to detect low levels of *C. jejuni* in artificially contaminated chicken meat, it did not provide satisfactory distinction between dead and viable *Campylobacter* cells, probably because 16S rRNA is an abundant and stable RNA species that can survive for a long period of time even in dead cells. A better way of differentiation between viable and dead *Campylobacter* cells may be accomplished using reverse transcriptase-PCR (RT-PCR) to amplify mRNA (Sails et al. 1998), which has a short half-life and disappears quickly after cells die. However, RT-PCR needs pure RNA preparations because DNA contamination in the reaction would result in misleading results. Purified bacterial RNAs are often digested with DNase prior to RT-PCR to remove contaminated DNA. In addition, critical controls are required for each sample to ensure that the amplicons are indeed from the target mRNA and not from the contaminated DNA. When choosing a target mRNA, one needs to consider the target carefully, because mRNAs that are nonconstitutive and highly unstable may yield false-negative results.

OTHER DETECTION AND IDENTIFICATION METHODS. Some nontraditional approaches recently have been reported for detection and identification of *Campylobacter* spp. such as matrix-assisted laser desorption/ionization time-of-flight mass spectrometry (MALTI-TOF MS) and biosensor-based methods (Ivnitski et al. 2001; Mandrell and Wachtel 1999; Winkler et al. 1999). In MALTI-TOF MS, proteins from a single bacterial colony are ionized in a matrix solution using laser irradiation. As a result, an ion profile specific for each pathogen (species or strain) is produced, which is then compared against the computerized database to determine the identity of the pathogen. This method has been shown to have enough discriminatory power to differentiate between *C. jejuni* and *C. coli* in recent studies (Mandrell and Wachtel 1999; Winkler et al. 1999). The usefulness of this technique is illustrated by the fact that this method correctly identified a hippuricase-negative *C. jejuni* isolate, which was originally identified as *C. coli* by conventional phenotypic methods (Mandrell and Wachtel 1999). Recently, Ivnitski et al. (2001) designed an ion-channel biosensor system for detecting *Campylobacter*. In this system, a biosensor was formed by coating a stainless-steel working electrode with artificial bilayer lipid membrane that served as a thin electrical insulator. *Campylobacter*-specific antibodies were embedded into bilayer lipid membranes and were used as channel-forming proteins. Selective interaction of *Campylobacter* cells with the membrane-bound antibodies resulted in changes in ion permeability of the bilayer lipid membrane and consequently the alteration of transmembrane ion-current that was detected amperometrically. The assay was highly sensitive and fast (ten minutes) when artificially contaminated water samples were tested. As with any other method utilizing antibodies, the sensitivity and specificity of this biosensor-based detection method are greatly affected by the quality of the employed antibodies as well as the design of the detection device. Although MALTI-TOF MS and the biosensor-based method showed great potential for analyzing pure *Campylobacter* cultures, their use for direct direction of *Campylobacter* in animal feces or processed food is practically questionable because of the great complexity of the samples.

Besides the DNA-based methods discussed in the previous section, many other molecular tools have been used for epidemiological studies of *Campylobacter* in animal reservoirs. The majority of these molecular methods are developed for typing or differentiation of *Campylobacter* isolates and are not suitable for detection purposes. Some commonly used molecular typing tools include pulsed-field gel electrophoresis (PFGE), random amplified polymorphic DNA (RAPD), amplified fragment length polymorphism (AFLP), and ribotyping. In addition, sequence-based typing methods involving the variable region of the *fla* gene (encoding the flagellin subunit) or the *cmp* gene (encoding the major outer membrane protein) have also been reported (Meinersmann et al. 1997; Zhang et al. 2000). A comprehensive review on various typing tools has been published recently by Wassenaar and Newell (2000).

DETECTION OF "VIABLE-BUT-NONCULTURABLE" (VBNC) *CAMPYLOBACTER*. It has been proposed that *C. jejuni* adopts a VBNC physiological state as a survival strategy against environmental stress and starvation. In this state, the organism cannot be recovered by conventional culture methods, but still maintains some metabolic activity and is able to cause infections in animals (Jones et al. 1991; Rollins and Colwell 1986; Tholozan et al. 1999). This theory has been controversial because conflicting results are reported with the resuscitation of putative VBNC *Campylobacter* cells (Bovill and Mackey 1997; Buswell et al. 1998). Some studies generated data supporting the presence of the VBNC state and its role in transmission of *Campylobacter* in animals (Cappelier et al. 1999; Jones et al. 1991; Pearson et al. 1993; Saha et al. 1991; Stern et al. 1994), whereas others found that the putative VBNC state is questionable and insignificant in the epidemiology of *Campylobacter* (Beumer et al. 1992; Hald et al. 2001; Medema et al. 1992; van de Giessen et al. 1996). Several methods have been used for detecting putative VBNC campylobacters. One of the commonly used methods is the double staining of *Campylobacter* cells with 5-cyano-2, 3-ditolyl tetrazolium chloride (CTC) and 4′-6 diamino-2 phenylindole (DAPI) (Cappelier et al. 1997), which allows the simultaneous detection of total and viable cells in a population by using an epifluorescent microscope. Passage in animals (Jones et al. 1991; Saha et al. 1991; Stern et al. 1994) or embryonated eggs (Cappelier et al. 1999) was also used to resuscitate VBNC campylobacters. In addition, RT-PCR can be potentially adapted to detect VBNC cells. Although several findings support the concept of the VBNC state, it has been difficult to differentiate the resuscitation of VBNC cells from the proliferation of low numbers of residual viable cells. Therefore, the role of VBNC state in the epidemiology and ecology of *Campylobacter* remains uncertain.

SUMMARY AND CONCLUSIONS. Sensitive and reliable detection methods are essential for understanding the epidemiology and ecology of *Campylobacter* spp. in animal reservoirs, and ultimately for the design of farm-based intervention strategies to control *Campylobacter* infections. Owing to the unique growth characteristics of *Campylobacter*, isolation of these organisms from field samples requires the use of special media and culture conditions, and is generally laborious and time-consuming. Despite these difficulties, isolation of *Campylobacter* has been successfully conducted in many epidemiological studies and is still the most commonly used method for detecting *Campylobacter* in animal reservoirs. Although a variety of different media and culture procedures have been developed, there is still no consensus regarding the best culture media and isolation protocols for the recovery of *Campylobacter* from various samples. Selection of a particular medium and an isolation procedure is often influenced by personal preference and experience and is dictated by the fact that different types of samples may need different culture procedures for optimal recovery of *Campylobacter*. Despite these problems, isolation of *Campylobacter* from animal feces can be achieved with high success rates. Recovery of *Campylobacter* from environmental samples can be challenging, because *Campylobacter* does not propagate in the environment and may exist in an "injured" or "VBNC" state. Therefore, it is important to consider the sample type before choosing a culture medium and isolation procedure for recovering *Campylobacter* spp.

The recent surge in the use of molecular detection methods has greatly facilitated the specific and rapid detection and identification of *Campylobacter* pathogens but has not replaced the traditional culture methods as the gold standard for detecting campylobacters. In fact, molecular tools are often combined with traditional culture methods in many epidemiological studies to provide improved detection speed and accuracy. With the rapid increases in bacterial genomic databases, it is possible to use bioinformatics to identify more suitable genetic markers for genus- or species-specific detection of *Campylobacter*. Ideally, these genetic markers should exist in multiple copies (for increased detection sensitivity) on the genome and should be unique for *Campylobacter*. In addition, genomics can help to identify those antigens that can induce specific and highly potent antibodies that can be used for detection and identification of *Campylobacter*. With the rapid advancement of molecular techniques and the continued improvement of culture-based methods, it is predictable that in the future, the detection of *Campylobacter* spp. from animal reservoirs will become automated and much quicker. These innovations will certainly increase our ability to monitor and control *Campylobacter* infections in farm animals.

REFERENCES

al Rashid, S.T., I. Dakuna, H. Louie, D. Ng, P. Vandamme, W. Johnson, V.L. Chan. 2000. Identification of *Campylobacter jejuni, C. coli, C. lari, C. upsaliensis, arcobacter butzleri,* and *A. butzleri*-like species based on the glyA gene. *J. Clin. Microbiol.* 38:1488–1494.

Allmann, M., C. Hofelein, E. Koppel, J. Luthy, R. Meyer, C. Niederhauser, B. Wegmuller, U. Candrian. 1995. Polymerase chain reaction (PCR) for detection of pathogenic microorganisms in bacteriological monitoring of dairy products. *Res. Microbiol.* 146:85–97.

Beumer, R.R., J. de Vries, F.M. Rombouts. 1992. *Campylobacter jejuni* non-culturable coccoid cells. *Int. J. Food Microbiol.* 15:153–163.

Bolton, F.J., D. Coates, P.M. Hinchliffe, L. Robertson. 1983. Comparison of selective media for isolation of *Campylobacter jejuni/coli*. *J. Clin. Pathol.* 36:78–83.

Bovill, R.A., B.M. Mackey. 1997. Resuscitation of 'non-culturable' cells from aged cultures of *Campylobacter jejuni*. *Microbiology* 143 (Pt 5):1575–1581.

Buswell, C.M., Y.M. Herlihy, L.M. Lawrence, J.T. McGuigan, P.D. Marsh, C.W. Keevil, S.A Leach. 1998. Extended survival and persistence of *Campylobacter* spp. in water and aquatic biofilms and their detection by immunofluorescent-antibody and –rRNA staining. *Appl. Environ. Microbiol.* 64:733–741.

Cappelier, J.M., B. Lazaro, A. Rossero, A. Fernandez-Astorga, M. Federighi. 1997. Double staining (CTC-DAPI) for detection and enumeration of viable but non-culturable *Campylobacter jejuni* cells. *Vet. Res.* 28:547–555.

Cappelier, J.M., J. Minet, C. Magras, R.R. Colwell, M. Federighi. 1999. Recovery in embryonated eggs of viable but nonculturable *Campylobacter jejuni* cells and maintenance of ability to adhere to HeLa cells after resuscitation. *Appl. Environ. Microbiol.* 65:5154–5157.

Chaiyaroj, S.C., T. Sirisereewan, N. Jiamwatanasuk, S. Sirisinha. 1995. Production of monoclonal antibody specific to *Campylobacter jejuni* and its potential in diagnosis of *Campylobacter* enteritis. *Asian Pac. J. Allergy Immunol.* 13:55–61.

Chuma, T., S. Hashimoto, K. Okamoto. 2000. Detection of thermophilic *Campylobacter* from sparrows by multiplex PCR: the role of sparrows as a source of contamination of broilers with *Campylobacter*. *J. Vet. Med. Sci.* 62:1291–1295.

Chuma, T., T. Yamada, K. Okamoto, H. Yugi, T. Ohya. 1993. Application of a DNA-DNA hybridization method for detection of *Campylobacter jejuni* in chicken feces. *J. Vet. Med. Sci.* 55:1027–1029.

Chum, T., T. Yamada, K. Yano, K. Okamoto, H. Yugi. 1994. A survey of *Campylobacter jejuni* in broilers from assignment to slaughter using DNA-DNA hybridization. *J. Vet. Med. Sci.* 56:697–700.

Chuma, T., K. Yano, H. Omori, K. Okamoto, H. Yugi.1997. Direct detection of *Campylobacter jejuni* in chicken cecal contents by PCR. *J. Vet. Med. Sci.* 59:85–87.

Collins, E., M. Glennon, S. Hanley, A.M. Murray, M. Cormican, T. Smith, M. Maher. 2001. Evaluation of a PCR/DNA probe colorimetric membrane assay for identification of *Campylobacter* spp. in human stool specimens. *J. Clin. Microbiol.* 39:4163–4165.

Corry, J.E., H.I. Atabay. 2001. Poultry as a source of *Campylobacter* and related organisms. *Appl. Microbiol.* 96S–114S.

Corry, J.E., D.E. Post, P. Colin, M.J. Laisney. 1995. Culture media for the isolation of campylobacters. *Int. J. Food Microbiol.* 26:43–76.

Day, W.A., Jr., I.L. Pepper, L.A. Joens. 1997. Use of an arbitrarily primed PCR product in the development of a *Campylobacter jejuni*-specific PCR. *Appl. Environ. Microbiol.* 63:1019–1023.

Denis, M., J. Refregier-Petton, M.J. Laisney, G. Ermel, G. Salvat. 2001. *Campylobacter* contamination in French chicken production from farm to consumers. Use of a PCR assay for detection and identification of *Campylobacter jejuni* and *C. coli*. *J. Appl. Microbiol.* 91:255–267.

Denis, M., C. Soumet, K. Rivoal, G. Ermel, D. Blivet, G. Salvat, P. Colin. 1999. Development of a m-PCR assay for simultaneous identification of *Campylobacter jejuni* and *C. coli*. *Lett. Appl. Microbiol.* 29:406–410.

Docherty, L., M.R. Adams, P. Patel, J. McFadden. 1996. The magnetic immuno-polymerase chain reaction assay for the detection of *Campylobacter* in milk and poultry. *Lett. Appl. Microbiol.* 22:288–292.

Doyle, M.P., D.J. Roman. 1981. Growth and survival of *Campylobacter fetus* subsp. *jejuni* as a function of temperature an pH. *J. Food Prot.* 44:596–601.

Endtz, H.P., C.W. Ang, N. van den Braak, A. Luijendijk, B.C. Jacobs, P. de Man, J.M. van Duin, A. Van Belkum, H.A. Verbrugh. 2000. Evaluation of a new commercial immunoassay for rapid detection of *Campylobacter jejuni* in stool samples. *Eur. J. Clin. Microbiol. Infect. Dis.* 19:794–797.

Eyers, M., S. Chapelle, G. Van Camp, H. Goossens, R. De Wachter. 1993. Discrimination among thermophilic *Campylobacter* species by polymerase chain reaction amplification of 23S rRNA gene fragments. *J. Clin. Microbiol.* 31:340–3343.

Fermer, C., E.O. Engvall. 1999. Specific PCR identification and differentiation of the thermophilic campylobacters, *Campylobacter jejuni, C. coli, C. lari,* and *C. upsaliensis*. *J. Clin. Microbiol.* 37:3370–3373.

Gonzalez, I., K.A. Grant, P.T. Richardson, S.F. Park, M.D. Collins. 1997. Specific identification of the enteropathogens *Campylobacter jejuni* and *Campylobacter coli* by using a PCR test based on the ceuE gene encoding a putative virulence determinant. *J. Clin. Microbiol.* 35:759–763.

Grennan, B., N.A. O'Sullivan, R. Fallon, C. Carroll, T. Smith, M. Glennon, M. Maher. 2001. PCR-ELISAs for the detection of *Campylobacter jejuni* and *Campylobacter coli* in poultry samples. *Biotechniques* 30:602–610.

Gun-Munro, J., R. P. Rennie, J.H. Thornley, H.L. Richardson, D. Hodge, J. Lynch. 1987. Laboratory and clinical evaluation of isolation media for *Campylobacter jejuni*. *J. Clin. Microbiol.* 25:2274–2277.

Hald, B., K. Knudsen, P. Lind, M. Madsen. 2001. Study of the infectivity of saline-stored *Campylobacter jejuni* for day-old chicks. *Appl. Environ. Microbiol.* 67:2388–2392.

Harmon, K.M., G. M. Ransom, I.V. Wesley. 1997. Differentiation of *Campylobacter jejuni* and *Campylobacter coli* by polymerase chain reaction. *Mol. Cell Probes* 11:195–200.

Hazeleger, W.C., J.A. Wouters, F.M. Rombouts, T. Abee. 1998. Physiological activity of *Campylobacter jejuni* far below the minimal growth temperature. *Appl. Environ. Microbiol.* 64:3917–3922.

Hindiyeh, M., S. Jense, S. Hohmann, H. Benett, C. Edwards, W. Aldeen, A. Croft, J. Daly, S. Mottice, K.C. Carroll. 2000. Rapid detection of *Campylobacter jejuni* in stool specimens by an enzyme immunoassay and surveillance for *Campylobacter upsaliensis* in the greater Salt Lake City area. *J. Clin. Microbiol.* 38:3076–3079.

Hoorfar, J., E.M. Nielsen, H. Stryhn, S. Andersen. 1999. Evaluation of two automated enzyme-immunoassays for detection of thermophilic campylobacters in faecal samples from cattle and swine. *J. Microbiol. Methods* 38:101–106.

Houng, H.S., O. Sethabutr, W. Nirdnoy, D.E. Katz, L.W. Pang. 2001. Development of a ceuE-based multiplex polymerase chain reaction (PCR) assay for direct detection and differentiation of *Campylobacter jejuni* and *Campylobacter coli* in Thailand. *Diagn. Microbiol. Infect. Dis.* 40:11–19.

Itoh, R., S. Saitoh, J. Yatsuyanagi. 1995. Specific detection of *Campylobacter jejuni* by means of polymerase chain reaction in chicken litter. *J. Vet. Med. Sci.* 57:125–127.

Ivnitski, D., E. Wilkins, H.T. Tien, A. Ottova. 2001. Electrochemical biosensor based on supported planar lipid bilayers for fast detection of pathogenic bacteria. *Electrochemistry Communications* 2:457–460.

Jacobs-Reitsma, W. 2000. *Campylobacter* in the Food Supply. In *Campylobacter*. I.Nachamkin, and M.Blaser, eds. Washington, D.C.: American Society for Microbiology, 457–460.

Jones, D.M., E.M. Sutcliffe, A. Curry 1991. Recovery of viable but non-culturable *Campylobacter jejuni*. *J. Gen. Microbiol.* 137(Pt 10):2477–2482.

Jones, K., S. Howard, J.S. Wallace. 1999. Intermittent shedding of thermophilic campylobacters by sheep at pasture. *J. Appl. Microbiol.* 86:531–536.

Kaino, K., H. Hayashidani, K. Kaneko, M. Ogawa. 1988. Intestinal colonization of *Campylobacter jejuni* in chickens. *Jpn. J. Vet. Sci.* 50:489–494.

Konkel, M.E., S.A. Gray, B.J. Kim, S.G. Garvis, J. Yoon. 1999. Identification of the enteropathogens *Campylobacter jejuni* and *Campylobacter coli* based on the cadF virulence gene and its product. *J. Clin. Microbiol.* 37:510–517.

Korolik, V., D.T. Friendship, T. Peduru-Hewa, D.A. Alfredson, B.N. Fry, P.J. Coloe. 2001. Specific identification, grouping and differentiation of *Campylobacter jejuni* among thermophilic campylobacters using multiplex PCR. *Epidemiol. Infect.* 127:1–5.

Lamoureux, M., A. MacKay, S. Messier, I. Fliss, B.W. Blais, R.A. Holley, R.E. Simard. 1997. Detection of *Campylobacter jejuni* in food and poultry viscera using immunomagnetic separation and microtitre hybridization. *J. Appl. Microbiol.* 83:641–651.

Lawson, A.J., J.M. Logan, G.L O'Neill, M. Desai, J. Stanley. 1999. Large-scale survey of *Campylobacter* species in human gastroenteritis by PCR and PCR-enzyme-linked immunosorbent assay. *J. Clin. Microbiol.* 37:3860–3864.

Lawson, A.J., M.S. Shafi, K. Pathak, J. Stanley. 1998. Detection of *Campylobacter* in gastroenteritis: comparison of direct PCR assay of faecal samples with selective culture. *Epidemiol. Infect.* 121:547–553.

Lilja, L., M-L. Hanninen. 2001. Evaluation of a commercial automated ELISA and PCR-method for rapid detection and identification of *Campylobacter jejuni* and *C.coli* in poultry products. *Food Microbiology* 18:205–209.

Linton, D., A.J. Lawson, R.J. Owen, J. Stanley. 1997. PCR detection, identification to species level, and fingerprinting of *Campylobacter jejuni* and *Campylobacter coli* direct from diarrheic samples. *J. Clin. Microbiol.* 35:2568–2572.

Logan, J.M., K.J. Edwards, N.A. Saunders, J. Stanley. 2001. Rapid identification of *Campylobacter* spp. by melting peak analysis of biprobes in real-time PCR. *J. Clin. Microbiol.* 39:2227–2232.

Madden, R.H., L. Moran, P. Scates. 2000. Optimising recovery of *Campylobacter* spp. from the lower porcine gastrointestinal tract. *J. Microbiol. Methods* 42:115–119.

Mandrell, R.E., M.R. Wachtel. 1999. Novel detection techniques for human pathogens that contaminate poultry. *Curr. Opin. Biotechnol.* 10:273–278.

Manzano, M, C. Pipan, G. Botta, G. Comi. 1995. Polymerase chain reaction assay for detection of *Campylobacter coli* and *Campylobacter jejuni* in poultry meat. *Zentralbl. Hyg. Umweltmed.* 197:370–386.

Martin, P.M., J. Mathiot, J. Ipero, M. Kirimat, A.J. Georges, M.C. Georges-Courbot. 1989. Immune response to *Campylobacter jejuni* and *Campylobacter coli* in a cohort of children from birth to 2 years of age. *Infect. Immun.* 57:2542–2546.

Medema, G.J., F.M. Schets, A.W. van de Giessen, A.H. Havelaar. 1992. Lack of colonization of 1 day old chicks by viable, non-culturable *Campylobacter jejuni*. *J. Appl. Bacteriol.* 72:512–516.

Meinersmann, R.J., L.O. Helsel, P.I. Fields, K.L. Hiett. 1997. Discrimination of *Campylobacter jejuni* isolates by fla gene sequencing. *J. Clin. Microbiol.* 35:2810–2814.

Metherell, L.A., J.M. Logan, J. Stanley. 1999. PCR-enzyme-linked immunosorbent assay for detection and identification of *Campylobacter* species: application to isolates and stool samples. *J. Clin. Microbiol.* 37:433–435.

Musgrove, M.T., M.E. Berrang, J.A. Byrd, N.J. Stern, N.A. Cox. 2001. Detection of *Campylobacter* spp. in ceca and crops with and without enrichment. *Poult. Sci.* 80:825–828.

Nachamkin, I., J. Engberg, F.M. Aarestrup. 2000. Diagnosis and antimicrobial susceptibility of *Campylobacter* species. In *Campylobacter* I. Nachamkin and M. Blaser, Ed., Washington, D.C.: American Society for Microbiology, 45–66.

Nielsen, E.M., J. Engberg, M. Madsen. 1997. Distribution of serotypes of *Campylobacter jejuni* and *C. coli* from Danish patients, poultry, cattle and swine. *FEMS Immunol. Med. Microbiol.* 19:47–56.

Nogva, H.K., A. Bergh, A. Holck, K. Rudi. 2000. Application of the 5′-nuclease PCR assay in evaluation and development of methods for quantitative detection of *Campylobacter jejuni*. *Appl. Environ. Microbiol.* 66:4029–4036.

O'Sullivan, N.A., R. Fallon, C. Carroll, T. Smith, M. Maher. 2000. Detection and differentiation of *Campylobacter jejuni* and *Campylobacter coli* in broiler chicken samples using a PCR/DNA probe membrane based colorimetric detection assay. *Mol. Cell Probes* 14:7–16.

Olsen, J.E., S. Aabo, W. Hill, S. Notermans, K. Wernars, P.E. Granum, T. Popovic, H.N. Rasmussen, O. Olsvik. 1995. Probes and polymerase chain reaction for detection of food-borne bacterial pathogens. *Int. J. Food Microbiol.* 28:1–78.

On, S.L. 1996. Identification methods for campylobacters, helicobacters, and related organisms. *Clin. Microbiol. Rev.* 9:405–422.

Oyofo, B.A., D.M. Rollins. 1993. Efficacy of filter types for detecting *Campylobacter jejuni* and *Campylobacter coli* in environmental water samples by polymerase chain reaction. *Appl. Environ. Microbiol.* 59:4090–4095.

Pearson, A.D., M. Greenwood, T.D. Healing, D. Rollins, M. Shahamat, J. Donaldson, R.R. Colwell. 1993. Colonization of broiler chickens by waterborne *Campylobacter jejuni*. *Appl. Environ. Microbiol.* 59:987–996.

Quentin, R., D. Chevrier, J.L Guesdon, C. Martin, F. Pierre, A. Goudeau. 1993. Use of nonradioactive DNA probes to identify a *Campylobacter jejuni* strain causing abortion. *Eur. J. Clin. Microbiol. Infect. Dis.* 12:627–630.

Rasmussen, H.N., J.E. Olsen, K. Jorgensen, O.F. Rasmussen. 1996. Detection of *Campylobacter jejuni* and *C. coli* in chicken faecal samples by PCR. *Lett. Appl. Microbiol.* 23:363–366.

Rice, B.E., C. Lamichhane, S.W. Joseph, D.M. Rollins. 1996. Development of a rapid and specific colony-lift immunoassay for detection and enumeration of *Campylobacter jejuni, C. coli*, and *C. lari*. *Clin. Diagn. Lab Immunol.* 3:669–677.

Rollins, D.M., R.R. Colwell. 1986. Viable but nonculturable stage of *Campylobacter jejuni* and its role in survival in the natural aquatic environment. *Appl. Environ. Microbiol.* 52:531–538.

Saha, S.K., S. Saha, S.C. Sanyal. 1991. Recovery of injured *Campylobacter jejuni* cells after animal passage. *Appl. Environ. Microbiol.* 57:3388–3389.

Sahin, O., P. Kobalka, Q. Zhang. 2001a. Detection and survivability of *Campylobacter jejuni* in chicken eggs. Poster presented at the 82nd Annual Meeting of CRWAD Proceedings, 11–13 November, St. Louis, MO.

Sahin, O., Q. Zhang, J.C. Meitzler. 2001b. Effects of anti-*Campylobacter* maternal antibody on the colonization of *Campylobacter jejuni* in poultry. Poster presented at the 101th General Meeting of American Society for Microbiology, 20–24 May, Orlando, FL.

Sahin, O., Q. Zhang, J.C. Meitzler, B.S. Harr, T.Y. Morishita, R. Mohan. 2001c. Prevalence, antigenic specificity, and bactericidal activity of poultry anti-*Campylobacter* maternal antibodies. *Appl. Environ. Microbiol.* 67:3951–3957.

Sails, A.D., F.J. Bolton, A.J. Fox, D.R.Wareing, D.L. Greenway. 1998. A reverse transcriptase polymerase chain reaction assay for the detection of thermophilic *Campylobacter* spp. *Mol. Cell Probes* 12:317–322.

Skirrow, M.B. 1977. *Campylobacter* enteritis: a "new" disease. *Br. Med. J.* 2:9–11.

Solomon, N.H., D.M. Hoover. 1999. *Campylobacter jejuni*: a bacterial paradox. *J. Food Safety* 19:121–136.

Stanley, K., R. Cunningham, K. Jones. 1998a. Isolation of *Campylobacter jejuni* from groundwater. *J. Appl. Microbiol.* 85:187–191.

Stanley, K.N., J.S. Wallace, J.E. Currie, P.J. Diggle, K. Jones. 1998b. Seasonal variation of thermophilic campylobacters in lambs at slaughter. *J. Appl. Microbiol.* 84:1111–1116.

Stanley, K.N., J.S. Wallace, J.E. Currie, P.J. Diggle, K. Jones. 1998c. The seasonal variation of thermophilic campylobacters in beef cattle, dairy cattle and calves. *J. Appl. Microbiol.* 85:472–480.

Steinbrueckner, B., G. Haerter, K. Pelz, M. Kist. 1999. Routine identification of *Campylobacter jejuni* and *Campylobacter coli* from human stool samples. *FEMS Microbiol. Lett.* 179:227–232.

Stern, N.J. 1992. Reservoirs for *C. jejuni* and approaches for intervention in poultry. In *Campylobacter jejuni: Current Status and Future Trends*, Ed. I. Nachamkin, M. Blaser, and L.S. Tompkins. Washington, D.C.: American Society for Microbiology, 49–60.

Stern, N.J., D.M. Jones, I.V. Wesley, and D.M. Rollins. 1994. Colonization of chicks by non-culturable *Campylobacter* spp. *Lett. Appl. Microbiol.* 18:333–336.

Stucki, U., J. Frey, J. Nicolet, A.P. Burnens. 1995. Identification of *Campylobacter jejuni* on the basis of a species-specific gene that encodes a membrane protein. *J. Clin. Microbiol.* 33:855–859.

Studer, E., J. Luthy, P. Hubner. 1999. Study of the presence of *Campylobacter jejuni* and *C. coli* in sand samples from four Swiss chicken farms. *Res. Microbiol.* 150:213–219.

Taylor, D.E., N. Chang. 1987. Immunoblot and enzyme-linked immunosorbent assays of *Campylobacter* major outer-membrane protein and application to the differentiation of *Campylobacter* species. *Mol. Cell Probes* 1:261–274.

Taylor, D.E., K. Hiratsuka. 1990. Use of non-radioactive DNA probes for detection of *Campylobacter jejuni* and *Campylobacter coli* in stool specimens. *Mol. Cell Probes* 4:261–271.

Tholozan, J.L., J.M. Cappelier, J.P. Tissier, G. Delattre, M. Federighi. 1999. Physiological characterization of viable-but-nonculturable *Campylobacter jejuni* cells. *Appl. Environ. Microbiol.* 65:1110–1116.

Tran, T.T. 1995. Evaluation of Oxyrase enrichment method for isolation of *Campylobacter jejuni* from inoculated foods. *Lett. Appl. Microbiol.* 21:345–347.

Uyttendaele, M., A. Bastiaansen, J. Debevere. 1997. Evaluation of the NASBA nucleic acid amplification system for assessment of the viability of *Campylobacter jejuni*. *Int. J. Food Microbiol.* 37:13–20.

van de Giessen, A.W., C.J. Heuvelman, T. Abee, W.C. Hazeleger. 1996. Experimental studies on the infectivity of non-culturable forms of *Campylobacter* spp. in chicks and mice. *Epidemiol. Infect.* 117:463–470.

van Doorn, L.J., A. Verschuuren-van Haperen, A. Burnens, M. Huysmans, P. Vandamme, B.A. Giesendorf, M.J. Blaser, W.G. Quint. 1999. Rapid identification of thermotolerant *Campylobacter jejuni, Campylobacter coli, Campylobacter lari*, and *Campylobacter upsaliensis* from various geographic locations by a GTPase-based PCR-reverse hybridization assay. *J. Clin. Microbiol.* 37:1790–1796.

Waage, A.S., T. Vardund, V. Lund, G. Kapperud. 1999. Detection of small numbers of *Campylobacter jejuni* and *Campylobacter coli* cells in environmental water, sewage, and food samples by a seminested PCR assay. *Appl. Environ. Microbiol.* 65:1636–1643.

Waegel, A., I. Nachamkin. 1996. Detection and molecular typing of *Campylobacter jejuni* in fecal samples by polymerase chain reaction. *Mol. Cell Probes* 10:75–80.

Waller, D.F., S.A. Ogata. 2000. Quantitative immunocapture PCR assay for detection of *Campylobacter jejuni* in foods. *Appl. Environ. Microbiol.* 66:4115–4118.

Wang, H., J.M. Farber, N. Malik, G. Sanders. 1999. Improved PCR detection of *Campylobacter jejuni* from chicken rinses by a simple sample preparation procedure. *Int. J. Food Microbiol.* 52:39–45.

Wassenaar, T.M., D.G. Newell. 2000. Genotyping of *Campylobacter* spp., *Appl. Environ. Microbiol.* 66:1–9.

Winkler, M.A., J. Uher, S. Cepa. 1999. Direct analysis and identification of *Helicobacter* and *Campylobacter* species by MALDI-TOF mass spectrometry. *Anal. Chem.* 71:3416–3419.

Winters, D.K., M.F. Slavik. 2000. Multiplex PCR detection of *Campylobacter jejuni* and *Arcobacter butzleri* in food products. *Mol. Cell Probes* 14:95–99.

Wonglumsom, W., A. Vishnubhatla, J.M. Kim, D.Y. Fung. 2001. Enrichment media for isolation of *Campylobacter jejuni* from inoculated ground beef and chicken skin under normal atmosphere. *J. Food Prot.* 64:630–634.

Yu, L.S., J. Uknalis, S.I. Tu. 2001. Immunomagnetic separation methods for the isolation of *Campylobacter jejuni* from ground poultry meats. *J. Immunol. Methods* 256:11–18.

Zhang, Q., J.C. Meitzler, S. Huang, T. Morishita. 2000. Sequence polymorphism, predicted secondary structures, and surface-exposed conformational epitopes of *Campylobacter* major outer membrane protein. *Infect. Immun.* 68:5679–5689.

21

IN VITRO AND *IN VIVO* MODELS USED TO STUDY *CAMPYLOBACTER JEJUNI* VIRULENCE PROPERTIES

MICHAEL E. KONKEL, MARSHALL R. MONTEVILLE, JOHN D. KLENA, AND LYNN A. JOENS

INTRODUCTION. Members of the genus *Campylobacter* are Gram-negative curved rods that range in size from 0.2 to 0.8 μm in width and 0.5 to 5 μm in length. They are nonsporeforming, nonsaccharolytic, motile bacteria that possess unipolar or bipolar flagella. These bacteria grow optimally at temperatures between 37 and 42°C under microaerophilic conditions. When stressed, these bacteria typically become spherical in shape (Boucher et al. 1994; Federighi et al. 1998). The genome of *C. jejuni* is approximately 1.6 to 1.7 Mbp, with a mol (G + C) content of approximately 30 percent (Chang and Taylor 1990; Karlyshev et al. 1998; Parkhill et al. 2000). To better understand the metabolic capacity and virulence properties of *C. jejuni*, the genome sequence of strain NCTC 11168 was determined (Parkhill et al. 2000). Protein coding regions account for 94.3 percent of the NCTC 11168 genome.

Although members of *Campylobacter* spp. were initially recognized to cause disease in sheep and cattle, *C. jejuni* was not recognized as a human pathogen until much later . The first reported isolation of *C. jejuni* from diarrheal stools of humans was in 1972, with researchers using a filtration technique designed for veterinary diagnostics (Dekeyser et al. 1972). Subsequently, a selective medium for *Campylobacter* isolation from diarrheal stools of animals and humans was published in 1979 (Butzler and Skirrow 1979). Now *Campylobacter* spp. have emerged as the leading cause of human gastroenteritis in developed countries (Tauxe 1992; Altekruse et al. 1999). For example, in 1980, campylobacteriosis became a notifiable disease in New Zealand. By the year 2000, New Zealand had one of the highest reported incidence rates (greater than 230 per 100,000) for campylobacteriosis in the developed world (Anonymous 2001; Savill et al. 2001). In the New Zealand context, this increase in reported cases does not appear to be caused by increased reporting or changes in methodologies for the detection of the pathogen (McNicholas et al. 1995). In other countries, the sharp rise in the number of reported cases of *C. jejuni* infections can be attributed to increased awareness of the disease by laboratory personnel and physicians, an increase in attempts to isolate the microorganism, better isolation methods, and improved reporting (Skirrow 1991; Cowden 1992; Kapperud and Aasen 1992). In the United States, the annual incidence of *C. jejuni* infection has been shown to be five to six per 100,000 persons (Tauxe et al. 1988). However, it has been estimated that 2.4 to 4 million cases of human campylobacteriosis occur every year in the United States (Tauxe 1992), suggesting that this disease is significantly underreported in the United States (Blaser 1997).

All ages are affected by *C. jejuni* infection, although in developed countries, a greater number of infections occur in children less than four years of age and in young adults (Allos and Blaser 1995). Infection caused by *C. jejuni* is more common in developing countries, but because of high-level exposure within the first five years of life and the development of protective immunity, symptomatic infections among adults are much less common (Blaser and Reller 1981). A slightly higher reporting rate exists among males than females (Skirrow 1987).

In developing countries, campylobacteriosis is usually a mild diarrheal disease with watery stools and little dysentery (Glass et al. 1983). In contrast, the presentation of disease in developed countries is more severe, characterized by bloody stools, abdominal pain, and fever (Blaser et al. 1979; Karmali and Fleming 1979; Svedhem and Kaijser 1980; Blaser et al. 1983). These differences probably reflect the repeated exposure of individuals in developing countries and may also reflect differences in the virulence among *C. jejuni* strains. *Campylobacter* infections can be fatal, with a case fatality rate for *Campylobacter* infection of 0.05 per 1,000 infections (Allos 2001). However, the case fatality rate increases in immunocompromised individuals; this is especially true of individuals infected by the human immunodeficiency virus (Altekruse et al. 1998; Altekruse et al. 1999).

Although most cases of campylobacteriosis are thought to be sporadic, some outbreaks have been associated with drinking improperly treated water and unpasteurized milk (Potter et al. 1983; Vogt et al. 1984; Humphrey and Hart 1988; Pebody et al. 1997). These outbreaks frequently occur in the spring and fall (Altekruse et al. 1998). In contrast, sporadic infections occur most commonly as a result of handling or consuming raw or undercooked meats, especially poultry (Hopkins et al. 1984; Harris et al. 1986; Deming et al. 1987; Adak, Cowden et al. 1995; Eberhardt-Phillips et al. 1997). In temperate climates, sporadic infections peak in the early summer months (Brieseman 1990; Skirrow 1991; Tauxe 1992; Allos and Blaser 1995).

The infective dose varies depending on the nature of the contaminated food, but as few as five hundred bacteria can cause disease (Robinson 1981; Black et al. 1988; Black et al. 1992). Although *Campylobacter*-mediated enteritis can begin as early as eighteen hours after exposure, the average onset of disease is 3.2 days. The duration of clinical disease, diarrhea with abdominal pain, is usually three to four days. Following recovery, relapses of diarrhea and cramps have been reported (Blaser et al. 1979; Drake et al. 1981). There have also been reports of continued shedding of *C. jejuni* in the feces of individuals for weeks to months beyond the resolution of symptoms (Karmali and Fleming 1979; Kapperud and Aasen 1992). Extraintestinal infections associated with *C. jejuni* disease are rarely reported (Blaser et al. 1986; Allos 2001).

Although most people fully recover from campylobacteriosis, serious post-infectious sequelae can occur. The most notable post-infectious processes are Guillain-Barré (GBS) and Miller-Fisher (MFS) Syndromes, autoimmune-mediated disorders of the peripheral nervous system (Salloway et al. 1996; Nachamkin et al. 1998). These disorders are associated with host antibodies generated against the bacterial surface antigens; these antibodies may alternatively react with the host myelin sheaths of the sensory nerve fibers. This demyelination results in a flaccid paralysis; in some cases this can be fatal, but most affected individuals either completely recover or experience chronic disease (Mishu et al. 1993).

Treatment of campylobacteriosis is not required for most individuals because the disease is self-limiting. For individuals who seek treatment, the macrolide erythromycin and fluoroquinolones are effective drug treatments for *C. jejuni* infections. However, emergence of antibiotic resistance has been reported in developed and developing countries (reviewed in (Engberg et al. 2001)). Erythromycin resistance appears to be chromosomally mediated, involving changes in the peptidyltransferase binding site (23S rRNA domain V) (Weisblum 1995; Jensen and Aarestrup 2001). Quinolone and fluoroquinolone resistance is mainly caused by alterations in DNA gyrase (*gyrA*) or, less commonly, topoisomerase IV (*parC*) (Gibreel et al. 1998). Use of antimicrobial agents such as quinolones and fluoroquinolones in animal feed appears to directly influence the emergence of resistant isolates in humans (Smith et al. 1999; van den Bogaard and Stobberingh 2000; Engberg et al. 2001). Azithromycin is recommended as an alternative treatment for travelers from areas where conventional therapies are compromised by the emergence of antimicrobial resistance (DuPont 1995; Kuschner et al. 1995).

This article focuses on *in vitro* and *in vivo* models used to study some of the pathogenic properties of *C. jejuni*. More specifically, we discuss the *in vitro* models used to examine the interactions of *C. jejuni* with nonprofessional and professional phagocytic cells, and the *in vivo* models to examine *C. jejuni* colonization and infection. Finally, we discuss advances in genetic manipulation techniques for *C. jejuni*; application of these techniques should permit a rapid increase in our understanding of the pathogenesis of this organism in the future.

INTERACTIONS OF *C. JEJUNI* WITH NONPROFESSIONAL PHAGOCYTIC CELLS. Investigators have utilized a variety of assays to assess the pathogenic properties of *C. jejuni* isolates *in vitro*. Here we discuss the model systems used to examine *C. jejuni* adherence, invasion, and translocation.

Adherence. The ability of enteropathogenic bacteria to bind to nonprofessional phagocytic cells is considered an important virulence determinant for bacteria that colonize the intestinal tract of a host because it prevents the organism from being swept away by mechanical cleansing forces such as peristalsis and fluid flow. In some instances, binding is also a prerequisite for entry into a host cell (Roberts 1990; Alrutz and Isberg 1998). Invading a host cell is advantageous to organisms that survive intracellularly, because they are protected from the host immune responses (Isberg and Tran Van Nhieu 1994). The ability of *C. jejuni* to bind to host cells is hypothesized to play an early role in the development of campylobacteriosis. This hypothesis is supported by the finding that *C. jejuni* isolates cultured from individuals with fever and diarrhea adhere to cultured cells at a greater efficiency than those isolates cultured from asymptomatic individuals (Fauchere et al. 1986).

In vitro adherence assays have been used extensively to assess the binding potential of *C. jejuni* isolates (Fauchere et al. 1986; McSweegan and Walker 1986; de Melo and Pechère 1990; Konkel et al. 1997; Pei et al. 1998). A typical assay involves inoculating a monolayer of undifferentiated eukaryotic cells of human or nonhuman origin with a bacterial suspension containing a known number of bacterial-colony-forming units (Fig. 21.1). Researchers have used a variety of cell lines when performing this assay, but the INT 407 (Henle, a human intestinal epithelial cell line) and Caco-2 (a human colonic cell line) cell lines are frequently used. These eukaryotic cells are considered to be reflective of those that the organism encounters *in vivo*. After inoculation of the eukaryotic cells with the bacteria and subsequent centrifugation to promote contact and synchronize infection, the inoculated cells are incubated in a 37°C humidifed, CO_2-enriched incubator to allow the bacteria to bind to the host cells. Following incubation, the monolayers are rinsed to remove the non-adherent bacteria, and epithelial cells are lysed with a detergent such as Triton-X 100 or sodium deoxycholate. The viable number of adherent bacteria is determined by plating serial dilutions of the lysates on a solid medium, incubating the cultures under the appropriate set of conditions, and determining the number of bacteria. Each isolate is usually tested in triplicate to obtain a mean and standard deviation for the number of viable bacteria bound to a cell monolayer. Advantages of this assay include its simplicity and cost (inexpensive); also, it allows for several parameters to be manipulated (for example, the temperature at which the bacteria are cultured prior to the assay, the number of bacteria that are used to inoculate a cell monolayer, the temperature at which the assay is performed, and the addition of drugs to inhibit various host cell processes).

The number of *C. jejuni* bound to host cells can also be determined by microscopy examination of infected cell monolayers using an immunofluorescence assay (Newell et al. 1985). After the infected cells are incubated, the monolayers are rinsed to remove the non-adherent bacteria. The cell monolayers are then treated with a fixative, such as methanol, rather than lysing the cells with a detergent. The inoculated cell monolayers are incubated with a primary antibody that reacts with the bound bacteria, followed by a fluorescence-labeled secondary antibody. An advantage of this assay is that the binding of nonviable bacteria to host cells can be quantitatively assessed. Other methods to visualize the binding of *C. jejuni* to host cells include scanning and transmission electron microscopy (Newell et al. 1985).

C. jejuni synthesize a set of surface-exposed molecules that facilitate binding to host cells. The molecules that promote the binding of *C. jejuni* to host cell receptors are collectively referred to as adhesins. Because metabolically inactive (heat-killed or sodium azide-killed) *C. jejuni* bind to cultured cells at levels equivalent to metabolically active organisms (Konkel and Cieplak 1992), the *C. jejuni* adhesins appear to be synthesized constitutively rather than induced upon the organism's co-cultivation with mammalian cells. It is currently unknown whether the expression of the genes encoding the adhesins is upregulated in response to changes in culture conditions.

The best-characterized *C. jejuni* adhesins include PEB1, CadF, and JlpA. PEB1 is 28 kDa protein that shares homology with membrane proteins that function in amino acid transport. The gene encoding this protein was initially identified by screening a *C. jejuni* genomic - λgt11 library with a hyperimmune antibody raised against the 28 kDa protein (Pei and Blaser 1993). This work was preceded by another study in which De Melo and Pechère (1990) identified four outer membrane proteins (omps), with apparent molecular masses of 28, 32, 36, and 42 kDa, which demonstrated adhesive properties as judged by a lig-and-binding assay (de Melo and Pechère 1990). Evidence indicating that PEB1 is an adhesin includes the finding that a *C. jejuni peb1A* null mutant exhibits a reduction in the duration of mouse intestinal colonization when compared to the *C. jejuni* parental isolate (Pei et al. 1998). CadF (*Campylobacter* adhesion to Fibronectin) is a *C. jejuni* 37 kDa omp (Konkel et al. 1997). CadF is conserved among all *C. jejuni* and *C. coli* isolates tested to date (Konkel et al. 1999). Whether the 36 kDa protein identified by De Melo and Pechère (de Melo and Pechère 1990) and the 37 kDa CadF protein represent the same protein is not known.

A *C. jejuni cadF* mutant lacks the ability to colonize the cecum of newly hatched leghorn chickens (Ziprin et al. 1999). Jin et al. (Jin et al. 2001) identified a 42.3 kDa lipoprotein termed JlpA (*jejuni* lipoprotein A), which mediates the binding of *C. jejuni* to HEp-2 cells. A mutation in the *jlpA* gene resulted in an 18 to 19.4 percent reduction in adherence when compared to the *C. jejuni* wild-type isolate, but had no effect on *C. jejuni* invasion. In addition, pretreatment of HEp-2 cells with recombinant JlpA reduced the binding of *C. jejuni* to the cells in a dose-dependent fashion. Other molecules proposed to act as adhesins include the flagellum (Pavlovskis et al. 1991; Wassenaar et al. 1991; Grant et al. 1993; Wassenaar et al. 1993), lipopolysaccharide (McSweegan and Walker 1986), the major outer membrane protein (MOMP, also called OmpE) (Moser et al. 1997; Schröder and Moser 1997), and a protein termed P95 (Kelle et al. 1998).

In summary, evidence supports the hypothesis that adhesins play a role in *C. jejuni* colonization. Future studies should examine whether *C. jejuni* utilizes the same set of adhesins in a human host versus that of cattle, chickens, and other reservoir hosts in which *C. jejuni* is considered commensal flora. In addition, future studies should focus on other factors that influence the ability of *C. jejuni* to colonize a host, including the organism's motility, chemotactic behavior, and surface charge.

Invasion. The ability of *C. jejuni* to enter, survive, and replicate in mammalian cells has been studied extensively using tissue culture models (Newell et al. 1985; de Melo et al. 1989; Konkel and Joens 1989; Konkel et al. 1990; Wassenaar et al. 1991; Everest et al. 1992; Konkel et al. 1992; Konkel et al. 1992; Grant et al. 1993; Oelschlaeger et al. 1993; Yao et al. 1994; Doig et al. 1996; Pei et al. 1998; Konkel et al. 1999). The most commonly used experimental assay for assessing invasion involves determining the number of bacteria protected from an aminoglycosidic antibiotic, such as gentamicin, that does not penetrate eukaryotic cell membranes (Fig. 21.1). Everest et al. (Everest et al. 1992) found a statistically significant difference in the level of HeLa and Caco-2 cell invasion between *C. jejuni* isolates taken from individuals with colitis versus those isolated from individuals with noninflammatory diarrhea. In colonic biopsy specimens obtained after inoculation of primates, *C. jejuni* were observed to be in the process of being taken up by epithelial cells (Russell et al. 1993). *C. jejuni* were also observed within vacuoles and within the cytoplasm of damaged cells. The investigators concluded that early mucosal damage, occurring prior to any inflammatory response, resulted from *C. jejuni* invading the colonic epithelial cells.

The relative ability of *C. jejuni* to invade cultured cells is strain-dependent (Newell et al. 1985; Konkel and Joens 1989; Everest et al. 1992). Newell et al. (1985) found that environmental isolates were much less invasive than clinical isolates as determined by

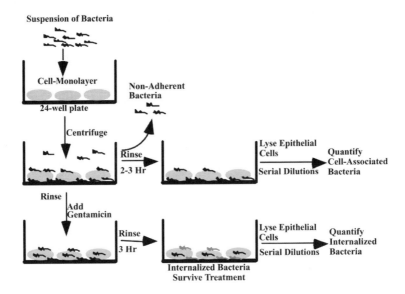

FIG 21.1—Diagram showing the adherence and invasion assay.

immunofluorescence and electron microscopy examination of *C. jejuni*-infected HeLa cells (Newell et al. 1985). The percent of the inoculum internalized for *C. jejuni* 81-176, a strain isolated from a milkborne outbreak of diarrheal illness, has been reported to range between 0.8 to 1.8 percent (Yao et al. 1994; Doig et al. 1996; Yao et al. 1997).

Parasite-directed endocytosis is a process in which a microorganism synthesizes the proteins required to promote internalization. The internalization of *C. jejuni* is inhibited by chloramphenicol, a specific inhibitor of bacterial protein synthesis (Konkel and Cieplak 1992; Oelschlaeger et al. 1993). In addition, metabolically inactive (sodium azide-killed) but intact *C. jejuni* are not internalized (Konkel and Cieplak 1992). One and two-dimensional electrophoretic analyses of metabolically labeled *C. jejuni* cultured in the presence and absence of epithelial cells have also demonstrated that a number of proteins are synthesized, exclusively or preferentially, in the presence of epithelial cells, whereas others are selectively repressed (Konkel and Cieplak 1992; Konkel et al. 1993). The newly synthesized proteins are distinct from those proteins induced by thermal stress (Konkel et al. 1998). Independently, Panigrahi et al. (1992) reported that *C. jejuni* synthesized a set of proteins when grown in rabbit ileal loops that were not produced when the organism was cultured on standard laboratory medium (Panigrahi et al. 1992). Two of the newly synthesized proteins, with apparent molecular masses of 84 and 47 kDa, were detectable using convalescent sera from *C. jejuni*-infected individuals. Combined, these results suggest that *C. jejuni* synthesize entry-promoting proteins upon exposure to host cells and that these proteins are part of a coordinated bacterial response, resulting from the regulation of expression of a defined subset of genes.

Studies in our laboratory have demonstrated that a subset of the *de novo* synthesized proteins, termed Cia(s) for *C*ampylobacter *i*nvasion *a*ntigens, are secreted upon co-cultivation of *C. jejuni* with intestinal cells. By differentially screening *C. jejuni* genomic DNA-phage expression libraries with antisera generated in rabbits against *C. jejuni* cultured in the presence of INT 407 cells (Cj+INT) and *C. jejuni* cultured in the absence of INT 407 cells (Cj–INT), plaques were identified that reacted only with the Cj+INT antiserum. This screening resulted in the identification of a gene termed *ciaB* (Konkel et al. 1999). *In vitro* assays revealed that a *C. jejuni ciaB* null mutant bound to INT 407 cells in numbers equal to or greater than the wild-type isolate, but exhibited a significant reduction (~100-fold) in internalization. Confocal microscopy studies using an anti-CiaB serum revealed an intense fluorescence signal in the cytoplasm of the *C. jejuni*-infected INT 407 cells but not in the cytoplasm of mock-infected INT 407 cells. Nonetheless, the specific function of CiaB is not known.

Other work has revealed that Cia protein synthesis and secretion are separable, and that secretion is the rate-limiting step in these processes (Rivera-Amill et al. 2001). Cia protein synthesis is induced in response to bile salts and various eukaryotic host cell components, whereas the latter is capable of also inducing Cia protein secretion. We propose that the incubation of *C. jejuni* in the presence of 0.1 percent of sodium deoxycholate, which is similar to the concentration of bile salts found in the lumen of the small intestine (Pope et al. 1995), provides an environment that partially mimics that which *C. jejuni* would encounter *in vivo*. Also noteworthy is that culturing *C. jejuni* on deoxycholate-supplemented medium retarded the inhibitory effect of chloramphenicol on epithelial invasion as judged by the gentamicin-protection assay (Rivera-Amill et al. 2001). These data, in combination with the observation that *C. jejuni* cultured with eukaryotic cells synthesize a subset of new proteins not synthesized by organisms cultured in the absence of eukaryotic cells (Konkel and Cieplak 1992; Konkel et al. 1993), suggest that the coordinate expression of the genes encoding the Cia proteins is subject to environmental regulation.

Preliminary data has been generated in our laboratory suggesting that the Cia proteins are secreted via the flagellar type III secretion apparatus (Konkel et al., unpublished observations). Mutations in the genes encoding components of the *C. jejuni* flagellar apparatus abolished Cia protein secretion. A caveat of this work is that *cia* and flagellar-structural genes may be co-regulated. The secretion of virulence proteins via the flagellar apparatus would seem to be beneficial to *C. jejuni*, eliminating additional energy expenditure for the synthesis of a distinct Cia secretory apparatus. There is precedent for protein secretion through the flagellar apparatus; *Yersinia* spp. secrete proteins termed flagellar outer proteins (Fops) from this apparatus (Macnab 1999; Young et al. 1999).

Bacon et al. (2000) reported the presence of a plasmid termed pVir in *C. jejuni* 81-176 (Bacon et al. 2000). Nucleotide sequence of four open reading frames located on the 35 kbp pVir plasmid showed similarity with *Helicobacter pylori* proteins (ranging from 49–60 percent similarity), and 27–34 percent similarity to proteins that comprise type IV secretion systems from a diverse range of Gram-negative bacteria. Mutations in two of the four genes (*comB3* and *virB11*) affected the virulence of *C. jejuni* strain 81-176 as judged by *in vitro* binding and internalization assays and *in vivo* assays utilizing the ferret diarrhea model. Independent mutations in each gene resulted in a reduction of adherence (29 percent of wild type with *comB3*, 13 percent of wild type with *virB11*) with a corresponding reduction (28 percent and 9 percent, respectively) in invasion into INT 407 cells. Compared to the wild type *C. jejuni* strain 81-176, the *virB11* mutant was significantly affected in its capacity to cause diarrhea in the ferret model at both high (9 × 10^{10} to 8 × 10^{11} bacterial cells) and low (8 × 10^9 to 8 × 10^{10} bacterial cells) doses. Hybridization experiments determined that 10 percent (six of fifty-eight samples) of *C. jejuni* clinical isolates obtained from Thailand contained the *virB11* gene. Based on this finding, the investigators concluded that distinct pathogenic mechanisms differentiate strains of *C. jejuni*. The relationship between the *virB11* system and the Cia secretion proteins from *C. jejuni* is currently undefined.

Translocation. Investigators have utilized a cell culture system to assess the ability of pathogens to migrate, or translocate, across a cell barrier (Finlay and

Falkow 1990; Cruz et al. 1994; Kops et al. 1996; Nataro et al. 1996). Translocation is considered an important virulence attribute for certain pathogens because it permits them access to underlying tissues and may allow for their dissemination throughout a host. The *in vitro* assay to assess bacterial translocation involves culturing eukaryotic cells on a permeable membrane contained within a plastic insert (Fig. 21.2). The inserts are placed in a plastic tray (for example, a twenty-four-well tissue culture plate), thereby establishing apical and basolateral chambers. Culture media is added to each chamber to promote cell growth and differentiation. Cell differentiation results in a polarized cell monolayer with distinct apical and basolateral surfaces. Depending on the cell line used, the apical cell surfaces are characterized by well-developed microvilli and brush borders. Human adenocarcinoma cell lines HT29, Caco-2, and T84 have all been used in this assay system, because these cells will form a polarized cell monolayer when cultured under appropriate conditions. The ability of bacteria to traverse cell monolayers is assessed after the addition of the bacteria to the apical chamber of the insert, followed by plating of the basolateral chamber medium on agar plates after various periods of time. This culture system enables investigators to monitor the integrity of a cell monolayer by measuring the transepithelial electrical resistance (TER). A decrease, or loss of TER, indicates disruption of cellular tight junctions (Finlay et al. 1988; Finlay and Falkow 1990).

Everest et al. (1992) noted that 86 percent of *Campylobacter* isolates from individuals with colitis were able to translocate across polarized Caco-2 cells, versus 48

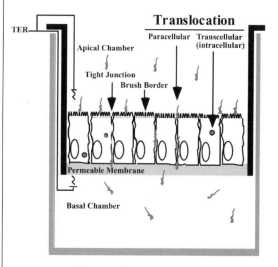

FIG. 21.2—Diagram showing cross-section of a polarized cell monolayer. The paracellular and transcellular routes of translocation are indicated (arrows). TER = Transepithelial electrical resistance.

percent of strains isolated from individual with non-inflammatory disease (Everest et al. 1992). Transloca-tion of *C. jejuni* cells across polarized Caco-2 cell monolayers was determined using the culture system described previously. It was also noted that six *C. jejuni* isolates, characterized as non-invasive, as judged by the gentamicin protection assay using Caco-2 cells, were able to translocate across the polarized cell monolayers. Harvey et al. (1999) compared four *C. jejuni* isolates with differing abilities to invade Caco-2 cells in their ability to translocate across polarized epithelial mem-branes (Harvey et al. 1999). In this report, invasiveness of *C. jejuni* did not quantitatively correlate with the ability to translocate across tissue cell culture mono-layers. Although the investigators detected fluctuations in the measurable TER with the different *C. jejuni* iso-lates over the course of the six-hour assay, cell culture monolayer integrity was maintained, and final TER val-ues were comparable to starting baseline values. Main-tenance of monolayer integrity, at least over a relatively short period of time (eight hours), has also been reported by others (Konkel et al. 1992; Brás and Ketley 1999). Noteworthy is that Brás et al. (1999) detected a loss in TER of Caco-2 cells inoculated with *C. jejuni* after twenty-four hours, indicating an eventual disrup-tion of cellular tight junctions. These investigators pro-posed that the loss in monolayer integrity was the result of long-term effects of translocation and/or invasion or the accumulation of a bacterial toxin(s). The aforemen-tioned studies argue that the genes encoding the prod-ucts responsible for invasion in *C. jejuni* are distinct from those that confer translocation ability.

The predominant route of *C. jejuni* translocation across polarized cell monolayers is unclear. The pres-ence of intracellular bacteria is supporting evidence for a transcellular (through a cell) route of passage. So is the observation that *C. jejuni*-cellular translocation is reduced at 20°C (Konkel et al. 1992). Temperatures of 18–22°C preferentially inhibit eukaryotic endocytic and phagocytic processes (Silverstein et al. 1977). In contrast, evidence exists that supports the paracellular (between cell) route of passage, including the observa-tion that *C. jejuni* can be recovered from the basolateral chamber as early as fifteen minutes post-inoculation of a polarized cell monolayer (Konkel et al. 1992). In addition, the invasiveness of *C. jejuni* isolates does not quantitatively correlate with translocation efficiency (Harvey et al. 1999). Studies have demonstrated that tight junctions temporally relax to allow regulated pas-sage of both solutes and neutrophils (Madara 1998). Cellular tight junctions also reseal following penetra-tion by certain pathogens (Takeuchi 1967). Therefore it is plausible that *C. jejuni* utilizes a paracellular route of passage, with only a transient change in cellular integrity. Future studies will be required to clarify the major route of *C. jejuni* translocation through the epithelium.

The *in vivo* relevance of *C. jejuni* translocation across the intestinal epithelium is not known. Whether *C. jejuni* reaches the lamina propria via migration across the epithelial cells lining the intestinal tract or exclusively via M cells is unclear. In the lamina pro-pria, *C. jejuni* would have access to different cellular receptors and professional phagocytic cells that are likely to play a role in infection. Nevertheless, the inci-dence of bacteremia in individuals with *C. jejuni* infec-tion is 0.4 percent (Allos and Blaser 1995), suggesting that the host's immune system is effective in preventing the spread of the infection. Movement of *C. jejuni* from the apical to basolateral epithelial cell surface is clearly important for pathogenesis. However, whether translo-cation through epithelial cells plays a more significant role than M cell sampling has yet to be determined.

INTERACTIONS OF *C. JEJUNI* WITH PROFES-SIONAL PHAGOCYTIC CELLS. Macrophages provide intestinal pathogens with an unoccupied space for acquiring nutrients and a shelter from immune sur-veillance. If a pathogen can survive within these cells, it is also possible for it to propagate and disseminate throughout a host. Subsequently, the pathogen may establish a carrier state in the host and serve as a source of infection for other individuals. As well as the tissue culture studies described in the preceding section, ani-mal studies have demonstrated that *C. jejuni* cells invade intestinal mucosal epithelial cells and translo-cate to the lamina propria and deeper submucosa of the host (Babakhani et al. 1993). Deeper tissue involve-ment results in a hemorrhagic necrosis in the lamina propria, formation of crypt abscesses, and infiltration of inflammatory cells (Blaser et al. 1980; Duffy et al. 1980). At this stage in a *C. jejuni* infection, the organ-ism is likely engulfed by macrophages.

Engulfment. Field et al. (1991) demonstrated a corre-lation between the uptake of *C. jejuni* by mouse peri-toneal macrophages and the organism's clearance from the bloodstream of mice (Field et al. 1991). In addition, avirulent strains of *C. jejuni*, defined by their inability to invade the chorioallantoic membrane of chicken embryos, were cleared *in vivo* and engulfed *in vitro* by peritoneal macrophages from Balb/c mice at signifi-cantly higher rates than virulent strains. Additional experiments revealed that these findings were inde-pendent of complement-mediated opsonization, sug-gesting that *C. jejuni* virulence was associated with resistance to phagocytosis (Field et al. 1991). Resis-tance to phagocytosis was also demonstrated by *C. jejuni* isolates exposed to guinea-pig resident peritoneal macrophages when compared to *C. coli* strains that were engulfed in significantly higher numbers (Banfi et al. 1986). These authors proposed that the difference in uptake of the two species of *Campylobacter* was caused by the presence of an antiphagocytic capsular-like material in *C. jejuni* that was absent in *C. coli*.

In 1997, Wassenaar et al. reported a linear correla-tion between infective dose and internalized *C. jejuni* in monocytes from human blood donors (Wassenaar et al. 1997). This experiment supported previous evidence obtained by Myszewski and Stern (1991), who found that *C. jejuni* was readily internalized by chicken macrophages within a thirty-minute incubation period with increased internalization when serum or macrophages from previously colonized hosts were added to the assay (Myszewski and Stern 1991). Recently, Day et al. (2000) demonstrated that the uptake of the clinical *C. jejuni* M129 isolate in mouse and porcine peritoneal macrophages, as well as a mouse macrophage cell line, occurred in the absence of serum and complement (Day et al. 2000) (Fig. 21.3). These data suggest that the internalization of *C. jejuni* by phagocytes is strain dependent. Interestingly, work from Guerry et al. (2000, 2002) strongly implicates sialylation of LOS as important for serum resistance (Guerry et al. 2000; Guerry et al. 2002). Guerry et al. (2002) created mutations in *neuC*1, a gene encoding an N-acetylglucosamine(GlcNAc)-6-phosphate 2-epimerase/GlcNAc-6-phosphatase, necessary for sialylation, as well as the LOS structural gene *cgtA* (a UDP-N-acetylgalactosaminyl transferase) (Guerry et al. 2002). Loss of sialylation impaired resistance to normal human sera but had no effect on invasion of INT407 cells. Alteration of the LOS structure enhanced invasion. This is clearly an area that will receive more attention in the future.

Survival. Wassenaar et al. (1997) examined the ability of sixteen clinical and laboratory-adapted isolates of *C. jejuni* to resist killing by activated human peripheral monocytes (Wassenaar et al. 1997). The investigators found that the majority of *C. jejuni* were killed by the monocytes within twenty-four to –forty-eight hours. However, the monocytes from approximately 10 percent of donor-individuals demonstrated normal uptake of *C. jejuni* but failed to kill the bacterium. Myszewski and Stern (1991) examined the ability of both a *C.*

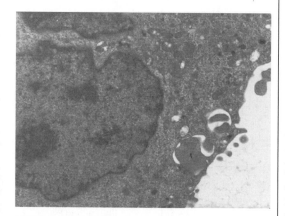

FIG 21.3—Transmission electron micrograph of J774A.1 macrophage infected with *C. jejuni* strain M129. The sample was processed six hours post-infection. After sectioning, the sample was stained with aqueous uranyl acetate and lead citrate, and examined with a JEOL electron microscope (model CX2) at 80V. Bar = 5 μm.

jejuni high-passage clinical isolate and a *C. jejuni* chicken isolate to resist killing by macrophages (Myszewski and Stern 1991). They found that both *C. jejuni* isolates were killed by chicken peritoneal macrophages within a six-hour period. These observations suggest that the macrophage cell type is important for *C. jejuni* intracellular survival.

Evidence suggesting that strain variation does have a significant role in intracellular survival has been gathered by several investigators. Joens et al. (unpublished data), in examining the survival of five environmental isolates of *C. jejuni* exposed to a mouse macrophage cell line (J774A.1), found that two of five isolates were inactivated at twenty-four hours. The remaining three isolates survived for seventy-two hours after engulfment. Day et al. (2000) examined the survival of a *C. jejuni* clinical isolate in various host macrophages (Day et al. 2000). The clinical isolate M129 was able to survive for seventy-two hours in porcine and murine peritoneal macrophages and in the J774A.1 macrophage cell line, although there was a noticeable reduction in the number of *C. jejuni* recovered from the three phagocytic cell types at seventy-two hours post-inoculation. Similar findings were reported by Kiehlbauch et al. (1985) when the survival of the *C. jejuni* 2964 clinical isolate was examined in different macrophage lineages (Kiehlbauch et al. 1985). This group was able to recover the clinical *C. jejuni* isolate from three phagocytic cell types over a six-day period. Together, these findings support the view that the particular *C. jejuni* isolate, as well as the macrophage cell type, are both important factors involved in intracellular survival.

Role of Phagolysosome Processing in Macrophage Survival. Although an intracellular existence provides bacteria with an unoccupied space and shelter from the host immune system, phagocytosed bacteria must be able to survive antigen processing in professional phagocytes. Reactive products such as the superoxide radical and hydrogen peroxide are produced in the phagolysosome in response to the respiratory burst and are extremely toxic to microorganisms. However, intracellular bacteria produce proteins that can effectively neutralize these toxic products (De Groote et al. 1997). To address the role of superoxide dismutase in *C. jejuni* intra-macrophage survival, assays were performed with a *C. jejuni sodB* mutant and the J774A.1 macrophage cell-line (Joens et al., unpublished data). A *sodB* mutant created in *C. jejuni* was able to survive in the J774A.1 macrophage cell-line at a level equal to that of the *C. jejuni* wild-type strain. In contrast, catalase production was shown to be important in *C. jejuni* intra-macrophage survival (Day et al. 2000). More specifically, a *katA* mutant of *C. jejuni* was found to be more susceptible to killing by J774A.1 cells than was the *C. jejuni* wild-type isolate. When the respiratory burst or production of nitric oxide were inhibited, the *C. jejuni katA* mutant and wild-type isolate were recovered in equal numbers from the J774A.1 cells following a seventy-two-hour period of incubation.

Many bacteria are able to circumvent antigen processing by altering the intracellular trafficking of the phagosome. *Mycobacteria* spp. prevent acidification of the phagosome, thereby preventing phagosome-lysosome fusion. *Salmonella enterica* also interfere with antigen processing and the endocytic pathway by preventing the maturation of the phagosome. Pitts et al. analyzed the intracellular trafficking of a *C. jejuni* clinical isolate in J774A.1 macrophages (Pitts et al., unpublished data). Macrophages were infected with the *C. jejuni* isolate at a multiplicity of infection of ten bacteria per macrophage and the phagosome examined for the presence of early and late protein markers over a seventy-two-hour time period. This experiment revealed that approximately 80–85 percent of the *C. jejuni* internalized co-localized with early and late proteins of the endocytic pathway, and that 90 percent of the phagosomes were acidified as determined by staining with lysotracker dye. Thus, the majority of the bacteria appear to be processed through the normal endocytic pathway. It is not known whether the remaining bacteria modify the phagosome in a manner that promotes their survival.

IN VIVO COLONIZATION MODELS.

Mice and chickens are the two animals most commonly used to study *C. jejuni* colonization. Oral inoculation of most inbred and outbred strains of mice with *C. jejuni* leads to intestinal tract colonization without significant clinical disease (Field et al. 1981; Blaser et al. 1983). However, diarrhea and clinical disease have been reported in scid mice orally infected with *C. jejuni* low passage clinical isolates (Hodgson et al. 1998). Systemic infection has also been reported in mice, with recovery of the organism from extraintestinal sites (Vuckovic et al. 1998). An early debate focused on whether *C. jejuni* adhere to the mouse intestinal mucus or directly to the intestinal epithelial cells. However, a reduction in the duration of mouse intestinal colonization was observed upon inoculation of mice with a *C. jejuni* cell-binding factor (*peb1A*) mutant when compared to the *C. jejuni* wild-type isolate (Pei et al. 1998). This finding suggests that *C. jejuni* are capable of binding to the intestinal cells of mice. Two of the advantages of the mouse model are that the system is relatively inexpensive to maintain and operate, and various mutant lines are available for study. In addition, numerous reagents (such as monoclonal antibodies) are available to study the host response following colonization with *C. jejuni*.

C. jejuni cells normally colonize chicks ten to twenty-one days after hatching, following the depletion of maternal antibody. Experimentally, chicks are colonized with 1×10^{10} *C. jejuni* per gram of fecal content within three days of oral inoculation (Wassenaar et al. 1993). Colonization occurs throughout the bird's intestinal tract, with the highest number of cells being found in the cecum. Similar to *C. jejuni*-colonized mice, chickens colonized with *C. jejuni* are most often asymptomatic (Shanker et al. 1990). It is unclear whether *C. jejuni* bind to the mucus or mucus-producing epithelial cells in the chick intestinal tract. However, Ziprin et al. (1999) noted that a *C. jejuni* CadF mutant did not colonize the cecum of newly hatched leghorn chickens (Ziprin et al. 1999). CadF is a *C. jejuni* 37 kDa outer membrane protein conserved among *C. jejuni* and *C. coli* isolates (Konkel et al. 1997). There are reports of extraintestinal recovery of *C. jejuni* in colonized chicks (Beery et al. 1988), but whether this dissemination results from the organism's translocation or direct invasion of intestinal epithelium is not known. Because poultry are naturally colonized with *C. jejuni*, this animal model seems ideal to study the mechanism of *Campylobacter* adherence. The major disadvantage of this system is the lack of reagents to study the effect of colonization within the host.

IN VIVO INFECTION MODELS.

As previously discussed, campylobacteriosis in humans is frequently characterized by one to three days of fever followed by intense abdominal pain and watery or hemorrhagic diarrhea (Price et al. 1979; Drake et al. 1981; Blaser et al. 1983; Allos and Blaser 1995). The severity of the disease is strain related and is usually self-limiting, with shedding of *C. jejuni* lasting from days to months (Blaser and Reller 1981; Black et al. 1988; Black et al. 1992). During the acute phase of the disease, lesions can be found throughout the intestine, but are more prominent in the terminal ileum and colon. Endoscopic lesions range from hyperemia, edema to frank hemorrhage with ucleration (Lambert et al. 1979; Colgan et al. 1980; Lambert et al. 1982; McKendrick et al. 1982; Mee et al. 1985). Histological lesions consist of acute inflammation of the mucosa with edema, inflammatory infiltrates into the lamina propria, and crypt abscesses (Van Spreeuwel et al. 1985).

Although *C. jejuni* has been isolated from the feces of many animals, most animals do not suffer from campylobacteriosis.

Small-Animal Models. The most frequently used laboratory animal model to study *Campylobacter* pathogenesis has been the young weanling ferret (Fox et al. 1987; Bell and Manning 1991). Oral inoculation with *C. jejuni* results in intestinal colonization lasting from two to twelve days with the presence of a mild to moderate diarrhea. The diarrhea, sometimes mucoid with occult blood, can last for two to three days (Fox et al. 1987). Histologically, the lesions seen in ferrets infected with *C. jejuni* suggest a mild colitis. This model has been used to assess the pathogenicity of a variety of *C. jejuni* mutants including *pspA*, *cheY*, *virB*11 and *astA*, as well as the *C. jejuni* NCTC 11168 sequence strain and the strain 81-176 (Doig et al. 1996; Yao et al. 1997; Bacon et al. 2000). A reduction in virulence (assessed by the absence of diarrhea) of *C. jejuni* *pspA*, *cheY*, and *virB*11 mutants was observed when compared to the parental *C. jejuni* 81-176 isolate. CheY is involved in *C. jejuni* chemotaxis (Yao et al. 1997). The function of the PspA protein is not known (Gaynor et al. 2001). The *virB*11 gene product has been linked to type IV secretion, based on a low level of sequence similarity as previously discussed in this review. NCTC 11168 and the mutation in *astA*, however, were found to be avirulent when examined in this model (Yao et al. 1997; Bacon et al. 2000). Although the relevance of the ferret model in terms of human disease has yet to be established, it is certain that large inocula, (10^8 to 10^{11} bacterial cells) are required to induce diarrhea. In addition, control animals may excrete feces containing green mucus, which is a symptom displayed by infected animals (Bell and Manning 1991). Nevertheless, this laboratory model appears to have potential in defining adherence mechanisms used by the bacterium to colonize the intestinal tract of mammals.

Large-Animal Models. Of the large, nonprimate models used to assess the pathogenicity of *Campylobacter* organisms, the use of piglets versus other animals has the advantage in that these animals develop symptoms of infection and disease similar to that observed in humans infected with *C. jejuni*. The reason for this is two-fold: 1) the human and porcine physiological system and gastrointestinal tract are similar and, 2) *Campylobacter*-induced enteritis occurs naturally in weaned pigs (Taylor and Olubunmi 1981). Taylor and Olubunmi (1981) observed inflammatory lesions and shortening of the villous epithelium in the small and large intestines of weaned piglets that resulted from a natural infection with *Campylobacter fetus* subspecies *coli* (Taylor and Olubunmi 1981). Histopathology results from this study were similar to those lesions described in human biopsies from campylobacter-induced enteritis (Price et al. 1979; Van Spreeuwel et al. 1985).

Gnotobiotic pigs have also been used to study the pathogenesis of *Campylobacter* organisms (Boosinger and Powe 1988). In a study by Boosinger and Powe (1988), *C. jejuni*-infected gnotobiotic pigs exhibited watery diarrhea that was yellowish in color two days post-inoculation (Boosinger and Powe 1988). The diarrhea lasted throughout the course of the twelve-day experiment. Lesions were present in the cecum and colon. The lesions were characterized by the sloughed epithelial cells, distended crypts, severe edema of the submucosa, and diffuse infiltration of inflammatory cells. Although the use of gnotobiotic pigs as a model system for campylobacteriosis is artificial because of the lack of intestinal flora, the pigs develop lesions consistent with those in humans. The main disadvantage of using these animals is that they are expensive and require specialized housing.

In 1993, Babakhani et al. described the use of a colostrum-deprived newborn piglet as a model to study the pathogenesis of *C. jejuni* (Babakhani et al. 1993). Although these piglets were free of maternal antibod-

ies, competing intestinal microflora were present. In the study, piglets were inoculated orally and observed for clinical signs over a nine-day period. Watery diarrhea with the presence of blood and mucus was observed one day post-infection. Lesions were detected primarily in the large intestine and consisted of a subacute, diffuse, mild to moderate, erosive colitis and typhlitis. Histologically, mucosal epithelial cells were rounded with exfoliation into the lumen, which resulted in a generalized necrosis of the lamina propria. Extensive infiltrates of inflammatory cells were found in the lamina propria and submucosa. The lumen and the crypts of pigs with severe lesions contained *Campylobacter* organisms.

More recently, the newborn pig has been used to examine the virulence of two different isogenic mutants (Joens, unpublished data). Piglets were infected with a *C. jejuni* M129 *katA* mutant and the isogenic wild-type isolate. The *katA* mutant was able to colonize the inoculated pigs but failed to produce intestinal lesions. In contrast, the *C. jejuni* wild-type isolate produced lesions in both the small and large intestines of the piglets. The lesions were characterized by epithelial cell sloughing, necrosis, inflammatory infiltrates, and occasional hemorrhage (Fig. 21.4). Konkel et al. tested the virulence of a *C. jejuni ciaB* mutant in the piglet model and noted fewer lesions in piglets inoculated with the mutant when compared to the piglets inoculated with the *C. jejuni* F38011 wild-type isolate and *C. jejuni ciaB* isolate complemented *in trans* (unpublished data). The clinical signs and lesions produced by the wild-type strains in these two experiments were similar to those reported in the original study (Babakhani et al. 1993). Collectively, these studies illustrate the usefulness of the piglet model in defining the pathogenesis of campylobacteriosis and confirm its ability to mimic the disease found in humans.

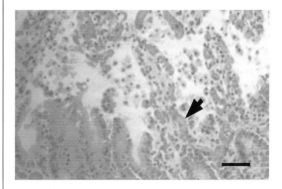

FIG. 21.4—Hematoxylin and eosin-stained section of the small intestine of a *C. jejuni*-inoculated piglet. Arrow indicates an area of luminal exudates and hyperemia. Also note the shortening of the villous epithelium caused by necrosis. Bar = 100 µm.

MOLECULAR APPROACHES USED IN *CAMPYLOBACTER* RESEARCH. Although the vectors used to generate defined *C. jejuni* mutants have been available to investigators for some time (Labigne-Roussel et al. 1987; Yao et al. 1993; Richardson and Park 1997), the tools to generate random mutations in *C. jejuni* have only recently become available (Bleumink-Pluym et al. 1999; Golden et al. 2000; Colegio et al. 2001; Hendrixson et al. 2001). The development of methods to genetically manipulate the organism and the accessibility of the *C. jejuni* NCTC 11168 genome sequence will likely lead to significant advances in the field of *Campylobacter* research. The availability of the *C. jejuni* genome sequence also coincides with advances in bioinformatics, protein chemistry, and microarray technologies. We refer the reader to additional papers on this topic to gain a broader understanding of the implications that sequencing the *C. jejuni* genome will have on future *Campylobacter* research (Dorrell et al. 2001; Wren et al. 2001).

Methods Used to Generate Defined *C. jejuni* Mutants. The method of allelic replacement is the most frequently used technique to generate defined *C. jejuni* mutants (see, for example, Yao and Guerry 1996; Bacon et al. 2000; Guerry et al. 2002; Pumbwe and Piddock 2002). This method employs the use of a *C. jejuni* suicide vector, which by definition can replicate in a heterologous host such as *E. coli* but cannot replicate in *Campylobacter*. The procedure involves ligation of a gene into a suicide vector, and then the disruption of this target gene by the insertion of an antibiotic resistance cassette that functions in *C. jejuni*. Various methods, including high-voltage electroporation, conjugation, or natural transformation, can be used to introduce the suicide vector or just the *Campylobacter* DNA harboring the modified gene into *C. jejuni* (Labigne-Roussel et al. 1988; Miller et al. 1988; Wang and Taylor 1990). If conjugation is to be used to introduce the vector from *E. coli* into *C. jejuni,* it is necessary to use a suicide vector containing additional genetic elements (Labigne-Roussel et al. 1988). Following introduction of the vector into *C. jejuni*, the bacteria are cultured under nonselective conditions to allow for expression of the antibiotic resistance gene. Transformants are then selected by transferring the electroporation or conjugation mixtures onto medium supplemented with the appropriate antibiotic. Mutation of the chromosomal gene is generated by homologous recombination via a double cross-over event, thereby resulting in incorporation of the allele containing the antibiotic resistance cassette into the chromosome. Disruption of the gene of interest can be confirmed by Southern hybridization analysis using a gene-specific probe or PCR using gene-specific primers.

Although allelic replacement has been used extensively to generate *C. jejuni* mutants, a disadvantage of the method is the use of an intermediate host such as *E. coli*, in which the gene of interest is cloned in its entirety. This has been demonstrated to be an obstacle in some instances because the *C. jejuni* gene may be toxic when expressed in *E. coli*. As an alternative to allelic replacement, one strategy to generate a defined mutant involves disrupting a gene by a second type of insertional inactivation (see, for example, Konkel et al. 1997; Konkel et al. 1998; Konkel et al. 1999). More specifically, the chromosomal gene is disrupted by homologous recombination via a single cross-over event between itself and an internal fragment of the target gene (constructed to have deletions at both the 5' and the 3' end of the coding region of the gene) on a suicide vector. In this strategy, the recombinant suicide vector is introduced into *C. jejuni* by electroporation or conjugation. A single cross-over event creating two defective alleles of the target gene is generated in the chromosome. Insertional mutants are initially identified by the acquisition of antibiotic resistance located on the suicide vector. The specific mutation can be confirmed by either Southern hybridization analysis using a gene-specific probe or PCR using gene-specific primers.

Assessment of gene function in *C. jejuni* is most often performed by comparing the phenotypes of a mutant with the wild-type isolate. However, insertion mutagenesis as described previously can result in disruption of downstream genes located in the same operon (polar effect). To alleviate this concern, investigators have complemented the modified gene by introducing a functional copy of the gene into the mutated isolate *in trans* using a shuttle vector. The term shuttle vector is used to define plasmids that are capable of replicating in *C. jejuni* and a heterologous host such as *E. coli*. The first *Campylobacter* shuttle vector was generated by Labigne-Roussel et al. (Labigne-Roussel et al. 1987). Today, *Campylobacter* spp. shuttle vectors are available that harbor a variety of antibiotic resistance genes and other markers (Wang and Taylor 1990; Yao et al. 1993; Yao and Guerry 1996; Park 1999; Miller et al. 2000). Some of these vectors also harbor the genes allowing for their introduction into *Campylobacter* organisms via conjugation.

An alternative technique for generating nonpolar, in-frame mutations has recently been described by Hendrixson et al. (2001) and has its basis in two previously published methods (Higuchi 1990; Pei and Blaser 1993; Skorupski and Taylor 1996). The technique was demonstrated by the disruption of *fliA* and *rpoN,* which encode specialized sigma factors*,* as well as in-frame deletions in *cetA* (Cj1190c) and *cetB* (Cj1189c), which are proposed to be involved in an energy taxis response. Several separate cloning steps were required for this method. Initially, a streptomycin-resistant mutant of *C. jejuni* strain NCTC 11168 was selected on a medium containing a gradient of streptomycin. The *rpsL* gene, conferring streptomycin-resistance, was amplified from the mutant by the polymerase chain reaction and cloned into the suicide vector pUC19. A *C. jejuni* 81-176 *rpsL*[Sm] mutant was generated by allellic exchange (double cross-over) of this fragment, selecting for streptomycin-resistant mutants. Following the generation of the *C. jejuni* 81-176 *rpsL*^{Sm}, DNA fragments containing the entire coding region of *fliA, rpoN, cetA,* and *cetB,* as well as 500 bp of upstream and downstream flanking DNA, were PCR amplified from *C. jejuni* 81-176 and cloned individually into a suicide plasmid (pUC19 or pBR322). Target genes were disrupted by insertion of a *cat-rpsL* (*cat,* chloramphenicol-resistance) cassette into an individual target gene harboured within a suicide plasmid. In addition to the chloramphenicol-resistance marker, each of these constructs had the NCTC 11168 wild-type (streptomycin-sensitive) *rpsL* gene. Chloramphenicol-resistant transformants were generated in 81-176 *rpsL*^{Sm} using each recombinant plasmid. Thus, through a double cross-over event, the wild-type target gene was replaced with the mutant allele. Transformants at this stage were referred to as "intermediate strains" with respect to their streptomycin sensitivity. Intermediate level of resistance to streptomycin is the result of the recessive nature of the *rpsL*^{Sm} mutation. The next step of the procedure involved generating additional suicide vectors harboring in-frame deletions of each of the target genes. The in-frame deletions were constructed using the SOEing reaction (splicing by overlap extension). The SOEing reaction involves amplifying 5′ and 3′ fragments of a gene by PCR, annealing the two amplified fragments, and a second round of PCR amplification to generate fusions of the upstream and downstream DNA segments of each gene (Higuchi 1990). Products of the SOEing reactions were cloned into a suicide plasmid, and the resultant recombinant plasmids electroporated into the corresponding *C. jejuni* 81-176 *rpsL* ^{Sm} intermediate strain. The transformants, in which allelic replacement had again occurred, were identified based on their resistance to streptomycin (loss of the NCTC 11168 *rpsL* gene) and sensitivity to chloramphenicol. Each mutation was confirmed by PCR using gene-specific primers.

Methods Used to Generate Random *C. jejuni* Mutants (Transposons). None of the previous methods described allow for the random mutagenesis of the *C. jejuni* genome. This lack has severely hampered the genetic analysis of pathogenesis in this organism. Golden et al. (2000) was the first to demonstrate random transposon mutagenesis in *C. jejuni* using a *mariner*-based transposon (Golden et al. 2000). The construction of pOTHM mini-transposon *mariner*-based transposon vector, which is unable to replicate in *C. jejuni*, involved replacement of the resident chloramphenicol resistance cassette with one that would be functional in both *E. coli* and *C. jejuni*. In addition, a *C. jejuni*-specific promoter was cloned upstream of the *Himar1* transposase gene to drive its expression in *C. jejuni*. Electroporation of *C. jejuni* with the pOTHM vector typically resulted in 37 ± 4 chloramphenicol-resistant colonies. Southern blot analysis of nineteen transformants concluded that random insertion within the chromosome of *C. jejuni* had indeed occurred. Additionally, twelve mutants were selected and the

mini-transposon chromosomal junctions were sequenced to map the sites of insertion of the *cat* gene. Transposon insertion appeared random in the *C. jejuni* genome without a site preference other than the invariant TA dinucleotide directing insertion.

Hendrixson et al. (2001) also generated random mutants in *C. jejuni* 81-176 using a derivative of the *mariner*-based transposon (Hendrixson et al. 2001). Two mini-transposons were constructed, termed pFalcon (Kan^r) and pEnterprise (Cm^r). In contrast to Golden et al., the transposition reactions were performed *in vitro* with the pFalcon and pEnterprise transposons and purified chromosomal DNA from *C. jejuni* strain 81-176, whereby the purified *Himar1* MarC9 transposase enzyme was added separately to each reaction. The transposed chromosomal DNA was subsequently introduced into *C. jejuni* 81-176 by biphasic natural transformation. Typically, 600–1,450 mutants were obtained per transposition reaction with both the pFalcon and pEnterprise transposons. The investigators also demonstrated the functionality of the pFalcon mutagenesis system by identifying mutants defective in motility and sequencing flanking regions within the chromosome by inverse PCR. The sequences obtained were used to search the *C. jejuni* NCTC 11168 genome to determine the location of transposition events. Mutations were identified in genes encoding flagellar structural components, motor proteins, chemotaxis proteins, and proteins involved in flagellar gene transcription (*fliA* and *rpoN*).

Colegio et al. (2001) developed an *in vitro* transposon system for the generation of random mutants in *C. jejuni* utilizing a strategy based on the *Staphylococcus aureus* transposon Tn*552* (Colegio et al. 2001). This transposon requires the coding of a single-subunit transposase and a single accessory protein for *in vivo* transposition. The Tn*552* transposon has an advantage in that a transposition event requires only the 48-bp terminal inverted repeats. Additionally, Tn*552* also displays no target site preference for insertion. A derivative of Tn*552* was constructed by replacing the *cat* cassette with an *aphA*-3 kanamycin resistance gene. The resultant plasmid, harboring Tn*552* with the new *aphA*-3 gene, was referred to as pSB1698. Transposition reactions were initially performed *in vitro* using a protocol similar to that used by Hendrixson et al. (2001). Purified chromosomal DNA from *C. jejuni* 81-176 was mixed with agarose gel-purified Tn552 harboring the *aphA*-3 gene from pSB1698 and the purified His-tagged TnpA transposase enzyme. The reaction mixture was allowed to incubate at 37°C for one hour, ethanol precipitated, and then introduced into *C. jejuni* 81-176 via electroporation. Approximately one hundred kanamycin-resistant mutants were recovered from the reaction. The relatively low recovery of mutants was hypothesized to be a result of restriction of the *E. coli*-grown pSB1698 transposon vector. Therefore, the Tn*552* transposon with *aphA*-3 kanamycin resistance gene was cloned into a *Campylobacter* shuttle plasmid, resulting in plasmid pSB1699. The pSB1699 shuttle plasmid was introduced into *C. jejuni* 81-176 via conjugation, and then purified. The transposition reactions were subsequently performed with *C. jejuni* 81-176 chromosomal DNA, agarose gel-purified Tn552 harboring the *aphA*-3 gene from pSB1699, and the purified His-tagged TnpA transposase enzyme. Approximately 3,000–8,000 *C. jejuni* kanamycin-resistant colonies were obtained per reaction. The insertions appeared at random sites within the *C. jejuni* genome as judged by sequencing and searching the *C. jejuni* NCTC 11168 chromosome with the sequences obtained. The investigators also identified nine nonmotile mutants by screening 205 of the kanamycin-resistant colonies. Sequence data from six of the nine nonmotile mutants revealed transposon insertions in genes known or thought to be associated with motility (*flaD*, *cheA*, *fliP*, *fliY*, *rpoN*, and *flgE*). The remaining three mutants had insertions in genes of unknown function.

Novel Molecular Approaches. As stated previously, the availability of the *C. jejuni* genome sequence will have a significant impact on our understanding of *Campylobacter* biology because it provides new investigative approaches. For example, a comparison of *C. jejuni* strains by whole genome-microarray analysis revealed extensive genetic diversity (Dorrell et al. 2001). Based on the comparison of eleven *C. jejuni* strains, 21 percent of the genes in the *C. jejuni* NCTC1168 sequence strain were proposed to be dispensable because they were either absent or highly divergent among the other isolates included in the study. Dorrell et al. (2001) also noted that many of the virulence genes identified to date are conserved in *C. jejuni* strains (Dorrell et al. 2001). Included among the strain-variable genes are those that are involved in iron acquisition, DNA restriction/modification, sialylation, flagellar biosynthesis, lipo-oligosaccharide biosynthesis, and capsular biosynthesis (Parkhill et al. 2000; Wren et al. 2001). The investigators proposed that the variable genes may encode factors that contribute to different disease presentations and that allow the organism to establish unique ecological niches.

SUMMARY. In the past few years, significant advances have been made in the tools used to generate random *C. jejuni* mutants. The ability to generate large numbers of random *C. jejuni* mutants will accelerate the rate of research progress in this area, ultimately resulting in a better understanding of the biology of *Campylobacter* organisms. It is also likely that our understanding of the pathogenic mechanisms utilized by *C. jejuni* organisms will increase given that researchers have access to the *C. jejuni* genome sequence and advances in bioinformatics, protein chemistry, and microarray technologies. Still needed is the development of a small-animal model that can be widely utilized by *Campylobacter* researchers. Given the prevalence of *C. jejuni* infections world-wide and the interest in understanding the biology of this organism, the next few years will certainly prove to be an exciting time in the field of *Campylobacter* research.

REFERENCES

Adak, G.K., J.M. Cowden, S. Nicholas, and H.S. Evans. 1995. The Public Health Laboratory Service national case-control study of primary indigenous sporadic cases of campylobacter infection. *Epidemiol. Infect.* 115: 15–22.

Allos, B.M. 2001. *Campylobacter jejuni* infections: update on emerging issues and trends. *Clin. Infect. Dis.* 32:1201–1206.

Allos, B.M., and M.J. Blaser. (1995). *Campylobacter jejuni* and the expanding spectrum of related infections. *Clin. Infect. Dis.* 20:1092–1099.

Alrutz, M.A., and R.R. Isberg. 1998. Involvement of focal adhesion kinase in invasin-mediated uptake. *Proc. Natl. Acad. Sci. USA* 95:13658–13663.

Altekruse, S.F., N.J. Stern, P.I. Fields, and D.L. Swerdlow. 1999. *Campylobacter jejuni*—an emerging foodborne pathogen. *Emerg. Infect. Dis.* 5(1):28–35.

Altekruse, S.F., D.L. Swerdlow, and N.J. Stern. 1998. Microbial food borne pathogens: *Campylobacter jejuni*. *Vet. Clinics of N. Am.: Food Anim. Pract.* 14(1):31–40.

Anonymous. 2001. Surveillance Data. *N. Zealand Publ. Hlth. Rep.* 8(2):6.

Babakhani, F.K., G.A. Bradley, L.A. Joens. 1993. Newborn piglet model for campylobacteriosis. *Infect. Immun.* 61(8):3466–3475.

Bacon, D.J., R.A. Alm, D.H. Burr, L. Hu, D.J. Kopecko, C.P. Ewing, T.J. Trust, and P. Guerry. 2000. Involvement of a plasmid in virulence of *Campylobacter jejuni* 81-176. *Infect. Immun.* 68:4384–4390.

Banfi, E., M. Cinco, and G. Zabucchi. 1986. Phagocytosis of *Campylobacter jejuni* and *C. coli* by peritoneal macrophages. *J. Gen. Microbiol.* 132:2409–2412.

Beery, J.T., M.B. Hugdahl, and M.P. Doyle.. 1988. Colonization of gastrointestinal tracts of chicks by *Campylobacter jejuni*. *Appl. Environ. Microbiol.* 54(10):2365–2370.

Bell, J.A., and D.D. Manning (1991). Evaluation of *Campylobacter jejuni* colonization of the domestic ferret intestine as a model of proliferative colitis. *Am. J. Vet. Res.* 52:826–832.

Black, R.E., M.M. Levine, M.L. Clements, T.P. Huges, and M.J. Blaser. 1988. Experimental *Campylobacter jejuni* infections in humans. *J. Infect. Dis.* 157(3):472–479.

Black, R.E., D. Perlman, M.L. Clments, M.M. Levine, and M.J. Blaser. 1992. Human volunteer studies with *Campylobacter jejuni*. In *Campylobacter jejuni*: current status and future trends. I. Nuchamkin, M.J. Blaser, and L. Tompkins, eds. Washington, D.C.: American Society for Microbiology, ,207–215.

Blaser, M. J. 1997. Epidemiology and clinical features of *Campylobacter jejuni* infections. *J. Infect. Dis.* 176 (Suppl 2):S103–105.

Blaser, M.J., I.D. Berkowitz, F.M. LaForce, J. Cravens, L.B. Reller, and W.L. Wang. 1979. Campylobacter enteritis:clinical and epidemiologic features. *Ann. Int. Med.* 91:179–185.

Blaser, M.J., R.B. Parsons, and W. WangL. 1980. Acute colitis caused by *Campylobacter fetus* subsp. jejuni. *Gastroenterol* 78(3):448–453.

Blaser, M.J., G. Perez Perez, P.F. Smith, C. Patton, F.C. Tenover, A.J. Lastovica, and W.-l WangL. 1986. Extraintestinal *Campylobacter jejuni* and *Campylobacter coli* infections: host factors and strain characteristics. *J. Infect. Dis.* 153(3):552–559.

Blaser, M.J., and L.B. Reller. 1981. *Campylobacter* enteritis. *N. Engl. J. Med.* 305(24):1444–1452.

Blaser, M.J., J.G. Wells, R.A. Feldman, R.A. Pollard, and J. Allen. 1983. *Campylobacter* enteritis in the United States. A multicenter study. *Ann. Int. Med.* 98:360–365.

Bleumink-Pluym, N.M.C., F. Verschoor, W. Gaastra, B.A.M. van der Zeijst, and B.N. Fry. 1999. A novel approach for the construction of a *Campylobacter* mutant library. *Microbiol.* 145(Part 8):2145–2151.

Boosinger, T.R., and T.A. Powe. 1988. *Campylobacter jejuni* infections in gnotobiotic pigs. *Am. J. Vet. Res.* 49(4):456–458.

Boucher, S.N., E.R. Slater, A.H. Chamberlain, and M.R. Adams. 1994. Production and viability of coccoid forms of *Campylobacter jejuni*. *J. Appl. Microbiol.* 77:303–307.

Brás, A.M., and J.M. Ketley. 1999. Transcellular translocation of *Campylobacter jejuni* across human polarised epithelial monolayers. *FEMS Microbiol. Lett.* 179(2):209–215.

Brieseman, M.A. (1990). A further study of the epidemiology of *Campylobacter jejuni* infections. *N. Zealand Med. J.* 103:207–209.

Butzler, J.P. and M.B. Skirrow. 1979. Campylobacter enteritis. *Clin. Gastroenterol.* 8:737–765.

Chang, N., and D.E. Taylor. 1990. Use of pulsed-field gel electrophoresis to size genomes of *Campylobacter* species and to construct a *Sal*I map of *Campylobacter jejuni* UA580. *J. Bacteriol.* 172(9):5211–5217.

Colegio, O.R., T.J.I. Griffin, N.D. Grindley, and J.E. Galen. 2001. In vitro transposition system for efficient generation of random mutants of *Campylobacter jejuni*. *J. Bacteriol.*183(7):2384–2388.

Colgan, T., J.R. Lambert, A. Newman, and S.C. Luk. 1980. *Campylobacter jejuni* enterocolitis. A clinicopathologic study. *Arch. Pathol. Lab. Med.* 104:571–574.

Cowden, J. 1992. Campylobacter:epidemiological paradoxes. *Brit. Med. J.* 305:132–133.

Cruz, N., Q. Lu, X. Alvarez, and E.A. Deitch. 1994. Bacterial translocation is bacterial species dependent: results using the human Caco-2 intestinal cell line. *J. Trauma* 36(5):612–616.

Day, W.A.J., J.L. Sajecki, T.M. Pitts, and L.A. Joens. 2000. Role of catalase in *Campylobacter jejuni* intracellular survival. *Infect. Immun.* 68:6337–6345.

De Groote, M.A., U.A. Ochsner, M.U. Shiloh, C. Nathan, J.M. McCord, M.C. Dinaure, S.J. Libby, A. Vasquez-Torres, Y. Xu, and F.C. Fang. 1997. Periplasmic superoxide dismutase protects *Salmonella* from products of phagocyte NADPH-oxidase and nitric oxidase synthase. *Proc. Natl. Acad. Sci. USA* 94(25):13997–14001.

de Melo, M.A., G. Gabbiani, and J.-C. Pechère. 1989. Cellular events and intracellular survival of *Campylobacter jejuni* during infection of HEp-2 cells. *Infect. Immun.* 57(7):2214–2222.

de Melo, M.A., and J.-C. Pechère. 1990. Identification of *Campylobacter jejuni* surface proteins that bind to eucaryotic cells *in vitro*. *Infect. Immun.* 58(6):1749–1756.

Dekeyser, P., M. Gossuin-Detrain, J.P. Butzler, and J. Sternon. 1972. Acute enteritis due to related vibrio: first positive stool cultures. *J. Infect. Dis.* 125:390–392.

Deming, M.S., R.V. Tauxe, P.A. Blake, S.E. Dixon, B.S. Fowler, T.S. Jones, E.A. Lockamy, C.M. Patton, and R.O. Sikes. 1987. *Campylobacter* enteritis at a university: transmission from eating chicken and from cats. *Am. J. Epidemiol.* 126(3):526–534.

Doig, P., N. Kinsella, P. Guerry, and T.J. Trust. 1996. Characterization of a post-translational modification of *Campy-*lobacter flagellin: identification of a sero-specific glycosyl moiety. *Molec. Microbiol.* 19(2):379–387.

Doig, P., R. Yao, D.H. Burr, P. Tuerry, and T.J. Trust. 1996. An environmentally regulated pilus-like appendage involved in *Campylobacter* pathogenesis. *Molec. Microbiol.* 20:885–894.

Dorrell, N., J.A. Mangan, K.G. Laing, J. Hinds, D. Linton, H. Al-Ghusein, B.G. Barrell, J. Parkhill, N.G. Stoker, A.V. Karlyshev, P.D. Butcher, and B.W. Wren. 2001. Whole genome comparison of *Campylobacter jejuni* human isolates using a low-cost microarray reveals extensive genetic diversity. *Genome Res.* 11:1706–1715.

Drake, A.A., M.J. Gilchrist, J.A.n. Washington, K.A. Huizenga, and R.E. Van Scoy. 1981. Diarrhea due to *Campylobacter fetus* subspecies *jejuni*. A clinical review of 63 cases. *Mayo Clinic Proc.* 56:414–423.

Duffy, M.C., J.B. Benson, and S.J. Rubin. 1980. Mucosal invasion in *Campylobacter* enteritis. *Am. J. Clin. Pathol.* 73(5):706–708.

DuPont, H.L. 1995. Antimicrobial-resistant *Campylobacter* species—a new threat to travelers to Thailand. *Clin. Infect. Dis.* 21:542–543.

Eberhardt-Phillips, J., N. Walker, N. Garrett, D. Bell, D. Sinclair, W. Rainger, and M. Bates. 1997. Campylobacteriosis in New Zealand: results of a case-control study. *J. Epidemiol. Commun. Hlth.* 51:686–691.

Engberg, J., F.M. Aarestrup, D.E. Taylor, P. Gerner-smidt, and I. Nachamkin. 2001. Quinolone and macrolide resistance in *Campylobacter jejuni* and *C. coli*: resistance mechanisms and trends in human disease. *Emerg. Infect. Dis.* 7(1):24–34.

Everest, P.H., H. Goossens, J.P. Butzler, D. Lloyd, S. Knutton, J.M. Ketley, and P.H. Williams. 1992. Differentiated Caco-2 cells as a model for enteric invasion by *Campylobacter jejuni* and *C. coli*. *J. Med. Microbiol.* 37:319–325.

Fauchere, J.L., A. Rosenau, M. Veron, E.N. Moyen, S. Richard, and A. Pfister. 1986. Association with HeLa cells of *Campylobacter jejuni* and *Campylobacter coli* isolated from human feces. *Infect. Immun.* 54(2):283–287.

Federighi, M., J.L. Tholozan, J.M. Cappelier, J.P. Tissier, and J. Jouve. (1998). Evidence of non-coccoid viable but non-culturable *Campylobacter jejuni* cells in microcosm water by direct viable count, CTC-DAPI double staining, and scanning electron microscopy. *Food Microbiol.* 15:539–550.

Field, L.H., J.L. Underwood, S.M. Payne, and L.J. Berry. (1991). Virulence of *Campylobacter jejuni* for chicken embryos is associated with decreased bloodstream clearance and resistance to phagocytosis. *Infect. Immun.* 59:1448–1456.

Field, L.H., J.L. Underwood, L.M. Pope, and L.J. Berry. 1981. Intestinal colonization of neonatal animals by *Campylobacter fetus* subsp. *jejuni*. *Infect. Immun.* 33:884–892.

Finlay, B.B. and S. Falkow (1990). *Salmonella* interactions with polarized human intestinal Caco-2 epithelial cells. *J. Infect. Dis.* 162:1096–1106.

Finlay, B.B., B. Gumbiner, and S. Falkow. 1988. Penetration of Salmonella through polarized Madin-Darby canine kidney epithelial cell monolayer. *J. Cell Biol.* 107:221–230.

Fox, J.G., J.I. Ackerman, N. Taylor, M. Claps, and J.C. Murphy. 1987. *Campylobacter jejuni* infection in the ferret: an animal model of human campylobacteriosis. *Am. J. Vet. Res.* 48:85–90.

Gaynor, E.C., N. Ghori, and S. Falkow. 2001. Bile-induced "pili" in *Campylobacter jejuni* are bacteria-independent artifacts of the culture medium. *Molec. Microbiol.* 39:1546–1549.

Gibreel, A., E. Sjörgren, B. Kaijser, B. Wretlind, and O. Sköld. 1998. Rapid emergence of high level resistance to quinolones in *Campylobacter jejuni* associated with mutational changes in *gyrA* and *parC*. *Antimicrobial Agents Chemother.* 42(12):3276–3278.

Glass, R.I., B.J. Stoll, M.I. Hug, M.J. Struelens, M.J. Blaser, and A.K.M.G. Kibrya. 1983. Epidemiologic and clinical features of endemic *Campylobacter jejuni* infections in Bangladesh. *J. Infect. Dis.* 148:292–296.

Golden, N.J., A. Camilli, and D.W.K. Acheson. 2000. Random transposon mutagenesis of *Campylobacter jejuni. Infect. Immun.* 68(9):5450–5453.

Grant, C.C.R., M.E. Konkel, W.J. Cieplak, and L.S. Tompkins. 1993. Role of flagella in adherence, internalization, and translocation of *Campylobacter jejuni* in nonpolarized and polarized epithelial cell cultures. *Infect. Immun.* 61(5):1764–1771.

Guerry, P., C.P. Ewing, T.E. Hickey, M.M. Prendergast, and A.P. Moran. 2000. Sialylation of lipooligosaccharide cores affects immunogenicity and serum resistance of *Campylobacter jejuni. Infect. Immun.* 68(12):6656–6662.

Guerry, P., C.M. Szymanski, M.M. Prendergast, T.E. Hickey, C.P. Ewing, D.L. Pattarini, and M.A. P. 2002. Phase variation of *Campylobacter jejuni* 81-176 lipooligosaccharide affects ganglioside mimicry and invasiveness in vitro. *Infect. Immun.* 70(2):787–793.

Harris, N.V., N.S. Weiss, and C.M. Nolan. (1986). The role of poultry and meats in the etiology of *Campylobacter jejuni/coli* enteritis. *Am J. Pub. Hlth.* 76:407–411.

Harvey, P., T. Battle, and S. Leach. 1999. Different invasion phenotypes of *Campylobacter* isolates in Caco-2 cell monolayers. *J. Med. Microbiol.* 48:461–469.

Hendrixson, D.R., B.J. Akerley, and V. J. Dirita. 2001. Transposon mutagenesis of *Campylobacter jejuni* identifies a bipartite energy taxis system required for motility. *Molec. Microbiol.* 40:214–224.

Higuchi, R. (1990). Recombinant PCR. PCR protocols: A guide to methods and applications. M.A. Innis, D.H. Gelfand, J.J. Sninsky, and T.J. White, eds. San Diego, CA: Academic Press, Inc., 177–183.

Hodgson, A.E., B.W. McBride, M.J. Hudson, G. Hall, and S.A. Leach. 1998. Experimental campylobacter infection and diarrhoea in immunodeficient mice. *J. Med. Microbiol.* 47:799–809.

Hopkins, R.S., R. Olmsted, and G.R. Istre. 1984. Endemic *Campylobacter jejuni* infection in Colorado: identified risk factors. *Am. J. Publ.Hlth.* 74(3):249–250.

Humphrey, T.J., and R.J. Hart. 1988. *Campylobacter* and *Salmonella* contamination of unpasteurized cows' milk on sale to the public. *J. Appl. Microbiol.* 65:463–467.

Isberg, R.R. and G. Tran Van Nhieu. 1994. Two mammalian cell internalization strategies used by pathogenic bacteria. *Ann. Rev. Genet.* 28:395–422.

Jensen, L.B. and F.M. Aarestrup. 2001. Macrolide resistance in *Campylobacter coli* of animal origin. *Antimicrobial Agents Chemother.* 45(1):371–372.

Jin, S., A. Joe, J. Lynett, E.K. Hani, P. Sherman, and V.L. Chan. 2001. JlpA, a novel surface-exposed lipoprotein specific to *Campylobacter jejuni*, mediates adherence to host epithelial cells. *Molec. Microbiol.* 39:1225–1236.

Kapperud, G., and S. Aasen. 1992. Descriptive epidemiology of infections due to thermotolerant *Campylobacter* spp. in Norway, 1979–1988. *Acta Patholog. Microbiol. Immunol. Scand.* 100:883–890.

Karlyshev, A.V., J. Henderson, J.M. Ketley, and B.W. Wren. 1998. An improved physical and genetic map of *Campylobacter jejuni* NCTC 11168 (UA580). *Microbiol.* 144:503–508.

Karmali, M.A., and P.C. Fleming. 1979. *Campylobacter* enteritis in children. *J. Ped.* 94:527–533.

Kelle, K., J.-M. Pages, and J.M. Bolla. 1998. A putative adhesin gene cloned from *Campylobacter jejuni. Res. Microbiol.* 149:723–733.

Kiehlbauch, J.A., R.A. Albach, L.L. Baum, and K.-P. Chang. 1985. Phagocytosis of *Campylobacter jejuni* and its intracellular survival in mononuclear phagocytes. *Infect. Immun.* 48:446–451.

Konkel, M.E., F. Babakhani, and L.A. Joens. 1990. Invasion-related antigens of *Campylobacter jejuni. J. Infect. Dis.* 162:888–895.

Konkel, M.E., and W. Cieplak, Jr. 1992. Altered synthetic response of *Campylobacter jejuni* to cocultivation with human epithelial cells is associated with enhanced internalization. *Infect. Immun.* 60(11):4945–4949.

Konkel, M.E., M.D. Corwin, L.A. Joens, and W. Ciplak, Jr. 1992. Factors that influence the interaction of *Campylobacter jejuni* with cultured mammalian cells. *J. Med. Microbiol.* 37:30–37.

Konkel, M.E., S.G. Garvis, S.L. Tipton, D.E. Anderson, Jr., and W. Cieplak, Jr. 1997. Identification and molecular cloning of a gene encoding a fibronectin-binding protein (CadF) from *Campylobacter jejuni. Molec. Microbiol.* 24(5):953–963.

Konkel, M.E., S.A. Gray, B.J. Kim, S.G. Garvis, and J. Yoon. 1999. Identification of the enteropathogens *Campylobacter jejuni* and *Campylobacter coli* based on the *cadF* virulence gene and its product. *J. Clin Microbiol.* 37(3):510–517.

Konkel, M.E., S.F. Hayes, L.A. Joens, and W. Cieplak, Jr. 1992. Characteristics of the internalization and intracellular survival of *Campylobacter jejuni* in human epithelial cell cultures. *Microbial Pathog.* 13:357–370.

Konkel, M.E., and L.A. Joens. 1989. Adhesion to and invasion of HEp-2 cells by *Campylobacter* spp. *Infect. Immun.*57:2984–2990.

Konkel, M.E., B.J. Kim, J.D. Klena, C.R. Young, and R. Ziprin. 1998. Characterization of the thermal stress response of *Campylobacter jejuni. Infect. Immun.* 66(8):3666–3672.

Konkel, M.E., B.J. Kim, V. Rivera-Amill, and S.G. Garvis. 1999. Bacterial secreted proteins are required for the internalization of *Campylobacter jejuni* into cultured mammalian cells. *Molec. Microbiol.* 32(4):691–701.

Konkel, M.E., D.J. Mead, and W. Cieplak, Jr. 1993. Kinetic and antigenic characterization of altered protein synthesis by *Campylobacter jejuni* during cultivation with human epithelial cells. *J. Infect. Dis.* 168:948–954.

Konkel, M.E., D.J. Mead, S.F. Hayes, and W. Cieplak, Jr. 1992. Translocation of *Campylobacter jejuni* across human polarized epithelial cell monolayer cultures. *J. Infect. Dis.* 166:308–315.

Kops, S.K., D.K. Lowe, W.M. Bement, and A.B. West. 1996. Migration of *Salmonella typhi* through intestinal epithelial monolayers: an *in vitro* study. *Microbiol. Immunol.* 40:799–811.

Kuschner, R.A., A.F. Trofa, R.J. Thomas, C.W. Hoge, C. Pitarangsi, S. Amato, R.P. Olfason, P. Echeverria, J.C. Sadoff, and D.N. Taylor. 1995. Use of azithromycin for the treatment of campylobacter enteritis in travelers to Thailand, an area where ciprofloxacin resistance is prevalent. *Clin. Infect. Dis.* 21:536–541.

Labigne-Roussel, A., P. Courcoux, and L. Tompkins. 1988. Gene disruption and replacement as a feasible approach for mutagenesis of *Campylobacter jejuni. J. Bacteriol.* 170(4):1704–1708.

Labigne-Roussel, A., J. Harel, and L. Tompkins. 1987. Gene transfer from *Escherichia coli* to *Campylobacter* species: development of shuttle vectors for genetic analysis of *Campylobacter jejuni*. *J. Bacteriol.* 169(11):5320–5323.

Lambert, M., E. Marion, E. Coche, and J.P. Butzler. 1982. *Campylobacter* enteritis and erythema nodosum. *Lancet* i:1409.

Lambert, M.E., P.F. Schofield, A.G. Ironside, and B.K. Mandal. 1979. *Campylobacter* colitis. *Brit. Med. J.* 1:857–859.

Macnab, R.M. 1999. The bacterial flagellum: reversible rotary propellor and type III export apparatus. *J. Bacteriol.* 181:7149–7153.

Madara, J.L. 1998. Regulation of the movement of solutes across tight junctions. *Ann. Rev. Physiol.* 60:143–159.

McKendrick, M.W., A.M. Geddes, and J. Gearty. 1982. Campylobacter enteritis: a study of clinical features and rectal mucosal changes. *Scand. J. Infect. Dis.* 14:35–38.

McNicholas, A.M., M. Bates, E. Kiddle, and J. Wright. 1995. Is New Zealand's recent increase in campylobacteriosis due to changes in laboratory procedures? A survey of 69 medical laboratories. *N. Zealand Med. J.* 108:459–461.

McSweegan, E., and R.I. Walker. 1986. Identification and characterization of two *Campylobacter jejuni* adhesins for cellular and mucous substrates. *Infect. Immun.* 53(1):141–148.

Mee, A.S., M. Shield, and M. Burke. 1985. *Campylobacter* colitis: differentiation from acute inflammatory bowel disease. *J. Royal Soc. Med.* 78:217–223.

Miller, J.F., W.J. Dower, and L.S. Tompkins. 1988. High-voltage electroporation of bacteria: genetic transformation of *Campylobacter jejuni* with plasmid DNA. *Proc. Natl. Acad. Sci. USA* 85:856–860.

Miller, W.G., A.H. Bates, S.T. Horn, M.T. Brandl, M.R. Wachtel, and R.E. Mandrell. 2000. Detection on surfaces and in Caco-2 cells of *Campylobacter jejuni* cells transformed with new *gfp*, *yfp*, and *cfp* marker plasmids. *Appl. Environ. Microbiol.* 66(12):5426–5436.

Mishu, B., A.A. Ilyas, C.L. Koski, F. Vriesendrop, S.D. Cook, F.A. Mithen, and M.A. Blaser. 1993. Serological evidence of previous *Campylobacter jejuni* infection in patients with the Guillain-Barré syndrome. *Ann. Int. Med.* 118:947–953.

Moser, I., W. Schroeder, and J. Salnikow. 1997. *Campylobacter jejuni* major outer membrane protein and a 59-kDa protein are involved in binding to fibronectin and INT 407 cell membranes. *FEMS Microbiol. Lett.* 157:233–238.

Myszewski, M.A., and N.J. Stern. 1991. Phagocytosis and intracellular killing of *Campylobacter jejuni* by elicited chicken peritoneal macrophages. *Avian Dis.* 35:750–755.

Nachamkin, I., B.M. Allos, and T. Ho. 1998. Campylobacter species and Guillain-Barré Syndrome. *Clin. Microbiol. Rev.* 11(3):555–567.

Nataro, J.P., S. Hicks, A.D. Phillips, P.A. Vial, and C.L. Sears. 1996. T84 cells in culture as a model for enteroaggregative *Escherichia coli* pathogenesis. *Infect. Immun.* 64:4761–4768.

Newell, D.G., H. McBride, and J.M. Dolby. 1985. The virulence of clinical and environmental isolates of *Campylobacter jejuni*. *J. Hyg.* 94:45–54.

Oelschlaeger, T.A., P. Guerry, and D.J. Kopecko. 1993. Unusual microtubule-dependent endocytosis mechanisms triggered by *Campylobacter jejuni* and *Citrobacter freundii*. *Proc. Natl. Acad. Sci. USA* 90:6884–6888.

Panigrahi, P., G. Losonky, L.J. DeTolla, and J.G. Morris. 1992. Human immune response to *Campylobacter jejuni* proteins expressed *in vitro*. *Infect. Immun.* 60(11):4938–4944.

Park, S.F. 1999. The use of *hipO*, encoding benzoylglycine amidohydrolase (hippuricase), as a reporter of gene expression in *Campylobacter coli*. *Lett. Appl. Microbiol.* 28(4):285–290.

Parkhill, J., B.W. Wren, K. Mungall, J.M. Ketley, C. Churcher, D. Basham, T. Chillingworth, R.M. Davies, T. Feltwell, S. Holroyd, K. Jagels, A.V. Karlshev, S. Mouse, M.J. Pallen, C.W. Penn, M.A. Quall, M.A. Rajandrean, K.M. Rutherford, A.H.M. Van Vliet, S. Whitehead, and B.G. Barrell. 2000. The genome sequence of the food-borne pathogen *Campylobacter jejuni* reveals hypervariable sequences. *Nature* 403:665–668.

Pavlovskis, O.R., D.M. Rollins, R.L.J. Haberberger, A.E. Green, L. Habash, S. Strocko, and R.I. Walker. 1991. Significance of flagella in colonization resistance of rabbits immunized with *Campylobacter* spp. *Infect. Immun.* 59:2259–2264.

Pebody, R.G., M.J. Ryan, and P.G. Wall. 1997. Outbreaks of *Campylobacter* infection: rare events for a common pathogen. *Comm. Dis. Rep. CDR Rev.* 7:R33–R37.

Pei, Z., and M.J. Blaser. 1993. PEB1, the major cell-binding factor of *Campylobacter jejuni*, is a homolog of the binding component in Gram-negative nutrient trans-port systems. *The J. Biol. Chem.* 268(25):18717–18725.

Pei, Z., C. Burucoa, B. Grignon, S. Baqar, X.-Z Huang, D.J. Kopecko, A.L. Bourgeois, J.-L. Fauchere, and M.J. Blaser. 1998. Mutation in the *peb1A* locus of *Campylobacter jejuni* reduces interactions with epithelial cells and intestinal colonization of mice. *Infect. Immun.* 66(3):938–943.

Pope, L.M., K.E. Reed, and S.M. Payne. 1995. Increased protein secretion and adherence to HeLa cells by *Shigella* spp. following growth in the presence of bile salts. *Infect. Immun.* 63:3642–3648.

Potter, M.E., M.J. Blaser, R.K. Sikes, A.F. Kaufmann, and J.G. Wells. 1983. Human *Campylobacter* infection associated with certified raw milk. *Am. J. Epidemiol.* 117:475–483.

Price, A.B., J. Jewkes, and P.J. Sanderson. 1979. Acute diarrhoea: *Campylobacter* colitis and the role of rectal biopsy. *J. Clin. Pathol.* 32:990–997.

Pumbwe, L., and L.J.V. Piddock. 2002. Identification and molecular characterisation of CmeB, a *Campylobacter jejuni* multidrug efflux pump. *FEMS Microbiol. Lett.* 206:185–189.

Richardson, P.T., and S.F. Park. 1997. Integration of heterologous plasmid DNA into multiple sites on the genome of *Campylobacter coli* following natural transformation. *J. Bacteriol.* 179(5):1809–1812.

Rivera-Amill, V., B.J. Kim, J. Seshu, and M.E. Konkel. 2001. Secretion of the virulence associated *Campylobacter* invasion antigens from *Campylobacter jejuni* requires a stimulatory signal. *J. Infect. Dis.* 183:1607–1616.

Roberts, D.D. 1990. Interactions of respiratory pathogens with host cell surface and extracellular matrix components. *Am. J. Resp. Cell Molec. Biol.* 3:181–186.

Robinson, D.A. 1981. Infective dose of *Campylobacter jejuni* in milk. *Brit. Med. J.* 282:1584.

Russell, R.G., M.O'Donnoghue, D.C. Blake, Jr., J. Zulty, and L.J. DeTolla. 1993. Early colonic damage and invasion of *Campylobacter jejuni* in experimentally challenged infant *Macaca mulatta*. *J. Infect. Dis.* 168:210–215.

Salloway, S., L.A. Mermel, M. Seamans, G.O. Aspinall, J.E.N. Shin, L.A. Kurjanczyk, and J.L. Penner. 1996. Miller-Fisher syndrome associated with *Campylobacter jejuni* bearing lipopolysaccharide molecules that mimic human ganglioside GD3. *Infect. Immun.* 64(8):2945–2949.

Savill, M.G., J.A. Hudson, A. Ball, J.D. Klena, P. Scholes, R.J. Whyte, R.E. McCormick and D. Jankovic. 2001. Enumeration of *Campylobacter* in New Zealand recreational and drinking waters. *Lett. Appl. Microbiol.* 91:38–46.

Schröder, W., and I. Moser. 1997. Primary structure analysis and adhesion studies on the major outer membrane protein of *Campylobacter jejuni. FEMS Microbiol. Lett.* 150:141–147.

Shanker, S., A.Lee, and T.C. Sorrell. 1990. Horizontal transmission of *Campylobacter jejuni* amongst broiler chicks: experimental studies. *Epidemiol. Infect.* 104:101–110.

Silverstein, S.C., R.M. Steinman, and Z.A. Cohn. 1977. Endocytosis. *Ann. Rev. Biochem.* 46:669–722.

Skirrow, M.B. 1977. *Campylobacter* enteritis: a new disease. *Brit. Med. J.* 2:9–11.

Skirrow, M.B. 1987. A demographic survey of campylobacter, salmonella and shigella infections in England. A Public Health Laboratory Service Survey. *Epidemiol. Infect.* 99:647–657.

Skirrow, M.B. 1991. Epidemiology of Campylobacter enteritis. *Int. J. Food Microbiol.* 12:9–16.

Skorupski, K., and R.K. Taylor. 1996. Positive selection vectors for allelic exchange. *Gene* 169:47–52.

Smith, K.E., J.M. Besser, C.W. Hedberg, F.T. Leano, J.B. Bender, J.H. Wicklund, B.P. Johnson, K.A. Moore, and M.T. Osterholm. 1999. Quinolone-resistant *Campylobacter jejuni* infections in Minnesota, 1992–1998. *N. Engl. J. Med.* 340(20):1525–1532.

Svedhem, A., and B. Kaijser. 1980. *Campylobacter fetus* subspecies *jejuni*: a common cause of diarrhea in Sweden. *J. Infect. Dis.* 142:353–359.

Takeuchi, A. 1967. Electron microscopic studies of experimental *Salmonella* infection. Penetration into the intestinal epithelium by *Salmonella typhimurium. Am. J. Pathol.* 50:109–136.

Tauxe, R.V. 1992. Epidemiology of *Campylobacter jejuni* infections in the United States and other industrialised nations. *Campylobacter jejuni*: Current Status and Future Trends. I. Nachamkin, M.J. Blaser and L. Tompkins, eds. Washington, D.C.: American Society for Microbiology, 9–19.

Tauxe, R.V., N. Hargrett-Bean, C.M. Patton, and I.K. Wachsmuth. 1988. *Campylobacter* isolates in the United States, 1982–1986. *MMWR* 37(SS-2):1–13.

Taylor, D.J., and P. A. Olubunmi. 1981. A re-examination of the role of *Campylobacter fetus* subspecies *coli* in enteric disease of the pig. *Vet. Rec.* 109:112–115.

van den Bogaard, A.E., and E.E. Stobberingh. 2000. Epidemiology of resistance to antibiotics. Links between animals and humans. *Int. J. Antimicrobial Agents* 14:327–335.

Van Spreeuwel, J.P., G.C. Duursma, C.J.L.M. Meijer, R. Bax, P.C.M. Rosekrans, and J. Lindeman. 1985. *Campylobacter* colitis: histological immunohostochemical and ultrastructural findings. *Gut* 26:945–951.

Vogt, R.L., A.A. Little, T. Barrett, R.A. Fedlman, R.J. Dickinson, and L. Witherell. 1984. Serotyping and serology studies of campylobacteriosis associated with the consumption of raw milk. *J. Clin. Microbiol.* 20(5):998–1000.

Vuckovic, D., M. Abram, and M. Doric. 1998. Primary *Campylobacter jejuni* infection in different mice strains. *Microbial Pathog.* 24:263–268.

Wang, Y. and D.E. Taylor. 1990. Natural transformation in *Campylobacter* species. *J. Bacteriol.* 172(2):949–955.

Wassenaar, T.M., N.M.C. Bleumink-Plum, and B.A. van der Zeijst. 1991. Inactivation of *Campylobacter jejuni* flagellin genes by homologous recombination demonstrates that *flaA* but not *flaB* is required for invasion. *EMBO J.* 10(8):2055–2061.

Wassenaar, T.M., M. Engelskirchen, S. Park, and A. Lastovica. 1997. Differential uptake and killing potential of *Campylobacter jejuni* by human peripheral monocytes/macrophages. *Med. Microbiol. Immunol.* 186:139–144.

Wassenaar, T.M., B.A. van der Zeijst, R. Ayling, and D.G. Newell. 1993. Colonization of chicks by motility mutants of *Campylobacter jejuni* demonstrates the importance of flagellin A expression. *J. Gen. Microbiol.* 139:1171–1175.

Weisblum, B. 1995. Erythromycin resistance by ribosome modification. *Antimicrobial Agents Chemother.* 39(3):577–585.

Wren, B.W., D. Linton, N. Dorrell, and A.V. Karlyshev. 2001. Post genome analysis of *Campylobacter jejuni. J. Appl. Microbiol.* 90:36S–44S.

Yao, R., R.A. Alm, T.J. Trust, and P. Guerry. 1993. Construction of new *Campylobacter* cloning vectors and a new mutational *cat* cassette. *Gene* 130:127–130.

Yao, R., D.H. Burr, P. Doig, T.J. Trust, H. Niu, and P. Guerry. 1994. Isolation of motile and non-motile insertional mutants of *Campylobacter jejuni*: the role of motility in adherence and invasion of eukaryotic cells. *Molec. Microbiol.* 14(5):883–893.

Yao, R., D.H. Burr, and P. Guerry. 1997. CheY-mediated modulation of *Campylobacter jejuni* virulence. *Molec. Microbiol.* 23(5):1021–1031.

Yao, R., and P. Guerry. 1996. Molecular cloning and site-specific mutagenesis of a gene involved in arylsulfatase production in *Campylobacter jejuni. J. Bacteriol.* 178(11):3335–3338.

Young, G.M., D.H. Schmiel, and V.L. Miller. 1999. A new pathway for the secretion of virulence factors by bacteria: the flagellar export apparatus functions as a protein-secretion system. *Proc. Natl. Acad. Sci. USA* 96:6456–6461.

Ziprin, R.L., C.R. Young, M.E. Hume, and M.E. Konkel. 1999. The absence of cecal colonization of chicks by a mutant of *Campylobacter jejuni* not expressing bacterial fibronectin-binding protein. *Avian Dis.* 43:586–589.

CAMPYLOBACTER: CONTROL AND PREVENTION

DIANE G. NEWELL AND HELEN C. DAVISON

INTRODUCTION. The genus *Campylobacter* comprises a number of species of characteristically short, s-shaped motile, microaerophilic, Gram-negative bacteria. Many of these species colonize mucosal surfaces, usually as commensals, but some can be pathogenic to man and other animals (Skirrow 1994). *Campylobacter jejuni* and its close relative *C. coli* are of particular interest because these are common causes of human acute bacterial enteritis and therefore are the focus of this chapter. Other campylobacters found in animal agriculture include *C. hyointestinalis, C. upsaliensis, C. lari, C. fetus* subspecies *fetus* and *C. fetus* subspecies *venerealis*, but overall, these appear to be relatively rare causes of human disease (Lastovica and Skirrow 2000) and therefore will not in general be considered here. Similarly, species of the *Arcobacter* and *Helicobacter* genera that are closely related to campylobacters are also considered infrequent zoonotic diseases (Lastovica and Skirrow 2000) and will be excluded from this chapter. Additionally, to reduce the number of references but provide sufficient information for further reading, we have quoted review articles where possible.

In most industrialized countries, the thermophilic campylobacters, *C. jejuni* and *C. coli,* are reported to be common, and sometimes the primary, infectious cause of intestinal disease (Food Standards Agency 2001; World Health Organisation 2002). The incidence of laboratory-reported campylobacteriosis has been estimated to be thirty to sixty cases per 100,000 people per year in the United States (Wesley et al. 2000), but is substantially higher in children under five years of age (up to three hundred cases per 100,000). The problem in children is worldwide (World Health Organisation 2002); however, few countries have national campylobacter surveillance systems that accurately monitor the true incidence of human campylobacteriosis, particularly in the community where the majority of cases are not detected (Food Standards Agency 2001). Nevertheless, internationally there is an apparent trend of increasing number of campylobacteriosis cases that is likely to be a result of both an improved awareness of the disease and a true increase in the number of cases. The reasons underlying this trend are not known. Many sources of human campylobacteriosis are recorded but their relative importance remains unclear. Although large outbreaks have been reported, usually associated with contaminated raw milk (Lerner 2000) or water (Friedman et al. 2000), the vast majority of clinical cases are sporadic. Nevertheless, increasing exposure to campylobacters acquired via the food chain are considered to have an important role in this disease (Smith et al. 2000).

Human campylobacteriosis generally presents as watery or bloody diarrhea, usually lasting five to seven days and accompanied by acute abdominal pain (Skirrow and Blaser 2000). Although the disease is generally self-limiting, recent sentinel studies in the United Kingdom indicate that up to 10 percent of clinical cases can require hospitalization (Communicable Disease Report Weekly 2000). Infection can result in bacteraemia, arthropathies, or abortion. In addition, prior *C.jejuni* infection is considered to be a predisposing factor in about 50 percent of cases of the severe polyneuropathy, Guillain-Barré syndrome (Allos 1998). Therefore, campylobacteriosis is a major social and economic public health problem worldwide. Recently this problem has been further enhanced by the increasing prevalence of antimicrobial resistance in campylobacters, particularly to macrolides and fluoroquinolones, which are used to treat human infections (Smith et al. 1999, 2000). This antimicrobial resistance is considered to be a consequence of the inappropriate use of antibiotics both in humans and food animal production.

OCCURENCE OF CAMPYLOBACTERS IN ANIMAL AGRICULTURE.

The thermophilic campylobacters, *C. jejuni* and *C. coli,* are common components of the gut flora of a wide range of mammalian and avian hosts, including common livestock (such as cattle, sheep, pigs), other domestic animals (such as cats, dogs, horses), poultry (such as chickens, turkeys, ducks, geese) and wildlife (such as hares, rabbits, wild boar, pheasants, quails). Generally in such animals, *C. jejuni/coli* act as gut commensals. However, campylobacter-associated enteritis has been reported in some young animals, for example calves and piglets, and both *C. jejuni* and *C. coli* can be recovered from aborted bovine and ovine fetuses (Skirrow 1994). Of considerable recent concern is the increasing proportion of antibiotic-resistant campylobacters recovered from livestock, particularly from poultry and pigs. For example, a high proportion of *C. coli* from pigs is resistant to macrolides (Aarestrup 2001), and campylobacters from poultry increasingly are resistant to fluoroquinolones (Smith et al. 1999).

Campylobacter jejuni/coli colonize the mucous layer overlying the intestinal epithelium, particularly of the lower small intestine, caecum, and colon. Infected animals shed organisms, intermittently or persistently, in their faeces. However, this fecal carriage may be difficult to detect when the numbers are low, and contaminating fecal flora are inhibitory to campylobacter growth. The prevalence of campylobacter carriage in different animal populations in many countries is unknown, and only a few national surveillance programs exist (such as in Denmark; see www.vetinst.dk/file/WEB-Annual%20Report.pdf). Prevalence estimates are affected by many factors, including the detection methods used, sample type and size, age of animal, season, management, and stage of production. Nevertheless, current data suggest that campylobacters are widely distributed throughout animal agriculture systems (see Table 22.1).

A transmission cycle of campylobacter appears to exist between animals and their environment. Under normal environmental conditions, *C. jejuni/coli* cannot grow but can survive for extended periods. Survival is best under cool moist conditions, and the organisms are sensitive to dessication, high temperatures, atmospheric oxygen, and lack of nutrients (Jones 2001). Nevertheless, survival can be sustained in agricultural environments such as soil, slurry, natural and artificial water systems, and damp bedding. Surrounding environments may then become infected from these sources, for example, through farm effluent contamination of water courses. The traffic of humans, animals, and even wildlife and wild birds (including migratory birds) can then transmit campylobacters within and between farms. The detection and tracking of such campylobacters in agricultural environments are difficult because of the ubiquitous nature of the organisms and the diversity of strains. Moreover, some campylobacters exist in the environment in a stressed state in which they are viable but their recoverability is poor. The role of such viable but not culturable forms as a source of infection for animals is unclear.

Poultry. The physiological characteristics of *C. jejuni/coli* suggest that these organisms have evolved to optimally colonize the avian gut. Chickens, ducks, turkeys and wild birds are frequently colonized with large numbers of these thermophilic campylobacters. In Europe, surveys indicate that the prevalence of campylobacter-positive broiler flocks is between 10 and 95 percent, with a lower prevalence found in the northernmost countries such as Sweden (Neilsen 1997). This prevalence can be seasonal, usually with a summer peak, and is dependent on the type of production system, for example reaching 100 percent in organic and free-range flocks (Heuer et al. 2001). About 90 percent of the strains isolated are *C. jejuni* and the remainder are *C. coli*. Infection is not usually detectable in broiler flocks until the birds are two to three weeks of age. The reasons for this so-called "lag phase" are unknown, but increased susceptibility at this age may reflect waning maternal antibody protection, changing gut flora, or both (Newell and Wagenaar 2000). After colonization of the flock occurs, the infection spreads rapidly, affecting up to 100 percent of birds within the flock and persisting until slaughter. Different campylobacter strains can be isolated from flocks (and occasionally in individual birds), but in flocks intensively reared in closed houses, usually only one to two strains are found. The number of campylobacters in the caeca can be extremely high: up to 10^{10} colony-forming units (cfu) per gram of cecal contents (Cawthraw et al. 1996). Under experimental conditions, this colonization is invariably asymptomatic. In older birds, such as layers, in-flock colonization levels and shedding is reduced over time, but birds rarely become cecal-culture negative. With such high numbers of organisms in broilers, contamination of carcasses from spilled gut contents, during slaughter and processing, is frequent. This results in the abattoir environment, in particular the immersion tank and defeathering machinery, becoming heavily contaminated,

Table 22.1—Occurrence of Campylobacters in Poultry in Different Countries

Host species	Country	*Campylobacter* species	Sampling location	Sample type	Number			Reference
					Farms	Samples/ animals tested	Samples/ animals positive	
Broiler chickens	Denmark	*C. jejuni, C. coli*	Abattoir	Feces	NK	1,217	*C. jejuni:* 36% *C. coli:* 3%	DANMAP 2000
Ostrich	USA	*Campylobacter* spp.	9 abattoirs	Carcass sponge	NK	191	10%	Ley et al. 2001
Broiler chickens	France	*Campylobacter* spp.	Farm	Fresh fecal dropping	75	75 flocks	42.7%	Refrégier-Petton et al. 2001
Broiler flocks	Denmark	*Campylobacter* spp.	10 abattoirs	Cloacal swab	8911 flocks	89,110	42.5%	Wedderkopp et al. 2001

NK: Not known

leading to the cross-contamination of subsequent flocks (Newell et al. 2001). In some countries, efforts to schedule the slaughtering of campylobacter-negative flocks before campylobacter-positive flocks are being made, together with improved diagnostic techniques for rapid identification of flock status.

The control of infection at the farm level is considered an important factor in the reduction or elimination of campylobacters from the human food chain. To date, control at the broiler farm has proved very difficult. The dose required to infect chickens is very low: as few as three organisms in experimental models (Cawthraw et al. 1996). However, this dose may be much higher in organisms stressed by environmental conditions. Nevertheless, *in vivo* passage of stressed organisms up-regulates colonization factors enhancing infectivity (Cawthraw et al. 1996). Effective transmission through a flock is mediated by coprophagic activity, enhanced by behavioral clustering of young chicks and fecal contamination of feed and water lines. This transmission can occur between groups of birds even in the absence of direct contact (Shreeve et al. 2000). Thus, prevention of the first bird becoming infected is important for successful control. Currently, two main strategies exist for control at the farm level: (1) biosecurity to exclude the organisms from the flock, and (2) reduction of bird susceptibility to infection by methods such as vaccination and competitive exclusion.

The challenge is to achieve practical and cost-effective long-term control of campylobacters on poultry farms. Empirical biosecurity measures are only partly successful in preventing flock colonization (Gibbens et al. 2001). Those measures applied so successfully to salmonella control in poultry have had no, or little, discernible effect on campylobacter colonization in the medium to long term. Current best practices for recommended control measures are given in Table 22.2. Applying such measures can delay but generally do not prevent the onset of colonization. In some countries, such as Sweden, the rigorous application of biosecurity measures are considered to possibly prevent colonization. However, the compliance required for countries with warmer climates and possibly greater environmental loads of this organism appears to be too great for such measures to be successful and cost-effective.

The sources of campylobacter infection in broiler flocks remain unclear. Although organisms are recoverable from the oviducts of layers (Camarda et al. 2001) and some evidence exists that vertical transmission can occur (Pearson et al. 1996), it is generally considered to have only a minor role, if any, in flock colonization. Campylobacters are ubiquitous in the environment in and around broiler farms, and horizontal transmission from these environmental sources appears to be a much more likely explanation. In Europe, the litter from previous flocks is removed and the houses cleaned and disinfected before the subsequent flock is housed. Campylobacters are rarely recovered from such cleaned and disinfected houses even using enrichment techniques (Evans and Sayer 2000). This is supported by the infrequent detection of carry-over of strains between sequential flocks. Dry conditions are lethal to campylobacters, so contaminated dry feed and fresh litter are unlikely sources. However, these organisms can survive in untreated water for long periods. The header tanks and water lines may harbor campylobacters tracked up from previously positive flocks, though this occurrence has yet to be demonstrated. The role of biofilms in such water survival is unknown. The most likely sources are outside the broiler houses. Organisms are routinely recovered from a number of potential sources in the farm environment, including from the feces of wild birds and other wild and domestic animals on-site and from puddles and soil around the broiler house (van der Giessen et al. 1996; van der Giessen et al. 1998; Jacobs-Reitsma 1998). Vehicles, catching machinery, and transport crates (Newell et al. 2001; Slader et al. 2002) travelling between poultry farms or from the abattoir to a farm are heavily contaminated potential sources. Such environmental contaminants may be transferred into the house via workers' boots and outer clothes or by vermin and wild birds overcoming inadequate biosecurity measures. Crop thinning, as practiced in many industrialized countries, is a clear risk factor (Hald et al. 2001), presumably as a consequence of catching crews and equipment entering the house partway through the flock growing period to remove a proportion of the birds. Because campylobacters can be widespread, sometimes in high numbers, in the environment, more stringent measures clearly are required to exclude these

Table 22.2—Empericial Biosecurity Measures to Prevent Campylobacter Colonization of Poultry Flocks

1.	All-in all-out policy.
2.	Disinfect building between flocks.
3.	Maintain buildings in good repair.
4.	Clean water and/or water treatment.
5.	Dispose of dead birds properly.
6.	Keep buildings free of vermin and wild birds.
7.	Change and disinfect boots.
8.	Change outer protective clothing.
9.	Avoid thinning (removal of birds) during production cycle.
10.	Restrict visitor access.
11.	Do not keep other domestic animals on the same site.

organisms. However, the measures and the compliance required to achieve adequate biosecurity are generally considered unrealistic.

Biosecurity measures are not fully effective and measures targeted at specific sources might be a more efficient control option in the future. However, this requires more accurate identification and differentiation of the relative contributions of the various sources and routes of transmission. For this purpose, methods of strain typing are essential. The recent development of molecular typing methods (Wassenaar and Newell 2000) for campylobacters has allowed the investigation of the roles of carry over between flocks (Shreeve et al. 2002) , bird-to-bird contact (Shreeve et al. 2000), vertical transmission (Camarda et al. 2000), and external environments and transport crates (Newell et al. 2001) as sources of bird and poultry product contamination.

Clearly, total prevention of flock exposure to campylobacters is unlikely to be achievable given the farm environment. Therefore, supportive measures to minimize or prevent colonization of the birds are required. For salmonella, vaccines and competitive exclusion agents have been developed. Currently, no such products specific to campylobacter exist. The competitive exclusion agents known to be effective against salmonella may have some effect on campylobacters under experimental conditions (Newell and Wagenaar 2000) but effectiveness, if any, has yet to be duplicated in field situations. The same is true for other probiotics with reported efficacy. Vaccine development is in very early stages. Systemic and mucosal antibody responses are induced in colonized birds (Cawthraw et al. 1994). Preliminary studies suggest that these responses can partly protect birds from colonization. However, killed and subunit vaccines have little demonstrable efficacy, and live vaccines, attenuated for human disease, have yet to be developed (Newell and Wagenaar 2000). Other potential control strategies include the breeding of genetically-resistant chicken strains and the use of bacteriophages. Some preliminary evidence indicates that some inbred chicken lines have a greater resistance to campylobacter colonization than others (P. Barrow, personal communication). Whether such traits can be identified and integrated into birds with the desired feed-conversion characteristics is not known. Some bacteriophages specific for campylobacters cause bacteriolysis under specified conditions. Oral treatment of birds with such bacteriophages at or close to slaughter may reduce cecal colonization levels, thus reducing the bacterial load entering the abattoir (Newell and Wagenaar 2001). All such strategies are far from product delivery. In the meantime, effective biosecurity remains the only achievable target. Further understanding of the ecology of the organism and the nature of avian gut colonization will be essential to the development of effective supporting control strategies.

Cattle. *C. jejuni*, *C. coli*, *C. hyointestinalis*. *C. fetus* subsp. *fetus*, and *C. lari* are all recoverable from the feces of healthy cattle. *C. fetus* subsp. *fetus* and *C. fetus* subsp. *venerealis* can cause bovine abortion and infertility, respectively, but are not associated with bovine enteric disease. Both *C. jejuni* and *C. coli* have been associated, sometimes with concurrent infections with other organisms, with diarrhea in calves (Skirrow 1994; Busato et al. 1999); however, their role as pathogens for cattle is not clearly defined. The fecal carriage of campylobacters in cattle is common, with up to 100 percent farms tested being positive (Table 22.3). *C. jejuni* is the most common *Campylobacter* species found in cattle, but *C. coli* and *C. hyointestinalis* also are frequently recovered. Colonization levels within individual cattle are substantially lower than in poultry, but cattle can shed campylobacters intermittently for several months or, in some cases, for the life of the animal. Some individuals shed continuously, possibly because of (unknown) host-specific factors or concurrent infections. The prevalence of different campylobacters reported in cattle sampled either on-farm or at slaughter varies with age and management, such as housing and diet (Busato, et al. 1999) (Table 22.3). Mixed campylobacter infections have been reported in calves, which can have a higher prevalence of infection than adult cattle. Milk can become infected either directly, as a result of a campylobacter infection of the udder, or by fecal contamination after milking (Jacobs-Reitsma 2000).

Some evidence exists of seasonality in shedding patterns, but peak periods of shedding vary between studies. For example, a higher prevalence has been observed in winter-housed cattle than in summer-grazing cattle, when longer daylight hours and higher temperatures would be expected to reduce the campylobacter burden on pastures, and stocking densities are lower. However, a higher prevalence was found in summer-grazing cattle that had access to a campylobacter-positive lake and were housed with a chlorinated source of water in winter in Finland (Hanninen et al. 1998). This study showed the possibility of a cow-water-cow cycle of infection. In general, specific campylobacter control strategies are not used in cattle, but measures aimed to break identified cycles of infection, for example by treatment of contaminated water sources, could be useful. Exclusion of infection at the farm or group level is difficult, because all-in all-out systems are rarely used and there is a lack of evidence-based approaches.

Sheep and Goats. Fecal carriage of campylobacters is common in sheep and goats. This carriage is usually asymptomatic, although *C. fetus* subsp. *fetus* is a cause of abortion in sheep and goats and *C. jejuni* has been associated with diarrhea in weaning lambs (Skirrow 1994). There have been few studies, and these investigated commensal campylobacters in sheep (Table 22.3). Jones et al. (1999) identified the majority of isolates recovered from flocks around northeast England as *C. jejuni* (90 percent) with some *C. coli* (8 percent) and *C. lari* (2 percent). However, in a national abattoir survey in Great Britain, approximately equal proportions of *C.jejuni* and *C.coli* were

Table 22.3. Occurrence of Campylobacters in Cattle and Sheep Production in Different Countries

Host species	Country	*Campylobacter* species	Sampling location	Sample type	Farms	Samples/ animals tested	% Samples/ animals positive (95 CIs)	Reference
Dairy and beef cattle	Japan	*C.jejuni; C. coli* (+ other spp.)	Farm	Rectal feces	6	34 calves + 60 cows	*C. jejuni* (calves): 61.8% *C. jejuni* (cows): 13.3% *C. coli* (calves): 2.9% *C. coli* (cows):0%	Giacoboni et al. 1993
Buffaloes	Pakistan	*C. jejuni/C. coli*	Farm	Fecal swab	NK	100	10.0%	Khalil et al. 1993
Dairy cows	Sweden	*C. jejuni/C. coli*	Farm	Feces	NK	80	2.5%	Khalil et al. 1993
Beef cattle	Switzerland	*C. jejuni/C. coli* (+ other spp.)	Farm	feces	67	395	[a] *C. jejuni:* 38.5% [a] *C. coli:* 3.4% [b] *C. jejuni:* 13.3% [b] *C. coli:* 1.7%	Busato et al. 1999
Beef cattle	Great Britain	*C. jejuni/C.coli/ C. lari*	Abattoir	Feces	NK	891	13.5%	Paiba et al. 2000
Cattle	Denmark	*C. jejuni, C. coli*	Abattoir	Feces	NK	89	*C. jejuni:* 56% *C. coli:* 1%	DANMAP 2000
Dairy cattle	USA	*C. jejuni, C. coli* (by PCR)	Farm	Feces	31	2085	*C. jejuni:* 37.7%	Wesley et al. 2000
Beef cattle	California,USA	*Campylobacter* spp.	Farm	Feces	17	401	5.0%	Hoar et al. 2001
Sheep	England	*Campylobacter* spp.	Farm	Feces	3	510	21.5–36.2%	Stanley et al. 1998
Sheep	England	*Campylobacter* spp.	Farm	Feces	3	510	21.5–36.2%	Stanley et al. 1998
Sheep	England	*Campylobacter* spp.	Abattoir	Small intestinal contents	NK	360	57–100%	Stanley et al. 1998
Sheep	Great Britain	*C. jejuni/C.coli/ C. lari*	Abattoir	Feces	NK	973	15.8%	Paiba et al. 2000

NK: Not known

215

recovered from ovine feces (Paiba et al. 2001). *Campylobacter hyointestinalis* also is a frequent commensal of the ovine gut. As with cattle, sheep tend to have a lower level of colonization than poultry. There is an apparent increase in shedding at lambing time and weaning, suggesting that shedding is stress associated, and lambs become colonized one to five days after birth (Jones, et al. 1999). Diet and types of grazing also can influence shedding rates. As with cattle, a sheep-water-sheep cycle may occur. In Norway, an outbreak of human campylobacteriosis was linked with contamination of a reservoir by run-off from a pasture where infected sheep grazed. Little data exists pertaining to campylobacters in goats, but goat milk consumption was a risk factor for human cases in the United States, and contaminated raw goat's milk has been confirmed as the source of infection for human enteritis cases (Harris et al. 1987).

Pigs. Surveys of pigs on farms, or in the abattoir, demonstrate a high prevalence of campylobacters in porcine intestinal and cecal contents or fecal samples (Table 22.4). High numbers of organisms can be recovered from positive samples. In a recent national survey, up to 87 percent of animals at slaughter were colonized with campylobacters in Great Britain. Similar figures for prevalence have been reported in Denmark (Neilsen 1997) and the United States (Young et al. 2000). The majority of these porcine campylobacters are *C. coli* (96 percent in Great Britain), but *C.jejuni* strains can be more common (Young et al. 2000). Most infected pigs are asymptomatic but *C. hyointestinalis,* which is often found in pigs, may be associated with disease such as proliferative enteritis. The contribution of porcine campylobacters to human intestinal infectious disease is unclear, and in most countries only about 10 percent of human infections are caused by *C.coli.* The absence of higher numbers of human infections may be a consequence of low pork product contamination or the lower survival rates of *C.coli* during processing. However, in countries with a higher consumption of pork products *C. coli*-related human infections are considered more frequent, and outbreaks of human illness have been associated with consumption of infected pork meat.

Campylobacter infection is acquired early in life and infection is reported to be a cause of diarrhea in piglets (Skirrow 1994), but in experimental infections only gnotobiotic or colostrum-deprived piglets develop symptoms (Newell 2001). The management practices for pig production are variable and involve either intensive or extensive production systems. Pigs bred outside are exposed to environmental sources of campylobacters, including wild birds and surface water, from birth and repeatedly thereafter. Immunity may be acquired rapidly; however, intestinal carriage remains chronic. Increases in fecal shedding are likely during periods of stress such as parturition or transport. Although prevention of exposure to campylobacters during extensive farming is unlikely to be feasible, control strategies taken during intensive rearing may be possible,

particularly in all-in all-out systems (Weijtens et al. 2000).

FOOD SAFETY. Humans can become infected with campylobacter either by direct contact with the feces of infected animals or, more commonly, by contact with fecally contaminated animal products. The latter may originate from an infected animal or have become contaminated during processing, food preparation, or both. Because the relative pathogenicity of different campylobacter strains is unknown, a precautionary approach should be adopted that all campylobacters in food are potentially zoonotic. However, preliminary evidence suggests that not all campylobacters are pathogenic. Studies are urgently required to determine the pathogenic potential of campylobacter strains so that the impact of different strains for public health can be more accurately assessed.

The role of the various potential sources of campylobacters from food-producing animals in human intestinal infectious disease is currently unknown. Such information requires appropriate strain-tracing techniques. There has been considerable difficulty with the typing of campylobacter strains for the purpose of comparing clinical and veterinary isolates. Phenotypic detection methods (for example, serotyping) are not suitable for the identification of animal campylobacters for epidemiological studies because a high number of isolates cannot be typed. The use of genotyping methods is increasing, including pulsed-field gel electrophoresis (PFGE) and amplified fragment length polymorphism (AFLP) (Wassenaar and Newell 2000). However, the wide diversity of strains, genomic instability over time and geographical distance, and the weak clonality of the population structure of *C.jejuni/coli* have precluded the use of typing techniques for assessment of the relative contributions of veterinary campylobacters to human disease. The recent development of multilocus sequence typing (MLST) (Dingle et al. 2001) is now enabling the population structure and evolutionary trends of campylobacters to be assessed, which may allow some understanding of the overlaps between human and veterinary isolates to be determined in the future.

The majority of human campylobacteriosis cases are assumed to be foodborne. However, because most cases are sporadic and outbreaks occur only rarely (even secondary cases within households are uncommon), identifying sources of infection is difficult. Case-control studies have identified several common risk factors associating human disease with food-producing animals, food, or water. These have included handling and consumption of raw or undercooked poultry meat; contact with poultry or poultry products; consumption of beef, pork, or lamb; consumption of contaminated milk, vegetables, or water, including recreational water (accidental ingestion); and contact with domestic animals, including livestock (Jacobs-Reitsma 2000; Rodrigues et al. 2001). However, some of these risk factors have been shown to have a protective effect

Table 22.4—Occurrence of Campylobacters in Pigs in Different Countries

Host species	Country	*Campylobacter* species	Sampling location	Sample type	Number		Samples/ animals positive	Reference
					Farms	Samples/ animals tested		
Pigs	The Netherlands	*C. jejuni/C. coli*	Abattoir	Gastric/ileal/ cecal contents	8	725	85%	Weijtens et al. 1993
Pigs	USA	*C. jejuni; C. coli, C. lari*	Abattoir	Cecal contents	4	595	*C. jejuni*: 0–76% *C. coli*: 20–100% *C. lari*: < 1%	Harvey et al. 1999
Pigs	USA	*C. jejuni; C. coli, C. lari*	Abattoir/ farm	Rectal swab/ Cecal contents	1	152	*C. jejuni*: 31.7–89.0% *C. coli*: 11–68.3% *C. lari*: 0–2.6%	Young et al. 2000
Pigs	England and Wales	*C. jejuni; C. coli, C. lari*	Abattoir	Cecal contents	NK	860	*C. jejuni*: 3.4% *C. coli*: 83.7% *C. lari*: 0%	Dalziel et al. 2001
Pigs	Denmark	*C. jejuni, C. coli*	Abattoir	Feces	NK	277	*C. jejuni*: 4% *C. coli*: 56%	DANMAP 2000

NK: Not known

217

(Adak et al. 1995), possibly as a consequence of acquired immunity. The sources of many cases remain unknown.

In contrast to salmonella, campylobacter does not multiply on food. Campylobacter numbers on and within carcasses can be reduced by freezing and eliminated by irradiation, but these methods are not widely accepted by consumers. Pasteurisation kills campylobacter, but post-pasteurization contamination must be avoided. Cheese and yoghurt are less likely to have viable campylobacter because of its sensitivity to acid. Quantitative risk assessments (QRA) have been developed to evaluate the level of exposure to campylobacters originating in broiler flocks and the subsequent risk to human health (see, for example, Harnett et al. 2001; www.fao.org/ES/ESN/pagerisk/riskpage.htm). Such QRAs will be useful to explore the potential impact of different control strategies along the food chain, and this is an approach currently being expoited in Europe (www.COST920.com).

CONCLUSIONS. Many government and intergovernment agencies are now requiring the reduction or elimination of campylobacters from the food chain to achieve targets to reduce human food poisoning. Because campylobacters are essentially normal gut flora constituents, few surveillance programs for these organisms exist. Thus, the extent of the problem at the farm level is largely unknown. It is proving extremely difficult to reduce the colonization with these organisms of animals reared in intensive and extensive production systems. With the move toward improved animal welfare and the concomitant increase in extensive farming practices, removing campylobacters from the food chain using biosecurity alone will be even more difficult. The development of supportive measures, such as vaccination or probiotics to reduce colonization at the farm level is required, particularly in poultry flocks. This will necessitate considerably more research into the ecology of the organisms. As colonization levels are reduced, control downstream will become more important. Pathogen reduction and hazard analysis and critical control point (HACCP) approaches are now being applied in poultry slaughterhouses and processing plants in the United States. In addition, educational campaigns are now being used to reduce cross-contamination within food retailers, including restaurants and within consumers' kitchens. Worldwide, government agencies are targetting foodborne pathogens to improve public health. To achieve such targets, evidence-based control strategies throughout the food chain will be essential.

REFERENCES

Aarestrup, F.M., and J. Engberg. 2001. Antimicrobial resistance of thermophilic *Campylobacter*. *Vet. Res.* 32:311–21.

Adak, G.K., J.M. Cowden, S. Nicholas, and H. S. Evans. 1995. The Public Health Laboratory Service national case-control study of primary indigenous sporadic cases of campylobacter infection. *Epidemiol. Infect.* 115:15–22.

Allos, B.M. 1998. *Campylobacter jejuni* infection as a cause of the Guillain-Barré syndrome. *Infect. Dis. Clinics N. Am.* 12:173–84.

Busato, A., D. Hofer, T. Lentze, C. Gaillard, andA. Burnens. 1999. Prevalence and infection risks of zoonotic enteropathogenic bacteria in Swiss cow-calf farms. *Vet. Microbiol.* 69:251–63.

Camarda, A., D.G. Newell, R. Nasti, and G. Di Modugnoa. 2000. Genotyping *Campylobacter jejuni* strains isolated from the gut and oviduct of laying hens. *Avian Dis.* 44:907–12.

Cawthraw, S., R. Ayling, P. Nuijten, T. Wassenaar, and D.G. Newell. 1994. Isotype, specificity, and kinetics of systemic and mucosal antibodies to *Campylobacter jejuni* antigens, including flagellin, during experimental oral infections of chickens. *Avian Dis.* 38:341–9.

Cawthraw, S.A., T.M. Wassenaar, R. Ayling, and D.G. Newell. 1996. Increased colonization potential of *Campylobacter jejuni* strain 81116 after passage through chickens and its implication on the rate of transmission within flocks. *Epidemiol. Infect.* 117:213–5.

Communicable Disease Report Weekly 2000. Public Health Laboratory Service 10:437

Dalziel, R.W., R.H. Davies, D.G. Newell, I.M. McLaren, C.J. Teale, P.J. Heath, G.A. Paiba, J.B.M. Ryan, and A. Mars. 2001. Prevalence of fecal carriage of foodborne pathogens in pigs at slaughter in Great Britain. Final report on Project FT 9119. Department of Environment, Food and Rural Affairs, Great Britain.

Dingle K.E., F.M. Colles, D.R. Wareing, R. Ure, A.J. Fox, F.E. Bolton, H.J. Bootsma, R.J. Willems, R. Unwin, and M.C. Maiden. 2001. Multilocus sequence typing system for *Campylobacter jejuni*. *J. Clin. Microbiol.* 39:14–23.

Evans S.J., and Sayers, A.R. 2000. A longitudinal study of campylobacter infection of broiler flocks in Great Britain. *Prevent. Vet. Med.* 46:209–23.

Food Standards Agency 2001. Report on the study of intestinal infectious disease in England. The Stationary Office, London. See also www.foodstandards.gov.uk/science/research/research_archieve/scientific/intestinal.

Friedman C.R., J. Neimann, H.C. Wegner, and R.V.Tauxe. 2000. Epidemiology of Campylobacter jejuni infections in the United States and other industrialized nations. In *Campylobacter 2nd Ed.*, I. Nachamkin and M.J. Blaser, eds. Washington, D.C.: ASM Press, 121–138.

Gibbens J.C., S.J.S. Pascoe, S.J. Evans, R.H. Davies, and A.R. Sayers. 2001. A trial of biosecurity as a means to control Campylobacter infection of broiler chickens. *Prev. Vet. Med.* 48:85–99.

Giacoboni, G.I., K. Itoh, K. Hirayama, E. Takahashi, and T. Mitsuoka. 1993. Comparison of fecal *Campylobacter* in calves and cattle of different ages and areas in Japan. *J. Vet. Med. Sci.* 55:555–559.

Grau, F.H. 1988. *Campylobacter jejuni* and *Campylobacter hyointestinalis* in the intestinal tract and on the carcasses of calves and cattle. *J. Food Prot.* 51:857–861.

Hanninen, M.L., M. Niskanen, and L. Korhonen. 1998. Water as a reservoir for *Campylobacter jejuni* infection in cows studied by serotyping and pulsed-field gel electrophoresis (PFGE). *Zentralbl. Veterinarmed.* 45:37–42.

Hartnett, E., L. Kelly, D. Newell, M. Wooldridge, and G. Gettinby. 2001. A quantitative risk assessment for the occurrence of campylobacter in chickens at the point of slaughter. *Epidemiol. Infect.* 127:195–206.

Harris, N.V., T.J. Kimball, P. Bennett, Y. Johnson, D. Wakely, and C.M. Nolan. 1987. *Campylobacter jejuni* enteritis associated with raw goat's milk. *Am. J. Epidemiol.* 126:179–86.

Harvey, R.G., C.R. Young, R.L. Ziprin, M.E. Hume, K.J. Genovese, R.C. Anderson, R.E. Droleskey, L.H. Stanker, and D.J. Nisbet. 1999. Prevalence of *Campylobacter* spp. isolated from the intestinal tract of pigs raised in an integrated swine production system. *J.A.V.M.A.* 215:1601–1604.

Hald, B., E. Rattenborg, and M. Madsen. 2001. Role of batch depletion of broiler houses on the occurrence of *Campylobacter* spp. in chicken flocks. *Lett. Appl. Microbiol.* 32:253–6.

Hoar, B.R., E.R. Atwill, C. Elmi, and T.B. Farver. 2001. An examination of risk factors associated with beef cattle shedding pathogens of potential zoonotic concern. *Epidemiol. Infect.* 127:147–155.

Heuer, O.E., K. Pedersen J.S. Andersen, and M. Madsen. 2001. Prevalence and antimicrobial suseptibility of thermophilic Campylobacter in organic and conventional broiler flocks. *Lett. Appl. Microbiol.* 33:269–274.

Jacobs-Reitsma, W.F. 1997. Aspects of epidemiology of Campylobacter in poultry. *Veterinary Quarterly* 19:113–7.

Jacobs-Reitsma, W. 2000. Campylobacter in the food supply. In *Campylobacter 2nd Ed.*, I. Nachamkin and M.J. Blaser, eds. . Washington, D.C.: ASM Press, 467–481.

Jones, K., S. Howard, and J.S. Wallace. 1999. Intermittent shedding of thermophilic campylobacters by sheep at pasture. *J. Appl. Microbiol.* 86:531–536.

Jones, K. 2001. Campylobacters in water, sewage and the environment. *J. Appl. Microbiol.* 90:68S–79S.

Jones, K., S. Howard, and J.S. Wallace. 1999. Intermittent shedding of thermophilic campylobacters by sheep at pasture. *J. Appl. Microbiol.* 86:531–6.

Khalil, K., G.B. Lindblom, K. Mazhar, E. Sjögren, and B. Kaijser. 1993. Frequency and enterotoxicity of *Campylobacter jejuni* and *C. coli* in domestic animals in Pakistan as compared to Sweden. *J. Trop. Med. and Hyg.* 96:35–40.

Lastovica, A.J., and M.B. Skirrow. 2000. Clinical significance of *Campylobacter* and related species other than *Campylobacter jejuni* and *C. coli*. In *Campylobacter 2nd Ed.*, I. Nachamkin and M.J. Blaser, eds. . Washington, D.C.: ASM Press, 89–120.

Lehner, A., C. Schneck, G. Feierl, P. Pless, A. Deutz, E. Brandl, M. Wagner. 2000. Epidemiologic application of pulsed-field gel electrophoresis to an outbreak of *Campylobacter jejuni* in an Austrian youth centre. *Epidemiol. Infect.* 125:13–16.

Ley, E.C., T.Y. Morishita, T. Brisker, and B.S. Harr. 2001. Prevalence of Salmonella, Campylobacter, and *Esherichia coli* on strich carcasses and the susceptibility of ostrich-origin *E. coli* isolates to various antibiotics. *Avian Dis.* 45:696–700.

Nielsen, E.M., J. Engberg, and M. Madsen. 1997. Distribution of serotypes of *Campylobacter jejuni* and *C. coli* from Danish patients, poultry, cattle and swine. *FEMS Immunol. Med. Microbiol.* 19:47–56.

Newell, D.G. 2001. Animal models of *Campylobacter jejuni* colonization and disease and the lessons to be learnt from similar *Helicobacter pylori* models. *J. Appl. Microbiol.* 90:57S–67S.

Newell, D.G., J.E. Shreeve, M. Toszeghy, G. Domingue, S. Bull, T. Humphrey, and G. Mead. 2001. Changes in the carriage of campylobacter strains by poultry carcasses during processing in abattoirs. *Appl. Environ. Microbiol.* 67:2636–40.

Newell, D.G., and J. A. Wagenaar. 2000. Poultry infections and their control at the farm level. In *Campylobacter, 2nd Ed.* I. Nachamkin and M.J. Blaser, eds. Washington, D.C.: ASM Press, 497–509.

Paiba, G.A., and J.C. Gibbens. 2000. The prevalence of fecal carriage of VTEC O157 and other foodborne pathogens by cattle and sheep at slaughter in Great Britain. Final report on Project FSZ 2500. Department of Environment, Food and Rural Affairs. Great Britain.

Pearson, A.D., M.H. Greenwood, R.K. Feltham, T.D. Healing, J. Donaldson, D.M. Jones, and R.R. Colwell. 1996. Microbial ecology of *Campylobacter jejuni* in a United Kingdom chicken supply chain: intermittent common source, vertical transmission, and amplification by flock propagation. *Appl. Environ. Microbiol.* 62:4614–20.

Rao, M.R., B.N. Abdollah, S.J. Savarino, R. Abu-Elyazeed, T.F. Wierzba, L.F. Peruski, I. Abdel-Messih, R. Frenck, and J.D. Clemens. 2001. Pathogenicity and convalescent excretion of *Campylobacter* in rural Egyptian children. *Am. J. Epidemiol.* 154:166–173.

Refrégier-Petton, J., N. Rose, M. Denis, and G. Salvat. 2001. Risk factors for *Campylobacter* spp. contamination in French broiler-chicken flocks at the end of the rearing period. *Prevent. Vet. Med.* 50:89–100.

Rodrigues, L.C., J.M. Cowden, J.G. Wheeler, D. Sethi, P.G. Wall, P. Cumberland, D.S. Tompkins, M.J. Hudson, J.A. Roberts, and P. Roderick. 2001. The study of infectious intestinal disease in England: risk factors for cases of infectious intestinal disease with Campylobacter jejuni infection. *Epidemiol. Infect.* 127:185–93.

Shreeve, J.E., M. Toszeghy, M. Pattison, and D.G. Newell. 2000. Sequential spread of Campylobacter infection in a multi-pen broiler house *Avian Dis.* 44:983–8.

Shreeve, J., M. Toszeghy, A. Ridely, and D.G. Newell. 2002. The carry-over of campylobacter isolates between sequential poultry flocks. *Avian Dis.*, in press.

Skirrow, M.B. 1994. Diseases due to *Campylobacter, Helicobacter* and related bacteria. *J. Comp. Pathol.* 111:113–149.

Skirrow, M.B., and M.J. Blaser. 2000. Clinical aspects of *Campylobacter* infection. In *Campylobacter 2nd Ed.*, I. Nachamkin and M.J. Blaser, eds. Washington, D.C.: ASM Press, 69–88.

Slader, J., G. Domingue, F. Jorgensen, K. McAlpine, R.J. Owen, F.J. Bolton, and T.J. Humphrey. 2002. Impact of transport crate reuse and of catching and processing on campylobacter and salmonella contamination of broiler chickens. *Appl. Environ. Microbiol.* 68:713–9.

Smith, K., J. Bender, and M. Osterholm. 2000. Antimicrobial resistance in animals and relevance to human infections. In *Campylobacter 2nd Ed.* I. Nachamkin and M. Blaser,eds. Washington, D.C.: ASM Press, 483–496.

Smith, K. E., J.M. Besser, C.W. Hedberg, F.T. Leano, J.B. Bender, J.H. Wicklund, B.P. Johnson, K. Moore, and M.T. Osterholm. 1999. Quinolone-resistant *Campylobacter jejuni* infections in Minnesota, 1992–1998. *N. Engl. J. Med.* 340:1525–32.

Stanley, K.N., J.S. Wallace, J.E. Currie, P.J. Diggle, and K. Jones. 1998. The seasonal variation of thermophilic campylobacters in beef cattle, dairy cattle and calves. *J. Appl. Microbiol.* 85:472–480.

Stanley, K.N., J.S. Wallace, J.E. Currie, P.J. Diggle, and K. Jones. 1998. Seasonal variation of thermophilic campylobacters in lambs at slaughter. *J. Appl. Microbiol.* 84:1111–1116.

van de Giessen, A.W., B.P. Bloemberg, W.S. Ritmeester, and J.J. Tilburg. 1996. Epidemiological study on risk factors and risk reducing measures for campylobacter infections in Dutch broiler flocks. *Epidemiol. Infect.* 117:245–50.

van de Giessen, A.W., J.J. Tilburg, W.S. Ritmeester, and J. van der Plas. 1998. Reduction of campylobacter infections in broiler flocks by application of hygiene measures. *Epidemiol. Infect.* 121:57–66.

Wassenaar, T. M., and D.G. Newell. 2000. Genotyping of *Campylobacter* spp. *Appl. Environ. Microbiol.* 66:1–9.

Weijtens, M.J., H.A. Urlings, and J. van der Plas. 2000. Establishing a campylobacter-free pig population through a top-down approach. *Lett. Appl. Microbiol.* 30:479–484.

Wesley, I.V., S.J. Wells, K.M. Harmon, A. Green, L. Schroeder-Tucker, M. Glover, and I. Siddique. 2000.

Fecal shedding of *Campylobacter* and *Arcobacter* spp. in dairy cattle. *Appl. Environ. Microbiol.* 66:1994–2000.

World Health Organization 2001. The increasing incidence of human campylobacteriosis. Report and procedings of a WHO Consultation of experts, Copenhagen, Denmark. November 21–25, 2000. WHO/CDS/CSR/APH 2001:7.

Young, C.R., R. Harvey, R. Anderson, D. Nisbet, and L.H. Stanker. 2000. Enteric colonization following natural exposure to *Campylobacter* in pigs. *Res. Vet. Sci.* 68:75–8.

PART

LISTERIA MONOCYTOGENS

23

EPIDEMIOLOGY OF LISTERIOSIS

YNTE H. SCHUKKEN, YRJO T. GROHN, AND MARTIN WIEDMANN

INTRODUCTION. Listeriosis has emerged as a human foodborne disease of major concern during the last two decades (Wesley 1999; Farber and Peterkin 1991). The emergence of listeriosis could be explained by factors such as changes in food production, handling and preparation practices, demographic changes in the population such as an increased proportion of elderly and immuno-compromised individuals, improved detection and diagnosis of listeriosis, and possibly the emergence of new bacterial strains (Rocourt and Cossart 1997). However, listeriosis is not a "new" disease. It was first recognized in rabbits in the 1920s (Murray et al. 1926) and in New Zealand sheep in 1931 (Gill 1931).

The genus *Listeria* contains seven species, only two of which are pathogenic: *L. ivanovii* and *L. monocytogenes* (Seeliger and Jones 1986). The former is mainly associated with disease in sheep, whereas the latter causes both human and animal disease. *Listeria monocytogenes* is thought to be ubiquitous. It has been found in many species of domestic and wild mammals, birds, fish, crustaceans, insects, and in the environment on six continents (Weis and Seeliger 1975). It can tolerate a variety of conditions, such as temperatures between 3 and 45° Celsius and a pH between 5.4 and 9.6. It can

even survive short periods of pasteurization temperatures and long periods of freezing (Cooper and Walker 1998). Appropriate pasteurization is effective in killing *L. monocytogenes.*

Because of it's the ubiquitous nature of *Listeria*, contamination of the environment is common, and proper farm management and food handling are essential. For example, on the farm, *Listeria* is found in many environmental sources, including water, bedding, and feed. Birds may contaminate feed and water troughs with their feces (Welshimer and Donker-Voet 1971). Farm animals may also be infected through contaminated silage. For example, it is suggested that animals fed silage from the bottom of a silo may be at risk of acquiring listeriosis caused by soil contamination (Seeliger 1961). Although contaminated raw agricultural commodities (such as milk) may be direct sources of human listeriosis cases, post-processing contamination from processing plant environments and equipment appears to be a more common source of food product contamination. *L. monocytogenes* appears to be able to survive in the environment of food processing plants and on the surface of processing equipment (Fig. 23.1), facilitating post-processing contamination. A variety of food products have been found contaminated with *Listeria spp.*, including raw and processed meats, uncooked seafood, and leafy vegetables (Gravani 1999; Harvey and Gilmour 1994). A recent USDA/FDA draft *L. monocytogenes* risk assessment provides detailed information on different food sources and their relative risk of causing human listeriosis cases (FDA 2001). Although such cases are rare, cases of listeriosis occurring through direct contact through cuts in the skin or through the eyes have been documented (Cooper and Walker 1998).

This chapter describes the epidemiology and molecular epidemiology of listeriosis, including foodborne outbreaks, and summarizes the implications of this knowledge for control of listeriosis.

LISTERIOSIS IN ANIMALS. A significant number of animals may be asymptomatic carriers of *L. monocytogenes,* often shedding the organism in fecal material (Husu 1990). *Listeria monocytogenes* is predominantly transmitted orally. Invasion generally occurs in the intestinal tract with subsequent hematogenous spread leading to septicemia and possibly transuterine infection (Rocourt and Cossart 1997). The pathogenesis of listerial encephalitis is still somewhat controversial (Gray and Killinger 1966). Some indications are that that *L. monocytogenes* may enter through the oral mucous membranes (possibly through preformed lesions) and subsequently migrate centripetally inside cranial nerves to the brain and particularly to the brain stem (Chakraborty 1999). For example, oral lesions in animals changing dentition may provide an entry port for *L. monocytogenes* (Low and Linklater 1985; Barlow and McGorum 1985; Otter and Blakemore 1989).

Sheep and Goats. A common manifestation of ovine listeriosis is encephalitis that is a result of brain stem lesions caused by *L. monocytogenes* (Low and Renton 1985). Common signs of encephalitis include fever,

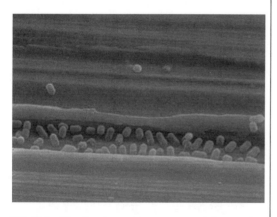

FIG. 23.1—*Listeria monocytogenes* on a scratch in a metal surface. (© Amy Lee Wong, licensed for use, ASM Microbe Library [linked to http://www.microbelibrary.org].)

depression, lack of feed and water intake, dullness, turning or twisting of the head to one side, and movement in circles. Listeriosis has been observed in lambs as young as four to eight months (Wesley 1999). The incubation period can be up to three weeks and in most cases of listerial encephalitis, death occurs within two to three days of the onset of clinical signs (Ladds et al. 1974).

Abortion is another common clinical manifestation of listeriosis in sheep. *Listeria monocytogenes* is transmitted to the fetus via the placenta, leading to a septic infection of the fetus. Clinical signs in the ewe are resolved following abortion of the fetus, which usually occurs as a stillbirth in the third trimester of pregnancy. However, a dam's subclinical infection can lead to the birth of an apparently healthy lamb that becomes septicemic and dies within a few days (Gronstol 1979). Septicemia is the most frequent infection type in neonates and lambs and has been documented to develop in lambs as young as five weeks old (Wesley 1999). Up to 5 to 10 percent of sheep exposed to *L. monocytogenes* during an outbreak have been reported to show clinical signs (Kimberling 1988). Many more are likely to have subclinical infections and to shed the bacteria in their feces (Low et al. 1993).

The clinical manifestations of listeriosis in goats closely resemble those seen in sheep: encephalitis, septicemia, and abortion. In cases of listerial encephalitis in goats, meningitis frequently develops secondarily to encephalitic brain stem lesions. As with sheep, goats may be asymptomatic carriers, shedding *L. monocytogenes* in the feces and milk. This shedding provides the potential for environmental contamination and infection of newborn kids housed with the does; the infection can be transmitted through the navel or by sucking on dirty teats.

Cattle. Some sources report that cattle account for about 80 to 90 percent of reported animal listeriosis cases in North America (Wesley 1999). This may not necessarily reflect a higher true incidence of listeriosis in cattle, but instead may reflect a lower reporting rate for listeriosis cases in small ruminants. Encephalitis, abortion, septicemia, and mastitis caused by *L. monocytogenes* have been documented in cattle (Cooper and Walker 1998). In an encephalitic infection, death usually occurs about two weeks after the appearance of the first clinical signs, in contrast to the very rapid death seen in sheep.

The same clinical signs seen in sheep and goats characterize bovine listerial septicemia. Listeriosis may also manifest as abortion in cattle. Less common manifestations of listeriosis in cattle include mastitis, kerato-conjunctivitis, and iritis (Cooper and Walker 1998). Listerial mastitis is characterized by shedding of *L. monocytogenes* in the milk, either accompanied by a general septicemia or without any other clinical signs. In mastitic cows the condition may be chronic, with the animal shedding the bacteria in her milk intermittently for up to twelve months. An interesting note is that healthy calves can be born to chronic carriers who shed *L. monocytogenes* in their milk. Listerial keratoconjunctivitis and iritis are thought to be caused by direct contact between the eye and contaminated silage, for example, when an animal reaches its head into the feed (Wiedmann and Evans 2002).

Diagnosis. Clinical diagnosis of listeriosis is generally difficult because a variety of other diseases can cause similar clinical signs. Definitive diagnosis of listeriosis can be achieved by bacterial culture, by histopathology, and sometimes by serology. In many cases, definitive diagnosis of listeriosis cannot be accomplished in a live animal, but requires a post-mortem pathological and histopathological examination, microbial culture, or both.

Because listeriosis is often a rapidly progressing disease, rapid diagnosis and treatment initiation immediately after onset of clinical signs are necessary for successful treatment. However, early diagnosis prior to death is rarely achieved, so treatment is often unsuccessful (Rebhun and Delahunta 1982).

EPIDEMIOLOGICAL DISEASE PATTERNS. The epidemiological aspects and pathogenesis of infection in ruminants remain poorly understood. Because listeriosis is not a reportable animal disease, a good estimate of the incidence of the disease is currently not available. Reports in the literature tend to be biased toward outbreaks (Wiedmann et al. 1994; Vandegraff et al. 1981), whereas low-incidence, sporadic disease occurrence is often neither reported nor diagnosed as listeriosis. The clinical disease patterns that are observed in animals would indicate that the majority of cases are single events, with occasional outbreaks of abortions and encephalitis (Wesley 1999).

Sporadic Listeriosis. Dutch data indicated that approximately 1.2 percent of all abortion cases that were submitted to a large diagnostic laboratory could be diagnosed as having been caused by *L. monocytogenes*. The incidence of listerial abortions was much higher (7 percent) in years when poor silage was common (Dijkstra 1986). Similar prevalences have been reported from South Dakota, Germany, and the midwestern United States (Wesley 1999). Abortion occurs in approximately 3 to 5 percent of pregnancies; therefore, the incidence of clinical *Listeria* infections can be estimated in this population as three to five cases per 10,000 cow-years at risk. *Listeria* may cause mastitis and may also be cultured from milk samples collected from cows (Jensen et al. 1996). In a Danish study, the percentage of cows that were cultured in the Danish mastitis control program and were infected with *L. monocytogenes* varied from 0.01 to 0.1 percent (mean .04 percent). The number of herds in this study that showed at least one infected cow varied between 0.2 to 4.2 percent (mean 1.2 percent) through the twenty-three-year period (Jensen et al. 1996).

Numerous species, including man, have been shown to shed *L. monocytogenes* without exhibiting any symptoms of disease (Farber and Peterkin 1991). The prevalence of these healthy shedders is not precisely known, but some studies in man suggest a prevalence as high as 10 percent. Studies in animal populations also show the presence of *L. monocytogenes* in healthy carriers. Healthy cows from herds without recent clinical cases of listeriosis showed a prevalence of approximately 2 percent, whereas in herds with recent cases of listeriosis a prevalence in healthy animals of approximately 7 percent was observed (Farber and Peterkin 1991). A large Finnish study observed a prevalence of healthy shedders of 9.2 percent in the housing season and 3.1 percent in the grazing season (Husu 1990). This study indicates that infection is present throughout the year, although infection prevalence may be somewhat higher in the housing season. It is difficult to differentiate shedding and infection from persistent exposure (for example, through contaminated silage), so these data need to be interpreted carefully.

Bulk milk contamination with *Listeria* spp. and *L. monocytogenes* has been investigated in several studies. Prevalences between 4 percent and 10 percent were reported (Wesley 1999). In most studies, seasonality in prevalence was observed, with a lower prevalence in the hot weather months and a higher prevalence in the colder months when cows are housed inside and fed silage. In a recent study from New York State, 404 dairy farms were investigated (Hassan et al. 2000). In-line milk filters were collected from each farm for bacteriological examination of *L. monocytogenes* and *Salmonella* spp. *Listeria monocytogenes* was isolated from fifty-one (12.6 percent) of the milk filters. Region-specific differences in the rate of farms with positive milk filters for this pathogen were observed. Although some variation appeared to exist in the seasonal prevalence, the variation was not found statistically significant. However, the prevalence in spring (18 percent) was higher compared to other seasons of the year (Fig. 23.2). The prevalence declined by 6 to 8 percent from spring to winter. *Listeria monocytogenes* was almost twice as likely to be isolated from samples in spring than in other seasons (odds ratio = 1.8).

***Listeria* Outbreaks in Animals.** Outbreaks of listeriosis described in the literature usually show a range of attack rates from 5 percent to 30 percent. The case-fatality rate varies from 20 percent to almost 100 percent (Wesley 1999; Vazquez-Boland et al. 1992; VandeGraaff et al. 1981; Wiedmann et al. 1997). From the published reports, outbreaks in sheep and goat herds appear to have higher attack rates (up to 30 percent) compared to cattle herds (up to 15 percent) (Low and Donachie 1997; Rebhun and Delahunta 1982).

An outbreak of listeriosis in a large sheep flock of approximately 650 sheep was described by Wiedmann et al. (1997). The incidence of mortality was approximately 2 percent; the outbreak curve is shown in Figure 23.3. Molecular evidence using ribotyping sug-

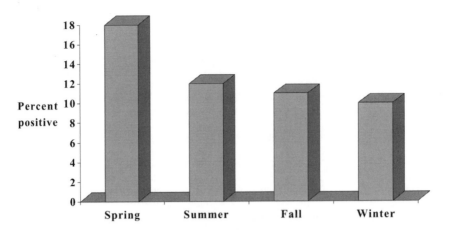

FIG. 23.2—The percentage of milk filters positive for *L. monocytogenes* sampled from New York State dairy herds. (From Hassan et al. 2000.)

gested that a tractor with front loader picked up the causal organism from a corn dump and then contaminated the silage that was fed to the sheep. The epidemiological data as depicted in Figure 23.3 also suggest that feeding silage was causally involved in the epidemic. Interestingly, none of the dairy cattle on the farm fed with the same front loader was affected with clinical listeriosis.

Risk Factors. Because most studies report a higher incidence of listeriosis in the housing season, risk factors present in the housing season and not in the grazing season likely are candidates for an association with clinical listeriosis. A case-control study involving 128 selected dairy farms was conducted by Sanaa et al. (1993) to assess the association of several suspected risk factors with the odds of contamination of raw milk by *L. monocytogenes*. Poor quality of silage (pH > 4.0), inadequate frequency of cleaning the exercise area, poor cow cleanliness, insufficient lighting of milking barns and parlors, and incorrect disinfection of towels between milkings were significantly associated with milk contamination by *L. monocytogenes* (Sanaa et al. 1993).

Although *L. monocytogenes* has been isolated from many locations in the environment of farm animals, most research indicates that silage has the highest prevalence. Silage, particularly of poor quality (usually determined by a relative high pH), may contain a large number of bacterial cells (>10^6 cfu per gram of material). In a recent Vermont study, 107 high-quality (pH < 4.0) corn silage samples accounted for eight *Listeria* spp. isolates, whereas five *Listeria* spp were isolated from twenty-two low-quality corn silage samples. In seventy low-quality hay silage (pH ≥ 4.0) samples, twenty *Listeria* spp were isolated, whereas one *Listeria* spp. was isolated from six high-quality hay silage samples (Ryser et al. 1997). Overall, poor quality silage

resulted in a three-times-higher isolation rate of *Listeria* spp. (Ryser et al. 1997).

Risk factors in animals include calving or lambing stress, concurrent diseases, changes in dentition, and viral or physical damage to the intestinal epithelial lining (Wesley 1999). Other stress-related events such as transport, changes in diet, adverse weather conditions, and parturition have also been associated with clinical disease. Administration of dexamethasone to subclinically infected animals increased the shedding levels up to 100-fold, providing more evidence for a relationship between listeriosis and stress (Wesley et al. 1989).

MOLECULAR EPIDEMIOLOGY. In the last few years, molecular subtyping tools have provided a more precise genetic comparison between *L. monocytogenes* isolates from different sources (Graves et al. 1999). These tools have increased the understanding of the epidemiology of *L. monocytogenes*. Studies that implicated animals as the main source of human infections have been refuted using such microbial genomic techniques (Farber and Peterkin 1991). In these cases, contamination of food was traced back to *L. monocytogenes* strains resident in packing plants, but not necessarily identical to the ones identified on the carcasses or finished products (Ryser et al. 1999). Such studies highlight the importance of precise isolate identification techniques in the determination of environmental sources that put animals and ultimately people at risk for infection (Jeffers et al. 2001).

For more in-depth coverage on molecular subtyping tools, the reader is referred to specific book chapters and reviews on this topic (Graves et al. 1999; Wiedmann 2002). Phenotypic and DNA-based subtyping methods allow for differentiation of *L. monocytogenes* beyond the species and subspecies level. Subtyping methods not only have improved the ability to detect

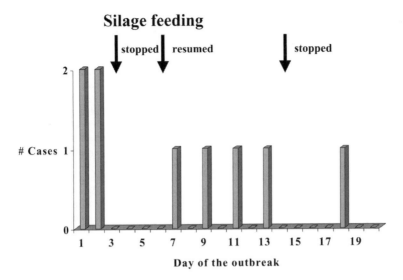

FIG 23.3—Outbreak of Listeriosis in a sheep farm (from Wiedmann et al. 1997). The arrows indicate the silage feeding procedures on the farm.

and track animal and human listeriosis outbreaks but also provide tools to trace sources of *L. monocytogenes* contamination throughout the food system. The last five years have provided tremendous advancements in the development of sensitive, rapid, and automated molecular subtyping methods for *L. monocytogenes* (Wiedmann 2002a). A comprehensive comparison of different subtyping methods for *L. monocytogenes* has been performed under the auspices of the World Health Organization (WHO), and the results from this study were published in 1996 (Bille and Rocourt 1996). Commonly used DNA-based subtyping approaches for *L. monocytogenes* include PCR-based methods (such as random amplification of polymorphic DNA, RAPD), Pulsed-Field Gel Electrophoresis (PFGE), and ribotyping. DNA-sequencing-based subtyping methods for *L. monocytogenes* are currently being developed and will likely find increasing use in molecular epidemiology (Wiedmann 2002a).

Serotyping represents one of the first phenotypic subtyping methods for *L. monocytogenes* (Seeliger and Hohne 1979). This method is based on the fact that different strains of bacteria differ in the antigens they carry on their surfaces. Antibodies and antisera can detect these surface antigens. Although serotyping has been used in epidemiological studies, this method shows poor discriminatory power, particularly as compared to many molecular subtyping methods (Graves et al. 1999). Specifically, serotyping differentiates thirteen different *L. monocytogenes* serotypes (Farber and Peterkin 1991), whereas some molecular subtyping methods yield more than one hundred different *L. monocytogenes* subtypes, allowing more sensitive strain discrimination and a higher level of standardization and reproducibility. Molecular subtyping methods

have mostly replaced serotyping as laboratory tools for epidemiological studies of *L. monocytogenes*. The rest of this section will thus focus on molecular subtyping methods.

Multilocus Enzyme Electrophoresis (MLEE). MLEE is a molecular subtyping method that differentiates bacterial strains by variations in the electrophoretic mobility of different constitutive enzymes (Boerlin and Piffaretti 1991). Cell extracts containing soluble enzymes are separated by size in non-denaturing starch gels, and enzyme activities are determined in the gel by using color-generating substrates (Graves et al. 1999). This method usually provides 100 percent typability but is difficult to standardize between laboratories. MLEE has been widely used for studies on the population genetics of many bacterial pathogens, including *L. monocytogenes* (Piffaretti et al. 1989). Although MLEE was used for epidemiological studies of human listeriosis in the early 1990s (Bibb et al. 1989; Harvey and Gilmour 1994), it appears to be less discriminatory than some DNA-based subtyping methods (Graves et al. 1999).

Pulsed-Field Gel Electrophoresis (PFGE). PFGE characterizes bacteria into subtypes (sometimes referred as *pulsotypes*) by generating DNA banding patterns after restriction digestion of the bacterial DNA. DNA banding patterns for different bacterial isolates are compared to differentiate distinct bacterial subtypes (or strains) from those that share identical (or very similar) DNA fragment patterns. Restriction enzymes commonly used for PFGE typing of *L. monocytogenes* include *Asc*I and *Apa*I. PFGE of a given isolate is often performed using different restriction enzymes in separate reactions to achieve improved discrimination (Brosch et al. 1994). PFGE shows a high level of sensitivity for discrimination of *L. monocytogenes* strains and is often considered the current gold standard for discriminatory ability (Autio et al. 1999). However, it is important to realize that PFGE as well as other subtyping methods may also sometimes detect small genetic differences that may not be epidemiologically significant (Tenover et al. 1995). Conversely, the detection of an identical PFGE type (or a subtype determined by another method) in two samples (such as a food sample and a sample from a clinically affected human) does not necessarily imply a causal relationship or a link between these two isolates. In outbreak investigations, molecular subtyping information needs to be analyzed in conjunction with epidemiological data to determine causal relationships between two or more isolates (Wiedmann 2002). The Centers for Disease Control and Prevention (CDC) and state health departments use PFGE in a national network (PulseNet) to exchange DNA subtypes for human and food isolates of foodborne pathogens in an effort to detect and track foodborne listeriosis outbreaks (Swaminathan et al. 2001).

Ribotyping. Ribotyping is another DNA-based subtyping method in which bacterial DNA is initially cut into fragments using restriction enzymes. Although PFGE uses restriction enzymes that cut the bacterial DNA into very few large pieces, the initial DNA digestion for ribotyping cuts DNA into many (>300–500) smaller pieces. These DNA fragments are separated by size through agarose gel electrophoresis. A subsequent Southern blot step uses DNA probes to specifically label and detect those DNA fragments that contain the bacterial genes encoding the ribosomal RNA (rRNA). The resulting DNA banding patterns are thus based only on those DNA fragments that contain the rRNA genes. The restriction enzyme *Eco*RI is commonly used for ribotyping of *L. monocytogenes* (Bruce et al.

1995; Jacquet et al. 1992). Although *Eco*RI ribotyping provides robust and sensitive differentiation of *L. monocytogenes* into subtypes that appear to correlate with phenotypic and virulence characteristics, the use of different restriction enzymes (for example, *Pvu*II) in separate reactions can provide increased strain discrimination (De Cesare et al. 2001; Gendel and Ulaszek 2000). A completely automated system for ribotyping (the RiboPrinter Microbial Characterization system) has been developed by Qualicon Inc. (Wilmington, DE) and is commercially available (Bruce 1996).

Use of ribotype analyses in epidemiological and population genetics studies confirmed and expanded findings from previous investigations using phenotype and MEE-based subtyping (Baxter et al. 1993). Ribotype data showed that on some farms a single subtype might be responsible for multiple listeriosis cases, whereas on other farms a variety of subtypes may cause infections of different animals during a listeriosis outbreak (Wiedmann et al. 1994, 1996, 1997, 1999). Figure 23.4 shows an example of ribotype results on isolates collected from clinical samples, healthy animals, and environmental sources on a single farm (unpublished data).

FOODBORNE OUTBREAKS AND ZOONOTIC LISTERIOSIS CASES.

Human *L. monocytogenes* infections can be manifested as a severe invasive disease, which predominantly occurs in the immuno-compromised, the elderly, and pregnant women. Typical symptoms of invasive listeriosis include septicemia, meningitis, and abortion. Human invasive *L. monocytogenes* infections are generally rare but are characterized by high case mortality rates between 20 and 40 percent. The incidence of invasive human listeriosis in developed countries is reported to vary between two and fifteen cases/million population (Farber and Peterkin 1991). Recent estimates by the researchers at

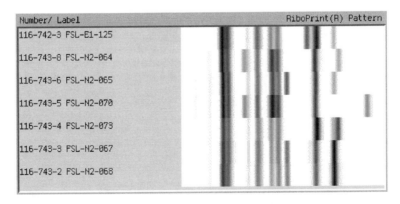

FIG. 23.4—*Eco*RI ribotype patterns for clinical and environmental *L. monocytogenes* isolates collected from a listeriosis case in a goat kid and from soil, water, and plant samples. The figure illustrates the diversity of *L. monocytogenes* subtypes found in farm environments.

CDC indicate that about 2,500 human listeriosis cases, including 500 deaths, occur in the United States on an annual basis (Mead et al. 1999). Infections of immuno-competent people are generally non-invasive and may manifest as mild, flu-like symptoms or as diarrheal disease. Some studies seem to indicate that the prevalence of *L. monocytogenes* in stool samples from healthy people and from people with diarrheal illnesses is around 1 percent to 2.5 percent. Considerably higher prevalences (up to 10 percent) have been reported for people without diarrhea who may be at higher risk of exposure to this organism, such as slaughterhouse workers, laboratory workers who have daily contact with *L. monocytogenes*, and household contacts of listeriosis patients (Slutsker and Schuchat 1999). A cutaneous form of human listeriosis has also been described (McLauchlin and Low 1994). Overall, no data appear to be available on the true incidence of these non-invasive manifestations of listeriosis (Slutsker and Schuchat 1999).

Although most human listeriosis infections traditionally have been thought to be sporadic, recent evidence indicates that more cases than generally assumed may represent case clusters and outbreaks. Surveillance of human listeriosis and detection of listeriosis clusters represent a particular challenge. Unlike diseases caused by many other foodborne pathogens, such as *E. coli* O157:H7 or *Salmonella*, foodborne listeriosis is characterized by long incubation periods (between seven and sixty days). Furthermore, only specific segments of the population (such as the immuno-compromised, the elderly, and pregnant women) are likely to develop clinical disease after exposure to contaminated foods. Thus, listeriosis outbreaks may occur over a wide geographical and temporal range and may not be detected by classical epidemiological approaches. The recent, more widespread application of molecular subtyping approaches, including the implementation of the PulseNet system (Swaminathan et al. 2001), has led to the detection of an increased number of human listeriosis outbreaks in the United States and other countries. Although only one human listeriosis outbreak was detected in the United States between 1990 and 1997 (Slutsker and Schuchat 1999), at least four human listeriosis outbreaks were detected between 1998 and 2000. One multistate outbreak in 1998-1999 involved more than one hundred cases, including twenty deaths, and was linked to consumption of contaminated hot dogs and deli meats (Anon. 1999). In 2000, one multistate outbreak involving at least twenty-nine cases was linked to consumption of contaminated sliced deli turkey (Anon. 2000), and another, single-state outbreak in 2000-2001 was linked to consumption of queso fresco (Anon. 2001).

Foodborne Transmission of *L. monocytogenes*. The majority of human listeriosis cases appear to be foodborne (Slutsker and Schuchat 1999). Some sources assume that as many as 99 percent of human cases are caused by foodborne infection (Mead et al. 1999). The minimum human infectious dose for *L. monocytogenes*

is not known and depends on many factors, including strain variation, environmental factors, and host susceptibility. In some countries, the presence of <100 CFU of *L. monocytogenes*/g is considered within regulatory limits for some ready-to-eat (RTE) foods, whereas other countries, such as the United States, enforce a "zero tolerance" for this organism in some or all RTE foods. It is believed that *Listeria* will always be present in the environment. Therefore the critical issue may not be how to prevent *Listeria* in foods, but how to control its survival and growth to minimize the potential risk (Donnelly 2001). In many foods, complete absence of *Listeria* may be unrealistic and probably not necessary; therefore, attempting to achieve this goal can limit trade without having an appreciable benefit to public health. A better understanding of the molecular epidemiology of *L. monocytogenes* may limit the need for zero tolerance to a relatively small number of more pathogenic strains. A relevant risk management option is to focus on foods that have historically been associated with human disease and support the growth of *Listeria* to high levels. The ability of *L. monocytogenes* to multiply at refrigeration temperatures represents an important consideration in the epidemiology of foodborne listeriosis. Even at temperatures less than 8°C, this organism may multiply from initially low numbers (for example, <1–10 CFU/g or ml) to reach numbers representing a considerable public health concern. This is a particular issue if products are stored under refrigeration for extended periods or if they are temperature-abused.

Animal sources can play an important role in animal-derived food products that are not processed before consumption (such as raw milk). For example, an outbreak involving forty-two cases in Nova Scotia in 1981 was linked to the consumption of coleslaw. This coleslaw was produced from cabbage harvested from fields fertilized with untreated sheep manure that was traced to a farm with a history of ovine listeriosis (Schlech et al. 1983).

Because *L. monocytogenes* is heat sensitive and is effectively destroyed by appropriate pasteurization procedures, contamination of most RTE foods seems to originate from post-processing environmental contamination rather than directly from animal sources. Although the prevalence of *L. monocytogenes* in most RTE foods ranges from <1–5 percent (Ryser 1999), in some surveys *L. monocytogenes* prevalence in certain RTE foods has been found to be higher than 10 percent (Nesbakken et al. 1996). The reservoirs of *L. monocytogenes* and the sources for the introduction of this organism into the food chain are still not completely understood. Nevertheless, no conclusive picture has emerged yet on the transmission dynamics and ecology of different *L. monocytogenes* subtypes. Further studies on the epidemiology, ecology, and population genetics of *L. monocytogenes* are needed to clarify the transmission pathways of this versatile pathogen and to better understand the importance of listeriosis infections in animals as a source and reservoir for strains causing human infections.

Direct Zoonotic Transmission of *L. Monocytogenes.* No evidence exists of direct animal-human transmission of *L. monocytogenes* leading to invasive human disease. At least two studies, though, report the occurrence of mild cutaneous listeriosis with symptoms, including papules and pustules, in veterinarians and farmers after contact with material from bovine abortions (Cain and McCann 1986; McLauchlin and Low 1994).

IMPLICATIONS FOR CONTROL. The described epidemiology and molecular epidemiology of *L. monocytogenes* have important applications for the control of this pathogen. In this section, the essential components of infection control are discussed.

Infection Control in Animals. True eradication or elimination of infection is not feasible with *L. monocytogenes.* This bacterium is generally present on the farm, but infection densities between farm premises appear to be different (Husu et al. 1990). The only feasible goal of control programs is to reduce exposure of animals and to reduce environmental contamination of products of animal origin (Hird and Genigeorigis 1990).

Because early diagnosis of listeriosis in animals is rarely achieved, appropriate treatment is often initiated late and unsuccessfully (Rebhun and Delahunta 1982). Treatment is therefore not a suitable method for reduction of shedding of infectious material in a herd or flock. Detection of carriers as a routine practice to reduce the environmental contamination is currently not feasible given the relatively high cost of diagnostics and the low incidence of disease. Whether detection of carriers and subsequent treatment would impact the overall exposure of animals in the herd is still questionable, because *L. monocytogenes* grows well in other areas of the environment, for example, ensilaged feed. Identification of risk factors and control methods focused on those risk factors (such as silage) are essential.

Control of Foodborne Contamination. To reduce the risk of foodborne infections, a complete process control is essential. Processes on the farm, in the processing plant, at the retailer and in the home affect the likelihood of bacterial contamination of edible products of animal origin (Husu 1990; Hassan et al., 2000; Sanaa et al., 1993; Bemrah et al., 1988). A schematic representation of the processes and the infection routes is presented in Figure 23.5. Without going into the detail of each of these processes, the on-farm processes will be discussed more thoroughly.

Harvesting. The process of milk harvesting is probably one of the most important steps in reducing the infection risk of this edible product. In a European study of 249 herds that were followed for twelve months, the prevalence of *L. monocytogenes* in milk reflected the prevalence in fecal material sampled at the same time, and did not reflect the prevalence in the

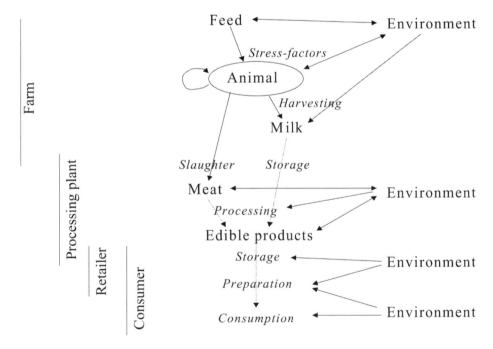

FIG. 23.5—Schematic presentation of the flow of products from farm to consumer (indicated by dotted arrows) and the possibilities for infection with *L. monocytogenes* (line arrows). Processes associated with risk of infection are indicated in italic.

main roughage used. Fecal material was considered the most important source of contamination of raw milk (Husu et al. 1990). Sanaa et al. (1993) also concluded that attention to good milking and barn hygiene are important for diminishing the risks of exogenous contamination of raw milk by *L. monocytogenes*. Studies showed that appropriate pre-milking cleaning procedures resulted in elimination of *L. monocytogenes* from the teat before milking (Husu et al. 1990). Hygienic milking equipment during all phases of the harvesting process is also essential for reduction of contamination risk (Husu et al. 1990; Nesbakken et al. 1996). Studies on milk filters indicated that these in-line filters are often soiled with fecal and environmental material. This may lead to contamination of the raw milk. Figure 23.6 shows how contamination of milking equipment with fecal material may be present. Hide contamination of animals before slaughter is also an important risk factor for carcasses being infected with *L. monocytogenes* (Gray and Killinger 1966; Nesbakken et al. 1996).

Post Processing. Pasteurization of milk is effective in killing all *Listeria* cells. However, heat treatments of meat and meat products help in reducing the bacterial load of edible products but do not necessarily eliminate all bacteria. Post-pasteurization contamination in processing plants has been documented (Norton et al. 2001; Linnan et al. 1988). Genotyping of isolates collected during the meat and milk processing has shown that *L. monocytogenes* strains present after processing and heat treatment may or may not be different from those found when the animal or the raw milk entered the processing plant (Wesley 1999). For example, case-control studies of an outbreak in California in 1985 implicated a Mexican-style soft cheese. Isolates collected from the patients, the cheese, and the environment of the processing plant all revealed identical phage types of *L. monocytogenes*, whereas no isolates

FIG. 23.6—A liner of a milking machine photographed between two milkings. (Courtesy of F. Welcome.)

were recovered from the raw milk of the farms supplying the processing plant (Linnan et al. 1988). Some cases have resulted in the identification of identical phage types of *L. monocytogenes* in pasteurized milk and the raw milk of farms supplying to the processing plant (Fleming et al. 1985). These studies suggest that the infection routes and processes as indicated in Figure 23.5 may all be responsible for contamination of the final edible product. Risk reduction programs are necessary along the entire production chain. No single process or step will result in elimination of *L. monocytogenes* contamination from edible products.

SUMMARY AND CONCLUSIONS. In this chapter we have discussed the epidemiological features of *L. monocytogenes* infections in animal populations and the risk for transmission of infection from an animal source to humans. The epidemiological pathways of infections have not all been illuminated yet (Foxman and Riley 2001). It is expected that further, precise descriptions of listeriosis using newly emerging molecular methods will shed light on the transmission cycles of *L. monocytogenes* in animal and human populations (Trott et al. 1993). In the vast majority of current descriptive studies, isolates are described at species level and hence the precise source and transmission routes are still not well understood. The development of new molecular techniques has improved the understanding of infection dynamics of *L. monocytogenes* in processing plants (Norrung and Skovgaard, 1993). For example it has shown that some of *L. monocytogenes* contamination cases were not caused by isolates from the raw farm product but from isolates residing in the processing facility (Lawrence and Gilmour 1995). Further studies on the dynamics of clone-specific transmission routes will further our understanding on the epidemiology of this important organism (vanBelkum et al. 2001).

The transmission dynamics of endemic listeriosis on farms is not yet fully understood. Still unclear is whether carrier animals are the key component, or whether continuous reinfection from the environment is responsible for endemic infections (Ryser et al. 1997). Herd-level studies on the dynamics of infection (again at the clonal level) are necessary to better understand the transmission cycle. This knowledge is also a prerequisite for the design of effective control programs. Current information would suggest that silage of poor quality combined with extra susceptibility in the host leads to outbreaks of clinical listeriosis (Arimi et al. 1997). On the other hand, observational studies have shown a high prevalence of healthy shedders, and these may indeed be able to maintain the infection in a population. Poor silage would in such a situation be a multiplier of existing infection pressure and not the primary cause (Husu 1990).

With the availability of modern diagnostic techniques, research into the epidemiology of *L. monocytogenes* has entered a new era. Many unanswered questions are expected to be tackled in the next five to ten years. This research may result in more definitive prevention programs for listeriosis in animals and humans.

REFERENCES

Anonymous. 1999. Update: multistate outbreak of listeriosis—United States, 1998–1999. *M.M.W.R.* 47:1117-1118.

Anonymous. 2000. Multistate outbreak of listeriosis—United States, 2000. *M.M.W.R* 49:1129–1130.

Anonymous. 2001. Outbreak of Listeriosis Associated With Homemade Mexican-Style Cheese—North Carolina, October 2000–January 2001. *M.M.W.R* 50:560–562.

Arimi, S.M., E.T. Ryser, T.J. Pritchard, and C.W. Donnelly. 1997. Diversity of *Listeria* ribotypes recovered from dairy cattle, silage, and dairy processing environments. *J. Food Prot.* 60:811–816.

Autio, T., S. Hielm, M. Miettinen, A.-M. Sjoberg, K. Aarnisalo, J. Bjorkroth, T. Mattila-Sandholm, and H. Korkeala. 1999. Sources of *Listeria monocytogenes* contamination in a cold-smoked rainbow trout processing plant detected by pulsed-field gel electrophoresis typing. *Appl. Environ. Microbiol.* 65:150–155.

Barlow, R.M., and B. McGorum. 1985. Ovine listerial encephalitis: analysis, hypothesis and synthesis. *Vet. Rec.* 116:233–236.

Baxter, F., F. Wright, R.M. Chalmers, J.C. Low, and W. Donachie. 1993. Characterization by multilocus enzyme electrophoresis of *Listeria monocytogenes* isolates involved in ovine listeriosis outbreaks in Scotland from 1989 to 1991. *Appl. Environ. Microbiol.* 59:3126–3129.

Bemrah, N, M. Sanaa, M.H. Cassin, M.W. Griffiths, O. Cerf. 1998. Quantitative risk assessment of human listeriosis from consumption of soft cheese made from raw milk. *Prev. Vet. Med.* 37:129–45.

Bibb, W.F., B. Schwartz, B.G. Gellin, B.D. Plikaytis, and R.E. Weaver. 1989. Analysis of *Listeria monocytogenes* by multilocus enzyme elctrophoresis and application of the method to epidemiologic investigations. *Int. J. Food Microbiol.* 8:233–239.

Bille, J., and J. Rocourt. 1996. WHO International Multicenter *Listeria moncoytogenes* subtyping study—rational and set-up of the study. *Int. J. Food Microbiol.* 32:251–262.

Boerlin, P., and J.-C.Piffaretti. 1991. Typing of human, animal, food, and environmental isolates of *Listeria monocytogenes* by multilocus enzyme electrophoresis. *Appl. Environ. Microbiol.* 57:1624–1629.

Brosch, R., J. Chen, and J.B. Luchansky. 1994. Pulsed-field fingerprinting of listeriae: identification of genomic divisions for *Listeria monocytogenes* and their correlation with serovar. *Appl. Environ. Microbiol.* 60:2584–2592.

Bruce, J. 1996. Automated system rapidly identifies and characterizes microorganisms in food. *Food Technol.* 50:77–81.

Bruce, J.L., R.J. Hubner, E.M. Cole, C.I. McDowell, and J.A. Webster. 1995. Sets of *Eco*RI fragments containing ribosomal RNA sequences are conserved among different strains of *Listeria monocytogenes. Proc. Natl. Acad. Sci. USA* 92:5229–5233.

Chakraborty, T. 1999. Molecular and cell biological aspects of infection by *Listeria monocytogenes. Immunobiol.* 201:155–163.

Cain, D.B., and V.L. McCann. 1986. An unusual case of listeriosis. *J. Clin. Microbiol.* 23:976–977.

Cooper, J., and R.D. Walker. 1998. Listeriosis. *Veterinary Clinics of North America: Food Animal Practice* 14:113–125.

De Cesare, A., J.L. Bruce, T.R. Dambaugh, M.E. Guerzoni, and M. Wiedmann. 2001. Automated ribotyping using different enzymes to improve discrimination of *Listeria monocytogenes*, with a particular focus on serotype 4b strains. *J. Clin. Microbiol.* 39:3002–3005.

Dijkstra, R.G. 1986. A fifteen year survey of isolations of *L. monocytogenes* out of animals and the environment in Northern Netherlands (1970–1984). In *Listeriose, Listeria, Listeriosis 1985–1986.* A.L. Courtieu, ed. Nantes, France: University of Nantes.

Donnelly, C.W. 2001. Listeria monocytogenes: a continuing challenge. *Nutr. Rev.* 59:183–194.

Edelson, B.T., and E.R.Unanue. 2000. Immunity to *Listeria* infection. *Current Opinion in Immunology* 12:425–431.

Farber, J.M., and P.I. Peterkin. 1991. *Listeria monocytogenes*, a food-borne pathogen. *Microbiol. Rev.* 55:76–511.

Farber, J.M., and P.I. Peterkin. 1999. Incidence and behaviour of *Listeria monocytogenes* in meat products. In *Listeria, listeriosis, and food safety.* E.T. Ryser and E.H. Marth, eds. New York: Marcel Dekker, Inc., 505–564.

FDA/USDA. 2001. Draft assessment of the relative risk to public health from foodborne *Listeria monocytogenes* among selected categories of Ready-to-Eat foods. Available at: http://www.foodsafety.gov/~dms/lmrisk.html.

Fleming, D.W., S.L. Cochi, K.L. MacDonald , J. Brondum , P.S. Hayes, B.D. Plikaytis, M.B. Holmes, A. Audurier, C.V. Broome, A.L. Reingold. 1985. Pasteurized milk as a vehicle of infection in an outbreak of listeriosis. *N. Engl. J. Med.* Feb. 14;312(7):404–7.

Foxman, B., and L. Riley. 2001. Molecular epidemiology: focus on infection. *Am. J. Epidemiol.* 153:1135–1141.

Gendel, S.M., and J. Ulaszek. 2000. Ribotype analysis of strain distribution in Listeria monocytogenes. *J. Food Prot.* 63:179–185.

Gill, D.A. 1931. Circling disease of sheep in New Zealand. *Vet J.* 87:60–74.

Gravani, R. 1999. Incidence and control of *Listeria* in food-processing facilities. In *Listeria, listeriosis and food safety.* E.T. Ryser and E.H. Marth, eds. New York: Marcel Dekker Inc., 657–709.

Graves, L.M., B. Swaminathan, and S.B. Hunter. 1999. Subtyping *Listeria monocytogenes*. In *Listeria, listeriosis, and food safety.* E. T. Ryser and E . H. Marth. eds. New York: Marcel Dekker Inc. 279–297.

Gray, M.L., and A.H. Killinger. 1966. *Listeria monocytogenes* and listeric infections. *Bacteriol. Rev.* 30:309–382.

Gronstol, H. 1979. Listeriosis in sheep. Listeria monocytogenes excretion and immunological state in sheep flocks with clinical listeriosis. *Acta Vet. Scand.* 20:417–428.

Harvey, J., and A. Gilmour. 1994. Application of multilocus enzyme electrophoresis and restriction length polymorphism analysis to the typing of *Listeria monocytogenes* strains isolated from raw milk, nondairy foods, and clinical and veterinary sources. *Appl. Environ. Microbiol.* 60:1547–1553.

Hassan, L., H.O. Mohammed, P.L. McDonough, and R.N. Gonzalez. 2000. A cross-sectional study on the prevalence of Listeria monocytogenes and Salmonella in New York dairy herds. *J. Dairy Sci.* 83:2441–7.

Husu, J.R. 1990. Epidemiological studies on the occurrence of Listeria monocytogenes in the feces of dairy cattle. *Zentralbl Veterinarmed* [B].Jun;37(4):276–82.

Husu, J.R., J.T. Seppanen, S.K. Sivela, and A.L. Rauramaa. 1990. Contamination of raw milk by Listeria monocytogenes on dairy farms. *Zentralbl Veterinarmed* [B].Jun;37(4):268–75.

Hird, D., and C. Genigeorgis. 1990. Listeriosis in food animals: Clinical signs and livestock as a potential source of direct (nonfoodborne) infections in humans. In: *Food-*

borne listeriosis. A.J. Miller, J.L. Smith, and G.A. Somkuti, eds.. Amsterdam: Elsevier, 31–39.

Jacquet, C., J. Bille, and J. Rocourt. 1992. Typing of *Listeria monocytogenes* by restriction polymorphism of the ribosomal ribonucelic acid gene region. *Zbl. Bakt.* 276:356–365.

Jeffers, G.T., J.L. Bruce, P. McDonough, J. Scarlett, K.J. Boor, and M. Wiedmann. 2001. Comparative genetic characterization of *Listeria monocytogenes* isolates from human and animal listeriosis cases. *Microbiol.* 147:1095–1104.

Jensen, N.E., F.M. Aarestrup, J. Jensen J, and H.C. Wegener. 1996. *Listeria monocytogenes* in bovine mastitis. Possible implication for human health. *Int. J. Food Microbiol.* 32:209–216.

Kimberling, C.V. 1988. Diseases of the central nervous system. In *Jensen and Swift's Diseases of Sheep*, C.V. Kimberling, ed. Philadelphia: Lea & Febiger, 195–199.

Ladds, P.W., S.M. Dennis, and C.O. Njoku. 1974. Pathology of listeric infection in domestic animals. *Vet. Bulletin* 44:67–74.

Lawrence, L.M. and A. Gilmour. 1995. Characterization of *Listeria monocytogenes* isolated from poultry products and from the poultry-processing environment by random amplification of polymorphic DNA and multilocus enzyme electrophoresis. *Appl. Environ. Microbiol.* 61:2139–2144.

Linnan, M.J., L. Mascola, X.D. Lou, V. Goulet, S. May, C. Salminen, D.W. Hird, M.L. Yonekora, P. Hayes, R. Weaver, A. Audurier, B.D. Plikaytis, S.L. Fannin, A. Kleks, and C.V. Broome. 1988. Epidemic listeriosis associated with Mexican-style cheese. *N. Engl. J. Med.* 319:823–828.

Low, J.C., and W. Donachie. 1997. A review of Listeria monocytogenes and listeriosis. *Vet. J.* 153:9–29.

Low, J.C., and K. Linklater. 1985. Listeriosis in sheep. *Practice* March 1985, 65–67.

Low, J.C., and C.P. Renton. 1985. Septicemia, encephalitis and abortions in a housed flock of sheep caused by *Listeria monocytogenes* type 1/2. *Vet. Rec.* 116:147–150.

Low, J.C., F. Wright, J. McLauchlin, and W. Donachie. 1993. Serotyping and distribution of *Listeria* isolates from cases of ovine listeriosis. *Vet. Rec.* 133:165–166.

McLauchlin, J., and J.C. Low. 1994. Primary cutaneous listeriosis in adults: an occupational disease of farmers and veterinarians. *Vet. Rec.* 135:615–617.

Mead, P., L. Slutsker, V. Dietz, L.F. McCaig, J.S. Bresee, C. Shapiro, P. Griffin, and R. V. Tauxe. 1999. Food-related illness and death in the United States. *Emerg. Infect. Dis.* 5:607–625.

Murray, E.G.D., R.A. Webb, and M.B.R. Swann. 1926. A disease of rabbits characterized by a large mononuclear leucocytosis caused by a hitherto undescribed *bacillus Bacterium monocytogenes.* *J. Pathol. Bacteriol.* 28:407–439.

Nesbakken, T., G. Kapperud, and D. Caugant. 1996. Pathways of Listeria monocytogenes contamination in the meat processing industry. *Int. J. Food Microbiol.* 31:161–171.

Norrung, B., and N. Skovgaard. 1993. Application of moltilocus enzyme electrophoresis in studies of the epidemiology of *Listeria monocytogenes* in Denmark. *Appl. Environ. Microbiol.* 59:2817–2822.

Norton, D., M. McCamey, K. Gall, J. Scarlett, K. Boor, and M. Wiedmann. 2001. Molecular studies on the ecology of *Listeria monocytogenes* in the smoked fish processing industry. *Appl. Environ. Microbiol.* 67:198–205.

Otter, A., and W.F. Blakemore. 1989. Observation on the presence of *Listeria monocytogenes* in axons. *Acta Microbiol. Hung.* 36:125–131.

Piffaretti, J.C., H. Kressebuch, M. Aeschenbacher, J. Bille, E. Bannerman, J.M. Musser, R.K. Seelander, and J. Rocourt. 1989. Genetic characterization of clones of the bacterium *Listeria monocytogenes* causing epidemis disease. *Proc. Natl. Acad. Sci. USA* 86:3818–3822.

Rebhun, W.C., and A. deLahunta. 1982. Diagnosis and treatment of bovine listeriosis. *J.A.V.M.A.* 180:395–398.

Rocourt, J., and P. Cossart. 1997. Listeria monocytogenes. In *Food Microbiology: Fundamentals and Frontiers.* M.P. Doyle, L.R. Beuchat, T.J. Montville, eds. Washington, D.C.: ASM Press.

Ryser, E.T. 1999. Incidence and behaviour of *Listeria monocytogenes* in cheese and other fermented dairy products. In *Listeria, listeriosis, and food safety.* E.T. Ryser and E.H. Marth, eds. New York: Marcel Dekker Inc. 411–503.

Ryser, E.T., S. M. Arimi, and C.W. Donnelly. 1997. Effects of pH on distribution of Listeria ribotypes in corn, hay, and grass silage. *Appl Environ Microbiol.* Sep;63(9):3695–7.

Sanaa, M., B. Poutrel, J.L. Menard, and F. Serieys. 1993. Risk factors associated with contamination of raw milk by *Listeria monocytogenes* in dairy farms. *J Dairy Sci.* 76:2891–8.

Schlech, W.F.I., P.M. Lavigne, R.A. Bortolussi, A.C. Allen, E.V. Haldane, A.J. Wort, A.W. Hightower, S.E. Johnson, S.H. King, E.S. Nicholls, and C.V. Broome. 1983. Epidemic listeriosis—evidence for transmission by food. *N. Engl. J. Med.* 308:203–206.

Seeliger, H.P.R., and K. Hohne. 1979. Serotyping of *Listeria monocytogenes* and related species. *Meth. Microbiol.* 13:31–49.

Seeliger, H.P.R. 1961. Listeriosis in Animals. In *Listeriosis.* New York: Hafner Publishing Co., Inc., 60–111.

Seeliger, H.P.R., and D. Jones. 1986. Genus Listeria. In *Bergey's Manual of Systematic Bacteriology* Volume 2. P.H.A. Sneath, N.S Mair, M.E. Sharpe, and J.G. Holt, eds. Baltimore: William and Wilkins, 1235–1245.

Slutsker, L., and A. Schuchat. 1999. Listeriosis in humans. In *Listeria, listeriosis, and food safety.* E. T. Ryser and E. H. Marth, eds. New York: Marcel Dekker Inc., 75–95.

Southwick, F.S., and D.L. Purich. 1996. Intracellular pathogenesis of listeriosis. *N. Engl. J. Med.* 334:770–776.

Swaminathan, B., T. Barrett, S. Hunter, and R. Tauxe. 2001. PulseNet: the molecular subtyping network for foodborne bacterial disease surveillance, United States. *Emerg. Infect. Dis.* 7:382–389.

Tenover, F.C., R.D. Arbeit, R.V. Goering, P.A. Mickelsen, B.E. Murray, D.H. Persing, and B. Swaminathan. 1995. Interpreting chromosomal DNA restriction patterns produced by pulsed-field gel electrophoresis: criteria for bacterial strain typing. *J. Clin. Microbiol.* 33:2233–2239.

Trott, D.J., I.D. Robertson, and D.J. Hampson. 1993. Genetic characterization of isolates of *Listeria monocytogenes* from man, animals, and foods. *J. Med. Microbiol.* 38:122–128.

van Belkum, A., M. Struelens, A. de Visser, H. Verbrugh, and M. Tibayrenc. 2001. Role of genomic typing in taxonomy, evolutionary genetics, and microbial epidemiology. *Clin. Microbiol. Rev.* 14:547–560.

Vandegraaff, R, N.A. Borland, and J.W. Browning. 1981. An outbreak of listerial meningo-encephalitis in sheep. *Aust. Vet. J.* 57:94–96.

Vazquez-Boland, J.A., L. Dominguez, M.Blanco, J. Rocourt, J.F. Fernandez-Garayzabal, C.B. Gutierrez, R.I. Tascon, and E.F. Rodriguez-Ferri. 1992. Epidemiologic investigation of a silage-associated epizootic of ovine listeric encephalitis, using a new Listeria-selective enumeration medium and phage typing. *Am. J. Vet. Res.* 53:368–371.

Weis, J., and H.P.R. Seeliger. 1975. Incidence of *Listeria monocytogenes* in nature. *Appl. Microbiol.* 30:29–32.

Welshimer, H.J., and J. Donker-Voet. 1971. *Listeria monocytogenes* in nature. *Appl. Microbiol.* 21:516–519.

Wesley, I.V. 1999. Listeriosis in Animals. In *Listeria, Listeriosis and Food Safety*. E.T. Ryser and E.H. Harth, eds. New York: Marcel Dekker, Inc., 39–73.

Wiedmann, M. 2002. Molecular subtyping methods for *Listeria monocytogenes*. *J. Assoc. Off. Anal. Chem.* (accepted).

Wiedmann, M., and K.E. Evans. 2002. Infectious diseases of dairy animals: Listeriosis. In *Encyclopedia of Dairy Science*. London: Academic Press (in press).

Wiedmann, M., T. Arvik, J.L. Bruce, F. del Piero, M.C. Smith, J. Hurley, H.O. Mohammed, and C.A. Batt. 1997. An outbreak investigation of listeriosis in sheep in New York state. *Am. J. Vet. Res.* 58:733–737.

Wiedmann, M., J.L. Bruce, R. Knorr, M. Bodis, E.M. Cole, C.I. McDowell, P.L. McDonough, and C.A. Batt. 1996. Ribotype diversity of *Listeria monocytogenes* strains associated with outbreaks of listeriosis in ruminants. *J. Clin. Microbiol.* 34:1086–1090.

Wiedmann, M., J. Czajka, N. Bsat, M. Bodis, M.C. Smith, T.J. Divers, and C.A. Batt. 1994. Diagnosis and epidemiological association of *Listeria monocytogenes* strains in two outbreaks of listerial encephalitis in small ruminants. *J. Clin. Microbiol.* 32:991–996.

Wiedmann, M., S. Mobini, J.R. Cole Jr., C.K. Watson, G. Jeffers, and K.J. Boor. 1999. Molecular investigation of a listeriosis outbreak in goats caused by an unusual strain of *Listeria monocytogenes*. *J.A.V.M.A.* 215:369–371.

DETECTION AND DIAGNOSIS OF *LISTERIA* AND LISTERIOSIS IN ANIMALS

IRENE V. WESLEY, MONICA BORUCKI, DOUGLAS R. CALL, DAVID LARSON, AND LINDA SCHROEDER-TUCKER

INTRODUCTION. Members of the genus *Listeria* are small (~0.5 x 0.5–2.0 mm) Gram-positive coccobacilli that can survive in a wide range of environments. For example, *L. monocytogenes* can grow at temperatures between 4°C and 45°C and between pH 4.5 and 9.6 (Fenlon 1999). There are six species of *Listeria*: *L. monocytogenes, L. ivanovii,* which are frank pathogens, as well as *L. seelegeri, L. murrayi, L. welshimeri,* and *L. innocua* (Rocourt 1999). Both humans and livestock are susceptible to infection with *L. monocytogenes,* especially the immunocompromised, including neonates and pregnant females. *L. ivanovii* is associated with abortion in livestock; human infections are rare. A disproportionately high number of *L. welshhimeri* were recovered from clinically healthy turkeys, suggesting that turkeys may be a preferred host (Genigeorgis et al. 1990). *L. innocua* is most closely related to *L. monocytogenes* (Tran and Kathariou 2002) and is generally regarded as a non-pathogen. Nevertheless, it has been infrequently recorded as a cause of listeric encephalitis in ruminants. Molecular methods, including sequencing of the variable region of the 16S rRNA gene as well as the absence of the *hly* and *prfA* genes, which are present in *L. monocytogenes,* verified *L. innocua* in brainstem tissues (Walker et al. 1994). Because of its public health significance in both veterinary and human medicine, we focus on listeriosis resulting from infection with *L. monocytogenes.*

Virtually all domestic animals are susceptible to listeriosis, with sheep, cattle goats, and, less frequently, poultry succumbing to infection (Gray et al. 1966; Rahlovich 1984; Seeliger et al. 1961; Wesley 1999). Sheep and goats may be more susceptible than cattle (Radostis 2000). The majority of *L. monocytogenes* infections are subclinical, resulting in healthy asymptomatic animals that shed *L. monocytogenes* in their feces. Clinical listeriosis in livestock can occur either sporadically or as epidemics and presents as encephalitis, abortions, and septicemia (Radostits 2000).

Listeriosis is reported most frequently in temperate and cold climates. Seasonal variation in the number of cases of animal listeriosis has often been observed especially during the fall, winter, and early spring (Kimberling 1988). In the Northern Hemisphere (England, Bulgaria, Hungary, United States, France, and Germany), listeriosis in domestic animals generally occurs from late November to early May, with the greatest incidence during February and March (Gray et al. 1966). Animals that are stressed by adverse weather, poor housing, and poor nutrition are more at risk to contract listeriosis, particularly if they are fed silage, which may contain ~10^6 *L.* monocytogenes /g (Radostits 2000). Other factors that alter susceptibility as well as the course of infection include genetic predisposition, sudden changes in feed, introduction of new animals into the flock, overcrowding, transportation, stress, concurrent viral disease, and climatic changes such as heavy rains. Physical or viral damage to the buccal cavity or epithelial lining of the digestive tract may facilitate entry of *L. monocytogenes* (Fenlon 1996; Green 1994; Killinger 1970; Kimberling 1988; Meredith 1984; Nash 1995; Reuter 1989).

L. monocytogenes resides in soil, water, decaying organic material (plant or animal), as well as in the gut of clinically normal humans and animals. Because it thrives in the environment, improperly prepared silage (pH >5.0) poses a special risk of listeriosis for livestock, especially when animals are stressed. To illustrate, in the Netherlands, most cases of listeric abortion in cattle occurred in winter between December and May, with approximately 40 percent of these cases linked to consumption of contaminated silage. In Great Britain, a significant association exists between silage feeding and development of ovine listeric encephalitais (relative risk of 3.8, Wilesmith 1986). Changes in silage production methods influence levels of *L. monocytogenes* (Fenlon 1999). In Great Britain, alterations in silage production may have preceded an increase in ovine listeriosis (Wilesmith 1986).

Although it does not result in reportable diseases, the prevalence of listeriosis in domestic animals has increased worldwide, especially in temperate climates. This increase may reflect changes in husbandry, fluctuations in the distribution of susceptible hosts, improved detection methods, or all of these factors (Anonymous 1988). To illustrate, in Canada, the majority of listeriosis cases occur in bovine (82 percent), with a smaller percentage in sheep (17 percent) and fewer still in pigs (Beauregard 1971). In marked contrast, from 1975 to 1984 listeriosis cases in Great Britain were primarily in sheep (63 percent) and cattle (32 percent), with pigs, goats, fowls and other species constituting less than 1 percent each of the total submissions (Wilesmith 1986). We analyzed listeriosis cases (n=253) submitted to the Iowa State University Veterinary Diagnostic Laboratory from 1993 to 2000. During this interval, the majority of clinical cases were in bovine (86 percent), with fewer in ovine (10 percent) and goats (4 percent). This result may reflect the host distribution of cattle (3.7 million) versus sheep (0.3 million) in the state. In addition, 88 percent of the listeriosis cases (305 of 346) submitted to the Iowa State University Veterinary Diagnostic Laboratory were diagnosed as encephalitis (Wesley et al. 2002).

In livestock—especially cattle and sheep—and humans, listeriosis presents with three clinical syndromes: encephalitis, abortion late in pregnancy, generalized infection (septicemia). It infrequently presents as ocular iritis and keratoconjunctivitis (Low 1997; Quinn et al. 1994; Radostits 2000).

DISEASES. In the following sections, the pathogenic, biological, and molecular characteristics of *L. monocytogenes* are discussed.

Encephalitis. *L. monocytogenes* can infect branches of cranial nerves, usually the trigeminal nerve, and then travel to the brain stem (pons and medulla) and anterior spinal cord. The lesions in the brainstem are not grossly visible, but can be seen with a light microscope. The most common lesion is a focal accumulation of mononuclear cells with or without neutrophils close to or around a blood vessel. In some cases, however, diffuse infiltration of inflammatory cells and microabscesses may occur in the brain stem with minimal tissue necrosis, and mononuclear leptomeningitis may be present. The affected animal has clinical signs that reflect the site of brain stem inflammation. Clinical signs include walking in circles (circling), head tilt, facial hypalgesia (decreased sensitivity), facial paralysis with drooping of the ear, dropped jaw caused by paresis of jaw muscles, paralysis of the lips, and ptosis (drooping of the upper eyelid). Ataxia with consistent falling to one side may be seen (Jones 1997). In the United Kingdom, meningo-encephalitis caused by *L. monocytogenes* is the most common bacterial infection of the central nervous system of adult sheep, although young lambs (~5 weeks) may develop the septicemic form (Scott 1993). Panophthalmitis with pus in the anterior chamber of one or both eyes may be observed in cattle.

Abortion. In cattle, *Listeria* abortion usually occurs in the last trimester (gestation period is ~280 days) without any clinical signs of infection in the cow. Fetuses usually die *in utero*, but stillbirths and neonatal deaths occur. Retention of the afterbirth is common and results in clinical illness with a high fever. Cattle abortions are usually sporadic, but the rate of abortion varies and may reach 15 percent in some herds (Radostits 2000).

In sheep, *Listeria* abortion may occur after the twelfth week of pregnancy (gestation period is ~21 weeks). The afterbirth is usually retained, and a bloody vaginal discharge occurs for several days after the abor-tion. Fetuses usually die *in utero*. Ewes that retain the dead fetuses may die from septicemia. The ewe abortion rate in a flock is usually low but may reach a rate of 20 percent in some outbreaks (Radostits 2000).

Listeria ivanovii can cause sporadic abortions in cattle with the same clinical features as *L. monocytogenes*. *L. ivanovii* can cause outbreaks in sheep characterized by abortion and stillbirth. Infected lambs do not survive (Gill et al. 1997).

Septicemia. Generalized listeriosis most commonly occurs in newborn calves, lambs, and monogastric animals (including human infants). Septicemia is not a common form of listeriosis in adult ruminants. Young lambs develop septicemia, whereas older feedlot lambs (four to eight months) typically manifest meningoencephalitis. The clinical signs associated with *L. monocytogenes* septicemia include nonspecific fever, weakness, depression, and emaciation. Some cases may also present with diarrhea with lesions of gastroenteritis evident at necropsy. Septicemia may cause retention of a dead fetus and placenta, and the nervous system normally is not involved. Corneal opacity, nystagmus, and opisthotonos can occur in septicemic neonatal calves less than ten days old, with ophthalmitis and serofibrinous meningitis evident at necropsy (Low 1997).

LABORATORY DIAGNOSIS. The protocol for identifying *L. monocytogenes* in a veterinary diagnostic laboratory is outlined in Figures 24.1A through 24.1C (Johnson et al. 1995). Even in cases of clinical meningoencephaitis, which represent the majority of listeriosis cases, usually no visible gross lesions occur in the brain at necropsy. Therefore, brains are prepared for histopathology as well as bacterial isolation as follows (Figure 24.1A). Hematoxylin-eosin (HE) stained sections of brainstem are examined for microscopic lesions typical of *L. monocytogenes*. These include cuffing of blood vessels by mostly mononuclear cells and a few neutrophils, microabscesses, and in some cases diffuse purulent inflammation of the neuropil of the brain stem. Nonsuppurative leptomeningitis may be evident. Special staining may reveal low numbers of small Gram-positive bacterial rods within microabscesses compatible with the diagnosis of encephalitic listeriosis (Jones 1997). Sections of brain stem that contain inflammatory lesions may be examined immunohistochemically for the presence of *Listeria* using a highly specific antiserum (Domingo et al. 1986). Polymerase chain reaction (PCR) assays may be adapted for detection of *L. monocytogenes* by *in situ* hybridization of tissue sections.

In aborted fetuses, multifocal necrotic lesions can be seen microscopically in the liver (Figure 24.1B). These foci of necrosis may also be seen in other organs such as the lung, spleen, and myocardium. The placenta may be necrotic, or inflamed and edematous, or all of these. Diagnosis can be made by identifying *Listeria* in the lesions of various organs and placenta with light micro-

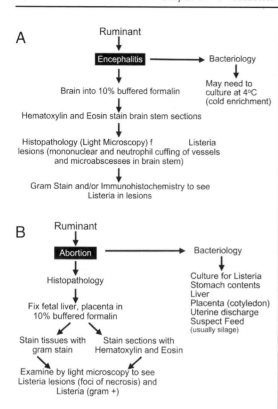

A

Ruminant

↓

Encephalitis ──────────→ Bacteriology

↓ ↓

Brain into 10% buffered formalin

May need to
culture at 4°C
(cold enrichment)

↓

Hematoxylin and Eosin stain brain stem sections

↓

Histopathology (Light Microscopy) f   Listeria
lesions (mononuclear and neutrophil cuffing of vessels
and microabscesses in brain stem)

↓

Gram Stain and/or Immunohistochemistry to see
Listeria in lesions

B

Ruminant

↓

Abortion ──────────→ Bacteriology

↓ ↓

Histopathology

Culture for Listeria
Stomach contents
Liver
Placenta (cotyledon)
Uterine discharge
Suspect Feed
(usually silage)

↓

Fix fetal liver, placenta in
10% buffered formalin

↓

Stain tissues with   Stain sections with
gram stain   Hematoxylin and Eosin

↓

Examine by light microscopy to see
Listeria lesions (foci of necrosis) and
Listeria (gram +)

C

Ruminant

↓

Septicemia
(visceral form)

↙   ↘

Fix viscera in 10%
buffered formalin

Bacteriology on liver and
other viscera for Listeria

↓

Light microscopy

↓

Typical Septicemic lesions
(foci or necrosis)

↙   ↘

Immunohistochemistry for
Listeria in lesions

Gram Stain for Listeria in
lesions

FIG. 24.1—Schematic of diagnostic procedure to recover *L. monocytogenes* from cases of encephalitis (A), abortion (B), and septicemia (C).

scopic examination of HE-stained tissues, the use of special bacterial stains such as Gram stain, or by immunohistochemistry as well as bacterial isolation (Jones 1997; Marco et al. 1988).

Adults with the septicemic form of *Listeria* have the foci of necrosis in the liver and other organs as do those of aborted fetuses (Figure 24.1C). *Listeria* can be identified in these lesions by microscopic examination of organs using HE and Gram stains, immunohistochemical staining of tissue sections [IHC and FA techniques], as well as bacterial culture (Jones 1997).

For bacterial isolation of *Listeria*, tissues, such as brain or fetal tissues, are plated to tryptose blood agar base supplemented with 5 percent defibrinated bovine blood and incubated in aerobic, anaerobic, and microaerobic (6 percent CO_2) environments (forty-eight hours, 37°C). Colonies grown on blood agar (1–2 mm in diameter after twenty-four hours) will be surrounded by a narrow zone of beta-hemolysis. *L. ivanovii* may be distinguished by a broad zone of hemoysis. Suspect *Listeria* colonies are Gram stained and identified by biochemical phenotype or by PCR. *L. monocytogenes* is differentiated from several other species of *Listeria* by production of acid from rhamnose (+), but not from mannitol and xylose. The CAMP test as well as pathogenicity in mice may also be used in identification (Donnelly 1999; Quinn 1994).

Contaminated specimens or specimens for which the number of *Listeria* organisms is low may require enrichment before any isolation can be made. In cold enrichment, tissues are homogenized and enriched in modified University of Vermont (UVM) *Listeria* enrichment broth (10 percent w/v) and incubated (4°C). Every week for up to twelve weeks, an aliquot is plated to either blood agar or to *Listeria* selective PAL-CAM agar (twenty-four to forty-eight hours, 37°C). Presumptive *Listeria* colonies, which appear as black colonies on the orange-red PALCAM agar, are transferred to either brain heart infusion enriched with 0.5 percent yeast extract or to blood agar slants. Up to five suspect colonies from each plate may be subcultured for serotyping and genetic analysis, especially in epidemiological investigations.

PCR assays may be used for screening of enrichment cultures as well as confirmation of presumptive *Listeria* colonies (Batt 1999) and are commercially available (Norton 2002). Second-generation fluorogenic 5′ nuclease PCR assays have been developed for the identification of *L. innocua* and *L. monocytogenes* (Bassler et al 1995; Batt 1999; Hein et al. 2001; Nogva et al. 2000; Norton 2002; Norton and Batt 1999). Undoubtedly, fluorogenic PCR assays combined with the advent of portable thermal cyclers may accelerate detection and quantitation of *Listeria*.

Serology. Although the humoral immune response may not have a major role in acquired resistance against listeriosis, serum antibodies are useful for the serodiagnosis of *Listeria* (Bhunia 1997). Heat-killed whole cells have been used in both direct agglutination and in the complement fixation formats. Antigens are commercially available (Difco, Detroit Mich.).

Because *Listeria* antibodies may cross-react with a variety of Gram-positive microbes, serosurveys of healthy livestock using heat-killed *Listeria* should be interpreted with caution. An alternative protocol involves testing for antibodies specific for listerolysin O (LLO) protein, which is unique to *L. monocytogenes*. Antibodies to LLO have been detected in human patients (Berche et al. 1990) as well as in experimentally infected lambs (Baetz et al. 1996), goats (Miettinen and Husu 1991), and cattle (Baetz et al. 1995). LLO antibodies have been detected in livestock by using ELISA or Western blot formats (Baetz et al. 1996). In our study, the agglutination test using heat-killed *L. monocytogenes* gave false-positive reactions when tested with serum from sheep experimentally infected with *L. innocua*. In contrast, only sera from *L. monocytogenes*–infected sheep gave a positive result in the highly specific LLO assay. The requirement for highly purified LLO is a major drawback to the widespread use of this test in serosurveys in livestock.

Serotyping and Phage Typing. Although thirteen serotypes are known, only three serotypes—4b, 1/2a, and 1/2b—account for the majority of veterinary and human cases (Tappero et al. 1995). Therefore, serotyping is of limited value in ecological and epidemiological investigations. Serotype 4b has caused the majority of human epidemics; sporadic cases of listeriosis are caused by serotypes 4b, 1/2a, and 1/2b. Likewise, ovine outbreaks are predominantly of 4b; sporadic ovine cases are mainly attributed to serovar 1/2 (Low 1993). Serotypes 4a and 4c may be restricted to animals because they are infrequently recovered from humans.

In the United States, Centers for Disease Control and Prevention and the Food and Drug Administration still maintain serotyping capabilities. These reagents are also commercially available (Accurate Scientific, Westbury, New York; Denka Seiken, Tokyo, Japan; Oxoid, Ogdensburg, New York; Toxin Technology, Sarasota, Florida). Alternatively, because 32 percent of clinical isolates are of serogroup 1 (1/2a, 1/2b, 1/2c) and 64 percent are attributed to serogroup 4 (4a, 4b, 4c), an inexpensive rapid slide agglutination test has been developed to broadly differentiate serogroups 1 and 4 (Difco Laboratories, Detroit, Michigan).

A limited number of reference laboratories perform phage typing, and this technique characterizes only ~60–80 percent of *L. monocytogenes* isolates (Audurier et al. 1986). Recent improvements to the system have incorporated phages for most serogroup 1/2 strains and the majority of serovar 4b isolates (Low et al. 1997). Nevertheless, dependence on reference laboratories for either serotyping or phage typing has accelerated the development of molecular formats to characterize *L. monocytogenes*.

Genetic Typing. DNA subtyping methods have been used to further characterize isolates of the same serotype, delineate the phylogeny of *L. monocytogenes*, trace the source of infection, and distinguish virulent from avirulent isolates. Genotyping analysis includes pulsed field gel electrophoresis (PFGE), ribotyping as well as PCR-based formats such as randomly amplified fragment length polymorphic DNA (RAPD), amplified fragment length polymorphism (AFLP), and amplification of enterobacterial repetitive intergenic consensus (ERIC) (Wiedmann 2002).

Macrorestriction enzyme analysis or PFGE uses rare cutting enzymes, such as *Asc*I and *Apa*I, that cleave DNA into large fragments (40–600 kb) that are then size-separated by pulsating electrical currents. The method is reliable and reproducible (range 79 percent to 90 percent) when compared with other protocols evaluated by the WHO Multicentre *L. monocytogenes* Subtyping Study (Graves et al. 1999). The use of standard conditions and availability of a national database recommend this technique for clonal identification. In the United States, CDC's PulseNet is based on PFGE as the preferred method to characterize clinical isolates of *L. monocytogenes*. PFGE has been used to differentiate serotype 4b isolates recovered from the major human listeriosis outbreaks (Brosch et al. 1994). Identity of PFGE profiles confirmed that contaminated chocolate milk was the source of a gastroenteritis outbreak in Illinois (Graves et al. 1999). In contrast, we studied a case of sporadic ovine meningoencephalitis that occurred on a sheep farm in the Midwest that also raised soybeans for the organic foods market. Distinctive PFGE profiles of *L. monocytogenes* isolates recovered from the sheep brain of the index case, from fecal samples of healthy sheep on the premises and from soybean compost piles, indicated that the index case was unrelated to other strains recovered on that farm (<50 percent homology). Although PFGE is the most reliable and discriminatory of existing subtyping techniques, it provides limited genetic characterization of isolates.

Ribotyping combines restriction enzyme digestion with Southern blot hybridization using probes to the genes encoding the 16S rRNA (Graves et al. 1999). A commercial system miniaturizes traditional restriction enzyme digestion with *Eco*RI, which generates numerous (>300–500) small fragments (2 to 20 kb), with Southern blot hybridization (RiboPrinter, Qualicon, Wilmington, Delaware). Uniform reagents, standardized methods, and access to a large database facilitate laboratory comparisons of patterns or ribotypes (Wiedmann 2002). Wiedmann et al. (1997a) evaluated seven clinical strains with isolates from grass and corn silage associated with an outbreak of ovine listeriosis. The ribotype identity coupled with epidemiological association suggested that sheep were probably infected by eating traces of contaminated corn silage left on farm equipment that was also used to feed cattle.

Random amplification of polymorphic DNA (RAPD) is a PCR-based method that uses low, nonstringent annealing temperatures so that the primers anneal at several sites to generate an array of DNA products. Identity of genetic profiles obtained by RAPD analysis linked *L. monocytogenes* serotype 1/2a present in the silage to an outbreak in sheep, whereas in another study, isolates from a goat brain and silage were shown to be clearly different by RAPD analysis (Vazquez-Boland et al. 1992; Wiedmann et al. 1994). Although RAPD is an inexpensive subtyping method, the WHO Multicentre *L. monocytogenes* Subtyping Study revealed inconsistent results between the six participating laboratories (average 86.5, range 0–100 percent, Graves et al. 1999).

Molecular analysis, using multilocus enzyme electrophoresis (MLEE), PFGE, allelic variation, sequence analysis, and AFLP, identified two main serotypic divisions within the *L. monocytogenes* species (Aarts et al. 1999; Bibb et al. 1990; Brosch et al. 1994; Nadon et al. 2001; Norton 2002; Piffaretti et al. 1989). Multilocus enzyme electrophoresis, which measured allelic variation for sixteen enzymes, again clustered *L. monocytogenes* into two major phylogenetic divisions (Piffaretti et al. 1989). By ribotyping, two divisions identical to those based on MLEE were also observed with the addition of new serotypes (Graves et al. 1994).

The key virulence genes for *L. monocytogenes* cell entry (*inlA*, *inlB*), cell invasion (*iap*), escape from the host phagocytic vacuole (*hlyA*), and cell-to-cell spread via polymerization of actin protein (*actA*) have been identified. Sequencing of *iap*, *hlyA*, and the *flaA*, which encodes flagellin proteins, likewise aligned serotypes into two major categories and assigned isolates of serotype 4a to a distinctive third cluster (Rasmussen et al. 1995). Based on these efforts, the major serotypes have been tentatively arranged into three clusters: I (1/2a, 1/2c, 3a, 3c), II (1/2b, 3b, 4b, 4d and 4e), and III (4a and 4c).

Virulence genes may be amplified and the resultant PCR products compared in multilocus sequence typing (MLST) (Wiedmann 2002) or cleaved with a specific restriction enzyme (PCR-RFLP). Combining ribotyping with PCR-RFLP analysis of the *hlyA* gene suggests that *L. monocytogenes* strains may differ in their pathogenicity for humans and livestock (Jeffers et al. 2001; Wiedmann et al. 1997b). Specifically, human isolates, including serotype 1/2b and epidemic 4b isolates, primarily cluster together in lineage I, whereas veterinary isolates cluster more frequently in another group, lineage III. Mismatch amplification mutation assay (MAMA-PCR) has also been used to classify isolates into putative evolutionary lineages and thereby estimate pathogenic potential (Jinneman et al. 2001).

Understanding the ecology of *L. monocytogenes* is key to identifying sources of both human and animal outbreaks and requires a means to differentiate between clonal lineages. Existing molecular subtyping techniques are adequate for strain typing in epidemiological studies but provide limited genotype information for individual isolates. Because the FDA maintains a "zero tolerance" policy for the presence of *L. monocytogenes* in ready-to-eat products, identification of genetic markers for serotypes and virulence genes may predict the relative health risks for different *L. monocytogenes* strains. Therefore, a subtyping technique that is accurate, discriminatory, genetically informative, and rapid would greatly aid in ecological and epidemiological investigations. The ideal genetic subtyping method would discriminate groups of epidemiologically related isolates with high resolution and provide insight into their virulence potential.

DNA microarrays are the latest technological advancement for multi-gene detection and diagnostics. In essence, a microarray is a reverse dot-blot that uses the same principles of hybridization (probes and targets) and detection that for many years have been used with established blotting methods such as Southern and Northern blots. Microarrays are typically composed of DNA "probes" (oligonucleotides of known sequence) that are bound to a solid substrate such as glass. Each microarray consists of lattice of spots of densely packed probes robotically imprinted onto the substrate. Probes are usually oligonucleotides or PCR products from cloned genes or gene fragments. For *E. coli* M54, (~4,639,000 bp genome) for example, one commercial genome array consists of probes (25-mer each) for ~4,200 open reading frames and ~2,700 intergenic regions. Each of the total of 219,360 oligonucleotide probes (25-mer each) is densely packed on a single chip (1.28 cm^2) (www.affymetrix.org). A comparable *L. monocytogenes* array to scan the ~2,944,528 bp genome (Glaser et al. 2001) would hypothetically accommodate ~147,220 probes. The "target" DNA (sequence to be analyzed) is labeled and applied to the microarray, where it hybridizes with its complementary probe. The hybridized targets are then detected using a fluorescent scanner or similar device.

DNA microarrays are used to identify differential mRNA expression for either eukaryotic or prokaryotic organisms. To illustrate, seventy-four up-regulated RNAs and twenty-three down-regulated RNAs were detected following *L. monocytogenes* infection of promyelocytic THP1 cell lines using a probe miroarray representing ~6,800 human genes (Cohen et al. 2000). Microarrays are also being used increasingly to examine differences in the composition of entire genomes. Arrays are constructed from either a single reference strain or from a mix of reference strains (Dong et al. 2001; Dorrell et al. 2001; Fitzgerald et al. 2001; Hakenbeck et al. 2001; Liang et al. 2001). The former method involves hybridizing genomic DNA from different bacterial strains to an array composed of all open reading frames from a single reference strain. This is a very powerful tool for identifying classes of genes that are "dispensable" versus genes that are the essence of the species itself (Boucher et al. 2001). Alternatively, mixed-genome microarrays function as parallel subtractive hybridization experiments. In this format, genomic DNA from a diversity of strains is pooled to construct a genomic library. The library is then used to construct an array composed of genes shared by all the original isolates and genes specific to different strains. Genomic hybridizations are then used to identify which of the gene probes are polymorphic between strains. When identified, the polymorphic probes from the array can be sequenced, and this information can be used to identify potential strain markers and potential functional genes related to virulence. Consequently, DNA microarrays can serve as a tool for subtyping while simultaneously serving as a tool for detection of specific markers and genes. After specific markers have been identified, PCR-based assays can be constructed for rapid strain typing. Both the DNA microarrays themselves and PCR assays with or without DNA

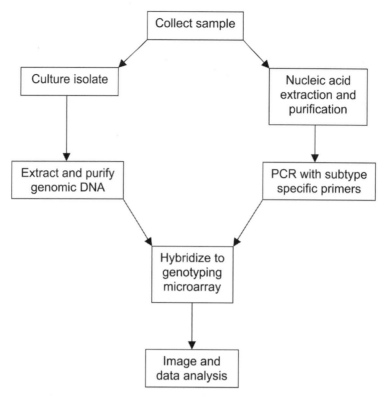

FIG. 24.2—Flowchart of *Listeria monocytogenes* subtyping using mixed-genome microarray technology. A mixed-genome microarray may be used to subtype laboratory isolates or used in conjunction with serotype/subtype-specific PCR primers to characterize *L. monocytogenes* DNA directly from a sample.

microarray detectors are amenable to standardization and automation (Figure 24.2).

Because microarray technology allows the rapid genetic characterization of a large number of strains, genetic diversity of a species can quickly be assessed in ecological and epidemiological studies. To investigate the utility of microarray technology for both epidemiological subtyping and gene discovery with *L. monocytogenes*, a 585-spot microarray was constructed from a mixed genomic library composed of DNA from ten different strains of *L. monocytogenes*. Twenty-nine strains of *L. monocytogenes* (six serotypes) were characterized using the microarray, and phylogenetic relationships were constructed by comparing the number of probes shared between strains (Borucki et al., in review) (Figure 24.3). Microarray analysis grouped isolates according to serotype lineage and, within serotype groupings, those with identical PFGE patterns clustered very closely (Figure 24.3: 12716E, 10867C, 10867D and 35568E, 32490G, 35568A). This preliminary 585-spot array assayed only a small portion of the entire coverage of the *L.* genome. Increasing the size of the array will enhance the discriminatory power of the array as well as identify additional potential markers for distinguishing important serogroups and subtypes of *L. monocytogenes*.

SUMMARY AND CONCLUSIONS. *L. monocytogenes* has been implicated as a cause of encephalitis, abortion, and septicemia in veterinary and human medicine. Because of its prominence as a human foodborne pathogen and as a major cause of food product recalls, intensive efforts have been made to improve its detection and diagnosis and to elucidate its epidemiology. Because of its high mortality rate (~25 percent of human clinical infections), molecular strategies, based in part on the publication of its entire genome sequence, will be critical in differentiating harmless from virulent *L. monocytogenes* strains. This is especially important in gauging the pathogenicity and therefore relative risk of environmental, food, and animal isolates.

REFERENCES

Anonymous. 1988. Veterinary Investigation Diagnosis and Analysis II. Ministry of Agriculture Fisheries and Food, Weybridge, England.

Aarts, H.J., L.E. Hakemulder, and A.M. Van Hoef. 1999. Genomic typing of *Listeria monocytogenes* strains by automated laser fluorescence analysis of amplified fragment length polymorphism fingerprint patterns. *Int. J. Food Microbiol.* 49:95–102.

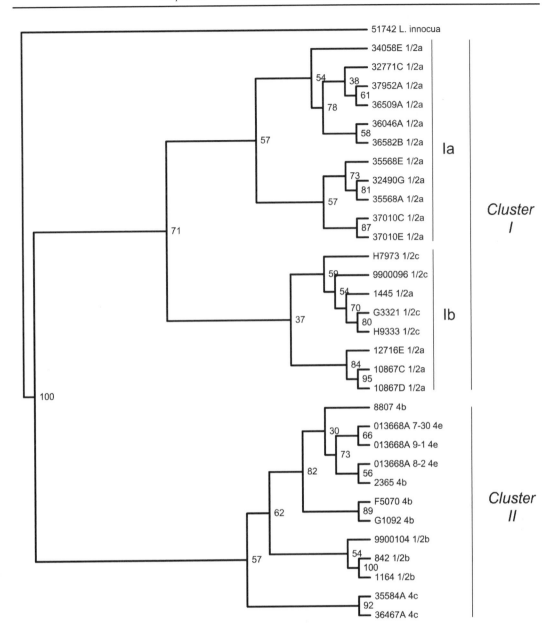

FIG 24.3—Dendogram (UPGMA and bootstrap; 1,000 replicates) showing the genetic relationship of *L. monocytogenes* isolates (Borucki at al., in review). Genomic DNA from 29 isolates was hybridized to a 585 probe mixed-genome microarray as described in the text. Isolates clustered according to serotype lineage. Cluster Ia is composed entirely of milk isolates (designated by a five-digit number followed by a letter). Clusters Ib and II include clinical and milk isolates. To check the reproducibility of the results, sample 013668A was prepared on three different dates and analyzed in three separate experiments. Additionally, samples 842 and 1164 are replicate samples.

Audurier, A., M. Gitter, and A. Raoult. 1986. Phage typing of *Listeria* strains isolated by the Veterinary Investigation Centres in Great Britain from 1981–1984. *Proceedings of the 9th International Symposium on the Problems of Listeriosis*. In A.L. Courtieu, E.P. Espaze, A.E. Reynaud, eds. France: University of Nantes, 410.

Baetz, A.L., and I.V. Wesley. 1995. Detection of antilisteriolysin O in dairy cattle experimentally infected with *Listeria monocytogenes*. *J. Vet. Diagn. Invest.* 7:82–86.

Baetz, A.L., I.V. Wesley, and M.G. Stevens. 1996. The use of listeriolysin O in an ELISA, a skin test and a lymphocyte blastogenesis assay on sheep experimentally infected with *Listeria monocytogenes*, *Listeria ivanovii* or *Listeria innocua*. *Vet. Microbiol.* 51:151–159.

Bassler, H.A., S.J.H. Flood, K.J. Livak,J. Marmaro, R. Knorr, and C.A. Batt. 1995. Use of a fluorogenic probe in PCR-based assay for the detection of *Listeria monocytogenes*. *Appl. Environ. Microbiol.* 61:3724–3728.

Batt, C.A. 1999. Rapid methods for detection of *Listeria*. In *Listeria, Listeriosis, and Food Safety, Second Edition*. E.T. Ryser and E.H. Marth, eds. New York: Marcel Dekker Inc., 261–278.

Beauregard, M., and K.L. Malkin. 1971. Isolation of *Listeria monocytogenes* from brain specimens of domestic animals in Ontario. *Can. Vet. J.* 12:221–223.

Berche, P., K.A. Reich, M. Bonnichon, J.L. Beretti, C. Geoffroy, J. Raveneau, P. Cossart, J.L. Gaillard, P. Geslin, H. Kreis, M. Veron. 1990. Detection of anti-listeriolysin O for the serodiagnosis of human listeriosis. *Lancet* 335:624–627.

Bhunia , A.K. 1997. Antibodies to *Listeria monocytogenes*. *Crit. Rev. Microbiol.* 23:77–107.

Bibb, W.F., B.G. Gellin, R. Weaver, B. Schwartz, B.D. Plikaytis, M.W. Reeves, R.W. Pinner, and C.B. Broome. 1990. Analysis of clinical and food-borne isolates of *Listeria monocytogenes* in the United States by multilocus enzyme electrophoresis and application of the method to epidemiologic investigations. *Appl. Environ. Microbiol.* 56:2133–2141.

Boucher, Y., C.L. Nesbo, and W.F. Doolittle. 2001. Microbial genomes: dealing with diversity. *Curr. Opin. Microbiol.* 4:285–289.

Brosch, R., J. Chen, and J.B. Luchansky. 1994. Pulsed-field fingerprinting of listeriae: identification of genomic divisions for *Listeria monocytogenes* and their correlation with serovar. *Appl. Environ. Microbiol.* 60:2584–2592.

Cohen, P., M. Bouaboula, M. Bellis, V. Baron, O. Jbilo, C. Poinot-Chazel, S. Galiegue, E.H. Habidi, and P. Casellas. 2000. Monitoring cellular responses to *Listeria monocytogenes* with oligonucleotide arrays. *J. Biol. Chem.* 275:11181–11190.

Donnelly, C.W. 1999. Conventional methods to detect and isolate *Listeria monocytogenes*. In *Listeria, Listeriosis, and Food Safety, Second Edition*. E.T. Ryser and E.H. Marth, eds. New York: Marcel Dekker Inc., 225–260.

Domingo, M., J.A. Ramos, L. Dominguez, L. Ferrer, and A. Marco. 1986. Demonstration of *Listeria monocytogenes* with PAP technique in formalin-fixed and paraffin-embedded tissues of experimentally infected mice. *J. Vet. Med. Ser. B.* 33:537–542.

Dong, Y., J.D. Glasner, F.R. Blattner, and E.W. Triplett. 2001. Genomic interspecies microarray hybridization: rapid discovery of three thousand genes in the maize endophyte, *Klebsiella pneumoniae* 342, by microarray hybridization with *Escherichia coli* K-12 open reading frames . *Appl. Environ. Microbiol.* 67:1911–1921.

Dorrell, N., J.A. Mangan, K.G. Laing, J. Hinds, D. Linton, H. Al Ghusein, B.G. Barrell, J. Parkhill, N.G. Stoker, A.V. Karlyshev, P.D. Butcher, and B.W. Wren. 2001. Whole genome comparison of *Campylobacter jejuni* human isolates using a low-cost microarray reveals extensive genetic diversity. *Genome Res.* 11:1706–1715.

Fenlon, D.R. 1999.*Listeria monocytogenes* in the natural environment. In *Listeria, Listeriosis, and Food Safety, Second Edition*. E.T. Ryser and E.H. Marth, eds. New York: Marcel Dekker Inc., 21–39.

Fitzgerald, J.R., D.E. Sturdevant, S.M. Mackie, S.R. Gill, and J.M. Musser. 2001. Evolutionary genomics of *Staphylococcus aureus*: insights into the origin of methicillin-resistant strains and the toxic shock syndrome epidemic. *Proc. Natl. Acad. Sci. USA* 98:8821–8826.

Genigeorgis, C.A., P. Oanca, and D. Dutulescu. 1990. Prevalence of *Listeria* spp. in turkey meat at the supermarket and slaughterhouse level. *J. Food Prot.* 53:282–288.

Gill, P.A., J.G. Boulton, G.C. Fraser, A.E. Stevenson, and L.A. Reddacliff. 1997. Bovine abortion caused by *Listeria ivanovii*. *Aust. Vet. J.* 75:214.

Glaser, P., L. Frangeul, C. Buchrieser, C. Rusniok, A. Amend, F. Baquero, P. Berche, H. Bloecker, P. Brandt, T. Chakraborty, et al. 2001. Comparative genomics of *Listeria* species. *Science* 294:849–852.

Gray, M. L., and A.H. Killinger. 1966. *Listeria monocytogenes* and listeric infections. *Bacteriol. Rev.* 30:309–382.

Graves, L.M., B. Swaminathan, M.W. Reeves, S.B. Hunter, R.E. Weaver, B.D. Plikaytis, and A. Schuchat. 1994. Comparison of ribotyping and multilocus enzyme electrophoresis for subtyping of *Listeria monocytogenes* isolates. *J. Clin. Microbiol.* 32:2936–2943.

Graves, L.M., B. Swaminathan, and S.B. Hunter. 1999. Subtyping *Listeria monocytogenes*. In *Listeria, Listeriosis, and Food Safety, Second Edition*. E.T. Ryser and E.H. Marth, eds. New York: Marcel Dekker Inc., 39–74.

Green, L.E., and K.L. Morgan. 1994. Descriptive epidemiology of listeria meningoencephalitis in housed lambs. *Prev. Vet. Med.* 18:79–87.

Hakenbeck, R., N. Balmelle, B. Weber, C. Gardes, W. Keck, and A. de Saizieu. 2001. Mosaic genes and mosaic chromosomes: intra- and interspecies genomic variation of *Streptococcus pneumoniae*. *Infect. Immun.* 69:2477–2486.

Hein, I., D. Klein, A. Lehner, A. Bubert, E. Brandl, and M. Wagner. 2001. Detection and quantification of the *iap* gene of *Listeria monocytenes* and *Listeria innocua* by a new real-time quantitative PCR assay. *Res. Microbiol.* 152:37–46.

Jeffers, G.T., J.L. Bruce, P.L. McDonough, J. Scarlett, K.J. Boor, and M. Wiedmann. 2001. Comparative genetic characterization of *Listeria monocytogenes* isolates from human and animal listeriosis cases. *Microbiol.* 147:1095–1104.

Jinneman, K.C., and W.E. Hill. 2001. *Listeria monocytogenes* lineage group classification by MAMA-PCR of the listeriolysin gene. *Curr. Microbiol.* 43:129–133.

Johnson, G.C., W.H. Fales, C.W. Maddox, and J.A. Ramos-Vara. 1995. Evaluation of laboratory tests for confirming the diagnosis of encephalitic listeriosis in ruminanats. *J. Vet. Diagn. Invest.* 7:223–228.

Jones, T.C., R.D. Hunt, and N.W. King. 1997. Diseases caused by bacteria. In *Veterinary Pathology, Sixth Edition*. Baltimore, MD: Williams and Wilkins, 461–463.

Killinger, A.H., and M.E. Mansfield. 1970. Epizootiology of listeric infection in sheep. *J. Amer. Vet. Med. Assoc.* 157:1318–1324.

Kimberling, C.V. 1988. Diseases of the central nervous system. In *Jensen and Swift's Diseases of Sheep, Third Edition*. Philadelphia, PA: Lea and Febiger, 195–199.

Liang, X., X.Q. Pham, M.V. Olson, and S. Lory. 2001. Identification of a genomic island present in the majority of pathogenic isolates of *Pseudomonas aeruginosa*. *J. Bacteriol.* 183:843–853.

Low, J.C., and W. Donachie. 1991. Clinical and serum antibody responses of lambs to infection by *Listeria monocytogenes*. *Res. Vet. Sci.* 51:185–192.

Low, J.C., and W. Donachie. 1997. A review of *Listeria monocytogenes* and listeriosis. *Vet. J.* 153:9–29.

Marco, A., J.A. Ramos, M. Dominguez, M. Domingo, and L. Gonzalez. 1988. Immunochemical detection of *Listeria monocytogenes* in tissue with the peroxidase-antiperoxidase technique. *Vet. Pathol.* 25:385–387.

Meredith, C., and D. Schneider. 1984. An outbreak of ovine listeriosis associated with poor flock management practices. *J. S. Afr. Vet. Med. Assoc.* 55:55–56.

Miettinen, A., and J. Husu. 1991. Antibodies to listeriolysin O reflect the acquired resistance of *Listeria monocytogenes* in experimentally infected goats. *FEMS Microbiol. Lett.* 77:181–186.

Nadon, C.A., D.L. Woodward, C. Young, F.G. Rodgers, and M. Wiedmann. 2001. Correlations between molecular subtyping and serotyping of *Listeria monocytogenes*. *Clin. Microbiol.* 39:2704–2707.

Nash, M.L., L.L. Hungerford, T.G. Nash, and G.M. Zinn. 1995. Epidemiology and economics of clinical listeriosis in a sheep flock. *Prev. Vet. Med.* 3:75–83.

Nogva, H.K., K. Ridi, K. Naterstad, A. Holck, and D. Lillehaug. 2000. Application of 5'nuclease PCR for quantitative detection *of Listeria monocytogenes* in pure cultures, water, skim milk, and unpasteurized whole milk. *Appl. Environ. Microbiol.* 66:4266–4271.

Norton, D. 2002. Polymerase chain reaction-based methods for detection of *Listeria monocytogenes*: toward real-time screening for food and environmental samples. *J. AOAC Int.* 85:505–515.

Norton, D., and C.A. Batt. 1999. Detection of viable *Listeria monocytogenes* with a 5' nuclease PCR assay. *Appl. Environ. Microbiol.* 65:2122–2127.

Piffaretti, J.C., H. Kressebuch, M. Aeschbacher, J. Bille, E. Bannerman, J.M. Musser, R.K. Selander, and J. Rocourt. 1989. Genetic characterization of clones of the bacterium *Listeria monocytogenes* causing epidemic disease. *Proc. Natl. Acad. Sci. USA* 86:3818–3822.

Quinn, P.J., M.E. Carter, B. Markey, and G.R. Cater. 1994. *Listeria* species. In *Clinical Veterinary Microbiology*. London: Wolfe Publishing Yearbook Europe Limited, 170–174.

Radostits, O.M., C.C. Gay, D.C. Blood, and K.W. Hinchcliff. 2000. Diseases caused by *Listeria* sp. In *Veterinary Medicine: A textbook of the diseases of cattle, sheep, pigs, goats and horses, Ninth Edition*. St. Louis, MO: WB Saunders Co., 736–741.

Ralovich, B. 1984. *Listeriosis Research—Present Situation and Perspective*. Budapest: Akademiai Kiado.

Rasmussen, O., P. Skouboe, L. Dons, L. Rossen, and J.E. Olsen. 1995. *Listeria monocytogenes* exists in at least three evolutionary lines: evidence from flagellin, invasive associated protein and listeriolysin O genes. *Microbiol.* 141:2053–2061.

Reuter, R., M. Bowden, and M. Palmer. 1989. Ovine listeriosis in south coastal Western Australia. *Aust. Vet. J.* 66:223–224.

Rocourt, J. 1999. The genus *Listeria* and *Listeria monocytogenes*: Phylogenetic position, taxonomy, and identification. In *Listeria, Listeriosis, and Food Safety, Second Edition*. E.T. Ryser and E.H. Marth, eds. New York: Marcel Dekker Inc., 1–20.

Scott, P.R. 1993. A field study of ovine listerial meningoencephalitis with particular reference to cerebrospinal fluid analysis as an aid to diagnosis and prognosis. *Br. Vet. J.* 149:165–170.

Seeliger, H.P.R. 1961. *Listeriosis*. New York: Hafner Publishing Co.

Tappero, J.W., A. Schuchat, K.A. Deaver, L. Mascola, and J.D. Wenger. 1995. Reduction in the incidence of human listeriosis in the United States. Effectiveness of prevention efforts? The Listeriosis Study Group. *JAMA* 273:1118–1122.

Tran, H.L., and S. Kathariou. 2002. Restriction fragment length polymorphisms detected with novel DNA probes differentiate among diverse lineages of serogroup 4 *Listeria monocytogenes* and identify four distinct lineages in serotype 4b. *Appl. Environ. Mirobiol.* 68:59–64.

Vazquez-Boland, H.J.A., L. Dominguez, M. Blanco, J. Rocourt, J.F. Fernandez-Garayzabal, C.B. Gutierrez, R.I. Tascon, and E.F. Rodriquez-Ferri. 1992. Epidemiologic investigation of a silage-associated epizootic of ovine listeric encephalitis, using a new *Listeria* selective enumeration medium and phage typing. *Am. J. Vet. Res.* 3:368–371.

Walker J.K., J.H. Morgan, J. McLauchlin, K.A. Grant, and J.A. Shallcross. 1994. *Listeria innocua* isolated from a case of ovine meningoencephalitis. *Vet. Microbiol.* 42:245–253.

Wesley, I.V. 1999. Listeriosis in animals. In *Listeria, Listeriosis, and Food Safety, Second Edition*. E.T. Ryser and E.H. Marth, eds. New York: Marcel Dekker Inc., 39–74.

Wesley, I.V., J.H. Bryner, and M.J. van der Maaten. 1989. Effects of dexamethasone on shedding of *Listeria monocytogenes* in dairy cattle. *Amer. J. Vet. Res.* 50:2009–2113.

Wesley, I.V., D.J. Larson, K.M. Harmon, J.B. Luchansky, and A.R. Schwartz. 2002. A case report of sporadic ovine listerial meningoencephalitis in Iowa with an overview of livestock and human cases. *Vet. Diag. Invest.* 14: (in press).

Wiedmann, M. 2002. Molecular subtyping methods for *Listeria monocytogenes*. *J. AOAC Int.* 85:524–531.

Wiedmann, M., T. Arvik, J. Bruce, J. Neubauer, F. del Piero, M.C. Smith, J. Hurley, H.O. Mohammed, and C.A. Batt. 1997. Investigation of a listeriosis epizootic in sheep in New York state. *Amer. J. Vet. Res.* 58:733–737.

Wiedmann, M., J.L. Bruce, C. Keating, A.E. , P.L. McDonough, and C.A. Batt. 1997. Ribotypes and virulence gene polymorphisms suggest three distinct *Listeria monocytogenes* lineages with differences in pathogenic potential. *Infect. Immun.* 65:2707–2716.

Wiedmann, M., J. Czajka, N. Bsat, N.M. Bodis, M.C. Smith, T.J. Divers, and C.A. Batt. 1994. Diagnosis and epidemiological association of *Listeria monocytogenes* strains in two outbreaks of listerial encephalitis in small ruminants. *J. Clin. Microbiol.* 32:991–996.

Wilesmith, J.W., and M. Gitter. 1986. Epidemiology of ovine listeriosis in Great Britain. *Vet. Rec.* 119:467–470.

25

FOODBORNE OUTBREAKS OF LISTERIOSIS AND EPIDEMIC-ASSOCIATED LINEAGES OF *LISTERIA MONOCYTOGENES*

SOPHIA KATHARIOU

INTRODUCTION. Human listeriosis is a rare but severe illness associated with infection by the facultative intracellular bacterium, *Listeria monocytogenes*. Individuals at primary risk are pregnant women, newborn babies, and individuals in several categories of immunosuppression, including organ transplantation recipients, HIV-infected patients, cancer patients and the elderly. A hallmark of human listeriosis is the severity of the symptoms (stillbirth and abortions for pregnant women; septicemia, meningitis and other central nervous system complications for other patients) and the high case fatality rate (20 to 30 percent, even with use of antibiotics). The clinical symptoms and epidemiology of human listeriosis have been reviewed extensively (Farber and Peterkin 1991; Schuchat et al. 1991; Jay 1996; Rocourt et al. 2000; Vazquez-Boland et al. 2001), as have the virulence mechanims and biology of the pathogen (Farber and Peterkin 1991; Jay 1996; Kathariou 2000; Vazquez-Boland et al. 2001).

The epidemiological and bacteriological investigations of the 1981 outbreak of listeriosis in the Maritime Provinces of Canada, provided, for the first time, documentation of the role of contaminated food (in this case, coleslaw) in transmission of the disease (Schlech et al. 1983). Subsequent epidemiological investigations have strongly suggested food-borne transmission in numerous outbreaks, and the pathogen has been frequently isolated from the implicated food vehicles (reviewed in Rocourt et al. 2000; Kathariou 2002; see also summary in Table 25.1). In contrast, only rarely has the mode of transmission been determined for sporadic cases. Although human listeriosis may be primarily foodborne, alternative, rarer modes of transmission may also operate. Nosocomial transmission among neonates has been documented on several occasions and has resulted in clusters of illness (McLaughlin and Hoffman 1989; Schuchat et al. 1991a; Pejaver et al. 1993). In addition, asymptomatic carriers may contribute to the epidemiology of listeriosis, although such involvement has not yet been characterized or documented (Rocourt et al. 2000; Kathariou 2002).

The capability of *L. monocytogenes* to contaminate food and cause severe foodborne illness has created substantial challenges both to public health and to the food industry. The organism is fairly ubiquitous and can colonize the food processing environment. It is also capable of growing at low temperature, thus becoming a problematic contaminant of cold-stored foods. In heat-processed foods in which competing microflora have been reduced or eliminated, environmental contamination by *Listeria* and subsequent refrigeration of the product can result in multiplication of the pathogen, which can reach numbers sufficient for infection. Although the infectious dose remains undetermined, it is now clear that food vehicles most likely to be commonly involved in human listeriosis are ready-to-eat foods, especially those that are highly processed, and foods that have been refrigerated for varying lengths of time and that are commonly consumed without substantial heating.

Numerous studies have addressed the general physiological, genetic, and virulence-related attributes of the pathogen that are of relevance to its contamination of food and its involvement in food-borne disease, and recent reviews of the subject have been published (Rocourt et al. 2000; Vazquez-Boland et al. 2001; Kathariou 2000, 2002). The special challenges posed by listerial contamination of the processing plant have also been recently reviewed (Tompkin et al. 1999; Tompkin 2002). In this review, focus is placed on a specific area of special relevance to human foodborne listeriosis: epidemic disease and the characteristics of the bacterial lineages that have become implicated in foodborne outbreaks of human listeriosis.

THE SPECIAL ROLE OF OUTBREAKS IN THE EPIDEMIOLOGY OF LISTERIOSIS. Epidemiologically, human listeriosis is characterized by outbreaks of the disease that occur periodically over a background of sporadic cases. Although much of human illness is sporadic, a disproportionate amount of attention has focused on outbreaks. From the food safety and scientific point of view, the reasons are evident and clearly valid: outbreaks provide not only a strong impetus but also a unique forum for the epidemiological and bacteriological studies needed to identify the agent, as well as the implicated food vehicle. From the economic, social, and public health point of view, there are equally compelling reasons for paying due attention to epidemics. In the food production and distribution systems prevalent in the United States and other industrialized nations, outbreaks often involve a product with wide distribution, and thus can cross state and national boundaries. Substantial burden is placed on the food industry by the associated nega-

243

Table 25.1—Selected Major Outbreaks of Foodborne Listeriosis[1], and Implicated Foods

Location	(No. of Cases)	Year	Serotype	Food vehicle	Reference
I. Dairy products					
Halle, former East Germany	(279)	1945	1/2a	Milk (raw)	Ortel 1968
Massachusetts (USA)	(49)	1983	4b	Milk (pasteurized)	Fleming et al. 1985
California (USA)	(142)	1985	4b	Mexican-style (Jalisco) cheese (improper pasteurization)	Linnan et al. 1988
Switzerland	(122)	1983–1987	4b	Vacherin Mont d'or cheese	Bille 1989
France	(37)	1995	4b	Soft cheese	Goulet et al. 1995
France	(14)	1997	4b	Soft cheese	Goulet et al. 1995
Finland	(25)	1999	3a	Butter	Lyytikäinen et al. 2000
N. Carolina (USA)	(12)	2000	4b	Mexican-style cheese (raw milk)	CDC 2001a
British Columbia (Canada)	(19)	2002	Not known	Specialty cheese (surface spray contamination)	BCCDC 2002
II. Meat products					
England	(300)	1989–1990	4b, 4b(X)	Paté	McLaughlin et al. 1991
France	(279)	1992	4b	Jellied pork tongue	Goulet et al. 1995
France	(39)	1993	4b	Potted pork	Goulet et al. 1993
USA (multistate)	(101)	1998–1999	4b	Hot dogs	CDC 1998, 1999
France	(10)	1999–2000	4b	Rillettes[2]	DeValk et al. 2001
France	(32)	1999–2000	4b	Jellied pork tongue	DeValk et al. 2001
USA (multistate)	(29)	2000	1/2a	Deli turkey meats	CDC 2001a
III. Other food products					
Nova Scotia, Canada	(41)	1981	4b	Coleslaw	Schlech et al. 1983
New Zealand	(4)	1992	1/2b	Smoked mussels	Brett et al. 1998
Sweden	(6–9)	1994–1995	4b	Gravad[3] rainbow trout	Ericsson et al. 1997

[1] Selected outbreaks of invasive disease, with abortions/stillbirths, septicemia, meningitis, or death as outcomes.
[2] Paté-like meat product.
[3] Product made from raw fish fillets rubbed with mixture of sugar, salt, and pepper; then covered with dill, placed in plastic bag, and left at refrigeration temperature for two days before becoming vacuum-packaged and released for consumption (Ericsson et al. 1997).

tive publicity, product recalls, and possible litigation. Outbreaks of listeriosis are associated with substantial disease burden because numerous patients are involved, with the high case fatality rates characteristic of human listeriosis (20 to 30 percent). Last but not least, timely recognition of outbreaks, along with associated product recalls and consumer alerts, have proven their value in the curtailing of the outbreak (CDC 1998, 1999, 2001; DeValk et al. 2001).

FOOD VEHICLES COMMONLY IMPLICATED IN FOODBORNE OUTBREAKS. Ready-to-eat (RTE) foods that are refrigerated before consumption and do not undergo any substantial treatment by the consumer have been most commonly implicated in outbreaks of listeriosis. The prevalent categories include dairy products, especially soft cheeses, processed RTE meats, and RTE seafood. In some outbreaks, a commodity has been implicated on epidemiological grounds alone, and not confirmed bacteriologically (Fleming et al. 1983; Devalk et al. 2001). In several well-publicized outbreaks, however, implication of a contaminated food product was based on both epidemiologic analysis and bacteriologic confirmation (Farber and Peterkin 1991; Jacquet et al. 1995; CDC 1998, 2000; Lyytikäinen et al. 2000; CDC 2001a). Following is a description of salient features of outbreaks associated with the different food commodities. These descriptions do not represent an exhaustive list of outbreaks; rather, they are selected on the basis of the extent to which these outbreaks were investigated, epidemiologically and bacteriologically, and on the basis of their public health impact.

Milk and Dairy Products. Contaminated milk (unpasteurized) may have been responsible for the first major outbreak of human listeriosis that has been reported, and which took place in Halle, former E. Germany, in 1945 (Ortel 1968). In 1983, contaminated pasteurized milk was epidemiologically implicated in an outbreak in Boston, Massachusetts (Fleming et al. 1985). The commonly used laboratory strain Scott A was a clinical isolate from this outbreak. In several subsequent outbreaks, it was possible to obtain bacteriological confirmation of the contamination of the dairy product by *L. monocytogenes* (Table 25.1). Of pivotal importance to our understanding of the epidemiology of foodborne listeriosis was the Jalisco cheese outbreak in Los Angeles, California (Linnan et al. 1987). The outbreak was the first to be investigated effectively by integrated approaches at the epidemiologic, bacteriologic, and food processing/environmental levels, which led to the conclusive identification of the source of contamination, and to consumer and product alerts. This was also the first documented major outbreak of listeriosis in the United States, and as such, it had major impact in the shaping of public health policy in this country. The Jalisco cheese outbreak was soon followed by active surveillance of listeriosis (Gellin et al. 1991) and led to enhanced awareness of the pathogen as foodborne agent in the dairy and other industries.

Soft cheeses have been more frequently implicated in outbreaks of listeriosis than have other dairy products, and in certain cases, contamination was traced to milk that was raw or improperly pasteurized (Linnan et al. 1987; CDC 2001a). In a recent outbreak in British Columbia, a specialty cheese was contaminated from a *Listeria*-contaminated fungal culture used to spray the surface of the product (BCCDC 2002). One outbreak was unusual not only because it is the only outbreak of listeriosis associated with butter but also because it involved serotype 3a, which is not commonly implicated in human infection (Lyytikäinen et al. 2000). The contaminant may have been established in the processing line of the butter manufacturer for several months prior to the outbreak (Maijala et al. 2001).

In addition to outbreaks of invasive illness, in 1995 an outbreak of febrile gastroenteritis was caused by chocolate milk heavily contaminated by *L. monocytogenes* (Table 25.2). Such atypical outbreaks with gastroenteritis symptoms are discussed later.

Meat Products. Meat had not been implicated in human listeriosis until 1988, when investigation of one case in Oklahoma led to the identification of contaminated turkey franks as the vehicle of infection (CDC 1989). In the 1990s, however, contaminated ready-to-eat meat products led to several outbreaks (Table 25.1), some of which were highly publicized and had profound epidemiogical impact.

The jellied pork tongue outbreaks in France can be seen as paradigms for the challenges associated with epidemiological investigations of listeriosis outbreaks, where a specialty food item is implicated and additional products (such as other deli products), may

Table 25.2—Febrile Gastroenteritis Outbreaks Caused by *Listeria Monocytogenes*

Location	(No. of cases)	Year	Food vehicle	Serotype	Reference
Italy	(39)	1992	Rice salad	1/2b	Salamina et al. 1996
Illinois, USA	(45)	1994	Chocolate milk	1/2b	Dalton et al. 1997
Finland	(5)	1997	Rainbow trout (cold-smoked)	1/2a	Miettinen et al. 1999
Italy[1]	(292)	1997	Corn salad	4b	Aureli et al. 2000

[1] Hospitalized individuals. A total of 1,566 of the individuals who were interviewed had symptoms, and *L. monocytogenes* was isolated from the stools of 123 individuals (Aureli et al. 2000).

become secondarily cross-contaminated. The additional challenges posed by the very low attack rate and wide geographical distribution (Jacquet et al. 1995) are typical of other foodborne outbreaks of listeriosis as well, including the U.S. multistate outbreaks traced to hot dogs and turkey deli products. The first jellied pork tongue outbreak in France (279 cases in 1992, between March and December) was also the first to clearly demonstrate the essential role of both phenotypic (based on serotyping and phage typing) and molecular (based on pulsed field gel electrophoresis, or PFGE) subtyping to identify outbreak-related strains and putative food vehicles. Molecular subtyping not only allowed the delineation of outbreak-associated cases and assisted in identification of the putative food vehicle but also allowed the exclusion of other food items. Thus, strains from cheeses could be excluded from implication in the outbreak because their PFGE patterns were distinct from the patient isolates (Jaquet et al. 1995). Delineation of outbreak strains is essential, considering that many listeriosis outbreaks that involve widely distributed products take place over a long time and large geographical areas, and the outbreak cases are necessarily superimposed upon a background of sporadic cases.

The 1998-1999 hot dog outbreak in the United States[1] involved multiple states and 101 cases (CDC 1998, 1999) and also can be viewed as an epidemiological paradigm. This was the first documented multistate outbreak in this country, and its impact on the public (including the media) and the food industry and food safety sectors was monumental. A large component of the impact was the fact that, since the Jalisco cheese outbreak in 1985, no common-source epidemics of foodborne listeriosis were documented in the United States, and the incidence of sporadic listeriosis itself appeared to be on the decrease (Schuchat et al. 1991; Tappero et al. 1995; CDC 2001b, 2002). The 1998-1999 outbreak also clearly demonstrated the potential for contamination and human disease involvement of a RTE meat product. Although a contaminated RTE meat product (turkey hot dogs) had been earlier shown to be implicated in a case of human listeriosis (CDC 1989), this was the first time that the potential of such products to cause a major outbreak was demonstrated. The epidemic also clearly demonstrated the ability and effectiveness of molecular subtyping, coordinated through PulseNet (Swaminathan et al. 2001), to identify the epidemic strain in clinical samples isolated from several different states, separated by large distances (CDC 1998, 1999). The identification of strains with the same PFGE type from the clinical isolates and from the implicated food validated the findings of the epidemiologic investigation and prompted the product recall that was instrumental in the curtailing of the outbreak.

The 1998–1999 hot dog outbreak, as well as the multistate outbreak that followed in 2000 (CDC 1998, 1999, 2000) also had major impact on our understanding of the bacterial strains that were involved. As with most other epidemic-associated strains, the strain implicated in the 1998-1999 outbreak was of serotype 4b, but it was of a molecular subtype that had not been commonly seen before. In an even more dramatic divergence from the previous trends, the strain in the multistate outbreak of 2000 was of serotype 1/2a, which was for the first time implicated in a major foodborne outbreak of listeriosis in the United States. The strains implicated in these outbreaks clearly revealed that the genotypic repertoire of the epidemic-associated lineages of the pathogen was much more diverse than previously thought.

Other Products (Vegetables and Seafood). A contaminated vegetable (coleslaw) was implicated in the Nova Scotia (Maritime Provinces) outbreak, in which the majority of cases were perinatal (Table 25.1). The outbreak was a landmark event. The epidemiologic investigation of this outbreak documented, for the first time, the role of contaminated food in the transmission of human listeriosis (Schlech et al. 1983).

No other major outbreaks of invasive illness with documented involvement of contaminated vegetables have been reported. An outbreak of febrile gastroenteritis caused by contaminated corn (Table 25.2) is discussed later, along with other, similar outbreaks.

Contaminated seafood has been implicated in several small outbreaks (Table 25.1). Of special interest is an outbreak in Sweden, in which for the first time contaminated trout ("gravad" trout) was implicated in invasive human illness (Ericsson et al. 1997). Several points in the RTE fish product chain have been shown to be prone to contamination by *L. monocytogenes* (Rorvik et al. 1995; Destro et al 1996; Autio et al. 1999; Johansson et al. 1999; Norton et al. 2001), and the incidence of epidemics may increase as RTE seafood products become more widely accessible and more commonly consumed. Data from Sweden, where gravad trout and other RTE seafood products are consumed at relatively high frequency, indeed suggest that such products may contribute significantly to human listeriosis (communication by W. Tham, XIV International Symposium on Problems of Listeriosis, Mannheim, Germany, May 2001).

STRAIN TYPING TOOLS OF SPECIAL VALUE IN OUTBREAK INVESTIGATIONS. Until the advent of standardized protocols for PFGE and implementation of PulseNet (Swaminathan et al. 2001) or equivalent systems, strain subtyping in outbreak investigations relied on phenotypic methods, primarily serotyping, phage typing, and multi-locus enzyme electrophoresis (MLEE). Genotypic methods primarily included ribotyping and detection of restriction fragment length polymorphisms (RFLPs) in selected genomic regions (Farber and Peterkin 1991; Schuchat et al. 1991). Currently, PFGE and automated ribotyping (riboprinting) are being used extensively in outbreak investigations as well as for routine surveillance of strains from clinical, food, or environmental sources.

The high resolution of PFGE and real-time transportability of molecular subtyping data made possible by PulseNet (and similar systems in other countries) have enabled a reversal of the course historically followed in outbreak investigations. Rather than the clinicians first noticing an outbreak (feasible in localized settings, but much more difficult in geographically dispersed outbreaks) and then the epidemiologists and bacteriologists confirming the common source, currently an outbreak can be noticed first by the occurrence of clinical isolates of the same molecular subtype. Examples are the multistate outbreaks in the United States, where identification of clinical isolates from different states with identical PFGE patterns and ribotypes provided the first indication of the occurrence of a common-source multistate outbreak. In each case, the outbreak became subsequently confirmed by epidemiologic and bacteriologic investigations (CDC 1998, 1999, 2000).

Of the phenotypic subtyping tools, serotyping based on the scheme of Seeliger and Hohne (1975) remains of substantial value, because it is rapid, inexpensive, and can be used to screen multiple isolates from patients and suspected foods. In the 1992 jellied pork tongue outbreak (serotype 4b) in France, for instance, serotyping allowed exclusion of serogroup 1/2 strains that commonly contaminate foods, and narrowed the group of candidate food vehicles to those harboring strains of serotype 4b. Serotyping, along with phage typing, thus allowed screening of thousands of strains during the outbreak investigation (15,000 strains screened in seven months) and facilitated the identification of the outbreak strains, which were defined and confirmed by PFGE (Jacquet et al. 1995). It is anticipated that for several applications, conventional serotyping may be replaced by DNA-based, "molecular serotyping" methods that can provide information relevant to the serotype of an isolate. DNA sequences specific for serotype 4b (and the related serotypes 4d and 4e) have already been described (Lei et al. 1997, 2001) and can be used to construct molecular serotyping tools (utilizing Southern blots, Polymerase Chain Reaction, DNA chips, or other formats). Similar tools can be constructed on the basis of sequences specific to other serotypes as well (S. Kathariou, P. Fields, and B. Swaminathan, unpublished results).

OUTBREAKS OF FEBRILE GASTROENTERITIS. In the 1990s, several outbreaks were noticed that were unusual in having predominantly gastroenteritis presenting symptoms, high attack rates, short incubation times, and in targeting previously healthy individuals (Salamina et al. 1996; Dalton et al. 1997; Miettinen et al. 1999; Aureli et al. 2000). These outbreaks have involved different food vehicles and strains of serotypes 1/2a, 1/2b, and 4b (Table 25.2). A common characteristic was the unusually high contamination levels in the food. In the chocolate milk outbreak, contamination was found to be ca.10^9 CFU/ml, resulting in an estimated dose of ca. 10^{11} bacteria for the infected individuals (Dalton et al. 1997). It remains unclear whether the unusual clinical picture of these infections was associated with specific virulence factors of the strains or with the very high inoculum that resulted in the observed intestinal inflammatory response.

During the chocolate milk outbreak, it was noticed that, in addition to the febrile gastroenteritis cases, the outbreak strain was also isolated from a small number of patients with invasive illness (Proctor et al. 1995; Dalton et al. 1997). Whether such invasive cases resulted from ingestion of a lower inoculum or had predisposing factors that obviated gastrointestinal symptoms and resulted in invasive illness instead is unclear.

OUTBREAKS OF LISTERIOSIS INVOLVING MULTIPLE STRAINS. On several occasions, apparent outbreaks have been found to involve multiple strains of *L. monocytogenes*. During the 1987 outbreak in Philadelphia, strains of different serotypes (for example, 4b, 1/2b) and of several distinct MLEE types were isolated from the patients. The results of the epidemiologic and bacteriologic investigation of this unusual outbreak led to the conclusion that multiple food vehicles were likely involved, and that individuals in the population had become susceptible to infection by *L. monocytogenes* that may have been present at background levels in several diverse foods. Susceptibility to listeriosis may have been enhanced by a co-infecting agent or other immunosuppressive factor in the community (Schwartz et al. 1989). A more recent example involves four cases of listeriosis, all of serotype 4b, identified within one month in the summer of 1999 in Hawaii, where listeriosis usually does not exceed one case per year. Interestingly, PFGE analysis of the isolates revealed that each strain had a distinct PFGE type, although two of the strains appeared to be genetically closely related (Fig. 25.1). The factors responsible for the observed clustering of the cases, or the possible food vehicles, were not identified. The identification of different strains is consistent with the hypothesis that more than one source was involved in this outbreak. However, in the absence of further epidemiological and bacteriological data, we cannot exclude the possibility that the implicated food(s) were from a common source but were contaminated by multiple strains. In the outbreak of listeriosis that was traced to contaminated rainbow trout, it was found that trout from the same producer were contaminated with three different clonal types of *L. monocytogenes*. Strains of one clonal type were isolated from three patients, whereas two other patients were infected each by one of the other two clonal types (Ericsson et al. 1997).

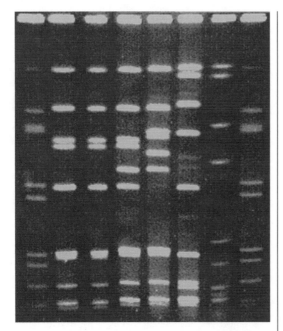

FIG. 25.1—PFGE of DNA of *Listeria monocytogenes* serotype 4b strains digested by *Asc*I. Lane M, molecular marker; lanes 1 and 2, California (Jalisco) outbreak isolates; lanes 3–6, clinical isolates from the State of Hawaii (summer 1999). Isolates shown on lane 3 and 4 have PFGE patterns similar to each other, and to the Jalisco epidemic strains, and possess ECI-specific RFLP and DNA methylation markers. The isolate in lane 5 has genetic markers typically seen in sporadic serotype 4b isolates. (Sample preparation and PFGE done by R. Y. Kanenaka, State of Hawaii Dept of Health.)

EPIDEMIC-ASSOCIATED STRAINS AND CLONAL GROUPS OF L. MONOCYTOGENES.

Starting with the Nova Scotia outbreak of 1981, most major outbreaks of food-borne listeriosis have involved strains of serotype 4b (Table 25.1). Serotype 4b strains contribute significantly to sporadic illness as well, along with serotypes 1/2a and 1/2b (Farber and Peterkin 1991; Schuchat et al. 1991; Rocourt et al. 2001). The remaining several serotypes are noticeably underrepresented among human infections (Schuchat et al. 1991; Rocourt et al. 2000; Kathariou 2000). Serotype 4b belongs to one of the two major genomic divisions of *L. monocytogenes*, which also includes serotypes 1/2b and 3b (Bibb et al. 1989, 1990; Piffaretti et al. 1989; Brosch et al. 1994; Graves et al. 1994; Rasmussen et al. 1995; Wiedmann et al. 1997; Aarts et al. 1999; reviewed in Kathariou 2000). The other major genomic division (serotypes 1/2a, 1/2c, 3a, 3c) has also contributed to outbreaks, involving serotypes 1/2a and 3a (Table 25.1), albeit much less frequently.

Early MLEE-based strain subtyping data showed that strains implicated in several major outbreaks were genetically closely related (Piffaretti et al. 1989; Bibb et al., 1990), and these findings were confirmed by ribotyping and virulence gene polymorphism analysis

(Wiedmann et al. 1997; De Cesare et al. 2001) and by PFGE (Buchrieser et al. 1993; Jacquet et al. 1995; Brosch et al. 1996). To date, most major outbreaks appear to have been caused by a rather small number of clonal groups (lineages) that consist of closely related strains.

The repeated involvement of some of these lineages in outbreaks has often prompted the speculation that epidemic-associated strains may have enhanced human virulence. However, several efforts using cell culture and animal models have failed to identify differences in virulence between serotype 4b epidemic-associated strains and sporadic clinical isolates of the same serotype (Pine et al. 1990, 1991; Del Corral et al. 1990; Brosch et al. 1993; Bula et al. 1995; Van Langendonck et al. 1998). Furthermore, case fatality rates in epidemics of listeriosis are generally similar to those observed among sporadic cases (20 to 30 percent), and the clinical picture is equally severe for sporadic and outbreak cases (Gellin et al. 1991; Jacquet et al. 1995; Rocourt et al. 2000). In addition to the possibility (remaining to be confirmed) that epidemic strains have enhanced human virulence, another possibility, also remaining to be confirmed, is that these strains have enhanced ability to establish themselves in foods and food processing environments.

The "clone" concept (Orskov and Orskov 1983) is useful in discussions of epidemic-associated strains. In this context, "epidemic clone" refers to a group of closely related strains implicated in one or more epidemics, and presumably of a common ancestry. Three epidemic-associated clonal groups (designated here as Epidemic Clone I, II, and III, and listed in Table 25.3) are discussed in more detail because they constitute epidemiological "landmarks" for food-borne listeriosis. Of these clonal groups, Epidemic Clone I has been repeatedly involved in outbreaks and has been most extensively investigated. Epidemic clones II and III, implicated in the 1998-1999 and 2000 multistate outbreaks, respectively, were identified relatively recently and have not yet been studied extensively.

Clearly, the potential exists for currently unidentified clonal groups to be implicated in outbreaks in the future. Increased understanding of the currently known clones, however, will strengthen our knowledge base for the biology, ecology, and genetics of these strains and will be instrumental for the evaluation of the attributes, and for possible later control strategies, of the emerging lineages of the future.

Epidemic Clone I (ECI). ECI appears to be a cosmopolitan epidemic clonal lineage, having become implicated in outbreaks in several countries in Europe and North America (Table 25.3). Genetic subtyping studies have indicated that the strains from the different populations are closely related (Piffaretti et al. 1989; Bibb et al. 1990; Schuchat et al. 1991; Buchrieser et al. 1993; Jacquet et al. 1995; Wiedmann et al. 1997; Kathariou 2000; Rocourt et al. 2000 De Cesare et al. 2001), even though each outbreak population is genetically distinct

Table 25.3—Major Epidemic Clones Implicated in Foodborne Outbreaks of Listeriosis

Epidemic Clone[1]	Location, year, implicated food[2]	Remarks
Epidemic Clone I (ECI)	Nova Scotia, 1981, coleslaw California, 1985, Mexican-style cheese Switzerland, 1983–1987, cheese Denmark, 1985–1987, vehicle unknown *France, 1992, jellied pork tongue	Serotype 4b; Distinct MEE types and ribotypes; DNA resistant to digestion by *Sau*3AI (Zheng and Kathariou 1997); Special genetic markers detected by Southern blot (Zheng and Kathariou 1995; Tran et al. 2002), genomic subtraction (Herd and Kocks 2001); genome sequence determined (California outbreak strain)
Epidemic Clone Ia (ECIa)[3]	Massachusetts, 1983, pasteurized milk Boston, 1985	Serotype 4b; Distinct MEE types and ribotypes
Epidemic Clone II (ECII)	Multistate, USA, 1998–1999 Hot dogs	Serotype 4b; Previously rare ribotype; Previously rare PFGE type ("type E", with limited variations)
Epidemic Clone III (ECIII)	Multistate, USA, 2000 Turkey deli meats	Serotype 1/2a; Identical PFGE type identified in implicated processing plant in 1988

[1] Epidemic clone: closely related strains implicated in one or more outbreaks.
[2] Additional information on these outbreaks (cases, references) is included in Table 25.1.
[3] The term "ECIa" was chosen to reflect the apparent close genetic relatedness between the epidemic-associated strains in this group and those of ECI, as evidenced by MEE (Piffaretti et al. 1989; Bibb et al. 1990).
*Denmark, 1989–1990, blue mould cheese

(Wesley and Ashton 1991; Buchrieser et al. 1993; Brosch et al. 1996). Within ECI, different degrees of relatedness among outbreak populations have been seen. Strains from the outbreaks in California (Jalisco cheese), Denmark (vehicle unknown), Switzerland (Vacherin Mont d'or cheese), and France (1992 jellied pork tongue outbreak) are very closely related to each other (Buchrieser et al. 1993; Jacquet et al. 1995) and distinct genotypically from the Nova Scotia outbreak strains (Buchrieser et al. 1993; Brosch et al. 1996).

Examination of strains from several of the ECI outbreak populations, including Nova Scotia, California, and Switzerland, identified a unique RFLP in a genomic region essential for low temperature (4°C) growth of *L. monocytogenes* (Zheng and Kathariou 1995). The strains from these outbreaks also share an additional unique RFLP, in a gene that may be mediating mannitol transport (Tran et al. 2002). Furthermore, these same strains appear to methylate cytosines at GATC sites in their DNA that renders the DNA resistant to digestion by the restriction enzyme *Sau*3AI (Zheng and Kathariou 1997).

Recently, genomic subtraction protocols were successfully employed to identify genomic fragments present in strains from the California outbreak but absent (or highly divergent) from the genome of other serotype 4b strains (Herd and Kocks 2001). Interestingly, certain fragments were conserved between the epidemic strains and strains of serotypes 1/2b and 3b. Genetic transfer between such strains would not be surprising, because strains of serotypes 4b, 1/2b, and 3b constitute one of the two major genomic divisions of *L. monocytogenes* and have significant genetic similarity, except in serotype-associated determinants (Kathariou 2000, 2002). The fact that certain sequences are shared between ECI and 1/2b or 3b strains (but not other

serotype 4b strains) would suggest that some of the genes unique to ECI may (1) have been acquired from (or transferred to) the lineage that includes strains of serotypes 1/2b and 3b, or (2) the genes may have been present in the original 4b-1/2b-3b lineage, and these genes were maintained in ECI but lost from other serotype 4b strains. Interestingly, the G+C content of many of these same fragments that are unique to the epidemic strains was in the range of 24 to 33 percent. This range is significantly lower than the average for *L. monocytogenes* (38 percent), suggesting the possibility that these genes were acquired by horizontal transfer from sources other than *Listeria*.

On the basis of sequence analysis, one of the ECI-specific genes is likely to encode a putative methylase that may be responsible for the observed resistance of the DNA of these strains to *Sau*3AI digestion, discussed previously. Putative functions of the other unique genes remain undetermined. In addition, these genes were shown to be present in strains from the California and the Swiss outbreaks, but their presence in strains from other epidemics involving ECI was not determined. Such determinations can be readily made and are expected to be forthcoming. In addition, it should be feasible to perform functional genetic studies to obtain further information on the roles of these genes in the physiology and possibly the virulence of the organism.

With funding from USDA-ARS, the DNA sequence determination of the entire genome of one serotype 4b ECI strain (implicated in the California Jalisco cheese outbreak) is currently near completion (www.tigr.org). Comparative genomic analysis between the genome of this strain and the already published genome of *L. monocytogenes* strain EGD (of serotype 1/2a) (Glaser et al. 2001) is expected to readily identify genes that are unique to each. Subsequent genomic studies can reveal which of the genes present in the California outbreak strain, but absent from EGD, are present in other ECI strains as well and are characteristic and unique to ECI as a whole (that is, present in ECI but absent from other serotype 4b lineages, including sporadic serotype 4b strains). It is also anticipated that genomic approaches will facilitate the identification of sequences that may be unique to different ECI outbreak populations (for example, present in the California outbreak strains but absent from the Nova Scotia outbreak population). Clearly, genomic data of this type will open a new era for our understanding of the evolution and unique attributes of ECI, as well as other epidemic lineages.

Epidemic Clone Ia (ECIa). DNA sequences and bacteriologic or genetic features that may be unique to strains of this lineage have not yet been reported. The Massachusetts outbreak strain was found to lack the genomic fragments found to be unique to ECI (Herd and Kocks 2001). In addition, it lacks the RFLPs and cytosine methylation at GATC sites that are characteristic of ECI (Zheng and Kathariou 1995, 1997; Tran et al. 2002).

Epidemic Clone II (ECII). In 1998-1999, a new genotype of *L. monocytogenes* serotype 4b was implicated in a multistate outbreak of listeriosis in the United States that involved contaminated hot dogs. Strains from this outbreak[2] had unique ribotype and PFGE patterns not commonly encountered in previous surveys (117, 118). At the physiological level, searches for possible differences in attributes such as heat tolerance have proven unsuccessful (Mazzotta and Gombas 2001). Genomic fragments unique to these strains have not yet been identified. However, Southern blots have identified a unique RFLP that can differentiate these strains from other isolates of serotype 4b, including ECI strains (Fig. 25.2).

Another recent finding is that strains from this outbreak appear to have undergone substantial divergence in certain of the genes that are conserved among other serotype 4b strains, thus failing to produce the expected product by PCR (Fig. 25.3). At this time, the forces that have driven this divergence are not known. However, at the applied level, these findings suggest that it is now possible to employ Southern blots, or PCR, to identify such strains as being of the ECII lineage.

Epidemic Clone III (ECIII). In 2000, a multistate outbreak of listeriosis in the United States involved contaminated turkey deli meat products and resulted in several cases of listeriosis and a massive product recall (CDC 2000). Unlike most other outbreaks, the implicated strain was serotype 1/2a. An especially interest-

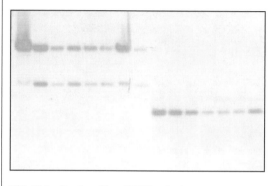

FIG. 25.2—Southern blot of ECII and other *L. monocytogenes* strains. DNA was digested with *Eco*RI and hybridized with digoxigenin-labeled probe constructed on the basis of genomic sequences of *L. monocytogenes* EGD. Lanes 1–8, *L. monocytogenes* ECII isolates; lanes 10–15, other *L. monocytogenes* strains, of serotypes 4b and 1/2b. (S. Bowen, B. Swaminathan, and S. Kathariou, unpublished results.)

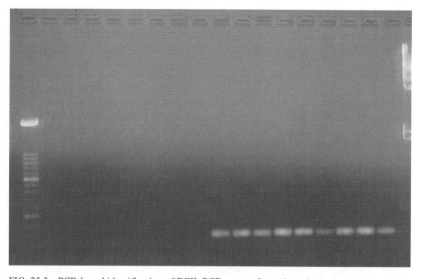

FIG. 25.3—PCR-based identification of ECII. PCR was performed employing primers derived from the genomic sequence of *L. monocytogenes* serotype 4b. Lane 1: 100 bp ladder, lanes 2–9: *L. monocytogenes* ECII isolates, lanes 10–18: other *L. monocytogenes* strains of serotype 4b; lane 19: λ DNA digested by *Hin*dIII, used as markers. ECII strains do not yield a product because of divergence in the gene used to construct the primers. (M. Evans, S. Kathariou, and B. Swaminathan, unpublished findings.)

ing finding was that the outbreak strain had a genotype (as determined by PFGE with several enzymes) identical to that of a strain that was implicated earlier in a human listeriosis case associated with consumption of contaminated turkey franks in 1988 (CDC 1989). The products implicated in the 2000 multistate outbreak were from the same food processing facility as the earlier isolate, suggesting that this strain had persisted in that facility for several years without detectable genotypic changes (Communications by B. Swaminathan and P. Mead, XIV International Symposium on Problems of Listeriosis, Mannheim, Germany, May 2001).

At this time, no genetic or bacteriological features unique to ECIII are known. ECIII is of serotype 1/2a, and finding features exclusive to this clonal group may pose some special challenges. Serotype 1/2a strains have been found to be significantly more diverse genetically than 4b (Bibb et al. 1989; Aarts et al. 1999; Buchrieser et al. 1993; Brosch et al. 1994; Moore and Datta 1994; Jersek et al. 1999). A large portion of this diversity is currently undefined, partly because the clonal groups associated with foods (where serotype 1/2a strains are quite common) have not yet been surveyed extensively, and none has been genetically characterized in detail. On the other hand, the published *L. monocytogenes* genome sequence is that of a serotype 1/2a strain (strain EGD) (Glaser et al. 2001), and the available genomic information is likely to greatly facilitate the employment of "genomotyping" schemes to identify ECIII-specific nucleotide sequences.

EPIDEMIC STRAINS: AMPLIFICATION AND RESERVOIRS.

In some cases, different outbreaks involve the same clonal type (or "epidemic clone") of *L. monocytogenes*, whereas in others an outbreak involves a clonal type rarely, if at all, encountered earlier. The periodic involvement of the same epidemic clone in different outbreaks suggests that the implicated strains have a reservoir during the often lengthy inter-outbreak intervals. Similarly, a reservoir is expected to exist for bacteria of previously rare types that emerge as the cause of an outbreak. In the latter case, the search for possible reservoirs is complicated by the cryptic nature of the clonal type during periods outside of the outbreak in which it was implicated. To understand the evolution of epidemic lineages, it is crucial to identify the reservoir and to characterize the environmental niches or hosts permitting amplification of the bacteria.

ECI: A Cosmopolitan Clonal Group with an Unknown Reservoir.

The periodic involvement of the same epidemic clone in different outbreaks suggests that the implicated strains have a reservoir during the often lengthy intervals between outbreaks. Widespread incidence and repeated involvement of ECI in outbreaks suggests that this may be a ubiquitous environmental lineage, which can become amplified in animals and humans (Boerlin et al. 1991). Animals (including

animals ill with listeriosis) and human carriers, or subclinically infected individuals, may also serve as reservoirs for this clone, although documentation to this effect is currently lacking. Data from France indicate that the strain implicated in the jellied pork tongue outbreak of 1992 was also encountered in previous years, although at low numbers (six to twenty-seven cases/yr), along with other sporadic strains that constituted the background levels of human listeriosis in that country (Jacquet et al. 1995). Evidence for ECI strains outside of common-source outbreaks has also been obtained in the United States; analysis of the cluster of non-common-source infections in Philadelphia in 1987 (Schwartz et al 1989) showed that some of the serotype 4b strains that were involved had the RFLP and DNA modification characteristic of ECI (Zheng and Kathariou 1997). Thus, in this complex outbreak that involved multiple strains, of different serotypes, ECI appeared to be present, albeit not predominant.

A similar situation was found during the investigation of a cluster of four cases, all four involving serotype 4b strains, in the state of Hawaii, in 1999. The strains were found to be genotypically different, as evidenced by PFGE (Fig. 25.1). Two of these strains, however, had the RFLPs and DNA modification characteristic of ECI. PFGE analysis suggested that the strains were closely related to prototype ECI strains (California outbreak) (Fig. 25.1). Such data suggest that the two strains in question belong to the ECI lineage. However, the source(s) for such apparent ECI strains that are found to cause infection during the periods between common-source outbreaks remain unidentified.

ECI strains are not commonly isolated from routine surveys of food (Boerlin et al. 1991; 1997; Ryser et al. 1996; Gendel and Ulaszek 2000.), suggesting that this lineage is not frequently established in the food processing environment. It is possible that establishment of such strains in food processing facilities may require specialized conditions or specific environmental niches.

Establishment of Outbreak Strains in the Food Manufacturing Facility.

In the case of ECIII, implicated in the 2000 turkey deli meat outbreak (CDC 2000), the outbreak strain appeared to have persisted in the processing plant for at least twelve years, suggesting that environmental habitat(s) in the manufacturing plant environment served as the reservoir for survival and possible multiplication of the pathogen. However, the prevalence of this strain in the plant before, during, and after the outbreak is not known; neither is there information on the specific environment that allowed such prolonged persistence of the bacteria. In addition, a persistent clone may not result in problematic (or even detectable) contamination of the product, if its population remains at very low levels.

Listeria contamination in the food processing plants commonly involves established, "resident" clones, characterized by their persistence, as well as transient strains, that may be prevalent at a specific time but do not persist. Some of the clones that are prevalent and persistent in food manufacturing plants consist of strains of serotype 1/2a and 1/2c that have not been shown to be commonly associated with epidemic illness (Autio et al. 1999; Giovanacci et al. 1999; Norton et al. 1999; Ojeniyi et al. 2000; Kathariou 2002). However, extensive environmental contamination by the outbreak strain (serotype 4b) was found in the investigation of the Jalisco cheese outbreak (Linnan et al. 1987), and a strain (serotype 4b) that was both persistent and prevalent in a trout processing plant was implicated in an outbreak of listeriosis (Ericsson et al. 1997). The investigation of the butter-associated epidemic in Finland reported sporadic isolation of the epidemic strain (serotype 3a) from the product several months before the actual outbreak (Maijala et al. 2001), suggesting that during that time, the manufacturing facility had been colonized by the outbreak strain. Finally, as mentioned previously, genotyping of the strains from the turkey deli meat outbreak (CDC 2000) also suggested lengthy persistence of the outbreak strain (serotype 1/2a) in the implicated manufacturing plant. Such findings suggest that establishment in the food manufacturing facility may be one of the key parameters in a strain's access to the final food product and potential involvement in epidemic illness.

GENETIC VARIATION WITHIN OUTBREAK POPULATIONS. In a common-source outbreak, special selective pressures in distinct microenvironments in the contaminated food, in the course of an infection, or both of these, may result in production of genetic variants. Depending on the phenotypes that are affected and the strain characterization methods that are employed, such variants may or may not be easily detectable genotypically and phenotypically. Identification of putative variants by necessity requires study of multiple strains from a specific outbreak population.

Variation within the outbreak population has indeed been observed. Twenty-seven percent of the patient-derived strains from the Nova Scotia outbreak lacked galactose substituents in the teichoic acid of the cell wall, which serves as major antigenic determinant for *L. monocytogenes*. Such strains were negative with serotype 4b-specific monoclonal antibodies and resistant to serotype 4b-specific phages. Similar antibody-negative, phage-resistant strains were identified in both the California and the Massachusetts outbreaks (Clark et al. 2001). Although we cannot exclude the possibility that these phenotypes became established during isolation (from clinical or food specimens), passage, or storage of the bacteria in the laboratory, an alternative possibility is that the observed variants were selected during infection of their human host, or in response to other selective pressures.

The existence of genetic variants within the outbreak population may account for certain atypical results reported in the literature. For instance, the Swiss outbreak (1983–1987) involved strains of one common MLEE type (Piffaretti et al. 1989; Bibb et al. 1990) but two closely related genotypes and phage types. In addition, identical genotypes were seen in strains that were either nontypable by phage typing or were of different phage types (Nocera et al. 1990). Inconsistencies with phage typing results have been reported in a more recent outbreak as well (DeValk et al. 2001), possibly because of the presence of genetic variants with alterations in phage receptors.

Investigation of the 1992 jellied pork tongue outbreak in France contains a detailed description of closely related genotypes (as detected by PFGE) identified in the course of the outbreak. Strains of the same phage type had twenty different but closely related PFGE profiles as seen following restriction by three different enzymes (*Apa*I, *Sma*I, *Not*I). Of the 289 human isolates, 247 (89 percent) had identical PFGE profiles, and such strains were designated "epidemic strain sensu stricto" by the authors (Jacquet et al. 1995). Possibly the other strains represented genetic variants of the "epidemic strain sensu stricto." Interestingly, a closely related PFGE profile (identical *Not*I and *Apa*I but distinct *Sma*I patterns) was identified among strains isolated from cheeses, but was totally absent from human isolates. It is possible that genotypic variants of the strain had been established in the cheeses, but were, for unknown reasons (possibly impaired virulence, low levels, or other), not involved in the outbreak.

High-resolution genotyping by PFGE (using two or more enzymes) or other tools might reveal similar genetic variation within the strains implicated in other outbreaks as well, and may provide clues as to the possible selective pressures encountered by the bacteria. One may speculate that strains that have colonized a manufacturing facility for a long time would have greater potential for localized adaptation and genotypic differentiation in the food manufacturing plant. In such a scenario, the final product may be contaminated by slightly different genotypes, which (presuming that virulence was not affected) would subsequently become isolated from the patients during the outbreak.

FUTURE PROSPECTS. With enhanced implementation for PulseNet or similar surveillance and subtyping systems, our ability to recognize and monitor outbreaks of listeriosis is expected to continue improving. Scientific study of the implicated strains will be essential for our understanding of the evolution and special attributes of the pathogen, including, but not limited to: potential for infection; adaptation to environmental stresses; ability to colonize food and environmental niches; ability to compete with other bacteria and with other *Listeria* strains. The epidemic clones that will be identified in the future may or may not be related to those known currently. Emerging lineages are expected, and it will be essential that they are investigated in the context of those that have caused epidemics or sporadic disease in the past so that the evolution of the pathogen is placed in proper perspective.

The area of virulence is of keen interest from the point of view of food safety and public health. In the case of epidemic strains, human virulence is not in question, because the strains have been proven to be virulent by their very association with invasive disease. It would be important, however, to determine whether such strains may have enhanced virulence potential to humans, in comparison to other *L. monocytogenes* strains such as those causing sporadic cases of illness, and whether this attribute contributes to their involvement in epidemic illness.

Commonly used animal and cell culture models have not been successful in identifying epidemiologically relevant differences in virulence. Selection of strains according to genotype (Norrung and Andersen 2000) and employment of alternative models for virulence, including recently described models that may better permit oral infection evaluations (Lecuit et al. 2001) may make novel contributions in the study of virulence. Laboratory systems that monitor gene expression relevant to infection and pathogenicity may also serve this area. The latter will be made increasingly feasible with the increasing availability of genome sequences for *L. monocytogenes*. The currently almost complete genome sequence of an epidemic strain (California outbreak) (www.tigr.org) along with the genome of *L. monocytogenes* strain EGD (Glaser et al. 2001) are already major resources for the relevant molecular tools for such studies.

In terms of the epidemic clones themselves, our understanding of their origin and evolution will be immensely aided by knowing what is special and unique to their genome and their gene expression and proteomic profiles. "Genomotyping" utilizing DNA arrays or other formats will greatly facilitate the identification of genes unique to distinct epidemic clones. Genomic approaches such as these have been used successfully with other pathogens (Salama et al. 2000; Dorrell et al. 2001) and will allow identification of unique genes, whose function can be studied in relevant genetic, physiological, or ecological systems and in virulence models. Advances in proteomics (such as MALDI-MS) are expected to contribute in major ways to new knowledge on the protein repertoire of these and other *Listeria* strains under conditions relevant to environmental and food contamination, infection, and disease.

SUMMARY AND CONCLUSIONS. From the earliest rigorous discourses on *Listeria* (Seeliger 1961; Gray and Killinger 1966) to the recently published report on the genomes of *L. monocytogenes* strain EGD and of *L. innocua* (Glaser et al.), it is abundantly clear that ecological adaptations, often unexpected and intriguing, feature prominently in the biology of *Listeria*, and they are likely to feature even more prominently in the epidemic clones. Ability to become established in the food manufacturing plants and to persist there, sometimes for years; and ability to survive and often to grow in foods, including under refrigeration, are some of the key known adaptations of the pathogen. Many others remain unknown, and for those that are known, the underlying mechanisms remain largely uncharacterized. Epidemic lineages may have unusual attributes in regard to such adaptations, and this is an area of great interest that is currently virtually unexplored.

Continued scientific progress in terms of our understanding of epidemic-associated clones of this pathogen will require the integrated efforts of members of the public health (State Health Departments, CDC), food industry (Manufacturing plant surveillance for *Listeria*), and basic microbiology sectors. Publications from the food industry (Tompkin et al. 1999; Tompkin 2002) are already making major contributions to our understanding of the pathogen in one of the settings of the greatest relevance in terms of food safety—the food processing plant. Much progress can be made by integrated, coordinated efforts among representatives from the public health, industry, and academic research sectors. Food safety and public health will clearly be promoted by our improved ability to detect and monitor the epidemic-associated lineages and to understand the factors that influence their survival and prevalence.

REFERENCES

Aarts, H.J., L.E. Hakemulder, and A.M. van Hoef. 1999. Genomic typing of *Listeria monocytogenes* strains by automated laser fluorescence analysis of amplified fragment length polymorphism fingerprint patterns. *Int. J. Food Microbiol.* 49:95–102.

Aureli, P., G.C. Fiorucci, D. Caroli, G. Marchiaro, O. Novara, L. Leone, and S. Salmaso. 2000. An outbreak of febrile gastroenteritis associated with corn contaminated by *Listeria monocytogenes*. *N. Engl. J. Med.* 342:1236–1241.

Autio, T., S. Hielm, M. Miettinen, A.M. Sjoberg, K. Aarnisalo, J. Bjorkroth, T. Mattila-Sandholm, and H. Korkeala. 1999. Sources of *Listeria monocytogenes* contamination in a cold-smoked rainbow trout processing plant detected by pulsed-field gel electrophoresis typing. *Appl. Environ. Microbiol.* 65:150–155.

BCCDC (British Columbia Centre for Disease Control), 2002. Listeriosis: British Columbia (Update). March 1, 2002. Infectious Diseases News Brief http://www.hc-sc.gc.ca/pphb-dgspsp/bid-bmi/dsd-dsm/nb-ab/2002/nb0902_e.html.

Bibb, W.F., B. Schwartz, B.G. Gellin, B.D. Plikaytis, and R.E. Weaver. 1989. Analysis of *Listeria monocytogenes* by multilocus enzyme electrophoresis and application of the method to epidemiologic investigations. *Int. J. Food Microbiol.* 8:233–239.

Bibb, W.F., B.G. Gellin, R. Weaver, B. Schwartz, B.D. Plikaytis, M.W. Reeves, R.W. Pinner, and C.V. Broome. 1990. Analysis of clinical and food-borne isolates of *Listeria monocytogenes* in the United States by multilocus enzyme electrophoresis and application of the method to epidemiologic investigations. *Appl. Environ. Microbiol.* 56:2133–2141.

Boerlin, P., and J.C. Piffaretti. 1991. Typing of human, animal, food, and environmental isolates of *Listeria monocytogenes* by multilocus enzyme electrophoresis. *Appl. Environ. Microbiol.* 57:1624–1629.

Boerlin, P., F. Boerlin-Petzold, E. Bannerman, J. Bille, and T. Jemmi. 1997. Typing *Listeria monocytogenes* isolates from fish products and human listeriosis cases. *Appl. Environ. Microbiol.* 63:1338–1343.

Brett, M.S., P. Short, and J. McLauchlin. 1998. A small outbreak of listeriosis associated with smoked mussels. *Int. J. Food Microbiol.* 43:223–229.

Brosch, R., B. Catimel, G. Millon, C. Buchrieser, E. Vindel, and J. Rocourt. 1993. Virulence heterogeneity of *Listeria monocytogenes* strains from various sources (food, human, animal) in immunocompetent mice and its association with typing characteristics. *J. Food Prot.* 56:296–301.

Brosch, R., J. Chen, and J.B. Luchansky. 1994. Pulsed-field fingerprinting of *Listeriae*: identification of genomic divisions for *Listeria monocytogenes* and their correlation with serovar. *Appl. Environ. Microbiol.* 60:2584–2592.

Buchrieser, C., R. Brosch, B. Catimel, and J. Rocourt. 1993. Pulsed-field gel electrophoresis applied for comparing *Listeria monocytogenes* strains involved in outbreaks. *Can. J. Microbiol.* 39:395–401.

Bula, C.J., J. Bille, and M.P. Glauser. 1995. An epidemic of food-borne listeriosis in western Switzerland: description of 57 cases involving adults. *Clin. Infect. Dis.* 20:66–72.

CDC. 1989. Listeriosis associated with consumption of turkey franks. *MMRW* 38:267–168.

CDC. 1998. Multistate outbreak of listeriosis—United States, 1998. *MMRW* 47:1085–1086.

CDC. 1999. Update: multistate outbreak of listeriosis—United States, 1998–1999. *MMRW* 47:1117–1118.

CDC. 2000. Multistate outbreak of listeriosis—United States, 2000. *MMRW* 49:1129–1130.

CDC. 2001a. Outbreak of listeriosis associated with homemade Mexican-style cheese—North Carolina, October 2000–January 2001. *MMRW* 50:560–562.

CDC. 2001b. Preliminary Food Net data on the incidence of foodborne illnesses—selected sites, United States, 2000. *MMRW* 50:241–246.

CDC. 2002. Preliminary Food Net data on the incidence of foodborne illnesses—selected sites, United States, 2001. *MMRW* 51:325–329.

Clark, E.E., I. Wesley, F. Fiedler, N. Promadej, and S. Kathariou. 2000. Absence of serotype-specific surface antigen and altered teichoic acid glycosylation among epidemic-associated strains of *Listeria monocytogenes*. *J. Clin. Microbiol.* 38:3856–3859.

Dalton, C.B, C.C. Austin, J. Sobel, P.S. Hayes, W.F. Bibb, L.M. Graves, B. Swaminathan, M.E. Proctor, and P.M. Griffin. 1997. An outbreak of gastroenteritis and fever due to *Listeria monocytogenes* in milk. *N. Engl. J. Med.* 336:100–105.

Del Corral, F., R.L. Buchanan, M.M. Bencivengo, and P.H. Cooke. 1990. Quantitative comparison of selected virulence associated characteristics in food and clinical isolates of *Listeria*. *J. Food Prot.* 53:1003–1009.

De Cesare, A., J.L. Bruce, T.R. Dambaugh, M.E. Guerzoni, and M. Wiedmann. 2001. Automated ribotyping using different enzymes to improve discrimination of *Listeria monocytogenes* isolates, with a particular focus on serotype 4b strains. *J. Clin. Microbiol.* 39:3002–3005.

Destro, M.T., M.F.Leitao, and J.M. Farber. 1996. Use of molecular typing methods to trace the dissemination of *Listeria monocytogenes* in a shrimp processing plant. *Appl. Environ. Microbiol.* 62:705–711.

DeValk, H., V. Vaillant, C. Jacquet, J. Rocourt, F. Le Querrec, F. Stainer, N. Quelquejeu, O. Pierre, V. Pierre, J.-C. Desenclos, and V. Goulet. 2001. Two consecutive nationwide outbreaks of listeriosis in France, October 1999–February 2000. *Am. J. Epidemiol.* 154:944–950.

Dorrell, N., J.A. Mangan, K.G. Laing, J. Hinds, D. Linton, H. Al-Ghusein, B.G. Barrell, J. Parkhill, N.G. Stoker, A.V. Karlyshev, P.D. Butcher, and B.W. Wren. 2001. Whole genome comparison of *Campylobacter jejuni* human isolates using a low-cost microarray reveals extensive genetic diversity. *Genome Res.* 11:1706–1715.

Ericsson, H., A. Eklow, M. L. Danielsson-Tham, S. Loncarevic, L.O. Mentzing, I. Persson, H. Unnerstad H, and W. Tham. 1997. An outbreak of listeriosis suspected to have been caused by rainbow trout. *J. Clin. Microbiol.* 35:2904–2907.

Farber, J.M., and P.I. Peterkin. 1991. *Listeria monocytogenes*, a food-borne pathogen. *Microbiol. Rev.* 55:476–511.

Fleming, D.W., S.L. Cochi, K.L. MacDonald, J. Brondum, P.S. Hayes, B.D. Plikaytis, M.B. Holmes, A. Audurier, C.V. Broome, and A.L. Reingold. 1985. Pasteurized milk as a vehicle of infection in an outbreak of listeriosis. *N. Engl. J. Med.* 312:404–407.

Gellin, B.G., C.V. Broome, W.F. Bibb, R.E. Weaver, S. Gaventa, L. Mascola, and the Listeriosis Study Group. 1991. The epidemiology of listeriosis in the United States, 1986. *Am. J. Epidemiol.* 133:392–401.

Gendel, S.M., and J. Ulaszek. 2000. Ribotype analysis of strain distribution in *Listeria monocytogenes*. *J. Food Prot.* 63:179–185.

Giovannacci, I., C. Ragimbeau, S. Queguiner, G. Salvat, J.L. Vendeuvre, V. Carlier, and G. Ermel. 1999. *Listeria monocytogenes* in pork slaughtering and cutting plants. Use of RAPD, PFGE and PCR-REA for tracing and molecular epidemiology. *Int. J. Food Microbiol.* 53:127–140.

Glaser, P., L. Frangeul, C. Buchrieser, C. Rusniok, A. Amend, F. Baquero, P. Berche, et al. 2001. Comparative genomics of *Listeria* species. *Science* 294:849–852.

Goulet, V., J. Rocourt, I. Rebiere, C. Jacquet, C. Moyse, P. Dehaumont, G. Salvat, and P. Veit. 1995. Listeriosis from consumption of raw-milk cheese. *Lancet* 345: 1581–1582.

Graves, L.M., B. Swaminathan, M.W. Reeves, S.B. Hunter, R.E. Weaver, B.D. Plikaytis, and A. Schuchat. 1994. Comparison of ribotyping and multilocus enzyme electrophoresis for subtyping of *Listeria monocytogenes* isolates. *J. Clin. Microbiol.* 32:2936–2943.

Gray, M. L., and A. H. Killinger. 1966. *Listeria monocytogenes* and listeric infections. *Bacteriol. Rev.* 30:309–382.

Herd, M., and C. Kocks. 2001. Gene fragments distinguishing an epidemic-associated strain from a virulent prototype strain of *Listeria monocytogenes* belong to a distinct functional subset of genes and partially cross-hybridize with other *Listeria* species. *Infect. Immun.* 69:3972–3979.

Ho, J.L., K.N. Shands, G. Friedland, P. Eckind, and D.W. Fraser. 1986. An outbreak of type 4b *Listeria monocytogenes* infection involving patients from eight Boston hospitals. *Arch. Intern. Med.* 146:520–524.

Jacquet, C., B. Catimel, R. Brosch, C. Buchrieser, P. Dehaumont, V. Coulet, A. Lepoutre, P. Veit, and J. Rocourt. 1995. Investigations related to the epidemic strain involved in the French listeriosis outbreak in 1992. *Appl. Environ. Microbiol.* 61:2242–2246.

Jay, J.M. 1996. Foodborne listeriosis. In Modern Food Microbiology, 5ᵗʰ Edition. D.R. Heldman, ed..New York: Chapman & Hall, 478–506.

Jersek, B., P. Gilot, M. Gubina, N. Klun, J. Mehle, E. Tcherneva, N. Rijpens, and L. Herman. 1999.Typing of *Lis-*

teria monocytogenes strains by repetitive element sequence-based PCR. *J. Clin. Microbiol.* 37:103–109.

Johansson, T., L. Rantala, L. Palmu, and T. Honkanen-Buzalski. 1999. Occurrence and typing of *Listeria monocytogenes* strains in retail vacuum-packed fish products and in a production plant. *Int. J. Food Microbiol.* 47:111–119.

Kathariou, S. 2000. Pathogenesisdeterminants of *Listeria monocytogenes*. In Microbial Foodborne Diseases, Mechanisms of Pathogenesis and Toxin Synthesis. J.W. Cary, J.E. Linz, and D. Bhatnagar, Eds.. Lancaster, PA: Technomics Publishing Co., Inc., 295–314.

Kathariou, S. *Listeria monocytogenes* virulence and pathogenicity, a food safety perspective. *J. Food Prot.*65:1811–1829.

Lecuit, M., S. Vandormael-Pournin, J. Lefort, M. Huerre, P. Gounon, C. Dupuy, C. Babinet, and P. Cossart. 2001. A transgenic model for listeriosis: role of internalin in crossing the intestinal barrier. *Science* 292:1722–1725.

Lei, X.H., N. Promadej, and S. Kathariou. 1997. DNA fragments from regions involved in surface antigen expression specifically identify *Listeria monocytogenes* serovar 4 and a subset thereof: cluster IIB (serotypes 4b-4d-4e). *Appl. Ennviron. Microbiol.*63:1077–1082.

Lei, X.H., F. Fiedler, Z. Lan, and S. Kathariou. 2001. A novel serotype-specific gene cassette (*gltA-gltB*) is required for expression of teichoic acid-associated surface antigens in *Listeria monocytogenes* of serotype 4b. *J. Bacteriol.* 183:1133–1139.

Lyytikäinen, O., T. Autio, R. Maijala, P. Ruutu, T. Honkanen-Buzalski, M. Miettinen, M. Hatakka, J. Mikkola, et al. 2000. An outbreak of invasive *Listeria monocytogenes* serotype 3a infections from butter in Finland. *J. Infect. Dis.* 181:1838–1841.

Maijala, R., O. Lyytikäinen, T. Johansson, T. Autio, T. Aalto, L. Haavisto and T. Honkanen-Buzalski. 2001. Exposure of *Listeria monocytogenes* within an epidemic caused by butter in Finland. *Intern. J. Food Microbiol.* 70:97–109.

McLauchlin, J., A. Audurier, and A.G. Taylor. 1986. Aspects of the epidemiology of human *Listeria monocytogenes* infections in Britain 1967–1984; the use of serotyping and phage typing. *J. Med. Microbiol.* 22:367–377.

McLauchlin, J., N. Crofts, and D.M. Campbell. 1989. A possible outbreak of listeriosis caused by an unusual strain of *Listeria monocytogenes*. *J. Infect.* 18:179–187.

McLauchlin, J., and P.N. Hoffman. 1989. Neonatal cross-infection from *Listeria monocytogenes*. *Comm. Dis. Rep.* 16:3–4.

McLauchlin, J., S.M. Hall, S.K. Velani, and R.J. Gilbert. 1991. Human listeriosis and paté: a possible association. *Br. Med. J.* 303:773–775.

Miettinen, M.K., A. Siitonen, P. Heiskanen, H. Haajanen, K.J. Bjorkroth, and H.J. Korkeala. 1999. Molecular epidemiology of an outbreak of febrile gastroenteritis caused by *Listeria monocytogenes* in cold-smoked rainbow trout. *J. Clin. Microbiol.* 37:2358–2360.

Moore, M.A., and A.R. Datta. 1994. DNA fingerprinting of *Listeria monocytogenes* strains by pulsed-field gel electrophoresis. *Food Microbiol.* 11:31–38.

Nocera, D., E. Bannerman, J. Rocourt, K. Jaton-Ogay, and J. Bille. 1990. Characterization by DNA restriction endonuclease analysis of *Listeria monocytogenes* strains related to the Swiss epidemic of listeriosis. *J. Clin. Microbiol.* 28:2259–2263.

Norrung, B., and J.K. Andersen. 2000. Variations in virulence between different electrophoretic types of *Listeria monocytogenes*. *Lett. Appl. Microbiol.* 30:228–232.

Norton, D.M., M.A. McCamey, K.L. Gall, J.M. Scarlett, K.J. Boor, and M. Wiedmann. 2001. Molecular studies on the ecology of *Listeria monocytogenes* in the smoked fish

processing industry. *Appl. Environ. Microbiol.* 67:198–205.

Ojeniyi, B., J. Christiansen, and M. Bisgaard. 2000. Comparative investigations of *Listeria monocytogenes* isolated from a turkey processing plant, turkey products, and from human cases of listeriosis in Denmark. *Epidemiol. Infect.* 125:303–308.

Ørskov, F., and I. Ørskov. 1983. From the national institutes of health. Summary of a workshop on the clone concept in the epidemiology, taxonomy, and evolution of the enterobacteriaceae and other bacteria. *J. Infect. Dis.* 148:346–357.

Ortel, S. 1968. Bakteriologische, serologische, und epidemiologische Unitersuchungen während eine *Listeriose Epidemie. Dtsch. Gesundheitswes.* 23:753–759.

Pejaver, R.K., A.H. Watson, and E.S. Mucklow. 1993. Neonatal cross-infection with *Listeria monocytogenes*. *J. Infect.* 26:301–303.

Piffaretti, J C., H. Kressebuch, M. Aeschbacher, J. Bille, E. Bannerman, J.M. Musser, R.K. Selander, and J. Rocourt. 1989. Genetic characterization of clones of the bacterium *Listeria monocytogenes* causing epidemic disease. *Proc. Natl. Acad. Sci. USA* 86:3818–3822.

Pine, L., G.B. Malcolm, and B.D. Plikaytis. 1990. *Listeria monocytogenes* intragastric and intraperitoneal approximate 50% lethal doses for mice are comparable, but death occurs earlier by intragastric feeding. *Infect. Immun.* 58:2940–2945.

Pine L., S. Kathariou, F. Quinn, V. George, J.D. Wenger, and R. E. Weaver. 1991. Cytopathogenic effects in enterocytelike Caco-2 cells differentiate virulent from avirulent *Listeria* strains. *J. Clin. Microbiol.* 29:990–996.

Proctor, M.E., R. Brosch, J.W. Mellen, L.A. Garrett, C.W. Kaspar, and J.B. Luchansky. 1995. Use of pulsed-field gel electrophoresis to link sporadic cases of invasive listeriosis with recalled chocolate milk. *Appl. Environ. Microbiol.* 61:3177–3179.

Rasmussen, O.F., P. Skouboe, L. Dons, L. Rossen, and J.E. Olsen. 1995. *Listeria monocytogenes* exists in at least three evolutionary lines: evidence from flagellin, invasion associated protein, and listeriolysin O. *Microbiology* 141:2053–2061.

Rocourt, J., C. Jacquet, and A. Reilly. 2000. Epidemiology of human listeriosis and seafoods. *Intern. J. Food Microbiol.* 62:197–209.

Rørvik, L.M., D.A. Caugant, and M. Yndestad. 1995. Contamination pattern of *Listeria monocytogenes* and other *Listeria* spp. in a salmon slaughterhouse and smoked salmon processing plant. *Int. J. Food Microbiol.* 25:19–27.

Ryser, E.T., S.M. Arimi, M.M.-C. Bunduki, and C.W. Connely. 1996. Recovery of different *Listeria* ribotypes from naturally contaminated, raw refrigerated meat and poultry products with two primary enrichment media. *Appl. Environ. Microbiol.* 62:1781–1787.

Salama, N., K. Guillemin, T.K. McDaniel, G. Sherlock, L. Tompkins, and S. Falkow. 2000. A whole-genome microarray reveals genetic diversity among *Helicobacter pylori* strains. *Proc. Natl. Acad. Sci. USA.* 97:14668–14673.

Salamina, G., E. Dalle Donne, A. Niccolini, G. Poda, D. Cesaroni, M. Bucci, R. Fini, M. Maldini, A. Schuchat, B. Swaminathan, W. Bibb, J. Rocourt, N. Binkin, and S. Salmaso. 1996. A foodborne outbreak of gastroenteritis involving *Listeria monocytogenes*. *Epidemiol. Infect.* 117:429–436.

Samuelsson, S., N.P. Rothgardt, A. Carvajal, and W. Frederiksen. 1990. Human listeriosis in Denmark 1981–1987 including an outbreak November 1985–March 1987. *J. Infect.* 20:251–259.

Schlech, W.F. III, P.M. Lavigne, R.A. Bortolussi, A.C. Allen, E.V. Haldane, A.J. Wort, A.W. Hightower, S.E. Johnson, S.H. King, E.S. Nicholls, and C.V. Broome. 1983. Epidemic listeriosis—evidence for transmission by food. *N. Engl. J. Med.* 308:203–206.

Schuchat, A., B. Swaminathan, and C.V. Broome. 1991. Epidemiology of human listeriosis. *Clin. Microbiol. Rev.* 4:169–183.

Schuchat, A., C. Lizano, C.V. Broome, B. Swaminathan, C. Kim, and K. Winn. 1991a. Outbreak of neonatal listeriosis associated with mineral oil. *Pediatr. Infect. Dis. J.* 10:183–189.

Seeliger, H.P.R. 1961. Listeriosis. Basel: Karger Verlag.

Seeliger, H.P.R., and K. Hoehne. 1979. Serotypes of *Listeria monocytogenes* and related species. *Methods Microbiol.* 13:31–49.

Swaminathan, B., T.J. Barrett, S.B. Hunter, R.V. Tauxe, and the CDC PulseNet Task Force. 2001. PulseNet: the molecular subtyping network for foodborne bacterial disease surveillance, United States. *Emerg. Infect. Dis.* 7:382–389.

Tappero, J.W., A. Schuchat, K.A. Deaver, L. Mascola, and J.D. Wenger. 1995. Reduction in the incidence of human listeriosis in the United States. Effectiveness of prevention efforts? The Listeriosis Study Group. *JAMA.* 273:1118–1122.

Tompkin, R.B., V.N. Scott, D.T. Bernard, W.H. Sveum, and K.S. Gombas. 1999. Guidelines to prevent post-processing contamination from *Listeria monocytogenes. Dairy Food Environ. Sanit.* 19:551–562.

Tompkin, R.B. 2002. Control of *Listeria monocytogenes* in the food-processing environment. *J. Food Prot.* 65:709–725.

Tran, H.L., and S. Kathariou. 2002. Restriction fragment length polymorphisms detected with novel DNA probes differentiate among diverse lineages of serogroup 4 *Listeria monocytogenes* and identify four distinct lineages in serotype 4b. *Appl. Environ. Microbiol.* 68:59–64.

Van Langendonck, N., E. Bottreau, S. Bailly, M. Tabouret, J. Marly, P. Pardon, and P. Velge. 1998. Tissue culture assays using Caco-2 cell line differentiate virulent from non-virulent *Listeria monocytogenes* strains. *J. Appl. Microbiol.* 85:337–346.

Vazquez-Boland, J.A., M. Kuhn, P. Berche, T. Chakraborty, G. Dominguez-Bernal, W. Goebel, B. Gonzalez-Zorn, J. Wehland, and J. Kreft. 2001. *Listeria* pathgenesis and molecular virulence determinanats. *Clin. Microbiol. Rev.* 14:584–640.

Wendlinger, G., M.J. Loessner, and S. Scherer. 1996. Bacteriophage receptors on *Listeria monocytogenes* cells are the N-acetyl-D-glucosamine and rhamnose substituents of teichoic acids or the peptidoglycan itself. *Microbiology* 142:985–992.

Wesley, I.V., and F. Aston. 1991. Restriction enzyme analysis of *Listeria monocytogenes* strains associated with food-borne epidemics. *Appl.Environ. Microbiol.* 57:969–975.

Wiedmann, M., J.L. Bruce, C. Keating, A.E. Johnson, P.L. McDonough, and C.A. Batt. 1997. Ribotypes and virulence gene polymorphisms suggest three distinct *Listeria monocytogenes* lineages with differences in pathogenic potential. *Infect. Immun.* 65:2707–2716.

Zheng, W., and S. Kathariou S. 1994. Transposon-induced mutants of *Listeria monocytogenes* incapable of growth at low temperature (4 degrees C). *FEMS Microbiol. Lett.* 121:287–291.

Zheng, W., and S. Kathariou. 1995. Differentiation of epidemic-associated strains of *Listeria monocytogenes* by restriction fragment length polymorphism in a gene region essential for growth at low temperatures, (4°C). *Appl. Environ. Microbiol.* 61:4310–4314.

Zheng, W., and S. Kathariou. 1997. Host-mediated modification of *Sau*3AI restriction in *Listeria monocytogenes*: Prevalence in epidemic-associated strains. *Appl. Ennviron. Microbiol.* 63:3085–3089.

26

LISTERIA MONOCYTOGENES

SCOTT E. MARTIN

INTRODUCTION. *Listeria monocytogenes* is the causative agent of the disease listeriosis and was discovered almost ninety years ago (Gray and Killinger 1966; Donnelly 1994). The organism can be found intracellularly within monocytes and neutrophils, and the name is derived from the observation that increased numbers of monocytes are found in the peripheral blood of infected animals. In 1929 the organism was isolated for the first time from the blood of a human with a mononucleosis-like illness (Gray and Killinger 1966). *L. monocytogenes* causes a high percentage of fatalities in foodborne disease outbreaks, exceeding even *Clostridium botulinum* and *Salmonella*. It has been suggested that listeriosis may be the leading fatal foodborne infection in the United States (Farber and Peterkin 1991). Consumption of foods contaminated with *Listeria* can cause both sporadic illness and foodborne disease epidemics. The annual rate of listeriosis incidence is between 0.2 and 0.7 cases per 100,000. The rate is three times higher for persons over seventy and is seventeen times higher for pregnant women (Southwick and Purich 1996). The overall fatality rate for recent outbreaks has been 33 percent (Morris 1986; Linnan et al. 1988). Although invasive listeriosis is a rare illness, exposure to *L. monocytogenes* is common. The reasons for this low incidence are unclear; early acquired immunity or the fact that most strains are only weakly virulent could be the explanation (Swaminathan 2001).

The number of recognized listerial species varies with the source of reference. Donnelly (1994) and Sneath et al. (1986) suggest that five species are recognized in the *Listeria* genus: *L. monocytogenes, L. ivanovii, L. seeligeri, L. innocua,* and *L. welshimeri*. In addition to these five species, Graham et al. (1997) include the species *L. grayi*. Riviera et al. (1993) added a seventh species: *L. murrayi*, which some authors now include with the species *L. grayi*. *L. monocytogenes* is pathogenic for man and animals, *L. ivanovii* is an animal pathogen, and the remaining species generally are considered to be nonpathogenic (Graham et al. 1997). However, both *L. innocua* and *L. seeligeri* have very rarely been reported to cause human illness (McLauchlin 1987; Riviera et. al. 1993; Donnelly 1994). *L. monocytogenes* is generally regarded as the most important disease-causing species in man. Biochemical characteristics of the six species are found in Table 26.1.

The genome sequences of both *L. monocytogenes* and *L. innocua* have recently been published (Glaser et al. 2001). *L. monocytogenes* was found to contain one circular chromosome of 2,944,528 base pairs (bp) with an average G+C content of 39 percent and contained 2,853 protein-encoding genes; the *L. innocua* chromosome had 3,011,209 bp and a G+C content of 37 percent, and contained 2,973 protein-encoding genes. These authors reported that the listerial-encoded proteins were very similar to those of *Bacillus subtilis,* another Gram-positive rod, suggesting a common origin for these three bacterial species.

L. monocytogenes are small (0.5–2.0 μm in length × 0.4-0.5 μm in diameter), Gram-positive, facultatively anaerobic, nonspore-forming coccoid-shaped rods. Older cultures have a tendency to stain variably between Gram-positive and Gram-negative. In addition, they may also occur as elongated rods and long filaments in older cultures or when grown under adverse conditions (Figure 26.1). They are motile by means of peritrichous flagella and exhibit end-over-end tumbling motility when grown below 25°C, but not at 37°C. When the bacterium is inoculated into motility agar and incubated at 25°C, an umbrella-like zone of growth occurs 3 to 5 millimeters below the surface. *L. monocytogenes* can grow over the temperature range of <1° to <50°C (optimum between 30° and 37°C) and survives freezing (Sneath et al. 1986). It grows over a pH range of 4.0 to 9.5, and in medium containing up to 10 percent NaCl. The minimum water activity for growth ranges from 0.90 to 0.97. They do not survive heating at 60°C for thirty minutes.

Fiedler and Ruhland (1987) examined the biochemistry of the cell surface of *L. monocytogenes* and found that it was typical of other Gram-positive bacteria, with the cell wall having a thick homogeneous layer surrounding the cytoplasmic membrane. The peptidoglycan layer accounts for about 35 percent of the cell wall dry weight. The glycosamine content is high. Teichoic acids are covalently linked to the glycan layers and constitute about 60 to 75 percent of the cell wall dry weight. Lipoteichoic acids are also present and resemble lipopolysaccharides of Gram-negative bacteria in both structure and biological function (Riviera et al. 1993). The antigenic composition of *L. monocytogenes* consists of O (somatic, designated with numbers and lowercase letters) and H (flagellar, designated with uppercase letters) antigens. Serovars exhibit the fol-

Table 26.1—Characteristics of *Listeria* Species

Characteristic	*L. monocytogenes*	*L. ivanovii*	*L. seeligeri*	*L. innocua*	*L. welshimeri*	*L. grayi*
β-hemolysis	+	+	+	−	−	−
CAMP-*S. aureus*	+	−	+	−	−	−
CAMP-*R. equi*	−	+	−	−	−	−
Fermentation of:						
Mannitol	−	−	−	−	−	+
Xylose	−	+	+	−	+	−
Rhamnose	+	−	−	+/−	+/−	−
Virulent (mouse)	+	+	−	−	−	−

lowing antigenic composition: 1/2a, 1/2b, 1/2c, 3a, 3b, 3c, 4a, 4ab, 4b, 4c, 4d, 4e, and serovar 7. Serotypes 1/2a, 1/2b and 4b predominate (92 percent) among isolates from man and animals. All nonpathogenic species, except *L. welshimeri,* share one or more somatic and flagellar antigens with *L. monocytogenes.* Studies have indicated that immunological cross-reactions occur between strains of *Streptococcus, Staphylococcus, Escherichia,* and some *Corynebacterium* species. Joklik et al. (1984) report no correlation between the various serotypes and any particular clinical syndrome or specific host. There is, however, a significant geographic distribution of the various serotypes, with 4b being predominant in the United States, Canada, and Europe (Farber and Peterkin 1991).

L. monocytogenes requires at least four B-vitamins (biotin, riboflavin, thiamin, lipoate) as well as six amino acids (arginine, methionine, cysteine, isoleucine, leucine, valine). Metabolic reconstruction indicated that the biosynthetic pathways for all four vitamins were missing or incomplete, whereas all amino acid biosynthesis pathways were present (Glaser et al. 2001). Carbohydrates are essential for growth, with glucose being the optimum. Under anaerobic conditions, only hexose and pentose sugars support growth; aerobically, maltose and lactose, but not sucrose, also supported growth (Farber and Peterkin 1991). The organism grows well in nutrient-rich media such as tryptic soy broth and brain heart infusion broth. Colonies of *L. monocytogenes* on nutrient agar are small (0.5–1.5 mm/24–48 hr), round, and translucent. The colonies appear bluish-gray under normal illumination but have a blue-green sheen when viewed using obliquely transmitted light. Good growth occurs on blood agar, upon which colonies are surrounded by a narrow band of β-hemolysis. However, strains isolated from fecal matter and environmental sources are frequently nonhemolytic on 5 percent sheep blood agar.

SOURCES. *L. monocytogenes* is a ubiquitous organism that has been isolated from a variety of natural sources including soil, water, and sewage. A primary source is decaying vegetation. It has been found associated with numerous species of mammals (at least thirty-seven including sheep, cattle, swine, dogs), poultry, sewage, milk, cheese, coleslaw, lettuce, and meat products. Silage is an important source of *L. monocy-* *togenes,* and a correlation has been found between the feeding of silage and outbreaks of listeriosis in ruminants (Donnelly 1994). The presence of *L. monocytogenes* in silage has been found to be strongly influenced by pH: silage with a pH 5.0 to 6.0 and above is far more likely to contain *L. monocytogenes* than is silage below pH 5.0. The organism has been observed in cattle and to cause udder infections. Between 11 to 52 percent of healthy animals have been found to be carriers of *L. monocytogenes,* whereas healthy human intestinal carriers occur at a rate of 1 to 5 percent of the total population.

The microorganism can be isolated from a number of sites in food processing plants. It gains entry into processing plants via a number of different ways: through soil on workers' shoes and clothing and on transport equipment; animal products having contaminated hides or surfaces; raw foods (plant or animal); and human carriers. *L. monocytogenes* is capable of growth in any site in a food manufacturing facility having favorable growth conditions (nutrient source, temperature, high humidity), which can include floor drains, stagnant water, floors, food residues, and processing equipment (Swaminathan 2001). The ability of *L. monocytogenes* to form biofilms may aid its growth in food processing plants and its capacity to survive cleaning and sanitation treatments. In dairies it has been found in raw milk, on cooler floors, in freezers, processing rooms, cases, and on mats (Klausner and Donnelly 1991). It has been isolated from various sites in meat and seafood processing facilities. In heat-processed foods, the presence of *L. monocytogenes* occurs via post-processing contamination. No evidence exists to suggest that it can survive heat-processing protocols used to render food safe (Swaminathan 2001).

Foods. *L. monocytogenes* has been found in a number of different types of foods and ingredients, both raw and processed. Ready-to-eat foods have been identified as a significant source of the pathogen. The United States Department of Agriculture, Food Safety and Inspection Service conducted a microbiological testing study of ready-to-eat meat and poultry products produced at approximately 1,800 federally inspected establishments (Levine et al. 2001). The cumulative ten-year prevalences for *L. monocytogenes* were as follows: jerky, 0.52 percent; uncured poultry products, 2.12 percent; large-diameter cooked sausages, 1.31 percent; small-

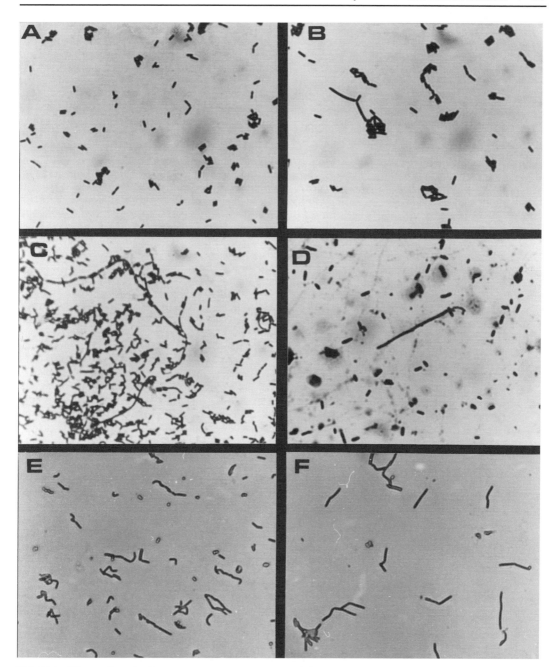

FIG. 26.1—Filament formation in *L. monocytogenes* grown under environmental stress. Cells were photographed (magnification 412.5×) after growth in tryptic soy broth: (A) without acid; (B) pH 6.0, citric acid; (C) pH 9.2, NaOH; (D) pH 9.4, NaOH; (E) 1200 mM NaCl; (F) 1500 mM NaCl (from Isom et al. 1995).

diameter cooked sausages, 3.56 percent; cooked beef, roast beef, and cooked corned beef, 3.09 percent; salads, spreads, and pates, 3.03 percent; and sliced ham and luncheon meat, 5.16 percent. The cumulative three-year prevalence for dry and semidry fermented sausage was 3.25 percent. Several species of *Listeria* (*L. mono-*

cytogenes, L. ivanovii, L. seeligeri, L. innocua, L. welshimeri, and *L. grayi)* were isolated from lettuce, spinach, and potato omelettes in Spain (Soriano et al. 2001). Several *Listeria* species have been isolated from cheese, being most frequently found in soft and semi-soft cheeses (Rodolf and Scherer 2001). *L. monocyto-*

genes has been found as a contaminant in seafood. The highest prevalence was found in cold-smoked fish (34 to 60 percent), with the lowest found in heat-treated and cured seafood (Jorgensen and Huss 1998). *L. monocytogenes* accounted for the greatest number of food product recalls in the United States because of microbial contamination in the period from October 1993 through September 1998 (Wong et al. 2000).

Animals. *L. monocytogenes* has been found to be associated with many animal types and to cause listeriosis in sheep, goats, cattle, swine, and fowl (chickens, turkeys, ducks, geese, and pheasants). In addition to *L. monocytogenes,* other listerial species have also been isolated from animals. Most listerial infections in animals are subclinical, although listeriosis can occur either sporadically or as epidemics that can lead to fatal forms of encephalitis, septicemia, and abortion (Wesely 1999). The most common syndrome of listeriosis in ruminants is encephalitis, leading to observations of nervous system involvement in cattle and sheep (Donnelly 1994). These animals become disoriented, causing them to circle in one direction. Listeriosis in ruminants is often referred to as circling disease. Transmission of *L. monocytogenes* from animals to humans occurs in two primary ways: direct contact with infected animals and by consumption of contaminated raw milk. Milk from animals (cows, sheep, or goats) having generalized listerial infections can cause mastitis in these animals, leading to contaminated milk. In swine, listeriosis primarily occurs as septicemia, with encephalitis and abortions rarely being observed. In fowl, listeriosis is uncommon and has been suggested to be a secondary infection. Septicemia is the most common manifestation of listeriosis in chickens and other domestic fowl. Evidence suggests that fish and shellfish are not normally carriers of *L. monocytogenes.*

GROWTH. *Listeria monocytogenes* is a hardy bacterial pathogen. It can survive and grow under conditions that are inhibitory to many other pathogenic bacteria. Some of the environmental factors that influence the growth of *L. monocytogenes* include: pH, temperature, water activity (salt concentration), and modified atmosphere.

pH. The influence of pH has been studied both in terms of absolute value and acid type, with special emphasis on dairy products. When the pH was lowered to 5.0, survival time of *L. monocytogenes* was significantly reduced. Listeriae can grow at pH 5.0 at temperatures >13°C. Ahamad and Marth (1989) established that acetic acid was most detrimental to *L. monocytogenes* growth, followed by lactic and citric acids. Conner et al. (1990) found the growth inhibitory pH to be as follows: 5.0 for propionic acid; 4.5 for acetic and lactic acids; and 4.0 for citric and hydrochloric acids at 4°C. It appears that pH, acid type, and temperature all influence the growth and survival of *L. monocytogenes.*

Temperature. *L. monocytogenes* is a psychrotrophic bacterium, able to grow at refrigeration temperatures, although optimal growth occurs between 30° and 37°C. The influence of temperature on the survival and growth of *L. monocytogenes* has been examined, frequently in conjunction with other environmental variables. Buchanan et al. (1989) studied temperature and its interactions with initial pH, atmosphere, sodium chloride content, and sodium nitrite concentration. They found that growth kinetics were dependent on the interaction of the five variables. Junttila et al. (1988) reported that the mean minimum growth temperature of *L. monocytogenes* was $+1.1 \pm 0.3°C$. They also found that hemolytic listerial isolates grew better than nonhemolytic strains under cold conditions. Doyle (1988) stated that generation times in various milk products ranged from 1.2 to 1.7 days at 4°C, 5.0 to 7.2 hr at 13°C, and 0.65 to 0.69 hr at 35°C. Walker et al. (1990) established that the listerial minimum growth temperatures ranged from −0.1 to −0.4°C in chicken broth, with generation times ranging from 13 to 24 hr at 5°C to 62 to 131 hr at 0°C.

Growth temperature has also been shown to influence the virulence of *L. monocytogenes.* Czuprynski et al. (1989) showed that growth at 4°C significantly increased the virulence of three clinical listerial isolates in mice when compared to cells grown at 37°C for intravenously, but not intragastrically, inoculated mice. In agreement, Stecha et al. (1989) showed that growth at 4°C significantly decreased the killing of test strains by human neutrophils. Myers and Martin (1994) found that *L. monocytogenes* strain 10403S did not increase in virulence when grown at low temperature or in increased levels of NaCl.

Salt. Sodium chloride is frequently added to food to improve flavor and to serve as an agent to reduce water activity. Doyle (1988) found that *L. monocytogenes* is tolerant to NaCl, being capable of growth in 10 percent NaCl and surviving for one year in 16 percent NaCl. As with pH and temperature, the influence of salt is often examined with other variables. Beuchat et al. (1986) found that the thermal inactivation of *L. monocytogenes* was not significantly influenced by the presence of up to 2 percent NaCl, but that heat-stressed cells had increased sensitivity to the salt. Cole et al. (1990) observed that low salt concentrations (4 to 6 percent) improved survival, whereas higher concentrations reduced survival of listerial cells at limiting pH values. The optimum a_w for listerial growth is ≥ 0.97. Although some strains can grow at a_w values as low as 0.90, the minimum a_w value for most strains is 0.93. *L. monocytogenes* can survive for long periods at a_w values as low as 0.83.

Modified Atmosphere. Modified atmosphere and its effects on *L. monocytogenes* have been examined by fewer investigators than have pH, temperature, or NaCl. Doyle (1988) reported that growth of *L. monocytogenes* was enhanced under decreased oxygen concentrations and when supplemented with carbon dioxide. Berrang et al. (1989) examined the survival and growth of *L. monocytogenes* on three fresh vegetables under controlled atmosphere storage (asparagus: 15 percent O_2, 6 percent CO_2, 79 percent N_2; broccoli: 11 percent O_2, 10 percent CO_2, 79 percent N_2; cauliflower: 18 percent O_2, 3 percent CO_2, 79 percent N_2). They found that the listerial cells increased in number during storage, and the type of atmosphere did not influence the rate of growth. Knabel et al. (1990) observed that strictly anaerobic conditions significantly increased the recovery of heat-injured listeriae when compared with aerobically incubated controls.

Flagella. *L. monocytogenes* cells are able to survive throughout the environment and undergo morphological changes to adapt to new surroundings. It is known that temperature and osmolarity of the growth medium have a role in flagellar expression. Therefore, flagella of *L. monocytogenes* may play a role in survival of the organism.

The degree of listerial flagellation and extent of motility at 37°C remain controversial. Several reports indicate that the production of flagella in *L. monocytogenes* is markedly reduced at 37°C, with motility repressed. Other reports indicate the presence of flagella and flagella-based motility at this temperature (Galsworthy et al. 1990). When grown between 20° and 25°C, *L. monocytogenes* are peritrichously flagellated, vigorously motile, and display a tumbling motility interspersed with smooth swimming. Listerial strains are vigorously motile via a peritrichous flagellar arrangement that causes the cell to exhibit a tumbling motility with bursts of rapid motion occurring in one direction for ten to twenty cell lengths in liquid medium. One test for motility in *L. monocytogenes* utilizes motility agar and involves stabbing a semisolid motility agar tube (about 1 cm deep) and incubation at 20°–25°C (Seeliger and Höhne 1979). Studies using soft agar in Petri dishes have also been conducted to correlate swarming with the degree of motility. The growth environment has been found to influence listerial flagellation and motility.

Conflicting reports exist as to the extent of flagellation and degree of motility at different temperatures. Peel et al. (1988) reported a characteristic temperature-dependent motility, where *L. monocytogenes* was motile when grown only between 20° and 25°C. Studies on chemotaxis showed that *L. monocytogenes* exhibited directional motility at 37°C, and it has been hypothesized that directional motility of *L. monocytogenes* may facilitate penetration of the intestinal epithelium. The small quantity of flagellin produced at 37°C was thought to be highly immunogenic upon infection; the largest single antigenic component of *L. monocytogenes* grown at 20°C is flagellin.

LISTERIOSIS. In 1983, forty-nine pregnant women and immuno-compromised adults in Massachusetts contracted listeriosis caused by *L. monocytogenes*. Fourteen people died as a result of this foodborne dis-

ease outbreak, including two unborn infants. The epidemiologically implicated source of the organisms was pasteurized whole and 2 percent low-fat milk produced by a single dairy plant. *L. monocytogenes* was the cause of one of the most deadly outbreaks of food poisoning in the history of the United States: the Jalisco Cheese episode in southern California. Following consumption of this soft Mexican-style cheese, 142 cases of listeriosis were recorded. Of ninety-three pregnant women infected, twenty-nine lost their babies; of forty-nine immuno-compromised adults, eighteen died. The overall fatality rate was 33 percent. An outbreak of listeriosis involving gastrointestinal symptoms occurred in San Giorgio di Piano, Italy, in 1993. Eighteen young, previously healthy, nonpregnant adults developed symptoms (mostly gastrointestinal) after ingestion of rice salad (Salamina et al. 1996). Table 26.2 lists several representative outbreaks of *L. monocytogenes*. Most cases of human listeriosis occur as sporadic illnesses. Retrospective epidemiological studies have identified a number of small foodborne outbreaks of listeriosis. In a twenty-five–month study, the Centers for Disease Control identified 301 cases of listeriosis (Schuchat et al. 1992). Underlying patient conditions included pregnancy, steroid therapy, cancer, renal or liver disease, diabetes, HIV infection, and organ transplant patients. Of the 301 cases, 32 percent were attributed to delicatessen foods.

Listeriosis can occur in persons of all ages but occurs most frequently in individuals who have an underlying condition leading to suppression of T-cell-mediated immunity. It is clinically defined when the organism is isolated from a normally sterile site such as blood or cerebrospinal fluid. The disease may take a number of complications, with the most common being meningitis in the elderly and in immunocompromised patients. In the average, healthy individual infections are usually asymptomatic or cause mild influenza-like symptoms (fever, fatigue, malaise, nausea, cramps, vomiting, and diarrhea). The estimated infective dose varies with the strain of *L. monocytogenes* and the host, but ingestion of fewer than one thousand cells is thought to be sufficient to cause disease. Time of onset of gastrointestinal symptoms is unknown but is probably more than twelve hours. Gastrointestinal symptoms have been epidemiologically associated with consumption of antacids or cimetidine.

Serious complications of human listeriosis include: septicemia, infectious mononucleosis-like syndrome, pneumonia, endocarditis, aortic aneurysm, localized abscesses, cutaneous lesions, conjunctivitis, hepatitis, and urethritis. Time of onset for the more serious forms of listeriosis may range from a few days to three weeks. The mortality rate among untreated patients is approximately 70 percent. The disease is most common in hosts whose immune systems are compromised because of preexisting disease or administration of immuno-suppressive agents (AIDS, cancer, diabetes, age, alcoholism). *L. monocytogenes* can survive and multiply within macrophages. Selective plating, monoclonal antibodies, polymerase chain reaction, and DNA

Table 26.2—Outbreaks of Listeriosis Associated with Food

Year	Location	Serotype	Outbreak and associated food
1979	Boston, MA	4b	Salads or pasteurized milk implicated in nosocomial outbreak among immunocompromised patients
1979	Nova Scotia, Canada	4b	Sixty-six cases in pregnant women, 18 deaths (infants); cabbage thought to be contaminated with sheep manure
1981	Auckland, New Zealand		Consumption of raw fish and oysters, epidemiologically linked to perinatal listeriosis
1983	Massachusetts	4b	Pasteurized milk linked to 49 cases with 14 deaths; mainly immunocompromised adults
1985	Los Angeles, California	4b	Mexican-style soft cheese; more than 100 cases and 30 deaths, mostly infants
1983-1987	Vaub, Switzerland	4b	Vacherin Mont D'Or cheese over a period of years; 31 deaths
1993	San Giorgio di Piano, Italy	1/2b	Rice salad, 18 victims with gastrointestinal symptoms; no deaths
2000	Winston-Salem, North Carolina		Homemade Mexican-style cheese, 5 stillbirths

probe techniques have been developed for detecting *Listeria* in foods. Clinical diagnosis is frequently initially accomplished by microscopic observation of infected fluids such as blood, cerebrospinal fluid, amniotic fluid, and genital tract secretions. Most clinical isolates are grown on conventional media such as brain-heart infusion agar containing 5 percent blood (from sheep, horse, or rabbit). Treatment with the antibiotics ampicillin or penicillin is usually very effective in treating listeriosis. Trimethoprim sulfamethoxazole may be used for patients allergic to penicillin.

A serious manifestation of *Listeria* infection is an intrauterine infection in pregnant women that can lead to abortion, stillbirth, or to delivery of an apparently healthy child who develops meningitis after birth. They occur most frequently during the third trimester, although they may occur any time during pregnancy (Mandell et al. 1995). The mother may have had a low-grade uterine infection and may have had few or no symptoms. Table 26.3 lists recommendations to prevent listeriosis in pregnant women and in other susceptible individuals.

Transplacental transmission causes a second category of infection unique to *L. monocytogenes,* granperinatal listeric septicemia (also known as granulomatosis infantiseptica; Davis et al. 1980). The immune system of the pregnant woman is altered so that the fetus is able to escape immunological attack by the maternal host defenses. Such changes in immune competence affect the defense of the pregnant host, leaving the woman and her fetus at high risk of infection. Two clinical forms of neonatal listeriosis are recognized: early- and late-onset (Farber and Peterkin 1991). In the early-onset form, symptoms occur within 1.5 days after birth. *L. monocytogenes* are found in numerous foci of necrosis throughout the infant body, especially in the liver, spleen, lungs, kidney, and brain (Mandell et al. 1995). The poorest prognosis appears to occur in the early-onset group.

In late-onset neonatal listeriosis, the mean onset of symptoms is 14.3 days after birth, with meningitis as the predominant form of the disease. Symptoms of late-

onset neonatal listeriosis include respiratory distress syndrome, rash, conjunctivitis, pneumonia, hyperexcitability, vomiting, cramps, shock, hematologic abnormalities, and either hyper- or hypothermia. Most neonatal deaths are from pneumonia and respiratory failure (Farber and Peterkin 1991). Prompt antibiotic therapy significantly lowers mortality, although the mortality rate is still about 36 percent. Because listeriosis has no distinctive clinical manifestations, isolation and identification of *Listeria* is an important aid to diagnosis.

L. monocytogenes Infection. *L. monocytogenes* is acquired by ingestion of contaminated food. Following listerial ingestion and survival of the acidity of the stomach, listerial cells attach to the intestinal mucosa. In mouse and guinea pig models, M cells or Peyer's Patches may be the primary intestinal invasion site. It is thought that β-D-galactose residues on the bacterial surface bind the β-D-galactose receptors on intestinal cells. Flagella may be an important virulence factor in the attachment in that they may aid the bacteria in moving to the intestinal target site of infection. Following translocation across the intestinal barrier, *L. monocytogenes* cells can be seen in phagocytic cells in the underlying lamina propria and in nonphagocytic cells (Sheehan et al. 1994; Swaminathan 2001). In mice, the bacteria become disseminated to the liver and spleen (Ireton and Cossart 1997). Although most listerial cells are killed by phagocytes in the liver, survivors multiply in hepatocytes. In cultured epithelial cells, host cell entry requires the presence of surface proteins InlA (internalin) and InlB (Ireton and Cossart 1997). The receptor for InlA is the cell adhesion molecule E-cadherin. InlB-mediated entry requires activation of host cell protein phosphoinositide 3-kinase. After it is inside the host cell, the *L. monocytogenes* cell is encased in a membrane vesicle (Figure 26.2). Escape from this vesicle is mediated by listeriolysin O and phospholipases C. A zinc metalloprotease is directly involved in the maturation of one of the phospholipases C (Schwarzkopf 1996). When free from the vesicle, the listerial cells are free in the host cell cytoplasm, where

Table 26.3—Recommendations to Protect against Listerial Infection

- Avoid raw/unpasteurized milk
- Keep raw and cooked foods separate at all times
- Wash hands, knives, and all food contact surfaces after handling uncooked foods
- Wash raw vegetables thoroughly before eating
- Thoroughly cook all food of animal origin, including eggs; cook raw meat and fish to an internal temperature of 160°F, raw poultry to 180°F; reheat leftovers thoroughly
- Read and follow label instructions to "keep refrigerated" and "use by" a certain date
- Keep hot food hot (above 140°F); do not keep foods at room temperature for longer than two hours
- Keep cold food cold (at or below 40°F); do not keep foods at room temperature for longer than two hours
- Refrigerate small portions of food so that they chill rapidly and evenly
- Keep refrigerator clean and the temperature at 34–40°F

Recommendations for high-risk individuals (pregnant women, the elderly, the immunosuppressed):

- Avoid soft cheeses such as Mexican-style, feta, Brie, Camembert and blue cheeses
- If soft cheeses are eaten during pregnancy, cook them until they boil
- Do no eat hard cheese made from unpasteurized milk; use only hard cheeses aged at least 60 days
- Reheat leftovers and ready-to-eat foods thoroughly until steaming hot before eating
- Reheat all meats purchased at deli counters

rapid multiplication occurs. The final steps of infection involve intra- and inter-cellular movement, utilizing host cell actin. Actin polymerization is catalyzed by a listerial surface protein of the *actA* gene ActA. Intracytoplasmic *L. monocytogenes* appear to be covered by a "cloud" of host cell actin filaments, which rearrange into a polymerized comet tail. As the comet tail grows, the listerial cell is propelled to the host cell membrane, forming protrusions that extend into adjacent cells. These protrusions are internalized by the neighboring host cell. Again with the aid of listeriolysin O and phospholipases C, the double membrane is lysed and the cell cycle begins again. Molecular genetic techniques have been used to analyze the factors responsible for the various steps of the infectious process. Virulence genes identified include: those involved in entry of *L. monocytogenes* in epithelial cells (*inlA* and *inlB*); escape from the phagosome (*hly* and *plcA*); intra- and intercellular movement (*actA*); and lysis of the two-membrane vacuoles (*plcB*) (Cossart 1994). All these genes are co-regulated and, except for the *inl* genes, are clustered on the same region of the chromosome (Figure 26.3). In addition, the enzymes catalase and superoxide dismutase may function as secondary virulence factors. These two enzymes act by degrading toxic oxygen free radicals produced during the respiratory burst following macrophage phagocytosis.

CONTROL. The complete elimination of *L. monocytogenes* on the farm and at the packing/processing level is very difficult, if not impossible. Because *L. monocytogenes* is ubiquitous in the environment, eliminating this bacterium from the farm would be virtually impossible. Because of the presence of *L. monocytogenes* in soil, water, and vegetation, it can be isolated from any equipment that has been in contact with, or contaminated by, these materials. Improperly fermented silage

is a frequent source of *L. monocytogenes* on the farm. It has been found in moldy silage samples and in silage with a pH both above and below 4.5.

Complete elimination of *L. monocytogenes* from food processing facilities also is a hard-to-achieve goal for food manufacturers. In food processing facilities, cells of *L. monocytogenes* are often found as biofilms. A biofilm is defined as a biologically active population of microorganisms that are attached to a surface and enclosed by an extracellular matrix. Bacteria in biofilms often are more resistant to antimicrobial substances than are unattached bacteria. This may be attributed to properties of the biofilm, such as: reduced ability of substances to diffuse into the biofilm, physiological changes to the bacteria because of reduced growth rate, and production of enzymes, which degrade antimicrobial substances. Even after proper cleaning, viable microorganisms can be recovered from treated

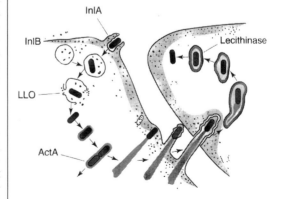

FIG. 26.2—The steps of *L. monocytogenes* infectious process (Ireton and Cossart 1997).

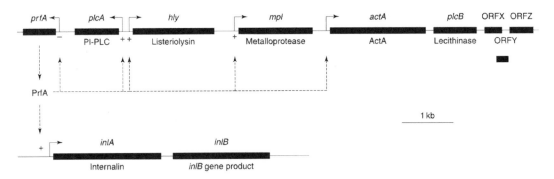

FIG. 26.3—Organization and regulatory control of some *L. monocytogenes* virulence genes that are clustered on the chromosome. *prfA*, regulatory protein (activator); genes activated by PrfA are indicted by arrows: *plcA*, PI-PLC, *plcB*, PC-PLC (lecithinase); *hly*, listeriolysin O; *actA*, protein involved in actin polymerization; *orfX,Y,Z* orfs of unknown function; probable locations of promoters are indicated by "P"; wavy lines indicated mRNA transcripts (Sheehan et al. 1994).

surfaces. Organisms remaining after cleaning may survive for extended periods and subsequently form new biofilms. *L. monocytogenes* have been shown to attach to stainless steel and are capable of developing an extensive extracellular matrix commonly associated with biofilm formation. Microorganisms existing in biofilms are less susceptible than suspended cells to heat and the presence of sanitizing agents. Attached microorganisms surviving cleaning and sanitizing procedures may desorb from food-contact surfaces and subsequently contaminate food products.

Four different categories exist of chemical sanitizers commonly used in food plants: chlorine, iodine, quaternary ammonium compounds, and acid-anionic surfactants. Chlorine (25–200 ppm), quaternary ammonium compounds (100–200 ppm), an iodophor (12.5–50 ppm), and an acid anionic surfactant (200 ppm) were all effective against *L. monocytogenes* cells at 25°C. Chlorine, in the form of sodium hypochlorite (200 ppm), caused a six log_{10} reduction in *L. monocytogenes* cells on inoculated brussels sprouts. A one log_{10} reduction in the number of listerial cells would be expected, regardless of the disinfectant used, for *L. monocytogenes* on fresh-cut vegetables. Current evidence suggests that chlorine washes are relatively ineffective in the elimination of *L. monocytogenes* from contaminated raw vegetables.

The use of high temperature to eliminate cells of *L. monocytogenes* can be employed as an alternative for, or used in addition to, chemical sanitizers in food processing plants. The heat resistance of *L. monocytogenes* is in question in the scientific literature. Some studies suggest that the pathogen can survive exposure to High Temperature-Short Time (HTST) pasteurization of milk at 71.7°C for fifteen seconds. As is the case for most sanitizing treatments, the higher the temperature or the longer exposure to santizing agents, the more effective is the elimination of *L. monocytogenes*.

Ozone (O$_3$) is one of the most powerful oxidizing agents. It is a 52 percent more potent oxidant than chlorine, acts more rapidly against a wide spectrum of microorganisms, and is a Generally Recognized As Safe (GRAS) substance for use in the food industry. Cells of *L. monocytogenes* were found to be more resistant to ozone than were cells of *Salmonella typhimurium* and *Escherichia coli* after exposure to 0.18 ppm ozone. Exposure of *L. monocytogenes* to ozone at 2.5 ppm for forty seconds caused five to six log_{10} decrease in viable cell counts.

Implementation of an effective Hazard Analysis and Critical Control Point (HACCP) program by food processors may help to reduce the total level of contamination by *L. monocytogenes*. The successful employment of an HACCP program can reduce contamination by *L. monocytogenes*.

SUMMARY AND CONCLUSIONS. *L. monocytogenes* has been recognized as a human pathogen for more than seventy years. *L. monocytogenes* are resistant to many environmental challenges and are one of the few foodborne pathogens able to grow at refrigeration temperatures. The bacterium is ubiquitous and can be isolated from a wide variety of sources, both the environment and food. Listeriosis is a potentially fatal illness for immunosuppressed victims. Serious complications of human listeriosis include meningitis and granperinatal listeric septicemia. *L. monocytogenes* cells are able to survive phagocytosis and multiply within host cells. Isolation and detection of listerial cells from foods require the use of enrichment and plating on antibiotic containing media.

REFERENCES

Ahamad, N., and E.H. Marth. 1989. Behavior of *Listeria monocytogenes* at 7, 12, 21 and 35°C in tryptose broth acidified with acetic, citric, or lactic acid. *J. Food Protect.* 52:688–695.

Bacteriological Analytical Manual, 7th Ed. 1992. Food and Drug Administration, Arlington, VA: AOAC International, 141–160.

Berrang, M.E., R.E. Brackett, and L.R. Beauchat. 1989. Growth of *Listeria monocytogenes* on fresh vegetables stored under controlled atmosphere. *J. Food Protect.* 52:702–705.

Beuchat, L.R., R.E. Brackett, D.Y. Hao, and D.E. Conner. 1986. Growth and thermal inactivation of *Listeria monocytogenes* in cabbage and cabbage juice. *Can. J. Microbiol.* 32:791–795.

Buchanan, R.L., H.G. Stahl, and R.C. Whitting. 1989. Effects and interactions of temperature, pH, atmosphere, sodium chloride and sodium nitrite on the growth of *Listeria monocytogenes. J. Food Protect.* 52:844–851.

Cole, M.B., M.V. Jones, and C. Holyoak. 1990. The effect of pH, salt concentration and temperature on the survival and growth of *Listeria monocytogenes. J. Appl. Bacteriol.* 69:63–72.

Conner, D.E., V.N. Scott, and D.T. Bernard. 1990. Growth, inhibition and survival of *Listeria monocytogenes* as affected by acidic conditions. *J. Food Protect.* 53:652–655.

Cossart, P. 1994. *Listeria monocytogenes:* Strategies for entry and survival in cells and tissues. *Baillère's Clin. Infect. Dis.* vol 1. no. 2, Chapt. 7, 285–304.

Czuprynski, C.J., J.F. Brown, and J.T. Roll. 1989. Growth at reduced temperatures increases the virulence of *Listeria monocytogenes* for intravenously but not intragastrically inoculated mice. *Microbiol. Pathogen.* 7:213–223.

Davis, B.D., R. Dulbecco, H.N. Eisen, and H.S. Ginsberg. 1980. *Microbiology,* 3rd edition. New York: Harper & Row, 799–800.

Donnelly, C.W. 1994. *Listeria monocytogenes.* In Foodborne Disease Handbook, vol. 1. Y.H. Hui, J.R. Gorham, K.D. Murrell, and D.O. Cliver, eds. New York: Marcel Dekker Inc., 215–252.

Doyle, M.P. 1988. Effect of environmental and processing conditions on *Listeria monocytogenes. Food Tech.* April:169–171.

Farber, J.M., and P.I. Peterkin. 1991. *Listeria monocytogenes,* a food-borne pathogen. *Microbiol. Rev.* 55:476–511.

Fiedler, F., and G.H. Ruhland. 1987. Structure of *Listeria monocytogenes* cell wall. *Bull. de l'Institut Pasteur* 85:287–300.

Galsworthy, S., S. Girdler, and S. Koval. 1990. Chemotaxis in *Listeria monocytogenes. Acta Microbiol. Hung.* 37:81–85.

Glaser, P., L. Frangeul, C. Buchrieser, C. Rusniok, A. Amend, F. Baquero, P. Berche, H. Bloecker, P. Brandt, T. Chakraborty, A. Charbit, F. Chetouani, E. Couvé, A. de Daruvar, P. Dehoux, E. Domann, G. Domínguez-Bernal, E. Duchaud, L. Durant, O. Dussurget, K.-D. Entian, H. Fsihi, F. Garcia-Del Portillo, P. Garrido, L. Gautier, W. Goebel, N. Gómez-López, T. Hain, J. Hauf, D. Jackson, L.-M. Jones, U. Kaerst, J. Kreft, M. Kuhn, F. Kinst, G. Kurapkat, E. Madueño, A. Maitourman, J. Mata Vincente, E. Ng, H. Nedjari, G. Nordsiek, S. Novella, B. de Pablos, J.-C. PéreDiaz, R. Purcell, B. Remmel, M. Rose, T. Schlueter, N. Simoes, A. Tierrez, J.-A Vázquez-Boland, H. Voss, J. Wehland, and P. Cossart. 2001. Comparative Genomics of *Listeria* species. *Science* 294:849–852.

Graham, T.A., E.J. Golsteyn-Thomas, J.E. Thomas, and V.P. Gannon. 1997. Inter- and Intraspecies Comparison of the 16S-23S rRNA Operon Intergenic Spacer Regions of Six *Listeria* spp. *Internat. J. System. Bacteriol.* 47:863–869.

Gray, M.L., and A.H. Killinger. 1966. *Listeria monocytogenes* and listeric infections. *Bacteriol. Rev.* 30:309–382.

Ireton, K., and Cossart, P. 1997. Host-pathogen interactions during entry and actin-based movement of *Listeria monocytogenes. Annu. Rev. Genet.* 31:113–138.

Isom, L.L., Z.S. Khambatta, J.L. Moluf, D.F. Akers, and S.E. Martin. 1995. Filament formation in *Listeria monocytogenes. J. Food Protect.* 58:1031–1033.

Joklik, W.K., H.P. Willett, and D.B. Amos, eds. 1984. *Zinsser Microbiology,* 18th Edition. Crofts, CN: Appleton-Century, 527–533.

Jorgensen, L.V., and H.H. Huss. 1998. Prevalence and growth of *Listeria monocytogenes* in naturally contaminated seafood. *Internat. J. Food Microbiol.* 42:127–131.

Junttila, R.R., S.I. Niemela, and J. Hirn. 1988. Minimum growth temperatures of *Listeria monocytogenes* and non-hemolytic listeria. *J. Appl. Bacteriol.* 65:321–327.

Klausner, R.B., and C.W. Donnelly. 1991. Environmental sources of *Listeria* and *Yersinia* in Vermont dairy plants. *J. Food Protect.* 54:607–611.

Knabel, S.J., H.W. Walker, P.A. Hartman, and A.F. Mendonca. 1990. Effects of growth temperature and strictly anaerobic recovery on the survival of *Listeria monocytogenes* during pasteurization. *Appl. Environ. Microbiol.* 56:370–376.

Levine, P., B. Rose, S. Green, G. Ransom, and W. Hill. 2001. Pathogen testing of ready-to-eat eat and poultry products collected at federally inspected establishments in the United States, 1990 to 1999. *J. Food Protect.* 64:1188–1193.

Linnan, M.J., L. Mascola, X.D. Lou, V. Goulet, S. May, C. Salminen, D.W. Hird, M.L. Yonekura, P. Hayes, R. Weaver, A. Audurier, B.D. Plikaytis, S. Fannin, A. Kleks, and C.V. Broome. 1988. Epidemic listeriosis associated with Mexican-style cheese. *New Eng. J. Med.* 319:823–828.

Mandell, G.L., J.E. Bennett, and R. Dolin. 1995. *Principles and Practice of Infectious Diseases.* 4th Edition. New York: Churchill Livingstone, 1880–1885.

McLauchlin, J. 1987. *Listeria monocytogenes,* recent advances in the taxonomy and epidemiology of listeriosis in humans. *J. Appl. Bacteriol.* 63:1–11.

Morris, C.E. 1986. Microbe tracking: The new dangers. *Food Eng.* June:64.

Myers, E.R., and S.E. Martin. 1994. Virulence of *Listeria monocytogenes* propagated in NaCl containing media at 4, 25, and 37° C. *J. Food Protect.* 57(6):475–478.

Peel, M., W. Donachie, and A. Shaw. 1988. Temperature-dependent expression of flagella of *Listeria monocytogenes* studied by electron microscopy, SDS-PAGE and Western Blotting. *J. Gen. Microbiol.* 134:2171–2178.

Riviera, L., F. Dubini, M.G. Bellotti. 1993. *Listeria monocytogenes* infections: the organism, its pathogenicity and antimicrobial drugs susceptibility. *Microbiologica* 16:189–204.

Rodolf, M., and S. Scherer. 2001. High incidence of *Listeria monocytogenes* in European red smear cheese. *Internat. J. Food Microbiol.* 63:91–98.

Salamina, G., E. Dalle Donne, A. Niccolini, G. Poda, D. Cesaroni, M. Bucci, R. Fini, M. Maldini, A. Schuchat, B. Swaminathan, W. Bibb, J. Rocourt, N. Binkin, S. Salmaso. 1996. A foodborne outbreak of gastroenteritis involving *Listeria monocytogenes. Edpidemiol. Infect.* 117:429–436.

Schuchat, A., K.A. Deaver, J.D. Wenger, B.D. Plikaytis, L. Mascola, R.W. Pinner, A.L. Reingold, and C.V. Broome, and the *Listeria* Study Group. 1992. Role of foods in sporadic listeriosis. 1. Case-control study of dietary risk factors. *JAMA* 267:2041–2045.

Schwarzkopf, A. 1996. *Listeria monocytogenes*—Aspects of Pathogenicity. *Path. Biol.* 9:769–774.

Seeliger, H., and K. Höhne. 1979. Serotyping of *Listeria monocytogenes* and related species. Meth. in Microbiol., Chapter 13. New York: Academic Press, 31–49.

Sheehan, B., C. Kocks, S. Dramsi, E. Gouin, A. D. Klarsfeld, J. Mengaud, and P. Cossart. 1994. Molecular and genetic determinants of the *Listeria monocytogenes* infectious process. *Curr. Top. Microbiol. Immuniol.* 192:187–216.

Sneath, P.H.A., N.S. Mair, M.E. Sharpe, and J.G. Holt. 1986. *Bergey's Manual of Systematic Bacteriology.* Baltimore, MD: Williams and Williams, 1235–1245.

Soriano, J.M., H. Rico, J.C. Molto, and J. Manes. 2001. *Listeria* species in raw and ready-to-eat foods from restaurants. *J. Food Protect.* 64:551–553.

Southwick, F.S., and D.L. Purich. 1996. Intracellular pathogenesis of listeriosis. *New Eng. J. Med.* 334:770–776.

Stecha, P.F., C.A. Heynen, J.T. Roll, J.F. Brown, and C.J. Czuprynski. 1989. Effects of growth temperature on the ingestion and killing of clinical isolates of *Listeria monocytogenes* by human neutrophils. *J. Clin. Microbiol.* 27:1572–1576.

Swaminathan, B. 2001. *Listeria monocytogenes.* In *Food Microbiol.* M.P. Doyle, L.R. Beuchat, and T.J. Montville, eds. Washington, D.C.: ASM Press.

Unanue, E.R. 1997. Inter-relationship among macrophages, natural killer cells and neutrophils in early stages of *Listeria* resistance. *Curr. Opin. Immunol.* 9:35–43.

USDA. 1992. Preventing foodborne listeriosis. http://vm.cfsan.fda.gov/~mow/FSISLIST.html.

Walker, S.J., P. Archer, and J.G. Banks. 1990. Growth of *Listeria monocytogenes* at refrigeration temperatures. *J. Appl. Microbiol.* 68:157–168.

Wesley, I.V. 1999. Listeriosis in Animals. In Listeria, *Listeriosis and Food Safety.* E.T. Ryser and E.H. Marth, eds. New York: Marcel Dekker Inc., 39–73.

Wong, S., D. Street, S. Delgado, and K. Klontz. 2000. Recalls of foods and cosmetics due to microbial contamination reported to the U. S. Food and Drug Administration. *J. Food Protect.* 63(8):1113–1116.

RISK ASSESSMENT

27

MICROBIAL RISK ASSESSMENT

A.S. AHL, D.M. BYRD, AND A. DESSAI

INTRODUCTION. For purposes of this chapter, risk is the probability that a person will develop an illness or die following exposure to an adverse event (Byrd et al. 2000). One of the roles of government, whether local, state, or federal, is to protect citizens from risks that the individual cannot control, including public health, environmental, workplace, and highway risks. Although the evaluation of risk and suggestions for mitigation have a long history, formal methods of risk assessment developed more recently. This chapter briefly describes the history of risk assessment in federal agencies, which began with radiation protection, chemical exposure, and carcinogenesis risk assessment, and contrasts this with the more recent and evolving microbial risk assessment methods.

ORIGINS OF HEALTH RISK ASSESSMENT. Initially, the U.S. Atomic Energy Commission estimated the risks of radiation, particularly to protect occupationally exposed persons. Cancer is a prominent outcome for radiation, and radiation biology still dominates many of the considerations about carcinogen risk assessment. Scientific evidence suggests that a single radiation event, such as the release of an alpha particle

from radon decay, can alter the genome of a cell and can induce a somatic mutation. In addition, most cancers originate monoclonally from a single mutated cell. These two observations suggest a model of carcinogenesis like the absorption of light by molecules in solution. A molecule will absorb a photon of light that hits it. Some photons do not collide with any molecule and pass through the solution. At low-incident light and in dilute solutions, absorption is proportional to the concentration of molecules. Similarly, the genetic material of a cell may absorb radiation particles that hit it.

Carcinogens appear to act through a complex process with multiple stages, each of which reflects "hit-like" damage. One of the early risk models, the Armitage-Doll, assumes that the effect of a substance on some stage of the carcinogenic process is additive to an effect induced by external stimuli at that stage (Armitage et al. 1954). The assumption of additivity for all stages results in the form of the Armitage-Doll model that regulatory agencies use to estimate cancer risk (Crump et al. 1976). This model also implies linearity at low exposures (Hartley et al. 1977). Although use of the model remains controversial, most agencies use linear, nonthreshold models (LNT) to interpolate carcinogenic risks to low exposure levels.

The use of LNT models to estimate exposure-response relationships, even for radiation, remains controversial (National Research Council 1999). Many experts prefer a mixed, linear-quadratic model. However, the U.S. Food and Drug Administration (FDA 1980) and, somewhat independently, the U.S. Environmental Protection Agency (EPA) (Albert 1994), applied various forms of LNT models to estimate carcinogenic risks of chemical substances. Most assessors believe that carcinogenic potency of a chemical substance would not exceed a LNT estimate and might, under some circumstances and for some substances, be the same. This belief leads to the idea that an LNT estimate is a plausible upper bound of risk.

Unlike the plausible upper bound approach of chemical risk assessment, microbial risk assessment has taken an entirely different direction, an attempt to forecast accurately the numbers of cases, hospitalizations, or deaths from an infectious agent. Assessment of microbial risks is intrinsically more difficult than the assessment of radiation or chemical risks. No scientific principles exist similar to conservation of energy or matter. After microbes infect a host, the microbes multiply or die. In addition, virulence may change, whereas neither radiation sources nor chemical substances change potency without the host acting on them.

For foods, both the Food and Drug Administration (FDA) Center for Food Safety and Applied Nutrition (CFSAN) and Center for Veterinary Medicine (CVM), as well as the U.S. Department of Agriculture (USDA) Food Safety and Inspection Service (FSIS), have a role in protecting the food supply. By the early 1990s, several microorganisms were associated with major outbreaks of foodborne disease, including *Escherichia coli* O157:H7 (Riley 1983) as well as *Salmonella enteritidis* (St. Louis et al. 1988) and *Listeria monocytogenes* (Schlec 1983). The urgency of dealing with newly recognized domestic threats to food safety hastened the development of microbial risk assessment and its application to problems of food safety by federal agencies. By 1995, FSIS proposed a new regulation to decrease microbial contamination of meats (FSIS 1996). This rule, termed the "HACCP Pathogen Reduction Act," proposed a systems approach (Hazard Analysis Critical Control Point) for testing and evaluating the risks posed by a particular carcass or meat process. This rule, evaluated by USDA Office of Risk Assessment and Cost-Benefit Analysis (ORACBA), was not based on risk assessment, but many aspects of the rule evidenced risk-based thinking (ORACBA 1996).

A few years before FSIS developed the HACCP rule, the Uruguay Round of the General Agreement on Tariffs and Trade (GATT) promulgated Sanitary and Phytosanitary (SPS) provisions (GATT 2002). GATT laid the foundation for the World Trade Organization (WTO), in which members have to accept agricultural exports, including foods, from other members unless the recipient country can show through science-based risk assessment that the commodity is unsafe for the health of its human population, crops, or livestock.

Thus, risk assessments became the basis for rejection of undesirable imports (WTO 2002). The development of risk assessment procedures to evaluate the safety of imported food became critical.

Passage of the U.S. Farm Bill of 1994 added to the urgency of the development of methods for microbial risk assessment. Congress required that a risk assessment and cost-benefit analysis accompany all USDA rules affecting human health, safety, or the environment and exceeding a cost of $100 million per year. A risk assessment, based on science, must show that a rule regulates an identified hazard in a cost-effective manner (ORACBA 1996). A conference sponsored by the USDA Economic Research Service (ERS) in 1995 brought the USDA, FDA, and other analysts together to forge approaches to microbial risk assessment (ERS 1995). Shortly afterward, FSIS began developing methods and approaches to assess all newly regulated risks. Because rules were already in place for residues, chemicals, and physical hazards, microbial risk assessment received the greatest attention.

As scientists documented more outbreaks of foodborne microbial illness, a coalition of FDA, USDA, the Centers for Disease Control and Prevention (CDC), and state and local public health departments created FoodNet in 1995. This surveillance system collects and monitors data from foodborne outbreaks in the United States (CDC 2002). It provides data to help identify the most important foodborne organisms and set priorities for risk assessments.

Right now, concern for microbial food safety and related risk assessment primarily addresses acute morbidity and mortality, but not long-term sequelae of infection. Yet viruses, bacteria, and prokaryotic microbes also induce cancer. In addition, infections with *Salmonella* spp. and other bacteria are associated with reactive arthritis (Smith et al. 1993), and *Campylobacter* is associated with some 40 percent of the cases of Guillain-Barré Syndrome (Ropper 1992). *Mycobacterium paratuberculosis*, the cause of Johne's Disease in cattle, may be associated with Crohn's Disease in humans (Collins et al. 2000). The science associated with microbial foodborne diseases is still at an early stage of development, but as knowledge in this field matures, microbial risk assessment will have to address these chronic illnesses.

MICROBIAL RISK ASSESSMENT. Regulatory risk assessment usually begins with the hazard identification, that is, specification of what can go wrong (NRC 1983). Risk denotes two factors: the likelihood of the adverse effect and the consequences of the effect (the outcome). These definitions and concepts are widely used in risk assessment but are subject to amplification for greater utility in food safety. Risk characterization integrates the two principal components of risk assessment, exposure assessment and dose-response assessment. The terminology associated with risk analysis is notoriously confusing. For food safety, the USDA,

FDA, Codex Alimentarius Commission (CAC), and others have developed a set of definitions and concepts that have general agreement in the field (http://www.foodriskclearinghouse.umd.edu). They are based on classic definitions but are modified for greater usefulness for food safety work. After outbreak reports identify a food hazard, assessors must specify the contacts between host and infectious organism for microbial risk assessment. Risk assessment defines this notion as exposure. In terms of food safety, microbes are foodborne. Classically, epidemiologic methods define the specific foods and provide estimates of how much affected persons consumed of these foods.

Given the exposure estimates, microbial growth modeling provides the dose for microbial risk assessments. This step involves estimating how many microorganisms are present in the food, and how fast they grow, given the substrate, time, and temperature at which the food was held before consumption (Whiting et al 1997). An evaluation of the consumer's health following foodborne pathogen ingestion completes the dose-response assessment. This assessment requires knowledge of whether the consumer sought medical care, whether the individual was hospitalized, survived, and other outcomes. FoodNet is a valuable source of information for both exposure and dose-response assessment. For each case of microbial foodborne illness, FoodNet identifies the microorganism causing the illness and the food vehicle, if possible, and records morbidity and mortality and information about the age and general health condition of the person affected. Epidemiological studies may also provide insight into the source and cause of the outbreak (Graham 1995). The USDA Agricultural Research Service (ARS) collects rather specific data on individual food intake, and this information can also be quite valuable in risk assessments (ARS 2002).

Risk characterization combines exposure and dose-response information from the hazard identification to complete the risk assessment. Risk characterization gives analysts and managers summary information to support the evaluation of a given pathogen and to compare foodborne risks created by different pathogens. Because foodborne microorganisms are part of nature and essentially ubiquitous in occurrence, the entire system of food production (from farm to table) must be engaged in tracking, evaluating, preventing, and mitigating the effects of foodborne microbial pathogens.

This expediency was recognized early in the development of food safety risk assessment (Roberts et al. 1995), and the first farm-to-table risk assessment was carried out by a team of scientists from the USDA and FDA for *Salmonella* Enteritidis (SE) in shell eggs (FSIS 1997). The multidisciplinary team included veterinarians, a physician, statistician, economists, microbiologists, and modelers. They invented new methods as they developed the risk assessment, and contributed enormously to the development of microbial risk assessment.

COMPARISON OF RISK ASSESSMENT FOR CHEMICALS AND MICROBES.

In the 1950s, the FDA developed approaches for ensuring the safety of food additives (Carrington et al. 2000). These procedures, generally described as safety assessment, and complementary methods for human health risk assessment associated with chemicals and toxins, have evolved rapidly since the formation of the Environmental Protection Agency (EPA) in 1972. The EPA and FDA have published many guidance documents and advisories since then. (See Byrd et al., 2000, for a review of these.) In contrast, microbial risk assessment is less than a decade into its development. The following sections provide a comparison and contrast of microbial and chemical risk assessments.

Hazard Identification. Hazard identification is the physical identification of the agent, chemical or biological, and the linkage of a particular hazard (or adverse effect) to the agent in a way that helps the risk assessment.

The methods for identifying chemicals and toxins are well developed, standardized, and largely automated so that qualitative and quantitative identification are rapid and accurate. Microbial identification methods are more problematic because they are less well developed and standardized. New methods and technologies such as microarrays, PCR, and blot tests are helping to make identification faster and more accurate. However, some microorganisms must still be grown in culture before identification can be made. Standardization of identification methods among laboratories has not yet been accomplished.

For chemicals and toxins, the identification and characterization of the hazards is never complete because new chemicals are invented, and new chemical studies on existing elements or compounds reevaluate their status. One example of this is the recent reassessment of a natural compound, arsenic, and its affects on human health from drinking water (EPA 2000). Hundreds of microorganisms are known, but more remain unknown. A major complication is that a microorganism known as a harmless commensal may suddenly emerge as a pathogen in food. Currently, predicting factors in genetics and natural selection that may result in the formation of a new pathogen is not possible. If a new chemical substance is developed in a laboratory, its function can be predicted, at least partially, on its structure.

The quantitative analysis of a chemical is straightforward. A total quantity is known although more specific calculations are necessary if the chemical is diluted or concentrated as it moves through the environment. In contrast, microorganisms are more difficult to quantify. Information on the total population is unknown. A microbe may replicate, die, or behave differently in different environments. One example is that dilution of a microbial population may actually amplify it by dispersing individuals and providing space and resources for additional population growth. This gives a few organisms an opportunity to develop into millions in a short time (Ray 1996).

Exposure Assessment. Exposure assessment describes contact between the agent and a receptor organism. It includes an evaluation of how contact will occur. For chemicals, exposure is a complex phenomenon that may occur by any or all modalities: ingestion, inhalation, dermal contact, or mucous membranes contact. Foodborne microorganisms are, by definition, encountered only by ingestion of food, water, or both. This fact makes it seem easier to assess exposure to microorganisms than to chemicals; however, microbial exposure is more complex. Each breath, each drink of water, each bite of food is accompanied by contact with multitudes of microorganisms, both known and as yet unidentified. The challenge is to find out which are pathogenic and which are a normal accompaniment of life. We cannot ensure that all food, water, or air are sterile, nor can we test each batch. Therefore, the number and kind of bacteria present in food or otherwise continuously entering our bodies are neither readily detectable nor quantifiable, nor is it practical to attempt such.

Chemicals may undergo transformation in the environment, and the changed chemical structure may have different toxicity. Chemists largely understand these modifications and their causes so that environmental fate and transportation are relatively predictable. Microbes are less predictable. They may pick up mobile genetic elements that alter pathogenicity, or they may mutate, which alters their identity. Because microbiological ecology is a new and developing field, the base knowledge for predicting environmental fate, survival, and transport of microbes is lacking.

Another difference between chemicals and microorganisms is their movement in the environment. Chemicals exhibit passive mobility by moving with air or water currents, mostly by diffusion, but do not move of their own accord. Microorganisms, on the other hand, actively move, and this mobility may allow them to spread very fast. This mobility is related to the fluidity of the medium in which they are located. For example, in very thick batches of food such as stuffing, pockets of pathogenic organisms can exist that infect the individuals who eat food from one area, whereas others who sample another part encounter no pathogens at all (CNN 1997). Tracking microorganisms from farm to table is an important element in microbial fate and transport. Before exposure can be assessed, the path of the microbes in the food production system needs to be understood. Pathogens can enter the system at many points, and the point of entry and subsequent time, temperature, and conditions can affect the parameters of exposure (Roberts et al. 1995).

Although chemical substances have an atomic structure, at analytically and biologically meaningful concentrations, they essentially behave as continuous variables. Risk assessment treats them as such mathematically. Microorganisms exist as discrete organisms, but for ease of computation in risk assessment, they, too, are treated as continuous variables. If the numbers of microorganisms are large, this is reasonable. However, when numbers are very small, this approach may not be optimal.

Dose-Response Assessment. Within biologically meaningful concentration ranges, a chemical substance provides a consistent response for a given dose. Dose-response relationships are predictable. Microorganisms have less predictable dose-response relationships. For example, sometimes response varies with strain, isolate of the same strain, or even culturing conditions. Consistent doses of the same microorganism produce variable responses within the same individual at different times or between two individuals at the same time. Variation of response may arise in part from difficulty in accurately calculating doses, but doses can change variably during gastrointestinal transit. For example, some microbes may survive transit through the upper part of the gastrointestinal tract, and during that transit, may increase or decrease in number, or under stress begin producing toxins that were not evident before their ingestion (Mekalanos 1992). Some differences relate to the amount and composition of food ingested. The composition of the food affects both microbial growth and survival as well as individual response to the pathogen (Ray 1996). The timing of the ingestion of the pathogen in relation to the timing of the meal may be influential. For example, a pathogen ingested early in a meal elicits a different effect than one ingested near the end of the meal (Schuchat et al. 1992).

Microbial dose-response relationships, which are highly idiosyncratic, relate to individual immune status, age, gender, preexisting diseases and conditions, genetic variability, prior exposure, and timing of exposure. Very young, very old, and immune-suppressed individuals have greater susceptibility to foodborne illness. Pregnant women are immune-suppressed as a condition of pregnancy and thus are more susceptible to foodborne pathogens. Individual variability in the genes that control immune response also may affect susceptibility to foodborne microbial response (Baird-Parker et al. 2000).

The human immune system is a dynamic one, and evidence is growing that an individual's prior exposure to a pathogen and its timing may also influence subsequent responses to exposure. Early exposure to certain microorganisms is thought to prime the immune system so that subsequent exposure is less likely to make a person ill (Institute of Food Technologists 2002). Additionally, the recent rise in microbial foodborne illness and in food allergies is postulated to be caused, at least in part, by lack of early exposure to certain microorganisms (Matricardi et al. 2000).

The medication status of the individual can also affect response to a given dose of foodborne pathogen. For example, the acid environment of the stomach is a potent deterrent to some bacterial growth, essentially killing the microbes in the stomach before they can damage the lower intestinal tract. An individual taking antacids or acid-lowering drugs is more susceptible to foodborne illness. Laxatives move food through the gastrointestinal tract faster than usual; this also can affect responses to foodborne pathogens. Antibiotics taken for infection not associated with the gastrointestinal tract often eliminate important commensal bacteria. This leaves resources of both space and food for invading pathogens, with no other competition. It is possible that even relatively harmless bacteria can colonize the gastrointestinal tract and cause serious health problems. The net effect of these sources of variation is that representation of a dose-response relationship as a parameter in a microbial risk assessment model has greater associated uncertainty than that in a chemical risk assessment, thus requiring more complex approaches.

Risk Characterization. Risk characterization brings together information from the exposure and dose-response assessments to estimate a number of expected cases. Some think that risk characterization should accomplish even more by essentially linking the risk assessment to the managerial decision-making process (NRC 1996). For microbial risk assessment, the ARS's ongoing survey of food intake provides some measure of the actual food intake (ARS 2002), and manufacturers' information about distribution of foods provides another source in supporting estimates. FoodNet provides a lower bound for past outbreaks. However, no completely satisfactory way of estimating population risk exists.

Historical/Cultural Context of Risk Assessment. Despite the biological differences in responses to chemical substances and microbial pathogens, and the consequent differences in assessment methods, perhaps the most important difference relates to the historical and cultural contexts in which the requirements that Congress placed on the federal agencies, mostly on the EPA after its formation in 1972. Risk assessment for chemical substances developed in the light of the "War on Cancer" and the fears about this group of serious, dreaded, and seemingly unpredictable set of diseases that were characterized by uncontrolled cell growth (Olin et al. 1997).

Scientific uncertainty and subsequent difficulty in making policy led to the plausible upper bound practices of chemical risk assessment that many decry. In essence, government accepted over-prediction of chemical risk, based on the idea that these practices "protect the public" while developing regulations about a single chemical substance. Thus, policy goals rather than scientific information drove chemical risk assessment methods (Wagner 1995; Powell 1999). Ultimately, these policies have distorted public health priorities because the public cannot compare the expected risks of different chemical substances, given standard regulatory practices. The extent of over-prediction, if

any, remains unknown. Although microbial risk assessment has a shorter history, its developers have aimed for accuracy of predictions.

Chemical risk assessments include many default assumptions, incorporated so that any errors are propagated throughout the risk assessment (Wilson 1998). In contrast, microbial risk assessments rarely use default assumptions. Instead they require a basis in data or expert opinion, even if by analogy to some similar microorganism or circumstance. Every step represents uncertainty through probability density functions that express scientific doubt, measurement error, and biological variation. The practices of microbial risk assessment carry this information throughout. Thus, microbial risk assessments explicitly state a most likely expectation of the number of deaths, hospitalizations, or cases and their associated uncertainty.

Currently, chemical risk assessments are not bounded in identifiable cases, whereas FoodNet enumerates morbidity and mortality data that provide lower bounds for risk estimation. Chemical risk assessments implicitly assume that exposure is an optional luxury; new chemicals or processes can bypass the risk. For foodborne pathogen risk assessment, some exposure is unavoidable if humanity continues to eat and thus to live.

The practice of microbial risk assessment grew directly from what was developed for plant and animal health in international agricultural trade (Kaplan 1990). International agricultural trade had long been restrictive, infamous among scientists for the creation of "regulatory diseases"; that is, diseases of crops and livestock that may be of small consequence, but served as a trade barrier. Sanitary and phytosanitary provisions require that agricultural trade in crops, livestock, and food more realistically portray risk, and comply with this requirement using transparent and science-based risk assessment (WTO 2002).

As regulators face difficult choices of which hazard(s) deserve the most immediate response, it would be useful to use risk assessments to rank risks by likelihood of occurrence and magnitude of effect. However, because risk assessment for chemicals and microorganisms has taken such different approaches, their results are not comparable. Thus, their use in policy decisions to target the worst risks first is not defensible.

SUMMARY AND CONCLUSIONS. Foodborne microbes provide many challenges. They move between food and the environment and pass through and among many hosts. This mobility gives them the opportunity to pick up genetic elements for resistance to antibiotics. As living organisms, microbes live, reproduce, die, and evolve in response to environmental pressures. Control of foodborne microbes is a moving target, making risk assessment to evaluate options for their control problematic and challenging.

In consideration of risk assessment for foodborne microbes, humanity has always lived with the foodborne pathogens. Historically, foodborne microbial risk assessment has grown from a system requiring that science guide policy and decision-making, not the other way around. This approach makes it difficult, if not impossible, to compare risk assessments for chemical substances and microbes as a way to choose which to target first.

Eating is both a necessity and, for many people, a major sensory pleasure. Attempts to diminish its importance will not make our food safer. Rather, we need to construct warnings about food safety carefully and place them in the right context. Protecting against one risk inevitably exposes consumers to other risks. Regulators must consider comparative microbial risk assessments in formulating policies for foodborne microorganisms.

REFERENCES

Agricultural Research Service (USDA/ARS). 2002. Food surveys research group—what we eat in America. Available at: http://www.barc.usda.gov/bhnrc/foodsurvey/home.htm.

Albert, R.E. 1994. Carcinogen risk assessment in the U.S. Environmental Protection Agency. *Crit. Rev. Toxicol.* 24:75–85.

Armitage, P., and R. Doll. 1954. The age distribution of cancer and a multistage theory of carcinogenesis. *Br. J. Cancer* 8:1–12.

Baird-Parker, C.T., and R.B. Tompkin. 2000. Risk and microbiological criteria. In *Microbiological Safety and Quality of Foods*. B.M. Lund, A.C. Baird-Parker and G.W. Gould, eds. Gaithersburg, MD: Aspen Publishers, Inc.

Byrd, D.M., and C.R. Cothern. 2000. *Introduction to Risk Analysis: A Systematic Approach to Science-Based Decision Making.* Dallas, TX: Government Institutes, 433.

Cable News Network (CNN). 2002. Officials say more may have been sickened by church dinner. Available at: http://www.cnn.com/US/9711/06/salmonella.folo/index.html.

Carrington, C., and M. Bolger. 2000. Safety assessment and risk assessment: sometimes more is less. *ORACBA News* 5(2); Spring.

Centers for Disease Control and Prevention (CDC). 2002. Emerging Infections Program, Foodborne Disease Active Surveillance Network (FoodNet). What is FoodNet? Available at: http://www.cdc.gov/foodnet/what_is.htm.

Codex Alimentarius Commission (CAC). 2000. Available at: http://www.codexalimentarius.net/default.htm.

Collins, M.T., G. Lisby, C. Moser, D. Chicks, S. Christensen, M. Reichelderfer, N. Hoiby, B.A. Harms, O. Thomsen, U. Skibsted, and K.V.Binder. 2000. Results of multiple diagnostic tests for Mycobacterium avium subsp. Paratuberculosis in patients with inflammatory bowel disease and in controls. *J. Clin. Microbiol.* 38:4373–4381.

Crump, K.S., D.G. Hoel, C.H. Langley, and R. Peto. 1976. Fundamental carcinogenic processes and their implications for low dose risk assessment. *Cancer Res.* 6:2973.

Economic Research Service (USDA/ERS). 1995. Conference on tracking foodborne pathogens from farm to table: data needs to evaluate control options, Washington, D.C., January 9–10, 188.

Environmental Protection Agency (EPA). 2000. National primary drinking water regulations: arsenic and clarifications to compliance and new source contaminants monitoring. *Federal Register* 65(121):38888–38983.

Food and Drug Administration (FDA), Food Safety Council. 1980. *Proposed System for Food Safety Assessment.* Washington, D.C., 160.

Food Safety and Inspection Service (USDA/FSIS). 1996. Pathogen Reduction: Hazard Analysis and Critical Control Point (HACCP) Systems; Final Rule. *Federal Register* 61(44):38805–38989. Thursday, July 25, 1996.

Food Safety and Inspection Service (USDA/ FSIS). 1997. *Salmonella* Enteritidis risk assessment. Available at: http://www.fsis.usda.gov/ophs/risk/contents.htm.

General Agreement on Tariffs and Trade (GATT). 2002. Available at: http://www.wto.org/english/tratop_e/tratop_e.htm.

Graham, J.D. 1995. *The Role of Epidemiology in Regulatory Risk Assessment.* Proceedings of the Conference on the Role of Epidemiology in Risk Analysis, Boston, MA, 13–14 October 1994. Amsterdam, The Netherlands: Elsevier Science B.V., 156.

Hartley, H.O., and R.L. Sielkin. 1977. Estimation of "safe doses" and carcinogenesis experiments. *Biometrics* 33:1.

Institute of Food Technologists. 2002. Expert report on emerging microbial food safety issues. February 20. Available at: http:///www.ift.org/govtrelations/microfs/webreport.pdf.

Kaplan, S. 1990. A preliminary probabilistic risk assessment (pra) on the importation of Chilean llamas into the United States. For the International Llama Association, Washington, D.C., March 15.

Matricardi, P.M., F. Romini, S. Riondino, M. Fortini, L. Ferrigno, M. Rapicetta and S. Bonini. 2000. Exposure to foodborne and orofecal microbes versus airborne viruses in relation to atopy and allergic asthma: epidemiological study. *Br. Med. J.* 320(7232):412–417.

Meklanos, J.J. 1992. Environmental signals controlling expression of virulence determinants in bacteria. *J. Bacteriol.* 174:1–7.

National Research Council (NRC). 1983. *Risk Assessment in the Federal Government: Managing the Process.* Washington, D.C.: The National Academy Press, 191.

National Research Council (NRC). 1996. *Understanding Risk: Informing Decisions in a Democratic Society.* Washington, D.C.: The National Academy Press, 249.

National Research Council (NRC). 1999. *Health Effects of Exposure to Radon.* Washington, D.C.: National Academy Press.

Office of Risk Assessment and Cost-Benefit Analysis (USDA/ORACBA). 1996. Review of the Pathogen Reduction, HACCP, Final Rule. March.

Olin, S.S., D.A. Neumann, J.A. Foran, and G.J. Scarano. 1997. Topics in cancer risk assessment. *Environ. Health Persp.* 105(Suppl. 1):117–126.

Powell, M.R. 1999. *Science at EPA: Information in the Regulatory Process.* Washington, D.C.: Resources for the Future, 433.

Ray, B. 1996. *Fundamental Food Microbiology.* New York: CRC Press.

Riley, L.W., R.S. Remis, S.D. Helgerson, H.B. McGee, J.G. Well, B.R. Davis, E.S. Hebert, L.M. Olcutt, L.M. Johnson, N.T. Hargett, P.A. Blake, and M.L. Cohen. 1983. Hemorrhagic colitis associated with a rare *Escherichia coli* serotype. *N. Engl. J. Med.* 308:681–685.

Roberts, T., A.S. Ahl, and R.M. McDowell. 1995. Risk assessment for foodborne microbial hazards. In *Tracking Foodborne Pathogens from Farm to Table: Data Needs to Evaluate Control Options.* T. Roberts, H. Jensen, and L. Unnevehr, eds. Washington, D.C.: USDA/ERS Publ. No. 1532, 95–115.

Ropper, A.H. 1992. The Guillain-Barré syndrome. *N. Engl. J. Med.* 326:1130–1136.

Schlec, W.F. 1983. Epidemic listeriosis—evidence for transmission by food. *N. Engl. J. Med.* 308:203–206.

Schuchat, A., K.A. Dearer, J.D. Wenger, B.D. Plikaytis, L. Mascola, R.W. Pinner, A.L. Reingold, and C.V. Broome. 1992. Role of foods in sporadic listeriosis. I. Case-control study of dietary risk factors. *JAMA* 267:2041–2045.

Smith, J.L., Palumbo, S.A., and Walls, I. 1993. Relationship between foodborne bacterial pathogens and the reactive arthritides. *J. Food Safety* 13:209–236.

St. Louis, M.E., D.L. Morse, M.E. Potter, T.M. DeMelfi, J.J. Guzewich, R.V. Tauxe, and P.A. Blake. 1988. The emergence of Grade A eggs as a major source of *Salmonella enteritidis* infections. *JAMA* 259(4):2103–2107.

Wagner, W. 1995. The science charade in toxic risk regulation. *Columbia Law Review* 95(7):1613–1723.

Whiting, R.C., and R.L. Buchanan. 1997. Development of a quantitative risk assessment model for *Salmonella enteritidis* in pasteurized eggs. *Intl. J. Food Microbiol.* 36:111–135.

Wilson, J.D. 1998. Default and Inference Options: Use in Recurrent and Ordinary Risk Decisions. Washington, D.C.: Resources for the Future, 24.

World Trade Organization (WTO). 2002. Available at: http://www.wto.org/english/tratop_e/sps_e/sps_e.htm.

28

SAMPLING TECHNIQUES FOR FOODBORNE PATHOGENS IN ANIMALS AND ANIMAL PRODUCTS

M. SALMAN, B. WAGNER, AND I. GARDNER

INTRODUCTION. Monitoring for foodborne pathogens in U.S. animal populations prior to harvest has relied primarily on clinical isolates submitted during an outbreak investigation or as part of a diagnostic work-up. Some sampling for these pathogens also has occurred as part of the National Animal Health Monitoring System (Dargatz et al. 2000). Post-harvest detection of foodborne pathogens has utilized clinical samples and samples from compliance monitoring undertaken by regulatory agencies such as the USDA's Food Safety Inspection Service (FSIS) and the Food and Drug Administration (FDA). Although there is value in linking pathogens with outbreak, diagnosis, and compliance events, there is also a need for increased efforts directed toward collection of reliable and representative samples for the precise and accurate assessment of pathogen levels in animals and in animal products. A well-designed monitoring system based on a sound sampling strategy makes it possible to make inferences concerning population prevalence or incidence levels and trends from a subset of the population. Associated risk factors and interventions can be assessed when the levels are reliably estimated.

SAMPLING METHODS. The purpose of this chapter is to review basic sampling techniques that can be used for estimating the prevalence or incidence of foodborne pathogens in animal or animal products. These techniques should account for the characteristics and distributions of the pathogens. Important considerations include expected prevalence, potential clustering or nonrandom distribution (geographic and biological), and pathogen level. Pathogens where the prevalence is extremely low, such as *Salmonella enteritidis* in eggs, may require a different sampling approach than a more common pathogen, such as *Campylobacter* spp. on processed poultry. Also, the phase of production can influence prevalence. *Escherichia coli* O157:H7, for example, is believed to occur in cattle on most feedlots (Chapman et al. 1993; Hancock et al. 1997) but is much rarer on processed beef (Chapman et al. 1993; Griffin 1995; Mahon et al. 1997). Sampling strategies also should address potential clustering of pathogens such as fecal contamination of beef carcasses. Last, the biological implications of pathogen levels in samples deserve special attention. For example, the presence of a few *E. coli* O157:H7 organisms on a food product can

have serious human health implications, whereas *Clostridium perfringens* requires much higher spore numbers before human health is of concern. Four primary categories of conventional sampling techniques exist: (1) simple random sampling; (2) stratified random sampling; (3) systematic sampling; and (4) cluster sampling. The remainder of this chapter focuses on the presentation of these four basic sampling techniques and their effectiveness and practicality for sampling foodborne pathogens. The discussion of each technique includes a description of important assumptions, characteristics, and application to foodborne pathogens. The material presented here is intended to be an introduction. Many texts (Schaeffer et al. 1986; Kish 1995) are available for more detailed and theoretical presentations of sampling techniques.

Simple Random Sampling (SRS). Simple random sampling (SRS) is the primary sampling technique, which also serves as the basis for other techniques. Simple random sampling occurs when a sample size of *n* (for example, selected subset of eggs) from a population of size *N* (for example, total number of eggs produced on a farm on a given day) is taken in such a way that every possible sample of size *n* has the same chance of being selected. Ordinarily, this is interpreted as the equal probability of selection of each element from the population, but this would omit simple random sampling with probabilities that are proportional to size (for example, more eggs selected from large houses on the farm). The use of SRS depends on having a list of all elements that are part of the population (called a list frame) from which to select the sample. The selection of the sample must be conducted using a random procedure. In a random procedure, each element of the population is assigned a unique identification. Random numbers are then chosen from a random number table or by a similar method. Then the sample is chosen by selecting the elements that are associated with the selected random numbers. For example, if the objective were to sample feces from ten cattle in a feedlot pen of one hundred cattle, we would start by creating a list frame of animals in the pen. Each animal could be assigned a unique number between one and one hundred. Ten unique numbers between one and one hundred would be selected from a random number table. The ten cattle represented by these random numbers would then comprise the sample.

Estimates of the proportion (for example, percentage of cattle shedding bacteria that are resistant to tetracycline) is made using the following formula:

$$\hat{p} = \frac{x}{n}$$

where $\hat{p}$ = the proportion of interest, x = the number of sampled elements that are positive for the condition of interest, and n = the sample size. So if four (x=4) of the ten (n=10) cattle sampled in this example are positive for resistance to tetracycline, then the estimated proportion of this resistance in this pen of cattle is 0.4.

Estimates from SRS are unbiased. Bias is the difference between the mean of the estimate that would be obtained after repetition of the process necessary to get this estimate and the true population value. Thus, if the estimate is unbiased, then on average the estimate will not differ from the population value. If bias is present, then the mean of the estimation process will differ from the population value. Simply stated, estimates that are unbiased are properly aimed at the population target value.

In the preceding example, if we had selected ten different cattle, we may have found a different number of cattle to be shedding resistant bacteria. Measurements should incorporate the variability that arises because all the elements of the population were not measured for the outcome of interest. This variability is called the variance of the estimated proportion. The formula for the variance of a proportion, V_p, under SRS is as follows:

$$V_p = \frac{\hat{p}(1-\hat{p})}{n}\left(\frac{N-n}{N}\right)$$

where N = population size. The second term in the equation is called the finite population correction factor (Scheaffer et al. 1986). If $n \leq 0.05N$, the finite population correction factor can be ignored. In the cattle feedlot pen example, $\hat{p}$=0.4 and n=10, then the estimated variance is 0.022. The standard error, also referred to as the square root of the variance, is 0.15. This standard error can be used to represent the variability associated with sampling when making an inference to the population. This procedure is called a confidence interval. A confidence interval is interpreted as the proportion of times that the true parameter would lie within the bounds of the interval if the sampling procedure were repeated many times. Assuming a normal distribution, the confidence interval under SRS would be:

$$CI = \hat{p} \pm z_{1-\alpha}se(\hat{p})$$

where CI = confidence interval, $z_{1-\alpha}$ = standard normal value for the α level of significance, and $se(\hat{p})$ = standard error of the proportion. To construct a 95 percent confidence interval for the cattle feedlot pen example, we use the $z_{1-\alpha}$=1.96 value to arrive at an interval of (0.11, 0.69). The product, $z_{1-\alpha}se(\hat{p})$, is referred to as the error bound (Scheaffer et al. 1986). The interval can be interpreted as follows: 95 percent of the time, the true proportion of cattle shedding tetracycline-resistant bacteria in the pen would be between 0.11 and 0.69 if we maintain the SRS sampling with a sample size of 10.

An important step in designing a SRS study is to determine the sample size needed to be able to estimate the prevalence with a desired error bound. The sample size formula is as follows:

$$n = \frac{Np(1-p)}{(N-1)\frac{B^2}{4}+p(1-p)}$$

where N = population size, p = the proportion being estimated, and B = bound on the error. For example, suppose that a proposed study objective is to estimate the percentage of dairy cattle shedding *Mycobacterium paratuberculosis* in a five hundred–head dairy herd. Prior studies estimate the within herd prevalence to be about 5 percent (p), and we want the error bound to be 2 percent (B). Using the preceding formula, the required sample size is 244. If no prior information is available regarding the proportion being estimated, a conservative approach would be to use a value of 0.5 for p.

Other parameters, such as mean (for example, number of colonies per plate) and total (for example, bacterial cell count in a gram of feces), also can be estimated from SRS. For example, baseline surveys by the USDA for *Clostridium perfringens* spores and cells in raw ground meat products are based on a direct plate count method in a 25-gram subsample of product (Marks and Coleman 1998). These samples are selected by randomly selecting beef processing plants and then randomly selecting a single sample of pound of ground beef from the plant. Because only one sample was taken from a plant, this is equivalent to SRS at the plant level. Count results were summarized as colony forming units (CFU)/g after adjustment for a 10-fold sample dilution. Of 453 samples, 389 (85.9 percent) were nondetectable for *C. perfringens.* Of the sixty-four positive samples, counts ranged from 10 to ≥100 CFU/g.

In addition to calculation of the percent of samples that have detectable organisms, the count data can be used to calculate the mean number of colony-forming units per gram. For purposes of calculation of the mean and standard deviation, we assume that the CFU/g measurements were 0 (n=389), 10 (30), 20 (14), 30 (4), 40 (3), 50 (1), 60 (4), 70 (1), 80 (3), 90 (1), and 100 (3).

The mean is calculated as:

$$\bar{x} = \frac{\sum_{i=1}^{n}x_i}{n} = \frac{\sum_{i=1}^{11}n_ix_i}{n} = \frac{1,180}{453} = 3.996 \, CFU/g$$

Standard deviation is calculated as:

$$\hat{S} = \sqrt{\frac{\sum_{i=1}^{n}(x_i - \bar{x})}{n-1}} = \sqrt{\frac{\sum_{i=1}^{11} n_i(x_i - \bar{x})}{n-1}} = \frac{888,867.99}{452}$$

$$= 14.022 CFU/g$$

where x_i = number of CFU units on a plate and n = number of samples (plates). The standard error is equal to the standard deviation divided by the square root of the sample size (SE= 0.659). The colony-forming unit counts often are transformed to meet the assumption of normality. A common method is to use the log transformation (natural log [x+1]). The log transformation of the count results in a mean of 0.432 and a standard deviation of 1.10 (SE=0.052). The 95 percent confidence interval for the transformed data is (0.330, 0.533). The confidence interval for the data at the original scale is obtained by first creating a confidence interval with the transformed mean and standard error and then transforming (exponentiating) the confi-

dence limits back to the original scale. The resulting confidence interval is (0.391, 0.704).

Advantages and disadvantages of SRS are listed in Table 28.1. Because elements are chosen at random, estimates of parameters for subgroups, such as cattle in different weight groups, can be calculated. The practicality of SRS is limited by a couple of issues, especially for large surveys for foodborne pathogens. Simple random sampling depends on the existence of a list frame. Typically, list frames do not exist for elements of any type of production animal or animal products, nor is it practical to produce such lists. The populations are simply too large and in such constant fluctuation that it is an impractical task. Even building a list of operations with the specified production animal is costly, time consuming, and would require constant updating. When there are substantial differences in means within groups of the data, there is no method within SRS to alter sampling to reduce variability.

Stratified Random Sampling (STRS). The second sampling technique is stratified random sampling (STRS). A stratified random sample is obtained when a

Table 28.1—Advantages and Disadvantages of the Four Primary Sampling Techniques

Sampling technique	Advantages	Disadvantages
SRS	Simplicity. Conventional formulas. Unbiased estimators.	Need a full list frame. Potentially very expensive. Doesn't reduce the effect of data variation on the error of estimation.
STRS	Unbiased estimators of subgroups. Sample size estimates relatively simple. Smaller variance than SRS for fixed sample size when measurements are homogeneous within strata—gain in reliability. Convenient when need estimates of subgroups (strata). Potential cost savings.	May produce larger variance than SRS if little variation between strata and more within strata. Subgroup estimates are biased estimates of population parameters. Sample size computations depend on within-strata variance, which might be difficult to obtain.
SS	Can post-stratify. Unbiased estimates. Estimation formulas are familiar to most. Easy and convenient (cheaper per unit cost than SRS). Can be implemented without a sampling list frame. Potentially better than SRS if correlation between pairs of elements is negative. Formulas for estimation are the same as SRS.	Performs poorly if there is systematic structure in the data (for example, linear trend or periodicity). Estimates can be biased, especially with small population sizes. The intraclass correlation can impact the variance.
CS	Cost effective in terms of lowered travel cost and list development. Can be done if there is an inadequate list frame of population elements. May be better, in terms of error, than SRS or STRS if measurements within a cluster are heterogeneous and cluster means (or other estimate) are similar. Can perform single-stage or multi-stage sampling.	High standard errors (variance) if measurements are similar within a cluster. Nontraditional estimators that, if ignored, could lead to improper estimates. Sample size determination and actual design can be complex. Need to estimate within cluster sample size and the number of clusters to sample.

SRS=Simple Random Sampling; STRS=Stratified Random Sampling; SS=Systematic Sampling; CS=Cluster or Two-Stage Sampling.

simple random sample is taken with distinct subgroups (strata) of the population. The principle behind stratified random sampling is to take advantage of similarities within strata and differences between strata to make both population and strata level estimates. Stratified random sampling may produce a more precise estimate than a simple random sample of the same size, especially if the measurements within strata are homogenous (Scheaffer et al. 1986). Suppose that the objective were to estimate pathogen prevalence in two regions of a country where one region is low prevalence and the other is relatively high. In STRS, random samples would be chosen for each region for determination of pathogen status. Estimates of prevalence could then be made for the regions individually and in combination. If there were relatively large within-strata variability and little between-strata variability, then the variance would tend to be larger than that of SRS with a fixed sample size. A measure of this difference (the ratio of the variance from a design relative to a SRS) is called the design effect.

Estimation of the population proportion in STRS is similar to that of SRS except that the formulas are essentially weighted by strata population size and sample size. The estimator of the population proportion is as follows:

$$\hat{p} = \frac{1}{N}\sum_{i=1}^{k} N_i \hat{p}_i$$

where N = whole population size, N_i = population size of the ith stratum, and $\hat{p}_i$ is estimate of the within stratum proportion (same equation as SRS). Returning to the pathogen in the two populations example, suppose that there were 5,000 and 10,000 animals in the first and second population, respectively (N=15,000). If the estimated prevalence were 5 percent and 40 percent in the first and second populations, the estimate of the prevalence for the whole population would be 0.283. The variance is as follows:

$$\hat{V}_{st} = \frac{1}{N^2}\sum_{i=1}^{k} N_i^2 \hat{V}_i(\hat{p}_i)\left(\frac{N_i-n_i}{n_i}\right)$$

where $\hat{V}_i$ = the variance for the ith stratum, based on an SRS formula, and k = number of strata. Confidence intervals can be constructed under the normality assumption, as was done with SRS with the use of this variance estimate. Sample size calculations are somewhat more complicated than those used for SRS because they depend on some knowledge of within-strata variability. Also, the sample can be allocated considering the cost of obtaining a sample in the strata. A researcher can then balance cost and variance to optimize the design. The formula is the following:

$$n = \frac{\sum_{i=1}^{k} N_i^2 p_i(1-p)_i/w_i}{N^2\frac{B^2}{4}+\sum_{i=1}^{k} N_i^2 p_i(1-p)_i}$$

where w_i = the fraction of the sample size allocated to the ith strata. An approximate allocation to minimize cost to put in the preceding equation can be obtained using the following formula:

$$n_i = \frac{N_i\sqrt{p_i(1-p)_i/c_i}}{\sum_{i=1}^{k} N_i\sqrt{p_i(1-p)_i/c_i}}$$

where c_i = the cost of obtaining a sample in strata i. To illustrate these equations, suppose that the preliminary estimate of prevalence in region 1 from the preceding example was 0.1 at a cost of $1.50 per sample, whereas in region 2 the estimated prevalence was 0.5 at a cost of $1.25. The total population remains the same (N_1=5,000 and N_2=10,000). The proportional allocation for region 1 would be 0.215 and region 2 would be 0.785. These are the w_i values for the prior equation. Thus, if we want to calculate a sample size with a bound on the error of 0.05, the total sample size is 295, of which 63 (or round up to 64) would be taken randomly from the first population and 232 would be taken randomly from the second population.

Stratified random sampling (STRS) has the advantages listed in Table 28.1. Estimates for the strata or subgroups as well as the population parameters are unbiased. Post-stratification is possible if the experimenter is aware of sampling issues that may lead to biased estimation (for example, overrepresentation of a group). Subgroup estimates, although they are unbiased estimates for the group they represent, are biased estimates of the population parameters.

Systematic Sampling (SS). Systematic sampling (SS) is the process that consists of obtaining a sample by randomly selecting an element from the first part of a frame or ordered group and then sampling every element that follows in the list at a specified initial interval. For example, if the interval was twenty, one of the first twenty elements in the frame would be randomly selected and then every twentieth element in the frame would be selected. The interval is determined by the sampling fraction, which is the ratio of the required sample size to the total population. For example, if the objective were to sample the specific site of a slaughtered animal from 100 carcasses in a slaughterhouse that processes 1,000/day, we would start by calculating the required interval for the SS (1000/100 = 10). The interval of ten indicates that every tenth carcass on that day will be sampled. The first sample carcass will be selected using a random procedure for those first ten carcasses that are processed on that day. At the end of the day, we should expect approximately 100 sampled carcasses.

Estimates of the proportion (for example, percentage of carcasses contaminated with bacteria that are resistant to tetracycline) and the variance for SS are calculated with the same formulas that are used for SRS. The confidence interval also is constructed in the same manner under the same assumption of a normal distribution. Similarly, the sample size calculation for SS is the same as it is for SRS.

Potentially, SS can have a lower variance than SRS if a negative correlation exists between elements, but SS will perform poorly if there is systematic structure to the data. This relationship can be seen by examining the following equation:

$$v(\bar{y}) = (\sigma^2/n)(1+(n-1)\rho)$$

where σ^2/n is the variance of SRS and ρ is intraclass correlation coefficient (a measure of the correlation between pairs of elements restricted between -1 and 1). If the correlation is positive, the variance will increase, whereas the opposite is true if the correlation is negative. If the population is in random order, then ρ will be close to zero; thus, there will be little impact of using SS. If the elements are ordered, then we would expect the observations to be heterogeneous and ρ to be less than zero, which means that SS would be more effective than SRS for a fixed sample size. If the population is arranged in a periodic fashion, it is possible for the measurements to be more homogenous than one would expect if the SRS was implemented; thus, ρ will be greater than zero. In production animal sampling, the researcher should be aware of the potential impact of clustering and evaluate it for the situation.

Systematic sampling (SS) is easy and convenient, but its biggest asset is probably the ability to be implemented without a list (see Table 28.1). For production animal sampling, this characteristic may be of key importance, especially for situations such as sampling animals going through a chute or along a processing line.

Cluster or Two-Stage Sampling (CS). In the context of foodborne pathogens in domestic animals, the elements of interest (for example, a steer, sow, or chicken) often occur in groups such as herds and flocks. Animals in these groups, also referred to as clusters, are typically managed similarly and may have similar exposures to pathogens of interest. Cluster or two-stage sampling (CS) can be used as a strategy to investigate groups of animals. The primary sampling unit in these types of sampling is the cluster, such as a herd. In cluster sampling, all the units within a cluster are sampled, whereas with two-stage sampling, only a randomly selected subset of the units in a cluster is chosen. Clusters can be randomly sampled, like the SRS, or subject to other strategies, such as STRS. Sampling clusters allows for estimation at both cluster level and within-cluster level for considering both herd and animal-level factors that may influence the presence of foodborne pathogens (McDermott and Schukken 1994).

Cluster sampling has the advantage of not needing a complete list of elements in the population. For animal production, this is an important advantage because of the inability to create an animal-level list, whereas it is possible, with work, to develop an oper-

ation-level list or some type of geographic cluster. If clusters are geographical in nature, like blocks in a city or operations in a county, then CS may result in substantial cost savings caused by reduced travel and interview time costs. Potentially, cluster sampling may have lower errors than SRS or STRS if measurements within a cluster are heterogeneous and cluster means are similar. In animal production, though, we would expect the opposite: within-cluster homogeneity and heterogeneity between clusters. Thus, the variance may be higher for CS than for SRS in many applications.

The measure of the similarity of diseased (or healthy animals) within a cluster is the intracluster correlation (McDermott and Schukken 1994; Donald 1993). If $\rho = 0$, then the disease is randomly distributed within a population, and animals from within a herd are no more likely to be diseased than any animal selected from the entire population. When $\rho = 1$, then all animals within a herd have the same status. Thus, when using cluster sampling, as ρ approaches 1, the herd is behaving more like an individual, and fewer animals within a herd would need to be evaluated to determine herd status. Under this scenario, sampling more herds and fewer animals per herd would be an appropriate approach. When ρ is close to zero, then animals are behaving as though they were randomly distributed throughout the population, and it may be more appropriate to sample fewer herds and more animals per herd to optimize the design. McDermott and Schukken (1994) reviewed a number of papers from the veterinary epidemiology literature and found that, where estimable, ρ varied between 0.0017 and 0.46.

Estimation of the proportion under CS sampling is similar to that of SRS. The formula is as follows:

$$\hat{p} = \frac{\sum_{i=1}^{l} x_i}{\sum_{i=1}^{l} m_i}$$

where x_i = the number of positive elements in the lth cluster, l = the number of clusters, and m_i = the total number of elements in the lth cluster. The estimated variance is the following:

$$\hat{V} = \left(\frac{N-n}{Nn\bar{M}^2}\right)\frac{\sum_{i=1}^{l}(x_i - \hat{p}m_i)^2}{n-1}$$

where $\bar{M}$ = mean number of elements in a cluster. The total number of elements to be sampled can be calculated with the following formula:

$$n = \frac{n\sigma_c^2}{N\frac{B^2\bar{M}^2}{4} + \sigma^2}$$

where σ^2 is approximated by:

$$s_c^2 = \frac{\sum\limits_{i=1}^{1}(x_i - \hat{p}m_i)^2}{n-1}$$

An example of the use of these formulae can be found in Scheaffer et al. (1986). Cluster and two-stage sampling and estimation can get very complex, especially in the presence of an imperfect diagnostic test. More detailed discussion of this topic can be found in Cameron and Baldock (1998a, 1998b), and Jordan and McEwen (1998).

SUMMARY AND CONCLUSIONS. Many statistical sampling tools are available to foodborne pathogen researchers. The discussion in this chapter demonstrates that, although only four methods were reviewed, sampling methods can be combined to meet study objectives and to address constraints such as list frame availability and resources. The complexities of pathogen distributions and the diversity of potential environments (for example, animal production systems, animal processing facilities, food, and humans) necessitates careful consideration of the sampling approach prior to study initiation to ensure that sampling effort and expense will provide interpretable results.

Study hypotheses should be carefully contemplated before a sampling technique is selected and the design is finalized. On the other hand, findings from a published study should be scrutinized in terms of the appropriateness of the sampling technique before the findings can be generalized. Although most of the common sampling techniques are presented in this chapter, investigators may use their judgment to consider selecting an alternative technique that can fit their situation. Regardless of the choice of techniques, extrapolation of the findings must be based on the implemented sampling technique.

REFERENCES

Cameron, A.R., and F.C. Baldock. 1998a. A new probability formula for surveys to substantiate freedom from disease. *Prev. Vet. Med.* 34:1–17.

Cameron, A.R., and F.C. Baldock. 1998b. Two-stage sampling in surveys to substantiate freedom from disease. *Prev. Vet. Med.* 34:19–30.

Chapman, P.A., C.A. Siddons, D.J. Wright, P. Norman, J. Fox, and E. Crick. 1993. Cattle as a possible source of vero-cytotoxin-producing Escherichia coli O157 infections in man. *Epidemiol. Infect.* 111:439–447.

Dargatz, D.A., P.J. Fedorka-Cray, S.R. Ladely, and K.E. Ferris. 2000. Survey of Salmonella serotypes shed in feces of beef cows and their antimicrobial susceptibility patterns. *J. Food Prot.* 63:1648–1653.

Donald, A. 1993. Prevalence estimation using diagnostic tests when there are multiple, correlated disease states in the same animal or farm. *Prev. Vet. Med.* 15:125–145.

Griffin, P.M. 1995. *Escherichia coli* O157:H7 and other enteropathogenic *E. coli*. In *Infections of the gastrointestinal tract*. M.J. Blaser, P.D. Smith, J.I. Ravdin, H.B.Greenberg, and R.I. Guerrant, eds. New York: Raven Press, Ltd., 739–761.

Hancock, D.D., D.H. Rice, L.A. Thomas, D.A. Dargatz, and T.E. Besser. 1997. Epidemiology of *Escherichia coli* O157 in feedlot cattle. *J. of Food Prot.* 60:462–465.

Jordan, D., and S.A. McEwen. 1998. Herd-level test performance based on uncertain estimates of individual test performance, individual true prevalence and herd true prevalence. *Prev. Vet. Med.* 36:187–209.

Kish, L. 1995. Survey sampling. New York: Wiley.

Mahon, B.E., P.M. Griffin, P.S. Mead, and R.V. Tauxe. 1997. Hemolytic uremic syndrome surveillance to monitor trends in infection with *Escherichia coli* O157:H7 and other shiga-producing *E. coli. Emerg. Infect. Dis.* (letter) 3(3):409–411.

Marks, H, and M. Coleman. 1998. Estimating distributions of numbers of organisms in food products. *J. of Food Prot.* 61:1535–1540.

McDermott, J.J., and Y.H. Schukken. 1994. A review of methods used to adjust for cluster effects in explanatory epidemiological studies of animal populations. *Prev. Vet. Med.* 18:155–173.

McDermott, J.J., Y.H. Schukken, and M.M. Shoukri. 1994. Study design and analytical methods for data collected from clusters of animals. *Prev. Vet. Med.*18:175–191.

Schaeffer, R.L., W. Mendenhall, and L. Ott. 1986. Elementary survey sampling. Third Edition. Boston: Prindle, Weber and Schmidt.

THE *SALMONELLA* ENTERIDITIS RISK ASSESSMENT

W.D. SCHLOSSER, E.D. EBEL, B.K. HOPE, A.T. HOGUE, R. WHITING, R. MORALES, R. McDOWELL, AND A. BAKER

INTRODUCTION. The occurrence of *Salmonella enterica* serotype Enteritidis (SE) in humans increased from 1,207 isolates identified in 1976 (0.6 isolates/100,000 population) to 10,201 in 1995 (4.0/100,000 population). Sporadic cases of SE infections showed an association with the consumption of raw or undercooked shell eggs (St. Louis et al. 1988; Hedberg et al. 1993; Passaro et al. 1996). In a study of SE human illness outbreaks, a vehicle was implicated in 45 percent of the outbreaks of SE, and shell eggs constituted 82 percent of this group (38 percent of total outbreaks) between 1985 and 1991 (Mishu et al. 1994). Furthermore, the results of a United States Department of Agriculture (USDA) survey of spent hens at slaughter and unpasteurized liquid eggs at breaker plants in 1991 and 1995 revealed an increase in the prevalence of SE isolates in most regions of the United States (Hogue et al. 1997a).

Because controls at the national level, including the SE traceback regulation (USDA 1991), and intensified efforts to educate food handlers and enforce safe food handling practices, had not reduced human isolates or the prevalence of SE in flocks or unpasteurized liquid eggs (Hogue et al. 1997b), the USDA's Food Safety and Inspection Service (FSIS) began a comprehensive risk assessment of SE in December 1996. The objectives of the risk assessment were to: (1) model from farm to table the unmitigated risk of foodborne illness caused by SE from the consumption of eggs and egg products; (2) identify target areas along the farm-to-table continuum for potential risk-reduction activities; (3) compare the public health benefits accruing from the mitigated risk of SE foodborne illness with the implementation of various intervention strategies; (4) provide information on risk effectiveness of mitigation to be used by the USDA for subsequent cost effectiveness and cost-benefit analysis; and (5) identify data gaps and guide future research and data-collection efforts. This quantitative risk assessment for shell eggs and egg products extended from pullet through production, processing, transportation, preparation, consumption, to human illness (production-to-consumption) (USDA 1998). Although contamination of egg products was evaluated in the risk assessment, only a partial exposure assessment and risk characterization was completed. Thus, contamination of egg products is not considered in this report.

The risk assessment focused on modeling the occurrence of human illness caused by the consumption of egg-containing meals prepared from SE-contaminated eggs. The baseline results provided a point of reference for identifying those mitigation strategies most likely to reduce the occurrence of SE infection in humans from eggs and egg products. Such a quantitative risk assessment can provide better guidance to risk managers than can traditional qualitative analyses, particularly when a lengthy, multistep food process is being evaluated.

In risk assessments for chemicals, a primary objective is to predict potential future public health outcomes even though it is difficult to validate their predictions because of the chronic nature of the outcomes (US EPA 1989). However, in the case of acute illnesses associated with microbial pathogens, surveillance data are available and can provide a best estimate of the magnitude of the public health problem. A primary objective of a risk assessment for microbial pathogens in food for which there are surveillance data is not, therefore, to predict risk but rather to provide a tool for analyzing how to more effectively mitigate risk. This model may serve as a useful tool for Pathogen Reduction and Hazard Analysis and Critical Control Point (HACCP) applications.

OVERVIEW OF THE *SALMONELLA* ENTERITIDIS RISK ASSESSMENT. The Codex Alimentarius Commission, which establishes international food standards, defines a microbial risk assessment as a scientifically-based process consisting of four steps: (1) *Hazard Identification* to identify, through collection and critical review of data and information, a pathogen that may be present in a particular food or groups of food and is capable of causing adverse health effects; (2) *Exposure Assessment* to qualitatively and/or quantitatively estimate, using food consumption and prevalence data, total pathogen intake; (3) *Hazard Characterization* to qualitatively or quantitatively (or both) evaluate the nature of adverse health affects associated with the identified pathogen, including a dose-response assessment to determine the relationship between the level of pathogen intake (dose) and the frequency of associated adverse health effects (response); and (4) *Risk Characterization* to integrate results of *Exposure Assessment* and *Hazard Characterization* to develop an estimate for the probability of occurrence and severity of adverse health effects in a given population, as well as the uncertainty associated with these estimates (Codex 1999).

Hazard Identification. *Salmonella* is the leading cause of bacterial foodborne disease outbreaks in the United States, accounting for 69 percent of all bacterial foodborne outbreaks between 1988 and 1992 (CDC 1997; Shapiro et al. 1999). An estimated one to four million cases of illness caused by nontyphoidal salmonellosis occur in the United States each year; about forty thousand of these are culture-confirmed cases reported to the Centers for Disease Control and Prevention (CDC) (CDC 1997; Shapiro et al. 1999). Approximately nine hundred deaths occur, and economic losses associated with human salmonellosis are estimated to range from $150 million to $870 million annually (Todd 1989; Buzby et al. 1996).

All human salmonella pathogens are serotypes within *S. enterica* subspecies *enterica* (Chin 2000). Salmonella is transmitted by ingestion of the organisms in food derived from infected animals or contaminated by feces of an infected animal or human (CDC 1997; Chin 2000). The spectrum of clinical illness ranges from acute enterocolitis, the most common symptomatic form, to bacteremia manifesting as septicemia or focal infections such as abscesses or septic arthritis. Certain subpopulations (that is, infants, elderly, immune-compromised individuals) may be more susceptible to infection and to experiencing clinical manifestations of disease (Levine et al. 1991; Lew et al. 1991; Gerber et al. 1996). Except in these susceptible populations, deaths are uncommon, but morbidity and associated health care costs may be high across all population groups (Buzby et al. 1996; Bopp et al. 1999). In most countries that maintain surveillance, two members of this subspecies, *S. enterica* serotype Typhimurium and *S. enterica* serotype Enteritidis (SE), are the most commonly reported (CDC 1997). The rel-

ative prevalence of the different serotypes varies from country to country. In the United States, the proportion of reported *Salmonella* isolates that were SE increased from 5 percent in 1976 to 26 percent in 1994, with the incidence of human infection highest in New England and the mid-Atlantic region (CDC 1997). From 1985 to 1995, state and territorial health departments reported 582 SE outbreaks to the CDC. SE was the serotype most frequently reported to the CDC in 1990, 1994, 1995, and 1996. This accounted for 24.5 percent of all *Salmonella* isolates reported in 1996 (CDC 1997).

Since SE from raw or lightly cooked shell eggs was first identified as an SE infection source in 1988 (St. Louis et al. 1988; Hedberg et al. 1993; Passaro et al. 1996), eggs have been a dominant vehicle (in 82 percent of outbreaks with a known vehicle) for this pathogen (Mishu et al. 1994). More than forty-six billion eggs are distributed, sold, and used as shell eggs yearly in the United States. SE is the predominant serotype isolated from the contents of intact eggs (Schlosser et al. 1999).

Exposure Assessment. As shown in Figure 29.1, the risk model consists of four modules, each representing both a stage in the farm-to-table continuum and a step in the microbial risk assessment process. Three modules, Shell Egg Production, Shell Egg Processing & Distribution, and Preparation & Consumption, address Exposure Assessment (Figures 29.2, 29.3, and 29.4), whereas one, Public Health Outcomes (Figure 29.5), integrates *Hazard Characterization* and *Risk Characterization*. Baseline results were generated by linking together the four modules within one spreadsheet program (Excel, Microsoft Corporation, Redmond, Washington). Inputs to each module were expressed as distributions rather

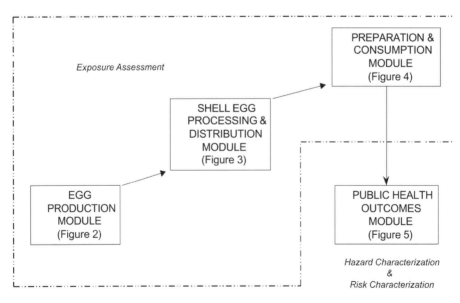

FIG. 29.1—Relationships between modules comprising the farm-to-table quantitative microbial risk assessment for SE in shell eggs and egg products.

than as point estimates to better express the uncertain estimation of their values. Inputs were parameterized using available data, which represented our best understanding of the ecology of SE in layer hens, shell eggs, and human behavior in the United States.

Monte Carlo simulations were performed using @Risk (Palisades Corporation, Newfield, NY) to generate results from the model. Each simulation consisted of 1,000 iterative spreadsheet calculations. Each iteration randomly selected a single value from each input distribution; all calculations were completed using these randomly selected values. Model results are not time specific, nor do they refer to any specific year's egg production or human illness incidence. However, data used to develop the module variables generally referenced the period from 1988 (when SE became a recognized problem in the United States) to the present.

SHELL EGG PRODUCTION. The Shell Egg Production module estimates the annual production of SE-contaminated eggs from U.S. flocks, beginning with the number of commercial egg production flocks in the United States. Figure 29.2 is a flowchart illustrating the principal inputs, processes, and outputs for this

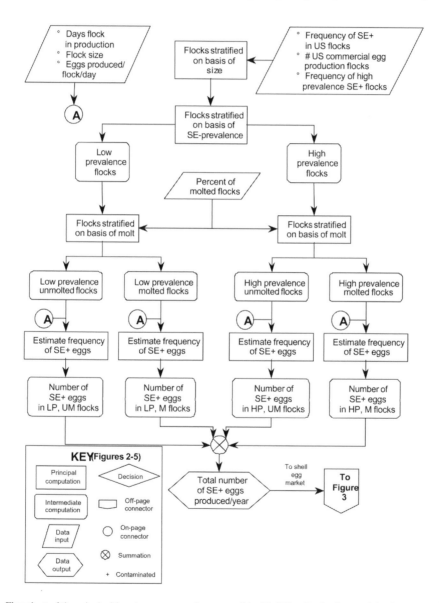

FIG. 29.2—Flowchart of the principal inputs, process, and outputs of the Shell Egg Production module. (Note that this flowchart, as well as those shown in Figures 29.3 through 29.5, are summaries and do not necessarily show all data elements and computations occurring in a given module.)

module. Flocks are first stratified into four size strata (to account for variability in egg production by various sized flocks) and then into SE-contaminated or non-contaminated categories based on the prevalence of SE in the flocks (Ebel et al. 1992; Hogue et al. 1997a). Contaminated flocks are further stratified into high- or low-prevalence flocks. Low-prevalence flocks produce SE-contaminated eggs at a very low rate (for example, 1/17,000 eggs laid), whereas high-prevalence flocks do so at a rate approximately ten times higher (for example, 1/1400 eggs laid). All flocks are again stratified into molted or not molted flocks (Henzler et al. 1998; Schlosser et al. 1999). For each type of infected flock, the number of SE-contaminated eggs produced per year is estimated, and then the number of contaminated eggs is summed to determine the total annual production of SE-contaminated eggs. The estimated number of SE-contaminated eggs is calculated as a final step in this module, and this estimate is a key input to Shell Egg Processing & Distribution module.

In contrast to some other microbial food safety problems, where the prevalence of contaminated commodities is influenced by processing steps (for example, slaughter of cattle and grinding of ground beef in the case of *E. coli* O157:H7), the prevalence of SE-contaminated eggs changes little between the time the eggs are laid and the time the eggs are prepared for consumption. Processing steps between production and consumption should affect only the level of contamination within contaminated eggs. Therefore, the fraction of eggs produced in the United States that are SE-contaminated is an important measurement in the exposure risk to U.S. consumers.

One output of the Shell Egg Production module is an uncertainty distribution for the fraction of eggs that are SE-contaminated during one year of production in the United States (Figure 29.6). This distribution's 5th, 50th, and 95th percentiles are 0.003 percent, 0.005 percent, and 0.008 percent, respectively (Ebel and Schlosser 2000). The mean is approximately one SE-affected egg in every twenty-thousand eggs annually produced.

We can compare the Shell Egg Production module's output to egg culturing data that were not used in developing this model. A random survey conducted in California found one SE-positive pooled egg sample in 1,416 pooled egg samples, where twenty eggs were pooled per sample (Gardner 1998). These results are assumed equivalent to finding one SE-positive egg in 28,320 eggs. To compare the California evidence to the module predictions, the model was simulated with the national prevalence of SE-affected flocks fixed at the regional prevalence representing California. Furthermore, uncertainty about the proportion of contaminated eggs from the California data was depicted using a beta(s+1, n-s+2) distribution, where s is the number of positive samples and n is the total eggs sampled (Vose 1996). The output of the production module is similar to the evidence independently derived from the California data (Figure 29.6). More uncertainty regarding the true fraction of contaminated

eggs is implied by the California data than by the model's output. Because prevalence and egg contamination frequency data in the model is substantial, increased certainty in the model's output, relative to a more limited sampling of eggs in California, is expected.

Using the mean of the distribution in Figure 29.6 as the best estimate of the fraction of U.S. eggs that are SE-contaminated, and assuming that roughly sixty-five billion (nonhatching) eggs are produced in the United States annually, an estimated 3.2 million contaminated eggs are produced per year. Not all of these eggs are consumed as table eggs. In 1999, approximately 30 percent of all shell eggs produced in the United States were marketed as egg products after undergoing some form of pasteurization. Assuming that eggs are marketed as egg products independent of the SE status of their flock of origin, then an estimated 2.2 million SE-contaminated eggs are marketed as table eggs. There are, however, quality-assurance programs in the United States requiring producers to divert eggs from known-infected flocks to pasteurization. It is difficult to precisely estimate what fraction of affected flocks operate in these programs, but it seems likely that there is some dependency between SE-status and the fraction of eggs marketed as egg products. Such dependency will influence the actual number of contaminated eggs marketed as table eggs. For example, if just 10 percent of infected flocks operate in quality-assurance programs that require diversion of their eggs to pasteurization, and the remaining 90 percent of of the eggs of infected flocks are marketed independent of their SE-status, then the estimate of the number of contaminated eggs marketed as table eggs is decreased to two million. Furthermore, assuming that these 10 percent of infected flocks are those infected flocks most easily detected, then the estimated number of contaminated table eggs is reduced to 1.9 million.

SHELL EGG PROCESSING AND DISTRIBUTION. The Shell Egg Processing and Distribution module simulates what happens to an SE-contaminated egg from the time it is laid until it is delivered to the consumer (Whiting et al. 2000). Figure 29.3 is a flow chart illustrating the principal inputs, processes, and outputs for this module. Shell eggs are stored, washed, packaged, and transported within this module. SE-contaminated eggs enter this module with a certain number of SE organisms (Humphrey 1994). The temperatures these eggs are exposed to, and the time these eggs experience different temperatures, determine whether SE grows in eggs. Ambient temperatures influence internal egg temperatures, and this effect is modeled through heat transfer equations that use cooling constants to account for different packaging materials and cooling environments (Czarick and Savage 1992).

SE growth within shell eggs cannot occur until the membrane surrounding the yolk breaks down enough to allow yolk nutrients to leak into the albumen of the egg (Humphrey 1994). These nutrients are considered

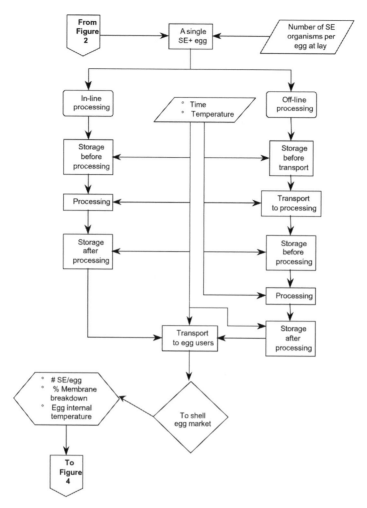

FIG. 29.3—Flowchart of the principal inputs, process, and outputs of the Shell Egg Production and Distribution module.

essential for SE to grow inside the egg. Yolk membrane breaks down according to a time-temperature relationship, and this module tracks the percent of yolk membrane remaining in each SE-contaminated egg during processing. When the egg reaches 100 percent yolk membrane breakdown, SE can begin to multiply within the egg (Humphrey 1994; Bradshaw et al. 1990; Schoeni et al. 1995). Equations were developed to estimate the internal temperature of the egg, the percent yolk membrane time remaining (or whether the egg would be able to support growth), and how much growth was taking place. Equations were developed from published data regarding when and how rapidly growth takes place in eggs (Whiting et al. 2000; Shapiro et al. 1999; Schoeni et al. 1995).

PREPARATION AND CONSUMPTION. The Preparation and Consumption module calculates the final number of SE organisms per meal served, and the total number

of servings at the various dose levels, after SE-contaminated shell eggs estimated in the Shell Egg Production module have been processed and distributed. This module considers the handling of a single SE-contaminated egg through storage, preparation, and cooking. An egg is assigned to one of sixteen different pathways depending on where it is used (home or institutional), how it is handled (pooled or not pooled), how it is used (as an egg or as an ingredient incorporated into another product), and to what extent it is cooked (fully or undercooked). All sixteen pathways are modeled simultaneously. Each pathway is associated with a variable probability of occurrence and a variable number of servings. Assumptions are made about the place of use, pooling, types of use, and cooking practices. With the exception of eggs in pools, post-production contamination (for example, by consumers) is not considered. Because the prevalence of SE-contaminated eggs is very low, only contaminated eggs were modeled, to avoid needless iterations involving hundreds of thousands of noncontaminated eggs.

Figure 29.4 illustrates the principal inputs, processes, and outputs for this module. Key inputs to this module are the percentage of yolk membrane breakdown, internal temperature of the egg, and the number of bacteria in the egg. The egg is either used in the home, in which case it is subjected to storage both at retail and in the home, or it is used in an institutional setting, where it is also subjected to storage. The egg is then either pooled or not pooled before use. If it is pooled before use, it is again subjected to storage. The egg is then either used and eaten as an egg, or it is used as an ingredient that is incorporated into a recipe. In either case, the product usually receives some type of

heat treatment through cooking. After the product is prepared, further short-term storage occurs before consumption. There are two key outputs from this module: the number of servings per SE-contaminated egg and the number of SE bacteria per serving.

Most eggs (70 percent) end up in only one serving, but some eggs are pooled and used to make products that serve more than one person. This both dilutes the number of organisms per serving and increases the number of people exposed, so that each SE-contaminated egg could contribute to an average of 4.4 servings. The Preparation and Consumption module predicts that contaminated eggs will contribute SE organisms to an average of ten million individual servings per year (that is, 2.3 million SE-contaminated

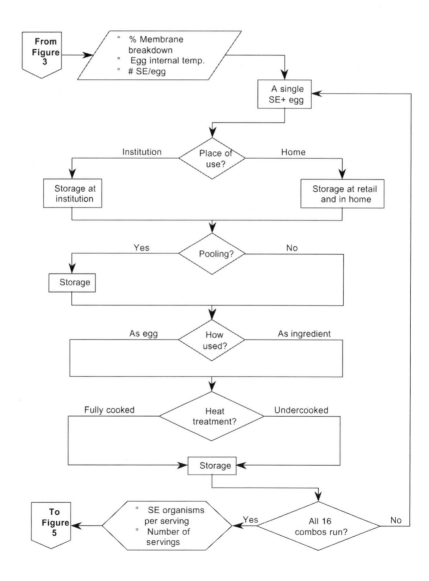

FIG. 29.4—Flowchart of the principal inputs, process, and outputs of the Preparation and Consumption module.

eggs/year × 4.4 servings/egg). Of these contaminated servings, 73 percent will contain no SE organisms after cooking, and 27 percent will contain from 1 to >1×10⁸ SE organisms. All human illness results from the 2.7 million servings per year that contain ≥1 SE organism.

Hazard Characterization and Risk Characterization. The Public Health Outcomes module combines *Hazard Characterization* and *Risk Characterization* by linking exposure to SE with the adverse health outcomes of morbidity and mortality that may arise from the ingestion of SE bacteria. It calculates the number of human illnesses resulting from exposure to meals containing varying levels of SE per serving. The consequences of illness may include physician visits, medical treatment, hospitalization, post-infection sequelae, and death. Figure 29.5 is a flow chart illustrating the prin-

cipal inputs, processes, and outputs of this module. The outcome from a single exposure to SE from shell eggs for an individual varies widely and is a function of the individual's age, health status, immune status, the number of SE bacteria consumed, and the fat content of the food vehicle (USDA 1998). The population exposed to SE is partitioned into two subpopulations: "normal" (that is, members of the general population assumed to be in good health) and "susceptible" (that is, persons who are at increased risk of illness from SE such as newborns, the elderly, and immunocompromised persons). The susceptible subpopulation is estimated to consist of about 20 percent of the U.S. population (Levine et al. 1991; Lew et al. 1991; Gerba et al. 1995).

The number of *Salmonella* organisms required to establish infection varies with serotype (Chin 2000). The probability of an individual person within a sub-

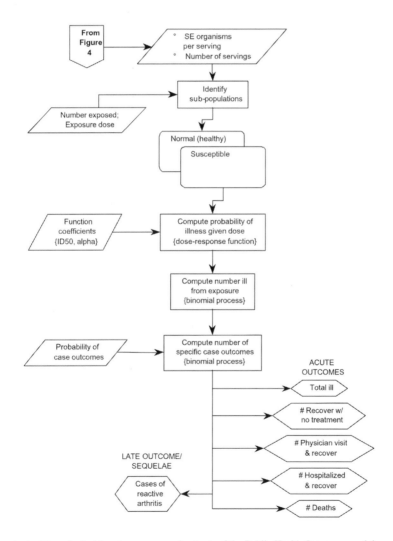

FIG. 29.5—Flowchart of the principal inputs, process, and outputs of the Public Health Outcomes module.

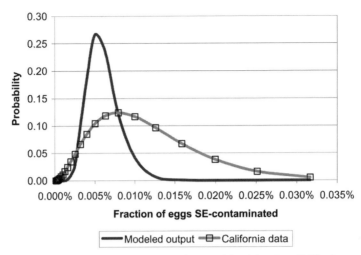

FIG. 29.6—Comparison of uncertainty distributions generated from model and data from California.

population becoming ill from exposure to a specified dose of SE was calculated using a stochastic dose-response function incorporating parameter variability. The distribution of the number of persons becoming ill in each subpopulation is estimated using the number of persons exposed and the computed probability of becoming ill from exposure to a specified dose of SE. However, not every person exposed to the dose of SE becomes ill. Results are presented on the basis of human illnesses, stratified into four categories: (1) illness and recovery without medical care; (2) illness with a physician visit; (3) illness with hospitalization; and (4) illness resulting in death. A specific case is assigned to one of these categories depending on the most severe outcome. Therefore, a case resulting in hospitalization is recorded only as such, even though it may have involved a physician visit as well. Additionally, long-term sequelae (for example, reactive arthritis) may be a sequela to cases who recovered without medical care, or from those who visited a physician or were hospitalized.

Public health impacts are estimated in terms of numbers of illnesses and specific case outcomes on an annual basis. Results from this module are the primary measure of the public health consequences of exposure to SE from shell eggs and may be used as the primary indicator of the public health benefits of specific risk mitigations introduced into other modules. The baseline model predicts a mean of nineteen human illnesses in the United States per year per million eggs consumed as shell eggs, with 5th and 95th percentiles of 4.0 and forty-six human illnesses, respectively. Alternatively, consumption of 2.3 million contaminated eggs could produce from 130,000 (5th percentile) to ≈1.7 million cases (95th percentile) of human illness per year, with a mean of 660,000 cases. The predicted average risk is 3.5 SE illnesses per one million egg-containing servings per year. The relative worth of specific

risk mitigation efforts to reduce SE exposures is measured in terms of these estimated health impacts. Although morbidity and mortality have measurable economic impacts, this module does not evaluate the economic costs of illness and the economic costs and benefits of mitigation activities.

Total illnesses in the normal and susceptible subpopulations occur at frequencies roughly consistent with the proportion of the population in each group. Using mean illnesses per million eggs consumed, approximately thirteen (68 percent) of the nineteen cases are predicted to occur in normal individuals and the remaining 32 percent of cases are predicted to occur in susceptible individuals. However, susceptible individuals experience more severe manifestations of SE infection than do normal individuals. In addition to their disproportionate contribution to deaths, susceptible individuals represent 51 percent of the hospitalized SE cases and 38 percent of cases requiring a physician visit. Among those surviving illness, almost three in every one hundred cases are predicted to experience post-infectious reactive arthritis. Susceptible individuals are not overrepresented in this group.

Most persons who become ill are predicted to recover without medical treatment, and the number of persons in each successive clinical outcome (physician visit and recovery, hospitalization and recovery, and death) declines about one order of magnitude. This pattern is consistent for the normal, susceptible, and total populations. Using the mean value as the reference point, ≈28 percent (610,000 of 2,100,000) of the exposed normal population, and ≈45 percent (270,000 of 600,000) of the susceptible population, are expected to become ill. More than 90 percent of ill people in both groups recover without medical treatment. However, a susceptible person is more likely than a normal person to see a physician, be hospitalized, or die. In the normal population, about 4.9 percent (30,000 of

610,000) of those who become ill are treated by physician and recover, whereas 0.35 percent are hospitalized and recover, and 0.02 percent die. In the susceptible population, 6.8 percent (18,000 of 270,000) of the ill are treated and recover, whereas 0.84 percent are hospitalized and recover, and 0.13 percent die. Thus, an ill susceptible person is 1.4 times (7 percent v. 4.9 percent) more likely to be treated by a physician, 2.4 times (0.84 percent v. 0.35 percent) more likely to be hospitalized, and 6.5 times (0.13 percent v. 0.02 percent) more likely to die than is a person in the normal subpopulation. The rate of reactive arthritis for persons who become ill is ≈3 percent in each group. The rate is slightly higher in the normal population because more of those who become ill survive than in the susceptible population; thus, a larger proportion are potentially able to develop post-illness sequelae such as reactive arthritis.

National public health surveillance statistics reported by the CDC were used as the basis for estimating the number of SE cases per year. An average of forty thousand *Salmonella* isolates are reported to CDC each year via the CDC's passive surveillance system; of these, 25 percent (ten thousand) are serotype *Salmonella* Enteritidis (CDC 1997). Although an average of ten thousand cases of SE are reported per year, this number represents only a proportion of all SE illnesses occurring per year, because of under-reporting intrinsic to passive surveillance systems. The probability of a case being reported was estimated in the range of 0.0028 to 0.026 (Aserkoff et al. 1970; Chalker and Blaser 1988; Todd 1989). Given the number of SE cases reported per year and the probability of a case being reported, a Pascal distribution was used to estimate a mean of 640,000 cases per year, with 5th and 95th percentiles of 250,000 and 1,200,000, respectively (Figure 29.7) (USDA 1998). This mean is less than that from the baseline risk assessment model (that is, 660,000), and the median of the distributions are very close in numerical value (630,000 for public health surveillance versus 680,000 for the risk assessment model). The modeled distribution is skewed to the right, which implies that the model predicts some probability of extremely high numbers of SE cases per year when compared to public health surveillance (Figure 29.7). Given the uncertain specifications of our model, this finding is not surprising.

EVALUATING MITIGATION SCENARIOS. Mitigation elasticity (ME) is defined as the ratio of the percent change in negative outcome (total human illness) to a fixed percent decrease in a model variable that resulted in the change. For example, if a 25 percent

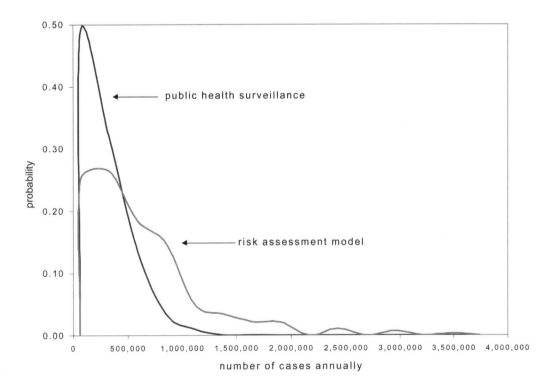

FIG. 29.7—Comparison of the number of human cases of SE from all causes predicted from public health surveillance to those predicted by the SE risk assessment for shell eggs.

decrease in an input variable resulted in a 30 percent decrease in the output variable, the ME would be 30/25 or 1.2, whereas the ME for a 25 percent input decrease associated with a 5 percent output decrease would be 5/25 or 0.20. ME is similar in concept to sensitivity analysis. However, because the complexity of this risk assessment precluded the use of traditional sensitivity analysis, ME was used to evaluate the effects of changing module variables on the baseline model output. It can be used to compare the likely benefit of several possible interventions. A complete cost-benefit analysis would provide additional information for selecting an optimal intervention strategy.

ME was calculated for variables that were considered possible useful mitigations or had been suggested as possible mitigations by producers, public health officials, or regulatory officials. Five mitigation scenarios were compared for their effect on the number of human illnesses. To allow for direct comparison of mitigations, the distribution of each example variable was modified to the same extent in order to reflect a 25 percent reduction in all values taken from the distribution. This change is not necessarily supported by research, but provides a reasonable assessment of the relative effects of the variables in the model.

Table 29.1 shows the expected total number of human illnesses (and associated ME values) after implementing each mitigation or each set of mitigations. For Scenario 1a, storage time of eggs in homes, institutions, and at retail were reduced by 25 percent; for Scenario 1b, storage temperatures in these three settings were similarly reduced. For Scenario 1c, both storage time and temperature were reduced. For Scenarios 2 and 3, respectively, the number of SE-contaminated flocks in the largest size strata and the number of high prevalence SE flocks were reduced by 25 percent. Scenario 4 examined the effect of diverting SE-contaminated eggs from the shell egg market to the egg products market, so as to result in 25 percent fewer SE-contaminated eggs available for shell egg consumers.

Scenario 5 combined 25 percent reductions in SE prevalence in the largest flocks and in storage time in homes, institutions, and at retail. None of the individual mitigations had an ME > 1; however, when separate mitigations were combined, an ME > 1 was obtained. This implies that a policy directed solely at one segment of the farm-to-table continuum will be less effective than a more broad-based approach. For example, a policy that encourages quality-assurance programs at the production level, cooling of eggs during processing and distribution, and proper food handling techniques is likely to be more effective than a policy that includes only one of these actions.

DISCUSSION. The objective of the baseline model was to describe the unmitigated risk of human illness caused by SE from eggs in the United States. Although model results correspond well with other estimates of the number of cases of human SE illnesses per year, the model is limited to describing SE illnesses caused by the consumption of eggs internally contaminated with SE and does not account for other sources of human illness due to SE in the United States. These other sources, if included, would increase the predicted annual SE cases. In addition, ongoing mitigation activities by producers, processors, and consumers are constantly changing the true incidence of illness caused by SE-contaminated eggs; these ongoing activities will cause this particular risk assessment (prepared in 1998) to become dated. In March 2002, FSIS began a comprehensive revision of the risk assessment. The goal of this revision is to incorporate new data, increase the efficiency of the model, and respond to comments generated by public and peer review.

Model results (Figure 29.7) imply that the distribution for illnesses predicted by the baseline model exceeds that predicted from public health reporting, although there remains considerable overlap of the

Table 29.1—Predicted Number of, and Reduction in, SE Cases for Five Different Mitigation Scenarios

Mitigation scenario	Mean number of SE cases	Reduction in SE cases	Percent reduction	ME
0 Baseline (i.e., no mitigation).	660,000			
1a Reduce storage time by 25%.	580,000	86,000	13%	0.5
1b Reduce temperature abuse occurrences by 25% in homes, institutions, and at retail.	580,000	77,000	12%	0.5
1c Reduce both time and temperature.	520,000	140,000	21%	0.8
2 Reduce prevalence of SE in flocks >100K by 25%.	560,000	100,000	15%	0.6
3 Reduce number of high-prevalence SE flocks by 25%.	570,000	94,000	14%	0.6
4 Divert 25% of all eggs from SE-contaminated flocks.	500,000	170,000	26%	1.0
5 Reduce prevalence of SE in flocks >100K by 25% and reduce storage by 25% in homes, institutions, and at retail.	450,000	210,000	32%	1.3

two distributions. Neither of these distributions can be verified. It is possible that predictions based on the public health data under- or overrepresent the annual occurrence of human SE illnesses per year. The baseline model may also inaccurately specify the production, processing, or preparation of SE-contaminated eggs, as well as the dose-response relationship for SE-contaminated meals. The fact that these distributions overlap suggests that the baseline model is a reasonable depiction of the farm-to-table continuum. This makes it a useful tool for evaluating the effect of mitigations, where the most important measurement is the resulting difference (reduction) in human cases. It becomes clear, through analysis of the model, which input variables are more important (sensitive) than others in their correlation with the number of human illnesses. Mitigation activities can be directed toward these more sensitive inputs, provided that they are candidates for intervention. For example, flock prevalence can possibly be influenced by policy decisions, but the inherent growth rate of SE in eggs cannot.

Although the goal was to make the model comprehensive, it has some important limitations. It is a static model and does not incorporate possible changes in SE over time as host, environment, or agent factors change. For many variables, data were limited or nonexistent. Some obvious sources of contamination, such as food handlers, restaurant environment, or other possible sites of contamination on or in the egg (such as the yolk), were not included. And, as complex as the model is, it still represents a simplistic view of the entire farm-to-table continuum. Finally, this analysis did not explicitly separate variability from uncertainty in specifying most of the model inputs. Distinguishing between variability and uncertainty provides an appreciation of how model outputs are affected by both ignorance of events occurring in the real population and by natural variability in the food-to-table system. The revision of the risk assessment will address these limitations.

SUMMARY AND CONCLUSIONS. One of the major benefits of conducting a quantitative risk assessment is that the discipline of the process highlights weaknesses in empirical data. When a variable is critical to the risk assessment, and few data are available to support estimation of the variable, the result is a wide-ranging probability distribution for that variable. This can suggest the need for more research to narrow our uncertainty about that variable. Preparation of this risk assessment has clarified the need to know more about the epidemiology of SE on farms, the bacteriology of SE in eggs, human behavior in food handling and preparation, and the variability of human responses to exposure to SE. These "data gaps" become priority target areas for future research efforts. The most prominent research needs were identified by module. Prioritization of needs across modules was not attempted,

because each module could significantly contribute to public health outcomes depending on a given pathway or scenario.

In conclusion, completion of this risk assessment depended on modeling a foodborne pathogen from its site of production through to the occurrence of human illness. Risk assessment, however, is an iterative process. Each risk assessment provides a starting point to focus future research and future assessments.

REFERENCES

Aserkoff, B., S.A. Schroeder, and P.S. Brachman. 1970. Salmonellosis in the United States—A five year review. *Am. J. Epidemiol.* 92:13–24.

Bopp, C.A., F.W. Brenner, J.G. Wells, and N.A. Strockbine. 1999. Escherichia, Shigella, and Salmonella. In *Manual of Clinical Microbiology, 7th Edition*. P.R. Murray, E.J. Barron, M.A Pfaller, F.C. Tenover, and R.H. Yolken, eds. Washington, D.C.: American Society for Microbiology, 467–471.

Bradshaw, J.G., D.B. Shah, E. Forney, and J.M. Madden. 1990. Growth of *Salmonella* Enteriditis in yolk of eggs from normal and seropositive hens. *J. Food Prot.* 53:1033–1036.

Buzby, J.C., T. Roberts, L.C.T. Jordan, and J.L. MacDonald. 1996, August. Bacterial Foodborne Disease: Medical Costs & Productivity Losses. *Agricultural Economic Report Number 741*. Washington, D.C.: Economic Research Service, U.S. Department of Agriculture.

Centers for Disease Control and Prevention. 1997. *CDC Salmonella Surveillance: Annual Tabulation Summary, 1996*. Washington, D.C.: U.S. Government Printing Office.

Chalker, R.B., and M.J. Blaser. 1988. A review of human salmonellosis: III. Magnitude of Salmonella infection in the United States. *Rev. Infect. Dis.* 10:111–124.

Chin, J. 2000. Control of Communicable Diseases Manual, 17th Edition. Washington, D.C.: American Public Health Association, 440–444.

Codex Alimentarius Commission. 1999. Codex Alimentarius Volume 1B, Codex Standards for General Requirements (Food Hygiene), CAC/GL 030-1999, Principles and Guidelines for the Conduct of Microbiological Risk Assessment. Available at: http://ftp.fao.org/codex/standard/volume1b/en/GL_030e.pdf.

Czarick, M., and S. Savage. 1992. Egg cooling characteristics in commercial egg coolers. *J. Appl. Poultry Res.* 1:389–394.

Ebel, E., and W. Schlosser. 2000. Estimating the annual fraction of eggs contaminated with *Salmonella* Enteriditis in the United States. *Intl. J. Food Microbiol.* 61:51–62.

Ebel, E.D., M.J. David, and J. Mason. 1992. Occurrence of *Salmonella* Enteriditis in the U.S. commercial egg industry: Report on a national spent hen survey. *Avian Dis.* 36:646–654.

Food and Drug Administration. 1997, May. Food Safety from Farm to Table: A National Food Safety Initiative. Available at: http://www.foodsafety.gov/~dms/fsreport.html.

Gardner, I. 1998. Personal communication. University of California, Davis.

Gast, R.K., and C.W. Beard. 1992. Detection and enumeration of *Salmonella* Enteriditis in fresh and stored eggs laid by experimentally infected hens. *J. Food Prot.* 55:152–156.

Gerba, C.P., J.B. Rose, and C.N. Haas. 1996. Sensitive populations: Who is at the greatest risk? *Intl. J. Food Microbiol.* 30:113–123.

Hedberg, C.W., M.J. David, K.E. White, K.L. MacDonald, and M.T. Osterholm. 1993. Role of egg consumption in sporadic *Salmonella enteritidis* and *S. typhimurium* infections in Minnesota. *J. Infect. Dis.* 167:107–111.

Henzler, D.J., D.C. Kradel, and W.M. Sischo. 1998. Management and environmental risk factors for *Salmonella* enteritidis contamination of eggs. *A.J.V.R.* 59:824–829.

Hogue, A.T., E.D. Ebel, L.A. Thomas, W.D. Schlosser, N.S. Bufano, and R.E. Ferris. 1997. Surveys of *Salmonella* Enteritidis in unpasteurized liquid egg and spent hens at slaughter. *J. Food Prot.* 60:1194–1200.

Hogue, A.T., P. White, J. Guard-Petter, W. Schlosser, R. Gast, E. Ebel, J. Farrar, T. Gomez, J. Madden, M. Madison, A.M. McNamara, R. Morales, D. Parham, P. Sparling, W. Sutherlin, and D. Swerdlow. 1997. Epidemiology and control of egg-associated *Salmonella* Enteritidis in the United States of America. *Review of Science and Technology, Office of International Epidemiology* 16:542–553.

Humphrey, T.J. 1994. Contamination of egg shell and contents with *Salmonella* Enteritidis: A review. *Intl. J. Food Microbiol.* 21:31–40.

Levine, W.C., J.F. Smart, D.L. Archer, N.H. Bean, and R.V. Tauxe. 1991. Foodborne disease outbreaks in nursing homes. *J.A.M.A.* 266:2105–2109.

Lew, J.R., R.I. Glass, R.E. Gangarossa, I.P. Cohen, C. Bern, and C.L. Mode. 1991. Diarrheal deaths in the United States, 1979–1987: A special problem for the elderly. *J.A.M.A.* 265:3280–3284.

Mead, P.S., L. Slutsker, V. Dietz, L.F. McCaig, J.S. Bresee, C. Shapiro, P.M. Griffin, and R.V. Tauxe. 1999. Food-related illnesses and death in the United States. *Emerg. Infect. Dis.* 5:607–625.

Mishu, B., J. Koehler, L.A. Lee, D. Rodrique, F.H. Brenner, P. Blake, and R.V. Tauxe. 1994. Outbreaks of *Salmonella* Enteritidis infections in the United States, 1985–1991. *J. Infect. Dis.* 169:547–552.

Passaro, D.J., R. Reporter, L. Mascola, L. Kilman, G.B. Malcolm, N. Rolka, S.B. Werner, and D.J. Vugia. 1996. Epidemic *Salmonella* Enteritidis (SE) infection in Los Angeles County, California; the predominance of phage type 4. *West. J. Med.* 165:126–130.

Schlosser, W., D. Henzler, J. Mason, D. Kradel, L. Shipman, S. Trock, H.S. Hurd, A. Hogue, W. Sischo, and E. Ebel. 1998. *The Salmonella Enteritidis Pilot Project.* In *Salmonella enterica serovar Enteritidis in Humans and Animals.* M. Saeed, ed. Ames, IA: Iowa State University Press.

Schoeni, J.L., K.A. Glass, J.L. McDermott, and A.C.L. Wong. 1995. Growth and penetration of *Salmonella* Enteritidis, *Salmonella* Heidelberg, and *Salmonella* Typhimurium in eggs. *J. Food Microbiol.* 24:385–396.

Shapiro, R., M.L. Ackers, S. Lance, M. Rabbani, L. Schaefer, J. Daugherty, C. Thelen, and D. Swerdlow. 1999. *Salmonella* Thompson associated with improper handling of roast beef at a restaurant in Sioux Falls, South Dakota. *J. Food Prot.* 62:118–122.

St. Louis, M.E., D.L. Morse, M.E. Potter, T.M. DeMelfin, J.J. Guzewich, R.V. Tauxe, and P.A. Blake. 1988. The emergence of grade A eggs as a major source of *Salmonella* Enteritidis infections: new implication for control of salmonellosis. *J.A.M.A.* 259:2103–2107.

Todd, E.C.D. 1989. Preliminary estimates to the cost of foodborne disease in the United States. *J. Food Prot.* 52:595–601.

U.S. Department of Agriculture. 1991. Chickens affected by *Salmonella* Enteritidis. Final Rule, 9 CFT Parts 71 & 82, Federal Register 56:3730–3743.

U.S. Department of Agriculture. 1998. *Salmonella* Enteritidis *Risk Assessment Shell Eggs and Egg Products*, Final Report. Available at: http://www.fsis.usda.gov/OPHS/risk/index.htm.

U.S. Environmental Protection Agency. 1989. Risk Assessment Guidance for Superfund (RAGS), Volume I—Human Health Evaluation Manual, Part A. (EPA/540/1-89/002). Washington, D.C.: Office of Emergency and Remedial Response.

Vose, D. 1996. Quantitative Risk Analysis: A Guide to Monte Carlo Simulation Modelling. Chichester, West Sussex, England: John Wiley & Sons Ltd.

Whiting, R.C., A. Hogue, W.D. Schlosser, E.D. Ebel, R.A. Morales, A. Baker, and R.M. McDowell. 2000. A quantitative process for *Salmonella* Enteritidis in shell eggs. *J. Food Sci.* 65(5):864–869.

30

CHARACTERIZING THE RISK OF ANTIMICROBIAL USE IN FOOD ANIMALS: FLUOROQUINOLONE-RESISTANT *CAMPYLOBACTER* FROM CONSUMPTION OF CHICKEN

MARY J. BARTHOLOMEW, KATHERINE HOLLINGER, AND DAVID VOSE

INTRODUCTION. The Center for Veterinary Medicine (CVM) is part of the Food and Drug Administration (FDA). The CVM's mission is to protect the public health by ensuring that animal drugs are safe and effective. One aspect of safety is human food safety. Thus, CVM ensures that edible products from food-producing animals treated with animal drugs are safe for humans.

Antimicrobial drugs are used in food animals to prevent, control, and treat diseases to enhance their health, production, and well being. Antimicrobial drug use in food animals has been implicated in the development of antimicrobial resistance in bacteria that are carried in and on food-producing animals. When animals are slaughtered for food, bacteria from their skin and gut can contaminate the meat intended for human consumption; some of these bacteria, such as *Salmonella* and *Campylobacter*, can cause illness that may require treatment in people. Bacteria of food animal origin that are carried on food consumed by humans may develop resistance during the animal's exposure to the antimicrobial drug. This can result in infections in people that are resistant to antimicrobial drugs needed to treat human foodborne illness (Endtz et al. 1991; Smith et al. 1999). When the transferred resistance confers resistance to an antimicrobial drug deemed important for the treatment of human illness, it poses a risk to public health.

Fluoroquinolones are considered an important class of antimicrobial drugs used for the empiric treatment of enteric illness in people. In 1995, fluoroquinolones were first approved for use in chickens for the treatment of colibacillosis. When chickens are treated, the fluoroquinolone exerts selection pressure, resulting in the selection of fluoroquinolone-resistant bacteria in the chickens. *Campylobacter* is a bacterial organism frequently isolated from broiler chickens that can cause campylobacteriosis, an enteric illness, in people exposed to the pathogen. When bacteria lose susceptibility through exposure to one fluoroquinolone, the bacteria exhibit cross-resistance to other drugs within the fluoroquinolone class. Consequently, individuals with fluoroquinolone-resistant *Campylobacter* infections acquired from consumption of chicken may have infections with decreased susceptibilities to any member of the fluoroquinolone class used for the empiric treatment of enteric illness in people.

The CVM approved fluoroquinolones for use in poultry as prescription-only medication and prohibited its extra-label use. It was expected that these measures would minimize the development of resistance. A nationally representative resistance monitoring system was established shortly after approval of the first poultry fluoroquinolone so that emerging resistance could be detected in either humans or animals.

The CVM became concerned about the potential public health impact of veterinary uses of the drug when, despite the precautionary use restrictions, emerging resistance to fluoroquinolones in *Campylobacter* was noted. This concern was reinforced by scientific literature reporting the emergence of fluoroquinolone resistance in human campylobacteriosis after approval of poultry fluoroquinolones in several foreign countries. CVM decided to assess, by means of a risk assessment model, the public health impact of fluoroquinolone resistance in *Campylobacter* that could be attributed to uses of fluoroquinolones in chickens. This chapter provides an overview of the risk assessment model. Although it is impossible to cover the entire assessment here, this chapter describes the modeling approach, provides a summary of the steps in the model, and discusses two of the assumptions used in the development of the model. The entire risk assessment entitled "Human Health Impact of Fluoroquinolone Resistant *Campylobacter* from the Consumption of Chicken" (FDA 2001) can be found on the Internet at http://www.fda.gov/cvm/antimicrobial/Risk_asses.htm.

BACKGROUND. Fluoroquinolones belong to a class of antimicrobial drugs with a broad spectrum of antimicrobial activity. They can be orally administered in humans and are very effective against most bacterial pathogens causing foodborne gastrointestinal illness (salmonellosis, campylobacteriosis, and others). Fluoroquinolones were first approved for use in people in 1986 in the United States and were commonly prescribed for the empiric treatment of travelers' diarrhea and for people with serious gastrointestinal infections. *Campylobacter* spp. are the most commonly isolated bacterial agents causing human gastroenteritis (Mead et al. 1999). Consumption of chicken has been identified as a risk for campylobacteriosis (Smith et al. 1999; Deming et al. 1987). Fluoroquinolone resistance among *Campylobacter* on chicken has raised concerns for the public health of humans with campylobacteriosis. In particular, there was concern that individuals with campylobacteriosis from consumption of chicken would be prescribed fluoroquinolones to treat fluoroquinolone-resistant infections and that the treatment would be ineffective.

Modern chicken-rearing practices involve housing large numbers of chickens (ten thousand to twenty thousand) in a single building for approximately forty to forty-seven days until they are taken to slaughter. Although broiler chickens are vaccinated against many diseases, they can succumb to respiratory infections that have the potential to spread through the flock. The most effective means for preventing the spread of an infection, after conducting the appropriate diagnostic tests to confirm a susceptible bacterial infection, is to medicate the flock using an antimicrobial drug in feed or water. Fluoroquinolone antimicrobials were first approved by FDA in 1995 for food-animal use in chickens with a respiratory disease called colibacillosis. Other countries had approved fluoroquinolones for this use in the 1980s. The emergence of fluoroquinolone-resistant *Campylobacter* infections in people was identified soon after the approval of the drug for use in chickens in the Netherlands (Endtz et al. 1991; Smith 1999). This temporal relationship between veterinary drug approval and emergence of resistant human infections was also observed in Spain and several other countries (Perez-Trallero et al. 1997; Velazquez et al. 1995). This raised questions about the public health implications of veterinary fluoroquinolone use and the potential magnitude of the risk in the United States.

Risk managers determined the risk assessment model and directed the development of a risk assessment that would estimate the public health impact of fluoroquinolone resistance in *Campylobacter* that was attributed to the use of fluoroquinolones in chickens. Work on the risk assessment began in 1998. CVM contracted with David Vose Consulting in France to develop the risk model.

The outcome measure used to estimate the public health impact was defined as the number of individuals with campylobacteriosis, who had fluoroquinolone-resistant infections, had sought care, and were prescribed a fluoroquinolone antibiotic to treat the infection. At the time the risk assessment was initiated, there were no food-animal approvals for fluoroquinolones except those in poultry. This simplified the modeling process because it was sufficient to identify other sources of resistance, estimate their contribution to the total pool of resistant *Campylobacter* infections, and remove them to determine those that could be attributed to uses of fluoroquinolones in chickens.

MODELING APPROACH. Predictive microbiology is used in microbial risk assessments to model the increases or decreases in microbial load under varied conditions of food processing or storage. Various microbial loads are then used as predictors in an assumed dose-response relationship to estimate the proportion of people who may become ill at any given level of microbial load. The model that was developed utilized information collected in surveillance systems identifying culture-confirmed cases of campylobacteriosis. This was important because the data describing the complex exposure, host-pathogen interactions necessary to recreate an accurate dose-response relationship were lacking. Additionally, measures that could indicate seasonal and annual changes in the level of contamination of food products at the point of exposure were lacking. An approach to the development of a risk assessment model that relied upon more robust assumptions and annually collected data was considered necessary. The Centers for Disease Control and Prevention (CDC) had established a foodborne disease surveillance system that collected information on the number of individuals who were ill and cultured positive for *Campylobacter* in a year, and susceptibility testing of isolates provided information on the level of fluoroquinolone resistance. The surveillance data were available on an annual basis, allowing the model to be updated with the current year's data. Several other studies provided modeling parameters that allowed an estimation of the total number of affected people. Then, instead of using predictive microbiology to estimate the public health impact, an estimate using data collected by the CDC FoodNet program of the annual number of people with campylobacteriosis was a starting point in the modeling process. The risk model provided a mechanism to express uncertainties in the estimated annual number of people with campylobacteriosis. Based upon the specificity of the risk assessment task assigned, the goal was to examine the potential link between the human health impact with fluoroquinolone-resistant *Campylobacter* on chicken product. In terms of formulating a model, this entailed the necessity to (1) estimate the expected annual number of cases seeking care that are fluoroquinolone-resistant and attributed to the use of fluoroquinolones in chickens; (2) estimate the quantity of consumed chicken with fluoroquinolone-resistant *Campylobacter*, V_i; and (3) estimate a parameter, K_{res}, linking the quantity of chicken with fluoroquinolone-resistant *Campylobacter* and λ. Figure 30.1 illustrates the link.

K_{res} is the proportionality constant which, when multiplied by the quantity of chicken consumed with fluoroquinolone-resistant *Campylobacter*, V_i, yields the average number of people impacted, λ. It is understood that the exposure and the consequence are measured over the same duration and over a complete period of seasonal variation. In the risk assessment, a year was used. The number of these people who are treated with a fluoroquinolone is determined by multiplying λ by the proportion of cases who seek care and are prescribed a fluoroquinolone.

SOURCES OF DATA. Step 1 required the estimation of the annual number of cases seeking care that are flu-

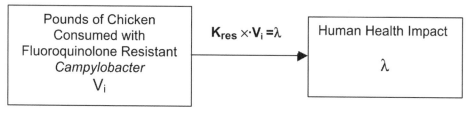

FIG. 30.1—K_{res} as the Proportionality Constant.

oroquinolone-resistant and attributed to the use of flu-oroquinolones in chickens. The starting point for this estimate was the annual number of cases of campy-lobacteriosis estimated by the CDC. The CDC had begun collecting prevalence data for campylobacterio-sis in humans in FoodNet in 1996. FoodNet is an active surveillance system for foodborne illness coordinated by the CDC with participating health departments reporting culture-confirmed cases. To produce an esti-mate of all cases of campylobacteriosis from these reportable cases, the CDC first estimated the fraction of cases that were not reported. Whether a case is reported depends on whether the case seeks care and whether the case submits a stool sample. Estimates for the pro-portion of times persons with diarrhea sought care and submitted stool samples were obtained from a tele-phone survey conducted by CDC (CDC 1998). In a telephone survey, approximately 150 persons within the FoodNet sites were called by random digit dialing each month and asked whether they or a person in the household had diarrhea in the past month. If the response was affirmative, the respondent was asked questions about whether they had sought care and, if so, whether a stool sample had been requested and sub-mitted for culture. Underreporting also depends on whether the stool culture actually yields *Campylobac-ter*; an estimate of the proportion of samples from per-sons with campylobacteriosis that actually yield *Campylobacter* was obtained from the literature (Ikram et al. 1994).

The CDC had begun collecting human *Campylobac-ter* isolates and testing for resistance in the National Antimicrobial Resistance Monitoring System (NARMS) in 1997 (CDC 1998; CDC 1999). Human isolates sent to the CDC for susceptibility testing con-stituted a statistically valid subset of the culture-con-firmed cases reported to FoodNet. The CDC was also in the midst of conducting a *Campylobacter* case-con-trol study (CDC 1998).

Step 2 required information about the amount of chicken consumed and the proportion of resistant iso-lates recovered from chicken carcasses. The amount of chicken consumed is available from the United States Department of Agriculture (USDA) Economic Research Service (ERS). USDA's Food Safety and Inspection Service (FSIS) collected chicken samples for susceptibility tests in 1998 as a part of NARMS;

USDA's Agricultural Research Service (ARS) did the testing for *Campylobacter* for NARMS (ARS 1998, 1999).

MODELING STEPS. The model was used to esti-mate two years of impact, 1998 and 1999. The three steps and their results are summarized in the following sections. At each step, the model output is the uncer-tainty distribution about the parameter estimated for that step. The mean of the uncertainty distribution is the average value one would expect to observe if it were possible to repeat each year many times.

Estimate λ. The parameter, λ, is a Poisson rate for a given year. When the rate is applied to the total U.S. population for that year, it yields an estimate of the number of campylobacteriosis cases seeking care who are prescribed a fluoroquinolone and who have fluoro-quinolone-resistant campylobacteriosis attributed to the use of fluoroquinolones in chickens.

For the purposes of this summary, the estimation of the number of campylobacteriosis cases seeking care who are prescribed a fluoroquinolone and who have fluoroquinolone-resistant campylobacteriosis attrib-uted to the use of fluoroquinolones in chickens consists of four subcomponent estimates. The estimates are for the total number of cases of campylobacteriosis; the number of campylobacteriosis cases attributed to chicken; the proportion of cases with resistant *Campy-lobacter* where the resistance is attributed to the use of fluoroquinolones in chickens; and the number of resist-ant cases attributed to chicken who seek medical care and are treated with fluoroquinolones.

TOTAL NUMBER OF CASES. The total number of cases of campylobacteriosis contracted in the United States annually is estimated by the CDC from the total num-ber of reported isolations in a year. The CVM used the process developed by the CDC and modeled the uncer-tainties in parameters used in the process of deriving the estimate. The CDC estimates the total number of cases by taking the observed number of culture-con-firmed cases within FoodNet sites and multiplying it by factors that account for undercounting and under-reporting (Mead et al. 1999). Input for these multiply-ing factors is from the population telephone survey

(CDC 1998a). A refined estimate is obtained by separately evaluating cases from three disease categories that differ with respect to their care choices: those with bloody diarrhea, nonbloody diarrhea, or invasive illness. Additionally, although most laboratories do test for *Campylobacter* in cultures, the ascertainment rate is about 75 percent (Ikram et al. 1994). Inability to detect, or ascertain, *Campylobacter* when it is truly in a sample is a function of the sensitivity of the laboratory method and of the non-uniformity of the bacteria in samples. The CVM model incorporates the uncertainty in the estimates of all the constituent proportions used to account for the undercounting: proportion of cases who seek care, proportion of cases who submit samples, and proportion of cases ascertained. The uncertainty distributions for these proportions are derived through Monte-Carlo simulation by drawing from Beta distributions. The parameters for a Beta distribution for a given proportion are the numbers of "successes" and the numbers of "failures" in the sample used to estimate that proportion. The model derives a distribution for the estimated mean number of cases summed over the three disease categories and extrapolated from the FoodNet sites to the entire United States. In the model, the FoodNet data were collected from the seven FoodNet sites. Catchment sizes of the sites vary and, therefore, the site estimates were weighted by fraction of the total catchment size to formulate the final estimates for FoodNet. A graph of the distribution of the estimated mean number of cases is shown in Figure 30.2.

NUMBER OF CASES THAT ARE ATTRIBUTED TO CHICKEN. The next step in the model derives the uncertainty distribution for the estimated mean number of cases of campylobacteriosis cases attributed to chicken. Two case-control studies from the literature were used for input values for determining the proportion of all campylobacteriosis cases attributable to chicken (Harris et al. 1986; Deming et al. 1987). In the Harris study, an attributable fraction of 45.2 percent was given; in the Deming study, the attributable fraction was 70 percent. The uncertainties surrounding these two attributable fractions were modeled using lognormal random variables. Walter (1975) describes the theory for attributable fraction uncertainty. For each iteration of the Monte-Carlo simulation, a value was drawn from each lognormal distribution. Then a value for the proportion of campylobacteriosis cases attributable to chicken was drawn from a uniform distribution defined between the two lognormal values, to recognize that significant uncertainty exists because of the difference between these two estimates.

PROPORTION OF CASES WITH RESISTANT *CAMPYLOBACTER* ARE ATTRIBUTED TO CHICKEN. To determine the proportion of domestically acquired chicken-associated cases that were drug resistant, it was necessary to eliminate the resistance attributed to other sources. Data from the *Campylobacter* case-control study assisted in the removal of proportions of resistance caused by other sources. Two other major sources

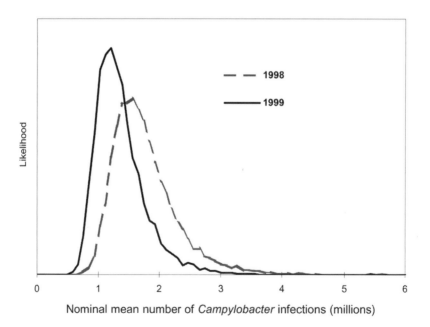

FIG. 30.2—Likelihood distributions for the mean numbers of *Campylobacter* infections in 1998 and 1999.
    Notes: The 5th percentiles are 1.1 million for 1998 and 0.9 million for 1999. The 95th percentiles are 2.8 million for 1998 and 2.2 million for 1999.

of fluoroquinolone-resistant *Campylobacter* in humans are foreign travel and human use of fluoroquinolone antimicrobials. The proportion of cases who had traveled outside the United States in the previous seven days was determined for those cases whose cultures yielded susceptible *Campylobacter* and for cases whose cultures yielded resistant *Campylobacter*. These cases were removed from the modeled estimates, leaving only domestically acquired cases. The proportion of cases who either had received a fluoroquinolone prior to submitting their stool samples or who were unsure when they had received fluoroquinolone in relation to when they had submitted their samples was also determined for those with susceptible and those with resistant *Campylobacter*. These cases were also removed from the estimates that were modeled. This information was collected as part of a FoodNet case-control study on reported cases of campylobacteriosis, and a sample of the collected isolates was tested for susceptibility.

By applying the proportion of travelers and prior fluoroquinolone users among the susceptible cases from the CDC case-control study to the susceptible cases from NARMS surveillance, it was possible to estimate the number of susceptible cases who would have traveled or used a fluoroquinolone. Subtracting them from the total number of NARMS-susceptible cases yields an estimate of susceptible cases that were designated domestically acquired. A similar process was carried out for the resistant cases. The proportion of travelers and prior fluoroquinolone users among resistant cases from the CDC case-control study (CDC 1998b) was applied to the resistant cases from NARMS surveillance. The result estimated the number of travelers and prior fluoroquinolone users who were subtracted from the total number of NARMS-resistant cases. Cases remaining were designated resistant cases that were domestically acquired. Adjusting for the effects of travel and prior human drug use, the remaining resistance among campylobacteriosis cases attributed to chicken was attributed to the use of fluoroquinolones in chickens. The proportion of resistance among domestically acquired cases was multiplied by the number of chicken-associated cases to estimate the mean number of cases resistant and attributed to resistance from chicken. The output distribution of the estimated mean number of cases resistant and attributed to resistance from chicken are shown in Figure 30.3.

NUMBER OF RESISTANT CASES WHO SEEK CARE AND ARE TREATED WITH THE DRUG. To estimate the mean number of resistant cases who had infections attributed to consumption of chicken who seek care and are treated with a fluoroquinolone and its associated

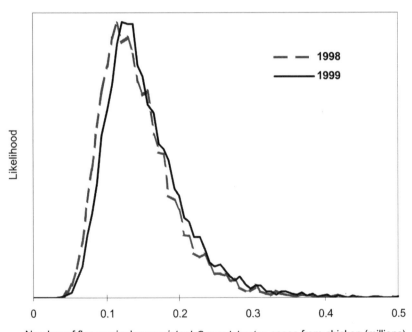

Number of fluoroquinolone resistant *Campylobacter* cases from chicken (millions)

FIG. 30.3—Likelihood distributions for the mean numbers of fluoroquinolone-resistant *Campylobacter* cases from chicken for 1998 and 1999.
   Notes: The 5th percentile estimates are 77,000 in 1998 and 84,000 in 1999. The 95th percentile estimates are 241,000 in 1998 and 258,000 in 1999.

uncertainty distribution, the mean number of cases with resistant campylobacteriosis attributed to chicken is multiplied by all the care-seeking proportions discussed previously. They are the proportion of cases who seek medical care, who submit a specimen for culture, and whose illness is ascertained by culture. In addition, the answer must be multiplied by the proportion of cases who receive antibiotics and by the proportion receiving a fluoroquinolone, given that they receive an antibiotic. The proportion of cases who receive antibiotics and the proportion who receive a fluoroquinolone were derived from the 1998–1999 CDC *Campylobacter* case control study. The distribution of the estimated mean number of fluoroquinolone-resistant *Campylobacter* cases from chicken who seek care and are prescribed fluoroquinolones is shown in Figure 30.4.

Estimate V_i. Estimation of the quantity of chicken with fluoroquinolone-resistant *Campylobacter* consumed and the uncertainties in the estimate are modeled next. Inputs to estimate the quantity of consumed chicken were taken from ERS with product sent for rendering, product diverted for pet food, exports, water added during processing, and imports subtracted (ERS 1999). The proportion of chicken with *Campylobacter* and the proportion of *Campylobacter* that were fluoroquinolone-resistant were determined from the samples

that FSIS and ARS analyzed as part of the NARMS project (ARS 1998, 1999). These estimates are likely to be underestimates of the true prevalence. NARMS susceptibility testing characterizes a single colony-forming unit (CFU) per sample. Chicken carcasses typically carry several different *Campylobacter* strains, each potentially exhibiting different levels of susceptibility to fluoroquinolones. Additionally, *Campylobacter* CFUs can be comprised of several strains, again each exhibiting a unique minimum inhibitory concentration. Estimates of the prevalence of contamination of poultry products with fluoroquinolone-resistant *Campylobacter* are therefore not accurate. More accurate estimates may be obtained by testing several isolates per sample. The resultant distribution for this step is shown in Figure 30.5.

Estimate K_{res}. Figure 30.1 showed K_{res} and its relationship to λ and V_i. K_{res} is formed as the ratio of the two variables. The numerator, λ, is the mean number of fluoroquinolone-resistant *Campylobacter* cases from chicken, shown in Figure 30.3. The denominator, V_i, is the amount of fluoroquinolone-resistant *Campylobacter*-contaminated chicken meat consumed, shown in Figure 30.5. The uncertainty distribution for the quantity K_{res} is shown in Figure 30.6. The interpretation of K_{res} is the probability that a pound of chicken meat with fluoroquinolone-resistant *Campylobacter* will result in

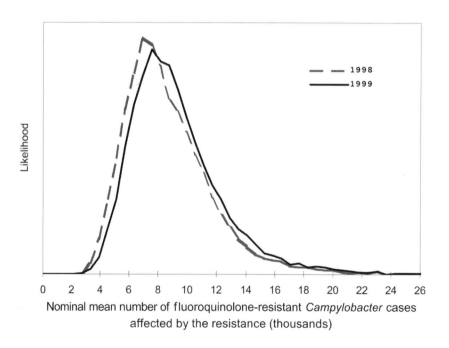

FIG. 30.4—Likelihood distribution for the mean numbers of campylobacteriosis cases that seek care and are treated with fluoroquinolones who have fluoroquinolone-resistant *Campylobacter* attributed to chicken, in 1998 and 1999.

Notes: The 5th percentile estimates are 4,760 in 1998 and 5,230 in 1999. The 95th percentile estimates are 14,370 in 1998 and 15,330 in 1999.

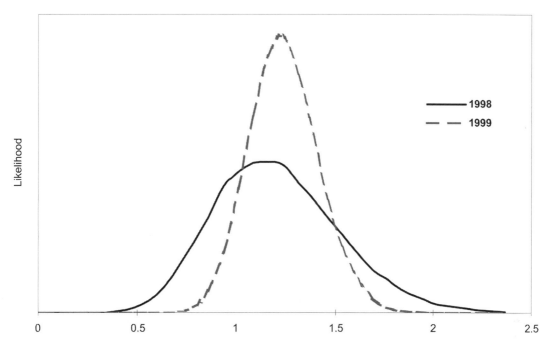

Fluoroquinolone-resistant *Campylobacter* contaminated chicken consumed
(billion lbs)

FIG. 30.5—Likelihood distributions for Vi, the quantities of chicken with fluoroquinolone-resistant *Campylobacter* consumed in 1998 and 1999.
Notes: The 5th percentile estimates are 0.7 billion pounds in 1998 and 1.0 billion pounds in 1999. The 95th percentile estimates are 1.8 billion pounds in 1998 and 1.5 billion pounds in 1999. Resistance data from chicken was collected during the entire year for 1999, but only in the latter part of the year for 1998, resulting in more uncertainty in the 1998 estimate than in the 1999 estimate.

a case of fluoroquinolone-resistant campylobacteriosis in a specific year.

ASSUMPTIONS. The CVM risk model attributes a proportion of all campylobacteriosis cases to chicken and a proportion of all fluoroquinolone-resistant campylobacteriosis cases in humans to the use of fluoroquinolones in chickens.

In general, tracing bacteria responsible for foodborne illness back to their source is a difficult task. However, sources of outbreaks of foodborne illnesses are sometimes found. In addition to poultry, water (Alary 1990; Sacks et al. 1986) or raw milk (Evans 1996; Headrick et al. 1998; Korlath et al. 1985) have been implicated as the source of *Campylobacter* in some outbreaks. In contrast, tracing back to the source is not typically accomplished for sporadic cases, that is, cases that are not outbreak related. Approximately 99 percent of campylobacteriosis cases are sporadic cases. Knowledge of the sources of the bacteria responsible for outbreaks helps identify potential sources for sporadic cases, but it does not define the universe of sources, nor does it provide an indication of the relative contribution of sources to the overall burden of sporadic campylobacteriosis cases. There have, however, been some epidemiologic investigations specifically conducted to identify sources of sporadic campylobacteriosis (Altekruse et al. 1999; Smith et al. 1999; Tauxe 1992). They indicate that chicken is a major source.

In 1995, about 88 percent of chicken carcasses yielded *Campylobacter* in the slaughterhouse, as reported by the FSIS (FSIS 1996a). This compares to 1–4 percent found on beef at slaughter in 1993 (FSIS 1996b). Additionally, the microbial load of *Campylobacter* is higher on chicken than on other foodanimal products. The CVM risk assessment calculated that approximately 50 pounds of domestically produced chicken per person was consumed in 1998–1999. This value was derived by subtracting off water weight and amounts sent for rendering and export from the pounds of chicken produced, as given by ERS. The combined information about *Campylobacter* contamination levels on chicken and the substantial exposure through consumption provide further credence to the epidemiologic study findings.

The CVM risk assessment used figures for fractions of campylobacteriosis attributable to chicken without

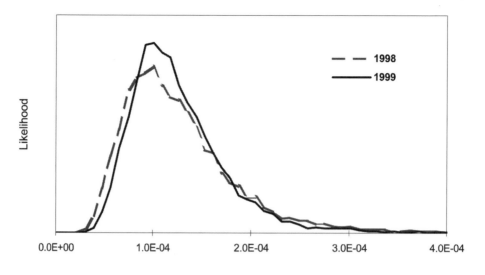

K for fluoroquinolone-resistant *Campylobacter*

FIG. 30.6—Likelihood distributions for K_{res} for 1998 and 1999.
   Notes: The 5th percentile estimates are 5.8E-05 casesin 1998 and 6.41E-05 cases in 1999. The 95th percentile estimates are 1.8 billion pounds in 1998 and 1.5 billion pounds in 1999.

amplifying the amount to account for cross-contamination from chicken to other foods in the kitchen or on the salad bar. Effectively, all types of cross-contamination are considered to be included in the risk measured by K_{res}.

As mentioned in the background section, the risk assessment began prior to FDA approvals for fluoroquinolone use in the United States in food-producing animals other than poultry. At least with respect to domestically acquired cases, tracking sources of resistance was limited then to the resistance attributed to the use of fluoroquinolones in chickens and in people. Fortunately, the 1998–1999 CDC case control study questioned the respondents about whether they had taken fluoroquinolones before submitting their stool specimens. The fraction of cases responding that they had taken human fluoroquinolones was factored out of those whose resistance was attributed to the use of fluoroquinolones in chickens in the United States.

Other countries also use fluoroquinolones in food-producing animals. This meant that U.S. travelers could acquire fluoroquinolone-resistant *Campylobacter* abroad that might erroneously be attributed to domestic use of fluoroquinolones. The risk assessment needed to eliminate that proportion of resistant cases whose resistance had come from a foreign source. Fortunately, the 1998–1999 CDC *Campylobacter* case control study questioned the respondents about whether they had traveled outside the country in the previous seven days. The fraction of cases responding that they had traveled in the past seven days was factored out of those whose resistance was attributed to

the use of fluoroquinolones in chickens in the United States.

Factoring out those who had reported taking a human fluoroquinolone before submitting their stool specimens and those who reported having traveled outside the United States may have resulted in an underestimate of the number of fluoroquinolone-resistant cases that could be attributed to the use of fluoroquinolones in chickens. Some of the cases with resistance who received the human fluoroquinolone or who had traveled outside the United States are likely to have acquired their fluoroquinolone-resistant *Campylobacter* from chicken in the United States regardless of the human drug use or travel. The fraction of cases with resistance attributed to the use of fluoroquinolones in chickens in the United States is likely to be higher than the value (mean of 14.2 percent for 1998) derived in the risk assessment.

SUMMARY AND CONCLUSIONS. The CVM risk assessment is an example of an antimicrobial resistance risk assessment. In microbial risk assessments, the object is to provide a scientifically based estimate of the impact of a particular microbial risk and/or an estimate of the effect of specified risk management actions, usually by tracing microbes from their sources through various pathways until ultimately humans are exposed to them. In contrast, antimicrobial resistance risk assessment traces resistance traits. To some extent, doing so may include tracing the bacteria carrying the resistance traits.

In the CVM risk assessment, an epidemiological approach was used to estimate the mean number of individuals impacted by fluoroquinolone-resistant campylobacteriosis attributable to chicken consumption. Individuals considered to be impacted were those with campylobacteriosis who had fluoroquinolone-resistant infections, had sought care, and were prescribed a fluoroquinolone antibiotic to treat the infections. In 1998, the mean number (the number of cases that would occur on average if 1998 were to be repeated many times) was estimated to be 8,678 between the 5th percentile, 4,758, and the 95th percentile, 14,369. This corresponds to a risk of one impacted individual in every 34,945 individuals in the United States.

The availability of surveillance data concerning the prevalence of campylobacteriosis, per se, as well as concerning the prevalence of resistance in both humans and in chickens was essential for the CVM risk assessment. This work would not have been possible without the data collection efforts of U.S. government agencies, notably the CDC, the USDA, and the Census Bureau. It was also facilitated by the research findings of numerous academic and foreign government researchers.

REFERENCES

Agricultural Research Service, USDA. 1998. Preliminary data: NARMS national antimicrobial resistance monitoring system: enteric bacteria—1998 animal *Campylobacter* isolate report. Athens, Georgia. Personal communication, Dr. P. Fedorka-Cray.

Agricultural Research Service, USDA. 1999. Preliminary data: NARMS national antimicrobial resistance monitoring system: enteric bacteria—1999 animal *Campylobacter* isolate report. Athens, Georgia. Personal communication, Dr. P. Fedorka-Cray.

Alary, M., and D.Nadeau. 1990. An outbreak of *Campylobacter enteriditis* associated with a community water supply. *Can. J. Public Health* 81(4):268–271.

Altekruse, S., N.Stern, P.Fields, and D. Swerdlow. 1999. *Campylobacter jejuni*—an emerging foodborne pathogen. *Emg. Infect. Dis.* 5(1):28–35.

Blaser, M. 1997. Epidemiologic and clinical features of *Campylobacter jejuni* infections. *J. Infect. Dis.* 176(Suppl 2):S103–5.

Centers for Disease Control and Prevention. 1998. Annual report, NARMS national antimicrobial resistance monitoring system: enteric bacteria. Atlanta, Georgia.

Centers for Disease Control and Prevention. 1999. Annual report, NARMS national antimicrobial resistance monitoring system: enteric bacteria. Atlanta, Georgia.

Centers for Disease Control and Prevention. 1999. FoodNet *Campylobacter* case control study 1998-1999, preliminary data. Atlanta, Georgia. Personal communication, Dr. Nina Marano.

Centers for Disease Control and Prevention. 2000. FoodNet Population (Telephone) Survey 1998-1999, Preliminary data. Atlanta, Georgia. personal communication Elizabeth Imhoff, M.P.H.

Deming, M., R. Tauxe, P. Blake, S. Dixon, B. Fowler, T. Jones, E. Lockamy, C. Patton, and R. Sikes. 1987. *Campylobacter enteriditis* at a university: transmission from eating chicken and from cats. *Am. J. Epidemiol.* 126(3):526–534.

Endtz, H., G.Ruijs, B.van Klingeren, W. Jansen, T. van der Reyden, and R. Mouton. 1991. Quinolone resistance in *Campylobacter* isolated from man and poultry following the introduction of fluoroquinolones in veterinary medicine. *J. Antimicrob. Chemo.* 27:199–208.

Evans, M., R. Roberts, C. Ribero, D. Gardner, and D. Kembrey. 1996. A milk-borne *Campylobacter* outbreak following an education farm visit. *Epidemiol. Infect.* 117:457–462.

Economic Research Service. USDA. April, 1999. Food consumption, prices and expenditures, 1970–97. Availble at: http://www.ers.usda.gov/publications/sb965/.

Food and Drug Administration. 2001. Human health impact of fluoroquinolone resistant *Campylobacter* from the consumption of chicken. Available at: http://www.fda.gov/cvm/antimicrobial/Risk_asses.htm.

Food Safety and Inspection Service, Science and Technology Microbiology Division. USDA. April 1996. Nationwide broiler chicken microbiological baseline data collection program, July 1994–June 1995. Available at http://www.fsis.usda.gov/OPHS/baseline/contents.htm.

Food Safety and Inspection Service, Science and Technology Microbiology Division. USDA. February 1996. Nationwide beef microbiological baseline data collection program: cows and bulls, December 1993–November 1994. Available at http://www.fsis.usda.gov/OPHS/baseline/contents.htm.

Harris, N., N. Weiss, and C.Nolan. 1986. The role of poultry and meats in the etiology of *Campylobacter jejuni/coli/enteriditis. Am. J. Publ. Health* 76(4):407–411.

Headrick, M., S. Korangy, N. Bean, F. Angulo, S. Altekruse, M. Potter, and K. Klontz. 1998. The epidemiology of raw milk-associated foodborne disease outbreaks reported in the United States, 1973 through 1992. *Am. J. Publ. Health* 88(8):1219–1221.

Ikram, R., S.Chambers, P. Mitchell, M. Brieseman, and O. Ikram. 1994. A case control study to determine risk factors for *Campylobacter* infection in Christ Church in the summer of 1992–3. *New Zealand Med. J.* 107:430–432.

Korlath, J., M.Osterholm, L. Judy, J. Forgang, and R. Robinson. 1985. A point-source outbreak of campylobacteriosis associated with consumption of raw milk. *J Infect. Dis.* 152(3):592–596.

Mead, P., L.Slutsker, V. Dietz, L. McCaig, J. Bresee, C. Sharpiro, P. Griffin, and R. Tauxe. 1999. Food-related illness and death in the United States. *Emerg. Infect. Dis.* 5(5):1–20.

Perez-Trallero, E., F.Otero, C. Lopez-Lopategui, M. Montes, J. Garcia-Arenzana, and M. Gomariz. 1997. High prevalence of ciprofloxacin resistant *Campylobacter jejuni/coli* in Spain. Abstracts of the 37th ICAAC, September 28–October 1, Toronto, Canada.

Sacks, J., S. Lieb, L. Baldy, S. Berta, C. Patton, M. White, W. Bigler, and J. Witte. 1986. Epidemic campylobacteriosis associated with a community water supply. *Am. J. Publ. Health* 76(4):424–428.

Smith, K., J. Besser, C. Hedberg, F. Leano, J. Bender, J. Wicklund, B. Johnson, K. Moore, and M. Osterholm. 1999. Quinolone-resistant *Campylobacter jejuni* infections in Minnesota, 1992–1998. *N. Engl. J. Med.* 340(20):1525–1532.

Walter, S. 1975. The distribution of Levin's measure of attributable risk. *Biometrika.* 62(2):371–374.

Velazquez, J., A. Jimenez, B. Chomon, and T. Villa. 1995. Incidence and transmission of antibiotic resistance in *Campylobacter jejuni* and *Campylobacter coli. J. Antimicrobiol. Chemother.* 35:173–178.

31

BOVINE SPONGIFORM ENCEPHALOPATHY: RISK ASSESSMENT AND GOVERNMENTAL POLICY

ANNE K. COURTNEY, MARY PORRETTA, JOSHUA T. COHEN, GEORGE M. GRAY, SILVIA KREINDEL, AND DANIEL L. GALLAGHER

BACKGROUND. Bovine Spongiform Encephalopathy (BSE) is a disease in cattle that has become of great concern due to its implications for human health. The following describes briefly what is understood about the disease itself as well as other diseases in the same class, methods currently available for testing, and government mitigations enacted to date to prevent the entry of BSE into the United States food supply as well as to prevent the transmission of the human variant (vCJD) of this disease.

Transmissible Spongiform Encephalopathies. BSE, commonly referred to as "Mad Cow Disease," is a slowly progressive degenerative disease that affects the central nervous system (CNS) of adult cattle. The average incubation period (the time from when an animal becomes infected until it first shows disease signs) for BSE is from four to five years. Following the onset of clinical signs, the animal's condition deteriorates until it either dies or is destroyed. This process usually takes from two weeks to six months. BSE is so named because of the spongy appearance of the brain tissue of infected cattle when sections are examined under a microscope.

BSE belongs to the family of diseases known as the transmissible spongiform encephalopathies (TSEs). Other TSEs include scrapie in sheep and goats, transmissible mink encephalopathy, chronic wasting disease (CWD) in deer and elk, and in humans, kuru, Creutzfeldt-Jakob Disease (CJD), Gerstmann-Straussler-Scheinker syndrome, and fatal familial insomnia. Other TSEs such as Feline Spongiform Encephalopathy and Variant Creutzfelt Jakob Disease (vCJD) are believed to be a result of cross-species transmission of a TSE (for example, vCJD arises from BSE).

The agent that causes BSE and other TSEs has yet to be fully characterized, but strong evidence collected over the past decade indicates that the agent is composed largely, if not entirely, of an abnormally shaped variant of a normal cellular protein referred to as a prion. This protein is a membrane-associated nerve cell protein presumed to be involved in cell signaling and is present in all mammals. The abnormal prion protein has a higher percentage of beta-sheet structure than does the naturally occurring isoform, which leads to a different tertiary structure (Pan et al. 1993; Stahl et al. 1993; Pergami et al. 1996; Prusiner 1998). The misfolded isoform of the protein, which is thought to cause the TSEs, is water insoluble and aggregates to form plaques in the central nervous tissue. The prion agent is highly resistant to heat, ultraviolet light, ionizing radiation, and common disinfectants that normally inactivate viruses or bacteria (Taylor 1993, 1996).

BSE was first diagnosed in 1986 in the United Kingdom (U.K.) and since then has been confirmed in native-born cattle in several other European countries and, most recently, in Japan. The disease is most likely spread by feeding rendered parts of cattle infected with BSE to other cattle in the form of meat and bone meal. Worldwide, more than 178,000 cases of BSE have been detected since the disease was first diagnosed, with more than 99 percent of those cases reported from the United Kingdom. No cases of BSE have been detected in the United States despite active surveillance for the disease since 1990. Other animal TSEs, such as scrapie and CWD, are present in the United States.

Testing. The distribution throughout the organs of a diseased animal and the total level of infectivity are dynamic throughout the disease process in cattle (Wells 1998). Animals clearly harbor infectivity prior to clinical signs, but there is no live-animal diagnostic test for BSE (Wells 1998; Peters 2000). Definitive diagnosis of BSE and vCJD requires testing of brain tissue samples to evaluate histology, immunohistochemical alterations, and protein changes that are indicative of the disease.

Variant Creutzfeldt-Jakob Disease (vCJD). In 1996, a newly recognized form of CJD was reported in ten patients in the United Kingdom. CJD is a chronic, neurodegenerative disease that affects humans. However, in contrast to the classic form of CJD, patients with this new form of CJD, now known as vCJD, experience early psychiatric symptoms, earlier loss of coordination, and later onset of dementia. Both forms of CJD are always fatal. Patients with vCJD tend to be younger than those with classic CJD (average age of onset twenty-nine years, as opposed to sixty-five years) and have a relatively longer duration of illness (median of fourteen months as opposed to 4.5 months). The classic form of CJD occurs sporadically (that is, with no known cause) at a rate of about one to two cases per one million people per year throughout the world, whereas vCJD has been detected only in persons who reside in countries in which BSE is known to exist.

Biochemical and epidemiological studies have linked vCJD to exposure to BSE, probably through human consumption of beef products contaminated with the agent that causes BSE. As with BSE, vCJD has never been detected in the United States.

From October 1996 to early December 2001, 113 confirmed or probable cases of vCJD had been reported in the United Kingdom, three in France, and a single case in the Republic of Ireland and Hong Kong. Figure 31.1 illustrates the number of cases of BSE and vCJD over time. Although the number of infected cattle in the United Kingdom has peaked and declined, the much smaller number of infected cattle in the rest of the European Union continues to increase, as does the number of cases of vCJD.

A number of studies have attempted to estimate the probable magnitude of vCJD infection and the associated fatalities (Valleron et al. 2001; Cousens et al. 1997; Ghani et al. 2000). The estimates have varied substantially, suggesting that the ultimate number of illnesses will range between 70 and 136,000. Unfortunately, it is already apparent that the lower estimates are incorrect but to what degree is not yet known.

Most models either presume that susceptibility is reduced nonlinearly with age or that persons exposed later in life simply have a longer incubation period. Addi-

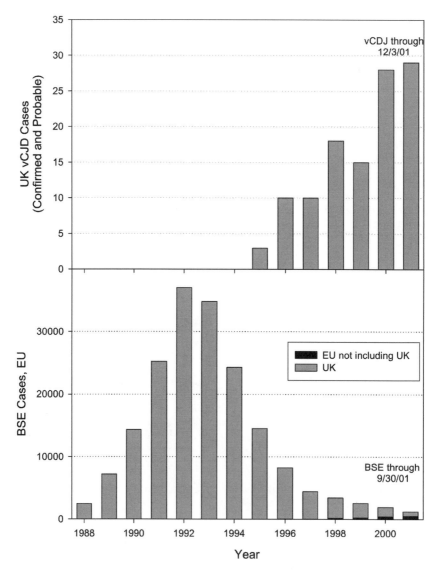

FIG. 31.1—Spread of BSE and vCJD in the United Kingdom. Sources: CJD data from http://www.cjd.ed.ac.uk/figures.htm, BSE data from http://europa.eu.int/comm/food/fs/bse/bse32_fr.pdf

tionally, to date, all victims of vCJD have had a specific genotype. vCJD susceptibility may depend on both strands of a person's DNA coding for amino acid 129 in the prion (codon 129 in the DNA) to be a methionine (Met). This genotype accounts for approximately 40 percent of the Caucasian population. No victims of vCJD thus far have had either of the other two possible genotypes: Methionine/Valine (Met/Val) or Val/Val. Models of the vCJD outbreak have presumed that only the population homogenous for Met at codon 129 are susceptible. In fact, the exclusion of Met/Val and Val/Val genotypes may be real or may imply extended incubation periods for these individuals, which also introduces uncertainty into estimates of risk of illness. In kuru, these genotypes were associated with a late age of onset, which is probably associated with an extended incubation period.

Mode of Infection. The method by which a prion infects an animal or person appears to vary for specific TSE diseases. BSE is believed to be transmitted primarily by consumption of infected tissue or organs from an infected animal. The BSE epidemic in the United Kingdom seems to have been initiated by the husbandry practice of feeding affected meat and bone meal (MBM) products (possibly scrapie-infected) to cattle as a protein supplement, and propagated by the recycling of the cattle-infected products back to cattle. Scrapie, which is different from BSE, appears to be transmissible between sheep that cohabitate (lateral transmission), through feeding of rendered MBM-containing infected tissues, through maternal transmission (Onodera et al. 1993; Pattison et al. 1972; Race et al. 1998), and possibly through contaminated vaccines (Consultation on Public Health and Animal TSEs: Epidemiology, Risk and Research Requirements, 1–3 December 1999). CWD is believed to transmit laterally and possibly through maternal transmission, although lateral transmission does not require cohabitation (Williams and Young 1992).

Laboratory studies have suggested a possibility of transmitting BSE in sheep via whole blood transfusions (Houston et al. 2000). However, epidemiological data regarding blood donations from sporadic CJD cases suggest that such iatrogenic transmission of the disease has never occurred. Similarly, blood donations from six victims of vCJD who have died have been traced, and the twelve recipients of those blood products remain healthy. (Consultation on Public Health and Animal TSEs: Epidemiology, Risk and Research Requirements, 1–3 December 1999).

Species Barrier. Interspecies transmission of TSEs is mitigated by a "species barrier" (Hill et al. 2000). This barrier represents the decreased efficiency with which TSEs are passed from one animal to a second animal of a different species, compared with the efficiency with which the TSE is passed among animals of the same species. It has been found that a much greater amount of infective material is necessary to infect an animal from a different species than is needed to pass the dis-

ease to an animal of the same species. In some cases, the species barrier prevents interspecies transmission; in others, it probably lengthens the nonclinical stage of the disease (latency period). For example, scrapie has never been demonstrated epidemiologically to affect human populations.

After a TSE disease crosses a species barrier, infection of the newly afflicted species by intraspecies (secondary) transmission is significantly more facile. It is believed that transmission is, in part, a function of the similarities between the prion proteins of the infected and the exposed animals. This explains why the disease is more readily transmitted after the initial species barrier is crossed; at such a time, the newly infected species is actually presenting its own prion as an infective agent and therefore the similarities are much greater, facilitating secondary transmission (Bartz et al. 1998).

A significant amount of work is currently being conducted to determine the susceptibility of sheep herds to BSE.

REGULATORY ACTIONS. The goal of the U.S. Department of Agriculture (USDA) with respect to Bovine Spongiform Encephalopathy is to prevent the introduction and spread of BSE in the United States and to prevent the introduction of the agent that causes BSE into the human food chain. To that end, the USDA and the Food and Drug Administration (FDA) have created a series of barriers. A major effect is the ban on importation of live ruminants and certain ruminant products from countries either where BSE is known to exist or where, because of widespread risk factors and inadequate surveillance, an undue risk of introducing BSE in the United States is present. Other actions include the ban on importation of rendered animal protein from all BSE-restricted countries, active surveillance of U.S. cattle for BSE, and the ban on the use of most mammalian protein in the manufacture of animal feeds given to cattle and other ruminants. These actions, for both the United States and the European Union, are summarized in the timeline in Figure 31.2.

Import Restrictions. Since 1989, the USDA's Animal and Plant Health Inspection Service (APHIS) has prohibited the import of live cattle and certain cattle products, including rendered protein products, from countries where BSE is known to exist. In 1997, because of concerns about widespread risk factors and inadequate surveillance for BSE in many European countries, these import restrictions were extended to include all the countries in Europe. As of December 7, 2000, APHIS has prohibited all imports of rendered animal protein products, regardless of species, from Europe because of concern that feed intended for cattle may have been cross-contaminated with the BSE agent.

BSE Surveillance in the United States. The United States has had an active surveillance program for

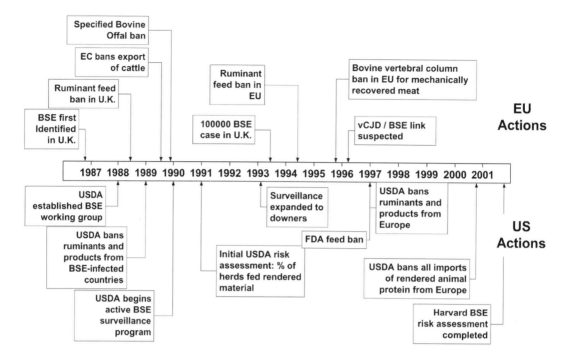

FIG. 31.2—Timeline of major actions taken to prevent spread of BSE and vCJD.

BSE in place since May 1990. Initial risk analyses were performed in 1991 to assess the risk factors associated with BSE. These risk factors, updated as necessary, have been used to identify the portion of the cattle population that would be at the highest risk of contracting BSE, and this at-risk population is where the majority of the active surveillance (brain submission and examination) has occurred. Specifically, this surveillance system samples field cases of cattle exhibiting signs of neurologic disease, cattle condemned at slaughter for neurologic reasons, rabies-negative cattle submitted to public health laboratories, neurologic cases submitted to veterinary diagnostic laboratories and teaching hospitals, and sampling of cattle that are nonambulatory ("downer cattle" or fallen stock).

As of October 31, 2001, 17,981 brains have been examined for TSE in cattle. The rate of surveillance in the United States for each of the last five years has been at least double the amount recommended by the Office International Des Epizooties (OIE). In 2001, the rate of surveillance was ten times that recommended by OIE. The number of samples analyzed each year by the National Veterinary Services Laboratory (NVSL), shown in Figure 31.3 (a smaller percentage of additional samples are analyzed by other labs), illustrates the increase in the sampling rate over time. The USDA plans to test more than 12,500 cattle in the next year. Figure 31.3 also illustrates the increasing focus on downer animals.

Feed Ban. In the European Union, certain materials known to contain the BSE agent in BSE-infected cattle are prohibited for use as human food. These banned materials are referred to as specified risk materials (SRMs) and include the skull (brain and eyes), tonsils, and spinal cord from cattle over twelve months of age and the intestine of all cattle regardless of age. Because most cases of BSE worldwide have been diagnosed in the United Kingdom, the list of materials designated as SRMs in the United Kingdom is more expansive and, in addition to the materials listed in the previous sentence, includes the entire head excluding the tongue but including the brains, eyes, trigeminal ganglia and tonsils; the thymus, spleen, and spinal cord of cattle over six months of age; and the vertebral column, including dorsal root ganglia (DRG), of cattle over thirty months of age. In addition to prohibiting certain materials for human food, the European Union has banned the use of bovine vertebral columns in the production of mechanically recovered meat and has banned the use of slaughter techniques that could introduce large pieces of brain into the circulatory systems of cattle. Generally, the European Union also prohibits healthy cattle over thirty months of age or at-risk cattle over twenty-three months of age from being used as human food unless such cattle have been tested for BSE and the test result does not indicate that they have BSE. However, some European countries (that is, Germany, France, and Italy) prohibit cattle over twenty-four months from being used as human food.

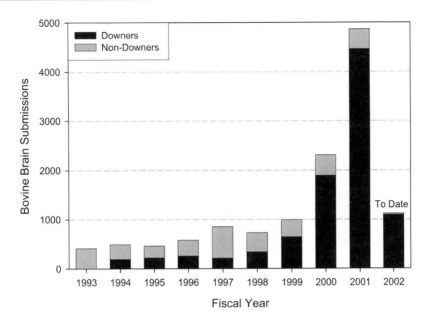

FIG. 31.3—BSE surveillance by APHIS.

The FDA implemented a final rule in 1997 that prohibits the use of most mammalian protein in feeds for ruminant animals. FDA began inspecting all renderers and feed mills for compliance with the feed ban. Firms initially out of compliance are later re-inspected. Firms have generally corrected noncompliance through better implementation of the rule or by eliminating prohibited materials from their operation. Of the 1,251 firms re-inspected to date, 106 (8 percent) were still found to be out of compliance. The current results of the feed ban inspections are shown in Table 31.1.

Other Actions to Prevent the Spread of vCJD in Humans. In August 1999, the FDA issued guidance that banned blood donations from individuals who had lived in the United Kingdom longer than six cumulative months between 1980 and 1996. Based on this guidance, the American Red Cross now restricts blood

Table 31.1—Compliance with the FDA Feed Ban[1]

		Feed Mills			
	Renderers	FDA Licensed	Non FDA Licensed	Other firms[2]	Total[3]
Estimated number of firms	264	1440	6000 – 8000	NA	——
Number of firms reporting	241	1176	4783	4094	9687
Number of firms handling materials prohibited for use in ruminant feed	183	435	1580	621	2653
Compliance problems at most recent inspection[4]					
Products not labeled as required	17 (9%)	47 (11%)	312 (20%)	84 (14%)	431 (16%)
Inadequate systems to prevent co-mingling	8 (4%)	45 (10%)	169 (11%)	25 (4%)	222 (8%)
Inadequate record keeping	3 (2%)	8 (2%)	85 (5%)	29 (5%)	112 (4%)
Overall number out of compliance[5]	25 (14%)	76 (17%)	421 (27%)	110 (18%)	591 (22%)

[1]As of June 12, 2001.
[2]Other firms include ruminant feeders, on-farm mixers, protein blenders, and distributors.
[3]Rows do not sum because firms may be involved in more than one industry segment..
[4]Percentages based on number of facilities handling materials prohibited for use in ruminant feed.
[5]Columns do not sum because facilities may be out of compliance in more than one aspect of the rule.

donations from anyone who has lived in the United Kingdom for longer than three months or who has lived in the rest of Europe in excess of six months during the period 1980 to the present (http://www.redcross.org/services/biomed/blood/supply/tse/bsepolicy.html).

The FDA has also requested that manufacturers discontinue using bovine products from countries known or suspected to harbor BSE in the manufacture of vaccines.

HARVARD RISK ASSESSMENT MODEL. In 1998, the USDA entered into a cooperative agreement with Harvard University's School of Public Health to conduct an analysis and evaluation of the agency's current measures to prevent BSE. The study (Cohen et al. 2001), referred to here as the Harvard risk assessment on BSE, reviewed current scientific information related to BSE and other TSEs, assessed the pathways by which BSE could potentially enter the United States, and identified any additional measures that could be taken to protect human and animal health in the United States. The risk assessment also identified pathways by which humans could potentially be exposed to the BSE agent through bovine materials or products. The full report is available at http://www.aphis.usda.gov/oa/bse/.

Model Description. Given the complexity of BSE transmission and the cattle slaughter, rendering, feed production, and feeding process, the Harvard authors decided to develop a quantitative computer model. The goal of the model was to describe the disposition of infectivity from an animal with BSE and the dynamics of the disease in the U.S. cattle population over time. A quantitative model has several advantages over a qualitative analysis. The model allows various potential scenarios to be evaluated in order to test the effectiveness of current or suggested actions to control BSE.

The impacts of such policies can be simulated, and the resulting spread of the disease from a hypothetical introduction can be evaluated.

The model was dynamic in that it tracked cattle and infectious material over time and took into account changes in conditions resulting from changing regulations, agricultural practices, and BSE prevalence. Model runs generally represented at least a twenty-year simulation period. The simulation used Monte Carlo simulation techniques to characterize the range of possible outcomes resulting from random events (for example, the possibility that a diseased animal is not selected for slaughter until after the amount of infectivity in its body grows substantially, or the possibility that the remains of diseased animal is rendered at a plant that uses technology that is relatively ineffective at reducing the level of infectivity in the processed material). By running the simulation one thousand times using a fixed set of assumptions about the probability of each of these outcomes, the model was able to characterize the range of possible results (for example, total number of cattle infected, potential human exposure to BSE-contaminated food products) consistent with those assumptions. The Harvard analysis also changed many of the assumptions, one at a time, to see how each assumption's uncertainty might potentially influence the results. Each time an assumption was changed, the simulation was again run one thousand times.

The structure of the simulation reflected beef industry practices and government regulations, including the slaughter of cattle, the rendering of their remains, cattle feed production, and the administration of feed to cattle. The major process steps simulated by the model are described next and shown in Figure 31.4.

The model quantified the amount of BSE infectivity in cattle, MBM, and feed in terms of cattle oral infectious dose$_{50}$ (ID$_{50}$). An ID$_{50}$ is defined to be the amount

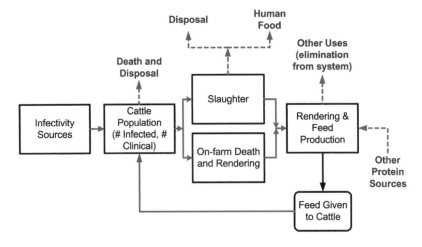

FIG. 31.4—Process flow diagram for the Harvard BSE model.

of infectious agent that would produce infection with 50 percent probability when administered to a bovine aged one to two months (older animals are generally far less susceptible to infection, a tendency the model also accounts for). The model is capable of describing the introduction of infectivity into the U.S. cattle population via the import of infected cattle, the introduction of infected meal from an exogenous source, or due to spontaneous development of disease. Following the introduction of infectivity, the simulation can model its spread via the recycling of bovine protein, bovine blood meal, and as the result of maternal transmission of disease to calves. The model keeps track of and reports the number of animals infected via each of these pathways, the flow of ID_{50}s through the slaughter, rendering, feed production, and feed processes, and how much various tissues contribute to the potential exposure of humans to BSE-contaminated products. Because no accepted dose-response model for vCJD exists, the model did not attempt to estimate the number of human illnesses that could arise under the various scenarios. Notably, European authorities suggest that humans may be ten to 100,000 times less susceptible to disease from bovine sources than cattle as a consequence of the species barrier (SSC 1999; SSC 2000).

Bovine Materials in Which the BSE Agent Has Been Detected. Each organ that might harbor infectivity was tracked separately during slaughter and rendering. The BSE agent has been detected only in certain tissues of cattle infected with BSE and, except in the distal ileum (gut), has been detected in those tissues only when the cattle are over twenty-four months of age, and most often in cattle over thirty-two months of age (Wells

1998). In cattle naturally infected with BSE, the BSE agent has been found only in brain tissue, spinal cord, and retina of the eye. In experiments, the BSE agent has also been detected in the brain, spinal cord, distal ileum, DRG, trigeminal ganglia, and, possibly, the bone marrow of deliberately infected cattle. Some tissues, such as brain and spinal cord, contain higher levels of BSE infectivity than others. The BSE agent has never been detected in the muscle tissue of BSE-infected cattle, regardless of the age of the animal; however, murine prions have recently been observed in the muscle tissue of infected mice (Bosque et al. 2002).

The level of infectious agent in each organ depends on the time since exposure. The Harvard model can vary both the organs and relative infectivity with age. The list of organs that harbor infectivity and their relative proportion of infectivity in the base case for mature cattle are shown in Figure 31.5. Other organs tracked by the model include the heart, lung, liver, kidney, ileum, muscle, and bone, as well as advanced meat recovery (AMR) meat.

As with the relative distribution of infectivity, the total amount of infectivity varies with time since exposure, as shown in Figure 31.6. Note that a log scale of infectivity is used to better illustrate the changes during the early onset of the disease.

The model assumes that the susceptibility of cattle to BSE infection varies with the age of the animal. In particular, the Harvard analysis assumed that susceptibility peaks at four months of age and declines exponentially thereafter.

Slaughtering and Rendering. Based on existing reporting data, animals are probabilistically selected for slaughter by age and type. The modeled slaughter-

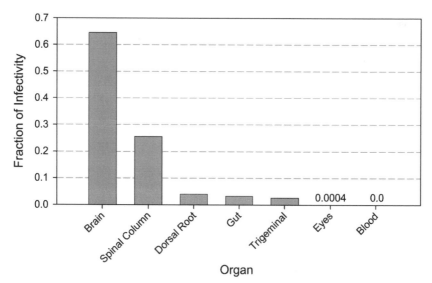

FIG. 31.5—Relative levels of BSE ID_{50}s for organs harboring infectivity at maturity as used in Harvard risk assessment model.

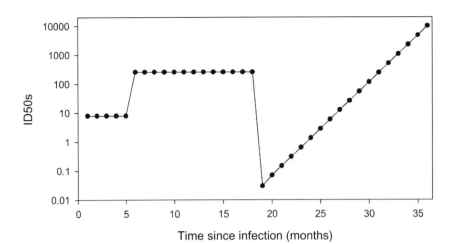

FIG. 31.6—Total amount of BSE infectivity per animal during the course of the disease as used in Harvard risk assessment model.

ing process is illustrated in Figure 31.7. "Out" refers to the removal of ID_{50}s from the bovine or human food chain. At the slaughter facility, the first step is ante-mortem inspection. Although this inspection cannot detect BSE except in the late clinical stages, some percentage of cattle are removed from the system for other reasons. The animals are then stunned and bled. The model can simulate the impact of different stunning technologies on the generation of emboli. These differences are potentially of interest because air-injected pneumatic stunning has been shown to introduce CNS emboli into other organs, whereas the captive bolt technology is much less likely to cause such contamination. After stunning, the head is removed and the carcass dressed and split longitudinally. Normally, after the vertebral column is split, the spinal cord is removed; however, the possibility of a missplit exists, which could leave spinal cord in the vertebral column. The carcass is then processed for human consumption. At some plants, AMR systems are used to recover residual meat from bones. It is at this step that ID_{50}s from miss-split spinal cord and dorsal root ganglia could enter the human food chain. In addition, some small percentage of possibly infected organs are consumed directly as food. The base model case (that is, the "best estimate" characterization of status quo conditions and practices developed as part of the Harvard analysis) assumes that 1 percent of brains directly enter the human food chain.

The remaining cattle tissues are then rendered. Historically, some of the resulting meat and bone meal was fed to cattle. This is now believed to be a major route for the spread of BSE and is banned in the United States and Europe. Under the current feed ban, three types of rendering plants exist: those that process material that is prohibited from being fed to ruminants, those that process material that may be fed to ruminants, and those that process both. Cross-contamina-

tion or generation of mislabeled MBM can occur in the latter type and facilitate future cattle infections from these food sources. Finally, prohibited feed may be inappropriately misfed to ruminants on the farm. These scenarios are the most likely to contribute to the spread of any BSE, after being introduced to the United States, and the model simulates each of these possibilities probabilistically.

MODEL RESULTS AND SENSITIVITY ANALYSIS. The model identified the following three pathways or practices as those that could contribute the most to the spread of BSE-infected tissues in the human food supply:

- Noncompliance with the FDA feed ban, including misfeeding on the farm and the mislabeling of feed and feed products prohibited for consumption by cattle.
- Rendering of downer cattle, including cattle that die on the farm.
- Inclusion of high-risk tissue, such as brain and spinal cord, in edible products.

Critical measures already implemented that reduce the spread of BSE include the ban on the import of live animals, ruminant meat, and bone meal from BSE infected countries, and the feed ban implemented by the FDA. Because the feed ban breaks the cycle of exposure to BSE-contaminated material, it greatly reduces the opportunity for BSE to spread from a sick animal back to other cattle.

Using the best data available, the model found that a likelihood exists of BSE spreading substantially in the United States. Using the base case scenario and assuming that ten infected cattle are introduced into the

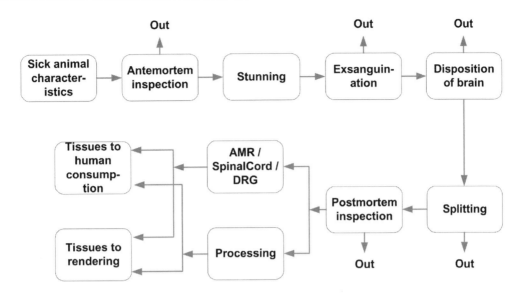

FIG. 31.7—Process flow diagram for cattle slaughter.

United States, the model predicted that on average only three new cases of BSE occur and that the disease is virtually certain to be eliminated from the country within twenty years of its introduction. The new cases of BSE arise primarily from the lack of compliance with the FDA feed ban.

Although some infectious material could potentially enter the human food chain, the level is small: approximately thirty-five cattle oral ID_{50}s. Human exposure to the BSE agent would most likely occur through the consumption of bovine brain and spinal cord and through the consumption of boneless beef product produced using advanced AMR systems. Other sources of potential human exposure to the BSE agent identified by the risk assessment include bone-in beef and distal ileum, although to a much lesser extent. The risk assessment also found that other materials, such as the blood, eyes, cheek meat, and organs contaminated with CNS tissue account for less than one percent of potential human exposure to the BSE agent.

Based on the model evaluation, two potential measures were identified to further reduce the already low likelihood that the disease could spread among cattle or contaminate human food. The first is a prohibition against rendering of animals that die on the farm. The second is a U.K.-style ban on specified risk material (such as spinal cords, brain) from both human food and animal feed. These options reduce the predicted average number of BSE cases by 77 percent and 80 percent, respectively.

Harvard also evaluated the possible historical introduction of BSE into the United States from the importation of British cattle between January 1981 and July 1989, prior to the importation ban implemented in July 1989. In reality, 496 such cattle entered the country;

however, these animals were primarily intended as breeding stock and not as beef or dairy production animals. Although the possibility exists that these animals might have introduced BSE into this country, the model determined that this risk was small. It is worth noting that APHIS has conducted a traceback effort to locate each of the animals, and many were disposed of in a manner that posed no risk to humans or other animals, suggesting that the disease likely did not spread even if was present in any of the imported animals.

Future Actions. The USDA will undertake a series of actions based on the BSE risk assessment. The risk assessment will undergo a peer-review by outside experts. The USDA will increase the annual number of cattle tested for BSE to more than 12,500—up from 5,000 in FY 2000. The USDA will propose prohibition of air-injected pneumatic stunning. The USDA will consider disposal requirements for dead and downer cattle on the farm. Finally, using the risk assessment model developed by Harvard, a series of scenarios will be simulated to evaluate the potential impact of possible regulatory actions that the USDA may implement to further minimize human exposure to materials that could potentially contain the BSE agent. These scenarios are likely to include prohibiting the use of brain and spinal cord from specified cattle in human food; prohibiting the use of central nervous system tissue in boneless beef products, including meat from advanced meat recovery systems; and prohibiting the use of the vertebral column from certain categories of cattle, including downed animals, in the production of meat from advanced meat recovery systems. These scenarios may vary the implementation of the rule by the age of the cattle.

SUMMARY AND CONCLUSIONS. No case of BSE or vCJD has ever been identified in the United States. The Harvard risk assessment has found that the U.S. policies currently in place would dramatically reduce the potential spread of the disease should it be introduced into domestic herds. The approach also demonstrated the value of a quantitative simulation approach to risk assessment.

REFERENCES

Bartz, J.C., R.F. Marsh, D.I. McKenzie, and J.M. Aiken. 1998. The host range of chronic wasting disease is altered on passage in ferrets. *Virology* 251:297–301.

Bosque, P.J., C. Ryou, G. Telling, D. Peretz, G. Legname, S.J. DeArmond, S.B. Prusiner. 2002. Prions in skeletal muscle. *PNAS* 99(6):3812–3817.

Cohen, J.T., K. Duggar, G.M. Gray, S. Kreindel, H. Abdelrahman, T. HabteMariam, D. Oryang, and B. Tameru. 2001. Evaluation of the potential for Bovine Spongiform Encephalopathy in the United States. Final Report to the USDA.

Cousens, S.N., E. Vynnycky, M. Zeidler, R.G. Will, and P.G. Smith. 1997. *Nature* 385:197.

Ghani, A.C., N.M. Ferguson, C.A. Donnelly, and R.M. Anderson. 2000. *Nature* 406:583.

Hill, A.F., S. Joiner, J. Linehan, M. Besbruslais, P.L. Lantos, and J. Collinge 2000. Species-barrier-independent prion replication in apparently resistant species. *Proc. Natl. Acad. Sci. USA* 97(18):10248–53.

Houston, F., J.D. Foster, A. Chong, N. Hunter, and C.J. Bostock. 2000. Transmission of BSE by blood transfusion in sheep. *Lancet* 356:999–1000.

Onodera, T., T. Ikeda, Y. Muramatsu, and M. Shinagawa. 1993. Isolation of the scrapie agent from the placenta of sheep with natural scrapie in Japan. *Microbiol. Immunol.* 37(4):311–316.

Pan, K.M., M. Baldwin, J. Nguyen, M. Gasset, A. Serban, D. Groth, I. Mehlhorn, Z. Huang, R.J. Fletterick, F.E. Cohen, and S.B. Prusiner. 1993. *Proc. Natl. Acad. Sci. USA* 90:10962–10966.

Pattison, I.H., M.N. Hoare, J.N. Jebbett, and W. Watson. 1972. Spread of scrapie to sheep and goats by oral dosing with foetal membranes from scrapie-affected sheep. *Vet Rec.* 90:465–468.

Pergami, P., H. Jaffe, and J. Safar. 1996. *Anal. Biochem.* 236:63–73.

Peters, J., J.M. Miller, A.L. Jenny, T.L. Peterson, and K.P. Carmichael. 2000. Immunohistochemical diagnosis of chronic wasting disease in preclinically affected elk from a captive herd. *J. Vet. Diagn. Invest.* 12:579–582.

Prusiner, S.B. 1998. Prions. *Proc. Natl. Acad. Sci. USA* 95:13363–13383.

Race, R., A. Jenny, and D. Sutton. 1998. Scrapie indectivity and proteinase K-resistant prion protein in sheep planceta, brain, spleen and lymph node: Implications for transmission and antemortem diagnosis. *J. Infect. Dis.* 178:949–953.

SSC. 1999. "Opinion of the Scientific Steering Committee on the Human Exposure Risk (HER) via food with respect to BSE—Adopted on 10 December 1999." Available at: http://europa.eu.int/comm/food/fs/sc/ssc/out67_en.html.

SSC. 2000. "Opinion—Oral exposure of Humans to the BSE agent: infective dose and species barrier adopted by the SSC at its meeting of 13–14 April 2000 following a public consultation via Internet between 6 and 27 March 2000." Available at: http://europa.eu.int/comm/food/fs/sc/ssc/out79_en.pdf.

Stahl, N., M.A. Baldwin, D.B. Teplow, L. Hood, B.W. Gibson, A.L. Burlingame, and S.B. Prusiner. 1993. *Biochemistry* 32:1991–2002.

Taylor, D.M. 1993. Inactivation of SE agents. *Brit. Med. Bull.* 49(4):810–821.

Taylor, D.M. 1996. Inactivation studies on BSE agent. *Brit. Food J.* 98(11):36–39.

Valleron, A.J., R.Y. Boelle, R. Will, and J.Y. Cesbron. 2001 Estimation of epidemic size and incubation time based on age characteristics of vCJD in the United Kingdom. *Science* 294:1726.

Wells, G.A.H., S.A.C. Hawkins, R.B. Green, A.R. Austin, I. Dexter, Y.I. Spencer, M.J. Chaplin, M.J. Stack, and M. Dawson. 1998. Preliminary observations on the pathogenesis of experimental bovine spongiform encephalopathy (BSE): an update. *Vet. Rec.* 1998:103–106.

Williams, E.S., and S. Young. 1992. Spongiform encephalopathies in Cervidae. *Rev. Sci. Tech. Off. Int. Epiz.* 11:551–567.

32

A RISK ASSESSMENT OF *ESCHERICHIA COLI* O157:H7 IN GROUND BEEF

E. EBEL, W. SCHLOSSER, K. ORLOSKI, J. KAUSE, T. ROBERTS, C. NARROD, S. MALCOLM, M. COLEMAN, AND M. POWELL

INTRODUCTION. The U.S. Department of Agriculture's Food Safety and Inspection Service (FSIS) has taken several regulatory measures to protect public health against *E. coli* O157:H7. In August 1994, FSIS declared ground beef contaminated with *E. coli* O157:H7 to be adulterated (that is, to contain a substance that may be injurious to public health). Under this policy, raw, chopped, or ground beef products that contained *E. coli* O157:H7 required further processing to destroy *E. coli* O157:H7. On October 17, 1994, FSIS initiated a microbiological testing program for *E. coli* O157:H7 in raw ground beef in meat plants and retail stores. The initial testing program was established and designed to test approximately five thousand raw ground beef samples, 50 percent from federally inspected plants and 50 percent from retail stores. In 1998, the sample size was increased from 25 grams to 325 grams to increase the likelihood of detecting this organism (FSIS Directive 10,010.1). In September 1999, FSIS modified laboratory testing for *E. coli* O157:H7 to include a selective capture step based on immunomagnetic separation (FSIS 2001).

Because of growing concerns, FSIS has conducted a risk assessment of *E. coli* O157:H7 to produce a baseline risk estimate that reflects current practices along the farm-to-table continuum and assesses the likelihood of human morbidity and mortality associated with the consumption of ground beef contaminated with *E. coli* O157:H7 in the United States. This chapter provides an overview of the risk assessment as well as results, limitations, and implications of the findings.

FSIS has based its risk assessment of *E. coli* O157:H7 in ground beef on a comprehensive review of the available literature and data. The baseline risk assessment follows the Codex Alimentarius Commission framework for microbial risk assessments with four primary components: (1) hazard identification, (2) exposure assessment, (3) hazard characterization, and (4) risk characterization (Codex 1999). Hazard identification provides information on the epidemiology, pathogenesis, transmission, and ecology of *E. coli* O157:H7. Exposure assessment describes the occurrence and number of *E. coli* O157:H7 bacteria from farm to table. Hazard characterization discusses the dose-response relationship for *E. coli* O157:H7. Risk characterization integrates the exposure assessment and dose-response relationship to estimate the risk of illness from *E. coli* O157:H7 in ground beef. The result

of the risk assessment is a computer model developed in Excel (Microsoft Corporation, Redmond, WA) using the add-in @Risk (Palisades Corporation, Newfield, NY). The model can be refined and updated for use in future risk assessments for ground beef products as new information and data become available. This risk assessment may assist FSIS in reviewing and refining its risk reduction strategy for *E. coli* O157:H7 in ground beef.

Hazard Identification. *E. coli* O157:H7 is a public health concern worldwide, causing reported outbreaks in the United States, Japan, Canada, Scotland, and Argentina (Michino et al. 1998; Spika et al. 1998; Ahmed and Donaghy 1998; Lopez et al. 1998). An estimated sixty-two thousand cases of symptomatic *E. coli* O157:H7 infections occur annually in the United States from foodborne exposures, resulting in an estimated 1,800 hospitalizations and fifty-two deaths (Mead et al. 1999). As many as three thousand cases may develop hemolytic uremic syndrome annually (Mead et al. 1999). Surveillance data indicate that the highest incidence of illness from *E. coli* O157:H7 occurs in children under five years of age (CDC 2000a).

In the United States, outbreaks of human illnesses caused by foodborne *E. coli* O157:H7 infection have often been linked to contaminated cattle-derived products such as ground beef or milk. In the first nationwide case-control study of sporadic *E. coli* O157:H7 infection, consumption of undercooked ground beef (described as "pink in the middle") was the only dietary factor independently associated with diarrhea in multivariate analysis (Slutsker et al. 1998). The population-attributable risk for this behavior was 34 percent. Other studies of sporadic cases of *E. coli* O157:H7 infection have also found an association between human *E. coli* O157:H7 illness and consumption of ground beef (Le Saux et al. 1993; Mead et al. 1997; MacDonald et al. 1988; Kassenborg et al. 2001). Interpretation of the evidence from outbreaks and sporadic infections indicate that consumption of ground beef is an important source of *E. coli* O157:H7 exposure.

Human infection with *E. coli* O157:H7 results in a mild to severe spectrum that includes asymptomatic infection, nonbloody diarrhea, bloody diarrhea, hemolytic-uremic syndrome (HUS), and thrombotic thrombocytopenic purpura (TTP). HUS is a condition that includes destruction of red blood cells, problems with blood clotting, and kidney failure. TTP is a condition that is similar to HUS but usually occurs over a longer period of time and may cause changes in mental status. Between 2 percent and 20 percent of patients develop HUS following *E. coli* O157:H7 infection; 3 percent to 5 percent of these patients die, and approximately 3 percent to 5 percent develop end-stage renal disease later in life (Mead and Griffin 1998). A recent study found that antiobiotic treatment of children may have contributed to the development of HUS in children infected with *E. coli* O157:H7 (Wong et al. 2000).

EXPOSURE ASSESSMENT. The purpose of the exposure assessment is to estimate the occurrence and number of *E. coli* O157:H7 organisms in servings of ground beef by modeling the processes involved from production of cattle on the farm to consumption of ground beef. The exposure assessment is divided into three modules: production, slaughter, and preparation. The production module estimates the herd and within-herd prevalence of *E. coli* O157:H7 infection in breeding cattle and feedlot cattle. Results from the production module are inputs for the slaughter module, which estimates the occurrence and number of *E. coli* O157:H7 organisms on carcasses and in beef trim. Results from the slaughter module are inputs for the preparation module, which estimates the occurrence and number of *E. coli* O157:H7 organisms in ground beef servings.

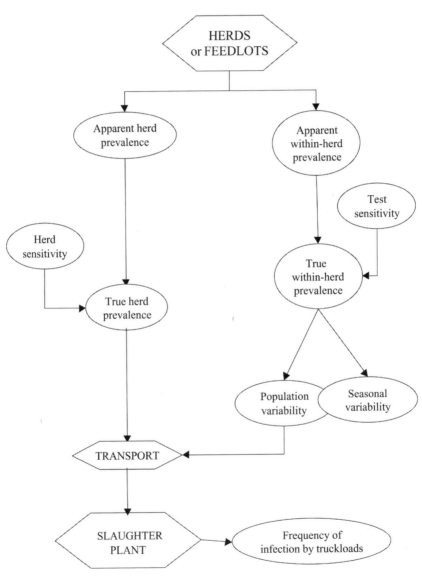

FIG. 32.1—Production module flowchart.

PRODUCTION. The production module estimates the prevalence of *E. coli* O157:H7-infected cattle entering U.S. slaughter plants (Figure 32.1). It models culled breeding cattle (cows and bulls) and feedlot cattle (steers and heifers) from their points of origin through transit to the slaughter plant. Culled breeding cattle and feedlot cattle are modeled separately in this risk assessment because the slaughter, processing, and distribution of meat from these types of cattle are different. It is assumed that imported beef originates from countries whose *E. coli* O157:H7 epidemiology is similar to that of the United States.

The prevalence of *E. coli* O157:H7-infected breeding herds and feedlots is estimated in the production module (Table 32.1). Herd prevalence is the proportion of all breeding herds that contain one or more infected cattle. Feedlot prevalence is defined similarly, but the reference population is U.S. feedlots. Herd sensitivity is the proportion of truly infected herds that are detected, and depends on the number of fecal samples collected per herd as well as the proportion of cattle in the herd shedding *E. coli* O157:H7 in their feces (Martin et al. 1992).

Most evidence on the occurrence and distribution of this organism in U.S. livestock comes from sampling cattle feces on farms and feedlots. National surveys (Dargatz et al. 1997; Garber et al. 1999) have not shown any geographic clustering of *E. coli* O157:H7 infection in cattle. Therefore, sampling data from multiple studies, representing different U.S. regions, are pooled to estimate national prevalence. Each study's results are adjusted for herd sensitivity before pooling of the results. The results suggest that we are 90 percent confident that the prevalence of affected breeding herds is between 55 percent and 72 percent. We are 90 percent confident that the prevalence of affected feedlots is between 78 percent and 97 percent.

The production module also estimates the proportion of *E. coli* O157:H7-infected cattle within breeding herds (within-herd prevalence) and feedlots (within-feedlot prevalence). Among the population of infected herds, within-herd prevalence is a variability distribu-

tion. Curves are fit to evidence from breeding herds (Hancock et al. 1997b; Garber et al. 1999) and feedlots (Dargatz et al. 1997) to model this herd-to-herd, and feedlot-to-feedlot, variability.

Within-herd and within-feedlot prevalence also varies systematically by season. *E. coli* O157:H7 fecal sampling evidence suggests that the greatest prevalence in cattle occurs during the summer months from June to September. It is thought that a summer rise in prevalence results from on-farm environmental conditions that permit increased transmission of *E. coli* O157:H7 among cattle (Hancock et al. 2001). For example, if feed and water are important in the transmission of *E. coli* O157:H7 to cattle within a herd, then summer ambient temperatures might induce substantial growth of *E. coli* O157:H7 in the feed and water that cattle ingest, thereby resulting in more infected cattle. A Canadian study found at least a 4-fold difference in *E. coli* O157:H7 fecal prevalence between samples collected in the winter and summer (Van Donkersgoed et al. 1999). The highest fecal prevalence was observed between June and August. Results from cross-sectional or cohort studies in the United States (Garber et al. 1999; Hancock et al. 1997a), the Netherlands (Heuvelink et al. 1998), and Italy (Bonardi et al. 1999) have shown similar seasonal patterns. Yet a yearlong study of ten cow-calf herds did not demonstrate any seasonal difference in prevalence (Sargeant et al. 2000).

Evidence concerning the apparent within-herd prevalence reported for each study was adjusted by test sensitivity (that is, the probability of detecting infected cattle). Estimation of test sensitivity for each study considered the quantity of fecal sample analyzed and the laboratory methods used (Sanderson et al. 1995). After estimating true within-herd prevalence (for example, true prevalence = apparent prevalence ÷ test sensitivity) for each study, seasonal population averages were estimated by pooling the studies that were conducted during June–September (considered hereafter as the high prevalence season) and during October–May (considered hereafter as the low prevalence

Table 32.1—Uncertainty about Low- and High-Prevalence Seasons' Estimated Average Cattle Prevalence

Result	5th percentile	Mean	95th percentile
Breeding herd prevalence[1]	55%	63%	72%
Feedlot prevalence[2]	78%	88%	97%
Low-prevalence season (October to May)			
Average within-herd prevalence[3]	2%	3%	4%
Average within-feedlot prevalence[4]	6%	9%	14%
High-prevalence season (June to September)			
Average within-herd prevalence	3%	4%	5%
Average within-feedlot prevalence	21%	22%	24%

[1] Prevalence based on: Hancock et al. 1997a; Hancock et al. 1997b; Rice et al. 1997; Hancock et al. 1998; Garber et al. 1999; Lagreid et al. 1999; Sargeant et al. 2000.
[2] Prevalence based on: Dargatz et al. 1997; Hancock et al. 1998; Smith 1999; Elder et al. 2000.
[3] Prevalence based on: Hancock et al. 1994; Besser et al. 1997; Rice et al. 1997; Garber et al. 1999; Sargeant et al. 2000.
[4] Prevalence based on: Dargatz et al. 1997; Hancock et al. 1998; Hancock et al. 1999; Smith 1999; Elder et al. 2000.

season). Table 32.1 shows the estimated uncertainty regarding average within-herd and within-feedlot prevalence in the United States. Generally, these results suggest that *E. coli* O157:H7 prevalence is higher for feedlot cattle than for breeding cattle and that for both types of cattle, prevalence is higher from June through September.

The production module simulates truckloads of cattle entering the slaughter process. Each truckload was assumed to contain forty cattle. The output of this module is the prevalence of infection within truckloads, which serves as the first input to the slaughter module. Prevalence at slaughter depends on both herd (or feedlot) prevalence and within-herd (or within-feedlot) prevalence. For breeding cattle, about 45 percent of truckloads are predicted to have no infected cattle during the low-prevalence season and 35 percent of truckloads are predicted to have no infected cattle during the high-prevalence season. For feedlot cattle, the frequency of truckloads with no infected cattle is about 32 percent in the low-prevalence season and 20 percent in the high-prevalence season (Figure 32.2).

Slaughter. The slaughter module estimates the occurrence and extent of *E. coli* O157:H7 contamination as live cattle transition to carcasses, then to meat trim, and finally to aggregates of meat trim in 2,000-pound combo bins (or 60-pound trim boxes) destined for ground beef production (Figure 32.3).

Culled breeding cattle and feedlot cattle are modeled separately. Slaughtering operations for the high- and low-prevalence seasons are also modeled separately.

The slaughter module includes seven steps: (1) arrival of live cattle at slaughter plant, (2) dehiding, (3) decontamination following dehiding, (4) evisceration, (5) final washing, (6) chilling, and (7) carcass fabrication (that is, creation of trim). Although there are other steps that are normally part of the slaughter process (for example, stunning, carcass splitting), these are not explicitly modeled. Generally, these other steps are incorporated into the seven steps of the model. This model assumes that contamination with *E. coli* O157:H7 can increase or decrease at different steps of slaughter (Galland 1997).

Truckloads of cattle are modeled in lots (step 1). A slaughter lot is defined as the number of cattle needed to fill a 2,000-pound combo bin of beef trim. Because slaughter lots may consist of multiple truckloads, the prevalence of *E. coli* O157:H7 infection in each truck is estimated in this step, and the total number of infected cattle in the lot is calculated.

Dehiding (step 2) is the transition from live cattle to carcasses. The process of removing the hide creates the first opportunity for surface contamination of the carcass with *E. coli* O157:H7 and other pathogenic and nonpathogenic microbes. The number of *E. coli* O157:H7 organisms that initially contaminate a carcass is variable and depends on the average density of *E. coli* O157:H7 per cm^2 of carcass surface area and the total surface area of a carcass that is contaminated. Limited data are available regarding the average density of contamination (FSIS 1994), and no research is available defining the size of contaminated surface area. Therefore, the initial amount of contamination on a carcass is uncertain, but may range from one

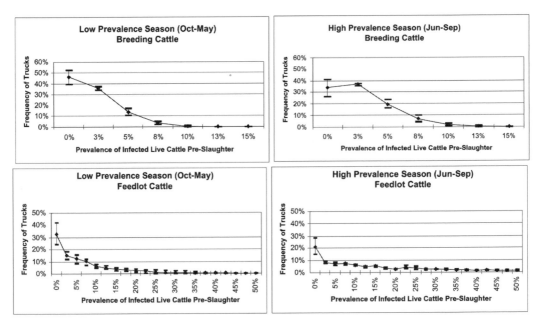

FIG. 32.2—Comparison of seasonal distributions for prevalence of infected cattle within truckloads of breeding and feedlot cattle sent to slaughter. Error bars show the 5th and 95th percentiles of uncertainty about frequency of trucks at each prevalence level.

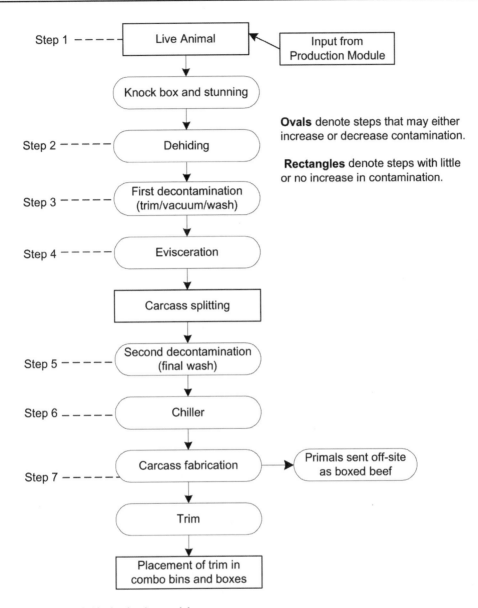

FIG. 32.3—Steps modeled in the slaughter module.

organism to more than three thousand organisms.

Contamination introduced during dehiding can be reduced by the first set of decontamination procedures (step 3). During decontamination, trimming, vacuuming, and/or washing of the carcass surface can reduce the number of organisms on contaminated carcass surfaces to a variable extent (Sheridan et al. 1992; Prasai et al. 1995; Dorsa et al. 1997; Galland 1997; Smeltzer et al. 1998; Bacon et al. 1999; Gill 1999). The available evidence suggests that the effectiveness of these decontamination procedures ranges from 0 to 1.2 logs (that is, 0 percent to 94 percent reduction in total contamination).

Evisceration (step 4) is another opportunity for contamination to be introduced. If any part of the gastrointestinal tract is perforated during the evisceration procedure, *E. coli* O157:H7 contamination of muscle tissue can occur. This is assumed to occur infrequently (that is, 0 percent to 2 percent of carcasses).

Final washing (step 5) follows evisceration and represents a second decontamination procedure applied to carcasses during slaughter. During final washing, carcasses are washed or steam pasteurized, and carcass contamination can be substantially reduced (Nutsch et al, 1997, 1998; Phebus et al. 1997; Gill 1998; Kastner 1998). The effectiveness of this decontamination step is variable. The available evidence suggests this effectiveness may range from 0 to 2.5 logs (that is, 0 percent to 99.7 percent reduction in total contamination).

Following final washing, the carcasses move to the chiller (step 6), where *E. coli* O157:H7 contamination may again increase or decrease depending on the effectiveness of carcass cooling (Gill and Bryant 1997). After chilling, the carcasses are separated into smaller units that are used to produce whole-muscle cuts of beef (step 7, fabrication). A by-product of this process for feedlot cattle is beef trim. Because carcasses from breeding cattle produce less valuable, whole-muscle cuts, greater proportions of these carcasses are converted to beef trim. The boneless meat trim from one animal is distributed based on fat content into multiple 2,000-pound combo bins or 60-pound boxes, where it is mixed with trim from other cattle. Fabrication can also result in new or additional contamination through cross-contamination of work surfaces.

For combo bins generated from the slaughter of breeding cattle, the model predicts that between 3 percent and 12 percent of combo bins contain one or more *E. coli* O157:H7 bacteria during the low-prevalence season. The range reflects uncertainties in the model inputs. This range is between 3 percent and 15 percent during the high-prevalence season. For combo bins generated from the slaughter of feedlot cattle, the model predicts that between 3 percent and 45 percent of combo bins contain one or more *E. coli* O157:H7 bacteria during the low-prevalence season. This range is between 17 percent and 58 percent during the high-prevalence season. Differences in contamination of combo bins between types of carcasses and season are largely reflective of differences in incoming live cattle prevalence. Because boxes of beef trim are modeled from combo bins of beef trim, differences in contamination by season and cattle type reflect those for combo bins, although the number of *E. coli* O157:H7 in 60-pound boxes is understandably lower than that for 2,000-pound combo bins.

Preparation. The preparation module simulates the annual consumption of approximately 18.2 billion ground beef servings. It considers the effects of storage and cooking on the amount of *E. coli* O157:H7 in contaminated servings.

Ground beef is consumed in many forms. Typical forms are hamburger patties, ground beef as a formed major ingredient (for examples, meatballs and meat loaf), and ground beef as a granulated ingredient (for example, ground beef in spaghetti sauce). The model focuses on the first two forms. Because granulated ground beef has a relatively large surface area compared to volume, the effect of cooking on this product is considered to be similar to intact beef products (for example, steaks), which are generally considered to be safe after cooking.

The preparation module consists of six steps (Figure 32.4). Five of these steps explicitly model growth,

decline, or dispersion of *E. coli* O157:H7 contamination: (1) grinding, (2) storage during processing by the retailer or distributor, (3) transportation home or to hotels, restaurants, and institutions, (4) storage at home and at hotels, restaurants, and institutions, and (5) cooking. Step 6 models the amount of ground beef consumed, which varies depending on the age of the consumer and the location where the meal was consumed (that is, home or away from home).

Grinding (step 1) transforms combo bins and boxes of meat trim into ground beef. Combo bins are processed in large commercial facilities. Boxes are typically processed in smaller settings such as grocery stores. Multiple combo bins or boxes are combined, mixed, and extruded to produce finished ground beef with a specific fat content. Because breeding and feedlot beef trim generally consist of different percentages of lean beef, combo bins generated from either type of cattle may be mixed together in a grinder. Although the numbers of *E. coli* O157:H7 in a grinder do not increase during the grinding process, contamination from a single combo bin or box can be dispersed during grinding to contaminate many individual ground beef servings.

Uncertainty about the prevalence of contaminated grinder loads was best estimated from FSIS ground beef sampling data (Table 32.2). These data were useful because they represented an entire year of sampling, were based on consistent sampling and testing methods, and resulted from improved culture methods. Assuming that the FSIS data fit a beta distribution (Vose 1996), we predict that the annual apparent prevalence is 0.52 percent with 90percent confidence bounds of 0.36 percent and 0.71 percent. The seasonal results demonstrate that there were significantly more positive samples in the high-prevalence season than in the low-prevalence season.

To calibrate the exposure assessment with these data, sampling from modeled grinder loads was simulated to mimic the FSIS methods by assuming 325-gram samples and the current FSIS culture methods. For each simulation of the model, the prevalence of positive samples from the simulated grinders was compared with the FSIS results. Model simulations that did not reasonably approximate this empiric evidence were considered infeasible and were not included in the model's results. The calibrated model predicts that between 40 percent (5th percentile) and 88 percent (95th percentile) of feasible grinder loads contained one or more *E. coli* O157:H7 in the low-prevalence season. In the high-prevalence season, between 61 percent (5th percentile) and 94 percent (95th percentile) of grinder loads contained one or more *E. coli* O157:H7.

Storage conditions at retail or wholesale (step 2) provide an opportunity for *E. coli* O157:H7 levels to increase as a result of time and temperature abuse (Walls and Scott 1996) or to decrease as a result of the effects of freezing ground beef (Sage and Ingham 1998; Ansay et al. 1999). Step 3 models the effects of

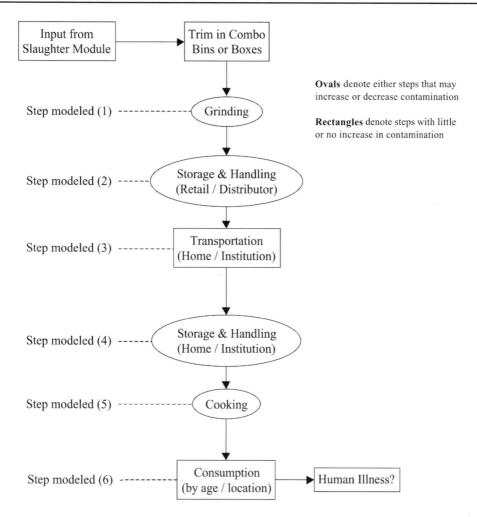

FIG. 32.4—Steps modeled in the preparation module.

Table 32.2—FSIS Ground Beef Sampling Results for Fiscal Year 2000 (October 1999 through September 2000)*

Season	Positive	Tested	5th Percentile	Mean	95th Percentile
Low prevalence (October–May)	10	3,139	0.20%	0.35%	0.54%
High prevalence (June–September)	13	1,447	0.59%	0.97%	1.42%
Annual	23	4,586	0.36%	0.52%	0.71%

*Based on FSIS-USDA sampling surveillance results after removing follow-up testing data conducted at processors following positive tests. These 325-gram samples were collected in federally inspected ground beef processing plants.

time and temperature during transportation on the level of *E. coli* O157:H7 after the ground beef is purchased. Step 4 models the storage of ground beef in the freezer or refrigerator prior to its preparation and consumption. This step provides another opportunity for increases or decreases in *E. coli* O157:H7 contamination in ground

beef servings. During steps 2–4, growth of *E. coli* O157:H7 is modeled based on predictive microbiology algorithms (Marks et al. 1998). These algorithms depend on time and temperature inputs whose distributions are generally estimated from consumer behavior surveys (FDA 1999).

Ground beef is usually cooked prior to consumption (step 5). Cooking can significantly reduce *E. coli* O157:H7 in ground beef servings. The model uses linear regressions estimated from a laboratory study (Jackson et al. 1996) to estimate the level of reduction in *E. coli* O157:H7 contamination in ground beef servings as a function of final internal product temperature. Temperature data from a commercial food temperature database were used to describe the variability in cooking practices (FDA 1999).

Step 6 models consumption of *E. coli* O157:H7-contaminated ground beef servings, taking into consideration the number of servings and serving size by age of the consumer and location where the meals were consumed. Survey data (ARS 1998) were used to characterize the U.S. population's ground beef consumption patterns.

The primary output from the preparation module is a variability distribution for the number of *E. coli* O157:H7 in cooked ground beef servings generated during low- and high-prevalence seasons. Figure 32.5 shows the results of one hundred simulations for exposure during the low- and high-prevalence seasons. Each simulation differs because of uncertainty about model inputs.

Very few ground beef servings are expected to contain *E. coli* O157:H7 organisms following cooking (Figure 32.5). The model predicts that the true fraction of servings contaminated after cooking is between one in thirty-six thousand and one in eight thousand in the low-prevalence season, and between one in fifteen thousand and one in three thousand in the high-prevalence season. Considerable uncertainty exists regarding the frequency of cooked ground beef servings that have one or more *E. coli* O157:H7 present.

HAZARD CHARACTERIZATION. Hazard characterization quantifies the nature and severity of the

adverse health effects (that is, illness or death) associated with the occurrence of *E. coli* O157:H7 in ground beef. For *E. coli* O157:H7, the precise relationship between the number of organisms consumed and the resulting adverse human health event is not known.

An *E. coli* O157:H7 dose-response function was derived using information from three sources: (1) the estimated annual number of symptomatic *E. coli* O157:H7 infections resulting from ground beef exposures (Powell et al. 2001); (2) the estimated number of contaminated ground beef servings from the exposure assessment; and (3) lower- and upper-bound dose-response curves derived using surrogate pathogens (Powell et al. 2000). The lower- and upper-bound dose-response curves describe the uncertainty about the probability of symptomatic illness at an ingested dose level based on bounding (minimum and maximum) estimates.

The estimated annual number of symptomatic *E. coli* O157:H7 infections resulting from ground beef exposures was derived from estimates of the baseline annual number of *E. coli* O157:H7 cases from all causes using the FoodNet national surveillance data (CDC 2000a). The baseline annual number of cases was first adjusted upward to account for recognized causes of underreporting—such as ill persons not seeking medical care, false-negative test results, and laboratories not culturing fecal samples for *E. coli* O157:H7—resulting in an estimate of the total annual number of symptomatic *E. coli* O157:H7 cases in the United States. The estimated total annual number of cases was then multiplied by estimates of the proportion of cases caused by ground beef exposure to arrive at the annual number of cases of symptomatic *E. coli* O157:H7 cases caused by ground beef exposure.

Total illnesses per year (T) depends on the total number of servings consumed per year (C), the frequency of exposure to various doses ($f(dose)$), and the proba-

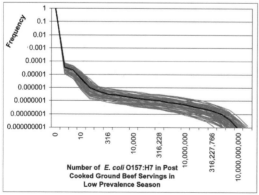

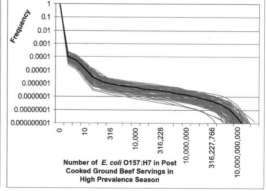

FIG. 32.5—Frequency of exposure to various levels of *E. coli* O157:H7 per cooked serving during the low-prevalence and high-prevalence seasons. The gray lines describe the uncertainty about the distribution. The dark line designates the average exposure distribution.

bility of becoming ill given various doses ($p(ill|dose)$) [that is, $T = C \int p(ill|dose) \times \int (dose) \, \rho \, dose$]. The dose-response function is symbolized as $p(ill|dose)$. Given our estimates of T, C, and $f(dose)$, the best-fitting *E. coli* O157:H7 dose-response function was estimated within the constraints of the upper and lower bounds.

The model estimated that a median of nineteen thousand cases of symptomatic *E. coli* O157:H7 infections would occur annually because of ground beef exposure in the United States. Considerable uncertainty exists about this estimate, as reflected by the 2.5th and 97.5th percentile values of 5,300 and 45,000 cases, respectively. The model also estimated the proportion of cases expected to develop severe clinical manifestations (for example, bloody diarrhea, HUS). The median annual number of patients with bloody diarrhea who sought medical care was eighteen hundred; of these approximately four hundred are hospitalized. Of the hospitalized patients, the model estimates that approximately 23 percent will develop HUS or TTP, and 11 percent of those with HUS or TTP will die (based on Cieslak et al. 1997; Hedberg et al. 1997).

RISK CHARACTERIZATION. Risk characterization integrates the results of the exposure assessment with the results of hazard characterization to estimate the risk of illness from *E. coli* O157:H7 in ground beef. Risk characterization also includes a sensitivity analysis to identify variables that influence the occurrence and extent of *E. coli* O157:H7 contamination in ground beef servings. These variables represent important areas where mitigation strategies could potentially be applied to reduce the risk of illness from *E. coli* O157:H7 in ground beef.

Population Risk of Illness. The risk of becoming ill from *E. coli* O157:H7 in ground beef depends both on the probability of being exposed to a specific number of *E. coli* O157:H7 in a ground beef serving *and* the probability of illness given that exposure. The model predicts the following average results per ground beef serving consumed; about one illness in each one million servings (9.6×10^{-7}), two hospitalized cases (that eventually recover) in each 100 million servings (2.0×10^{-8}), 4.2 HUS cases (that eventually recover) per one billion servings (4.2×10^{-9}), and six deaths per ten billion servings (5.9×10^{-10} per serving). These per-serving estimates are based on the central tendencies (median) of our uncertainty about both the exposure distribution and dose-response functions. Uncertainty about illnesses per serving ranges from about one illness in every three million ground beef servings to about two illnesses in every one million ground beef servings.

Effect of Season. The risk of illness from *E. coli* O157:H7 in ground beef was about three times higher during June to September than during the rest of the year. About one in every 600,000 ground beef servings

consumed during June through September is predicted to result in illness, whereas about one in every 1.6 million servings consumed during October to May is expected to result in illness based on the median results.

The hypothetical linkage between live cattle, ground beef, and human *E. coli* O157:H7 illnesses is strongly supported by these seasonal findings. Of the 18.2 billion ground beef servings consumed annually, it is assumed that one-third are consumed during the high-prevalence season and two-thirds are consumed during the remainder of the year. Combining this consumption pattern with the seasonal per-serving risk estimates implies that 58 percent of illnesses occur during the high-prevalence season, whereas 42 percent occur during the low-prevalence season. This finding is consistent with public health surveillance information. Of cases reported by FoodNet sites, 70 percent occurred during June through September for the years 1996 to 1998 (Bender et al. 2000). In 1998, 1,710 (54.1 percent) of 3,161 cases of *E. coli* O157:H7 reported to CDC occurred during those months (CDC 1999). Outbreaks also occur more frequently in the summer, with fifty (58.8 percent) of eighty-five foodborne outbreaks occurring during June through September for the period 1993 to 1997 (CDC 1999). During 1998 to 1999, twenty-one (50.0 percent) of forty-two foodborne outbreaks occurred during June through September (CDC 2000b, 2001).

Consistency in seasonal occurrence of human illness between the model and public health data is noteworthy because the model accounts for seasonality only in live cattle. Therefore, without any adjustment for seasonal differences in ground beef storage or cooking, these results imply that seasonal changes in prevalence on the farm subsequently influence levels of *E. coli* O157:H7 in combo bins, grinder loads, and servings, and predict changes in illnesses in a manner consistent with human health surveillance data.

SENSITIVITY ANALYSIS. A sensitivity analysis of this risk assessment serves to identify factors that most influence the occurrence or extent of *E. coli* O157:H7 contamination in ground beef. The occurrence and extent of *E. coli* O157:H7 contamination in beef trim and subsequent grinder loads was most influenced by feedlot and within-feedlot prevalence, probability of carcass contamination following dehiding, initial numbers of *E. coli* O157:H7 on contaminated carcasses, effectiveness of decontamination procedures, and carcass chilling.

The effect of these factors on the occurrence and extent of *E. coli* O157:H7 in combo bins and grinder loads varied by season and type of cattle (breeding or feedlot). For example, the initial amount of contamination on carcasses was correlated with the amount of *E. coli* O157:H7 in feedlot cattle combo bins during the high-prevalence season but not with the prevalence of *E. coli* O157:H7-contaminated combo bins during this season. In contrast, the initial amount of contamination on carcasses was not correlated with either the number of *E. coli* O157:H7 organisms in breeding cattle combo bins or the prevalence of *E. coli* O157:H7-contaminated breeding cattle combo bins for the low-prevalence season.

The occurrence of *E. coli* O157:H7 contamination in cooked ground beef servings was understandably influenced by the occurrence of *E. coli* O157:H7 contamination in beef trim and subsequent grinder loads. It was also greatly influenced by the proportion of ground beef that is frozen, the maximum population density of *E. coli* O157:H7 in ground beef, storage temperatures, and cooking.

Perhaps the most important finding was that consumers can still be exposed to *E. coli* O157:H7 in ground beef even if servings are cooked to effect a 5 log reduction, because extensive growth of *E. coli* O157:H7 can occur if servings are improperly stored prior to cooking.[1] Both proper storage and adequate cooking are necessary to prevent illness from *E. coli* O157:H7 in ground beef.

SUMMARY AND CONCLUSIONS. The baseline risk assessment described in this document models the occurrence of *E. coli* O157:H7 in cattle on the farm to the occurrence and extent of contaminated servings of cooked ground beef. The exposure assessment concludes that feedlot cattle are likely to have a higher prevalence of *E. coli* O157:H7 infection than are culled breeding cattle. Although only a fraction of infected live cattle results in contaminated carcasses, up to thousands of pounds of meat trim from carcasses are combined in the grinding process. Consequently, although the concentration of *E. coli* O157:H7 in these grinder loads may be quite low, the proportion of grinder loads that contain one or more *E. coli* O157:H7 organism is expected to be high. The effects of storage, holding times, chilling, and cooking were included throughout the model to account for organism growth or decline.

Finally, FSIS is releasing a draft report documenting the baseline risk assessment on *E. coli* O157:H7 in ground beef for public comment and scientific peer review. Thus, the risk assessment is a "work in progress" that will likely be continually revised based on improvements in science and risk assessment methodology, as well as the needs of users of such analysis.

REFERENCES

Agricultural Research Service, U.S.Department of Agriculture. 1998.The 1994–1996 continuing survey of food intakes by individuals and 1994–1996 diet and health knowledge survey and technical support databases. National Technical Information Service, Springfield, VA. 1998.

Ahmed, S., and M. Donaghy. 1998. An outbreak of *E. coli* O157:H7 in Central Scotland. In *E. coli O157:H7 and Other Shiga Toxin-Producing E. coli Strains*. J.B. Kaper and A.D. O'Brien eds. Washington, D.C.: ASM Press, 59–65.

Ansay, S.E., K.A. Darling, and C.W. Kaspar. 1999. Survival of *Escherichia coli* O157:H7 in ground-beef patties during storage at 2, −2, 15 and then −2 degrees C, and −20 degrees C. *J. Food Prot.* 62(11):1243–1247.

Bacon, R.T., J.N. Sofos, K.E. Belk, and G.C. Smith. 1999. Evaluation of multiple-hurdle microbiological interventions—to include pre-evisceration spray-washing and other technologies—for beef carcass decontamination. Final Report. Submitted to National Cattlemen's Beef Association, March 3, 1999.

Bender, J., K. Smith, A. McNees, T. Fiorentino, S. Segler, M. Carter, N. Spina, W. Keene, T. Van Gilder, and the EIP FoodNet Working Group. 2000. Surveillance for *E. coli* O157:H7 infections in FoodNet sites, 1996–1998: No decline in incidence and marked regional variation. 2nd International Conference on Emerging Infectious Diseases, Atlanta, Georgia, July.

Besser, T.E., D.D. Hancock, L.C. Pritchett, E.M. McRae, D.H. Rice, and P.I. Tarr. 1997. Duration of detection of fecal excretion of *Escherichia coli* O157:H7 in cattle. *J. Infect. Dis.* 175:726–729.

Bonardi, S., E. Maggi, A. Bottarelli, M.L. Pacciarini, A. Ansuini, G. Vellini, S. Morabito, and A. Caprioli. 1999. Isolation of verocytotoxin-producing Escherichia coli O157:H7 from cattle at slaughter in Italy. *Vet. Microbiol.* 67(3) 203–211.

Centers for Disease Control and Prevention (CDC). 1999. Summary of notifiable diseases, United States, 1998. *M.M.W.R.* 47(53):1–94.

Centers for Disease Control and Prevention (CDC). 2000a. FoodNet surveillance report for 1999. Final Report. November.

Centers for Disease Control and Prevention (CDC). 2000b. Surveillance for foodborne-disease outbreaks—United States, 1993–1997. *M.M.W.R.* 49(SS-1):1–62.

Centers for Disease Control and Prevention (CDC). 2001. Surveillance for outbreaks of *Escherichia coli* O157:H7 infection. Summary of 1999 Data. Report from the National Center for Infectious Diseases, Division of Bacterial and Mycotic Diseases to CSTE. June 15, 2000.

Cieslak, P.R., S.J. Noble, D.J. Maxson, L.C. Empey, O. Ravenholt, G. Legarza, J. Tuttle, M.P. Doyle, T.J. Barrett, J.G. Wells, A.M. McNamara, and P.M. Griffin. 1997. Hamburger-associated *Escherichia coli* O157:H7 infection in Las Vegas: A hidden epidemic. *Am. J. Publ. Health* 87(2):176–180.

Codex Alimentarius Commission. 1999. Codex Alimentarius Volume 1B, Codex Standards for General Requirements (Food Hygiene), CAC/GL 030-1999, Principles and Guidelines for the Conduct of Microbiological Risk Assessment. Available at: http://ftp.fao.org/codex/standard/volume1b/ en/GL_030e.pdf.

Dargatz, D.A., S.J. Wells, L.A. Thomas, D. Hancock, and L.P. Garber. 1997. Factors associated with the presence of *Escherichia coli* O157 in feces in feedlot cattle. *J. Food Prot.* 60:466–470.

Dorsa, W., C. Cutter, and G. Siragusa. 1997. Effects of acetic acid, lactic acid and trisodium phosphate on the microflora of refrigerated beef carcass surface tissue inoculated with *Escherichia coli* O157:H7, *Listeria Innocua*, and *Clostridium Sporogenes*. *J. Food Prot.* 60:619–624.

Elder, R.O., J.E. Keen, G.R. Siragusa, G.A. Barkocy-Gallagher, M. Koohmaraie, and W.W. Laegreid. 2000. Correlation of enterohemorrhagic *Escherichia coli* O157:H7 prevalence in feces, hides, and carcasses of beef cattle during processing. *Proc. Natl. Acad. Sci. USA* 97(7):2999–3003.

Food and Drug Administration. 1999. Audits International/FDA Home Cooking Temperature Interactive Database. Available at: http://www.foodriskclearinghouse.umd.edu/ColdFusion/ Cooking/Audits International.

Food Safety and Inspection Service (FSIS), U.S. Department of Agriculture. 1994. Nationwide beef microbiological baseline data collection program: steers and heifers—October 1992–September 1993.

Food Safety and Inspection Service (FSIS), U.S. Department of Agriculture. 2001 revision. Microbiology Laboratory Guidebook, 3rd edition, 1998. Chapter 5. Detection, isolation and identification of E. coli O157:H7 and O157:nm (nonmotile) from meat products. (Revision # 2; 2-23-01, W.C. Cray Jr., D.O. Abbott, F.J. Beacorn, S.T. and Benson.)

Galland, J.C. 1997. Risks and prevention of contamination of beef carcasses during the slaughter process in the United States of America. Contamination of animal products: Prevention and risks for public health. *Reveu Scientifique et Technique, Office International Des Epizooties.*

Garber, L., S. Wells, L. Schroeder–Tucker, and K. Ferris. 1999. Factors associated with the shedding of verotoxin-producing *Escherichia coli* O157 on dairy farms. *J. Food Prot.* 62(4):307–312.

Gill, C.O., and J. Bryant. 1997. Assessment of the hygienic performances of two beef carcass cooling processes from product temperature history data or enumeration of bacteria on carcass surfaces. *Food Microbiol.* 14:593–602.

Gill, C.O. 1998. Apparatus for Pasteurizing Red Meat Carcasses. Agriculture and Agri-Food Canada, Research Branch Technical Bulletin 1998-5E.

Gill, C.O. 1999. HACCP by guesswork or by the numbers? *Food Quality* Jan/Feb:28–32.

Hancock, D.D., T.E. Besser, M.L. Kinsel, P.I. Tarr, D.H. Rice, and M.G. Paros. 1994. The prevalence of *Escherichia coli* O157 in dairy and beef cattle in Washington State. *Epidemiol. and Infect.* 113:199–207.

Hancock, D.D., T.E. Besser, D.H. Rice, D.E. Herriott, and P.I. Tarr. 1997a. A longitudinal study of *Escherichia coli* O157 in fourteen cattle herds. *Epidemiol. and Infect.* 118:193–195.

Hancock, D.D., D.H. Rice, D.E. Herriot, T.E. Besser, E.D. Ebel, and L.V. Carpenter. 1997b. Effects of farm manure handling practices on *Escherichia coli* O157 prevalence in cattle. *J. Food Prot.* 60:363–366.

Hancock, D.D., T.E. Besser, D.G. Rice, E.D. Ebel, D.E. Herriott, and L.V. Carpenter. 1998. Multiple sources of *Escherichia coli* O157 in feedlots and dairy farms in the Northwestern USA. *Prev. Vet. Med.* 35:11–19.

Hancock, D.D., D.H. Rice, and T.E. Besser. 1999. Prevalence of *E. coli* O157:H7 in feedlot cattle at slaughter plants. Study funded by National Cattlemen's Association. Washington State University.

Hancock, D., T. Besser, J. Lejeune, M. Davis, and D. Rice. 2001. The control of VTEC in the animal reservoir. *Intl. J. Food Microbiol.* 66:71–78.

Hedberg, C., F. Angulo, J. Townes, J. Hadler, D. Vugia, and J. Garley. 1997. Differences in *Escherichia coli* O157:H7 annual incidence among FoodNet active surveillance sites. 5th International VTEC Producing *Escherichia coli* Meeting. Baltimore, MD, July.

Heuvelink, A.E., F.L. Van den Biggelaar, J. Zwartkruis–Nahuis, R.G. Herbes, R. Huyben, N. Nagelkerke, W.J. Melchers, L.A. Monnens, and E. de Boer. 1998. Occurrence of verocytotoxin-producing *Escherichia coli* O157 on Dutch dairy farms. *J. Clin. Microbiol.* Dec:3480–3487.

Jackson, T.C., M.D. Hardin, and G.R. Acuff. 1996. Heat resistance of *Escherichia coli* O157:H7 in a nutrient medium and in ground beef patties as influenced by storage and holding temperatures. *J. Food Prot.* 59:230–237.

Kassenborg, H., C. Hedberg, M. Hoekstra, M.C. Evans, A.E. Chin, R. Marcus, D. Vugia, K. Smith, S. Desai, L.

Slutsker, P. Griffin, and the FoodNet Working Group. 2001. Farm visits and undercooked hamburgers as major risk factors for sporadic *Escherichia coli* O157:H7 infections—data from a case-control study in five FoodNet sites. Manuscript in preparation.

Kastner, C. 1998. Research Highlights. Kansas State University Food Safety Consortium Annual Report.

Lagreid, W.W., R.O. Elder, and J.E. Keen. 1999. Prevalence of *E. coli* O157:H7 in range calves at weaning. *Epidemiol. and Infect.* 123(2):291–298.

Le Saux, N., J.S. Spika, B. Friesen, I. Johnson, D. Melnychuck, C. Anderon, R. Dion, M. Rahman, and W. Tostowaryk. 1993. Ground beef consumption in noncommercial settings is a risk factor for sporadic *Escherichia coli* O157:H7 infection in Canada. *J.Infect. Dis.* 167:500–502 (letter).

Lopez, E.L., M.M. Contrini, and M.F. De Rosa. 1998. Epidemiology of shiga toxin-producing *E. coli* in South America. In *E. coli O157:H7 and Other Shiga Toxin-Producing E. coli Strains*. J.B. Kaper and A.D. O'Brien. eds. Washington, D.C.: ASM Press, 30–73.

MacDonald, K.W., M.J. O'Leary, M.L. Cohen, P. Norris, J.G. Wells, E. Noll, J.M. Kobayashi, and P.A. Blake. 1988. *Escherichia coli* O157:H7, an emerging gastrointestinal pathogen: Results of a one-year, prospective, population-based study. *J.A.M.A* 259(24):3567–3570.

Marks, H.M., M.E. Coleman, C.T.J. Lin, and T. Roberts. 1998. Topics in risk assessment: Dynamic flow tree process. *Risk Analysis* 18:309–328.

Martin, S.W., M. Shoukri, and M.A. Thorburn. 1992. Evaluating the health status of herds based on tests applied to individuals. *Prev. Vet. Med.* 14(2):33–43.

Mead, P.S., L. Finelli, M.A. Lambert-Fair, D. Champ, J. Townes, L. Hutwagner, T. Barrett, K. Spitalny, and E. Mintz. 1997. Risk factors for sporadic infection with *Escherichia coli* O157:H7. *Arch. Intern. Med.* 157:204–208.

Mead, P.S., and P.M. Griffin. 1998. *Escherichia coli* O157:H7. *Lancet* 352:1207–1212.

Mead, P.S., L. Slutsker, V. Dietz, L.F. McCaig, J.S. Bresee, C. Shapiro, P.M. Griffin, and R.B. Tauxe. 1999. Food-related illness and death in the United States. *Emerg. Infect. Dis.* 5(5):607–625.

Michino, H., K. Araki, S. Minami, T. Nakayam, Y. Kjima, K. Hiroe, H. Tanaka, N. Fujita, S. Usami, M. Yonekawa, K. Sadamoto, S. Takaya, and N. Sakai. 1998. Recent outbreaks of infections caused by *E. coli* O157:H7 in Japan. In *E. coli O157:H7 and Other Shiga Toxin-Producing E. coli Strains*. J.B. Kaper and A.D. O'Brien, eds. Washington, D.C.: ASM Press, 73–82.

Nutsch, A.L., R.K. Phebus, M.J. Riemann, J.S. Kotrola, D.E. Schaefer, J.E. Boyer, Jr., R.C. Wilson, J.D. Leising, and C.L. Castner. 1997. Evaluation of a steam pasteurization process in a commercial beef processing facility. *J. Food Prot.* 60:485–492.

Nutsch, A.L., R.K. Phebus, M.J. Riemann, J.S., Kotrola, R.C. Wilson, J.E. Boyer, Jr., and T.L. Brown. 1998. Steam pasteurization of commercially slaughtered beef carcasses: Evaluation of bacterial populations at five anatomical locations. *J. Food Prot.* 61:571–577.

Phebus, R.K., A.L. Nutsch, D.E. Schafer, R.C. Wilson, M.J. Riemann, J.D. Leising, C.L. Kastner, J.R. Wolf, and R.K. Prasai. 1997. Comparison of steam pasteurization and other methods for reduction of pathogens on surfaces of freshly slaughtered beef. *J. Food Prot.* 60:476–484.

Powell, M., E. Ebel, W. Schlosser, M. Walderhaug, and J. Kause. 2000. Dose-response envelope for *Escherichia coli* O157:H7. *Quant. Microbiol.* 2(2):141–163.

Powell, M., E. Ebel, and W. Schlosser. 2001. Considering uncertainty in comparing the burden of illness due to foodborne microbial pathogens. *Intl. J. Food Microbiol.* 69:209–215.

Prasai, R., R.K. Phebus, C.M. Garcia Zepeda, C.C. Kastner, A.E. Boyle, and D.Y.C. Fung. 1995. Effectiveness of trimming and/or washing on microbiological quality of beef carcass. *J. Food Prot.* 58:1114–1177.

Rice, D.H., E.D. Ebel, D.D. Hancock, D.E. Herriott, and L.V. Carpenter. 1997. *Escherichia coli* O157 in cull dairy cows on farm, and at slaughter. *J. Food Prot.* 60:1–2.

Sage, J.R., and S.C. Ingham. 1998. Survival of *Escherichia coli* O157:H7 after freezing and thawing in ground beef patties. *J. Food Prot.* 61(9):1181–1183.

Sanderson, M.W., J.M. Gay, D.D. Hancock, C.C. Gray, L.K. Fox, and T.E. Besser. 1995. Sensitivity of bacteriologic culture for detection of *Escherichia coli* O157:H7 in bovine feces. *J. Clin. Microbiol.* 33(10):2616–2629.

Sargeant, J.M., J.R. Gillespie, R.D. Oberst, R.K. Phebus, D.R. Hyatt, L.K. Bohra, and J.C. Galland. 2000. Results of a longitudinal study of the prevalence of *E. coli* O157:H7 on cow-calf farms. *A.J.V.R.* 61(11):1375–1379.

Sheridan, J., B. Lynch, and D. Harrington. 1992. The effect of boning plant cleaning on the contamination of beef cuts in commercial boning hall. *Meat Sciences* 32:155–164.

Slutsker, L., A.A. Ries, K. Maloney, J.G. Wells, K.D. Greene, P.M. Griffin, and the *Escherichia coli* O157:H7 Study Group. 1998. A nationwide case-control study of *Escherichia coli* O157:H7 infection in the United States. *J. Infect. Dis.* 177:962–966.

Smeltzer, T., B. Peel, and G. Collins. 1998. The role of equipment that has direct contact with the carcass in the spread of *Salmonella* in beef abattoir. *Austr. Vet. J.* 55:275–277.

Smith, D.R. 1999. Designing strategies for food safety pathogen containment in cattle feedyards. Presented at Annual Meeting of U.S. Animal Health Association, San Diego, CA.

Spika, J.S., R. Khakhria, P. Michel, D. Milley, J. Wilson, and J. Waters. 1998. Shiga toxin-producing *Escherichia coli* infections in Canada. In *Escherichia coli O157:H7 and Other Shiga Toxin-Producing E. coli Strains.* J.B. Kaper and A.D. O'Brien, eds. Washington, D.C.: ASM Press, 23–29.

Van Donkersgoed, J., T. Graham, and V. Gannon. 1999. The prevalence of vertoxins, *E. coli* O157:H7, and *Salmonella* in the feces and rumen of cattle at processing. *Can. Vet. J.* 40:332–338.

Vose, D.J. 1996. Quantitative Risk Analysis: A Guide to Monte Carlo Simulation Modeling. West Sussex, England: John Wiley and Sons, Ltd.

Walls, I., and V.N. Scott. 1996. Validation of predictive mathematical models describing the growth of *Escherichia coli* O157:H7 in raw ground beef. *J. Food Prot.* 59:1331–1335.

Wong, C.S., S. Jelacic, R.L. Habeeb, S.L. Watkins, and P.I. Tarr. 2000. The risk of the hemolytic-uremic syndrome after antibiotic treatment of *E. coli* O157:H7 infections. *N. Engl. J. Med.* 342(26):1930–1936.

NOTES

1. The maximum population density for of *E. coli* O157:H7 in ground beef servings can vary depending on food matrix characteristics (for example, pH, water activity) and competitive microflora. The most likely maximum population density for *E. coli* O157:H7 in ground beef is between 5 and 10 logs. The importance of proper storage versus adequate cooking depends on the maximum population density.

33

BOVINE SPONGIFORM ENCEPHALOPATHY

WILLIAM D. HUESTON

INTRODUCTION. Bovine spongiform encephalopathy (BSE) became a global food safety issue with the recognition of the new human disease, variant Creutzfeldt-Jakob Disease (vCJD), in March 1996 (Will et al. 1996). Compelling evidence now exists that BSE and vCJD are caused by the same agent (Bruce et al. 1997; Collinge et al. 1996). The link between BSE and vCJD was hypothesized initially on the basis of their temporal and geographic parallels combined with the obvious differences between vCJD and classical Creutzfeldt-Jakob Disease (CJD) in terms of clinical signs, diagnostic test results, course of clinical disease, and histopathology (Will et al. 1996). Furthermore, no recognized CJD risk factors were identified among the initial vCJD cases. Epidemiologic studies to date suggest that the most likely route of human exposure to BSE was meat products contaminated with infective cattle tissues.

BSE emerged in 1986 in the United Kingdom (U.K.) as a new neurologic disease of cattle (Wells et al. 1987). Additional cases were soon diagnosed throughout the United Kingdom in all breeds of cattle. The epidemic gained momentum through the late 1980s and early 1990s, peaking in January 1993 at one thousand new suspect cases each week. The epidemic toll reached approximately 180,000 cases in the United Kingdom alone by the end of 2001. As of March 2002, the International Office of Animal Health (OIE) reports that BSE has been diagnosed in native cattle in twenty countries, including all of the European Union countries except Sweden, much of Eastern Europe, and Japan (OIE 2002a).

Assuring that food is safe from BSE has become a global priority. This chapter focuses on the critical importance of pre-harvest strategies for prevention, control, and eradication of BSE. The farmer plays a crucial role in preventing potential transmission of BSE on the farm through animal feed. The farmer is also key to disease surveillance by initiating diagnostic evaluation of ill cattle showing clinical signs compatible with BSE. The prompt detection of emergent cases of BSE on the farm will facilitate effective national response should the disease occur. Global eradication of BSE is imperative for removing infected cattle as a potential source of contaminated human foods.

ETIOLOGY. Bovine spongiform encephalopathy and vCJD join the growing family of transmissible spongiform encephalopathies (TSEs). Scrapie of sheep and goats, chronic wasting disease of deer and elk, and transmissible mink encephalopathy represent other animal TSEs, whereas Creutzfeldt-Jakob Disease, Kuru, and Gerstmann-Straussler-Shenker syndrome are the most well-known human TSEs.

All the TSE diseases are characterized by a long incubation period of months to years, progressive neurologic disease, degeneration of the central nervous system leading to death, the absence of host immune responses, lack of any effective treatment, and the ability to transmit the disease naturally and experimentally via infected central nervous system (CNS) tissue. Finally, all the TSE diseases are marked by the accumulation of prions, abnormally shaped isoforms of a normal host cell-surface glycoprotein (Pruisner 1982).

All mammals appear to have the normal cellular form (PrP^C) of prion protein (PrP), which is sensitive to protease digestion. The disease form of the prion (PrP^{Res}) results from a post-translational shape change and is resistant to protease digestion. In animals exposed to PrP^{Res}, the abnormally shaped prion protein catalyzes the conversion of normal host protein, PrP^C, into the aberrant isoform. The rate of conversion depends on the route of exposure and the degree of homology between the PrP^{Res} and the host PrP^C.

Accumulations of PrP^{Res} in CNS tissues are observed in all TSE cases. However, the scientific debate continues as to whether PrP^{Res} alone is the cause of TSE diseases or simply a product of the disease process. Three schools of thought predominate (Rubenstein 2000): (1) the prion protein itself is the agent; (2) the agent is an unusual virus not yet detected; and (3) the agent is a virion, a short segment of noncoding nucleic acid shielded by host protein. The unusual characteristics of TSE fuel this debate. The disease can be transmitted by brain extracts from TSE-affected animals that contain no evidence of nucleic acid, suggesting that the protein only may be the agent. Furthermore, the TSE agent is remarkably resistant to heat and disinfectants that readily denature nucleic acid. However, the observation of TSE strain differences is more compatible with a viral hypothesis than protein-only.

Regardless of the final scientific conclusion as to the exact nature of the TSE agent, the epidemiology of BSE has been described sufficiently to guide prevention and control programs.

EPIDEMIOLOGY. Epidemiologic investigations of the early BSE cases identified only one shared risk factor among affected farms: the feeding of proprietary cattle feeds (Wilesmith et al. 1988). Subsequent examination of calf feed formulations found meat and bone meal, a rendered animal protein, to be a statistically significant risk factor for BSE (Wilesmith et al. 1992a).

Until the emergence of BSE, the rendering process was considered sufficiently rigorous to destroy all known pathogens. Rendering was considered to be a safe and ecologically sound method for recycling animal protein.

Laboratory experiments now confirm that the rendering process fails to completely inactivate BSE infectivity (Taylor et al. 1995). Epidemiologic studies also have documented significant changes in the United Kingdom rendering industry in the 1970s and 1980s, including industry consolidation and adoption of newer technology (Wilesmith et al. 1991). Epidemiologic studies and challenge experiments have established oral exposure as the major, if not exclusive, route of natural transmission of BSE. The global BSE epidemic represents a large-scale animal feedborne epidemic, spread around the world through the distribution and feeding of rendered animal proteins of ruminant origin containing BSE infectivity.

The transmission of BSE has been studied extensively in the United Kingdom, with much of the research sponsored by the Department of Environment, Food and Rural Affairs (DEFRA). BSE can be transmitted experimentally through very small exposures to infective tissues. Calves exposed orally to 1 gm of BSE brain suspension have succumbed to the disease, and further challenge experiments are under way, with oral doses as low as 0.001 gm (DEFRA 2002a). The incubation period of BSE is negatively correlated with the exposure dose; for example, the smaller the exposure, the longer the incubation. Most naturally occurring cases of BSE have been observed in cattle between three and five years of age, with the youngest documented naturally occurring case being twenty months old and the oldest nineteen years (DEFRA 2002b). Taken together, these epidemiologic studies and challenge experiments suggest that most of the BSE cases observed in the United Kingdom and Europe were exposed to relatively small amounts of the BSE agent early in their lives.

Little evidence exists of direct BSE transmission from animal to animal. An epidemiologic study suggested a slightly higher risk of BSE in calves born within three days of a subsequently BSE-affected cow calving (Hoinville et al. 1995). More than 50 percent of the BSE-affected herds in the United Kingdom had cases in only a single birth cohort, again pointing to the likelihood of a contaminated batch of feed and little to no risk for cow-to-cow transmission (DEFRA 2002c). No cases of secondary spread of BSE have been identified in beef suckler herds after the purchase of an animal subsequently diagnosed with BSE (Wilesmith and Ryan 1997).

The accumulated epidemiologic evidence for BSE also does not support windborne or waterborne transmission of the agent or a role for environmental contamination as a source of later exposure of susceptible animals. To date, no BSE infectivity has been identified in normal body discharges such as urine, feces, or milk.

The potential risk of maternal transmission appears to be small. A prospective case-control study identified a 9.6 percent higher risk of BSE among calves born to cows with BSE compared to those born to unaffected dams (Wilesmith et al. 1997). However, BSE infectivity has not been detected in feti, fetal fluids, blood, or placenta, so it is unclear how the offspring might be exposed if in fact they are at a higher risk. Because the data are still unclear, current control strategies have taken a conservative approach, advocating the destruction of both affected animals and their offspring to insure against further cases. Transmission studies of germplasm suggest that transmission through semen and embryos is highly unlikely (Wrathall et al. 2002).

Indirect transmission of BSE through contaminated fomites such as farm equipment or veterinary instruments has not been observed. However, indirect transmission through contaminated neurosurgical instruments has been incriminated with human TSEs (Brown et al. 2000). Serendipitously, neurosurgery or other invasive surgical procedures targeting organs and tissues known to accumulate high levels of BSE infectivity (brain, spinal cord, distal ileum) are not routinely practiced in the veterinary medicine for cattle.

Theoretically, BSE also might be transmitted through other contaminated animal products, such as vaccines. Two examples demonstrate the iatrogenic dissemination of other TSEs. Scrapie was spread to susceptible sheep through vaccines produced with scrapie-affected tissues, one vaccine for louping ill (Gordon 1946) and another vaccine for contagious agalactia (Agrimi et al. 1999). Human TSEs also have been transmitted through tissue transplants and hormone extracts from CJD-affected persons (Brown et al. 2000). Thankfully, BSE infectivity appears to have a more limited tissue distribution in cattle than do many other TSEs. The fact that the infectivity has been detected only in the brain, spinal cord, and retina of clinical BSE cases reduces the likelihood of iatrogenic transmission in cattle. Furthermore, no BSE-infected cell lines have been identified to date.

The potential for spontaneous occurrence of BSE, albeit hypothetical, also must be considered. One theory of the origin of BSE suggests that cattle PrP malformations occur spontaneously, following the same pattern as observed with CJD in humans. These spontaneous cattle TSEs have been suggested as the cause of transmissible mink encephalopathy caused by the practice of feeding nonambulatory cattle carcasses (Marsh et al. 1991). Although no spontaneous cattle TSEs have been identified, theoretically these occurrences would be expected to occur quite rarely and be seen in older cattle.

PATHOGENESIS. The pathogenesis of BSE has been studied through a series of challenge studies. A group of four-month-old calves was orally challenged with a suspension of 100 gm of BSE brain and were sequentially sacrificed to follow the progression of the disease (Wells et al. 1998). Post experimental oral exposure, BSE infectivity was first detected in the distal ileum at the site of the Peyer's Patches at six months after challenge. No infectivity was found in any other tissue until thirty-two months post exposure, when infectivity could be detected in spinal cord, brain, and dorsal root ganglia. Clinical signs appeared at thirty-five to thirty-seven months after exposure. No infectivity was found outside the central nervous system and distal ileum except for a single isolation from bone marrow at thirty-eight months post infection, a finding that has not been replicated and thus must be interpreted cautiously.

INACTIVATION OF BSE. All the TSE agents are remarkably resistant to inactivation. Infectivity can be transmitted after ashing brain samples at 500°C and after multiple wash and autoclave cycles of neurosurgery instruments. Sodium hydroxide autoclaving appears most effective in eliminating infectivity.

Commercial rendering practices, previously considered to inactivate all animal and human pathogens, do not fully inactivate BSE. Although rendering procedures can reduce the infectivity titer by two to three logs, sufficient residual infectivity may remain to transmit the disease if high-risk tissues (brain, spinal cord, intestines) from BSE-infected cattle have been used as raw material. Current international guidelines for the production of meat and bone meal dictate that raw materials must be reduced to a particle size no greater than 50 mm and processed under saturated steam conditions to achieve a temperature of at least 133°C at an absolute pressure of at least 3 bar for twenty minutes or more (OIE 2001a). International guidelines for BSE-affected countries recommend the complete destruction of all confirmed cattle (OIE 2002b).

DIAGNOSTICS FOR BSE. No live animal diagnostic test for BSE is currently available. Exposed and affected cattle demonstrate no immune response to the BSE agent. Therefore, serology for BSE is of no use. Currently, diagnosis requires postmortem examination of tissues potentially containing high concentrations of PrP[Res], such as the brain and spinal cord. Neurohistopathology and demonstration of PrP[Res] by immunohistochemistry (IHC) remain the gold standard BSE diagnostic tests. Several rapid protein screening tests have become available over the past few years to screen for PrP[sc], including an ELISA and Western blot, but all these require confirmation, usually by IHC.

Current diagnostic tests focus on tissues with high levels of PrP[Res] and can detect cattle only in the clinical phase or the end stage of disease. No tests exist to determine whether cattle have been exposed to BSE or whether cattle may be in the early stages of the disease, prior to the point that infectivity can be found in the CNS.

Therefore, diagnosis of BSE on the farm depends on early recognition of clinical signs compatible with BSE, monitoring of suspect cattle, and testing of brain tissues after the animal's death. The clinical presentation of BSE involves an insidious onset of nonspecific changes in mentation, sensation, and locomotion (Wilesmith et al. 1992b). The first changes observed by most farmers involve the mental ability of the cow. Affected animals become more apprehensive and nervous, and may appear confused by their surroundings. These behavioral changes progress and animals may show reluctance to enter doorways; exaggerated response to touch, sounds, or light; and ataxia, frequently noticed first in the hind limbs. The gradual progression of behavior changes is often accompanied by weight loss or reduced milk production.

No successful treatment exists for BSE. Therefore, clinical signs gradually worsen and disease progression may be intensified by calving or other stressful events such as hauling. Some BSE suspects, subsequently confirmed as cases, have been identified during antemortem inspection at slaughter plants, although no abnormalities were observed on the farm prior to shipping. Because the clinical signs are nonspecific and demonstrate a gradual onset over time, many of the early cases of BSE would not be identified by cursory examination of a herd by the casual observer or even a trained veterinarian. Early diagnosis requires careful scrutiny on the part of the animal owner or herdsman and recognition of worsening clinical signs compatible with BSE.

Recognition of BSE suspects presents particular challenges for beef cattle producers or situations in which cattle are not closely observed on a regular basis. Insidious behavior changes may be missed without regular, close examination. Consequently, BSE must be considered one of the differential diagnoses when finding a disabled or nonambulatory animal (downer animal) or watching a nonresponsive course of clinical treatment of cattle exhibiting neurologic or locomotary abnormalities compatible with BSE. The only BSE case in Canada was identified after extra diagnostic scrutiny of an imported beef cow suffering from a broken leg.

RISK FACTORS ON THE FARM. Bovine spongiform encephalopathy enters the farm one of two ways: purchase of exposed or affected animals, or purchase of contaminated feed.

Imported Animals. In the United Kingdom, 32 percent of the confirmed BSE cases were purchased; therefore, their initial exposure to BSE occurred on a different farm from where the disease was diagnosed (DEFRA 2002d). Animals exported from the United Kingdom have developed BSE in a number of countries in Europe and elsewhere. Exported cases are the only recognized occurrence of BSE in the Middle East (Oman) and the Americas (Canada and the Falkland/Malvinas islands) (OIE 2002b). Importation of animals from BSE-affected countries or from countries with no evidence of adequate BSE surveillance and control programs represent the highest risk for cattle producers.

The importation of germplasm represents little to no risk for the introduction of BSE even for embryos or semen harvested from clinically-affected animals (Wrathall et al. 2002). Embryos should be washed according to the International Embryo Society (IETS) protocols (IETS 1998).

Establishing the history of potential BSE exposure for imported animals is very difficult because accurate feeding records and complete knowledge of feed ingredients are rarely available. Even cattle with known feeding histories at the farm of origin may be exposed in transit because of feeding during shipping or lairage. Imported purebred cattle represent a greater risk than imported young stock intended to be fed to slaughter weight. Purebred cattle are more likely to have multiple owners, so the origin may be obscured. Additionally, breeding animals are more likely to live to an old enough age that clinical disease can appear if the animal was exposed prior to purchase. Feeder cattle intended for slaughter, on the other hand, seldom live long enough to develop BSE even if they are exposed to contaminated feedstuffs.

Contaminated Animal Feeds. The second major risk factor for BSE exposure on the individual farm is contaminated animal feeds. The source of contamination is rendered animal protein products derived from BSE-affected cattle. Rendered products produced from raw materials that include brain, spinal cord, and intestinal tract of affected cattle represent the highest risk for contamination. If BSE-affected animals are rendered, infectivity will contaminate the resulting protein products, because no known rendering method can completely inactivate infectivity.

The international movement of rendered animal protein products as bulk commodities confounds global prevention efforts (WHO/FAO/OIE 2001). Bulk commodities quickly lose the identity of their source and details of their composition. Some imported protein supplements labeled as fish meal have been found to contain mammalian protein, leading to new United States Department of Agriculture (USDA) requirements for certification of imported fish protein meals. BSE exposure in Japan is thought to be linked to widespread incorporation of "enhanced" fish meal, which subsequently was found to contain mammalian protein.

Most rendered mammalian proteins currently are banned from ruminant feeds in the United States; however, they remain legal for pig, poultry, and other nonruminant feeds. Therefore, feeding pig and poultry feeds to cattle represents another potential risk. Exposure to BSE infectivity may also occur in situations in which cattle feeds are cross-contaminated with rendered mammalian proteins during processing, hauling, or storage. In addition, on-farm feed mixing may represent a potential risk if ruminant protein products are utilized, such as salvage pet food or processed plate waste.

RISK FACTORS FOR HUMAN EXPOSURE. Epidemiological investigations conducted by the National Creutzfeldt-Jakob Disease Surveillance Unit in the United Kingdom (NCJDSU) have examined potential risk factors for variant Creutzfeldt-Jakob cases in humans (NCJDSU 2000). No evidence currently exists to suggest that any particular occupation carries a higher risk of vCJD. Direct contact with BSE-affected cattle is not considered to represent a human health risk because no BSE infectivity has been identified in cattle secretions or excretions. The most likely risk factor for human vCJD remains consumption of high-risk tissues from BSE-affected cattle or meat products contaminated with high-risk tissues during preparation.

PREVENTION STRATEGIES. Pre-harvest prevention strategies for BSE relate directly to the two major risk factors for disease introduction, purchase of exposed or affected animals and purchase of contaminated feed. Therefore, successful BSE prevention strategies can rest on careful scrutiny of potential purchases of animals and feed, safe handling of feed to prevent exposure to high-risk ingredients (that is, ruminant-derived meat and bone meal), aggressive diagnostic work-up of cattle demonstrating compatible clinical signs, and ongoing education of farm workers, management, and veterinarians. On-farm prevention strategies must be complemented by national BSE surveillance programs and full compliance by the rendering and feed industry with regulations on the use of animal proteins in ruminant feeds.

Pre-Purchase Screening. The lack of live animal diagnostic tests hampers screening of potential purchases of breeding stock. Knowledge of the source of purchased animals and the conditions under which they have been raised provides the best assurance of low risk for BSE that is generally available, although these measures cannot provide complete assurance of BSE freedom. Cattle producers in the United States need to pay particular attention to animals imported from abroad. Although international guidelines recommend restrictions on the movement of breeding cattle from BSE-affected countries, cattle can change ownership repeatedly, obscuring their birthplace and country of origin. For example, several U.K. cattle were found to have transited through Canada on their way to U.S. farms and ranches in the 1980s.

The European Commission (EC) has characterized the overall risk of BSE of specific countries, a process called geographic BSE risk assessments (EC 2000). Countries were invited to submit data on existing BSE prevention, control, and surveillance practices to an international panel of experts. Each submitting country was assigned to one of four risk categories: (1) highly unlikely, (2) unlikely but cannot be excluded, (3) likely or exists at low levels, and (4) BSE known to exist at high levels. These categories provide some guidance but must be used with caution. Several countries considered unlikely to experience BSE (Category 2) have recently identified BSE cases in native cattle. In addition, these country-level risk assessments cannot characterize the range of risk in individual herds within the country. Undoubtedly, BSE will be identified in additional countries around the world because of the widespread dissemination of contaminated feedstuffs.

Accurate record keeping of the origin and parentage of cattle and the husbandry practices utilized by the producer provides additional data on which to evaluate potential risk. Furthermore, good records support rapid response should BSE be identified in U.S. herds. Good individual cow records will facilitate identifying the cattle at greatest risk and help protect valuable breeding stock at low risk. In the absence of good farm records, regulatory officials will tend to take more aggressive actions in culling cattle, given the uncertainty as to identification, origin, and management practices.

Assuring Feed Safety. Avoiding contamination of ruminant feeds with PrP^{Res} is the single most important factor in BSE prevention. Assuring the safety of ruminant feeds requires constant vigilance.

Farmers may not know the exact ingredients in proprietary feeds they purchase. Labeling requirements in the United States allow the use of some generic terms, such as "animal protein," that can lead to confusion because they cover a multitude of ingredients. Some animal protein ingredients such as milk, whey proteins, and blood represent little to no BSE risk and currently are allowed to be used in cattle rations. Animal proteins such as meat and bone meal of ruminant origin represent the highest potential risk for BSE exposure and must be excluded from cattle rations.

Many cattle feeders prepare total mixed rations on-farm, purchasing commodity ingredients and formulating their own rations using computer software packages.

Designers of these software programs may not be knowledgeable of all aspects of disease risk; therefore, specific commodities containing ruminant proteins may unwittingly be specified as recommended feed ingredients based on their nutritional value and price. Producers must be constantly vigilant to preclude accidental and unintentional contamination of rations. Anecdotal evidence of the use of low-cost ingredients such as salvage pet foods in the formulation of cattle rations provides a cautionary note for all cattle feeders.

Farms raising multiple species, such as cattle and poultry, face additional hazards for accidental contamination of cattle rations. Precautions must be exercised to prevent mixing of feed intended for different species in order to preclude potential exposure of cattle to high-risk feed ingredients.

No diagnostic tests currently exist for on-farm

scrutiny of cattle feed or ingredients for the presence of ruminant proteins. Therefore, identifying an ethical and trustworthy feed mill or ingredient supplier is a key component of pre-harvest BSE control. Feed industry quality-assurance and third-party certification programs have emerged in recent years to help identify specific companies in full compliance with protein feeding rules. Given U.S. federal government restrictions on importation of live animals for more than a decade, feeding practices represent the highest remaining risk for accidental exposure of cattle to BSE infectivity.

Education. Forewarned is forearmed. Successful pre-harvest prevention of BSE requires an ongoing commitment to education for producers, veterinarians, and feed manufacturers. BSE is a complicated disease that has generated extensive media coverage, some of which is misleading or even false. Therefore, active dissemination of scientifically sound and accurate information on the clinical signs of BSE, principles of early detection, and keys to prevention must be emphasized for all farm workers. Turnover in farm labor can set the stage for accidental exposure of cattle to potentially high-risk materials because of ignorance on the part of farm workers assigned responsibility for feeding or feed mixing.

Recognition of BSE Suspects. Every cattle producer has an obligation to know the clinical signs of BSE and to establish and maintain a good working relationship with a veterinarian to assure the health of the herd. Producers need to work with their veterinarian and other animal health professionals to conduct an on-farm risk assessment for BSE, including review of feed handling, storage and mixing, ration formulation, individual cow identification and records, purchasing procedures, and treatment protocols for ill animals.

Each farm and every veterinary practice needs an established protocol for diagnostic follow-up of neurologic cases and other cattle demonstrating clinical signs suspicious of BSE. Follow-up of clinical suspects is critical both to the individual farmer and to the nation as a whole. Early detection of BSE is key to control of the potential disease spread and minimizing impact on the farm if BSE is identified. One case of BSE in the United States would have an impact on every cattle owner in the nation. If BSE occurs in the United States, public confidence in the beef and dairy industries will be stronger if the disease is discovered early and aggressive responses implemented sooner rather than later.

National Surveillance. The investigation of suspects that ultimately are determined *not* to be BSE is just as important for the national surveillance system. The credibility of the U.S. surveillance system depends on the nation's ability to document the total number of cattle that have been examined, even if all of those exam-ined are negative. The recent implementation of mandatory testing of all slaughter cattle over thirty months of age in the European Union raises the international expectation for aggressive surveillance as evidenced by high numbers of cattle tested. The U.S. surveillance strategy to date has emphasized testing high-risk cattle rather than screening all older slaughter cattle. High-risk cattle would include those showing clinical signs compatible with BSE, rabies suspects, and nonambulatory cattle, because debilitation or the inability to rise may obscure recognition of clinical signs suggestive of BSE.

Renderers and Feed Manufacturers. The rendering and feed industries are key components of pre-harvest BSE prevention, because of their seminal role in the major risk factor for disease introduction, contaminated feedstuffs. The rendering industry serves a valuable function that is not widely understood. Rendering is a cooking process. Cooking of animal tissues releases water, thereby reducing the volume of the material, and allows for the separation of fats and oils from proteins. Edible rendering processes animal tissues fit for human consumption, such as animal fat to create a wide range of products used in human food such as lard. The raw materials for inedible rendering include excess tissues not destined for edible rendering and animal tissues deemed unfit for human consumption, including condemned carcasses or animal tissues including dead, nonambulatory, or diseased animals. Much of the commercial rendering in the United States is directly linked to large abattoirs that handle a single species only, such as poultry processing plants. Those renderers that handle older cattle (over two years of age) represent the most important component of the BSE prevention program because they process those cattle at the highest risk of BSE.

Feed compounding represents a critical control point for preventing the contamination of cattle feeds with rendered protein products of ruminant origin. Accurate labeling and prevention of cross-contamination represent important components of BSE prevention at the feed mill. The Food and Drug Administration (FDA) regulates renderers and feed mills and has promulgated regulations related to mammalian protein in cattle feed. Evidence of compliance with FDA regulations provides added assurance of the safety of cattle feed and ingredients for the producer.

SUMMARY AND CONCLUSIONS. BSE is a zoonotic disease of major importance to both animal and human health. Effective prevention and control of BSE require coordinated efforts by national governments, the rendering and feed industries, veterinarians, and cattle producers. National governments implement import restrictions, promulgate regulations restricting high-risk behaviors such as the feeding of ruminant-derived rendered animal proteins, establish surveillance programs, support educational programs, and establish preparedness plans in case BSE is diagnosed. However, national government actions alone are insufficient to prevent BSE from occurring or to control BSE after it has been identified. Successful prevention and control require the full cooperation of agribusiness, veterinarians, and cattle raisers. Actions implemented on the farm by cattle raisers are an essential component of both prevention and control.

The critical pre-harvest BSE prevention actions relate to two major risk factors for introduction of the disease: purchase of exposed or BSE-affected cattle and the exclusion of potentially contaminated rendered animal proteins from cattle rations. The lack of live-animal diagnostic tests for the early detection of BSE exposed or affected cattle requires more attention to be placed on the identification and documentation of cattle purchases. Individual animal identification and good record keeping provide crucial documentation for potential BSE suspects and support successful risk management by buyers.

Cattle feeding is the cornerstone of BSE prevention on the farm. The most likely source of exposure to BSE for U.S. cattle is through contaminated feedstuffs containing infective material from BSE-affected cattle. Imported bulk feed ingredients, cross-contamination, and access of cattle to feeds intended for nonruminants represent potential sources of cattle feed risks. Attention needs to be directed toward identifying trustworthy and credible sources of high-quality feed ingredients and protecting against accidental contamination of cattle feed on-farm with potentially infected rendered animal proteins.

Identification and diagnostic evaluation of cattle demonstrating clinical signs compatible with BSE represent a national imperative to protect the United States both from the disease and from adverse public opinion. Aggressive surveillance for BSE and thorough documentation of all tests conducted on BSE suspects, including negative test results, are necessary for maintaining the credibility of our national BSE surveillance program and access to global markets for U.S. animal products.

Finally, effective BSE prevention, surveillance, and prompt response in the event of BSE identification in the United States require education of farm workers, agribusiness, and veterinarians. All those involved in animal agriculture need to understand the basic epidemiology of the disease, the clinical signs and progression, and the rudiments of successful prevention and control.

REFERENCES

Agrimi, U., G. Ru, F. Cardone, M. Pocchiari, and M. Caramelli. 1999. Epidemic of transmissible spongiform encephalopathy in sheep and goats in Italy. *Lancet* 353:560–561.

Brown, P., M. Preece, J.P. Brandel, T. Sato, L. McShane, I. Zerr, A. Fletcher, R.G. Will, M. Pocchiari, N.R. Cashman, J.H. d'Aignaux, L. Cervenakova, J. Fradkin, L.B. Schonberger, S.J. Collins. 2000. Iatrogenic Creutzfeldt-Jakob disease at the millennium. *Neurology* 55:1075–1081.

Bruce, M.E., R.G. Will, J.W. Ironside, I. McConnell, D. Drummond, A. Suttie, L. McCardle, A. Chree, J. Hope, C. Birkett, S. Cousens, H. Fraser, and C. J. Bostock. 1997. Transmissions to mice indicate that a new variant CJD is caused by the BSE agent. *Nature* 389:498–501.

Collinge, J., K.C.L. Sidle, J. Meads, J. Ironside, and A.F. Hill. 1996. Molecular analysis of prion strain variation and the aetiology of "new variant" CJD. *Nature* 383:685–690.

DEFRA. 2002a. Infectious Dose Experiment. Accessed 4/1/2002 in the science section of the BSE Web site under research (pathogenesis). Available at: http://www.defra.gov.uk/animalh/bse/index.html.

DEFRA. 2002b. Youngest and oldest cases by year of onset. Accessed 4/1/2002 in the statistics section of the BSE Web site at http://www.defra.gov.uk/animalh/bse/index.html.

DEFRA. 2002c. Horizontal transmission. Accessed 4/1/2002 in the epidemiology section of the BSE Web site at http://www.defra.gov.uk/animalh/bse/index.html.

DEFRA. 2002d. General Statistics. Accessed 4/1/2002 in the statistics section of the BSE Web site at http://www.defra.gov.uk/animalh/bse/index.html.

European Commission. 2000. Geographical Risk of Bovine Spongiform Encephalopathy (GBR). Scientific Opinion adopted by the Scientific Steering Committee at its meeting of 6 July 2000. Accessible at http://europa.eur.int/comm/food/fs/sc/ssc/out113_en.pdf.

Gordon, W.S. 1946. Advances in veterinary research: louping-ill, tick-borne fever and scrapie. *Vet. Rec.* 58:516–*f*518.

Hoinville, L.J., J.W. Wilesmith, and M.S. Richards. 1995. An investigation of risk factors for cases of bovine spongiform encephalopathy born after the introduction of the "feed ban." *Vet. Rec.* 136:312–318.

IETS (International Embryo Transfer Society). 1998. Manual of the International Animal Embryo Transfer Society: A procedural guide and general information for the use of embryo transfer technology, emphasizing sanitary procedures. 3rd ed. D.A. Stringfellow, S.M. Seidel, eds. Savoy, IL: IETS.

Marsh, R.F., R.A. Bessen, S. Lehmann, and G.R. Hartsough. 1991. Epidemiological and experimental studies on a new incident of transmissible mink encephalopathy. *J. Gen. Vir.* 72:589–594.

NCJDSU (National Creutzfeldt-Jakob Surveillance Unit). 2000. Creutzfeldt-Jakob Disease Surveillance in the UK, Ninth Annual Report. Edinburgh, Scotland. Accessed 4/1/2002 at http://www.cjd.ed.ac.uk/2000rep.html#ccs.

OIE (International Office of Animal Health). 2002a. Number of reported cases of bovine spongiform encephalopathy (BSE) worldwide. Accessed 4/1/02 at http://www.oie.int/eng/info/en_esbmonde.htm.

OIE (International Office of Animal Health). 2002b. Countries/Territories having reported cases of BSE in imported animals only. Accessed 4/1/02 at http://www.oie.int/eng/info/en_esbimport.htm.

OIE (International Office of Animal Health). 2001a. Appendix 3.6.3. Transmissible spongiform encephalopathy agents inactivation procedures. In International Animal Health Code (2001). OIE. Paris: OIE Press, 2001. Also available at: http://www.oie.int/eng/normes/mcode/A_00146.htm.

OIE (International Office of Animal Health). 2001b. Chapter 2.3.13. Bovine Spongiform Encephalopathy. In International Animal Health Code (2001). OIE. Paris: OIE Press, 2001. Also available at: http://www.oie.int/eng/normes/mcode/A_00066.htm.

Pruisner, S.B. 1982. Novel proteinaceous infectious particles cause scrapie. *Science* 216:135–144.

Rubenstein, R. 2000. Nature of the Infectious Agent. In *Transmissible Spongiform Encephalopathies in the United States*. Council for Agricultural Science and Technology. Ames, IA, 29–30.

Taylor, D.M., S.L. Woodgate, and M.J. Atkinson. 1995. Inactivation of the bovine spongiform encephalopathy agent by rendering procedures. *Vet. Rec.* 137:605–610.

Wells, G.A.H., A.C. Scott, C.T. Johnson, R.F. Gunning, R.D. Hancock, M. Jeffrey, M. Dawson, and R. Bradley. 1987. A novel progressive spongiform encephalopathy in cattle. *Vet. Rec.* 121:419–420.

Wells, G.A.H., S.A.C. Hawkins, R.B. Green, A.R. Austin, I. Dexter, Y.I. Spencer, and M. J. Chaplin. 1998. Preliminary observations on the pathogenesis of experimental bovine spongiform encephalopathy (BSE): an update. *Vet. Rec.* 142:103–106.

WHO/FAO/OIE. 2001. Joint Technical Consultation on BSE: Public health, animal health and trade. Accessed 4/1/02 at http://www.who.int/emc-documents/tse/docs/whocd-scsraph20018.pdf.

Wilesmith, J.W., G.A.H. Wells, M.P. Cranwell, and J.B.M. Ryan. 1988. Bovine spongiform encephalopathy: Epidemiology studies. *Vet. Rec.* 123:638–644.

Wilesmith, J.W., J.B.M. Ryan, and M.J. Atkinson. 1991. Bovine spongiform encephalopathy: epidemiology studies on the origin. *Vet. Rec.* 128:199–203.

Wilesmith, J.W., J.B.M. Ryan, and W.D. Hueston. 1992a. Bovine spongiform encephalopathy: case-control studies of calf feeding practices and meat and bone meal inclusion in proprietary concentrates. *Res. Vet. Sci.* 52:325–331.

Wilesmith, J.W., L.J. Hoinville, J.B. Ryan, and A.R. Sayers. 1992b. Bovine Spongiform Encephalopathy: Aspects of the clinical picture and analyses of possible changes 1986–1990. *Vet. Rec.* 130:197–201.

Wilesmith, J.W., G.A.H. Wells, J.B.M. Ryan, D. Gavierwiden, and M.M. Simmons. 1997. A cohort study to examine maternally-associated risk factors for bovine spongiform encephalopathy. *Vet. Rec.* 141:239–243.

Wilesmith, J.W., J.B.M. Ryan. 1997. Absence of BSE in the offspring of pedigree suckler cows affected by BSE in Great Britain. *Vet. Rec.* 141:250–251.

Will, R.G., J.W. Ironside, M. Zeidler, S.N. Cousens, K. Estibeiro, A. Alperovitch, S. Poser, M. Pocchiari, A. Hofman, and P.G. Smith. 1996. A new variant of Creutzfeldt-Jakob Disease in the UK. *Lancet* 347:921–925.

Wrathall, A.E., K.F.D. Brown, A.R. Sayers, G.A.H. Wells, M.M. Simmons, S.S.J. Farrelly, P. Bellerby, J. Squirrell, Y.I. Spencer, M. Wells, M.J. Stack, B. Bastiman, D. Pullar, J. Scatcherd, L. Heasman, J. Parker, D.A.R. Hannam, D.W. Helliwell, A. Chree, and H. Fraser. 2002. Studies of embryo transfer from cattle clinically affected by bovine spongiform encephalopathy (BSE). *Vet. Rec.* 150:365–379.

34

CALICIVIRUSES AND OTHER POTENTIAL FOODBORNE VIRUSES

M. GUO, J. VINJÉ, AND L.J. SAIF

INTRODUCTION. Foodborne diseases remain a global public health problem despite major advances and improvements in hygiene, the quality of food, water, and sanitation, detection of foodborne pathogens, and the surveillance systems during the last century. They not only cause millions of cases of gastroenteritis annually in individual countries but also are a social and economic burden. A recent estimate of foodborne diseases indicates that they are responsible for about seventy-six million illnesses, 325,000 hospitalizations, and five thousand deaths in the United States each year (Mead et al. 1999), costing ~$50 billion annually. Known pathogens including bacteria, viruses, fungi, and parasites cause an estimated fourteen million illnesses, sixty-thousand hospitalizations and eighteen hundred deaths through foodborne transmission each year in the United States, whereas the etiologic agents for most cases and hospitalizations (~80 percent) remain unknown. Among the known pathogens, Norwalk-like viruses or, using their more recent designation, noroviruses (NLVs) cause an estimated twenty-three million cases of acute, epidemic gastroenteritis in the United States annually, accounting for an estimated 66.6 percent of all cases caused by known foodborne pathogens, and 33 percent of annual hospitalizations caused by known foodborne agents (Mead et al. 1999), although most cases used for estimation lacked identifiable pathogens. All other foodborne viruses such as hepatitis A virus (HAV), rotavirus, and astrovirus are suggested to account for only ~0.6 percent of all cases caused by known pathogens. Thus, this chapter focuses mainly on the NLVs. In the Netherlands, the United Kingdom, Japan, and Australia, NLVs have emerged as the most common pathogen in foodborne outbreaks of acute gastroenteritis (Lewis et al. 1997; Vinjé et al. 1997; Wright et al. 1998; Inouye et al. 2000). Recognition of NLVs as the most important foodborne virus is largely attributable to improvements in molecular diagnostics in recent years, permitting detection of the many diverse NLV strains. In developing countries where foodborne disease surveillance is usually limited and the quality of food and water and sanitation may be compromised, many food-related disease outbreaks remain either unreported or unidentified because of inadequate monitoring and surveillance. Thus, the social and economic impact of foodborne diseases may be greatly underestimated. In this chapter, we summarize recent informa-

tion on foodborne viruses with an emphasis on NLVs and discuss implications for control and intervention in outbreaks of foodborne viral disease.

COMMON CHARACTERISTICS OF FOODBORNE VIRUSES. Foodborne viruses are generally transmitted by the fecal-oral route and person-to-person contact (direct transmission) or by fomites and contaminated food or water (indirect transmission). Infection by these viruses through common foods or water frequently occurs during outbreaks, with primary cases consisting of people who ingest contaminated food or water. Secondary person-to-person spread also occurs following primary infection. Unlike foodborne bacteria and fungi, viruses do not grow in foods or produce toxins, and their numbers in contaminated foods are relatively low. Induction of disease depends on their ability to survive in foods for long periods of time, and following ingestion, it depends on their survival in the acidic environment of the stomach and the bile salts of the upper small intestine, thereafter initiating infection in intestinal epithelial cells or other target cells (for example, hepatocytes for HAV and hepatitis E virus, HEV). However, foodborne viruses have adapted to survive in adverse environments for prolonged periods and to infect humans, perhaps exclusively, although animal reservoirs for some foodborne viruses (HEV and enteric caliciviruses) may exist. The remarkable stability of foodborne viruses in foods or water may be related to their common physical properties: nonenveloped, spherical, 25 to 75 nm in diameter and icosahedral symmetry. Despite the low numbers of these viruses in foods or water, after they are ingested, they are capable of inducing infection and subsequent illness, although asymptomatic infections often occur.

VIRUSES ASSOCIATED WITH FOODBORNE DISEASE. Foodborne viruses (Table 34.1) can be divided into three groups based on their prevalence of foodborne transmission and disease induction in the gastrointestinal tract: (1) established foodborne, enteropathogenic viruses such as NLVs; (2) potential foodborne, enteropathogenic viruses including rotaviruses, astroviruses, and adenoviruses that are associated with acute gastroenteritis; (3) nonenteropathogenic viruses such as HAV, HEV, and tick-borne encephalitis (TBE) virus. The HAV and HEV water- and foodborne viruses invade the host via the intestinal tract, but these hepatotropic viruses replicate mainly in the liver and are shed in feces (Cromeans 2001). Tick-borne encephalitis caused by an arbovirus is perhaps the only clearly established viral zoonosis that is transmitted via food (Dumpis et al. 1999). It occurs mainly in Eastern Europe because of limited vector (the ticks *Ixodes ricinus* and *I. persulcatus*) or risk factor distribution. Goats, sheep, and cattle are infected after being bitten by TBEV-infected ticks, and TBE virus is then shed in their milk. Consumption of raw milk or products made with unpasteurized milk contaminated with TBE virus can cause infection, but this rarely occurs. In nature, TBE is transmitted predominantly via the bite of ixodid ticks (Dumpis et al. 1999). Another group, the human enteroviruses including polioviruses, coxsackieviruses, echoviruses, and other enteroviruses (serotypes 68 to 71) replicate in the intestinal tract and are shed in large numbers in feces, but they cause poliomyelitis, myocarditis, and other diseases instead of gastroenteritis (Melnick 1996). Enteroviruses are commonly detected in sewage effluents and in bivalve shellfish, but they are only rarely associated with foodborne illnesses or outbreaks despite their fecal-oral transmission and widespread distribution in the environment (Lees 2000).

Enteropathogenic foodborne viruses cause both endemic and epidemic gastroenteritis with different distributions in different age groups. The illness is manifested primarily as diarrhea and vomiting and is generally self-limiting and less severe than other food-related diseases. But these viruses can cause severe disease with dehydration in infants and young children, the elderly, and immunocompromised patients (Glass et al. 2001; Green et al. 2001). They infect and destroy the mature enterocytes lining the mucosal surface of the intestinal tract, resulting in malabsorption, increased water retention caused by osmotic effects and crypt cell secretion, and consequently diarrhea. Common intestinal lesions include villous blunting and atrophy, villous fusion, crypt hyperplasia, or even loss of villi in severe disease, depending on the viruses and their virulence and dose.

Epidemiologic Considerations. Foods can be contaminated with viruses at any point during the course of food production (pre-harvest) and processing or thereafter (post-harvest, such as contamination by infected food handlers); they also can be contaminated as a result of virus infections (for example, TBE) or virus retention and accumulation in shellfish that are harvested as food. The most frequently contaminated pre-harvest food is shellfish; thus, shellfish consumption represents a major risk factor for foodborne illnesses. Filter-feeding oysters, clams, cockles, and mussels can efficiently concentrate virus (up to 100-fold) from a large volume of water (Lees 2000; Cromeans 2001). Pre-harvest contamination of shellfish often results from human sewage contamination of offshore waters

where bivalve shellfish grow and are harvested. Other agricultural products, such as fruits and vegetables, can be contaminated at the source either from infected agricultural workers, or from sewage water and reclaimed wastewater used for irrigation (Glass et al. 2000; Cromeans 2001). Water contamination frequently occurs from human sewage because of floods or insufficient sanitation treatment. For the post-harvest contamination of food, infected food handlers may be the major source of contamination because foodborne viruses may be shed in feces for a prolonged period (1~2 weeks or longer) and because asymptomatic infections often occur. In this case, poor personal hygiene can be a major risk factor for food contamination. Contaminated water used for washing and preparing food has been linked to outbreaks. In addition, virus contamination of drinking water and recreation water from feces or sewage also leads to outbreaks of acute gastroenteritis (NLVs) and outbreaks of hepatitis (HAV and HEV) (Fankhauser et al. 1998; Glass et al. 2000; Cromeans 2001). Understanding the potential sources of virus contamination has important implications for food handling and for the prevention and control of outbreaks of foodborne diseases.

The fecal-oral route and person-to-person contacts are thought to be the most common modes of transmission for foodborne viruses (Lüthi 2001). Airborne transmission (aerosols) has been suggested for NLV infections as well (Green et al. 2001; Marx et al. 1999; Marks et al. 2000). Different modes of transmission reflect the exposure sources. For outbreak control and prevention, identification of transmission mode and the contamination source is critical, but it is difficult given the low virus load in foods, water, and other environmental samples and the potential diagnostic void. However, major molecular breakthroughs in the past decade have led to development and application of new molecular assays that have rendered diagnosis practical and feasible. By using new, sensitive diagnostic assays for virus detection and typing, outbreaks can be traced to the suspected foods, water, or other sources, and small and local outbreaks can then be linked nationally and internationally (Noel et al. 1999; Ponka et al. 1999; Glass et al. 2000).

Like NLV, the HAV and HEV also are able to cause large-scale outbreaks, but differ in that HAV is usually spread by food, whereas HEV is transmitted predominantly via contaminated water.

FOODBORNE VIRUSES OTHER THAN THE HUMAN CALICIVIRUSES. Other viruses are occasionally reported as pathogens in outbreaks of foodborne illness, such as rotavirus, astrovirus, and adenovirus (Table 34.1) (Lees 2000; Cromeans 2001; Glass et al. 2001; Lüthi 2001). Their role in causing foodborne illness requires further evaluation based on outbreak investigations. In this section, we briefly review the properties of these other viruses and their role in foodborne disease.

Table 34.1—Characteristics of Foodborne Viruses

Viruses	Family	Size (nm) & buoyant density[a]	Morphology	Genome and serotypes[a]	Growth in cell culture	Example	Disease	Transmission vehicle	Reservoir
Established foodborne viruses–acute gastroenteritis									
Calicivirus	*Caliciviridae* "NLV" genus	28–35 nm, 1.36–1.41	Spherical, amorphous surface, ragged outline	ss[b],[c]RNA, 7.4–7.7 kb, 3 ORFs[d]	No	Norwalk virus	Epidemic acute gastroenteritis in all ages	Food, water, fomites	Humans Animals (?)
	"SLV" genus	32–40 nm, 1.37–1.38	Prominent surface cuplike hollows, "Star of David" configuration	ss(+)RNA, 7.3–7.42 kb, 2 ORFs	No, except PEC/Cowden[e]	Sapporo virus	Endemic acute gastroenteritis in young children	Food, water, fomites	Humans Animals (?)
Established foodborne viruses–systemic infections									
Hepatitis A virus (HAV)	*Picornaviridae*	27–32 nm, 1.33–1.34	Spherical, smooth outer edge	ss(+)RNA, 7.5 kb	Yes, but no CPE[g]	HAV	Acute hepatitis	Food (esp. shellfish), water	Humans
Hepatitis E virus (HEV)	Unclassified	32–34 nm, ~1.29	Amorphous surface, ragged outline	ss(+)RNA, 7.5 kb, 3 ORFs	No	Human HEV	Acute hepatitis	Mostly water	Humans, Pigs (?)
Tickborne encephalitis virus	*Flaviviridae*	40–50 nm	Spherical, nonenveloped	ss(+)RNA, 11 kb, 6 subtypes	Yes	Tickborne encephalitis virus	Encephalitis	Ticks, contaminated milk	Rodents, Insects
Potential foodborne viruses–acute gastroenteritis									
Rotavirus	*Reoviridae*	60–75 nm, 1.36–1.38	Double shelled, wheel-like capsids, nonenveloped	ds[f]RNA, 11 segments, 18.5 kb, 6 groups	Yes	Group A rotavirus	Severe diarrhea in infants and young children	Food, water, fomites	Humans, Animals (?)
Astrovirus	*Astroviridae*	27–30 nm, 1.33–1.42	Spherical, 5–6 pointed surface star	ss(+)RNA, 6.8–7.9 kb, 2 ORFs, 8 serotypes	Yes	Human astrovirus	Acute gastroenteritis	Food, fomites	Humans
Aichi virus	*Picornaviridae*	~30 nm	Spherical, small round virus	ss(+)RNA, ~8.2 kb	Yes	Aichi virus	Acute gastroenteritis	Food (oyster), fomites	Humans
Adenovirus	*Adenoviridae*	70–90 nm, ~1.345	Nonenveloped, icosahedral	dsDNA, 30–42 kb, enteric serotypes 40, 41	Yes	Human adenovirus	Acute gastroenteritis	Fomites, seafood (?)	Humans

[a] Buoyant density determined in a CsCl gradient (g/cm^3) except HEV whose buoyant density was determined in a potassium tartrate/glycerol gradient (g/ml).
[b] ss, single-stranded.
[c] (+), plus sense.
[d] ORF, open reading frame.
[e] PEC/Cowden, porcine enteric calicivirus Cowden strain.
[f] ds, double-stranded.
[g] CPE, cytopathic effect.

Rotavirus. Rotaviruses are nonenveloped viruses with a triple-layered capsid. They comprise seven sero-groups (A to G), each of which has distinct group anti-gens. Group A rotavirus is the most common cause of severe diarrhea in infants and young children world-wide, accounting for 9 percent of the 1.5 billion diar-rheal episodes and >800,000 deaths each year (Dessel-berger et al. 2001). The seroprevalence to group A rotavirus is nearly 100 in children by five years of age. Common symptoms of rotavirus infection include watery diarrhea, anorexia, vomiting, and dehydration. They have been detected in seafoods (such as shell-fish), vegetables, sewage effluents, and drinking water, but only a few acute gastroenteritis outbreaks have been linked directly to rotavirus following food con-sumption (Lees 2000; Le Guyader et al. 2000; Sattar et al. 2001). However, water as a vehicle for rotavirus spread has been linked to a number of outbreaks (Sat-tar et al. 2001). Rotaviruses are also associated with diarrhea in a wide range of animal species, inducing similar clinical features and intestinal lesions to the ill-ness in humans. Rotaviruses of human origin can infect several animal species, and zoonotic transmission of animal rotaviruses to humans is suspected although not fully documented (Desselberger et al. 2001; Glass et al. 2001).

Astrovirus. Human astoviruses cause mostly spo-radic and occasionally epidemic acute gastroenteritis in infants, young children, and the elderly worldwide (Monroe et al. 2001). Clinical symptoms include infrequent vomiting and mild diarrhea of a short dura-tion. Using more sensitive molecular assays (RT-PCR) for virus detection, astroviruses have been asso-ciated with 3–9 percent of cases of childhood diarrhea (Monroe et al. 2001) and have been implicated in a few food- and waterborne acute gastroenteritis out-breaks (Oishi et al. 1994; Lees 2000). Although astro-viruses have been detected in sewage, water, and from shellfish (Pinto et al. 2001; Le Guyader et al. 2000), their role in causing foodborne outbreaks requires fur-ther investigation.

Hepatitis A Virus (HAV). The HAV causes acute hep-atitis in humans, with a wide spectrum of symptoms including malaise, nausea, fever, headache, vomiting, diarrhea, abdominal pain, arthritis, and jaundice (Cromeans 2001). The illness is generally self-limiting, lasts for two to six weeks, and rarely leads to death. Infected young children often have mild or no apparent illness, in contrast to more severe and overt hepatitis in older children and adults. Recovery is complete in most cases and, consequently, long-term immunity is acquired. The HAV infects and replicates in hepato-cytes and also in Kupffer cells, and virus is released from liver cells into bile and then shed in stool. Virus shedding in stools is detectable by immunoassays within a week upon onset of jaundice and lasts for three weeks or longer.

The HAV infections are distributed worldwide, but the numbers of disease cases and outbreaks are closely related to levels of sanitation and personal hygiene (Cromeans 2001). In developing countries, less than 90 percent of children may be infected by six years of age, although most of them remain asymptomatic. Conse-quently, in developed countries such as the United States, only a small proportion of foodborne illnesses are attributed to HAV, and large outbreaks are rarely observed. The virus is transmitted by contaminated food, water, or fomites, but the most common mode of transmission is person-to-person contact (Cromeans 2001). In the United States, food- and waterborne out-breaks account for approximately 5–7 percent of the reported cases of hepatitis A. The HAV are especially resistant to low pH, heat, and drying and are extremely stable in foods, water, and under other environmental conditions. A major public health concern is that HAV can cause large-scale outbreaks that are frequently linked to contaminated foods, especially raw or lightly cooked seafood (clams, oysters, shellfish, and others), but also fruit and vegetables that can be contaminated by irrigation with wastewater or by infected agricul-tural workers (Cromeans 2001).

Hepatitis E Virus (HEV). The HEV causes both epi-demic and sporadic acute hepatitis in older children, young adults, and pregnant women. The ingestion of fecally contaminated drinking water is the primary source of infection (Cromeans 2001). Hepatitis E shares similar symptoms with hepatitis A and has been identified as the major cause of non-A, non-B hepatitis in developing countries, particularly in Asia and Africa (Cromeans 2001), causing both outbreaks and sporadic cases. A major concern is the high mor-tality rate (17–33 percent) among pregnant women with hepatitis E. Potential transmission by foods other than shellfish has not yet been determined, but sewage contamination of shellfish areas could be a potential source of transmission. Recently, an HEV-like virus was detected in swine (Cromeans 2001) and this swine HEV is closely related genetically and anti-genically to human HEV, raising public health con-cerns for potential animal reservoirs and cross-species transmission of HEV.

Adenoviruses. Adenoviruses have a large number of serotypes and cause respiratory, ocular, gastroenteritis, and other infections in humans, birds, or other animals (Lees 2000). There are at least forty-nine human aden-ovirus serotypes and many of them can replicate in the intestine, but only enteric types 40 and 41 are associ-ated with gastroenteritis (Lees 2000). The fastidious enteric adenoviruses cause about 10 percent of infantile gastroenteritis cases and are a common cause of gas-troenteritis in children. Illness is less severe but lasts slightly longer than that of rotavirus infections. Aden-oviruses are detected in sewage effluents, surface water, and from shellfish (Chapron et al. 2000; Lees

2000), but there have been no reports of adenovirus-associated outbreaks following seafood consumption.

Aichi Virus. Aichi virus is a newly recognized picornavirus, 30 nm in diameter and morphologically similar to astrovirus (Yamashita et al. 1991, 1998). It was first identified in 1989 from gastroenteritis outbreaks in humans following consumption of oysters (Yamashita et al. 1991). Aichi virus was detected in human stool samples from a few oyster-associated diarrhea outbreaks (Yamashita et al. 1993), but the source of virus was not identified in those outbreaks. The association of Aichi virus with foodborne diarrhea outbreaks in humans requires further studies.

ENTERIC CALICIVIRUSES. Enteric caliciviruses comprise a wide spectrum of genetically diverse, enteropathogenic caliciviruses infecting humans and animals. Since the first cloning and sequencing of the RNA genome of Norwalk virus in 1990, human caliciviruses (HuCVs) including Norwalk and related viruses (mostly NLVs) are being increasingly recognized as a leading cause of foodborne outbreaks of gastroenteritis in all ages worldwide. Meanwhile, enteric caliciviruses of animals are emerging pathogens associated with gastroenteritis in their respective animal host, and these viruses are closely related genetically to their human counterparts. In this section, we focus on the HuCVs and related foodborne illnesses, highlight progress, and discuss the food safety aspect of these viruses and related infections.

History and Classification. Caliciviruses, members of the *Caliciviridae* family, are small, nonenveloped viruses, 27–40 nm in diameter (Green et al. 2001). They have a broad host range and cause a wide spectrum of disease syndromes in their respective hosts, including gastroenteritis (humans, pigs, calves, chickens, cats, dogs, and mink), vesicular lesions and reproductive failure (pigs, sea lions, and other marine animal species), respiratory disease (cats and cattle), and a fatal, hemorrhagic disease (rabbits) (Bridger 1990; Ohlinger et al. 1993; Smith et al. 1998; Green et al. 2000, 2001; Guo et al. 2001a).

Members of the family *Caliciviridae* can be divided into four genera: (1) *Vesivirus*, (2) *Lagovirus*, (3) "Norwalk-like viruses" or noroviruses (NLV), and (4) "Sapporo-like viruses" or sapoviruses (SLV) (Green et al. 2000). *Vesiviruses* include feline calicivirus (FCV), vesicular exanthema of swine virus (VESV), and San Miguel sea lion virus (SMSV), which are well-characterized, "classical" animal caliciviruses with distinct, cup-shaped surface depressions. *Lagoviruses* are comprised of two closely related viruses, rabbit hemorrhagic disease virus (RHDV) and European brown hare syndrome virus (EBHSV) (Green et al. 2000). The RHDV causes a highly contagious, fatal liver necrosis and systemic hemorrhage in rabbits over three months

of age (Ohlinger et al. 1993). It has been used successfully as a biologic agent to control wild rabbit populations in Australia and New Zealand. Enteric caliciviruses from both humans and animals belong to either NLVs or SLVs. NLVs are spherical particles with indistinct surface structure but a ragged outline. SLVs consist of enteric caliciviruses with classic calicivirus morphology with distinctive, cuplike surface depressions arranged in a "Star of David" configuration. Examples include Sapporo, Manchester, Plymouth, and Parkville viruses, porcine enteric calicivirus (PEC), and mink enteric calicivirus (MEC) (Green et al. 2001; Guo et al. 1999; Guo et al. 2001a). Norwalk virus (NV), the prototype NLV, was first discovered in 1972 in stool samples from human volunteers fed fecal filtrates from children who were affected by an outbreak of acute gastroenteritis at an elementary school in Norwalk, Ohio (Kapikian et al. 2000). Discovery of both NV and rotavirus (Bishop et al. 1973) as enteric pathogens by using immune electron microscopy (IEM) stimulated and ultimately led to the discovery of a number of enteric viruses, including ones previously identified as "small round-structured viruses" (SRSV) or NLVs, "classical" enteric caliciviruses now known as "Sapporo-like" viruses (SLV), and astrovirus in the 1970s and 1980s (Appleton & Higgins 1975; Madeley & Cosgrove 1976; Chiba et al. 2000). Like NV, many other enteric caliciviruses were designated by the geographic locations from which they were first detected. A cryptogram has been used for denotation of caliciviruses as follows: host species/genus abbreviation/species abbreviation/strain name/year of occurrence/country of origin (Green et al. 2001).

Genetic, Structural, and Biological Properties. Recognition of the importance of NLVs in foodborne illness is largely attributed to advances during the last decade in an understanding of their molecular and biological properties, and development of molecular diagnostic assays. Deciphering of the RNA genome of numerous HuCVs has led to improvements in diagnostic tests and shed light on the genetic heterogeneity and evolutionary lineage of enteric caliciviruses. Knowledge of NV structure provides pivotal insights into viral capsid assembly and architecture. Thus, in spite of the lack of a cell culture system or animal model, substantial progress has been made on the genetic, structural, and biological characterization of NLVs as reviewed in the following section.

GENOMIC ORGANIZATION AND VIRAL PROTEINS. Caliciviruses possess a single-stranded, positive-sense RNA genome of 7.3 to 8.3 kb in length, excluding the 3' poly(A)+ tail, which carries a small protein (virion protein, VPg) covalently attached to its 5' end (Green et al. 2001). In general, there are two different genomic organizations of ORFs in the caliciviruses (Fig. 34.1) (Clarke & Lambden 2000; Green et al. 2001). In NLVs and vesiviruses, the RNA genome is comprised of three

ORFs, coding for a polyprotein (ORF1), a single capsid protein (ORF2) of 56 to 71 kDa, and a minor structural protein of 15 to 23 kDa, respectively. The RNA genome of the SLVs and lagoviruses is composed of two ORFs; ORF1 encodes both the capsid protein and the polyprotein, and the capsid protein is fused to and contiguous with the polyprotein in the same ORF. ORF2 is similar to the ORF3 of NLVs. The polyprotein requires proteolytic cleavage for release and maturation of nonstructural proteins (2C-like helicase, 3C-like

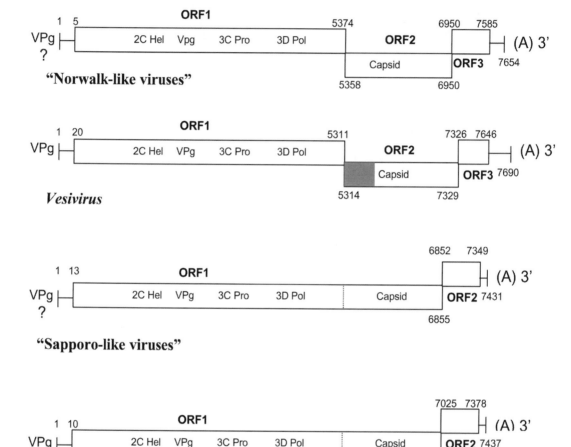

FIG. 34.1—Schematic representation of the genome organization of caliciviruses (Green et al., 2001). Genomic organization and reading frame usage are shown for representative strains in genera "Norwalk-like viruses" (NLVs, Hu/NLV/I/Norwalk/1968/US, Genbank accession no. M87661), *Vesivirus* (Fe/VV/FCV/F9/1958/US, M86379), "Sapporo-like viruses" (SLVs, Hu/SLV/Manchester/1993/UK, X86560), and *Lagovirus* (Ra/LV/RHDV/GH/1988/GE, M67473). Predicted helicase (2C Hel), protease (3C Pro), RNA polymerase (3D Pol) are indicated. Shaded region of ORF2 of vesiviruses illustrates leader sequence (~125 amino acids in length) of capsid precursor protein. The subgenomic RNA overlaps the genomic 3' terminal region from the predicted 5' end of capsid to 3' poly(A)+ tail, which has been detected in vesiviruses and a lagovirus (RHDV) but not the NLVs and SLVs.

protease, and 3D RNA-dependent RNA polymerase, RdRp), which are mediated by a caliciviral 3C-like cysteine protease. The polyprotein gene order of Southampton virus (SHV), a genogroup I(GI) NLV, has been determined as: NH2-p48-p41 (2C helicase)-p22-p16 (VPg)-p19 (3C protease)-p57(3D RdRp) (Clarke and Lambden 2000). Recently, the recombinant SHV 2C-like helicase protein (p41) was shown to be a nucleic acid-independent nucleoside triphosphatase with no detectable helicase activity (Pfister and Wimmer 2001).

Caliciviruses are unusual animal viruses because they have only a single major capsid protein that is capable of spontaneously self-assembling into virus-like particles (VLPs) when expressed in baculovirus (Jiang et al. 1992b; Clarke and Lambden 2000). The NV capsid protein is proteolytically cleaved by trypsin at amino acid K227/T (NV numbering), releasing a 32 kDa soluble protein as detected in NV-positive stools (Greenberg et al. 1981; Hardy et al. 1995). This cleavage by trypsin may partially explain why infected adults shed large amounts of soluble antigen in stools, although intact virions are resistant to trypsin cleavage. The 3′ terminal ORF protein has been identified as a minor structural protein in RHDV, NV, and FCV (Glass et al. 2000; Green et al. 2001).

STRUCTURAL PROPERTIES. Recently, the x-ray crystallographic structure of the rNV VLPs has been determined by using electron cryomicroscopy (Prasad et al. 1999), providing insights into the structure and relevant functions of the NV capsid. The capsid is composed of 90 dimers of the capsid protein, each forming an arch-like capsomere. Each capsid protein has two distinct domains, a shell (S) and a protruding (P) domain, which are connected by a flexible hinge. The S domain (aa 1 to 225 for NV) is folded into an eight-stranded β-barrel, forming the inner structure of the capsid. The P domain is further divided into two subdomains, P1 (two discrete regions, aa 226–278 and aa 406 to 520) forms the sides of the arch of the capsomer and P2 (aa 279 to 405) is located on top of the arch (at the exterior of the capsid) and is formed by the hypervariable capsid region 2. It is likely that P2 subdomain accounts for cell attachment, host specificity, and antigenic types of the NLVs (Prasad et al. 1999). The preceding NV capsid structure is similar to that of the classical animal caliciviruses. However, the morphologically typical caliciviruses (SLV and vesiviruses) have a longer P2 subdomain and a different shape at the top of the arch, resulting in their distinctive, morphological appearance by negative stain EM, in contrast to the feathery appearance of the NLVs.

ANTIGENIC SEROTYPES AND GENETIC DIVERSITY. Serotypes of NLVs and SLVs remain undefined because of the lack of a cell culture system. Four antigenic types of NLVs evaluated by cross-challenge studies in human volunteers, IEM, or a solid-phase IEM were identified: NV, Taunton, Hawaii (HV), and Snow Mountain viruses (SMV). So far, only one SLV antigenic type, Sapporo virus, is described (Kapikian et al. 1996; Jiang et al. 2000; Green et al. 2001). It is assumed that such antibody reactivity reflects the reactivity of neutralizing antibodies, but this may not be the case. By using ELISA based on hyperimmune antisera to VLPs, NV, Desert Shield virus (DSV), Mexico virus (MxV), Grimsby virus (GRV), HV and Sapporo viruses represent distinct antigenic types. Antigenic types of animal enteric caliciviruses have not been characterized, but multiple serotypes may exist, based on the genetic divergence observed.

GENETIC DIVERSITY OF NLVS AND SLVS. The genomes of both NLVs and SLVs show great genetic diversity and can be divided into genogroups that can differ up to 50 percent from each other (Ando et al. 2000; Vinjé et al. 2000a; Vinjé et al. 2000b; Schuffenecker et al. 2001; Green et al. 2001). Each genogroup can be subdivided into several genetic clusters or genotypes, which provisionally are classified based on >80 percent amino acid similarity of their complete capsid protein (Ando et al. 2000; Green et al. 2000). The NLVs can be divided into at least three genogroups (Fig. 34.2) with NV, DSV, and SHV placed in genogroup GI, whereas SMV, HV, Toronto virus (TV), Bristol virus (BV), and Lordsdale virus (LV) are included in GII. The BEC Newbury agent-2 and Jena virus constitute a new genogroup: GIII (Ando et al. 2000). Further analysis revealed that NLVs comprise at least seventeen genotypes (Ando et al. 2000). The SLVs can be divided into at least three different genogroups (Fig. 34.2), with Sapporo and Manchester viruses placed in SLV genogroup I (GI), London/92 viruses in GII, and PEC/Cowden viruses in GIII (Schuffenecker et al. 2001). The recently descibed enteric calicivirus from mink (MEC) may potentially be a cluster of yet another genogroup (GIV) (Guo et al. 2000a). This genetic classification scheme in general has correlated with antigenic diversity (Noel et al. 1997; Jiang et al. 1997; Vinjé et al. 2000b). However, it is unclear what percentage sequence identity will accurately reflect serotype differences.

BIOLOGICAL PROPERTIES. Both NLVs and SLVs survive in extreme environments for prolonged periods of time. The NV remains infectious for human volunteers following treatment with acid (at pH 2.7), ether, heating, and chlorine (3.75 to 6.25 mg/L) (Green et al. 2001), rendering virus elimination from contaminated water, foods, or the environment difficult. In contrast, poliovirus type 1, human rotavirus (Wa), simian rotavirus (SA11), and f2 bacteriophage are inactivated by similar concentrations of chlorine. However, NV is inactivated by treatment with 10 mg/L of chlorine.

The NLVs and most SLVs (except PEC/Cowden) remain refractory to propagation in cell culture, and no susceptible animal model is yet available for HuCVs. The NV VLPs specifically bind to a variety of human

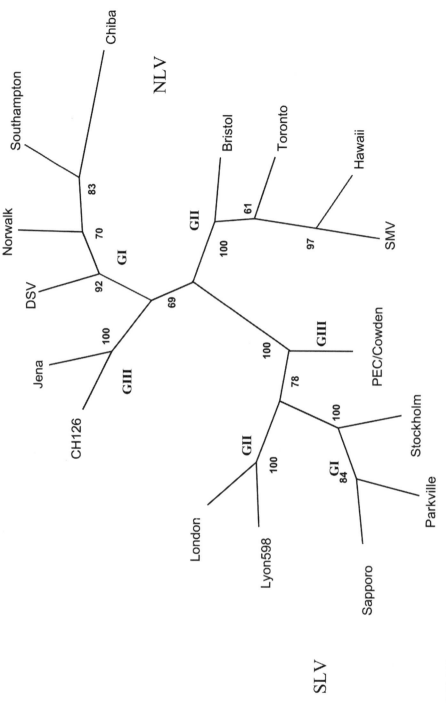

FIG. 34.2.—Unrooted phylogenetic tree based on entire capsid amino acid sequences from 10 NLV strains (Norwalk; M87661, Southampton; L07418, Desert Shield (DSV); U04469, Chiba; AB022679, Hawaii; U07611, Snow Mountain (SMV); U75682, Toronto; U02030, Bristol; X76716, Jena; AJ011099, CH126; AF320625) and 6 SLV strains (Sapporo; U65427, Parkville; U73124, Stockholm; AF194182, London; U95645, Lyon598; AJ271056, PEC/Cowden; AF182760). Consensus phylogenetic tree was constructed with ProML (Phylip 3.6) and bootstrap values are indicated as percentage of 100 replicates.

and animal cells *in vitro*, but they were internalized with low efficiency, which may partially explain their failure to grow in cell culture (White et al. 1996). The PEC Cowden strain, belonging to the SLV genus, is the only cultivable enteric calicivirus, but its growth requires addition of an intestinal contents preparation (ICP) from uninfected gnotobiotic pigs in the medium (Flynn & Saif 1988a; Parwani et al. 1991). To date, the identity of the virus growth promotion factors in ICP remains elusive. Interestingly, PEC cell culture adaptation and passage led to its virulence attenuation (Guo et al. 2001b), which may be associated with three clustered amino acid substitutions in the hypervariable capsid region that forms the externally located P2 subdomain on the virus surface and corresponds to the binding region of the NV (rVLPs) to human and animal cells *in vitro* (White et al. 1996; Guo et al. 1999; Prasad et al. 1999).

The NLVs and SLVs have a relatively broad host range, infecting humans, calves, pigs, mink, chickens, and other animal species. However, the viruses from humans and each animal species may be host-specific, although little is known about cross-species transmission among enteric caliciviruses. Enteric caliciviruses of animals, including PEC and MEC, are emerging pathogens associated with diarrhea in their respective animal hosts and are genetically related to SLVs infecting humans (Saif et al. 1980; Bridger 1990; Guo et al. 1999; Guo et al. 2001a). In addition, viruses genetically related to NLVs in humans have been detected in swine and cattle in the UK (Bridger 1990; Dastjerdi et al. 1999), Germany (Liu et al. 1999), Japan (Sugieda et al. 1998) and the Netherlands (van der Poel et al. 2000), raising public health concerns for potential cross-species transmission to humans from an animal reservoir. However, more data is needed to assess the zoonotic potential of these animal caliciviruses, including their genetic and antigenic similarity to NLV and SLV and their possible interspecies transmission.

REPLICATION STRATEGY. One striking feature of the replication strategy of caliciviruses is that vesiviruses and RHDV produce a 3′ co-terminal subgenomic RNA that encodes the capsid and the 3′ terminal ORF protein (Estes et al. 1997; Green et al. 2001). This subgenomic RNA is packaged into virions in RHDV and may facilitate the high-level capsid expression at a later stage of viral replication. Because the NLVs are uncultivable in cell culture, it is unknown whether a subgenomic RNA is produced in infected cells. However, an RNA molecule of 2.3 kb was detected in stools from an NV-infected volunteer, suggesting that subgenomic RNAs are produced during NLV replication (Jiang et al. 1993). The nonstructural polypeptides of caliciviruses are similar to those of picornaviruses in terms of their sequence and function similarities. But, unlike picornaviruses, caliciviruses have no ribosome entry sites in the 5′ terminal genomic region. Another property of cultivable caliciviruses is that the VPg (~15 kDa) is covalently linked to the 5′ end regions of genomic and subgenomic RNA molecules (Green et al. 2001). Although its definitive function remains unknown, VPg may function in initiation of transcription, translation, or both of viral genomic and subgenomic RNAs.

Clinical Features and Pathogenesis. Symptoms of NLV infections include nausea, vomiting, diarrhea, fever and abdominal cramps (Kapikian et al. 1996; Green et al. 2001). The illness is generally mild and self-limiting, featuring the sudden onset of vomiting, diarrhea, or both. Symptoms last twelve to twenty-four hours with an incubation time from twelve to sixty hours. The spectrum of illness may vary widely in individual patients, and subclinical infections and secondary transmission may be common. Prolonged virus shedding has been reported (Graham et al. 1994; Okhuysen et al. 1995), and a higher rate of asymptomatic infections facilitate virus spread and have implications for intervention strategies during outbreaks. Symptoms of gastroenteritis caused by SLVs were indistinguishable from those caused by NLVs (Estes et al. 1997; Green et al. 2001). However, SLVs induced diarrhea predominantly in affected young children. The predominance of symptoms may be related to the infectious dose and the age of the patient (Estes et al. 1997; Green et al. 2001). Clinical manifestations of PEC infection in pigs and BEC infections in calves are similar to NLV and SLV infections in humans, for example, mild to moderate diarrhea, short incubation period (two to three days) and illness duration (three to seven days) (Flynn et al. 1988b; Bridger 1990; Guo et al. 2001b). The human enteric caliciviruses preferably infect and destroy the mature villous enterocytes, but not the crypt enterocytes of the proximal small intestine, resulting in villous atrophy, reduced digestion, and malabsorptive diarrhea (Estes et al. 1997; Green et al. 2001). It is unknown whether other portions of the small intestine are involved in infections, and the site of virus replication has not been defined.

The PEC and BEC induced similar intestinal lesions to those reported in humans in the duodenum and jejunum of orally inoculated gnotobiotic (Gn) pigs and calves, respectively (Flynn et al. 1988b; Bridger 1990; Guo et al. 2001b). The duodenum and jejunum were the major replication sites for PEC and BEC. Furthermore, typical diarrhea and intestinal lesions were induced in pigs following intravenous (IV) inoculation with wild type (WT) PEC/Cowden and viremia occurred following oral or IV PEC infection (Guo et al. 2001b). The PEC infection of pigs may be a useful animal model for studies of the comparative pathogenesis of human NLV and SLV infections.

Detection and Diagnosis: General Procedures. Acute, infantile gastroenteritis caused by enteric viruses such as rotavirus, astrovirus, and enteric adenovirus may manifest similar or indistinguishable clinical signs, but acute gastroenteritis outbreaks caused by NLVs present some distinctive signs. A provisional diagnosis of illness caused by NLVs can be made during outbreaks of gastroenteritis if the following criteria are met: (1) absence of bacterial or parasitic pathogens; (2) vomiting in more than 50 percent of the cases; (3) duration of illness from twelve to sixty hours; and (4) an incubation period of twenty-four to forty-eight hours (Kaplan et al. 1982). However, a definitive diagnosis requires laboratory detection of viral antigen, genomic RNA, or antibody responses by using appropriate diagnostic assays. Following the cloning and sequencing of the NV and SHV genomic RNAs (Jiang et al. 1990; Lambden et al. 1993), first-generation RT-PCR assays were developed. These first assays appeared to be too specific (Jiang et al. 1992a; Moe et al. 1994), prompting the development of more refined primers based on the growing number of sequence information. To date, RT-PCR assays are widely used to detect NLVs and SLVs in stools and other clinical, environmental, and food samples (Atmar & Estes 2001). Later, enzyme immunoassays (EIAs) were developed using recombinant VLPs from NLV capsid expression in baculovirus (Jiang et al. 2000). Some of the available diagnostic assays for NLVs and other foodborne viruses are listed in Table 34.2, whereas the currently available baculovirus-expressed NLV capsid antigens and related assays are listed in Table 34.3.

ANTIGEN DETECTION. Immune electron miscroscopy (IEM) was the first diagnostic test for NV infection (Kapikian et al. 2000). It detects both virus in stools and serum antibodies, and is still widely used for the detection of NLVs in fecal samples worldwide. However, IEM is laborious, time-consuming, and relatively insensitive, having a detection limit of approximately 10^5 virus particles/ml. Direct EM remains a valuable diagnostic test for viral gastroenteritis because it is rapid and able to detect a wide range of enteric viruses in stool samples. However, because of the relative insensitivity of the method, it has limited value for the detection of NLVs or SLVs in foods.

For antigen detection using ELISA, a major advance came with the production and use of VLPs following NV capsid expression in baculovirus (Jiang et al. 2000; Atmar and Estes 2001), providing antigen (VLP) for antibody detection and for hyperimmune antiserum production. To date, VLPs have been produced for NV, SMV, HV, DSV, TV, MxV, GRV, SHV, LV, SV, and PEC, and EIAs utilizing hyperimmune antiserum have been developed for detection of each respective homologous virus in stools (Atmar and Estes 2001; Guo et al. 2001c). However, these new antigen ELISAs tend to be less cross-reactive, detecting only native viruses or closely related viruses (some but not all in the same genogroup) and giving rise to lower virus detection rates (~20 percent) in specimens from outbreaks (Jiang et al. 2000; Atmar & Estes 2001). The EIAs using monoclonal antibodies to native NV, native SMV, or NV VLPs have been shown to be more sensitive (Atmar & Estes 2001). Identification of a common epitope for GI NLVs may lead to development of a broadly reactive EIA for antigen detection (Hale et al. 2000).

ANTIBODY DETECTION. The EIAs using VLPs as coating antigen have been developed to detect IgG, IgA, or IgM antibody in serum samples for NV, MxV, TV, SHV, LV, SV, and PEC seroprevalence, immunity evaluation, or diagnosis (Atmar and Estes 2001; Guo et al. 2001c). These tests are more sensitive than assays using human reagents and are broadly reactive, but detect heterologous seroresponses at lower frequency and magnitude in comparison with the homologous seroresponse (Jiang et al. 2000; Atmar and Estes 2001). The antibody-detection EIAs have been widely used for seroprevalence studies, revealing the high seroprevalence of NLV infections in most age groups in both developed and developing countries (Jiang et al. 2000, Atmar and Estes 2001).

NUCLEIC ACID DETECTION. Using the appropriate broadly reactive RT-PCR primers (Ando et al. 1995; Green et al. 1995; Jiang et al. 1999; Vinjé and Koopmans 1996; Vinjé et al. 2000a), a wide range of distantly related SLVs and NLVs, and even SLVs and NLVs infecting calves and swine (van der Poel. 2000; Guo et al. 1999), can be detected, rendering the RT-PCR assay as the routine diagnostic method of choice. In addition, automatic sequencing of PCR products following amplification provides precise genetic information about the viruses detected, making it possible to trace the common exposure sources, to link small and local outbreaks to national or international ones, and to assess evolutionary trends and potential recombination of circulating NLV strains or to detect and differentiate NLVs and SLVs from humans and animals (Dowell et al. 1995; Noel et al. 1999; Berg et al. 2000; Glass et al. 2000). The majority of primers were designed based on the most conserved sequence in the RNA polymerase region (Ando et al. 2000; Atmar and Estes 2001). Modifications such as lower annealing temperature, nested PCR, or seminested PCR may increase the sensitivity of the assay, but the potential for carryover contamination needs to be eliminated or controlled. A combination of RT-PCR with hybridization assays (dot blot or slot blot hybridization, Southern blot hybridization, or liquid hybridization) increases the specificity and sensitivity of virus detection and can be used for NLV and SLV genotyping (Atmar and Estes 2001). Some new detection formats include a reverse line blot hybridization for simultaneous detection and genotyping (Vinjé and Koopmans 2000) and an RT-PCR-DNA enzyme immunoassay for confirmation of RT-PCR amplification (Schwab et al. 2001). These assays simplify postamplification analysis, increase the assay sensitivity and specificity, and decrease the time for sample examination.

Detection and Diagnosis in Foods/Shellfish. A recent estimate indicates that more than 60 percent of all foodborne disease outbreaks in the United States are caused by NLV (Mead et al. 1999; Parashar and Monroe 2000). Until recently, however, implicated food items were rarely tested for these viruses because of the complexity of different food commodities with respect

Table 34.2—Detection Limits of Diagnostic Methods for Foodborne Viruses

Viruses	Quantity in clinical specimens (particles/g)	EM or IEM[a] (particles/g)	Enzyme Immunoassay[b] Antigen detection	Enzyme Immunoassay[b] Antibody detection	RT-PCR[b] Detection in clinical sample	RT-PCR[b] Detection limits in foods	Hybridization[b] (particles/ml)
Caliviruses			0.025 ng antigen		(10 molecules of RNA	5–200 PCR units/1.5 g	~10^4 genome
Norwalk-like viruses	≤ 10^{7-8}	+ (10^{5-6})	++++	++++	++++	ND[c]	++
Sapporo-like viruses	≤ 10^{7-8}	+ (10^{5-6})	+++	++++	++++		
Group A rotavirus	≤ 10^{8-12}	++++ (10^{5-6})	+++	++++	++++	100 FFU[d]/g (nested)	2.5×10^5 or ~10^3 PFU[e]
Astrovirus	10^{7-8}	+ (10^{5-6})	+++	+++		ND	ND
Adenovirus	10^{7-8}	++ (10^{5-6})	+++	+++	++++ (PCR)	ND	ND
Hepatitis A virus	10^{7-8}	++ (10^{5-6})	+++	++++	++++	0.2–20 PFU/g	~10^4
Hepatitis E virus	10^{6-7}	+ (10^{5-6})	+++	+++	~10^3 particles	ND	ND

[a] EM or IEM, electron microscopy or immune electron microscop; +, detectable with low sensitivity; ++, moderate sensitivity; +++, easily detected.
[b] For enzyme immuno assay, RT-PCR and nucleic acid hybridization, +, low sensitivity; ++, moderate sensitivity; +++, high sensitivity; ++++, the highest sensitivity.
[c] ND, not determined.
[d] FFU, focus-forming unit.
[e] PFU, plaque-forming unit.

343

Table 34.3—Currently Available Baculovirus-Expressed Capsids of NLVs and SLVs and Related Enzyme Immunoassays (Modified from Jiang et al. 2000)

Genus[a]	Genotype	Recombinant capsid	Enzyme immunoassay		IgM	IgA	Reference
			Antigen	Antibody			
NLV	Norwalk	Norwalk	+	+	+	+	Jiang et al. 1992b
	Mexico	Mexico	+	+	+	+	Jiang et al. 1995
		Toronto	–	+	–	–	Leite et al. 1996
	Lordsdale	Lordsdale	–	+	–	–	Dingle et al. 1995
		Grimsby	+	+	–	–	Hale et al. 1999
	Hawaii	Hawaii	+	+	–	–	Green et al. 1997
SLV	Sapporo	Sapporo	–	–	–	–	Numata et al. 1997
		Houston/86	–	+[b]	–	–	Jiang et al. 1999
	Houston/90	Houston/90	+[b]	+[b]	–	–	Jiang et al. 1999
	PEC	PEC/Cowden	+	+	–	–	Guo et al. 2001c

[a] NLV, Norwalk-like virus; SLV, Sapporo-like virus; PEC, porcine enteric calicivirus.
[b] Methods are under development (Jiang et al. 2000).

to food sample processing and the presence of food-related RT-PCR inhibitors. Testing of implicated foods for NLV on a broader scale will give us a better understanding of the true burden of NLV foodborne outbreaks. The ideal method for the detection of NLVs in food and environmental samples would comprise the detection of infectious virus. However, because NLVs cannot yet be propagated in cell culture, the application of RT-PCR to the detection of NLV has emerged as the most important method to detect these viruses in implicated foods. Because of the historical association of shellfish consumption with viral foodborne disease, most of the methods developed for virus concentration and detection have been done using shellfish as the model food (Jaykus et al. 2000). Numerous methods have been described, focusing on different methods for the efficient recovery of viruses and the effective removal of PCR inhibitors from bivalve molluscan shellfish (oysters, calms, and mussels) that in most cases are expected to be contaminated with low levels of viruses (Shieh et al. 2000). The RT-PCR inhibitors, such as acidic polysaccharides, glycogen, and lipids, can be present in varying amounts, and the only method to monitor their effective removal is the use of an internal RNA standard (Atmar and Estes 2001). Most recently, methods for the successful detection of NLV in foods other than shellfish were described (Schwab et al. 2000; Leggitt and Jaykus 2000). Disadvantages of these methods include multiple sample manipulation steps, which may result in lower recovery and possible degradation of viral RNA, and the limitations that RT-PCR–based methods fail to discriminate between viable and inactivated viruses. Yet, tracing of foodborne viral contamination using the newest methods for food processing and RT-PCR may give us improved strategies for their prevention and control.

Epidemiology: General. Our understanding of the epidemiology of NLV infections has relied largely on the application of newly developed sensitive and specific diagnostic tools to investigate acute gastroenteritis outbreaks. Previously (from the 1970s to 1980s) using

less sensitive diagnostic tests for outbreak investigations, only 42 percent of acute gastroenteritis outbreaks were attributed to NLVs in the United States (Kaplan et al. 1982). However, in recent years sensitive molecular assays have been applied for outbreak investigations, showing that NLVs have emerged as the most common cause of outbreaks of foodborne gastroenteritis worldwide (Lewis et al. 1997; Vinjé et al. 1997; Fankhauser et al. 1998; Wright et al. 1998; Glass et al. 2000). The outbreaks occur year-round, but have a cold weather seasonality (Mounts et al. 2000; Green et al. 2001).

SEROPREVALENCE. Seroprevalence studies indicate that almost all children under five years of age have antibodies to the NLVs both in developed and developing countries (Glass et al. 2000), correlating with frequent infection of young children by NLVs (Pang et al. 2000). Seroprevalence is higher in almost every age group in developing countries than in developed countries (Jiang et al. 2000). In the United States, most children acquired serum antibodies to NV from five to fifteen years of age. In developing countries, infants had a high seroprevalence (90–99 percent) to NV and MxV at birth, which gradually decreased during their first seven to eleven months of life (Jiang et al. 2000). The seroprevalence in young children then increased sharply after one to three years of age, approaching 100 percent at eight to nine years of age and remaining high (89–98 percent) lifelong. High antibody prevalence in both children and adults suggests widespread NLV infections in humans worldwide. Because of a lack of diagnostic reagents, little is known about the prevalence of enteric calicivirus infections in animals. Recently, a high antibody prevalence (80–100 percent) to PEC in pigs of all ages was detected by ELISA using PEC VLPs as coating antigens (Guo et al. 2001c) in a PEC-infected Ohio swine herd (Guo and Saif 2001, unpublished data).

INCIDENCE OF NLVs. Specific incidence data for sporadic cases of gastroenteritis caused by NLVs in the United States is lacking. However, family studies have shown that nearly every family member experiences more than one episode of gastroenteritis each year in the United States, and severe or fatal illness often occurs in young children and the elderly. It is estimated that every child experiences 1.5–2.5 episodes per year, accounting for two million physician visits and 160,000 hospitalizations (Glass et al. 2000). In the Netherlands, NLVs accounted for 11 percent of the community cases of gastroenteritis during a year (de Wit et al. 2001) and 5 percent of the cases with acute gastroenteritis consulting a physician (Koopmans et al. 2000).

IMPORTANCE OF NLVS IN OUTBREAKS OF ACUTE GASTROENTERITIS. The NLVs have been established as the most important cause of outbreaks of acute gastroenteritis in the United States, United Kingdom, the Netherlands, Australia, and Japan. Early studies suggested that NLVs were responsible for ~42 percent of acute gastroenteritis outbreaks in the United States (Kaplan et al. 1982). Recently, more than 90 percent of reported outbreaks of viral gastroenteritis in the United States and 87 percent of all outbreaks in the Netherlands were caused by NLVs (Glass et al. 2000; Vinjé et al. 1997). The major public health concern about NLVs is that they are able to cause large-scale outbreaks in group settings. Studies of ninety outbreaks of nonbacterial gastroenteritis reported to the Centers for Disease Control and Prevention indicated that more outbreaks occurred in nursing homes and hospitals (43 percent) than in restaurants (26 percent), schools (11 percent), vacation settings (11 percent), or after oyster consumption (6 percent) (Fankhauser et al. 1998). Foodborne spread (21 percent) was the most common mode of transmission, followed by person-to-person contact (11 percent), oyster consumption (6 percent), or waterborne (3 percent).

The NLVs are also responsible for most acute gastroenteritis outbreaks in military personnel during field or shipboard maneuvers, affecting hundreds of soldiers each time. Recently, NLVs have been recognized as an important cause of endemic acute gastroenteritis in young children and the elderly (Pang et al. 2000; de Wit et al 2001). In Finland, NLVs were detected as frequently as rotaviruses (29 percent) in young children (<2 years of age) with diarrhea (Pang et al. 2000). Based on the high seroprevalence of NLV antibodies in children worldwide, childhood infections may be universal and repeated infections may be common, although the frequency of infection remains unknown, and usually only short-term or incomplete immunity is induced following infection (Jiang et al. 2000; Matsui and Greenberg 2000). Nosocomial infections with NLVs may be common, and infection of immunocompromised patients may induce severe and long-lasting illness (Glass et al. 2000).

TRANSMISSION. The NLVs are usually transmitted by the fecal-oral route and person-to-person contact or by fomites and contaminated food or water. Airborne transmission has been confirmed in some nosocomial infections and natural outbreaks (Marx et al. 1999; Marks et al. 2000; Green et al. 2001). Secondary attack rates (>50 percent) of family members or close contacts are high. The NLVs are highly infectious, and very few viruses (<100) are needed to induce infection (Glass et al. 2000). Infections spread rapidly, with an attack rate of 50–90 percent of individuals in group settings. In addition, NLVs are very stable in water, under adverse environmental conditions and likely in food, facilitating their spread.

Epidemiology of Foodborne Infections. Exposure to foodborne pathogens such as NLVs can occur via drinking water (associated with fecal contamination), seafood (wastewater disposal), fresh produce (irrigated or processed with contaminated water), or food items prepared by an infected foodhandler. Of these, seafood has been recognized as one of the most common vehicles for NLV gastroenteritis (Shieh et al. 2000). Other foods that have been implicated include salads, sandwiches, bakery products, orange juice, fruits, and vegetables (Fleet et al. 2000; Parashar and Monroe 2001). The source of contamination for these foods is through an infected foodhandler or, indirectly, via contaminated waters or soils. The application of RT-PCR for the detection of NLV genome in clinical specimens has shown that NLVs are among the most important causes of outbreaks of acute gastroenteritis worldwide (Glass et al. 2000). In several surveillance studies in developed countries with available transmission data, the proportion of outbreaks where food commodities were implicated ranged from 13 percent in Australia to 27 percent in the United States and 60 percent in Japan (Fleet et al. 2000; Parashar and Monroe 2001). However, it is likely that the true burden of foodborne illness is far greater because most cases are sporadic and often not reported.

Prevention, Control, and Immunity. Prevention of food and water contamination by enteric viruses either pre-harvest or post-harvest is one of the most efficient intervention strategies to reduce foodborne viral outbreaks (Richards 2001). The most effective and reliable approach to control shellfish-related viral diseases is to harvest them from areas with good water quality and not from waters contaminated with feces or sewage (Lees 2000). A less effective approach is the practice of depuration and relaying (Lees 2000; Richards 2001). However, the currently used fecal coliform indicator of water quality may not always reflect the presence or absence of enteric viruses. Commercial depuration may fail to satisfactorily remove enteric viruses because these viruses are eliminated from shellfish at slower rates than fecal coliforms, and the infectious dose is very low. Thus, proper heating, boiling, or cooking should be done before shellfish consumption. Another water quality indicator, the F+ RNA-bacteriophage, has been shown to be a better indication for fecal pollution. When possible, untreated feces or sewage sludge should not be used to grow foods such as vegetables and fruits. Fecally polluted waters should not be used for irrigation or food washing. Potential cross-contamination of foods with viruses should be avoided during harvesting, storage, and transportation by adopting good manufacturing practices (Richards 2001).

For food handlers, good personal hygiene practices such as handwashing and exclusion of sick persons from food handling, as well as prompt, rigorous cleaning of the area if vomiting occurred, are important to reduce virus transmission. However, prolonged virus shedding in stools and asymptomatic infections may increase the risk of exposure. In this regard, immunization of food handlers with enteric virus vaccines may prove beneficial. An important strategy is to educate the food industry, food handlers, and consumers to implement the food safety rules. It is important to implement the hazard analysis critical control point (HACCP) system in food handling and processing, food service, and retail.

Foodborne viral outbreaks often occur in group settings, and person-to-person contact, aerosols, and fomites can easily spread viruses. Ill persons should be isolated and the contaminated materials or environment should be cleaned and disinfected. After being identified, the ongoing common source of infection (contaminated foods, water, or foodhandlers) should be removed. One difficulty in outbreak management is the identification of asymptomatically infected food handlers or hospital staff who may shed viruses in stool up to two weeks after exposure (Estes et al. 1997; Green et al. 2001).

Currently, no vaccines are available to prevent human enteric calicivirus infections. Their development is hindered by the lack of a cell culture system to propagate virus for vaccines and to evaluate neutralizing antibody responses after infection or vaccination. Furthermore, the host factors that influence susceptibility to disease and the correlates of protective immunity to NLV are unclear. It is possible that the former impediment to vaccine development can be overcome by generation of calicivirus capsid VLPs for vaccines, as reported for a number of NLV strains. This approach has been explored using NLV VLPs administered orally to adult volunteers in phase I clinical trials (Ball et al. 1999). However, the genetic diversity of NLVs complicates vaccine development and will likely influence the effectiveness of vaccines produced. Results of human volunteer studies confirmed a lack of cross-protection between distinct NLV genotypes (Wyatt et al. 1974). However, protective immunity was evident after challenge with homologous NLV or antigenically similar NLV or repeated infections with NV, although some individuals remained susceptible after re-exposure to homologous virus (Parrino et al. 1977; Johnson et al. 1990). The latter finding has led to speculation that both genetic and immunologic factors may influence host susceptibility to NLV infections. Both the development of an animal model to study immunity to NLV and additional human volunteer studies are needed to advance this important area of research.

SUMMARY AND CONCLUSIONS. Tremendous progress has been made on studies of NLVs in the past decade, considering the lack of a cell culture system or a practical animal model for their propagation. These include the sequencing of the entire genome of the uncultivable NLVs (Jiang et al. 1990; Lambden et al. 1993) and using this information to develop specific primers and RT-PCR assays for the sensitive detection of NLVs and production of VLPs (Jiang et al. 2000) as substitute reagents for detection of NLVs in ELISA assays and for antiserum production. These developments have resulted in a new generation of reagents and sensitive tests for the detection of NLVs in foods, water, and clinical specimens, leading to their recognition as a potential major foodborne pathogen (Mead et al. 1999).

However, numerous challenges remain to advance our knowledge of foodborne calicivirus infections. The great diversity in NLV strains and their low concentrations and presence of RT-PCR inhibitors in foods suggest that many NLVs remain undetectable using current technology, contributing to the diagnostic void in diagnosis of foodborne illnesses. Simplified and sensitive virus extraction procedures from foods and additional assays and broadly reactive primers for RT-PCR are needed to improve diagnostic tests. In addition, the recent recognition that several animal enteric caliciviruses are genetically related to human NLVs and SLVs (Sugieda et al. 1998; Dastjerdi et al 1999; Guo et al 1999; Liu et al 1999; Guo et al 2000a; van der Poel 2000) and can be detected with the same generic primers complicates diagnosis of human caliciviruses from food and water sources.

The lack of a cell culture system for NLVs prevents propagation of virus for serotype analysis, vaccine studies, or diagnostic reagents and impedes our understanding of the cellular replication strategies of NLVs, an important prerequisite for the design of antiviral reagents and preventive approaches. However, a cultivable porcine enteric calicivirus (PEC/Cowden) related to human SLV exists that requires an intestinal contents preparation (ICP) in the medium for growth (Flynn and Saif, 1988). Further efforts to identify the growth-promoting factors in the ICP may provide new insights for cultivation of NLVs and SLVs. In addition, the virulence of the cell-adapted PEC is attenuated in pigs (Guo et al. 2001b) and, based on genetic analysis, the attenuated PEC has only five amino acid sequence changes from the virulent PEC/Cowden strain that does not grow in cell culture (Guo et al. 1999). Elucidation of the genetic basis for the cell culture adaptation and attenuation of PEC/Cowden by generation of infectious clones and use of reverse genetics could greatly enhance our understanding of these aspects of calicivirus replication and host virulence.

Finally the lack of an animal model to study human

NLV and SLV infections compromises our understanding of SLV and NLV pathogenesis and host immunity. Although human volunteer studies have been valuable to explore such parameters, they are of necessity limited in scope and costly to undertake. Important aspects of NLV and SLV infections, including the question of serotype-specific immunity and its duration, as well as the role of host genetic factors in susceptibility to disease, are unresolved. However, it is anticipated that additional studies of PEC and BEC infections in pigs and calves, respectively, will enhance our general knowledge of pathogenesis and immunity to enteric caliciviruses and serve as a surrogate model for human NLV and SLV infections. In addition, studies are ongoing to develop animal models (perhaps genetically modified hosts) for human NLV and SLV infections. Development of a cell culture system and an animal model for human NLV and SLV are likely to be the next major breakthroughs to significantly advance the field of enteric calicivirus research.

REFERENCES

Ando, T., S.S. Monroe, J.R. Gentsch, Q. Jin, D.C. Lewis, and R.I. Glass. 1995. Detection and differentiation of antigenically distinct small round-structured viruses (Norwalk-like viruses) by reverse transcription-PCR and Southern hybridization. *J. Clin. Microbiol.* 33:64–71.

Ando, T., J.S. Noel, and R.L. Fankhauser. 2000. Genetic classification of "Norwalk-like viruses. *J. Infect. Dis.* 181(Suppl 2):S336–S348.

Appleton, H., and P.G. Higgins. 1975. Viruses and gastroenteritis in infants. *Lancet* 1:1297.

Atmar, R.L., and M.K. Estes. 2001. Diagnosis of noncultivable gastroenteritis viruses, the human caliciviruses. *Clin. Microbiol. Rev.* 14:15–37.

Ball, J.M., D.Y. Graham, A.R. Opekun, M.A. Gilger, R.A. Guerrero, and M.K. Estes. 1999. Recombinant Norwalk virus-like particles given orally to volunteers: Phase I study. *Gastroenterol.* 117:40–48.

Berg, D.E., M.A. Kohn, T.A. Farley, and L.M. McFarland. 2000. Multi-state outbreaks of acute gastroenteritis traced to fecal-contaminated oysters harvested in Louisiana. *J. Infect. Dis.* 181(Suppl 2):S381–S386.

Bishop, R.F., G.P. Davidson, I.H. Holmes, and B.J. Ruck. 1973. Virus particles in epithelial cells of duodenal mucosa from children with acute non-bacterial gastroenteritis. *Lancet* 2:1281–1283.

Bridger, J.C. 1990. Small viruses associated with gastroenteritis in humans. In *Viral Diarrheas of Man and Animals.* L.J. Saif and K.W. Theil, eds. Boca Raton: CRC Press, 123–145.

Chapron, C.D., N.A. Ballester, J.H. Fontaine, C.N. Frades, and A.B. Margolin. 2000. Detection of astroviruses, enteroviruses, and adenovirus types 40 and 41 in surface waters collected and evaluated by the information collection rule and an integrated cell culture-nested PCR procedure. *Appl. Environ. Microbiol.* 66:2520–2525.

Chiba, S., S. Nakata, K. Numata-Kinoshita, and S. Honma. 2000. Sapporo virus: history and recent findings. *J. Infect. Dis.* 181(Suppl 2):S303–S308.

Clarke, I.N., and P.R. Lambden. 2000. Organization and expression of calicivirus genes. *J. Infect. Dis.* 181(Suppl. 2):S309–S316.

Cromeans, T.L. 2001. Hepatitis A and E viruses. In *Foodborne Disease Handbook,* 2nd Edition, Volume 2. Y.H. Hui, S.A. Sattar, K.D. Murrell, W. Nip, and P.S. Stanfield, eds. New York: Marcel Dekker Inc., 23–76.

Dastjerdi, A.M., J. Green, C.I. Gallimore, D.W. Brown, and J.C. Bridger. 1999. The bovine Newbury agent-2 is genetically more closely related to human SRSVs than to animal caliciviruses. *Virology* 254:1–5.

de Wit, M.A.S., M.P.G. Koopmans, L.M. Kortbeek, W.J.B. Wannet, J. Vinjé, F. van Leusden, A.I.M. Bartelds, and van Y.T.H.P. Duynhoven. 2001. Sensor, a population-based cohort study on gastroenteritis in the Netherlands: incidence and etiology. *Am. J. Epidemiol.* 154:666–674.

Desselberger, U., M. Iturriza-Gómara, and J.J. Gray. 2001. Rotavirus epidemiology and surveillance. *Novartis Fdn. Symp.* 238:125–152.

Dingle, K.E., P.R. Lambden, E.O. Caul, and I.N. Clarke. 1995. Human enteric Caliciviridae: the complete genome sequence and expression of virus-like particles from a genetic group II small round structured virus. *J. Gen. Virol.* 76:2349–2355.

Dowell, S.F., C. Groves, K.B. Kirkland, H.G. Cicirello, T. Ando, Q. Jin, J.R. Gentsch, S.S. Monroe, C.D. Humphrey, and C. Slemp. 1995. A multistate outbreak of oyster-associated gastroenteritis: implications for interstate tracing of contaminated shellfish. *J. Infect. Dis.* 171:1497–1503.

Dumpis, U., D. Crook, and J. Oksi. 1999. Tick-borne encephalitis. *Clin. Infect. Dis.* 28:882–90.

Estes, M. K., R.L. Atmar, and M.E. Hardy. 1997. "Norwalk and related diarrhea viruses." In *Clinical Virology,* 3rd Edition. Douglas D. Richman, ed., New York: Churchill Livingstone, 1073–1095.

Fankhauser, R.L., J.S. Noel, S.S. Monroe, T. Ando, and R.I. Glass. 1998. Molecular epidemiology of "Norwalk-like viruses" in outbreaks of gastroenteritis in the United States. *J. Infect. Dis.* 178:1571–1578.

Fleet, G.H., P. Heiskanen, I. Reid, and K.A. Buckle. 2000. Foodborne viral illness—status in Australia. *Int. J. Food Microbiol.* 59:127–136.

Flynn, W.T., and L.J. Saif. 1988. Serial propagation of porcine enteric calicivirus-like virus in primary porcine kidney cell cultures. *J. Clin. Microbiol.* 26:206–212.

Flynn, W.T., L.J. Saif, and P.D. Moorhead. 1988. Pathogenesis of porcine enteric calicivirus-like virus in four-day-old gnotobiotic pigs. *Am. J. Vet. Res.* 49:819–825.

Glass, R.I., J. Bresee, B. Jiang, J. Gentsch, T. Ando, R. Fankhauser, J. Noel, U.D. Parashar, B. Rosen, and S.S. Monroe. 2001. Gastroenteritis viruses: an overview. *Novartis Fdn. Symp.* 238:5–19.

Glass, R.I., J. Noel, T. Ando, R. Fankhauser, G. Belliot, A. Mounts, U.D. Parashar, J.S. Bresee, and S.S. Monroe. 2000. The epidemiology of enteric caliciviruses from humans: a reassessment using new diagnostics. *J. Infect. Dis.* 181(Suppl 2):S254–S261.

Glass, P.J., L.J. White, J.M. Ball, I. Leparc-Goffart, M.E. Hardy, and M.K. Estes. 2000. Norwalk virus open reading frame 3 encodes a minor structural protein. *J. Virol.* 74:6581–6591.

Graham, D.Y., X. Jiang, T. Tanaka, A.R. Opekun, H.P. Madore, and M.K. Estes. 1994. Norwalk virus infection of volunteers: new insights based on improved assays. *J. Infect. Dis.* 170:34–43.

Green, J., C.I. Gallimore, J.P. Norcott, D. Lewis, and D.W.G. Brown. 1995. Broadly reactive reverse transcriptase polymerase chain reaction for the diagnosis of SRSV-associated gastroenteritis. *J. Med. Virol.* 47:392–398.

Green, J., J. Vinjé, C.I. Gallimore, M. Koopmans, A. Hale, and D.W.G. Brown. 2000. Capsid diversity among Norwalk-like viruses. *Virus Genes* 20:227–236.

Green, K.Y., R.M. Chanock, and A.Z. Kapikian. 2001. Human caliciviruses. In *Fields Virology*, 4th Edition. D.M. Knipe, P.M. Howley, D.E. Griffin, et al., eds. Philadelphia: Lippincott-Raven, 841–874.

Green, K.Y., T. Ando, M.S. Balayan, T. Berke, I.N. Clarke, M.K. Estes, D.O. Matson, S. Nakata, J.D. Neill, M.J. Studdert, and H.-J. Thiel. 2000. Taxonomy of the caliciviruses. *J. Infect. Dis.* 181(Suppl 2):S322–S330.

Green, K.Y., A.Z. Kapikian, J. Valdesuso, S. Sosnovtsev, J.J. Treanor, and J.F. Lew. 1997. Expression and self-assembly of recombinant capsid protein from the antigenically distinct Hawaii human calicivirus. *J. Clin. Microbiol.* 35:1909–1914.

Greenberg, H.B., J.R. Valdesuso, A.R. Kalica, R.G. Wyatt, V.J. McAuliffe, A.Z. Kapikian, and R.M. Chanock. 1981. Proteins of Norwalk virus. *J. Virol.* 37:994–999.

Guo, M., K.O. Chang, M.E. Hardy, Q. Zhang, A.V. Parwani, and L.J. Saif. 1999. Molecular characterization of a porcine enteric calicivirus genetically related to Sapporo-like human caliciviruses. *J. Virol.* 73:9625–9631.

Guo, M., J.F. Evermann, and L.J. Saif. 2001a. Detection and molecular characterization of cultivable calicivirus from clinically normal mink and enteric caliciviruses associated with diarrhea in mink. *Arch. Virol.* 146:479–493.

Guo, M., J.R. Hayes, K.-O. Cho, A.V. Parwani, L.M. Lucas, and L.J. Saif. 2001b. Comparative pathogenesis of tissue culture-adapted and wild-type Cowden porcine enteric calicivirus (PEC) in gnotobiotic pigs and induction of diarrhea by intravenous inoculation of wild-type PEC. *J. Virol.* 75:9239–9251.

Guo, M., Y. Qian, K.O. Chang, and L.J. Saif. 2001c. Expression and self-assembly in baculovirus of porcine enteric calicivirus capsids into virus-like particles and their use in an enzyme-linked immunosorbent assay for antibody detection in swine. *J. Clin. Microbiol.* 39:1487–1493.

Hardy, M.E., L.J. White, J.M. Ball, and M.K. Estes. 1995. Specific proteolytic cleavage of recombinant Norwalk virus capsid protein. *J. Virol.* 69:1693–1698.

Hale, A.D., S.E. Crawford, M. Ciarlet, J. Green, C. Gallimore, D.W. Brown, X. Jiang, and M.K. Estes. 1999. Expression and self-assembly of Grimsby virus: antigenic distinction from Norwalk and Mexico viruses. *Clin. Diagn. Lab. Immunol.* 6:142–145.

Hale, A.D., T.N. Tanaka, N. Kitamoto, M. Ciarlet, X. Jiang, N. Takeda, D.W. Brown, and M.K. Estes. 2000. Identification of an epitope common to genogroup 1 "Norwalk-like viruses." *J. Clin. Microbiol.* 38:1656–1660.

Inouye, S., K. Yamashita, S. Yamadera, M. Yoshikawa, N. Kato, and N. Okabe. 2000. Surveillance of viral gastroenteritis in Japan: pediatric cases and outbreak incidents. *J. Infect. Dis.* 181(Suppl. 2):S270–S274.

Jaykus, L-A. 2000. Detection of human enteric viruses in foods. In *Foodborne Disease Handbook*, 2nd Edition. Y.H. Hui, S.A. Sattar, K.D. Murell, W. Nip, and P.S. Stanfield, eds. , New York: Marcel Dekker Inc., 137–163.

Jiang, X., D.Y. Graham, K.N. Wang, and M.K. Estes. 1990. Norwalk virus genome cloning and characterization. *Science* 250:1580–1583.

Jiang, X., P.W. Huang, W.M. Zhong, T. Farkas, D.W. Cubitt, and D.O. Matson. 1999. Design and evaluation of a primer pair that detects both Norwalk- and Sapporo-like caliciviruses by RT-PCR. *J. Virol. Meth.* 83:145–154.

Jiang, X., D.O. Matson, G.M. Ruiz-Palacios, J. Hu, J. Treanor, and L.K. Pickering. 1995. Expression, self-assembly, and antigenicity of a snow mountain agent–like calicivirus capsid protein. *J. Clin. Microbiol.* 33:1452–1455.

Jiang, X., M. Wang, D.Y. Graham, and M.K. Estes. 1992a. Detection of Norwalk virus in stool by polymerase chain reaction. *J. Clin. Microbiol.* 30:2529–2534.

Jiang, X., M. Wang, D.Y. Graham, and M.K. Estes. 1992b. Expression, self-assembly, and antigenicity of the Norwalk virus capsid protein. *J. Virol.* 66:6527–6532.

Jiang, X., M. Wang, K. Wang, and M.K. Estes. 1993. Sequence and genomic organization of Norwalk virus. *Virology* 195:51–61.

Jiang, X., W.D. Cubitt, T. Berke, W. Zhong, X. Dai, S. Nakata, L.K. Pickering, and D.O. Matson. 1997. Sapporo-like human caliciviruses are genetically and antigenically diverse. *Arch. Virol.* 142:1813–1827.

Jiang, X., N. Wilton, W.M. Zhong, T. Farkas, P.W. Huang, E. Barrett, M. Guerrero, G. Ruiz-Palacios, K.Y. Green, J. Green, A.D. Hale, M.K. Estes, L.K. Pickering, and D.O. Matson. 2000. Diagnosis of human caliciviruses by use of enzyme immunoassays. *J. Infect. Dis.* 181(Suppl. 2):S349–S359.

Jiang, X., W. Zhong, M. Kaplan, L.K. Pickering, and D.O. Matson. 1999. Expression and characterization of Sapporo-like human calicivirus capsid proteins in baculovirus. *J. Vir. Meth.* 78:81–91.

Kapikian, A.Z. 2000. The discovery of the 27-nm Norwalk virus: an historic perspective. *J. Infect. Dis.* 181 (Suppl. 2):S295–S302.

Kapikian, A.Z., M.K. Estes, and R.M. Chanock. 1996. Norwalk group of viruses. In *Fields Virology*, 3rd Edition. B.N. Fields, D.M. Knipe, P.M. Howley et al., eds.. Philadelphia, PA: Lippincott-Raven, 783–810.

Kaplan, J.E., R. Feldman, D.S. Campbell, C. Lookabaugh, and G.W. Gary. 1982. The frequency of a Norwalk-like pattern of illness in outbreaks of acute gastroenteritis. *Am. J. Pub. Hlth.* 72:1329–1332.

Koopmans, M., J. Vinjé, M. de Wit, I. Leenen, W. van der Poel, and Y. van Duynhoven. 2000. Molecular epidemiology of human enteric caliciviruses in The Netherlands. *J. Infect. Dis.* 181(Suppl. 2):S262–S269.

Lambden, P.R., E.O. Caul, C.R. Ashley, and I.N. Clarke. 1993. Sequence and genome organization of a human Small Round-Structured (Norwalk-like) virus. *Science* 259:516–519.

Lees, D.N. 2000. Viruses and bivalve shellfish. *Int. J. Food Microbiol.* 59:81–116.

Leggit, P.R., and L. Jaykus. 2000. Detection methods for human enteric viruses in representative foods. *J. Food. Prot.* 63:1738–1744

Leite, J.P., T. Ando, J.S. Noel, B. Jiang, C.D. Humphrey, J.F. Lew, K.Y. Green, R.I. Glass, and S.S. Monroe. 1996. Characterization of Toronto virus capsid protein expressed in baculovirus. *Arch. Virol.* 141:865–875.

Le Guyader, F.L. Haugarreau, L. Miossec, and M. Pommepuy. 2000. Three-year study to assess human enteric viruses in shellfish. *Appl. Environ. Microbiol.* 66:3241–3248.

Lewis, D.C., A. Hale, X. Jiang, R. Eglin, and D.W. Brown. 1997. Epidemiology of Mexico virus, a small round-structured virus in Yorkshire, United Kingdom, between January 1992 and March 1995. *J. Infect. Dis.* 175:951–954.

Liu, B.L., P.R. Lambden, H. Gunther, P. Otto, M. Elschner, and I.N. Clarke. 1999. Molecular characterization of a bovine enteric calicivirus: relationship to the Norwalk-like viruses. *J. Virol.* 73:819–825.

Lüthi, T.M. 2001. "Epidemiology of foodborne viral infections." In *Foodborne Disease Handbook,* 2nd Edition, Volume 2. Y.H. Hui, S.A. Sattar, K.D. Murrell, W-K. Nip, and P.S. Stanfield, eds., . New York: Marcel Dekker Inc., 183–204.

Madeley, C.R., and B.P. Crosgrove. 1976. Letter: Calicivirus in man. *Lancet* 1:199–200.

Marks, P.J., I.B. Vipond, D. Carlisle, D. Deakin, R.E. Fey, and E.O. Caul. 2000. Evidence for airborne transmission of

Norwalk-like virus (NLV) in a hotel restaurant. *Epidemiol. Infect.* 124:481–487.

Marx, A., D.K. Shay, J.S. Noel, C. Bragem, J.S. Bresee, S. Lipsky, S.S. Monroe, T. Ando, C.D. Humphrey, E.R. Alexander, and R.I. Glass. 1999. An outbreak of acute gastroenteritis in a geriatric long-term-care facility: combined application of epidemiological and molecular diagnostic methods. *Infect. Cont. Hosp. Epidemiol.* 20:306–311.

Matsui, S.M., and H.B. Greenberg. 2000. Immunity to calicivirus infection. *J. Infect. Dis.* 181(Suppl. 2):S331–S335.

Mead, P.S., L. Slutsker, V. Dietz, L.F. McCaig, J.S. Bresee, C. Shapiro, P.M. Griffin, and R.V. Tauxe. 1999. Food-related illness and death in the United States. *Emerg. Infect. Dis.* 5:607–625.

Melnick, J.L. 1996. Enteroviruses: polioviruses, coxsackieviruses, echoviruses and newer enteroviruses. In *Fields Virology*, 3rd Edition. B.N. Fields, D.M. Knipe, and P.M. Howley et al., eds. Philadelphia, PA: Lippincott-Raven, 655–714.

Moe, C.L., J. Gentsch, T. Ando, G.S. Grohmann, S.S. Monroe, X. Jiang, J. Wang, M.K. Estes, Y. Seto, C. Humphrey, S. Stine, and R.I. Glass. 1994. Application of PCR to detect Norwalk virus in fecal specimens from outbreaks of gastroenteritis. *J. Clin. Microbiol.* 32:642–648.

Monroe, S.S., J.J. Holmes, and G.M. Belliot. 2001. Molecular epidemiology of human astroviruses. *Novartis Fdn. Symp.* 238:237–245.

Mounts, A.W., T. Ando, M. Koopmans, J.S. Bresee, J. Noel, and R.I. Glass. 2000. Cold weather seasonality of gastroenteritis associated with Norwalk- like viruses. *J. Infect. Dis.* 181(Suppl. 2):S284–S287.

Noel, J.S., R.L. Fankhauser, T. Ando, S.S. Monroe, and R.I. Glass. 1999. Identification of a distinct common strain of "Norwalk-like viruses" having a global distribution. *J. Infect. Dis.* 179:1334–1344.

Noel, J.S., T. Ando, J.P. Leite, K.Y. Green, K.E. Dingle, M.K. Estes, Y. Seto. S.S. Monroe, and R.I. Glass. 1997. Correlation of patient immune responses with genetically characterized Small Round-Structured Viruses involved in outbreaks of nonbacterial acute gastroenteritis in the United States, 1990–1995. *J. Med. Virol.* 53:372–383.

Numata, K., M.E. Hardy, S. Nakata, S. Chiba, and M.K. Estes. 1997. Molecular characterization of morphologically typical human calicivirus Sapporo. *Arch. Virol.* 142:1537–1552.

Ohlinger, V.F., B. Haas, and H.-J. Thiel. 1993. Rabbit hemorrhagic disease (RHD): characterization of the causative calicivirus. *Vet. Res.* 24:103–116.

Oishi, I., K. Yamazaki, T. Kimoto, Y. Minekawa, E. Utagawa, S. Yamazaki, S. Inouye, G.S. Grohmann, S.S. Monroe, S.E. Stine, et al. 1994. A large outbreak of acute gastroenteritis associated with astrovirus among students and teachers in Osaka, Japan. *J. Infect. Dis.* 170(2):439–443.

Okhuysen, P.C., X. Jiang, L. Ye, P.C. Johnson, and M.K. Estes. 1995. Viral shedding and fecal IgA response after Norwalk virus infection. *J. Infect. Dis.* 171:566–569.

Pang, X.L., S. Honma, S. Nakata, and T. Vesikari. 2000. Human caliciviruses in acute gastroenteritis of young children in the community. *J. Infect. Dis.* 181(Suppl. 2):S288–S294.

Parashar, U.D., and S.S. Monroe. 2001. "Norwalk-like viruses" as a cause of foodborne disease outbreaks. *Rev. Med. Virol.* 11:243–252.

Parwani, A.V., W.T. Flynn, K.L. Gadfield, and L.J. Saif. 1991. Serial propagation of porcine enteric calicivirus in a continuous cell line: Effect of medium supplementation with intestinal contents or enzymes. *Arch. Virol.* 120:115–122.

Pfister, T., and E. Wimmer. 2001. Polypeptide p41 of a Norwalk-like virus is a nucleic acid-independent nucleoside triphosphatase. *J. Virol.* 75:1611–1619.

Pinto, R.M., C. Villena, F. Le Guyader, S. Guix, S. Caballero, M. Pommepuy, and A. Bosch. 2001. Astrovirus detection in wastewater samples. *Water Sci. Technol.* 43:73–76.

Ponka, A., L. Maunula, C.H. von Bonsdorff, and O. Lyytikainen. 1999. An outbreak of calicivirus associated with consumption of frozen raspberries. *Epidemiol. Infect.* 123:469–474.

Prasad, B.V., M.E. Hardy, T. Dokland, J. Bella, M.G. Rossmann, and M.K. Estes. 1999. X-ray crystallographic structure of the Norwalk virus capsid. *Science* 286:287–290.

Richards, G.P. 2001. Enteric virus contamination of foods through industrial practices: a primer on intervention strategies. *J. Indust. Microbiol Biotechnol.* 27:117–125.

Saif, L.J., E.H. Bohl, K.W. Theil, R.F. Cross, and J.A. House. 1980. Rotavirus-like, calicivirus-like, and 23-nm viruslike particles associated with diarrhea in young pigs. *J. Clin. Microbiol.* 12:105–111.

Sattar, S.A., S.V. Springthorpe, and J.A. Tetro. 2001. Rotavirus. In *Foodborne Disease Handbook*, 2nd Edition, Volume 2. Y.H. Hui, S.A. Sattar, K.D. Murrell, W-K. Nip, and P.S. Stanfield, eds. New York: Marcel Dekker, 99–125.

Schuffenecker, I., T. Ando, D. Thouvenot, B. Lina, and M. Aymard. 2001. Genetic classification of "Sapporo-like viruses." *Arch. Virol.* 146:2115$e–2132.

Schwab, K.J., F.H. Neill, R.L. Fankhauser, N.A. Daniels, S.S. Monroe, D.A. Bergmire-Sweat, M.K. Estes, R.L. Atmar. 2000. Development of methods to detect "Norwalk-like viruses" (NLVs) and hepatitis A virus in delicatessen foods: application to a food-borne NLV outbreak. *Appl. Environ. Microbiol.* 66:213–218.

Schwab, K.J., F.H. Neill, F. Le Guyader, M.K. Estes, and R.L. Atmar. 2001. Development of a reverse transcription-PCR-DNA enzyme immunoassay for detection of "Norwalk-like" viruses and hepatitis A virus in stool and shellfish. *Appl. Envrion. Microbiol.* 67:742–749.

Shieh, C.Y.-S., S.S. Monroe, R.L. Fankhauser, G.W. Langlois, W. Burkhardt III, and R.S. Baric. 2000. Detection of Norwalk-like virus in shellfish implicated in illness. *J. Infect. Dis.* 181(Suppl. 2): S360–366.

Smith, A.W., D.E. Skilling, N. Cherry, J.H. Mead, and D.O. Matson. 1998. Calicivirus emergence from ocean reservoirs: zoonotic and interspecies movements. *Emerg. Infect. Dis.* 4:13–20.

Sugieda, M., H. Nagaoka, Y. Kakishima, T. Ohshita, S. Nakamura, and S. Nakajima. 1998. Detection of Norwalk-like virus genes in the caecum contents of pigs. *Arch. Virol.* 143:1215–1221.

van der Poel, W.H.M., J. Vinjé, R. van der Heide, M.-I. Herrera, A. Vivo, and M.P. Koopmans. 2000. Norwalk-like calicivirus genes in farm animals. *Emerg. Infect. Dis.* 6:36–41.

Vinjé, J., and M.P.G. Koopmans. 1996. Molecular detection and epidemiology of small round-structured viruses in outbreaks of gastroenteritis in The Netherlands. *J. Infect. Dis.* 174(3):610–615.

Vinjé, J., S.A. Altena, and M.P.G. Koopmans. 1997. The incidence and genetic variability of small round-structured viruses in outbreaks of gastroenteritis in The Netherlands. *J. Infect. Dis.* 176:1374–1378.

Vinjé, J., H. Deijl, R. van der Heide, D. Lewis, K.O. Hedlund, L. Svensson, and M.P.G. Koopmans. 2000a. Molecular detection and epidemiology of Sapporo-like viruses. *J. Clin. Microbiol.* 38:530–536.

Vinjé, J., J. Green, D.C. Lewis, C.I. Gallimore, D.W. Brown, andM.P.G. Koopmans. 2000b. Genetic polymorphism across regions of the three open reading frames of "Norwalk-like viruses." *Arch. Virol.* 145:223–241.

Vinjé, J., and M.P.G. Koopmans. 2000. Simultaneous detection and genotyping of "Norwalk-like viruses" by oligonucleotide array in a reverse line blot hybridization format. *J. Clin. Microbiol.* 38:2595–2601.

White, L.J., J.M. Ball, M.E. Hardy, T.N. Tanaka, N. Kitamoto, and M.K. Estes. 1996. Attachment and entry of recombinant Norwalk virus capsids to cultured human and animal cell lines. *J. Virol.* 70:6589–6597.

Wright, P.J., I.C. Gunesekere, J.C. Doultree, and J.A. Marshall. 1998. Small round-structured (Norwalk-like) viruses and classical human caliciviruses in southeastern Australia, 1980–1996. *J. Med. Virol.* 55:312–320.

Wyatt, R.G., R. Dolin, N.R. Blacklow, H.L. Dupont, R.F. Buscho, T.S. Thornhill, A.Z. Kapikian, and R.M. Chanock. 1974. Comparison of three agents of acute infectious nonbacterial gastroenteritis by cross-challenge in volunteers. *J. Infect. Dis.* 129:709–714.

Yamashita, T., S. Kobayashi, K. Sakae, S. Nakata, X. Chiba, Y. Ishihara, and S. Isomura. 1991. Isolation of cytopathic small round viruses with BS-C-1 cells from patients with gastroenteritis. *J. Infect. Dis.* 164:954–957.

Yamashita, T., K. Sakae, Y. Ishihara, S. Isomura, and E. Utagawa. 1993. Prevalence of newly isolated, cytopathic small round virus (Aichi strain) in Japan. *J. Clin. Microbiol.* 31:2938–2943.

Yamashita, T., K. Sakae, H. Tsuzuki, Y. Suzuki, N. Ishikawa, N. Takeda, T. Miyamura, and S. Yamazaki. 1998. Complete nucleotide sequence and genetic organization of Aichi virus, a distinct member of the Picornaviridae associated with acute gastroenteritis in humans. *J. Virol.* 72:8408–8412.

35

PARATUBERCULOSIS: A FOOD SAFETY CONCERN?

WILLIAM P. SHULAW AND ALECIA LAREW-NAUGLE

INTRODUCTION. It is important to emphasize at the outset of this chapter that no consensus exists that *Mycobacterium avium* subspecies *paratuberculosis* (MAP), the causative agent of paratuberculosis or Johne's disease (JD) in ruminants, is causally associated with Crohn's disease (CD) in humans. Although some researchers have publicly stated their belief that such a relationship exists, others just as strongly express their opinion that one does not, citing evidence that they believe refutes the case for causality. Several extensive reviews of the potential association of MAP with CD have been published (Chiodini 1989; Thompson 1994; Van Kruiningen 1999; Hermon-Taylor 2000; El-Zaatari 2001; Hermon-Taylor 2001). Consequently, the issue of whether this organism constitutes a food safety risk is also debated (Hermon-Taylor 2001; Quirke 2001).

Nevertheless, research findings and concern about a possible causal role for MAP in CD during the last fifteen years have resulted in recent reports issued by the Food Standards Agency of the United Kingdom and the Scientific Committee on Animal Health and Animal Welfare of the European Union Directorate on possible links between the two diseases (Anonymous 2000; Commission 2000; Rubery 2001). In early 2001, the International Dairy Federation published a bulletin on MAP, a significant portion of which was devoted to the enumeration of viable MAP in milk and the efficiency of destruction of it in milk and milk products. In the United States, the National Institute of Allergy and Infectious Diseases of the National Institutes of Health held a workshop in 1998 "to review the current state of knowledge relevant to a microbial etiology of Crohn's disease" and, in particular, "to review evidence for and against the hypothesis that the bacterium, *Mycobacterium avium* subspecies *paratuberculosis* (Map) is the cause of CD, and to define needed research that could shed light on the etiology and pathogenesis of this chronic disease" (http://www.niaid.nih.gov/dmid/meetings/crohns.htm).

It is our intent here to briefly review the developments of the past few years that have created concern that MAP may constitute a food safety issue.

OVERVIEW OF JOHNE'S DISEASE AND CROHN'S DISEASE. A brief review of the epidemiology and clinical presentation of these two diseases

may provide a background for understanding why these concerns about MAP have arisen.

Johne's Disease. Johne's disease is a chronic, granulomatous enteritis of domestic, and occasionally wild, ruminants . Although MAP was first isolated from cases of JD in 1910, progress in improving cultural methods to isolate this organism continues through today (Whipple 1991; Kalis 1999; Whitlock 2000). The clinical disease is usually characterized by chronic weight loss progressing to emaciation and death. Diarrhea is a common clinical manifestation of the disease in cattle, but in some species, such as sheep and goats, diarrhea is not a consistent finding (Dubash 1996; Davies 1997). Infected animals may shed tens of billions of MAP daily in their feces whether diarrhea is present or not (Chiodini 1984; Whittington 2000).

The infection is usually transmitted by the fecal-oral route, and there is an age-related resistance to infection, with young animals in the first few months of life most susceptible (Chiodini 1984; Kreeger 1991). Often the bacteria are acquired when young animals nurse from contaminated teats and udders or explore a contaminated environment. MAP may be found in the udder secretions because it has also been recovered from aseptically collected milk and colostrum samples and from supramammary lymph nodes from subclinically infected cows, making infection at the first meal a possibility regardless of teat end contamination (Taylor 1981; Sweeney 1992; Koenig 1993; Streeter 1995). The organisms have also been recovered from fetuses, but the practical significance of this infection is not known (McQueen 1979; Seitz 1989; Sweeney 1992). Although MAP has been recovered from the semen and reproductive organs of bulls with JD, transmission by natural service or by artificial insemination is thought to be unlikely (Larsen 1970; Larsen 1981; Merkal 1982). Clinical disease is usually the culmination of a lifelong infection, and the incubation period may be from one to several years. Subclinically infected animals may shed bacteria in their feces for long periods before developing clinical signs of the infection. There is no satisfactory treatment.

The potential for tremendous contamination of the environment by animals shedding MAP in their feces and the ability of MAP to survive in the environment for up to a year makes control difficult (Lovell 1944; Jorgensen 1977; Eamens 2001). Control programs usually focus on reducing the environmental burden at the farm level by identification and removal of infected animals, and on reducing the risk of infection of young stock by removal and artificial rearing or by sanitation of the birthing area and relocation of dam-offspring pairs to clean pastures. The relatively recent discovery that MAP may be harbored in a wild-animal reservoir, such as rabbits, adds new complexity and concern in designing effective control programs that limit transmission to susceptible animals (Beard 2001).

Reliable estimates of the prevalence of JD in livestock populations are few. Recent reports produced by the National Animal Health Monitoring System of the USDA in the United States indicated that about 22 percent of dairy cattle herds and 8 percent of beef cattle herds were infected (Ott 1999; Dargatz 2001). Similar estimates for dairy herds in Belgium have been reported (Boelaert 2000). A serologic study estimated that about 55 percent of dairy herds were infected in the Netherlands, and an ELISA-based study of bulk milk tank samples suggested that about 47 percent of herds were infected in Denmark (Muskens 2000; Nielsen 2000; Nielsen 2000). Each of these studies has methodological limitations and should not be used for comparison purposes. However, they do indicate that this infection is probably very widespread.

Crohn's Disease. Crohn's disease is a chronic, incurable inflammatory disease of humans most commonly affecting the terminal ileum and colon, but it may affect any part of the gastrointestinal tract (Chiodini 1989; Thompson 1994; Van Kruiningen 1999). The disease usually strikes people in the fifteen– to twenty-four–year-old age group with a smaller secondary peak incidence in those forty-five to fifty five years of age. The greatest prevalence of CD is in Western Europe and North America. Estimates of annual incidence vary from 3.1 per 100,000 population in the United Kingdom and 4.3 per 100,000 in the United States (Langman 1997; Sawczenko 2001). There are regional differences and evidence of familial and nonfamilial clusters. Some researchers believe that the incidence is increasing in certain regions.

The disease causes the wall of the intestine to become thickened and inflamed. The thickening may lead to narrowing of the lumen with eventual obstruction, and the inflammation may progress to ulceration, fistula formation, and perforation. It is characterized by intermittent episodes of relapse and remission, and surgical intervention to repair obstructions, fistulae, and abscesses is eventually necessary in a significant proportion of patients. Patients with CD have variable signs, including reduced appetite, abdominal pain, diarrhea (often bloody), fatigue, and reduced growth rate.

In contrast to JD, extra-intestinal manifestations of CD are relatively common (Van Kruiningen 1999). They include arthritis, spondylitis, iritis, and skin lesions. Aphthoid ulcers of the mouth and intestine as well as perianal fissures and fistulae are often seen in CD patients. Ulceration, fistulae, and fissures are not seen in JD. The wide spectrum of lesions seen in CD have led some clinicians and researchers to propose that CD may not be a single-entity disease (Thompson 1994). Some investigators have proposed that there are two forms of CD, an aggressive form characterized by fistulae formation and an indolent, obstructive form (Mishina 1996). Evidence for this classification is currently lacking.

Recently, the potential contribution of genetics to the development of CD has been pursued. Several studies have reported greater than expected agreement with the clinical presentation of CD and age at onset among familial cases as opposed to sporadic cases (Polito 1996; Satsangi 1998; Satsangi 2001). Several reports have described association between susceptibility to CD and specific regions on chromosomes 12 and 16. These data strongly suggest a genetic component to this disease that probably involves several genes. It now appears that a combination of a genetic predisposition, an abnormal immune response, and environmental factors including bacteria, and perhaps dietary factors, are necessary for development of the disease.

ATTEMPTS TO ASSOCIATE CROHN'S DISEASE WITH JOHNE'S DISEASE AND MAP. The debate concerning a possible association between CD and JD has a long history. Dalziel first suggested a possible connection between JD in cattle and a specific form of ileitis, later known as CD, in humans in a 1913 article in the British Medical Journal (Dalziel 1913). This proposed connection originated from the clinical and pathological similarities noted for the two diseases. In 1932, Crohn, Ginzburg, and Oppenheimer published a case series report that described the pathology and clinical signs associated with regional ileitis in fourteen patients . This description was the first to identify clinical and diagnostic criteria for the specific form of regional ileitis subsequently referred to as CD. An etiology was not identified in this early report. Over the next several decades, the similarities between the two diseases caused a number of investigators to search for a mycobacterial etiology for CD.

In 1984, isolation of MAP from tissues of CD patients again focused attention on the potential relationship between CD and JD. Chiodini et al successfully isolated an unidentified mycobacterial species from resected intestinal tissue of a CD patient (Chiodini 1984; Chiodini 1984). This isolate was later shown to be MAP by the use of molecular techniques (McFadden 1987). One of these isolates produced granulomatous lesions when administered orally to a pygmy goat (Chiodini 1984). Subsequently, cell-wall-deficient forms of MAP were identified in resected tissues from other CD patients (Markesich 1988). In 1987, McClure et al. reported the isolation of MAP from twenty-nine of thirty-eight animals in a colony of stump-tailed macaques and presented evidence that this organism was the cause of diarrhea and unthriftiness in these monkeys. These reports further intensified the search for a mycobacterial cause of CD and reinitiated the debate regarding the role of MAP as a causative agent of CD.

Other attempts to culture MAP from tissues of CD patients were not very successful. It was reported that culture had been attempted in more than two hundred clinical specimens, with only six successful isolations of organisms confirmed to be MAP by molecular techniques (Fidler 1997). A similar review of the situation was published in 1999 (Van Kruiningen 1999). Several investigators have isolated cell-wall-deficient forms (spheroplasts) of MAP using artificial media (Markesich 1988; El-Zaatari 2001). Some have proposed that the organism exists in human tissue in this form in very low numbers, and this is offered as an explanation as to why the organisms have not been seen with regular histologic techniques and acid-fast staining (Hermon-Taylor 2000; El-Zaatari 2001). Efforts to improve the efficiency of isolation of MAP from animals have been reported in the past few years, and recent reports of improved methods of culture of MAP from humans have been published, but this has not yet been repeated by others (Naser 2000; Schwartz 2000).

Serological techniques have been used in an attempt to show an association between MAP and CD. Most reports using patients and controls have not provided convincing evidence. However, serological diagnosis of subclinical MAP infections in cattle is not particularly sensitive, perhaps because antibody production tends to be a late-stage manifestation of the infection. In addition, animals may be exposed to a number of organisms in their environment that share antigens with MAP, making development of highly sensitive, and yet highly specific, assays difficult. Most efforts examining human serum have used partially purified antigen preparations similar to those used in JD diagnosis. However, recent efforts to identify a humoral response in humans with CD using unique recombinant antigens of MAP or a novel mycobacterial antigen named HupB have suggested that MAP, or an unidentified mycobacterial species, may be involved in the serologically reactive patients identified .

Bacteria may acquire bits of DNA in the form of insertion elements and transposons. Insertion elements are thought to modify pathogenicity, virulence, and antibiotic resistance as well as possibly other characteristics. In 1989, an insertion sequence consisting of about 1,450 base pairs, IS900, was identified in MAP, and it appeared to be highly specific for this organism (Green 1989; Vary 1990). This finding has assisted diagnosis of JD in animals and facilitates epidemiologic studies and strain typing of isolates. A number of studies using polymerase chain reaction (PCR) techniques have been performed, on DNA extracted from both fresh and paraffin-embedded tissue specimens, in attempts to identify IS900 in tissues from CD patients.

In 1999, Van Kruiningen reviewed the results of such studies through 1997 and concluded that the evidence presented from studies using PCR was conflicting and has done nothing to confirm the presence of mycobacterial DNA in CD tissues (Van Kruiningen 1999). Variation in the techniques of sampling the tissues and performing PCR, insufficient stringency of the reactions, presence or absence of granulomas, use of paraffin-embedded tissue (DNA may be fragmented in extraction) in some studies, and possibly laboratory contamination have been cited as possible reasons for the variable results. Using primers to detect IS900, PCR techniques also have been used to confirm that the cell-wall-deficient isolates from cultures from CD patients were MAP. In situ hybridization techniques using labeled DNA probes have been used to detect the IS900 fragment in cell-wall-deficient forms of mycobacteria in tissues from animals with JD and CD patients (Hulten 2000; Sanna 2000; Hulten 2001; Sechi 2001). In one of these, the IS900 sequence was demonstrated in 40 percent of tissues from CD patients with granulomas, 4.5 percent of CD patients without granulomas, 9.5 percent of patients with ulcerative colitis, and in none of twenty-two noninflammatory bowel disease controls.

Although IS900 was initially thought to be highly specific for MAP and is still used as a diagnostic criterion for diagnosis of JD, new data have cast some doubt on this assumption. One report has indicated that 15/26 (57.6 percent) *Mycobacterium avium* subspecies *avium* isolates from AIDS patients were strongly positive for IS900 or an IS900-related sequence. Another study has reported the isolation of *Mycobacterium* spp. other than MAP that were positive by PCR using IS900-derived primers from the feces of three animals in Australia (Cousins 1999). These isolates appeared to be most closely related to *M. scrofulaceum*, and the authors "recommend the adoption of restriction endonuclease analysis of IS900 PCR product as a routine precaution to prevent the reporting of false positive IS900 PCR results." Future studies of CD must confirm positive IS900 data using either the methods suggested by these authors, or other methods, to present convincing evidence for a role of MAP in CD.

IF MAP CAUSES DISEASE IN HUMANS, HOW ARE THEY EXPOSED? If MAP is indeed an etiologic factor in the development of CD, how would people without direct exposure to infected animals acquire this organism? An indirect route of transmission between animals and humans has not been definitively identified. However, several possible routes of human exposure to MAP have been proposed. These include foodborne exposures via milk, milk products, meat, and environmental sources such as water.

Considerable attention has been focused on the potential role of retail milk supplies as a source of MAP. Several published reports have described attempts at isolation of viable MAP from milk and colostrum, spiked with varying numbers of organisms, following laboratory experiments designed to simulate pasteurization (Chiodini 1993; Grant 1996; Meylan 1996; Stabel 1997; Grant 1998; Keswani 1998; Grant 1999; Pearce 2001; Stabel 2001). Several methods of heat treatment were used in these studies, including the standard holder method, lab scale pasteurizers, double boilers, and capillary tubes in water baths. In these studies, the heat treatments used to simulate pasteurization were either 63°C for thirty minutes, when using holder methods, or 72°C for fifteen seconds with a lab-scale commercial pasteurizer or the double-boiler methods. It appears from this work that the standard holder method may not reliably destroy all MAP in milk. Thus, questions regarding the efficacy of commercial pasteurization in eliminating viable MAP have initiated considerable debate regarding retail milk as a potential source of MAP exposure.

It is important to note that three research groups have failed to obtain viable MAP following heat treatment at 72°C for fifteen seconds, the industry standard (Stabel 1997; Keswani 1998; Pearce 2001). Two of these reports represent work done with equipment simulating the turbulent flow conditions present during commercial milk processing . However, one of these has been criticized because the MAP cells were frozen and sonicated to break up clumps prior to inoculation in milk samples, possibly resulting in sublethally injured cells that may have been more sensitive to heat treatments (Hermon-Taylor 2000).

In addition to the results of these controlled, laboratory experiments, two studies have addressed the presence of viable MAP in pasteurized retail milk. In one British study, 7 percent of retail, pasteurized milk samples obtained over a 1.5 year period were positive by PCR for the insertion sequence, IS900. Interestingly, seasonal variation in the presence of MAP existed with as many as 25 percent of samples found test-positive during certain times of the year. The use of PCR techniques did not allow for direct assessment of organism viability. However, long-term broth cultures of the same samples resulted in the growth of acid-fast bacilli and positive PCR identification of MAP in 16 percent of the cultures that had originally been PCR-negative and in 50 percent of those samples originally PCR-positive. Even though these researchers were not able to isolate the organisms on solid media, they cite their successful culture of acid-fast organisms in broth, with PCR confirmation, as evidence of the viability of MAP in these retail milk samples. In 1999, another study was begun in the United Kingdom and is currently being completed. A total of 258 dairy processing plants participated in the study, and 830 samples of raw or pasteurized milk were collected. Thus far, viable MAP has been detected in 1.9 percent of raw milk samples and 2.1 percent of pasteurized milk samples analyzed as part of that investigation .

The available research has identified two important determinants for MAP's ability to survive pasteurization: heat tolerance and concentration in the milk. It appears that MAP may be more heat tolerant than both *Mycobacterium bovis* and *Coxiella burnetti*, the zoonotic organisms that are the targets of current milk pasteurization methods. Additionally, variations in heat tolerance may also exist between MAP strains. Several studies have suggested that the concentration of viable organism in the raw milk is an important determinant of the effectiveness of pasteurization against MAP (Chiodini 1993; Grant 1996; Stabel 1997; Grant 1998; Keswani 1998). Evidence suggests that as the concentration of MAP in the raw milk increases, the number of organisms surviving pasteurization also increases. However, two studies have concluded that survival of MAP following pasteurization at 72°C for fifteen seconds may occur even when low numbers of organisms, between 10^1 and 10^3 CFU/ml, are initially present (Grant 1998; Sung 1998).

Excretion of MAP into the colostrum and milk by both clinically and subclinically infected cows has been well documented. Fecal contamination of milk with MAP during its harvest is also possible. One report has suggested that average fecal contamination of milk is 10 mg of feces per liter. Thus, it is possible that fecal contamination of milk could result in significant levels of MAP in raw milk, given that the concentration found in the feces of infected cows and sheep can be at least 10^6 colony-forming units (cfu) per g. Unless the ability of pasteurization to eliminate all viable MAP from milk is verified, retail milk may continue to be considered a potential source of MAP. Raw milk or cheese products derived from unpasteurized or contaminated milk have also been identified as potential foodborne sources of MAP, but it would appear from limited studies that the combination of acids, salt concentration, and aging time may reduce the potential for survival of MAP if present at levels of 10^3 per gram or less (Collins 1997; Sung 2000; Anonymous 2001; Spahr 2001).

In addition to milk and milk products, meat has been identified as another potential source for human exposure to MAP. Although JD in ruminants primarily involves the gastrointestinal system, a systemic infection also occurs. This systemic infection is evidenced by the finding of MAP in the milk and mammary gland lymph nodes, reproductive organs and fetuses, blood, lymph, and tissue fluids (Larsen 1970; Sweeney 1992; Sweeney 1992; Koenig 1993). It is possible that meat obtained from infected animals could contain MAP as a result of a systemic infection or fecal contamination of the carcass or meat products. The thermal tolerance of MAP could facilitate human exposure under certain cooking conditions. However, no published studies have addressed the potential role for exposure of humans to MAP via meat products.

Exposures to MAP from environmental sources, specifically as a result of contamination of water supplies, have also been proposed (Hermon-Taylor 1993, 2001). Runoff from agricultural land contaminated with MAP as a result of infected cattle grazing pasture or the use of manure as fertilizer could result in contamination of water supplies. Other mycobacterial species within the *Mycobacterium avium* complex have been isolated from municipal water supplies. Potable drinking water has been documented as a source of nontuberculous mycobacterial infections in immunocompromised people and animals. A recent report demonstrated that water contaminated with more than 10^6 cfu/ml of MAP was not adequately disinfected after thirty minutes of contact time with 2 μg/ml of chlorine. However, it is unlikely that drinking water would be contaminated to this degree, and no data currently exists to show that MAP is present in potable water supplies or that it is there in this high concentration.

The role of wildlife species as a reservoir of MAP for humans or domestic animals and an additional source of environmental contamination has received considerable attention in the recent scientific literature. Sporadic reports of natural MAP infections in wild ruminants such as deer, elk, bison, bighorn sheep, and mountain goats have been reported in the United States (Riemann 1979; Williams 1979; Jessup 1981; Chiodini 1983). Infections in wild red deer in the Western Alps and in wild roe deer and fallow deer in the Czech Republic have been documented. More interestingly, natural MAP infections in nonruminant wildlife species have recently been reported in Scotland. Natural MAP infections were first confirmed by histopathological examination and tissue culture in wild rabbits taken from farms with a history of JD. Positive tissue cultures and histopathological lesions consistent with MAP infection were subsequently identified in tissues from two carnivore species, the fox and the stoat, and other wild species were found to harbor MAP as well, including the crow, rook, jackdaw, rat, wood mouse, hare, and badger (Beard 2001).

The potential role for wildlife, or other organisms, in the transmission of MAP to food animals or to food is unclear but needs further elucidation. For example, one report suggests that infected rabbits may be a source of MAP infection for cattle because cattle fail to avoid grazing pastures heavily contaminated with rabbit feces and may actually consume them in the course of grazing. Finally, MAP and several other species of mycobacteria were isolated from several *Diptera* species taken from contaminated environments, including a cattle slaughter facility (Fischer 2001).

RECENT DEVELOPMENTS. A recent report describes the therapy of a boy diagnosed with cervical adenitis and subsequent development of CD. He was apparently cured of CD following specific antimicrobial treatment aimed at MAP. Retrospectively, acid-fast organisms were seen on the original histopathologic examination of the lymph nodes, and archived lymphoid tissue was shown to have "abundant" MAP by DNA extraction and IS900-based PCR techniques (Hermon-Taylor 1998). A clinical report describing the isolation of MAP from the breast milk of two recent mothers with CD, but not from five healthy control mothers, has also recently been published (Naser 2000). These findings suggest that MAP may cause a systemic infection in humans similar to that seen in animals. However, these reports have not addressed the concerns about the potential lack of specificity of the IS900 in identification of MAP.

Two recent reports have identified variability in the *nod2* gene located on chromosome 16 that appears to be associated with susceptibility to CD (Hugot 2001; Ogura 2001). The *nod2* codes for a cytosolic protein involved in the innate immune response to bacterial components, including those associated with the bacterial cell wall. Results from these studies strongly indicate that increased susceptibility to CD is associated with mutations in the *nod2* (McGovern 2001). The possible mechanisms of immune dysfunction proposed by the authors vary between the studies. However, both studies provide evidence for the role of both bacteria and host genetic factors in the development of CD. It is important to recognize that although these studies favor a role for enteric bacteria in the development of CD, they do not implicate any single bacterial species as an etiologic agent.

SUMMARY AND CONCLUSIONS. Despite decades of speculation and research concerning the potential relationship between MAP and CD, the scientific community is still divided as to whether there is a role for MAP in the causation of CD. Recent reports issued by the Food Standards Agency of the United Kingdom, the Scientific Committee on Animal Health and Animal Welfare of the European Union Directorate, and the National Institute of Allergy and Infectious Diseases of the National Institutes of Health in the United States all conclude that insufficient evidence exists to either confirm or disprove a causative relationship between MAP and CD at the present time. That there is a genetic susceptibility in some people now seems certain. Positive IS900 data obtained in future studies must be confirmed by other techniques that conclusively identify MAP in tissues or cultures from humans.

Although the available scientific information should not be overlooked, additional research that more specifically addresses causation is needed. All three reports cite specific research priorities that, when considered together, would provide greater insight into this potential relationship. These priorities extend to both human and animal populations as well as requesting more basic research on the organism itself. With respect to food safety concerns, identifying temperature and time combinations for pasteurization of milk that will eliminate viable MAP has been recommended. Additionally, the role of water supplies in the transmission of MAP and the organism's ability to survive in water has been identified as a research priority. Although not specifically recommended in these reports, the lack of information concerning the potential presence of MAP, and its survival, in meat and meat products suggest these are other areas for study.

REFERENCES

Angus, K.W. 1990. Intestinal lesions resembling paratuberculosis in a wild rabbit (Oryctolagus cuniculus). *J. Comp. Pathol.* 103(1):101–105.

Anonymous. 2000. Report of the Scientific Committee on Animal Health and Animal Welfare: Possible links between Crohn's disease and Paratuberculosis. European Commission: 1–76.

Anonymous. 2001. Bulletin 362/2001: Mycobacterium paratuberculosis. Brussels, Belgium, International Dairy Federation: 1–61.

Beard, P.M., M.J. Daniels, D. Henderson, A. Pirie, K. Rudge, D. Buxton, S. Rhind, A. Greig, M.R. Hutchings, I. McKendrick, K. Stevenson, and J.M. Sharp. 2001. Paratuberculosis infection of nonruminant wildlife in Scotland. *J. Clin. Microbiol.* 39(4):1517–1521.

Beard, P.M., S.M. Rhind, D. Buxton, M.J. Daniels, D. Henderson, A. Pirie, K. Rudge, A Greig, M.R. Hutchings, K. Stevenson, and J.M. Sharp. 2001. Natural paratuberculosis infection in rabbits in Scotland. *J. Comp. Pathol.* 124(4):290–299.

Boelaert, F., K. Walravens, P. Biront, J.P. Vermeersch, D. Berkvens, and J. Godfroid. 2000. Prevalence of paratuberculosis (Johne's disease) in the Belgian cattle population. *Vet. Microbiol.* 77(3–4):269–281.

Chiodini, R.J. 1989). Crohn's disease and the mycobacterioses: a review and comparison of two disease entities. *Clin. Microbiol. Rev.* 2(1):90–117.

Chiodini, R.J., and J. Hermon-Taylor. 1993. The thermal resistance of Mycobacterium paratuberculosis in raw milk under conditions simulating pasteurization. *J. Vet. Diagn. Invest.* 5(4):629–631.

Chiodini, R.J., and H.J. Van Kruiningen. 1983. Eastern whitetailed deer as a reservoir of ruminant paratuberculosis. *J. Am. Vet. Med. Assoc.* 182(2):168–169.

Chiodini, R.J., H.J. Van Kruiningen, and R.S. Merkal. 1984. Ruminant paratuberculosis (Johne's disease): the current status and future prospects. *Cornell Vet.* 74(3):218–262.

Chiodini, R.J., H.J. Van Kruiningen, R.S. Merkal, W.R. Thayer, Jr., and J.A. Coutu. 1984. Characteristics of an unclassified Mycobacterium species isolated from patients with Crohn's disease. *J. Clin. Microbiol.* 20(5):966–971.

Chiodini, R.J., H.J. Van Kruiningen, W.R. Thayer, R.S. Merkal, and J.A. Coutu. 1984. Possible role of mycobacteria in inflammatory bowel disease. I. An unclassified Mycobacterium species isolated from patients with Crohn's disease. *Dig. Dis. Sci.* 29(12):1073–1079.

Collins, M. T. 1997. Mycobacterium paratuberculosis: a potential food-borne pathogen? *J. Dairy Sci.* 80(12):3445–3448.

Commission, E. 2000. Report of the Scientific Committee on Animal Health and Animal Welfare: Possible links between Crohn's disease and Paratuberculosis. European Commission: 76.

Cousins, D.V., R. Whittington, I. Marsh, A. Masters, R.J. Evans, and P. Kluver. 1999. Mycobacteria distinct from Mycobacterium avium subsp. paratuberculosis isolated from the faeces of ruminants possess IS900-like sequences detectable IS900 polymerase chain reaction: implications for diagnosis. *Mol. Cell Probes* 13(6): 431–442.

Crohn, B., Ginsberg, L., and G. Oppenheimer. 1932. Regional ileitis, a pathological and clinical entitiy. *J.A.M.A.* 99:1323–1329.

Dalziel, T.K. 1913. Chronic intestinal enteritis. *Brit. Med. J.* ii:1068–1070.

Daniels, M.J., N. Ball, M.R. Hutchings, and A. Greig. 2001. The grazing response of cattle to pasture contaminated with rabbit faeces and the implications for the transmission of paratuberculosis. *Vet. J.* 161(3):306–313.

Dargatz, D.A., B.A. Byrum, S.G. Hennager, L.K. Barber, C.A. Kopral, A. Wagner, and S.J. Wells. 2001. Prevalence of antibodies against Mycobacterium avium subsp. paratuberculosis among beef cow-calf herds. *J.A.V.M.A.* 219(4):497–501.

Davies, H.L. 1997. Ovine Johne's disease. *Aust. Vet. J.* 75(11):799.

Dubash, K., W.P. Shulaw, S. Bech-Nielsen, H.F. Stills, Jr., and R.D. Slemons. 1996. Evaluation of an agar gel immunodiffusion test kit for detection of antibodies to Mycobacterium paratuberculosis in sheep. *J.A.V.M.A.* 208(3):401–403.

Eamens, G., S. Spence, and M. Turner. 2001. Survival of Mycobacterium avium subsp. paratuberculosis in amitraz cattle dip fluid. *Aust. Vet. J.* 79(10):703–706.

El-Zaatari, F.A., S.A. Naser, K. Hulten, P. Burch, and D.Y. Graham. 1999. Characterization of Mycobacterium paratuberculosis p36 antigen and its seroreactivities in Crohn's disease. *Curr. Microbiol.* 39(2):115–119.

El-Zaatari, F.A., M.S. Osato, and D.Y. Graham. 2001. Etiology of Crohn's disease: the role of Mycobacterium avium paratuberculosis. *Trends Mol. Med.* 7(6):247–252.

Fidler, H.M., and J.J. McFadden 1997. Infective gents—mycobacteria. In *Inflammatory Bowel Diseases*. R.N. Allan, J.M. Rohdes, S.B. Hanauer, S.R.B. Keighley, J. Alexander-Williams, and V.W. Fazio., eds. New York: Churchill Livingstone, 125–132.

Fischer, O., L. Matlova, L. Dvorska, P. Svastove, J. Bartl, I Melicharek, R.T. Weston, and I. Pavlik. 2001). Diptera as vectors of mycobacterial infections in cattle and pigs. *Med. Vet. Entomol.* 15(2):208–211.

Grant, I. R. 1998). Does Mycobacterium paratuberculosis survive current pasteurization conditions? *Appl. Environ. Microbiol.* 64(7):2760–2761.

Grant, I.R., H.J. Ball, S.D. Neill, and M.T. Rowe. 1996. Inactivation of Mycobacterium paratuberculosis in cows' milk at pasteurization temperatures. *Appl. Environ. Microbiol.* 62(2):631–636.

Grant, I.R., H.J. Ball, and M.T. Rowe. 1998. Effect of high-temperature, short-time (HTST) pasteurization on milk containing low numbers of Mycobacterium paratuberculosis. *Lett. Appl. Microbiol.* 26(2):166–170.

Grant, I.R., H.J. Ball, and M.T. Rowe. 1999. Effect of higher pasteurization temperatures, and longer holding times at 72 degrees C, on the inactivation of Mycobacterium paratuberculosis in milk. *Lett. Appl. Microbiol.* 28(6):461–465.

Green, E.P., M.L. Tizard, M.T. Moss, J. Thompson, D.J. Winterbourne, J.H. McFadden, and J. Hermon-Taylor. 1989. Sequence and characteristics of IS900, an insertion element identified in a human Crohn's disease isolate of Mycobacterium paratuberculosis. *Nucleic Acids Res.* 17(22):9063–9073.

Hermon-Taylor, J. 1993. Causation of Crohn's disease: the impact of clusters. *Gastroenterol.* 104(2):643–646.

Hermon-Taylor, J. 2001. Mycobacterium avium subspecies paratuberculosis is a cause of Crohn's disease. *Gut* 49(6):755–757.

Hermon-Taylor, J., N. Barnes, C. Clarke, and C. Finlayson. 1998. Mycobacterium paratuberculosis cervical lymphadenitis, followed five years later by terminal ileitis similar to Crohn's disease. *B.M.J.* 316(7129):449–453.

Hermon-Taylor, J., T.J. Bull, J.M. Sheridan, J. Cheng, M.L. Stellakis, and N. Sumar. 2000. Causation of Crohn's disease by Mycobacterium avium subspecies paratuberculosis. *Can. J. Gastroenterol.* 14(6):521–539.

Hugot, J.P., M. Chamaillard, H. Zouali, S. Lesage, J.P. Cezard, J. Belaiche, S. Almer, C. Tysk, C.A. O'Morain, M. Gassull, V. Binder, Y. Finke., A. Cortot, R. Modigliani, P. Laurent_Puig, C. Gower-Rosseau, J. Macry, J.F. Colombel, M. Sahbatou, and G. Thomas. 2001. Association of NOD2 leucine-rich repeat variants with susceptibility to Crohn's disease. *Nature* 411(6837):599–603.

Hulten, K., H.M. El-Zimaity, T.J. Karttunen, A. Almashhrawi, M. R. Schwartz, D.Y. Graham, and F.A. El-Zaatari. 2001. Detection of Mycobacterium avium subspecies paratuberculosis in Crohn's diseased tissues by in situ hybridization. *Am. J. Gastroenterol.* 96(5):1529–1535.

Hulten, K., T.J. Karttunen, H.M. El-Zimaity, S.A. Naser, M.T. Collins, D.Y. Graham, and F.A. El-Zaatari. 2000. Identification of cell wall deficient forms of M. avium subsp. paratuberculosis in paraffin embedded tissues from animals with Johne's disease by in situ hybridization. *J. Microbiol. Methods* 42(2):185–195.

Jessup, D.A., B. Abbas, and D. Behymer. 1981. Paratuberculosis in tule elk in California. *J.A.V.M.A.* 179(11):1252–1254.

Jorgensen, J.B. 1977. Survival of Mycobacterium paratuberculosis in slurry. *Nord Vet. Med.* 29(6):267–270.

Kalis, C.H., J.W. Hesselink, E.W. Russchen, H.W. Barkema, M T. Collins, and I.J. Visser. 1999. Factors influencing the isolation of Mycobacterium avium subsp. paratuberculosis from bovine fecal samples. *J. Vet. Diagn. Invest.* 11(4):345–351.

Keswani, J., and J.F. Frank 1998. Thermal inactivation of Mycobacterium paratuberculosis in milk. *J. Food Prot.* 61(8):974–978.

Koenig, G.J., G.F. Hoffsis, W.P. Shulaw, S. Bech-Nielsen, D M. Rings, and G. St-Jean. 1993. Isolation of Mycobacterium paratuberculosis from mononuclear cells in tissues, blood, and mammary glands of cows with advanced paratuberculosis. *Am. J. Vet. Res.* 54(9):1441–1445.

Kreeger, J.M. 1991. Ruminant paratuberculosis—a century of progress and frustration. *J. Vet. Diagn. Invest.* 3(4):373–382.

Langman, M.J.S. 1997. Epidemiological overview of inflammatory bowel disease. *Inflammatory Bowel Diseases.* R.N. Allan, J.M. Rohdes, S.B. Hanauer, S.R.B. Keighley, J. Alexander-Williams, and V.W. Fazio., eds. New York: Churchill Livingstone, 35–39.

Larsen, A.B., and K.E. Kopecky. 1970. Mycobacterium paratuberculosis in reproductive organs and semen of bulls. *Am. J. Vet. Res.* 31(2):255–258.

Larsen, A.B., O.H. Stalheim, D.E. Hughes, L.H. Appell, W.D. Richards, and E.M. Himes. 1981. Mycobacterium paratuberculosis in the semen and genital organs of a semen-donor bull. *J.A.V.M.A.* 179(2):169–171.

Lovell, R., M. Levi, and J. Francis. 1944. Studies on the survivial of Johne's bacillus. *J. Comp. Pathol.* 54:120–129.

Markesich, D.C., D.Y. Graham, and H.H. Yoshimura. 1988.

Progress in culture and subculture of spheroplasts and fastidious acid-fast bacilli isolated from intestinal tissues. *J. Clin. Microbiol.* 26(8):1600–1603.

McClure, H.M., R.J. Chiodini, D.C. Anderson, R.B. Swenson, W.R. Thayer, and J.A. Coutu. 1987. Mycobacterium paratuberculosis infection in a colony of stumptail macaques (Macaca arctoides). *J. Infect. Dis.* 155(5):1011–1019.

McFadden, J.J., P.D. Butcher, R. Chiodini, and J. Hermon-Taylor. 1987. Crohn's disease-isolated mycobacteria are identical to Mycobacterium paratuberculosis, as determined by DNA probes that distinguish between mycobacterial species. *J. Clin. Microbiol.* 25(5):796–801.

McGovern, D.P., D.A. Van Heel, T. Ahmad, and D.P. Jewell. 2001. NOD2 (CARD15), the first susceptibility gene for Crohn's disease. *Gut* 49(6):752–754.

McQueen, D.S., and E.G. Russell. 1979. Culture of Mycobacterium paratuberculosis from bovine foetuses. *Aust. Vet. J.* 55(4):203–204.

Merkal, R.S., J.M. Miller, A.M. Hintz, and J.H. Bryner. 1982. Intrauterine inoculation of Mycobacterium paratuberculosis into guinea pigs and cattle. *Am. J. Vet. Res.* 43(4):676–678.

Meylan, M., D.M. Rings, W.P. Shulaw, J.J. Kowalski, S. Bech-Nielsen, and G.F. Hoffsis. 1996. Survival of Mycobacterium paratuberculosis and preservation of immunoglobulin G in bovine colostrum under experimental conditions simulating pasteurization. *Am. J. Vet. Res.* 57(11):1580–1585.

Millar, D., J. Ford, J. Sanderson, S. Withey, M. Tizard, T. Doran, and J. Hermon-Taylor. 1996. IS900 PCR to detect Mycobacterium paratuberculosis in retail supplies of whole pasteurized cows' milk in England and Wales. *Appl. Environ. Microbiol.* 62(9):3446–3452.

Mishina, D., P. Katsel, S.T. Brown, E.C. Gilberts, and R.J. Greenstein. 1996. On the etiology of Crohn's disease. *Proc. Natl. Acad. Sci. USA* 93(18):9816–9820.

Muskens, J., H.W. Barkema, E. Russchen, K. van Maanen, Y.H. Schukken, and D. Bakker. 2000. Prevalence and regional distribution of paratuberculosis in dairy herds in The Netherlands. *Vet. Microbiol.* 77(3–4):253–261.

Naser, S.A., J. Felix, H. Liping, C. Romero, N. Naser, A. Walsh, and W. Safranek. 1999. Occurrence of the IS900 gene in Mycobacterium avium complex derived from HIV patients. *Mol. Cell Probes* 13(5):367–372.

Naser, S.A., K. Hulten, I. Shafran, D.Y. Graham, and F.A. El-Zaatari. 2000. Specific seroreactivity of Crohn's disease patients against p35 and p36 antigens of M. avium subsp. paratuberculosis. *Vet. Microbiol.* 77(3–4):497–504.

Naser, S.A., D. Schwartz, and I. Shafran. 2000. Isolation of Mycobacterium avium subsp paratuberculosis from breast milk of Crohn's disease patients. *Am. J. Gastroenterol.* 95(4):1094–1095.

Nauta, M.J., and J.W. van der Giessen. 1998. Human exposure to Mycobacterium paratuberculosis via pasteurised milk: a modelling approach. *Vet. Rec.* 143(11):293–6.

Nebbia, P., P. Robino, E. Ferroglio, L. Rossi, G. Meneguz, and S. Rosati. 2000. Paratuberculosis in red deer (Cervus elaphus hippelaphus) in the western Alps. *Vet. Res. Commun.* 24(7):435–443.

Nielsen, S.S., S.M. Thamsborg, H. Houe, and V. Bitsch. 2000. Bulk-tank milk ELISA antibodies for estimating the prevalence of paratuberculosis in Danish dairy herds. *Prev. Vet. Med.* 44(1–2):1–7.

Nielsen, S.S., S.M. Thamsborg, H. Houe, and V.V. Bitsch. 2000. Corrigendum to bulk-tank milk ELISA antibodies for estimating the prevalence of paratuberculosis in Danish dairy herds. *Prev. Vet. Med.* 46(4):297.

Ogura, Y., D.K. Bonen, N. Inohara, D.L. Nicolae, F.F. Chen, R. Ramos, H. Britton, T. Moran, R. Karaliuskas, R.H.

Duerr, J.P. Achkar, S.R. Brant, T.M. Bayless, B.S. Kirschner, S.B. Hanauer, G. Nunez, and J.H. Cho. 2001. A frameshift mutation in NOD2 associated with susceptibility to Crohn's disease. *Nature* 411(6837):603–606.

Ott, S.L., S.J. Wells, and B.A. Wagner. 1999. Herd-level economic losses associated with Johne's disease on U.S. dairy operations. *Prev. Vet. Med.* 40(3–4):179–192.

Pavlik, I., J. Bartl, L. Dvorska, P. Svastova, R. du Maine, M. Machackova, W. Yayo Ayele, and A. Horvathova. 2000. Epidemiology of paratuberculosis in wild ruminants studied by restriction fragment length polymorphism in the Czech Republic during the period 1995–1998. *Vet. Microbiol.* 77(3–4):231–251.

Pearce, L.E., H.T. Truong, R.A. Crawford, G.F. Yates, S. Cavaignac, and G.W. de Lisle. 2001. Effect of turbulent-flow pasteurization on survival of Mycobacterium avium subsp. paratuberculosis added to raw milk. *Appl. Environ. Microbiol.* 67(9):3964–3969.

Polito, J.M., 2nd, R.C. Rees, B. Childs, A.I. Mendeloff, M.L. Harris, and T.M. Bayless. 1996. Preliminary evidence for genetic anticipation in Crohn's disease. *Lancet* 347(9004):798–800.

Quirke, P. 2001. Mycobacterium avium subspecies paratuberculosis is a cause of Crohn's disease. *Gut* 49(6):757–760.

Riemann, H., M.R. Zaman, R. Ruppanner, O. Aalund, J.B. Jorgensen, H. Worsaae and D. Behymer. 1979. Paratuberculosis in cattle and free-living exotic deer. *J.A.V.M.A.* 174(8):841–843.

Rubery, E. 2001. A review of the evidence for a link between exposure to Mycobacterium Paratuberculosis (MAP) and Crohn's Disease in humans. Cambridge, Food Standards Agency, United Kingdom.

Sanna, E., C.J. Woodall, N.J. Watt, C.J. Clarke, M. Pittau, A. Leoni, and A.M. Nieddu. 2000. In situ-PCR for the detection of Mycobacterium paratuberculosis DNA in paraffin-embedded tissues. *Eur. J. Histochem.* 44(2):179–184.

Satsangi, J. 2001. Genetics of inflammatory bowel disease: from bench to bedside? *Acta. Odontol. Scand.* 59(3):187–192.

Satsangi, J., D. Jewell, M. Parkes, and J. Bell. 1998. Genetics of inflammatory bowel disease. A personal view on progress and prospects. *Dig. Dis.* 16(6):370–374.

Sawczenko, A., B.K. Sandhu, R.F. Logan, H. Jenkins, C.J. Taylor, S. Mian, and R. Lynn 2001. Prospective survey of childhood inflammatory bowel disease in the British Isles. *Lancet* 357(9262):1093–1094.

Schwartz, D., I. Shafran, C. Romero, C. Piromalli, J. Biggerstaff, N. Naser, W. Chamberlin, and S.A. Naser. 2000. Use of short-term culture for identification of Mycobacterium avium subsp. paratuberculosis in tissue from Crohn's disease patients. *Clin. Microbiol. Infect.* 6(6):303–307.

Sechi, L.A., M. Manuela, T. Francesco, L. Amelia, S. Antonello, F. Giovanni, and Z. Stefania. 2001. Identification of Mycobacterium avium subsp. paratuberculosis in biopsy specimens from patients with Crohn's Disease identified by in situ hybridization. *J. Clin. Microbiol.* 39(12):4514–4517.

Secretariat. 2000). Preliminary results from the national study on the microbiological quality and heat processing of cow's milk: *Mycobacterium avium* subspecies *paratuberculosis* (MAP), Advisory Committee on the Microbiological Safety of Food, Food Standards Agency, 1–6.

Seitz, S.E., L.E. Heider, W.D. Heuston, S. Bech-Nielsen, D.M. Rings, and L. Spangler. 1989. Bovine fetal infection with Mycobacterium paratuberculosis. *J.A.V.M.A.* 194(10):1423–1426.

Spahr, U., and K. Schafroth. 2001. Fate of Mycobacterium avium subsp. paratuberculosis in Swiss hard and semi-hard cheese manufactured from raw milk. *Appl. Environ. Microbiol.* 67(9):4199–4205.

Squier, C., V.L. Yu, and J.E. Stout. 2000. Waterborne nosocomial infections. *Curr. Infect. Dis. Rep.* 2(6):490–496.

Stabel, J.R. 2001. On-farm batch pasteurization destroys Mycobacterium paratuberculosis in waste milk. *J. Dairy Sci.* 84(2):524–527.

Stabel, J.R., E.M. Steadham, and C.A. Bolin. 1997. Heat inactivation of Mycobacterium paratuberculosis in raw milk: are current pasteurization conditions effective? *Appl. Environ. Microbiol.* 63(12):4975–4977.

Streeter, R.N., G.F. Hoffsis, S. Bech-Nielsen, W.P. Shulaw, and D.M. Rings. 1995. Isolation of Mycobacterium paratuberculosis from colostrum and milk of subclinically infected cows. *Am. J. Vet. Res.* 56(10):1322–1324.

Sung, N., and M.T. Collins. 1998. Thermal tolerance of Mycobacterium paratuberculosis. *Appl. Environ. Microbiol.* 64(3):999–1005.

Sung, N., and M.T. Collins. 2000. Effect of three factors in cheese production (pH, salt, and heat) on Mycobacterium avium subsp. paratuberculosis viability. *Appl. Environ. Microbiol.* 66(4):1334–1339.

Sweeney, R.W., R.H. Whitlock, and A.E. Rosenberger. 1992. Mycobacterium paratuberculosis cultured from milk and supramammary lymph nodes of infected asymptomatic cows. *J. Clin. Microbiol.* 30(1):166–171.

Sweeney, R.W., R.H. Whitlock, and A.E. Rosenberger. 1992. Mycobacterium paratuberculosis isolated from fetuses of infected cows not manifesting signs of the disease. *Am. J. Vet. Res.* 53(4):477–480.

Taylor, T.K., C.R. Wilks, and D.S. McQueen. 1981. Isolation of Mycobacterium paratuberculosis from the milk of a cow with Johne's disease. *Vet. Rec.* 109(24):532–533.

Thompson, D.E. 1994. The role of mycobacteria in Crohn's disease. *J. Med. Microbiol.* 41(2):74–94.

Van Kruiningen, H.J. 1999. Lack of support for a common etiology in Johne's disease of animals and Crohn's disease in humans. *Inflamm. Bowel Dis.* 5(3):183–191.

Vary, P.H., P.R. Andersen, E. Green, J. Hermon-Taylor, and J.J. McFadden. 1990. Use of highly specific DNA probes and the polymerase chain reaction to detect Mycobacterium paratuberculosis in Johne's disease. *J. Clin. Microbiol.* 28(5):933–937.

Whan, L.B., I.R. Grant, H.J. Ball, R. Scott, and M.T. Rowe. 2001. Bactericidal effect of chlorine on Mycobacterium paratuberculosis in drinking water. *Lett. Appl. Microbiol.* 33(3):227–231.

Whipple, D.L., D.R. Callihan, and J.L. Jarnagin. 1991. Cultivation of Mycobacterium paratuberculosis from bovine fecal specimens and a suggested standardized procedure. *J. Vet. Diagn. Invest.* 3(4):368–373.

Whitlock, R.H., S.J. Wells, S.J. Wells, R.W. Sweeney, and J. Van Tiem. 2000. ELISA and fecal culture for paratuberculosis (Johne's disease): sensitivity and specificity of each method. *Vet. Microbiol.* 77(3–4):387–398.

Whittington, R.J., L.A. Reddacliff, I. Marsh, S. McAllister, and V. Saunders. 2000. Temporal patterns and quantification of excretion of Mycobacterium avium subsp. paratuberculosis in sheep with Johne's disease. *Aust. Vet. J.* 78(1):34–37.

Williams, E.S., T.R. Spraker, and G.G. Schoonveld. 1979. Paratuberculosis (Johne's disease) in bighorn sheep and a Rocky Mountain goat in Colorado. *J. Wildl. Dis.* 15(2):221–227.

36

TOXOPLASMA GONDII

DOLORES E. HILL AND J.P. DUBEY

INTRODUCTION. Infection with the protozoan parasite *Toxoplasma gondii* is one of the most common parasitic infections of humans and other warm-blooded animals (Dubey and Beattie 1988). It has been found worldwide from Alaska to Australia. Nearly one-third of humanity has been exposed to this parasite (Dubey and Beattie 1988). In most adults, it does not cause serious illness but can cause blindness and mental retardation in congenitally infected children and devastating disease in immunocompromised individuals. Consumption of raw or undercooked meat products are a major risk factor associated with *Toxoplasma gondii* infection.

CLASSIFICATION. *Toxoplasma gondii* is a coccidian parasite with cats as the definitive host and warm-blooded animals as intermediate hosts (Frenkel et al. 1970). It is one of the most important parasites of animals. It belongs to:

Phylum: Apicomplexa; Levine 1970
Class: Sporozoasida; Leukart 1879
Subclass: Coccidiasina; Leukart 1879
Order: Eimeriorina; Leger 1911
Family: Toxoplasmatidae, Biocca 1956

There is only one species of *Toxoplasma, T. gondii.*
Coccidia in general have complicated life cycles. Most coccidia are host-specific and are transmitted via a fecal-oral route. Transmission of *Toxoplasma gondii* occurs via the fecal oral route as well as through consumption of infected meat and by transplacental transfer from mother to fetus (Dubey and Beattie 1988; Frenkel et al. 1970).

STRUCTURE AND LIFE CYCLE. The name *Toxoplasma* (toxon = arc, plasma = form) is derived from the crescent shape of the tachyzoite stage (Fig. 36.1). *T. gondii* has three infectious stages: the tachyzoites (in groups), the bradyzoites (in tissue cysts), and the sporozoites (in oocysts).

The tachyzoite is often crescent-shaped and is approximately the size (2–6mm) of a red blood cell (Fig. 36.1A). The anterior end of the tachyzoite is pointed and the posterior end is round. It has a pellicle (outer covering), several organelles including subpel-licular microtubules, mitochondrium, smooth and rough endoplasmic reticulum, a Golgi apparatus, apicoplast, ribosomes, a micropore, and a well-defined nucleus. The nucleus is usually situated toward the posterior end or in the central area of the cell.

The tachyzoite enters the host cell by active penetration of the host cell membrane and can tilt, extend, and retract as it searches for a host cell. After entering the host cell, the tachyzoite becomes ovoid in shape and becomes surrounded by a parasitophorous vacuole. *Toxoplasma gondii* in a parasitophorous vacuole is protected from host defense mechanisms. The tachyzoite multiplies asexually within the host cell by repeated divisions in which two progeny form within the parent parasite, consuming it (Fig. 36.1A). Tachyzoites continue to divide until the host cell is filled with parasites.

After a few divisions, *T. gondii* form tissue cysts. Tissue cysts vary in size from 5–70 μm and remain intracellular (Figs. 36.1B,C). The tissue cyst wall is elastic, thin (< 0.5 μm), and may enclose hundreds of the crescent-shaped, slender *T. gondii* stage known as bradyzoites (Fig. 36.1C). The bradyzoites are approximately 7×1.5 μm. Bradyzoites differ structurally only slightly from tachyzoites. They have a nucleus situated toward the posterior end, whereas the nucleus in tachyzoites is more centrally located. Bradyzoites are more slender than are tachyzoites and are less susceptible to destruction by proteolytic enzymes than are tachyzoites. Although tissue cysts containing bradyzoites may develop in visceral organs, including lungs, liver, and kidneys, they are more prevalent in muscular and neural tissues (Fig. 36.1B), including the brain (Fig. 36.1C), eye, skeletal, and cardiac muscle. Intact tissue cysts probably do not cause any harm and can persist for the life of the host.

All coccidian parasites have an environmentally resistant stage in their life cycle, called the oocyst. Oocysts of *T. gondii* are formed only in cats, probably in all members of the Felidae (Fig. 36.2). Cats shed oocysts after ingesting any of the three infectious stages of *T. gondii*, that is, tachyzoites, bradyzoites, and sporozoites (Dubey and Frenkel 1972, 1976; Dubey 1996). Prepatent periods (time to the shedding of oocysts after initial infection) and frequency of oocyst shedding vary according to the stage of *T. gondii* ingested. Prepatent periods are three to ten days after ingesting tissue cysts and twenty-one days or more after ingesting tachyzoites

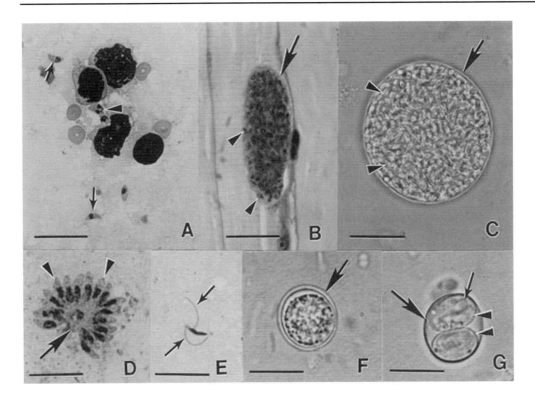

FIG. 36.1—Stages of *Toxoplasma gondii*. Scale bar in A–D = 20 μm, in E–G = 10 μm. **A.** Tachyzoites in impression smear of lung. Note crescent-shaped individual tachyzoites (arrows), dividing tachyzoites (arrowheads) compared with the size of host red blood cells and leukocytes. Giema stain. **B.** Tissue cysts in section of muscle. The tissue cyst wall is very thin (arrow) and encloses many tiny bradyzoites (arrowheads). Hematoxylin and eosin stain. **C.** Tissue cyst separated from host tissue by homogenization of infected brain. Note tissue cyst wall (arrow) and hundreds of bradyzoites (arrowheads). Unstained. **D.** Schizont (arrow) with several merozoites (arrowheads) separating from the main mass. Impression smear of infected cat intestine. Giemsa stain. **E.** A male gamete with two flagella (arrows). Impression smear of infected cat intestine. Giemsa stain. **F.** Unsporulated oocyst in fecal float of cat feces. Unstained. Note double-layered oocyst wall (arrow) enclosing a central undivided mass. **G.** Sporulated oocyst with a thin oocyst wall (large arrow), two sporocysts (arrowheads). Each sporocyst has four sporozoites (small arrow), which are not in complete focus. Unstained.

or oocysts (Dubey and Frenkel 1972, 1976; Dubey 1996). Less than 50 percent of cats shed oocysts after ingesting tachyzoites or oocysts, whereas nearly all cats shed oocysts after ingesting tissue cysts (Dubey and Frenkel 1976).

After the ingestion of tissue cysts by cats, the tissue cyst wall is dissolved by proteolytic enzymes in the stomach and small intestine. The released bradyzoites penetrate the epithelial cells of the small intestine and initiate development of numerous generations of asexual and sexual cycles of *T. gondii* (Dubey and Frenkel 1972). *Toxoplasma gondii* multiplies profusely in intestinal epithelial cells of cats (entero-epithelial cycle) and these stages are known as schizonts (Fig. 36.1D). Organisms (merozoites) released from schizonts form male and female gametes. The male gamete has two flagella (Fig. 36.1E), which swim to and enter the female gamete. After the female gamete is fertilized by the male gamete (Fig 36.1E), oocyst wall formation

begins around the fertilized gamete. When oocysts are mature, they are discharged into the intestinal lumen by the rupture of intestinal epithelial cells.

In freshly passed feces, oocysts are unsporulated (noninfective). Unsporulated oocysts are subspherical to spherical and are 10–12 μm in diameter (Fig. 36.1F). They sporulate (become infectious) outside the cat within one to five days depending upon aeration and temperature. Sporulated oocysts contain two ellipsoidal sporocysts (Fig. 36.1G). Each sporocyst contains four sporozoites. The sporozoites are 2×6 to 8 μm in size.

As the entero-epithelial cycle progresses, bradyzoites penetrate the lamina propria of the feline intestine and multiply as tachyzoites. Within a few hours after infection of cats, *T. gondii* may disseminate to extra-intestinal tissues. *Toxoplasma gondii* persists in intestinal and extra-intestinal tissues of cats for at least several months, and possibly for the life of the cat.

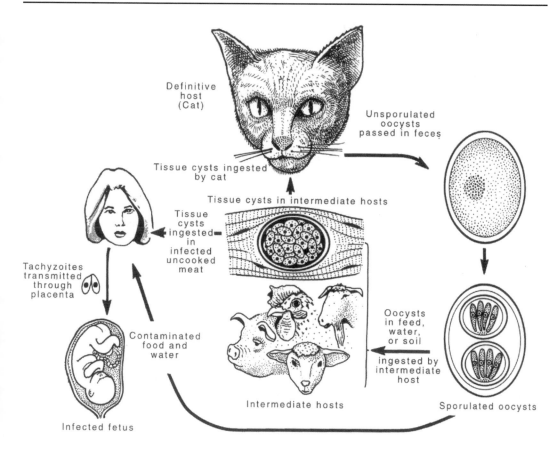

FIG. 36.2—Life cycle of *Toxoplasma gondii*.

MOLECULAR BIOLOGY AND GENETICS. The *T. gondii* nucleus is haploid except for the zygote, in the intestine of the cat (Pfefferkorn 1990). Sporozoites result from a miotic division followed by mitotic divisions, and genetic segregation seems to follow classical Mendelian laws. The total haploid genome contains approximately 8×10^7 base pairs. There is also a 36 kb circular mitochondrial DNA. Nine chromosomes have been identified by pulsed field gel electrophoresis.

Unlike many other microorganisms and in spite of a wide host range and worldwide distribution, *Toxoplasma gondii* has a low genetic diversity. *Toxoplasma gondii* strains have been classified into three genetic types (I, II, III), based on antigens, isoenzymes, and restriction fragment length polymorphism (RFLP) (Howe and Sibley 1995; Guo and Johnson 1996; Darde et al. 1988; Terry et al. 2001). Type I strains have been isolated predominantly from clinical cases of human toxoplasmosis and are considered highly virulent in outbred laboratory mice. Types II and III predominate in animals and are considered avirulent for mice (Howe and Sibley 1995; Howe et al. 1997; Mondragon et al. 1998; Owen and Trees 1999). However, most of the *T. gondii* strains genetically typed were from Europe and North America, and human-derived isolates were from clinical cases, whereas *T. gondii* strains from animals were from asymptomatic individuals (Howe and Sibley 1995; Literak et al. 1998; Fuentes et al. 2001). In a recent study, seventeen of twenty-five *T. gondii* isolates obtained from asymptomatic chickens from rural areas surrounding São Paulo, Brazil, were type I, the first report of isolation of predominantly type I strains of *T. gondii* from a food animal (Dubey et al. 2002). Although type I strains have been isolated from patients with symptoms, a definitive association of virulence with the type has not been established. Recently, Grigg et al. (2001a) suggested that type I strains or recombinants of types I and III are more likely to result in clinical ocular toxoplasmosis. Of twelve patients with ocular toxoplasmosis, three had type I strains and five had recombinant strains of *T. gondii* (Grigg et al. 2001a, 2001b).

HOST-PARASITE RELATIONSHIP. *Toxoplasma gondii* can multiply in virtually any cell in the body.

How *T. gondii* is destroyed in immune cells is not completely known (Renold et al. 1992). All extracellular forms of the parasite are directly affected by antibody but intracellular forms are not. Cellular factors, including lymphocytes and lymphokines, are believed to be more important than humoral factors in immune-mediated destruction of *T. gondii* (Renold et al. 1992).

Immunity does not eradicate infection. *Toxoplasma gondii* tissue cysts persist several years after acute infection. The fate of tissue cysts is not fully known. Whether bradyzoites can form new tissue cysts directly without transforming into tachyzoites is not known. It has been proposed that tissue cysts may at times rupture during the life of the host. The released bradyzoites may be destroyed by the host's immune responses, or there may be formation of new tissue cysts.

In immunosuppressed patients, such as those given large doses of immunosuppressive agents in preparation for organ transplants and in those with AIDS, rupture of a tissue cyst may result in transformation of bradyzoites into tachyzoites and renewed multiplication. The immunosuppressed host may die from toxoplasmosis unless treated. It is not known how corticosteroids cause relapse, but it is unlikely that they directly cause rupture of the tissue cysts.

Pathogenicity of *T. gondii* is determined by the virulence of the strain and the susceptibility of the host species. *Toxoplasma gondii* strains may vary in their pathogenicity in a given host. Certain strains of mice are more susceptible than others, and the severity of infection in individual mice within the same strain may vary. Certain species are genetically resistant to clinical toxoplasmosis. For example, adult rats do not become ill, whereas young rats can die of toxoplasmosis. Mice of any age are susceptible to clinical *T. gondii* infection. Adult dogs, like adult rats, are resistant, whereas puppies are fully susceptible to clinical toxoplasmosis. Cattle and horses are among the hosts more resistant to clinical toxoplasmosis, whereas certain marsupials and New World monkeys are highly susceptible to *T. gondii* infection (Dubey and Beattie 1988). Nothing is known concerning genetically determined susceptibility to clinical toxoplasmosis in higher mammals, including humans.

INFECTIONS IN HUMANS. *Toxoplasma gondii* infection is widespread in humans although its prevalence varies widely from place to place. In the United States and the United Kingdom, it is estimated that 16 to 40 percent of people are infected, whereas in Central and South America and continental Europe, estimates of infection range from 50 to 80 percent (Dubey and Beattie 1988). Most infections in humans are asymptomatic, but at times the parasite can produce devastating disease. Infection may be congenitally or postnatally acquired. Congenital infection occurs only when a woman becomes infected during pregnancy. Congenital infections acquired during the first trimester are

more severe than those acquired in the second and third trimester (Desmonts et al. 1974; Remington et al. 1995). Although the mother rarely has symptoms of infection, she does have a temporary parasitemia. Focal lesions develop in the placenta and the fetus may become infected. At first there is generalized infection in the fetus. Later, infection is cleared from the visceral tissues and may localize in the central nervous system. A wide spectrum of clinical diseases occur in congenitally infected children (Desmonts et al. 1974). Mild disease may consist of slightly diminished vision, whereas severely diseased children may have the full tetrad of signs: retinochoroiditis (inflammation of the inner layers of the eye), hydrocephalus (big head), convulsions, and intracerebral calcification. Of these, hydrocephalus is the least common but most dramatic lesion of toxoplasmosis. By far the most common sequela of congenital toxoplasmosis is ocular disease (Desmonts et al. 1974; Remington et al. 1995).

The socioeconomic impact of toxoplasmosis in human suffering and the cost of care of sick children, especially those with mental retardation and blindness, are enormous (Roberts and Frenkel 1990; Roberts et al. 1994). The testing of all pregnant women for *T. gondii* infection is compulsory in some European countries, including France and Austria. The cost benefits of such mass screening are being debated in many other countries (Remington et al. 1995).

Postnatally acquired infection may be localized or generalized. Oocyst-transmitted infections may be more severe than tissue cyst-induced infections (Teutsch et al. 1979; Benenson et al. 1982; Dubey and Beattie 1988; Smith 1993; Bowie et al. 1997; Burnett et al. 1998). Enlarged lymph nodes are the most frequently observed clinical form of toxoplasmosis in humans (Table 36.1). Lymphadenopathy may be associated with fever, fatigue, muscle pain, sore throat, and headache. Although the condition may be benign, its diagnosis is vital in pregnant women because of the risk to the fetus. In a British Columbia outbreak, of one hundred people who were diagnosed with acute infection, fifty-one had lymphadenopathy and twenty had retinitis (Aramini et al. 1998, 1999).

Encephalitis is the most important manifestation of toxoplasmosis in immunosuppressed patients, because it causes the most severe damage to the patient (Dubey and Beattie 1988; Luft and Remington 1992). Infection may occur in any organ. Patients may have headache, disorientation, drowsiness, hemiparesis, reflex changes, and convulsions, and many become comatose. Encephalitis caused by *T. gondii* is now recognized with great frequency in patients treated with immunosuppressive agents.

Toxoplasmosis ranks high on the list of diseases that lead to death of patients with acquired immunodeficiency syndrome (AIDS); approximately 10 percent of AIDS patients in the United States and up to 30 percent in Europe are estimated to die from toxoplasmosis (Luft and Remington 1992). Although in AIDS patients any organ may be involved, including the testis, dermis, and the spinal cord, infection of the brain is most frequently reported. Most AIDS patients suffering from toxoplasmosis have bilateral, severe, and persistent headache that responds poorly to analgesics. As the disease progresses, the headache may give way to a condition characterized by confusion, lethargy, ataxia, and coma. The predominant lesion in the brain is necrosis, especially of the thalamus (Renold et al. 1992).

INFECTION IN ANIMALS OTHER THAN HUMANS. *Toxoplasma gondii* is capable of causing infection and severe disease in animals other than humans (Dubey and Beattie 1988). Toxoplasmosis causes great losses in sheep and goats. *Toxoplasma gondii* may cause embryonic death and resorption, fetal death and mummification, abortion, stillbirth, and neonatal death in these animals. Disease is more severe in goats than in sheep. Outbreaks of toxoplasmosis in pigs have been reported from several countries, especially Japan (Dubey 1986; Nogami et al. 1999). Mortality in young pigs is more common than mortality in adult pigs. Pneumonia, myocarditis, encephalitis, and placental necrosis have been reported to occur in infected pigs. Sporadic and widespread outbreaks of toxoplasmosis occur in rabbits, mink, birds, and other domesticated and wild animals. Animals that survive infection harbor tissue cysts and can therefore transmit *T. gondii* infection to human consumers.

EPIDEMIOLOGY. Toxoplasmosis may be acquired by ingestion of oocysts or by ingestion of tissue-inhabiting stages of the parasite. Contamination of the environment by oocysts is widespread because oocysts are shed by domestic cats and other members of the Felidae (Dubey and Beattie 1988; Frenkel et al. 1970). Domestic cats are probably the major source of contamination because oocyst formation is greatest in domestic cats, and cats are extremely common. Widespread natural infection of the environment is possible because a cat may excrete millions of oocysts after ingesting as few as one bradyzoite or one tissue cyst, and many tissue cysts may be present in one infected mouse (Frenkel et al. 1970; Dubey 2001). Sporulated oocysts survive for long periods under most ordinary environmental conditions and even in harsh environments for months. They can survive in moist soil, for example, for months and even years (Dubey and Beattie 1988). Oocysts in soil can be mechanically transmitted by invertebrates such as flies, cockroaches, dung beetles, and earthworms, which can spread oocysts onto human food and animal feeds.

Infection rates in cats are determined by the rate of infection in local avian and rodent populations because cats are thought to become infected by eating these animals. The more oocysts in the environment, the more likely it is that prey animals would be infected, and this in turn would increase the infection rate in cats. In certain areas of Brazil, approximately 60 percent of six- to eight-year-old children have antibodies to *T. gondii* linked to the ingestion of oocysts from the environment heavily contaminated with *T. gondii* oocysts (Bahia-Oliveira et al. 2001). Infection in aquatic mammals indicates contamination and survival of oocysts in sea water (Cole et al. 2000). The largest recorded outbreak of clinical toxoplasmosis in humans was epidemiologically linked to drinking water from a municipal water reservoir in British Columbia, Canada (Aramini et al. 1998, 1999). This water reservoir was thought to be contaminated with *T. gondii* oocysts excreted by cougars (*Felis concolor*). Although attempts to recover *T. gondii* oocysts from water samples in the British Columbia outbreak were unsuccessful, methods to detect oocysts were reported (Isaac-Renton et al. 1998). At present, no commercial reagents are available to detect *T. gondii* oocysts in the environment.

In the United States, infection in humans is probably most often the result of ingestion of tissue cysts contained in undercooked meat (Dubey and Beattie 1988; Lopez et al. 2000; Roghmann et al. 1999), though the exact contribution of foodborne toxoplasmosis versus oocyst-induced toxoplasmosis to human infection is currently unknown. *Toxoplasma gondii* infection is common in many animals used for food, including sheep, pigs, goats, and rabbits. In one study, viable *T. gondii* tissue cysts were isolated from 17 percent of one thousand adult pigs (sows) from a slaughter plant in Iowa (Dubey et al. 1995). Serological surveys of pigs from Illinois pig farms indicate an infection rate of about 3 percent in market-weight animals and 20 percent for breeding pigs, suggesting that age is a factor for pigs acquiring *Toxoplasma* infection (Weigel et al. 1995). Serological surveys of pigs on New England farms revealed an overall infection rate of 47 percent (Gamble et al. 1999). Infection in cattle is less prevalent than is infection in sheep or pigs in the United States; however, recent surveys in several European countries using serology and PCR to detect parasite DNA have shown that infection rates in pigs and horses are negligible, whereas infection in sheep and cattle ranges from 1 to 6 percent (Wyss et al. 2000; Tenter et al. 2000). Serological surveys in eastern Poland revealed that 53 percent of cattle, 15 percent of pigs, and 0 to 6 percent of chickens, ducks, and turkeys were positive for *Toxoplasma* infection; nearly 50 percent of the people in the region were also serologically positive for toxoplasma infection (Sroka 2001). Dubey et al. (2002) recently isolated type I strains of *T. gondii* from 30 percent of eighty-two domestic free-range chickens from rural areas of Brazil, the first report of isolation of predominantly type I strains of *T. gondii* from a food animal. The prevalence of *Toxoplasma* infection of chickens in the United States has not been investigated; however, most chicken in the United States is cooled to near freezing or is completely frozen at the packing plant (Chan et al. 2001), which would kill organisms in tissue cysts (Kotula et al. 1991). The relative contributions of undercooked pork, beef, and chicken to *T. gondii* infection in humans is unknown; a nationwide retail meat survey is currently being conducted to determine the risk to U.S. consumers of purchasing beef, pork, or chicken containing viable *T. gondii* tissue cysts at the retail level (Dubey et al. unpublished).

Toxoplasma gondii infection is also prevalent in game animals. Among wild game, *T. gondii* infection is most prevalent in black bears and in white-tailed deer. Serological surveys of white-tailed deer in the United States have demonstrated seropositivity of 30 to 60 percent (Lindsay et al. 1991; Humphreys et al. 1995; Vanek et al. 1996). A recent study reported the occurrence of clinical toxoplasmosis and necrotizing retinitis in deer hunters with a history of consuming undercooked or raw venison (Ross et al. 2001). Approximately 80 percent of black bears are infected in the United States (Dubey et al. 1995), and about 60 percent of raccoons have antibodies to *T. gondii* (Dubey et al. 1995; Dubey and Odening 2001). Because raccoons and bears scavenge for their food, infection in these animals is a good indicator of the prevalence of *T. gondii* in the environment.

Virtually all edible portions of an animal can harbor viable *T. gondii* tissue cysts, and tissue cysts can survive in food animals for years. The number of *T. gondii* in meat from food animals is very low. It is estimated that as few as one tissue cyst may be present in 100 grams of meat. Because it is not practical to detect this low level of *T. gondii* infection in meat samples, digestion of meat samples in trypsin or pepsin is used to concentrate *T. gondii* for detection (Dubey 1988). Digestion in trypsin and pepsin ruptures the *T. gondii* tissue cyst wall, releasing hundreds of bradyzoites. The bradyzoites survive in the digests for several hours. Even in the digested samples, only a few *T. gondii* are present, and their identification by direct microscopic examination is not practical. Therefore, the digested material is bioassayed in mice (Dubey 1988). The mice inoculated with digested material have to be kept for six to eight weeks before *T. gondii* infection can be detected reliably; this procedure is not practical for mass scale samples. The detection of *T. gondii* DNA in meat samples by PCR has been reported (Warnekulasuriya et al. 1998), but no data exist on specificity and sensitivity of this method to detect *T. gondii*. A highly sensitive method using a Real-Time PCR and fluorogenic probes was found to detect *T. gondii* DNA from as few as four bradyzoites in meat samples (Jauregue et al. 2001). This method is now being tested to detect *T. gondii* in meat samples obtained from slaughtered animals. Cultural habits of people may affect the acquisition of *T. gondii* infection (Cook et al. 2000). For example, in France, the prevalence of antibody to *T. gondii* is very high in humans. Though 84 percent of pregnant women in Paris have antibodies to *T. gondii*, only 32 percent in New York City and 22 percent in London have such antibodies (Dubey and Beattie 1988). The high incidence of *T. gondii* infection in humans in France appears to be related in part to the French habit of eating some of their meat products undercooked or uncooked. In contrast, the high prevalence of *T. gondii*

infection in Central and South America is probably caused by high levels of contamination of the environment with oocysts (Dubey and Beattie 1988; Glasner et al. 1992; Neto et al. 2000). This said, however, it should be noted that the relative frequency of acquisition of toxoplasmosis from eating raw meat and that caused by ingestion of food or water contaminated by oocysts from cat feces is very difficult to determine and, as a result, statements on the subject are at best controversial. No tests exist at the present time to determine the source of infection in a given person.

There is little, if any, danger of *T. gondii* infection by drinking cow's milk and, in any case, cow's milk is generally pasteurized or even boiled; infection has, however, followed from drinking unboiled goat's milk (Dubey and Beattie 1988). Raw hens' eggs, although an important source of *Salmonella* infection, are extremely unlikely to transmit *T. gondii* infection.

DIAGNOSIS. Diagnosis is made by biologic, serologic, or histologic methods or by some combination of these. Clinical signs of toxoplasmosis are nonspecific and are not sufficiently characteristic for a definite diagnosis. Toxoplasmosis in fact mimics several other infectious diseases. Detection of *T. gondii* antibody in patients may aid diagnosis. Numerous serologic procedures are available for detection of humoral antibodies; these include the Sabin-Feldman dye test, the indirect hemagglutination assay, the indirect fluorescent antibody assay (IFA), the direct agglutination test, the latex agglutination test, the enzyme-linked immunoabsorbent assay (ELISA), and the immunoabsorbent agglutination assay test (IAAT). The IFA, IAAT, and ELISA have been modified to detect IgM antibodies (Frenkel et al. 1970; Remington et al. 1995). The IgM antibodies appear sooner after infection than the IgG antibodies, and the IgM antibodies disappear faster than IgG antibodies after recovery (Remington et al. 1995).

TREATMENT. Sulfadiazine and pyrimethamine (Daraprim) are two drugs widely used for treatment of toxoplasmosis (Guerina et al. 1994; Chirgwin et al. 2002). Although these drugs have a beneficial action when given in the acute stage of the disease process when there is active multiplication of the parasite, they usually will not eradicate infection. These drugs are believed to have little effect on subclinical infections, but the growth of tissue cysts in mice has been restrained with sulfonamides. Certain other drugs—diaminodiphenylsulfone, atovaquone, spiramycin, and clindamycin—also are used to treat toxoplasmosis in difficult cases.

PREVENTION AND CONTROL. To prevent infection of human beings by *T. gondii*, the hands of people handling meat should be washed thoroughly with soap and water before they go to other tasks (Dubey and Beattie 1988; Lopez et al. 2000). All cutting boards, sink tops, knives, and other materials coming in contact with uncooked meat should be washed with soap and water also. Washing is effective because the stages of *T. gondii* in meat are killed by contact with soap and water (Dubey and Beattie 1988).

Toxoplasma gondii organisms in meat can be killed by exposure to extreme cold or heat. Tissue cysts in meat are killed by heating the meat throughout to 67°C (Dubey et al. 1990). *Toxoplasma gondii* in meat is killed by cooling to −13°C (Kotula et al. 1991). *Toxoplasma* in tissue cysts are also killed by exposure to 0.5 kilorads of gamma irradiation (Dubey et al. 1994). Meat of any animal should be cooked to 67°C before consumption, and tasting meat while cooking or while seasoning should be avoided. Pregnant women, especially, should avoid contact with cats, soil, and raw meat. Pet cats should be fed only dry, canned, or cooked food. The cat litter box should be emptied every day, preferably not by a pregnant woman. Gloves should be worn while gardening. Vegetables should be washed thoroughly before eating because they may have been contaminated with cat feces. Expectant mothers should be aware of the dangers of toxoplasmosis (Foulon et al. 1994, 2000). At present, no vaccine is available to prevent toxoplasmosis in humans.

SUMMARY AND CONCLUSIONS. Infection by the protozoan parasite *Toxoplasma gondii* is widely prevalent in humans and animals. Although it causes asymptomatic infection in immune-competent adults, *T. gondii* can cause devastating disease in congenitally infected children and those with depressed immunity. To prevent human infection, all meat should be cooked well before consumption and gloves should be worn while gardening to prevent exposure to soil contaminated with *T. gondii* oocysts excreted in cat feces.

REFERENCES

Aramini, J.J., C. Stephen, J.P. Dubey. 1998. *Toxoplasma gondii* in Vancouver Island cougars (*Felis concolor vancouverensis*): serology and oocyst shedding. *J. Parasitol.* 84:438–440.

Aramini, J.J., C. Stephen, J.P. Dubey et al. 1999. Potential contamination of drinking water with *Toxoplasma gondii* oocysts. *Epidemiol. Infect.* Apr;122(2):305–315.

Bahia-Oliveira, L.M.G., A.M. Wilken de Abreu, J. Azevedo-Silva, et al. 2001. Toxoplasmosis in southeastern Brazil: an alarming situation of highly endemic acquired and congenital infection. *Int. J. Parasitol.* Feb;31(2):133–136.

Benenson, M.W., E.T. Takafuji, S.M. Lemon, R.L. Greenup, and A. Sulzer. 1982. Oocyst-transmitted toxoplasmosis associated with ingestion of contaminated water. *N. Engl. J. Med.* Sep 9;307:666–669.

Bowie, W.R., A.S. King, D.H. Werker, J.L. Isaac-Renton, A. Bell, S.B. Eng, and S.A. Marion. 1997. Outbreak of toxoplasmosis associated with municipal drinking water. The BC Toxoplasma Investigation Team. *Lancet* July 19;350:173–177.

Burnett, A.J., S.G. Shortt, J. Isaac-Renton, A.King, D. Werker, and W.R. Bowie, 1998. Multiple cases of acquired toxoplasmosis retinitis presenting in an outbreak. *Ophthalmology* 105:1032–1037.

Chan, K.F., H. Le Tran, R.Y. Kanenaka, and S. Kathariou. 2001. Survival of clinical and poultry-derived isolates of *Campylobacter jejuni* at a low temperature (4 degrees C). *Appl. Environ. Microbiol.* 67:4186–4191.

Chirgwin, K., R. Hafner, C. Leport, J. Remington, J. Andersen, E.M. Bosler, C. Roque, N. Rajicic, V. McAuliffe, P. Morlat, D.T. Jayaweera, J.L. Vilde, and B.J. Luft. 2002. Randomized phase II trial of atovaquone with pyrimethamine or sulfadiazine for treatment of toxoplasmic encephalitis in patients with acquired immunodeficiency syndrome: ACTG 237/ANRS 039 Study. *Clin. Infect. Dis.* 34:1243–1250.

Cole, R.A., D.S. Lindsay, D.K. Howe, C.L. Dubey, N.J. Thomas, L.A. Baeten. 2000. Biological and molecular characterizations of *Toxoplasma gondii* strains obtained from southern sea otters (Enhydra lutris nereis). *J. Parasitol.* Jun;86:526–530.

Cook, A.J., R.E. Gilbert, W. Buffolano, J. Zufferey, E. Petersen, P.A. Jenum, W. Foulon, A.E. Semprini, and D.T. Dunn. 2000. Sources of *Toxoplasma* infection in pregnant women: European multicentre case control study. European Research Network on Congenital Toxoplasmosis. *Br. Med. J.* Jul 15;321:142–147.

Darde, M.L., B. Bouteille, and M. Pestre-Alexandre. 1988. Isoenzyme characterization of seven strains of *Toxoplasma gondii* by isoelectrofocusing in polyacrylamide gels. *Am. J. Trop. Med. Hyg.* 39:551–558.

Desmonts, G., and J. Couvreur. 1974. Congenital toxoplasmosis. A prospective study of 378 pregnancies. *N. Engl. J. Med.* 290:1110–1116.

Dubey, J.P. 1986. A review of toxoplasmosis in pigs. *Vet. Parasitol.* 19:181–223.

Dubey, J.P. 1988. Refinement of pepsin digestion method for isolation of *Toxoplasma gondii* from infected tissues. *Vet. Parasitol.* 74:75–77.

Dubey, J.P. 1996. Infectivity and pathogenicity of *Toxoplasma gondii* oocysts for cats. *J. Parasitol.* 82:957–961.

Dubey, J.P. 2001. Oocyst shedding by cats fed isolated bradyzoites and comparison of infectivity of bradyzoites of the VEG strain *Toxoplasma gondii* to cats and mice. *J. Parasitol.* 87:215–219.

Dubey, J.P., and C.P. Beattie. 1988. *Toxoplasmosis of Animals and Man.* Boca Raton, FL: CRC Press.

Dubey, J.P., and J.K. Frenkel. 1972. Cyst-induced toxoplasmosis in cats. *J. Protozool.* 19:155–177.

Dubey, J.P., and J.K. Frenkel. 1976. Feline toxoplasmosis from acutely infected mice and the development of *Toxoplasma* cysts. *J. Protozool.* 23:537–546.

Dubey, J.P., D.H. Graham, C.R. Blackston, T. Lehmann, S.M. Gennari, A.M. Ragozo, S.M. Nishi, S.K. Shen, O.C. Kwok, D.E. Hill, and P. Thulliez. 2002. Biological and genetic characterisation of *Toxoplasma gondii* isolates from chickens (*Gallus domesticus*) from Sao Paulo, Brazil: unexpected findings. *Int. J. Parasitol.* 32:99–105.

Dubey, J.P., A.W. Kotula, A.K. Sharar, C.D. Andrews, and D.S. Lindsay. 1990. Effect of high temperature on infectivity of *Toxoplasma gondii* tissue cysts in pork. *J. Parasitol.* Apr;76:201–204.

Dubey, J.P., and K. Odening. *Parasitic Diseases of Wild Mammals.* Ames, IA: Iowa State University Press, 2001.

Dubey, J.P., and D.W. Thayer. 1994. Killing of different strains of *Toxoplasma gondii* tissue cysts by irradiation under defined conditions. *J. Parasitol.* 80:764–767.

Dubey, J.P., P. Thulliez, and E.C. Powell. 1995a. *Toxoplasma gondii* in Iowa sows: comparison of antibody titers to isolation of *T. gondii* by bioassays in mice and cats. *J. Parasitol.* 81:48–53.

Dubey, J.P., R.M. Weigel, A.M. Siegel, P. Thulliez, U.D. Kitron, M.A. Mitchell, A. Mannelli, N.E. Mateus-Pinilla, S.K. Shen, O.C. Kwok. 1995b. Sources and reservoirs of *Toxoplasma gondii* infection on 47 swine farms in Illinois. *J. Parasitol.* Oct;81:723–729.

Foulon, W., A. Naessens, and M.P. Derde, 1994. Evaluation of the possibilities for preventing congenital toxoplasmosis. *Am. J. Perinatol.* 11:57–62.

Foulon, W., A. Naessens, and D. Ho-Yen. 2000. Prevention of congenital toxoplasmosis. *J. Perinat. Med.* 28:337–45.

Frenkel, J.K., J.P. Dubey, and N.L. Miller. 1970. *Toxoplasma gondii* in cats: fecal stages identified as coccidian oocysts. *Science* 167:893–896.

Fuentes, I., J.M. Rubio, C. Ramirez, and J. Alvar., 2001. Genotypic characterization of *Toxoplasma gondii* strains associated with human toxoplasmosis in Spain. Direct analysis from clinical samples. *J. Clin. Microbiol.* 39:1566–1570.

Gamble, H.R., R.C. Brady, and J.P. Dubey. 1999. Prevalence of *Toxoplasma gondii* infection in domestic pigs in the New England states. *Vet. Parasitol.* 82:129–36.

Glasner, P.D., C. Silveira, D. Kruszon-Moran, M.C. Martins, M. Burnier Jr., S. Silveira, M.E. Camargo, R.B. Nussenblatt, R.A. Kaslow, and R. Belfort Jr. 1992. An unusually high prevalence of ocular toxoplasmosis in southern Brazil. *Am. J. Ophthalmol.* 114:136–44.

Grigg, M.E., J. Ganatra, J.C. Boothrooyd, and T.P. Margolis. 2001a. Unusual abundance of atypical strains associated with human ocular toxoplasmosis. *J. Infect. Dis.* 184:633–639.

Grigg, M.E., S. Bonnefoy, A.B. Hehl, Y. Suzuki, and J.C. Boothroyd. 2001b. Success and virulence in *Toxoplasma* as the result of sexual recombination between two distinct ancestries. *Science* 294:161–165.

Guerina, N.G., H.W. Hsu, H.C. Meissner, J.H. Maguire, R. Lynfield, B. Stechenberg, I. Abroms, M.S. Pasternack, R. Hoff, and R.B. Eaton. 1994. Neonatal serologic screening and early treatment for congenital *Toxoplasma gondii* infection. The New England Regional *Toxoplasma* Working Group. *N. .Engl. J. Med.* Jun 30;330:1858–1863.

Guo, Z.G., and A.M. Johnson. 1996. DNA polymorphisms associated with murine virulence of *Toxoplasma gondii* identified by RAPD-PCR. *Curr. Top. Microbiol. Immunol.* 219:17–26.

Howe, D.K., S. Honore, F. Derouin, and L.D. Sibley. 1997. Determination of genotypes of *Toxoplasma gondii* strains isolated from patients with toxoplasmosis. *J. Clin. Microbiol.* 35:1411–1414.

Howe, D.K., and L.D. Sibley. 1995. *Toxoplasma gondii* comprises three clonal lineages: Correlation of parasite genotypes with human disease. *J. Infect. Dis.* 172:1561–1566.

Humphreys, J.G., R.L. Stewart, and J.P. Dubey. 1995. Prevalence of *Toxoplasma gondii* antibodies in sera of hunter-killed white-tailed deer in Pennsylvania. *Am. J. Vet. Res.* 56:172–173.

Isaac-Renton, J., W.R. Bowie, A. King, G.S. Irwin, C.S. Ong, C.P. Fung, M.O. Shokeir, and J.P. Dubey. 1998. Detection of *Toxoplasma gondii* oocysts in drinking water. *Appl. Enviro.* Jun;64:2278–2280.

Jauregui, L.H., J.A. Higgins, D.S. Zarlenga, J.P. Dubey, and J.K. Lunney. 2001. Development of a real-time PCR assay for the detection of *Toxoplasma gondii* in pig and mouse tissues. *J. Clini. Microbiol.* Jun;39:2065–2071.

Kotula, A.W., J.P. Dubey, A.K. Sharar, C.D. Andrews, and D.S. Lindsay. 1991. Effect of freezing on infectivity of *Toxoplasma gondii* tissue cysts in pork. *J. Food Prot.* 54:687–690.

Lindsay, D.S., B.L. Blagburn, J.P. Dubey, and W.H. Mason. 1991. Prevalence and isolation of *Toxoplasma gondii* from white-tailed deer in Alabama. *J. Parasitol.* 77:62–64.

Literák, I., I. Rychlík, V. Svobodová, and Z. Pospíil. 1998. Restriction fragment length polymorphism and virulence of Czech *Toxoplasma gondii* strains. *Int. J. Parasitol.* 28:1367–1374.

Lopez, A., V.J. Dietz, M. Wilson, T.R. Navin, and J.L. Jones. 2000. Preventing congenital toxoplasmosis. *M.M.W.R.* 49:59–75.

Luft, B.J., and J.S. Remington. 1992. Toxoplasmic encephalitis in AIDS. *Clin. Infect. Dis.* 15:211–222.

Mondragon, R., D.K. Howe, J.P. Dubey, and L.D. Sibley. 1998. Genotypic analysis of *Toxoplasma gondii* isolates from pigs. *J. Parasitol.* 84:639–641.

Neto, E.C., E. Anele, R. Rubim, A. Brites, J. Schulte, D. Becker, T. Tuuminen. 2000. High prevalence of congenital toxoplasmosis in Brazil estimated in a 3-year prospective neonatal screening study. *Int. J. Epidemiol.* 29:941–947.

Nogami, S., A. Tabata, T. Moritomo, and Y. Hayashi. 1999. Prevalence of anti-*Toxoplasma gondii* antibody in wild boar, *Sus scrofa riukiuanus*, on Iriomote Island, Japan. *Vet. Res. Commun.* 23:211–214.

Owen, M.R., and A.J. Trees. 1999. Genotyping of *Toxoplasma gondii* associated with abortion in sheep. *J. Parasitol.* 85:382–384.

Pfefferkorn, E.R. 1990. *Cell Biology of* Toxoplasma gondii. New York, NY: Freedman.

Remington, J.S., R. McLeod, and G. Desmonts. 1995. Toxoplasmosis. In *Infectious Disease of the Fetus and Newborn Infant*. Philadelphia, PA: WB Saunders.

Renold, C., A. Sugar, J.P. Chave, L. Perrin, J. Delavelle, G. Pizzolato, P. Burkhard, V. Gabriel, and B. Hirschel. 1992. *Toxoplasma* encephalitis in patients with the acquired immunodeficiency syndrome. *Medicine* 71:224–239.

Roberts, T., and J.K. Frenkel. 1990. Estimating income losses and other preventable costs caused by congenital toxoplasmosis in people in the United States. *J. Am. Vet. Med. Assoc.* 196:249–256.

Roberts, T., K.D. Murrell, and S. Marks. 1994. Economic losses caused by foodborne parasitic diseases. *Parasitol. Today* 10:419–423.

Roghmann, M.C., C.T. Faulkner, A. Lefkowitz, S. Patton, J. Zimmerman, and J.G. Morris, Jr. 1999. Decreased seroprevalence for *Toxoplasma gondii* in Seventh Day Adventists in Maryland. *Am. J. Trop. Med. Hyg.* 60(5):790–792 .

Ross, R.D., L.A. Stec, J.C. Werner, M.S. Blumenkranz, L. Glazer, and G.A. Williams. 2001. Presumed acquired ocular toxoplasmosis in deer hunters. *Retina.* 21:226–229.

Smith J.L. 1993. Documented outbreaks of Toxoplasmosis: Transmission of *Toxoplasma gondii* to humans. *J. Food Prot.* 56:630–639.

Sroka, J. 2001. Seroepidemiology of toxoplasmosis in the Lublin region. *Ann. Agric. Environ. Med.* 8:25–31.

Tenter, A.M., A.R. Heckeroth, L.M. Weiss, 2000. *Toxoplasma gondii*: from animals to humans. *Int. J. Parasitol.* 30:1217–1258.

Terry, R.S., J.E. Smith, P. Duncanson, and G. Hide, G. 2001. MGE-PCR: a novel approach to the analysis of *Toxoplasma gondii* strain differentiation using mobile genetic elements. *Int. J. Parasitol.* 31:155–161.

Teutsch, S.M., D.D. Juranek, A. Sulzer, J.P. Dubey, and R.K. Sikes. 1979. Epidemic toxoplasmosis associated with infected cats. *N. Engl. J. Med.* Mar 29;300:695–699.

Vanek, J.A., J.P. Dubey, P. Thulliez, M.R. Riggs, and B.E. Stromberg. 1996. Prevalence of *Toxoplasma gondii* antibodies in hunter-killed white-tailed deer (*Odocoileus virginianus*) in four regions of Minnesota. *J. Parasitol.* 82:41–44.

Warnekulasuriya, M.R., J.D. Johnson, and R.E. Holliman. 1998. Detection of *Toxoplasma gondii* in cured meats. *Int. J. Food Microbiol.* 45:211–215.

Weigel, R.M., J.P. Dubey, A.M. Siegel, D. Hoefling, D. Reynolds, L. Herr, U.D. Kitron, Shen, S.K. Thulliez, and R. Fayer. 1995. Prevalence of antibodies to *Toxoplasma gondii* in swine in Illinois in 1992. *J. Am. Vet. Med. Assoc.* Jun 1;206:1747–1751.

Wyss, R., H. Sager, N.Muller, F. Inderbitzin, M. Konig, L. Audige, and B. Gottstein, B. 2000. The occurrence of *Toxoplasma gondii* and *Neospora caninum* as regards meat hygiene. *Schweiz Arch Tierheilkd.* 142:95–108.

37

AQUACULTURE AND PRE-HARVEST FOOD SAFETY

JAY F. LEVINE

INTRODUCTION. During the early 1970s, fish and shellfish were associated with more than 48 percent of all reported food-related outbreaks in the United States (Hughes et al. 1977; Bryan 1980). In contrast, an annual summary of food-related episodes in Canada conducted during 1977 noted seafood as the vehicle in only 6.6 percent of the outbreaks (Todd 1982). A marked reduction in the number of episodes associated with fish and shellfish in the United States was evident between 1993–1997; seafood products played a role in 19 percent of food-related outbreaks for which a vehicle was identified (Olsen et al. 2000). Differences in surveillance strategies, reporting, diagnostic techniques, and consumer awareness were probably reflected in the difference observed in the proportion of outbreaks involving seafood in the 1977 and 2000 reports. In the United Kingdom (U.K.), the Communicable Disease Surveillance Center investigated 1,425 outbreaks of gastrointestinal disease during 1992–1999 (Gillespie et al. 2001). Fish and shellfish were involved in 148 (10 percent) of the investigated food-related episodes. Fish-associated episodes in the United Kingdom occurred predominately in the summer (47 percent), and scromboid fish poisoning was the most frequent diagnosis. Viral pathogens in oysters and other mollusks accounted for 36 percent of the seafood-related episodes. Similar trends are noted globally in industrialized countries with healthy economies. However, the occurrence of seafood-related illnesses may be markedly higher in impoverished regions that lack the infrastructure and monetary resources to provide potable water and waste handling facilities, enforce sanitary and environmental regulations, and inform consumers about appropriate food-safety practices. Local differences observed in the specific agents or foods involved reflect the geographic, sociocultural, economic, and environmental diversity of human civilization.

Seafood-associated episodes of human illness are generally related to the consumption of raw or inadequately prepared fish and shellfish products, improper handling of seafood products during processing, or improper storage (Ahmed 1991; Rippey 1994). Many of these illnesses could be prevented by effective cold storage, careful processing, and thorough cooking. Consequently, the U.S. Food and Drug Administration's (FDA) Seafood Initiative, established in 1995 and implemented in December 1997, focused on the post-production aspects of the seafood industry. The U.S. Department of Agriculture's Food Safety Inspection Service (USDA-FSIS) introduced similar Hazard Analysis Critical Control Point programs (HACCP) in 1998 to improve the safety of meat and poultry products (Dyckman et al. 2001). These efforts focus on identifying post-harvest activities that perpetuate or introduce chemical or microbial agents into foods and make them unsafe for human consumption. A 2001 General Accounting Office review, however, reported that the HACCP guidelines established in support of the Seafood Initiative may not be effective ensuring the safety of seafood products reaching retail shelves (Dyckman et al. 2001).

Seafood harvested from natural stocks supplies approximately 74 percent of the aquatic species consumed worldwide (Wijkstrom et al. 2000). But many of the planet's traditional commercial fish species are either overexploited or approaching threshold levels of productivity. Aquaculture reared fish and shellfish are a viable alternative to the aquatic bounty from commercial harvests, and the diversity and number of species being explored as food products is continually expanding. A prime example is the rearing of various species of flounder (*Paralichthy* spp.) in "mariculture" systems (Bengston 1999). Harvests of flounder from natural stocks have declined, and research efforts have been initiated to culture this important seafood product (Burke et al., 1999).

The breadth of rearing techniques, seafood products, social issues, and marketing strategies that define aquaculture markedly exceeds the scope of what could conceivably be reviewed in a single chapter. Aquaculturists are supported by a wealth of extension materials (for example, U.S. Trout Farmers, 1994) and texts (such as Billard 1999) written in support of fish and shellfish farming. In addition, numerous excellent reviews (for example, Jensen and Greenlees 1997; Durborrow 1999) and books (such as Ahmed 1991) have been written that focus on the various toxins and pathogens associated with seafood and how post-harvest activities impact the wholesomeness of seafood products. Accordingly, this chapter is intended as an introduction to the complex seafood industry, aquaculture, and the pre-harvest challenges facing fish and shellfish farmers' efforts to market a safe product while simultaneously sustaining a profit.

THE SEAFOOD INDUSTRY. The term *seafood* is used collectively to refer to all aquatic species harvested for human consumption (FAO 1999), but from a consumer perspective it refers predominately to finfish and shellfish. Globally, seafood industries employ more than 35 million people. In 1999, approximately 125.2 million metric tons of seafood were harvested from natural stocks and aquaculture farms (Wijkstrom et al. 2000). More than 75 percent of the total harvest was sold for human consumption; the remainder was incorporated into fishmeal, fertilizer, and other fish-based products. Internationally, the majority of seafood purchased by consumers is either fresh (not frozen or preserved) (45.9 percent) or frozen (27.3 percent); the remainder is either cured (12 percent) or canned (14 percent) (Wijkstrom et al. 2000).

Seafood is an international commodity, and in many countries, demand markedly exceeds domestic seafood production (Wijkstrom et al. 2000). Consumers in industrialized countries on average consume more than 28 kg/person each year. Regional differences in seafood consumption reflect regional economic and social cultural differences. The relationship is somewhat intuitive; per capita consumption of seafood products is greatest in countries with coastal resources. For example, in the European community, per capita consumption of seafood in Portugal is almost 60 kg/person/per year, in contrast to 10 kg/person/year in landlocked Austria.

COMMERCIAL FISHING. Our hunter-gather roots defined the early use of fish and shellfish as components of human nutrition (Pillay 1993). However, the basic process of harvesting aquatic species remains the same; fishing vessels with crews armed with nets or hook and line harvest fish from natural populations, keep them on ice or in coolers during transit, and then carry their catch to port. At one time in the United States, seafood was often sold directly to the public or auctioned at the dock. However, the majority of fish and shellfish harvested in the United States today are sold to wholesale houses, where they are sorted and then relayed to retail markets. Similar trends are evident in Europe and other developed countries. In the United States, the market is increasingly vertically integrated. Companies often own their own fishing fleets or contract for a fisherman's entire catch. These companies then process and package the product for the retail market. In addition, during the past decade a vast array of value-added processed products (such as marinated fillets) have appeared on retail shelves (Meyers-Samuel 1995).

Diseases that impact fish and mollusk populations have contributed to declines in fisheries resources. Indeed, the oyster fishery along the East Coast of the United States serves as a clear example. Diseases associated with the protozoan pathogens, *Perkinsus mari-*

nus and *Haplosporidium nelsoni*, have severely reduced landings of the American oyster (*Crassostrea virginica*) (Andrews 1987, 1988). Similar health problems have impacted bivalve populations in Europe (Goulletquer and Heral 1997) and finfish populations in numerous locations. An additional burden is imposed by coastal pollution that prohibits commercial harvests in many locations.

The combined impact of fishing, disease, and pollution-associated mortality has depleted many natural fisheries stocks. Some species are severely imperiled, and the commercial fishing catch has become increasingly variable because of changes in the availability of individual species (Kosow et al. 1996). The depletion of natural stocks of species such as Atlantic sturgeon (*Acipenser oxyrhynchus*), Chinook salmon (*Oncorhynchus tshawytscha*), and sandbar shark (*Carcharhinus plumbeus)* have forced commercial fisherman to redirect their fishing effort toward other species. Species that were previously discarded are being harvested for human consumption because landings of traditionally more popular fish have declined. Consumers have adjusted their palates accordingly and now purchase species that were previously less popular or were available only in other countries before the advent of global shipment. However, some of these "in vogue" species, such as orange roughy (*Hoplostethus atlanticus*), have a longer length of time to reproduction. Consequently, they cannot accommodate the increased fishing pressure and can be rapidly depleted (Clark 2001). Nevertheless, millions of metric tons of fish are thrown back annually as by-catch (noncommercially relevant species) each year (Hall et al. 2000).

Surface waters that support fishing in the United States and many other countries are considered a "public trust," available to all via a right of citizenship. Consequently, various competing interests make use of surface waters. Hook and line recreational fishermen compete with commercial fisherman for access to the same fish stocks. The recreational viewpoint argues that recreational fishing and trophy fishing are more sustainable and do not impact the viability of natural populations. Aquaculturists actively argue that specific zones should be established and leased for fish and shellfish farming, and that commercial and recreational fishing should be restricted or prohibited in these zones. Zoning issues impact not just the location of the farming lease but also whether farming rights are granted for rearing aquaculture crops in the water column or on the bottom. Coastal and riparian land development further complicates the problem. Changes in land-use patterns and associated coastal pollution have greatly impacted nursery areas for many species of commercial fish. Agencies responsible for sustaining natural fisheries resources must juggle the disparate needs of the fishery, human nutrition, commercial fishing, recreational fishing and boating, municipalities, landowners, small businesses, and heavy industry.

AQUACULTURE. Aquaculture is the farming of aquatic plants and animals. Fish, shellfish, and aquatic plants have been consumed since the advent of human civilization and have been farmed for thousands of years. Aquaculture's origins trace back more than three thousand years to China (Li 1994). Aquaculture can be either marine or freshwater based and encompasses the culture of livestock used for human consumption, ornamental use, sport fishing, and the replenishment of stocks of depleted aquatic species. Today, approximately 15 percent of the seafood consumed in the United States and 26 percent consumed internationally are derived from aquaculture (Wijkstrom et al. 2000).

Aquaculturists rear more than 460 species of aquatic animals (Jhingran and Gopalakrishnan 1974). Carp (for example, *Cyprinus carpio*) dominate finfish aquaculture harvests, with the majority being reared in China and other Southeast Asian countries (Wijkstrom et al. 2000). Multiple species of carp are often grown in a single pond to exploit different trophic layers (Yang 1994). The Pacific oyster, *Crassostrea gigas,* is the most heavily produced bivalve (Wijkstrom et al. 2000). It has proven resistant to protozoan pathogens that plagued native oysters in North America and Europe. Consequently, it has been introduced as an alternative aquaculture crop to sustain oyster harvests where native indigenous stocks have been depleted (Goulletquer and Heral 1997). Shrimp and prawn culture dominate crustacean aquaculture (Wijkstrom et al. 2000).

Although the business aspects of running an aquaculture "farm" are somewhat similar to the business of farming other livestock, there are some key differences that slow the growth of aquaculture enterprises and aquaculture-based livestock industries. The first is direct competition with the commercial fishing industry. Hog and poultry farmers do not have to compete with commercial poultry or hog gatherers. There is no wild poultry or wild hog industry. Yet aquaculturists must compete with the commercial fishing industry. The wholesale price of an individual species (such as shrimp) can be driven up or down based on fish stock availability and the success of commercial fishermen. Aquaculture comprises everything from seeding estuarine acreage with young oysters to rearing fish in closed recirculating systems established in climatically controlled indoor facilities. Unless the facility is enclosed indoors, the producer has little control of the environment in which the livestock are being reared. From a food-safety perspective, factors and circumstances that are beyond the producer's control often alter the safety of the product. For example, a grower with a small clam lease would not be able to market animals from the lease if water quality is impacted by effluent from a municipal sewer system after a major storm. A third key difference relates to market share and consumer perception. Aquaculturists have had to fight an uphill battle informing consumers about aquaculture-grown products. Many producers, however, have begun to capital-ize on public concerns about the wholesomeness of seafood products by promoting the comparative safety of aquaculture products (Johnsen 1991).

Aquaculture Economics. Total global aquaculture production in 1999 was estimated at 32.9 million metric tons (Wijkstrom et al. 2000), with an estimated economic value of $47.9 billion dollars. The annual harvest of seafood from aquaculture is growing approximately 10 percent each year, and a total production of more than 47 million metric tons is anticipated by 2010 (Muir and Nugent 1999). By the year 2030, aquaculture production is expected to exceed harvests from capture fisheries (Wijkstrom et al. 2000). China is the world's leader in aquaculture production, and produced more than 22.7 million metric tons of aquaculture products during 1999, more than 68 percent of the world's total aquaculture production (Wijkstrom et al. 2000).

Aquaculture is the fastest-growing livestock industry in the United States (NASS 2001), with an estimated annual product value of more than $833 million. Nevertheless, U.S. aquaculture production comprises only a small fraction (< 3 percent) of the world's aquaculture harvest (Wijkstrom et al. 2000). The U.S. National Agricultural Statistics Service (NASS) recently reported that there are more than four thousand aquaculture farms nationwide (NASS 2001). The majority of farms are located in thirteen southern states that produce more than two-thirds of the aquaculture harvest. Aquaculture supports more than 200,000 jobs in the United States. The catfish (*Ictalurus punctatus*) industry is producing more than 593 million pounds of catfish each year, with an approximate market value of $500 million. More than 190,000 acres of catfish ponds are in production. The U.S. trout industry lies a distant second to catfish sales with an economic value of $75.8 million in 2000.

Salmon, tilapia, and other species are also grown in the United States. However, importation of salmon and tilapia far exceeds actual U.S. production of these commodities (NASS 2001). Clam, oyster, and scallop culture is active and has had some regionally specific success. A strong oyster mariculture industry along the Pacific Northwestern Coast and the clam industry along the Florida are notable examples. Other newer aquaculture commodities, including hybrid-striped bass (*Morone saxatilis* × *M. chrysops*), yellow perch *(Perca flavescens)*, and several species of flounder *(Paralichthys* spp.) are in their infancy. These new aquaculture products reflect the expanding role that aquaculture plays in attempting to accommodate consumer demand for seafood products. However, demand for seafood in the United States continues to exceed domestic production. During 2000, the combined value of imported shrimp, salmon, and tilapia was approximately $4.6 billion, exceeding the total value of hog and broiler exports during the same period (NASS 2001).

Aquaculture Livestock-Rearing Techniques. Aquaculture is truly a form of farming, and fish and shellfish farmers must consider many of the same factors that impact hog, poultry, or cattle farmers. Aquaculturists juggle capital, labor, and space resources with basic husbandry needs to produce, harvest, and deliver a product to market. They rear or purchase brood stock with high fecundity and strong genetic lines that can support rapid growth, feed efficiency, and disease resistance. Hatcheries supplying producers attempt to produce disease-free stock that yield a product of uniform size and "quality." Aquaculturists refine stocking densities, diets, husbandry practices, and environmental conditions to ensure livestock health and maximize livestock growth. Producer organizations help guide aquaculturists by providing quality-assurance guidelines and supporting the development of HACCP plans to sustain product safety from farm to table. Unlike other forms of livestock agriculture, for which the basic choices are limited to a small number of domesticated species, hundreds of species of fish and shellfish are being farmed with markedly different husbandry requirements.

The breadth of aquaculture systems reflect the husbandry needs of the species being reared and the topography, climate, financial resources, cultures, and traditions of the communities in which they are maintained. Open flow-through systems are generally located adjacent to surface waters that can be partitioned or diverted into a series of earthen ponds, tanks, containment ditches, or raceways (Pillay 1993). Flow-through raceway culture systems are used extensively to rear salmonids and are the infrastructural backbone of the U.S. trout industry. Raceway culture also supports the rearing of game fish in hatcheries reared for restocking, bivalves for sale to aquaculturists, soft-shell crab, and marine fish such as flounder and turbot (*Scophthalmus maximus*) (Strand and Oiestad 1997). Water quality is maintained by the continual input of water from the water source, and water is redirected back into the water source after passing through livestock holding sections. Nitrogenous waste in each rearing unit (for example, raceway) is continually diluted by the influx of fresh water. Water-quality problems arise when livestock density and waste production exceed the capacity of the water source to dilute nitrogenous wastes and sustain oxygen levels. Water-quality problems are often a problem in the lower raceways of stair-step type systems such as those used extensively by the salmonid industry (Fig. 37.1). Fish in the lower tiers live in water that has supported fish reared in the higher raceway sections. Consequently, ammonia and other waste products are comparatively elevated in lower raceways, and supplemental oxygen injection is often used to minimize periods of hypoxia during warm summer months (Hinshaw and Losordo 1993)

Cage and pen culture have been practiced in Asia for centuries (Pillay 1993). A broad array of cage and pen sizes, construction materials, and designs are used to

FIG. 37.1—Concrete stair-step raceways used for rearing trout.

hold fish in surface waters (Fig. 37.2). Catfish (*I. punctatus*), carp (*Cyprinus* spp.), salmon (*Oncorhynchus* spp.), and grouper (*Epinephelus* spp.) comprise a short list of species that aquaculturists have reared using cage and pen-based techniques. Cage culture offers a relatively inexpensive means of holding fish and making use of the flow and productivity of natural waterways.

Biosecurity has been a continual problem for pen-based aquaculture. The salmon net-pen industry and the recent detection of infectious salmon anemia virus in pen-reared salmon in Maine have highlighted the perils of conducting aquaculture in natural systems.

FIG. 37.2—Net-pen used for rearing salmon. (Image provided by Larry Hammell, Atlantic Veterinary College, Prince Edward Island, Canada)

Problems such as sea lice (*Lepeophtheirus salmonis*) acquired from free-ranging stock increase the susceptibility of salmon to pathogens (Nolan et al. 1999). The use of pesticides to eliminate sea lice has generated concern about the release of pesticides into surface waters and the presence of residues in marketed product (Haya et al. 2001). Environmentalists have also expressed concern about the accidental release of nonindigenous species from damaged net-pens (Anderson 1999) and the volume of organic and nitrogenous waste generated by high-density culture. The net-pen industry has also wrangled commercial fishing interests and coastal land-development entrepreneurs. The debate has led to restrictions on the expansion of net-pen industries in British Columbia and other regions.

The majority of aquaculture-reared finfish are grown in ponds. A wide variety of pond shapes and sizes is used to accommodate the type and intensity of the aquaculture enterprises and the various stages of the production process. Pond culture covers a broad range of producer management needs, from extensive systems providing relatively low productivity and requiring little management (such as rearing sturgeon for roe production) to intensive systems requiring feed supplementation, active water quality monitoring, and, often, supplemental aeration to maintain dissolved oxygen levels (Egna et al. 1997). Ponds either are dug into the flat landscape or created by dam or dike construction to take advantage of local topographical features (Pillay 1993). The bottom soil type, construction materials, slope, and depth are equally variable and constructed to accommodate local conditions, water sources, available capital, labor resources, and the culture of specific species, and they often reflect local traditions and expertise (Kelly and Kohler 1997). In the United States, pond culture is used extensively for rearing catfish, pike (*Esox* spp.), sturgeon, tilapia, crayfish, and, internationally, for rearing a lengthy list of finfish and crustacean species (Pillay 1993).

Water quality in mismanaged aquaculture ponds can degrade rapidly. Feeding and fertilizing increase pond nitrogen and phosphorous loads (Boyd and Tucker 1998), and fuel phytoplankton growth. Phytoplankton growth increases organic deposition to the pond bottom and the use of oxygen for bacterial degradation. Fish and benthic invertebrate respiration and chemical degradation further deplete the oxygen available for fish respiration (Wang 1980). Reduced bottom oxygen levels can decrease growth, increase susceptibility to pathogens, and cause mortality (Stoskopf 1992). Carbon dioxide from fish respiration and reducing pH can further degrade water quality and adversely impact fish respiration and health.

Closed systems continually recirculate water used for fish rearing. They are inherently capital intensive (Losordo et al. 1998), and their livestock products must often compete with livestock reared under less capital-intensive systems. Consequently, the margin of profit/pound is often small. Closed recirculating systems are most profitable when the species being reared have a high dollar value, high feed efficiency, and can tolerate high stocking densities. By placing the systems within a building, closed recirculating systems provide a more controlled environment for fish culture (Fig. 37.3) (Timmons and Losordo 1994). Environmental control is maximized when indoor temperatures are controlled with a building heating and air-conditioning system. Water quality in the holding tanks is maintained by organic particle removal and biofiltration that can accommodate the treatment of large quantities of nitrogenous waste. Tank water circulates through a closed loop, which generally includes some type of particulate trap or active removal system and high surface-area biofiltration (Losordo et al. 1998). Particle removal and biofiltration are often supplemented by carbon filtration, sand filtration, chemical treatment, ozonation, UV light, or protein skimming. Closed recirculating systems are dependent on high-yield wells with good-quality water that can be used to fill and replenish water levels in the holding tanks to sustain water quality. Because of the large capital investment, profitable closed recirculating aquaculture systems are designed for high-density fish culture.

Tilapia and other species with a high feed conversion ratio have recently been successfully grown in closed recirculating systems (Losordo et al. 1998). Although closed loop recirculating systems provide more control of the rearing environment, they heighten the level of care needed by aquaculturists to sustain water quality. Biosecurity is also a paramount concern, because fish reared in high density have a narrow margin of tolerance for changes in water quality and may be more susceptible to health problems (Bebak et al. 1997).

Bag culture is used in mariculture to suspend and hold bivalves in the water column or on the bottom of leased acreage. Young stock (seed) either are collected from natural nursery areas or purchased from hatcheries. Bivalve seed are then placed in a variety of bags, baskets, suspended lanterns, and racks and held at different levels of the water column within leased parcels

FIG. 37.3—Earthen pond (with aerator) used for rearing channel catfish.

(Fig. 37.4). The bags facilitate management of the crop and provide protection from crabs (for example, *Callinectes sapidus*) and other predators (Krantz and Chamberlin 1978). There are marked regional and local differences in the type of culture techniques used. After the stock is placed in the bags, various levels of maintenance are required to remove fouling organisms and redistribute the stock into groups of similar size and planting density. Bottom culture is an alternative to suspended-bag culture. Bivalve seed are sowed directly onto the bottom and then later harvested when the animals reach market size.

Aquaculture in the United States and Europe is predominately monoculture based, in contrast to the polyculture systems that are used extensively in Southeast Asia (Pillay 1993). Monoculture systems accommodate mechanization but heighten livestock disease risks. Polyculture systems maximize the productivity per unit area, but multicropping reduces product uniformity and is generally more labor intensive. Carp culture in Southeast Asia frequently includes several different species of carp, each utilizing a different trophic level in farm ponds (Yang 1994). Integrated systems are being developed that combine the use of aquatic plants, algae, bivalves (Stuart et al. 2001) and row crop agriculture as a means of handling nitrogenous wastes generated by aquaculture facilities. Shpigel and coworkers (1993) in Israel have developed polyculture systems that combine fish culture, the functional capacity of algae to remove nitrogenous waste, and the ability of bivalves to convert algae into bivalve biomass. Others have employed the co-culture of fish and rice (dela-Cruz 2001), fish and ducks (Szabo 1993), and shrimp and mussels (*Perna viridis*) (Lin et al. 1993). In the United States, partitioned, integrated culture raceway systems are being developed for the catfish industry to improve waste handling and reduce waste-product-associated impacts on fish growth, fish morbidity, and mortality (Drapchl and Brune 2000).

Aquaculture's Impact on the Environment. The impact of intensive aquaculture operations on native fauna and flora can be profound. In Southeast Asia, the shrimp industry has been associated with aquifer depletion, wetland damage, diminished water quality, and seawater incursion (Dierberg and Kiattismkul 1999; Paez-Osuna 2001). In Ecuador, shrimp farming has impacted mangrove forests (Kaiser 2001). Trout farming in flow-through raceways in the United States has raised concern about the release of nitrogenous waste and heightened organic sediment deposition in mountain streams. Environmentalists have challenged the ecologic viability of salmon net-pen farming due to organic particle deposition, changes to benthic communities, and the introduction of nonindigenous species (Anderson 1999).

Aquaculture practices can also pose a risk to public health. Aquaculture ponds are a potential breeding habitat for mosquitoes, snails, and other invertebrates. These unintended pond residents can serve as vectors of malaria, yellow fever, Dengue, schistosomiasis, and other human pathogens. From a food safety perspective, snails and other invertebrates may serve as intermediate hosts for human and zoonotic helminths (Deardorff and Overstreet 1991).

Bivalve aquaculture can have beneficial environmental effects. As suspension feeders, they remove particulate material from the water column and reduce overall particle deposition. Bivalves help improve water column clarity, which is important for sustaining the growth of submerged aquatic vegetation. Bivalve crops also attract an array of other fauna and flora that serve as food for other species and add to the diversity of organisms processing particulate and waste materials.

The future of aquaculture is dependent on its ability to maximize profitability without impacting the environment. Aquaculturists continue to refine and adapt new technologies to maximize the health and growth of

FIG. 37.4—Table culture used for rearing oysters held in bags.

FIG. 37.5—Closed recirculating system used for rearing juvenile salmon (smolts). (Image provided by Larry Hammell, Atlantic Veterinary College, Prince Edward Island, Canada)

livestock as well as maximize profitability. Refined diets and feeding regimes along with aquaculture system modifications that rapidly remove particulate waste and treat nitrogenous wastes are being developed to improve livestock health. These refinements are compatible with efforts to reduce the impact of aquaculture on the environment and improve product quality, uniformity, and seafood product safety. Improved health management of aquaculture livestock also reduces food residues by reducing the use of chemotherapeutic agents.

SEAFOOD SAFETY CONCERNS. Toxins and pathogens associated with fish and shellfish are naturally present in surface waters, introduced as a product of fecal contamination or other pollutant input, or the product of worker contamination. A wealth of HACCP guides and quality-assurance programs are available to assist processors and retailers attempting to prevent the introduction of contaminants into seafood products. Similarly, numerous resources are available for restaurants and consumers for minimizing contaminant introduction and proliferation during food preparation and storage. This review complements these efforts by focusing on pre-harvest considerations for aquaculturists.

Water Quality. Although a great variety of aquaculture rearing systems is available, there is a single universal environmental requirement of all aquaculture species: water. Water quality is the single most important pre-harvest aquaculture food safety issue. Water used must be available in sufficient quantity, accessible, and free of chemical or biologic agents that adversely impact animal or consumer health. Fish grown at high densities produce large quantities of nitrogenous and particulate waste (Losordo et al. 1998). Aquaculture feeds contribute large quantities of nitrogen and phosphorous and add to the particulate load. Removal of these waste materials must be sufficient to sustain water quality at levels that will not affect livestock growth or health. Discharges from aquaculture facilities into surface waters must be sufficiently treated to minimize any adverse impact on local aquatic life.

All aquatic ecosystems, including those maintained in aquaculture, are dynamic and subject to changes in the basic parameters that define the descriptor "water quality." Temperature, salinity, dissolved oxygen, hardness, carbon dioxide, suspended materials, pH, nitrogenous waste, bacteria, and other factors embody the environment in which aquaculturists are attempting to rear fish or shellfish (Stoskopf 1992). A change in one parameter will generally induce a change in one or more additional variables. Additional factors that define the exposure history and character of an aquaculture crop, such as the prior presence of a specific disease, fish age, species, type of production system, and spatial proximity to natural populations, alter the risk of disease in aquaculture stock (Bebak et al. 1997). To be successful financially, aquaculturists must manage this chemical soup and sustain conditions that maximize growth at the highest feasible stocking density, yet in a manner that does not negatively impact fish health or end-product quality and safety.

FECAL COLIFORM POLLUTION. Millions of acres of coastal habitat suitable for aquaculture and harvesting fish are not available for shellfish harvesting or culture because of fecal coliform contamination. Fecal contamination is derived from numerous sources. Human waste often enters surface waters from inadequate municipal wastewater treatment system outfalls, community wastewater treatment systems, leaking septic tanks, and watercraft. Agricultural sources include livestock units that spread fecal waste on pasture and cropland, leaking livestock waste management lagoons, and companion animal rearing facilities. Wildlife also contribute to the total fecal coliform load. In one study, Canada geese (*Branta canadensis*) were identified as a source of fecal coliforms in surface waters (Hussong et al. 1979). Although extensive research is being conducted to develop improved methods for differentiating the source of the contamination, consistently reliable techniques for differentiating between the origin of the fecal waste have not been adopted by shellfish sanitation agencies.

Specific Pathogens and Toxins. Various bacteria, protozoa, viruses, helminthes, and biotoxins contribute to seafood-related illnesses. Although the origin of a specific agent reflects local environmental, economic, and socio-cultural differences, the end product is the same: human morbidity and a loss of consumer confidence in seafood products. The review that follows is intended as an introduction to the breadth of pathogens and agents that can be associated with the occurrence of health problems after the consumption of fish and shellfish.

BACTERIA. Established indigenous gastrointestinal flora in fish will generally reflect the availability of specific bacteria in the environment (Hansen and Olafsen 1999). Similarly, indigenous bacterial microflora of the gills and shells of freshwater and marine crustaceans vary spatially and temporally with the waters in which they live (Cobb and Vanderzant 1976). However, the occurrence, distribution, and density of bacteria can be quite variable and may include human pathogens such as *Shigella* spp. and *Aeromonas hydrophilia* (Chittick et al. 2001).

VIBRIOS SPP. *Vibrios* spp. are ubiquitous in marine and estuarine sediments and pose a health problem for individuals that consume raw oysters or other bivalves (Pollack and Fuller 1999). *Vibrios* are particularly a problem for children, the elderly, and immunocompromised individuals (Gholami et al. 1998; Potasman et al. 2002). Daniels and co-authors (2000a) reported that during 1973–1998, 88 percent of the individuals diagnosed with *V. parahaemolyticus*–related illnesses had consumed raw oysters. In one outbreak in California, 209 people developed *V. parahaemolyticus* infections after consuming raw oysters. Serotypes of *V. parahaemolyticus* have been detected in areas approved for shellfish harvesting (Daniels et al. 2000b). Another *Vibrio* spp., *V. vulnificus,* has been associated with cutaneous wound infections and systemic illnesses (Shapiro et al. 1998). Shapiro and co-workers (1998) found that oysters harvested in the Gulf of Mexico were associated with *V. vulnificus* infections diagnosed in 422 patients. The organism was detected in 45 percent of patients with wound infections and 43 percent of patients with primary *V. vulnificus* septicemias. More than 60 percent of patients with primary septicemia died. *Vibrio* spp. have also been associated with crayfish consumption (Bean et al. 1998).

The wrath of *V. cholerae* has been recognized for centuries, and major pandemics have sporadically spread through Southeast Asia, Africa, Europe, Central America, and the United States. In 1994, more than twenty thousand cholera-related deaths were observed in the Democratic Republic of the Congo (Goma Epidemiology Group 1995). Cholera has been associated with the consumption of a variety of seafood, including: poorly cooked shrimp and mussels (*Perna viridis*) (Lim-Quizon et al. 1994) in the Philippines; raw oysters and cooked crab in Equador (Weber et al. 1994); dried fish in Tanzania (Acosta et al. 2001); and crab in the United States (Finelli et al. 1992). During 1995–2000, sixty-one cases of cholera were diagnosed in the US; fourteen (23 percent) were associated with undercooked seafood (Steinberg et al. 2001). In one episode in Miami, Florida, raw oysters that were harvested off the coast of Louisiana were identified as the vehicle (Klontz et al. 1987). Contaminated seawater in restaurant tanks used to keep fish alive was associated with twelve cases of cholera in Hong Kong. The outbreak was halted by instituting measures to improve the quality of water used in holding tanks and prohibiting the unlicensed sale of seafood (Kam et al. 1995). Global shipment of seafood from areas where cholera is still endemic may contribute to the cases that are observed in the United States and other developed countries each year (Blake 1993).

LISTERIA SPP. The FDA has a zero tolerance policy for *Listeria* spp. in processed food products (Jinnerman 1999). During 1987–1998 in the United States, the presence of *Listeria* resulted in more than one hundred class one recalls of seafood products. More than 250,000 pounds of ready-to-eat (for example, cooked shrimp) domestic or imported seafood products were recalled. The FDA tested 7,158 samples from seafood and processing plants during 1991–1998. *Listeria* was isolated from 8.7 percent of the samples.

Listeria monocytogenes is relatively ubiquitous in nature and is found almost anywhere that decaying plant material is present (Fenlon 1999). The pathogen can grow at low temperature and in both fresh (<0.5 ppt) and brackish water at salinities less than 10 ppt. In California, Colburn and coworkers (1990) isolated *Listeria* spp. from more than 60 percent of the fresh and low salinity sites they sampled. Although survival at individual sites will vary based on environmental conditions, *Listeria* spp. can survive for more than seven hundred days in pond or river water and for more than two hundred days in some types of stored feed (Fenlon 1999). The organism's ability to survive during cold storage (Rocourt et al. 2000) makes it both a pre-harvest and post-harvest concern.

The list of published accounts of *L. monocytogenes* in seafood products is extensive (Jinnerman 1999) and the presence of the organism in fish and shellfish is a global problem. *Listeria* has been isolated from the feces of healthy fish and crustaceans as well as numerous wildlife (such as seagulls) (Fenlon 1999). Humans can serve as carriers of *L. monocytogenes*, and the organism has been detected in human feces (Fenlon 1999). *Listeria* isolates have been obtained from oysters and shrimp (Motes 1991). Leung and coworkers (1992) found *L. monocytogenes* on the skin and in the viscera of catfish reared in ponds in Alabama. The pathogen has also been isolated from the skin (33 percent) and feces (40 percent) of rainbow trout in Switzerland, but it was detected on only 6 percent of the finished product. In another study it was not found in the skin, gills, intestines, tank water, or diet of striped bass raised in closed recirculating tanks (Nedoluha and Westhoff 1997). Van Wagner (1989) found *Listeria* on the exoskeleton of shrimp but didn't detect the pathogen within their gastrointestinal tract.

Although *Listeria* spp. are common residents of surface waters and have been found in numerous seafood products, episodes of illness associated with fish and shellfish are markedly less than those associated with other commodities (Jinnerman 1999). *L. monocytogenes* has been isolated from gravid (salt-cured raw fish) (Ericsson et al. 1997) and cold smoked rainbow trout (*O. mykiss*) (Tham et al. 2000). *Listeria* spp. were obtained from fish processed in thirty-seven of 141 fish processing plants, and in 14 of twenty-eight states from which fish were examined (Heinitz 1998). Ready-to-eat fish products, oysters, and other seafood consumed raw, poorly cooked, or lightly preserved (Rocourt et al. 2000) pose the greatest risk to consumers. Predictive models have been developed to assess the growth and survival of *Listeria* in fishery products, and dose-response relationships have been developed for *L. monocytogenes* in smoked fish (Ross et al. 2000). However, the predictive value of current modeling efforts is restricted by the need for additional information about *Listeria* growth and survival under storage conditions.

SALMONELLA SPP. *Salmonella enterica* serovar Typhi, the bacterium associated with typhoid fever, may be present in surface waters where the disease is endemic and human waste handling is less than optimal. *S. enterica* serovar Typhi infections are relatively rare in the United States and other developed countries with contemporary waste-handling infrastructures. However, *S. enterica* serovar Typhi infections are not unusual in impoverished countries (Sumaya 1991) and in regions affected by war or natural disasters. Raw consumption of bivalves harvested from surface waters contaminated with serovar Typhi has been associated with outbreaks of typhoid fever (Torne et al. 1988). Although cooking may be effective in reducing the threat of serovar Typhi infection, the bacterium remains viable during cold storage (Nishio et al. 1981).

Other *Salmonella* spp. have been associated with seafood-associated illnesses. In Berlin, Germany *S. enterica* serovar Blockley infections were associated with farm-reared eel (Fell et al. 2000). In Hong Kong, fecal samples were obtained from 5,718 patients hospitalized with diarrhea and from surrounding surface waters and shellfish (Yam et al. 1999). *Salmonella* was the most frequently identified pathogen in patient stool samples. Although *S. enterica* serovar Typhimurium and *S. enterica* serovar Enteritidis were the most prevalent serotypes in clinical samples, they were rare in environmental and shellfish samples. Only *S. enterica* serover Derby was found in similar proportions in clinical (16 percent) and environmental and shellfish samples (13.7 percent). The FDA examined more than 12,000 samples of seafood during 1990–1998. *Salmonella* spp. were present in 2.8 percent of raw and 0.47 percent of ready-to-eat domestic seafood, and in 10 percent of raw and 2.6 percent of ready-to-eat imported seafood (Heinitz et al 2000). *Salmonella* spp. have been isolated from cockles (Greenwood et al. 1998), mussels (*M. galloprovincialis*) (Ripabelli et al. 1999), oysters, clams, and various fish species (Heinitz et al. 2000). Although surface water contamination is associated with *Salmonella* spp.–related outbreaks, worker contamination during processing and preparation contributes to many seafood-related *Salmonella* infections.

AEROMONAS SPP. *Aeromonas hydrophilia* is a fish pathogen that has occasionally been associated with human episodes of gastrointestinal illness related to the consumption of seafood products (Greenlees et al. 1998). Rahim and Aziz (1994) purchased freshwater prawns (*Macrobrachium malcolmsonii*) from a fish market in Dhaka, Bangladesh, and isolated *Aeromonas* spp. from shrimp muscle and other tissues. Gobat and Jemmi (1993) isolated three species of *Aeromonas* from fresh and ready-to-eat seafood products in Switzerland. *Aeromonas* spp. were found in more than 10 percent of the smoked fish and gravad salmon that were examined. Hanninen and coworkers (1997) detected *A. hydrophilia* in shrimp, fish, fish-eggs, and freshwater samples. Children and immunocompromised individuals appear to be at greatest risk of developing *Aeromonas*-related illnesses (Juan et al. 2000). But strains isolated from children with diarrhea (*A. veronii, biovar sobria*) have not consistently been isolated from fish and shellfish (Hanninen et al. 1997).

CLOSTRIDIA SPP. During 1993–1997, *Clostridium botulinum* was associated with thirteen reported outbreaks of food-related illness investigated by the CDC (Olsen et al. 2000). Fish were identified as the vehicle in one outbreak. However, Mead and coworkers (2000) estimate that approximately fifty-eight cases occur each year in the United States. Although botulism is now comparatively uncommon in the United States, episodes of illness associated with the toxin in seafood products occur sporadically each year. In 1987, botulinum toxin contributed to outbreaks in both New York and Israel that were associated with the consumption of kapchunka, which are salted, air-dried, uneviscerated whitefish, from the same processor (Telzak et al. 1989). In a study conducted by Hielm and coworkers in Finland (1996) *C. botulinum* type E was detected by PCR in 5 percent of samples obtained from vacuum-packed freshwater fish and roe, and in 3 percent of the air-packed seafood sampled. Pullela and co-workers (1998) sampled the intestinal tracts of aquaculture-reared rainbow trout (*O. mykiss*), tilapia (*Oreochromis* spp.), hybrid striped bass (*M. saxatilis* × *M. chrysops*), and pacu (*Piaractus mesopotamicus*) and found *C. botulinum* in the gastrointestinal tracts of the trout and bass. The organism is frequently resident in sediment samples (Hielm et al. 1996) obtained in systems that experience periods of bottom anoxia.

OTHER BACTERIAL PATHOGENS. Bacterial fish pathogens of the genus *Edwardsiella* spp. have been associated with human gastroenteritis, wound infections, and meningitis (Janda and Abbott 1993). The pathogens have been isolated from numerous aquatic species and other wildlife, including waterfowl, crocodiles, and frogs (Wyatt et al. 1979). Wakabayashi and Egusa (1973) isolated *E. tarda* from pond-reared eels. In a study conducted by Wyatt and coauthors (1979), *E. tarda* was isolated from the skin and viscera of pond-reared catfish. *E. tarda* was also isolated from invertebrates, frogs, water samples, and sediment found in the catfish-rearing ponds.

Streptococcus iniae has been isolated from cutaneous wounds of patients that have handled live or recently killed tilapia in both the United States and Canada (MMWR 1996). Shoemaker and coworkers (2001) examined more than 1,543 hybrid striped bass (*M. saxatilis* X *M. chrysops*), channel catfish (*I. punctatus*), and tilapia (*Oreochromis* spp.) harvested from fish farms. The organism was isolated from tilapia and hybrid striped bass but not catfish. Several episodes of seafood-related *Shigella* spp. infections have been associated with gastrointestinal illness. Although worker contamination (Hackney and Potter 1994) and improper storage routinely play a role in many seafood-related shigellosis outbreaks, oysters are occasionally implicated as a vehicle (Rippey 1991). In one outbreak in Boston that affected twenty-four people, *Shigella sonnei* was detected in oysters that were shipped to eight restaurants. The use of buckets for toilets on the boat used to harvest the oysters was identified as the origin of the pathogen. The buckets were cleaned by emptying them in the water from which the oysters were obtained (Reeve et al. 1990). The episode could have been avoided if onboard regulatory inspection of sanitation procedures on fishing vessels was required and universally implemented.

Various *Escherichia coli* strains have been isolated from oysters and the surface waters and sediments from which they were harvested (Ogawa 1980). Isolates have also been obtained from mussels (*M. galloprovincialis*) collected from areas found suitable for shellfish harvesting (Ripabelli et al. 1999). Rahim and Aziz isolated *E. coli* from prawns (*Macrobrachium malcolmsonii*) and noted an increase in bacterial counts while the prawns were stored at 4°C. Although seafood-related enteropathogenic or enterotoxigenic *E. coli* associated illnesses are relatively rare in the United States, episodes of gastrointestinal illness have been sporadically observed in other countries. In one episode, more than 40 percent of patrons who consumed raw oysters at a restaurant became ill with *E. coli* 027:H7 (Hackney and Potter 1994). Another outbreak in Japan was associated with the consumption of salmon roe (Makino et al. 2000).

Campylobacter jejuni–associated gastroenteritis is a global problem and the illness is frequently food related (Butzler and Oosterom 1991). Although poultry are a common vehicle, *Campylobacter* spp. have been detected in raw mussels and oysters (Endtz et al. 1997) and surface waters (Bolton et al. 1987). Ripabelli and co-workers detected *C. jejuni* in mussels (*M. galloprovincialis*) collected from areas in the Adriatic Sea that had been approved for shellfish harvesting. Wilson and Moore (1996) also isolated *C. jejuni* from bivalves and cockles obtained from areas open to shellfish harvesting. In one episode in the state of Washington, a human case was associated with oyster consumption, and *Campylobacter* spp. was isolated from oysters, seawater, and waterfowl (Abeyta et al. 1993).

PREVENTING CONTAMINATION WITH BACTERIAL PATHOGENS. All seafood products will contain bacteria at the time of harvest. Bacteria cannot be eliminated from the livestock-rearing environment. Recreational and commercial fishermen can reduce the likelihood of pathogen presence by restricting fish and shellfish harvests to areas approved for fishing, and by using appropriate cold-storage procedures. The primary preharvest practices that aquaculturists can use to reduce bacterial pathogen presence are as follows: (1) careful site selection, avoiding areas subject to human or livestock waste contamination; (2) educating employees to use sound sanitation and biosecurity practices; (3) providing an optimum physical environment that reflects the needs of the livestock species being reared; and (4) effective water-quality monitoring. Adherence to quality-assurance production guidelines and the development of an HACCP-based plan for sustaining product quality can aid efforts to reduce the presence of pathogens in the rearing environment (O'Reilly and Kaferstein 1997).

VIRUSES. The consumption of raw bivalves is frequently associated with enteric illnesses in humans (Lees 2000). Shellfish- and fish-adapted viruses are not considered zoonoses, and the viruses do not replicate in seafood products (Clarke and Lambden 1997). Shellfish-related viral enteric illnesses indicate that the shellfish were either obtained from a location contaminated with human waste (Kilgen and Cole 1991) or that worker contamination occurred during processing (Cliver 1988). Human fecal effluent from waste-handling systems is replete with enteric viral pathogens (Melnick et al. 1978). Enteric viruses shed in human feces can survive the dynamic environments of freshwater ecosystems (Jaykus et al. 1994).

Oysters and other suspension-feeding bivalves, as well as various species of crustaceans, can accumulate human enteric viruses when they live in water contaminated by human waste (Jaykus et al. 1994). Virus accumulation levels can exceed that of the water from which molluscan shellfish were harvested (Hoff and Becker 1969). The rate and efficiency of uptake and accumulation appear to be driven by the type of virus, species of bivalve, virus concentration in the water, temperature, and other environmental variables (Pancorbo 1992; Jaykus et al. 1994).

The calicivirus, Norwalk virus, is a routine culprit in shellfish-associated outbreaks of human gastroenteritis (Shieh et al. 2000). A 1999 report from the CDC noted that Norwalk-like viruses were identified as a contributing factor in 67 percent of food-related illnesses for which a pathogen was isolated (Mead et. al 1999). Seven percent of the food-related deaths and 33 percent of the food-related hospitalizations were associated with Norwalk-like viruses. Another calicivirus, Snow Mountain agent (Truman et al. 1987), has also been detected in oysters and mussels (Le Guyader et al. 2000) and in the feces and vomitus of patients who experienced gastroenteritis after the consumption of shellfish (Ohyama et al. 1999; Shieh et al. 2000). Outbreaks of hepatitis A infection have also been associated with the consumption of shellfish (Desenclos et al. 1991; Bosch et al. 2001) from areas contaminated with human waste. The virus is stable in the environment (Conaty et al. 2000), can survive in feces for one month, resists very low pH (Scholz 1989), and persists during depuration (Sobsey et al. 1987). Bivalves can accumulate hepatitis A virus to levels one hundred times those observed in environmental samples (Enriquez et al. 1992). In one episode in China, more than 300,000 cases were associated with the consumption of shellfish (Yao 1991). Non-A non-B hepatitis virus (Torne et al. 1988) and other viruses have also been associated with illnesses after the consumption of bivalves. Strains of poliovirus used to produce modified-live vaccines are routinely isolated from shellfish in the United States, but pose no health risk (Kilgen and Cole 1991).

Regional and cultural differences in the preparation of bivalves can play a role in the level of risk posed by bivalve consumption. Although the potential risk posed by consuming raw shellfish harvested from contaminated surface waters is obvious, many enteric viruses are relatively resistant to steaming (Kirkland 1996), moderate cooking temperatures (60°C) (Peterson 1978) and cold storage (DiGirolamo et al. 1970). Shellfish sanitation guidelines attempt to prevent the harvest of bivalves from contaminated waters. However, current techniques routinely used to quantify total or fecal coliforms are inadequate indicators of the presence of many viruses (Jaykus et al. 1994; Lees 2000). Pathogenic viruses may be present in bivalves when routine total or fecal coliform monitoring suggest that shellfish harvested from a site may be safe for human consumption (Jaykus et al. 1994). Episodes of human illness have occurred after the consumption of oysters and other bivalves harvested from approved waters (Portnoy et al. 1975; Lees 2000).

To reduce the risk posed by human enteric viruses, seafood must be harvested from areas that have not been contaminated with human waste. Recreational and commercial fishermen must avoid areas closed by shellfish sanitation-monitoring agencies. Aquaculturists must select sites at which contamination with human waste is unlikely.

PROTOZOA. Livestock, wildlife, and human waste entering surface waters can introduce pathogenic protozoa into aquatic habitat. Bivalves filtering water contaminated with *Cryptosporidium parvum* or *Giardia lamblia* can concentrate cysts (Slifko et al. 2000). *C. parvum* has been detected in oysters harvested from Chesapeake Bay (Fayer et al. 1999). The pathogen has also been detected in mussels (*Mytilus galloprovincialis*), cockles (*Cerastoderma edule*) (Gomez-Bautista et al. 2000), clams (*Dosinia exoleta, Venerupis pullastra, V. rhomboideus*), and oysters (*Ostrea edulis*) harvested in Spain (Freire-Santos et al. 2000), clams (*Ruditapes philippinarum*) harvested in Italy (Freire-Santos et al. 2000), and mussels harvested in Ireland (Lowery et al. 2001). Laboratory studies indicate that the pathogen remains viable after ingestion by bivalves (Graczyk et al. 1998; Freire-Santos et al. 2001), and field studies have documented its ability to survive in seawater (Fayer et al. 1998; Tamburrini and Pozio 1999). *Cryptosporidium* spp. have also been detected in tilapia and other fish species (Durborrow 1991).

Seafood-related outbreaks associated with *G. lamblia* have been attributed to worker contamination (Slifko et al. 2000). In one study, viable oocysts were recovered from clams (*Ruditapes philippinarum*) (Mazas 2001). One outbreak in Minnesota was associated with canned salmon; however, the woman who prepared the salmon had changed the diaper of a child from whom *Giardia* was isolated (Osterholm et al. 1981).

G. lamblia or *C. parvum* could potentially be present in seafood products when fish or shellfish are harvested from surface waters that are frequented by wildlife or have been contaminated with human waste. Recreational and commercial fishermen can do little to reduce the risk of wildlife-related contamination. However, they can avoid fishing in areas that have been closed because of the presence of fecal coliform contamination. In marine coastal areas, mariculturists should avoid selecting sites that could potentially be contaminated with human or livestock waste. Aquaculturists rearing their livestock in ponds, raceways, or tanks can prevent contamination of their livestock holding waters by restricting wildlife access to their farms.

HELMINTHS. More than fifty species of helminths found in seafood impact the health of humans. Social-cultural factors often play an important role in determining the actual risk to consumers (Deardorff and Overstreet 1991). Residents of impoverished rural communities that use "night soil" for adding nutrients to fish ponds are at greatest risk of acquiring human helminth parasites (Ling 1993). Snails and other invertebrates inhabiting fish ponds often support the varied life cycles of helminths. Aquaculture-reared species harvested from farms that adhere to quality-assurance practices appear to pose a lower risk of infection with some helminths (Deardorff and Kent 1989).

TREMATODES. Trematode metacercariae in fish and snails that potentially pose a risk to human health are a global problem (Deardorff and Overstreet 1991). The consumption of raw or poorly cooked salmon (for example, *Oncorhynchus* spp.) is the primary factor resulting in human infection with the parasitic flatworm *Nanophyetus salmincola* (Eastburn et al. 1987). Abdominal discomfort, nausea, and diarrhea accompanied by an eosinophilia generally characterize the illness (Harrell and Deardorff 1990). *N. salmincola* can also be acquired when handling fresh seafood. The digenean fluke serves as a source of the rickettsia *Neorickettsia helminthoeca,* which can contribute to severe or fatal infection in canids ("salmon poisoning in dogs") (Millemann and Knapp 1970).

In Southeast Asia, more than eight million people are infected with *Opisthorchis viverrini.* The trematode is acquired when eating raw or poorly cooked carp (*Puntius gonionotus*) (Khamboonruang et al. 1997) and other species of freshwater fish. In one study, the fluke was found in more than 50 percent of residents examined who lived in two villages in Laos where the consumption of raw fish is routine. Eighty-percent of residents aged thirty-five to fifty-four years were infected (Kobayashi et al. 2000). Waikagul (1998) examined fish purchased from markets in Thailand for *O. viverrini.* The parasite was identified in fish purchased in three of fourteen provinces. In one district, Aranyaprathet, *O. viverrini* was identified in 25–28 percent of the fish examined.

Clonorchiasis, associated with the ingestion of the oriental liver fluke (*Clonorchis sinensis*), is also prevalent in Southeast Asia. Infection with the trematode often results in the painful inflammation and obstruction of bile ducts. Consumption of raw, partially cooked, pickled, salted, and smoked fish has been associated with *C. sinensis* infections (Chan et al. 2002). In one study conducted in the Kim Son district of Vietnam, more than 13 percent of 306 randomly selected residents were found to be infected with *C. sinensis* (Kino et al. 1998). *C. sinensis* metacercariae were identified in approximately 56–100 percent of *Hypophthalmichtys molitrix* reared in earthen ponds in the district. Cercariae of the trematode were found in more than 13 percent of *Melanoides tuberculatus,* an aquatic snail that was found in the fish ponds.

Nam and Sohn (2000) reported finding seven species of trematode metacercariae in pond smelts (*Hypomesus olidus*) in Korea. The fish were either purchased from a wholesaler or harvested by cast-net. *C. sinesis, Holostephansus nipponicus, C. orientalis, Diplostomum* sp., and *Metorchis orientalis* were observed in the fish examined. In Thailand, *Stellantchasmus falcatus* and five other species of tremaodes were detected in fish purchased in markets in twelve of fourteen provinces (Waikagul 1998). During another study, six genera of metaceraria were identified in more than three thousand fish representing thirty-two species in Thailand (Wongsawad et al. 2000). Similarly, Sukontason and coworkers (1999) documented the broad distribution of *O. viverrini* and other metacercariae in nine species of cyprinoid fish harvested from surface waters. *Metorchis conjunctus* has been observed in native Canadian aboriginal peoples (Behr et al. 1998) and in patients in Montreal, Canada, after the consumption of raw white sucker (*Catostomus commersoni*) (MacLean et al. 1996). Human infection with *Paragonimus* spp., a lungworm, has been diagnosed in Asia after the consumption of raw freshwater crab (Cui et al. 1998).

CESTODES. Humans are the definitive host for *Diphyllobothrium latum.* They acquire the parasite when they consume fish that serve as an intermediate host for a metacestode stage of the tapeworm (Deardorff and Overstreet 1991). Fish acquire the parasite when they consume infected copepods. Pike (*Esox* spp.), burbot (*Lota lota*), perch (*Perca* spp.), and other freshwater fish serve as the primary source of infection for humans in the United States (Ruttenber et al. 1984). Human infections have also been observed after the consumption of raw or poorly cooked salmon (*Oncorhynchus masou*) in Japan (Hutchinson et al. 1997), Hawaii (Hutchinson et al. 1997), and Korea (Lee et al. 2001), and after the consumption of trout (*Oncorhynchus mykiss*) from Scottish waters (Wooten and Smith 1979). Other cestodes of the genus *Diphyllobothriuim* also infect humans (Margolis et al. 1973; Curtis and Bylund 1991). *Diphyllobothrium nihonkaiense* infections have been associated with the consumption of raw salmon (*Oncorhynchus* sp.) in Japan (Anodo et al. 2001). Burbot and other carnivorous freshwater fish sometimes consume anadromous fish such as smelt (*Osmerus mordax dentex*). When the burbot are consumed, species of *Diphyllobothrium* spp. generally found in marine mammals and marine fish are sometimes observed in humans (Rausch and Adams 2000).

NEMATODES. Fish products are also implicated in Anisakiasis, human infection with the third-stage juveniles of one of several species of *Anisakis* nematodes. Anisakiasis is considered a global problem (Sakanari and McKerrow 1989). Patients infected with the nematode often experience intense abdominal pain. A case of anisakiasis in Norway highlights the potential severity of *A. simplex* infections: The patient required abdominal surgery to alleviate an obstructed duodenum (Eskesen et al. 2001). Anaphylaxis caused by an immediate hypersensitivity reaction to *A. simplex* (Dominguez-Ortega et al. 2001) reflects recently focused concern about the role the parasite plays as an allergin. Patients in numerous locations have been diagnosed with acute hypersensitivity reactions that are generally characterized by intensely pruritic urticaria, and some patients experience respiratory difficulty after consuming fish containing the parasite (Gracia-Bara et al. 2001). Although marine mammals serve as the definitive host, humans become infected when they consume the juveniles embedded in the flesh of fish that serve as an intermediate host. In the United States, *A. simplex* infections occur sporadically after consumption of raw or poorly cooked salmon (*Oncorhynchus* spp.) (Deardorff et al. 1991), yellowfin (*Thunnus albacores*) or skipjack tuna (*Euthynnus pelamis*), and rockfish (*Sebastes* spp.) (Bouree 1995). The light-colored juveniles are difficult to observe in fish fillets. In one study, Deardorff and Kent (1989) detected *A. simplex* in wild-caught salmon (*Oncorhynchus nerka*) but not in pen-reared salmon (*O. kisutch, O. tshawytscha, Salmo salar*). Feed supplementation with fresh herring (*Clupea harengus*) or other fish could introduce *A. simplex* into net-pen operations. In contrast to *A. simplex*, *Pseudoterranova decipiens*, a common nematode found in Atlantic cod (Bouree 1995), produces a reddish nodule that is more easily identified and usually results in the rejection of cod fillets before they reach consumers.

Another nematode, *Eustrongylides* spp., also infects humans (Wittner et al. 1989) and utilizes various species of waterfowl as their definitive hosts. Oligocheates serve as an intermediate host and fish as a paratenic host. Humans are infected when they consume raw or poorly cooked fish (Wittner et al. 1989). The consumption of raw prawns (*Macrobrachium* sp.) has been associated with *Angiostrongylus cantonensis* infections (Alicata and Brown 1962). The parasite, a lungworm of rats, can cause severe and sometimes fatal eosinophilic meningitis in humans. Shrimp serve as paratenic hosts. A recent outbreak in Kaohsiung Taiwan of *A. cantonensis* infection in eight patients who developed meningitis was associated with the con-

sumption of the intermediate host snails (Tsai et al. 2001). *Gnathostoma* spp. and *Capillaria* spp. have also contributed to human infections following the consumption of seafood products (Deardorff and Overstreet 1991). In the Philippines, an outbreak of infection with *C. philippinensis* was associated with the consumption of kinilaw or raw fish, raw shrimp, crab, and snails. Fish-eating birds were suspected to have spread the parasite to the Compostela Valley where the cases were diagnosed (Belizario et al. 2000).

The aquatic lifestyle of many gastropods warrants their consideration as seafood. Although cultured land snails (*Helix* spp.) are sold as esgargot, aquatic abalone (*Haliotis* spp.), perwinkles (*Littorina littorea*), and conch (*Strombus* spp.) are also eaten by people of various cultures. Snails serve as intermediate hosts for numerous helminths. Fish and waterfowl often serve as secondary intermediate hosts or definitive hosts and support the life cycle of the helminths. Echinostomiasis is a snail-associated, food-related zoonosis (Graczyk and Fried 1998). Sixteen species of digenean trematodes have been associated with the illness. Raw or poorly cooked fish, mollusks, amphibians, and crustaceans have all been associated with the disease. The disease occurs primarily in rural impoverished communities of Southeast Asia where "night soil" has been used to fertilize fish ponds. In the Philippines, *Echinostoma malayanum* infections have been associated with the local tradition of consuming its intermediate host, the snail *Lymnaea cumingiana*. The snail is consumed raw or as a partially cooked salted paste made from fish, "bagoong" (Tangtrongchitr and Monzon 1991). The prevalence in some areas is high, approaching 50 percent of the susceptible population. Ironically, fish have been introduced in some areas as a form of biologic control to reduce snail populations (Carney 1991), but fish can serve as a source of *Echinostoma* spp. infection.

Reducing the morbidity associated with helminths infections acquired from seafood has proven a difficult task. Readily visual helminths such as *P. decipiens* can be removed at the time of processing or food preparation. However, the majority of helminths are not readily visible to consumers. Cooking and freezing are effective in destroying most helminth parasites (Durborrow 1999). But traditional consumption of raw or lightly preserved foods plays an important role in sustaining helminth-associated morbidity in some regions. Although safe-food-handling education programs can be introduced, efforts to change consumer behavior have frequently proven ineffective. Consequently, effective reduction may be obtained only prior to harvest at the farm.

In many regions, "night soil" and animal waste are still used to fertilize aquaculture ponds (Ling et al. 1993). Elimination of waste-based pond enrichment practices could reduce the presence of helminths such as the cestode *D. latum* or the fluke *C. sinesis* (Durborrow 1999). However, alternative processed feeds are generally too expensive to import into impoverished rural areas where "night soil" use is a common practice.

Snail and crustaceans serve as intermediate hosts for many helminths (Deardorff and Overstreet 1991), and their control in ponds poses a complex paradox. The invertebrates play a beneficial role in pond ecosystems by degrading organic materials on the pond bottom. They also serve as a source of food for some aquatic species. Because they are prolific, manual removal is generally not feasible. Chemical molluscacides are also a problem, because they may be hazardous to the aquaculture crop and introduce unwanted residues into the final product. Another alternative, the use of antihelminthic agents such as praziquantel, is available to aquaculturists (Durborrow 1999). But these pharmaceuticals are expensive and beyond the financial reach of most aquaculturists in Southeast Asia, where seafood-associated helminth parasitism is most prevalent.

Adherence to quality-assurance production guidelines and the development of HACCP plans can be effective tools in the effort to market a helminth-free seafood product. In one study, HACCP-oriented strategies were introduced to attempt to limit the presence of *O. viverrini* in a pond. Water supply, the source of juvenile fish for stocking, fish feed, and pond conditions were considered critical control points. Hazards were identified and analyzed, control measures were introduced, critical limits were identified, monitoring conducted, and responses to monitoring actions were documented (Hongpromyart et al. 1997). Although the pilot study was limited in scope, the authors concluded that HACCP-based procedures could be employed for effective control of the trematode.

MARINE AND FRESHWATER TOXINS. In 1953, Bruce Halstead identified 518 species in ninety-five families of marine fauna that were considered to have toxins that could be associated with human health problems (Dack 1956). Today we identify many of these toxins as marine or freshwater bioxins, produced by dinoflagellates, diatoms, and cyanobacteria, which are associated with clinical illnesses. The toxins have been categorized based on chemical structure and associated pharmacologic action. Numerous reviews (Halstead 2001) and texts (Botanna 2000) have addressed the relevance of these agents to seafood safety; consequently, they are just briefly noted in the following text as an introduction.

Massive blooms of the saxitoxin-producing dinoflagellate, *Ptychodiscus breves (Gymnodinium breve)*, tint the water red and have been appropriately labeled by coastal residents as the "red tide" (Steidinger 1993). The blooms occur sporadically in the Gulf of Mexico and off the southeastern Atlantic Coast of the United States. Saxitoxins are heterocyclic guanidines (Llewellyn 2001) produced predominately by dinoflagellates in the marine genera *Gymnodinium, Alexandrium,* and *Pyrodinium,* as well as the freshwater cynaobacteria (Steidinger 1993). Saxitoxins interfere with sodium transport (Llewellyn 2001) at neuromuscular and neuronal synapses (Van Dolah 2000). They have profound cardiovascular and central nervous system effects, which comprise the syndrome commonly referred to as "paralytic shellfish poisoning" (PSP) (Llewellyn 2001). Clinical signs include tingling and numbness of the lips and mouth, incoherent speech, and, potentially, respiratory paralysis (Lagos and Andrinolo 2000). The toxin is also aerosolized by wave action, and commercial fisherman, biologists, and coastal residents who are exposed sometimes experience upper respiratory symptoms. Paralytic shellfish poisoning is usually associated with the consumption of filter-feeding bivalves, oysters, clams, scallops, and mussels (Llewellyn 2001). However, snails, crabs, and fish have been identified as vehicles in PSP episodes (Kodama 2000).

Acidic polyether toxins, including okadaic acid, dinophysis toxins, pectenotoxins, and yessotoxin are produced by dinoflagellates and associated with "diarrheic shellfish poisoning" (DSP) (James et al. 2000). The toxins inhibit protein phosphatase activity that plays a role in many cellular processes (Llewellyn 2001). In laboratory animal studies, pectenotoxins have been demonstrated to be tumor promoters (Burgess and Shaw 2001). Affected individuals develop mild to severe gastrointestinal discomfort characterized by nausea, abdominal pain, vomiting, diarrhea, and chills, often accompanied by headache and fever (Llewellyn 2001). Mussels (*Mytilus* spp.) and other bivalves have been associated with DSP outbreaks. Another polyether toxin has been associated with "neurotoxic shellfish poisoning." The toxin is produced by *P. brevis*, which is also associated with PSP toxin. Affected individuals describe both gastrointestinal and neurologic symptoms, which resemble those observed with ciguatera (described later) (Llewellyn 2001). Oysters have been identified as a vehicle (Ishida et al. 1996). Multiple toxins are often produced by dinoflagellates, and the presence of one toxin can potentiate the activity and uptake of other, simultaneously ingested toxins (Tripuraneni et al. 1997).

The diatom *Pseudonitzschia pungens* spp. and other members of the genus *Pseudoonitzschia* produce domoic acid (Llewellyn 2001). The risk posed by shellfish containing domoic acid was first recognized in 1987, when 250 people on Prince Edward Island, Canada, became ill after consuming blue mussels (*M. edulis*) (Perl et al. 1987). Many were hospitalized and four affected individuals died. The toxin induces accumulation of glutamate at neuronal synapses, disrupting cell ion homeostasis and causing cellular swelling and cell death (Niijar and Niijar 2001). Domoic acid induces focal lesions in the hippocampus as well as a broad range of gastrointestinal and neurologic manifestations. Affected individuals generally experience motor and cognitive dysfunction, and suffer a loss of short-term memory. The toxin has most frequently been associated with the consumption of blue mussels (Llewellyn 2001).

Freshwater blue-green algae, cyanobacteria of several genera (such as *Anabaena* spp.), and some marine species produce anatoxins that block acetylcholine receptors at the neuromuscular junction. Affected individuals develop neurologic problems. Other species produce microcystins and nodularians that target hepatic tissue and interfere with liver protein phosphatase. These toxins have been associated with intrahepatic hemorrhage, hypovolemic, and hepatic neoplasia (Yu 1989).

Ciguatoxins and maitotoxins are polyethers produced predominantly by tropical coral reef–associated dinoflagellates (Van Dolah 2000). Ciguatoxin, produced by *Gambierdiscus toxicus,* is associated with "Ciguatera poisoning," which possibly affects more than fifty thousand each year (Ting and Brown 2001). Ciguatera poisoning is observed after the consumption of snapper and other reef fish, and the toxin has been isolated from more than four hundred fish species (Randal 1980). Clinical signs include itching, and tingling of the lips, hands, and feet (Van Dolah 2000). During an outbreak in Melborne, Australia, forty-six people developed "ciguatera poisoning" after eating restaurant entrees prepared from a single fish (Ng and Gregory 2000).

Various species of fish of economic relevance to the commercial fishing industry, such as tuna, and mahi mahi, have been associated with "scromboid fish" poisoning. A fairly lengthy list of bacteria, including *Morganella morganii, Clostridium perfringens, and Klebsiella pneumonia*, use the enzyme histidine decarboxylase to convert histidine in fish flesh to histamine (Taylor et al. 1979). Hallmark clinical signs include rapidly developing erthyema and urticaria. Pruritis is severe, and bronchospasm, and associated severe respiratory distress can be life threatening. Ensuring the proper cold storage of fish after capture can prevent the occurrence of scromboid fish poisoning. When the histamine is present, however, it is not eliminated by cold storage or cooking.

From a pre-harvest perspective, the primary means of preventing illness associated with marine and freshwater biotoxins is preventing the harvest of shellfish from areas where the toxins are present. Vigilant active surveillance for the presence of the toxins by local, state, or federal agencies can be used to guide commercial fishermen and aquaculturists. Active surveillance is practiced in a number of countries, and standards have been developed to guide monitoring agencies, but the actual surveillance activities used are markedly variable. In the United States, the National Shellfish Sanitation Program has established standards for classifying specific harvesting and aquaculture areas (CFSANOS 1999).

The isolation of new toxins in greenshell mussels (*Perna canaliculus*) (Morohashi et al. 1999) reflects the continually magnifying complexity of dinoflagellate toxin chemistry, pharmacology, and biology. Consequentially, dinoflagellate toxin surveillance and identification are continually becoming more complex. In addition, successful surveillance activities must be effectively complemented with local legislation that reduces the influx of nutrients that support phytoplankton growth in shellfish growing areas. Vegetative and wetland buffers that serve as natural filters for nutrients must be preserved and in many locations reestablished to prevent algal and dinoflagellate blooms. Municipal, commercial, residential, and agricultural nutrient discharges must be effectively monitored to quickly identify, curtail, and prevent bloom formation.

PESTICIDES, METALS AND OTHER CONTAMINANTS. Fish and shellfish live in a chemical soup, and the list of compounds of anthropogenic origin that have been identified in fish is extensive. Many fish of commercial or recreational importance are predators, and contaminants (for example, organochlorine pesticides) can be bioaccumulated as they move up the food chain (Kearney et al. 1999). Pesticides have been detected both in free-ranging fish and aquaculture-reared species (Santerre et al. 2000). Suspension-feeding bivalves have a fairly limited ability to metabolize organic contaminants (Stegeman 1985) and can develop pollutant concentrations (such as polyaromatic hydrocarbons) that reach or exceed a steady state with the water column or surrounding sediments (for example, non-DDT pesticides) (Sericano et al. 1990). In fish and shrimp, heavy metals are concentrated in visceral organs, gills, and exoskeleton (Vanderzant et al. 1970). Some metals, such as cadmium and zinc in oysters, and other compounds are metabolized by hepatopancreas metal-binding polypeptides such as metallothioneins (Unger et al. 1991).

Co-existing species can have markedly different contaminant levels. In one study, arsenic levels in eels (*Gymnothorax undulatus*), copper in crab (*Grapsus tenuicrustatus*), and lead in coral were many times higher than levels observed in other tested species (Miao et al. 2001). Metals and other chemical contaminants found in seafood can have profound human health effects. Recognition of the relationship between Minamata disease, mercury, fish consumption, and environmental contamination provided a graphic example of the health risk posed by consuming fish from polluted waters (Harada 1995). In studies conducted in Sweden, sisters of fishermen who consistently consumed fish contaminated with organochlorines were more likely to have children of low birthweight (Rylander et al. 2000). Cultures (such as Inuit) that use seafood as their basic source of protein are particularly at risk (Harada 1995).

Maximum tissue-residue limits have not been established for many compounds (Jensen and Greenlees 1998) (for example, trichlorethylene). But contaminants are found both in wild fish (Amaraneni et al. 2001) and fish reared in aquaculture (Amodio-Coccieri et al. 2000). However, fish harvested from aquaculture routinely display contaminant loads below those observed in game fish and most commercial fishery stocks (Santerre 2000, 2001). Commercially prepared feeds and the short life span of fish reared for aquaculture reduce total accumulated body burdens.

The primary pre-harvest preventative measure that can be taken by aquaculturists to prevent the introduction of pesticides, metals, and other contaminants is the selection of sites that have no reported history of contaminant presence. Site selection should include a review of all adjacent land-use and historic environmental reports.

CHEMOTHERAPEUTIC DRUG USE AND RESIDUES. In the United States, two antimicrobial preparations have been available to aquaculturists for placement in livestock feed: oxytetracycline (DePaola et al. 1995) and a sulfadimethoxine and ormetoprim combination. The drugs have been approved only for use in lobsters, catfish, and salmonids (Durborrow and Francis-Floyd 1996). An anesthetic, tricaine methanesulfonate (MS 222), and formalin to control gill parasites have also been approved for use in finfish. Formalin residues and antibiotic residues have been observed in aquaculture-reared flounder (*Paralichtys olivaceus*), rockfish (*Sebasesschlegeli* sp.) (Jung et al. 2001), rainbow trout (*O. mykis*) (Sahagun et al. 1997), and other aquatic species (L'Abee-Lund and Sorum 2001; Schmidt et al. 2001).

Antimicrobials added to a culture system will alter resident bacterial flora in the holding system as well as the gastrointestinal tract of the species being reared. In this manner, antimicrobial use may impact the early growth and development of aquaculture species (Hansen and Olafsen 1999). The use of antibiotics by aquaculturists may also pose a potential threat to human health because of the development of antibiotic-resistant strains (D'Aoust 1994). In studies conducted by Gonzalez and co-workers (1999), multidrug, antimicrobial-resistant strains of *Salmonella* sp. and *Plesiomonas shigelloides* were found in the gastrointestinal tract of a pike (*Esox* spp.) and in the surface waters from which the fish was harvested. Cordano and Virgilio (1996) isolated drug-resistant *S. enterica* serovar Panama from shellfish and other food products. The strains isolated were similar to those observed in human patients. Ruiz and coworkers (1994) isolated *S. enterica* serovar Typhimurium from fish obtained in India, and identified microbial resistance to quinolones and other antibiotics.

Public health and environmental concerns have fueled efforts to reduce the use of antimicrobials for growth promotion and disease therapy by aquaculturists. Vaccine use promotes the social-political sustainability of aquaculture by reducing antimicrobial use (Gudding et al. 1999). The introduction of vaccines against *A. salmonicidia*, the agent associated with furunculosis in salmon, was associated with a decline in antibacterial use by the Norwegian salmon industry. However, the decline was short-lived because growers again increased their use of antibacterials to minimize the impact of a resurgence of vibriosis (Markestad and Grave 1999), which was not prevented by the *A. salmonicidia* vaccine. New vaccine strategies are needed to maximize the protection of aquatic livestock and reduce the use of anti-microbials in aquaculture.

Fish diets influence the gastrointestinal flora of aquaculture species (Ringo and Olsen 1999). In studies conducted in Arctic char (*Salvelinus alpinus*), dietary differences influenced the species composition of gastrointestinal flora. Diet alternation (Diez-Gonzalez et al. 1998; Buchko et. al 2000) has been suggested as one means of potentially reducing the presence of enteric pathogens in the gastrointestinal tracts of cattle. Dietary changes could be of similar value in aquaculture species and provide an alternative to the use of antimicrobials.

Minimizing the Presence of Pathogens and Other Contaminants in Seafood. Efforts to develop an international set of standards for food quality were initiated by the Food and Agricultural Organization in Association with the World Health Organization in 1961. The effort spawned the Codex Alimentarius Commission, charged with developing the Codex Alimentarius (FAO 1999), which defines the standards comprising the descriptor "food quality." The Codex serves as an international standard code for action levels of drug and pesticide residues and other contaminants in food products, as well as sanitary practices, labeling, product sampling and testing, and almost anything that potentially impacts "food quality." The Codex Commission serves as a forum for intergovernmental cooperation, education, and consensus building. Standards for most seafood commodities are articulated in the Codex. Product definitions and standards of practice for fishing vessels, including cold storage, storage containers, evisceration procedures, onboard sanitation, and processing and other variables that may impact product quality and safety are articulated in the guide (Codex 2000). Similar standards for aquaculture have been established to guide site selection, water quality, feed and feed storage, harvesting of aquaculture crops, personnel training, the use of chemotherapeutic agents, and record keeping. Numerous nation states (such as Canada) have used the Codex as a guide to establish their own seafood safety initiatives. Additional standards have been established by the European Union (EC No. 2065/2001).

Federal and State Regulatory Oversight in the United States. The seafood industry in the United States is a regulatory quagmire. Numerous federal departments, including the U.S. Departments of Agriculture, Human Services, Interior, and Commerce as well as the Environmental Protection Agency all play a role in supporting aquaculture and the seafood industry (Ahmed 1992). More than twenty federally legislated acts (for example, the National Aquaculture Improvement Act, 1985) define the strategies and regulations that impact aquaculture as an agribusiness. Regulatory oversight in many other countries with strong aquaculture industries is comparatively much less complex. However, product quality and safety may be more variable in countries without similar food quality and environmental oversight.

The FDA Center for Food Safety and Nutrition (CFSAN) is the primary agency responsible for ensuring the safety of seafood products and driving implementation of the U.S. Seafood Safety Initiative. It has primary regulatory responsibility for the harvesting, processing, packaging, and distribution of both domestic and imported seafood. The FDA Center for Veterinary Medicine (FDA-CVM) is responsible for the development, testing, and marketing of new therapeutic products used by aquaculturists. The FDA is authorized to perform product testing and conduct inspections of both domestic and foreign processing plants. It also plays an important role in the dissemination of information to consumers.

The Environmental Protection Agency (EPA) is the primary agency responsible for implementing the Clean Water Act. The EPA works closely with the U.S. Geological Survey and other federal and state agencies to monitor water quality. Comprehensive planning and site selection for mariculture coordinated by NOAA and directed by the Coastal Zone Management Act relies heavily on water-quality data collected by the EPA and other state and federal agencies.

The USDA is the lead agency for providing and disseminating information in support of aquaculture. Programs sponsored by the USDA support aquaculture-oriented research, provide diagnostic assistance for health assessments, regulate the development of new veterinary biologics for fish, and develop aquatic animal health disease surveillance programs. The USDA/Animal and Plant Health Inspection Service (APHIS) is charged with preventing the importation of foreign aquatic animal diseases, preventing wildlife damage (for example, from birds) to aquaculture crops, and ensuring the welfare of captive animals. The secretary of the USDA chairs the Joint Subcommittee on Aquaculture established by the National Aquaculture Development Act of 1980. The committee is responsible for implementing the National Aquaculture Development Plan, which articulates strategies for developing U.S. aquaculture and defines the roles of supporting agencies and departments.

The National Marine Fisheries Service (NMFS), a division of the National Oceanographic and Atmospheric Administration, is responsible for managing our marine fisheries and preserving marine species protected by the Endangered Species Act. They monitor commercial and recreational fishing harvests and bycatch, and play an important role in promoting mariculture. Their responsibilities include building international trade opportunities for U.S. fish and shellfish farmers, extension, and targeted training in mariculture

enterprises. The NMFS attempts to integrate aquaculture with stock management, supports environmental stewardship, and conducts pilot programs and research to explore new and environmentally friendly mariculture technologies and permitting. The NMFS also conducts and supports mariculture research and plays a vital role as a source of training and information in support of aquaculture and fisheries.

The U.S. Seafood Initiative. In 1995, the FDA established guidelines for the creation of a mandatory HACCP-based program to ensure the wholesomeness of seafood products. All processors of seafood were directed to conduct a hazard analysis and identify processing activities that may contribute to the adulteration of their seafood products. Processors were then required to develop an HACCP plan to correct the identified problems. Importers of seafood products were required to ensure that distributors of the imported products adhered to HACCP program guidelines. The HACCP plan must be updated annually, and designated records must be maintained during plant operation (CFSANOS 1998). The seafood industry initiative also established a zero toxin tolerance for *C. botulinum* that must be addressed by all processors distributing smoked fish.

Additional requirements were established for molluscan shellfish that complement the activities of the National Shellfish Sanitation Program (NSSP) (CFSANOS 1999). The sanitation program, established in 1984 in cooperation with the Interstate Shellfish Sanitation Conference, is an interstate cooperative program coordinated and administered by the FDA and supported by twenty-three states. Each participating state identifies a state agency that establishes criteria for ensuring the quality of fresh or frozen shellfish products available to consumers in their state. The state agencies are responsible for monitoring water quality and classifying areas where shellfish can be grown and harvested. They provide oversight for all phases of shellfish farming, harvesting, processing, packing, and distribution. In support of the NSSP, the FDA maintains a list of certified shellfish shippers that are in compliance with state licensure requirements (CFSANOS 2001). The NSSP Guide for the Control of Molluscan Shellfish articulates guidelines for NSSP program participants, and foreign countries exporting shellfish to the United States must participate in the program and be certified by the FDA (CFSANOS 1999). The FDA evaluates state and foreign shellfish sanitation programs, ensures standardization of laboratory procedures, and coordinates shellfish research.

Industry-producer associations (for example, Catfish Farmer of America), with the cooperation of extension agents and state and federal resource personnel, have established aquaculture quality-assurance programs for producers. The programs are pre-harvest oriented and suggest standard operating practices for producers that can help improve production efficiency and provide the public with high-quality seafood products (Warren 1998; USTFA 1994). Existing quality-assurance programs provide guidelines for site and water source selection, record keeping, and life-stage–specific feeding guidelines to maximize growth, fish health, and profitability. They also define suggested drug and chemical use practices and biosecurity measures to prevent the introduction of pathogens, and describe alternatives for waste management. By integrating HAACP-based production-hazard-identification strategies into quality-assurance programs, producers can ensure the safety of their products (Jensen and Greenlees 1997) and use participation in the program as a value-added identifier of product quality. The pre-harvest quality-assurance effort complements the federally mandated post-harvest–oriented Seafood Initiative. However, the producer-based quality-assurance programs are voluntary, and producer participation is far from universal.

Additional support for the Seafood Initiative has come from NOAA's Sea Grant program, the Extension Service, various state agencies, universities, and community colleges that have established training programs to aid processors, shippers, and retailers in their effort to comply with Seafood Initiative HACCP guidelines.

In 2001, at the request of the Senate, the GAO provided a review of the U.S. Seafood Initiative (Dyckman et al. 2001). Plans and procedures at processing facilities in Seattle, New England, and Florida were evaluated. They also compared the Seafood Initiative with HACCP programs in place in Chile and Canada. The GAO report noted that:

1. Numerous seafood processing facilities and commercial vessels that process seafood while at sea did not have HACCP plans.
2. Processing procedures for many seafood products had not been inspected by the FDA because approximately 48 percent of seafood products were not being processed at the time the FDA made its initial inspections. Serious hazards were identified at 50 percent of the processing facilities, and the FDA had not conducted follow-up inspections at many plants.
3. Warning letters noting significant violations were not sent to seafood processors in a timely manner.
4. Objective, quantifiable measures to evaluate the success of the seafood initiative had not been developed.
5. The FDA had not completed HACCP compliance agreements with foreign countries that export seafood to the United States. Less than one-third of importers had the needed documentation noting that their imported products originated at plants with HACCP-compliant practices. In addition, port of entry inspections had not been effective in identifying potential hazards in seafood. Less than 1 percent of seafood imported into the United States in 1999 was inspected by the FDA (Dyckman et al. 2001).

Although the FDA provided a response to the GAO report findings, it was apparent that more diligent enforcement and possibly some revision of guidelines would be needed to ensure the safety of U.S. seafood products.

Depuration. Bivalves are frequently held under controlled aquatic conditions after harvest to improve their safety before distribution to commercial markets (Dressel and Snyder 1991). The process, known as depuration, makes use of the natural clearing capabilities of bivalves (Jaykus et al. 1994). A similar rationale supports the relaying of oysters from "polluted" to "nonpolluted" habitat (Roderick and Schneider 1994; Ho and Tam 2000). Although some clearing of grit, bacteria, and other pathogens does occur, the microbial flora of bivalves is not completely eliminated by depuration, and many pathogens persist through the process (Pancorbo 1992). The persistence of biotoxins (Arino and Herrera 1993), hepatitis A virus (Sobsey 1987), and other pathogens has been noted in numerous studies. Results are highly variable because of differences in holding conditions, species of bivalve, and the type of pathogen (Jaykus et al 1994). Many metals (Yang et al. 1995), petroleum hydrocarbons (Fossato and Canzonier 1976), and other toxins are not readily eliminated by depuration (Blanco et al. 2002). Chlorine, ultraviolet irradiation, ozone, iodophors, and other compounds have been used to improve the efficacy of depuration (Roderick and Schneider 1994). However, no universally effective method of depuration has been identified.

Source Tracking. Identification of the type of food associated with a food-related illness is a key component of any food safety initiative. After the food-vehicle is identified, investigators generally attempt to identify its origin by tracing backward from the consumer to the point of harvest. The task is particularly arduous with seafood, and frequently isn't completed. Investigators must confront the great diversity of fish and shellfish species, seafood products, and the vast array of cultural differences in storage and preparation. Although the trail from processor to retail counter may be similar, pre-harvest considerations for products derived from the commercial fishing industry are markedly different. Product identification and product records that document the chain of custody from pre-harvest to table are the keys to successful outbreak source tracking. Consequently, record keeping and livestock identification are important components of the FDA's HAACP-based seafood initiative and the pre-harvest quality-assurance programs promoted by aquaculture commodity groups (for example, U.S. Trout Farmers Association).

Fish tagging is used extensively for monitoring the harvest of species in decline for which catch limits have been established (such as rockfish). A well-documented chain of custody has also been a consistent hallmark of the U.S. National Shellfish Sanitation Program. It has played an important role in investigating the origin of shellfish that have been associated with human illnesses. Chain-of-custody documents facilitate tracing products back to the point of harvest and help identify potential water-quality problems associated with the release of human waste into surface waters. A European Economic Union directive implemented in 2002 requires that all fish to be sold to wholesalers or retailers must be tagged in a manner that facilitates traceback to the point of harvest. Tagging is complicated by the morphologic and physical differences of finfish species (for example, flounder vs. swordfish). Håstein and co-authors (2001) provide an excellent review of current seafood identification alternatives that can be used to trace finfish, crustaceans, and bivalve mollusks. Improvements in finfish tagging and product package labeling have been facilitated by bar coding, microchipping and transponder identification, and genetic tracing. Genetic markers are now effectively used to facilitate the comparison of pathogen strains using multiplex polymerase chain reaction and other molecular techniques (Brasher et al. 1998). Both internal and external tagging can now be conducted, and procedures such as microsatellite analysis and amplified fragment length polymorphism analysis can potentially be used to determine the origin of individual fish fillets

Standardized monitoring and sampling protocols have been developed for many compounds, and a host of in situ techniques (for example, semi-permeable membranes) have been developed to accommodate routine monitoring of industrial and commercial point source discharges. However, the presence of recognized nonpoint source contaminants (such as NH_4) and in particular human or animal waste present more challenging problems. Adjacent land-used patterns provide some potential clues, but actual source tracking is difficult. In developed countries, municipal, state (provincial), or federal laboratories conduct routine nutrient analysis; however, the testing at times is not conducted with enough frequency to account for the natural variation that occurs in aquatic ecosystems. In addition, routine analyses for nutrient input does not facilitate actual source tracking unless it is conducted adjacent to discharge sites.

More than 80,000 metric tons of human feces are released in the Western Hemisphere each day (Blake 1993). Differentiating between human and animal waste presents complex problems for many municipalities. Most agencies conducting routine water-quality monitoring for the presence of human waste generally use the traditional, standardized total coliform or fecal coliform multiple-tube dilution techniques to identify the presence of fecal contamination. However, limitations of these most probable number (MPN) enumeration techniques are well-documented (Kator and Rhodes 1991). Human pathogens associated with human fecal waste such as hepatitis A may be present when total or fecal coliform testing suggests that the water column is safe for recreation or for harvesting bivalves for human consumption. Unless an obvious event such as a municipal waste-handling-plant spill or an overflowing livestock lagoon has been identified, the source of contamination cannot be readily identified. Consequently, hundreds of thousands of coastal acres in the United States are closed for the harvesting of shellfish for human consumption.

A variety of potential alternative procedures and indicators have been examined as potential replacements or adjuncts to the routine fecal MPN testing conducted by municipal and state agencies (Kator and Rhodes 1991; Pancorba 1992). Various aerobic and anaerobic bacteria, including fecal streptococci, other enterococci, bifidobacteria spp., bacteroides spp., *C. perfringes*, *Rhodococcus* spp., and bacteriophages, have been evaluated as alternative indicators of human fecal waste contamination. Chemical indicators such as coprostanol, a product of human caffeine consumption, and other sterols, and the presence of antibiotics in the water have been suggested as alternative indicators of fecal waste. Specialized incubators have also been developed that are combined with impedance analyzers to monitor changes in conductivity measurements obtained from inoculated standardized media. The changes in conductivity can be used to identify the growth of specific enteric pathogens and reduce the time required to examine a water or bivalve tissue sample. Characterization of isolates using DNA fingerprinting is being used to examine the potential relationship between isolates obtained from human patients, livestock, and environmental sources. No single technique has evolved as the "gold standard." However, there may be great opportunity in combining the use of several techniques to develop a profile of assay results that can be used to identify the presence of human waste.

Both active and passive monitoring programs are used to monitor for point and nonpoint source contaminant input into surface waters. Standard operating practices for sample collection, as well as laboratory procedures and "action levels" based on laboratory results generally guide collection of environmental data in the United States and many developed countries. However, regulatory agencies are often unable to conduct monitoring activities with the frequency and coverage needed to account for the natural variability in aquatic ecosystems. In addition, many developing countries lack the legislation, administrative infrastructure, and economic resources to engage in aquatic monitoring efforts to maintain water quality and ensure the safety of seafood products.

SUMMARY AND CONCLUSIONS.

Safe seafood products will continue to play an increasingly important role in achieving the optimistic goal of ensuring food security for the planet's population. As natural stocks of fisheries continue to be exploited, aquaculture will have to pick up the slack and play a prominent role in meeting the demand for seafood products. The mantra of sustainable aquaculture (aquaculture that can be perpetuated without adversely impacting the environment or depleting natural resources [Black 2001]) seems to define aquaculture's future as a livestock industry. The list of species being propagated and the locations at which aquaculture is practiced continue to increase. Training opportunities for fish and shellfish farmers have been heightened, and aquaculture investment capital is more readily available in some countries. To continue this growth, however, consumer demand for seafood products must increase, and for demand to increase, consumer confidence in seafood products must be improved.

From a local perspective, differences observed in the specific agents or foods involved in seafood-associated illnesses reflect the geographic, sociocultural, economic, and environmental diversity of human civilization. Internationally, however, the basic issues that must be addressed by aquaculture to ensure the quality of marketed products are universal. As the dialogue about pre-harvest food safety in aquaculture is heightened, a core group of specific pre-harvest challenges most be addressed:

1. The use of chemotherapeutic agents must be more closely controlled and monitored.
2. Water sources used for aquaculture production must be more closely scrutinized, and local environmental regulation and oversight strengthened.
3. Husbandry practices must be refined to enhance biosecurity and livestock health.
4. Improved diet formulations for carnivorous species (such as flounder) must be developed that reduce dependence on fish-based ingredients.
5. Genetic improvement in aquatic livestock is needed to enhance disease resistance, growth, and high livestock density tolerance.
6. Pre-harvest food safety HAACP-based guidelines need to be established.
7. An international consensus about product safety, quality, and standard practices must be developed.

REFERENCES

Abeyta, C. Jr, F.G. Deeter, C.A. Kaysner, R.F. Stott, and M.M. Wekell. 1993. *Campylobacter jejun* in a Washington State shellfish growing bed associated with illness. *J. Food Prot.* 56:323–325.

Acosta, C.J., C.M. Galindo, J. Kimario, K. Senkoro, H. Urassa, C. Casals, M. Corachan, N. Eseko, M. Tanner, H. Mshinda, F. Lwilla, J. Vila, and P.L. Alonso. 2001. Cholera outbreak in southern Tanzania: risk factors and patterns of transmission. *Emerg. Infect. Dis.* 7:583–587.

Ahmed, F.E. 1991. Executive Summary. *Seafood Safety. Committee on Evaluation of the Safety of Fishery Products, Food and Nutrition Board, Institute of Medicine.* Washington, D.C.: National Academy Press, 1–20.

Ahmed, F.E. 1992. Programs of Safety Surveillance and Control of Fishery Products. *Reg. Toxicol. Pharmacol.* 15:14–31.

Alicata, J.E., and R.W. Brown. 1962. Observations on the method of human infection with *Angiostrongylus cantonensis* in Tahiti. *Can. J. Zoology* 40:755–760.

Amaraneni, S.R., and R.R. Pillala. 2001. Concentrations of pesticide residues in tissues of fish from Kolleru Lake in India. *Environ. Toxicol.* 16:550–556.

Amodio-Coccieri, R.T. Cirillo, M. Amorena, M. Cavaliere, A. Lucisano, and U. Del Prete. 2000. Alkyltins in farmed fish and shellfish. *Intl. J. Food Sci. Nutr.* 51:147–151.

Anderson, R. 1999. Atlantic salmon escape into sound from pens. *Seattle Times.* June 15, 1999.

Andrews, J.D. 1987. Epizootiology of haplosporidium disease affecting oysters. *Comp. Pathobiol.* 7:243–269.

Andrews, J.D. 1988. Epizootiology of the disease caused by the oyster pathogen *Perkinsus marinus* and its effects on the oyster industry. *American Fisheries Society Special Publication* 18:47–63.

Anodo, K., K. Ishikura, T. Nakakugi, Y. Shimono, T. Tamai, M. Sugawa, W. Limviroj, Y. Chinzei. 2001. Five cases of *Diphyllobothrium nihonkaiense* infection with discovery of plerocercoids from an infective source, *Oncorhynchus masou ishikawae.* *J. Parasitol.* 87:96–100.

Arino, M., and M. Herrera. 1993. Biotoxins in marine foods: II. Shellfish poisonings. *Alimentaria* 30:43–47.

Bean, N.H, E.K. Maloney, M.E. Potter, P. Korazemo, B. Ray, J.P. Taylor, S. Siegler, and J. Snowden Jr. 1998. Crayfish: a newly recognized vehicle for *Vibrio* infections. *Epidemiol. and Infect.* 121:269–273.

Bebak, J., M. Baumgarten, and G. Smith. 1997. Risk factors for bacterial gill disease in young rainbow trout (*Oncorhynchus mykiss*) in North America. *Prev. Vet. Med.* 32:23–34.

Behr, M.A., T.W. Gyorkos, E. Kokoskin, B.J. Ward, and J.D. MacLean. 1998. North American liver fluke (*Metorchis conjunctus*) in a Canadian aboriginal population: a submerging human pathogen? *Can. J. Publ. Health* 89:258–259.

Belizario, V.Y, W.U. de Leon, D.G. Esparar, J.M. Galang, J. Fantone, and C. Verdadero. 2000. Compostela Valley: a new endemic focus for *Capillariasis philippinensis. The Southeast Asian J. Trop. Med. Publ. Health* 31:478–81.

Bengston, D.A., 1999. Aquaculture of summer flounder (*Paralichthys dentatus*): status of knowledge, current research, and future research priorities. *Aquaculture* 176:39–49.

Billard, R. 1999. On growing in ponds. In *Carp Biology and Culture.* R. Billard,ed. Chichester, U.K.: Springer, Praxis Publishing, 157–215.

Black, K.D. 2001. Sustainability of aquaculture. In *Environmental Impacts of Aquaculture.* K.D. Black, ed. Boca Raton: CRC Press, 199–212.

Blake, P.A. 1993. Epidemiology of cholera in the Americas. *Gastroenterology Clinics of North America* 22:639–660.

Blanco, J., C.P. Acosta, M. Bermudez dela Puente, and C. Salgado. 2002. Depuration and anatomical distribution of the amnesic shellfish poisoning (ASP) toxin domoic acid in the king scallop, *Pectin maximus. Aqual. Toxicol.* 6:111–121.

Bolton, F.J., D. Coates, D.N. Hutchinson, and A.F. Godgree.

1987. A study of thermophilic *Campylobacters* in a river system. *J. Appl. Bacteriol.* 62:167–76.

Botana, L.M. 2000. Seafood and Freshwater Toxins Pharmacology, Physiology, and Detection. New York:Marcel Dekker Inc. 798 pp.

Bosch, A., G. Sanchez, L. Le Guyader, H. Vanaclocha, L. Haugarreau, and R.M. Pinto. 2001. Human enteric viruses in Coquina clams associated with a large hepatitis A outbreak. *Water Science and Techn.* 43:61–65.

Bouree, P, A. Paugam, and J.C. Petithory. 1995. Anisakidosis: report of 25 cases and review of the literature. *Comp. Immunol. Microbiol. Infect. Dis.* 18:75–84.

Boyd, C.E., and C.S. Tucker. 1998. *Pond Aquaculture Water Quality Management* Boston: Kluwer Academic Publishers, 700 pp.

Brasher, C.W., A. DePaola, D.D. Jones, and A.K. Bej. 1998. Detection of microbial pathogens in shellfish with multiplex PCR. *Current Microbiol.* 37:101–107.

Bryan, F.I. 1980. Epidemiology of foodborne diseases transmitted by fish, shellfish, and marine crustaceans in the United States, 1970–1976. *J. Food Prot.* 43:859–876.

Buchko, S.J., R.A. Holley, W.O. Olson, V.P. Gannon, and D.M. Veira. 2000. The effect of different grain diets on fecal shedding of *Escherichia coli* O157:H7 by steers. *J. Food Prot.* 63:1467–1474.

Burgess, V., and G. Shaw. 2001. Pectenotoxins and issue for public health: a review of their comparative toxicology and metabolism. *Environ. Intl.* 27:275–283.

Burke, J.S., T. Seikai, Y. Tanaka, and M. Tanaka. 1999. Experimental intensive culture of summer flounder, *Paralichthys dentatus.* *Aquaculture* 176:135–144.

Butzler, J.P., and J. Oosterom. 1991. *Campylobacter:* pathogenicity and significance in foods. *Intl. J. Food Microbiol.* 12:1–8.

Carney, W.P. 1991. Echinostomiasis—a snail-borne intestinal trematode zoonosis. *The Southeast Asian J. Trop. Med. Publ. Health* 22 Supplement:206–211.

Center for Food Safety and Applied Nutrition Office of Seafood. 1998. *Fish and Fishery Products Hazards and Controls Guide* 2nd ed. U.S. Food and Drug Administration. Docket 93N-0195.

Center for Food Safety and Applied Nutrition Office of Seafood. 1999. *National Shellfish Sanitation Program (NSSP), Guide for the Control of Molluscan Shellfish, Model Ordinance.* Food and Drug Administration. 139 pp.

Center for Food Safety and Applied Nutrition Office of Seafood. 2001. *Interstate Certified Shellfish Shippers List.*

Chan, H.H., K.H. Lai, G.H. Lo, J.S. Cheng, J.S. Huang, P.I. Hsu, C.K. Lin, and E.M. Wang. 2002. The clinical and cholangiographic picture of hepatic clonorchiasis. *J. Clin. Gastroenterology* 34:183–186.

Chittick, B., M. Stoskopf, N. Heil, J. Levine, and M. Law. 2001. Evaluation of sandbar shiner as a surrogate for assessing health risks to the endangered Cape Fear shiner. *J. Aquatic Animal Health* 13:86–95.

Clark, M. 2001. Are deepwater fisheries sustainable? The example of orange roughy (*Hoplostethus atlanticus*) in New Zealand. *Fisheries Res.* 51:123–135.

Clarke, I.N., and P.R. Lambden. 1997. Viral zoonoses and food of animal origin: calciviruses and human disease. *Arch. Virol.* Supplement 13:141–152.

Cliver, D.O. 1988. Virus transmission in foods. A scientific status summary by the Institute of Food Technologists'-Expert Panel on Food Safety and Nutrition. *Food Techn.* 42:241–247.

Cobb, B.F., C. Vanderzant, M.O. Hanna, and C.-P.S. Yeh.

1976. Effect of ice storage on microbiological and chemical changes in shrimp and melting ice in a model system. *J. Food Sci.* 41:29.

Codex Committee on Fish and Fishery Products. 2000. Proposed draft code of practice for fish and fishery products. *Joint Food and Agricultural Organization of the United Nations/World Health Organization Food Standards Program* CX/FFP00/4. 148. pp.

Colburn, K.G., C.A. Kaysner, C. Abeyta, and M.M. Wekell. 1990. *Listeria* species in a California coast estuarine environment. *Appl. Environ. Microbiol.* 56:2007–2011.

Conaty, S., P. Bird, G. Bell, E. Kraa, G. Grohmann, J.M. McAnulty. 2000. Hepatitis A in New South Wales, Australia from consumption of oysters: the first reported outbreak. *Epidemiol. and Infect.* 124:121–130.

Cordano, A.M., and R. Virgilio. 1996. Evolution of drug resistance in *Salmonella panama* isolates in Chile. *Antimicrob. Agts. and Chemotherapy* 40:336–341.

Cui, J., Z.Q. Wang, F. Wu, and X.X. Jin. 1998. An outbreak of paragonimiasis in Zhenghou city, China. *Acta Tropica* 70:211–216.

Curtis, M.A., and G. Bylund. 1991. Diphyllobothriasis: fish tapeworm disease in the circumpolar north. *Artic Med. Res.* 50:18–24.

Dack, D.M. 1956. *Food Poisoning.* Chicago: University of Chicago Press, 251 pp.

Daniels, N.A., L. MacKinnon, R. Bishop, S. Altekruse, R.B. Ray, R.M. Hammond, S. Thompson, S. Wilson, N.H. Bean, P.M. Griffin, and L. Slutsker. 2000. Vibrio parahaemolyticus infections in the United States, 1973–1998. *J. Infect. Dis.* 18:1661–1666.

Daniels, N.A., B. Ray, A. Easton, N. Marano, E. Kahn, A.L. McShan 2nd, L. Del Rosario, T. Baldwin, M.A. Kingsley, N.D. Puhr, J.G. Wells, and F. J. Angulo. 2000b. Emergence of a new *Vibrio parahaemolyticus* serotype in raw oysters: A prevention quandary. *J.A.M.A.* 284:1541–1545.

D'Aoust, J.Y. 1994. *Salmonella* and the international food trade. *Intl. J. Food Microbiol.* 24:11–31.

Deardorff, T.L., and M.L. Kent. 1989. Prevalence of larval *Anisakis simplex* in pen-reared and wild caught salmon (Salmonidae) from Puget Sound, Washington. *J. Wildlife Dis.* 25:416–419.

Deardorff, T.L., and R.M. Overstreet. 1991. Seafood-transmitted zoonoses in the United States: the fishes, the dishes, and the worms. In *Microbiology of Marine Food Products* D.R. Ward and C. Hackney, eds. New York: Van Nostrand Reinhold, 211–266.

Deardorff, T.L., S.G. Kayes, and T. Fukumura. 1991. Human anisakiasis transmitted by marine food products. *Hawaii Med. J.* 50:9–16.

Deirberg, F.E., and W. Kiattisimkul. 1996. Issues, Impacts, and Implications of Shrimp, Aquaculture in Thailand. *Environ. Mgmt.* 20:649–666.

dela-Cruz, C., R.C. Sevilleja, and J. Torres. 2001. Rice-fish system in Guimba, Nueva Ecija, Philippines. FAO Fisheries Technical Paper 407:85–89.

DePaola, A., J.T Peeler, and G.E. Rodrick. 1995. Effect of oxytetracycline-mediated feed on antibiotic resistance of gram-positive bacteria in catfish ponds. *Appl. Environ. Microbiol.* 61:2335–2340, 61:3513.

Desenclos, J.A., K.C. Klontz, M.H. Wilder, O.W. Nainan, H.S. Margolis, and R.A. Gunn. 1991. A multistate outbreak of hepatitis A caused by the consumption of raw oysters. *Am. J. Publ. Health* 81:1268–1272.

Diez, F., T.R. Callaway, M.G. Kizoulis, and J.B. Russell. 1998. Grain feeding and the dissemination of acid-resistant *Escherichia coli* from cattle. *Science* 281:1578–1579.

DiGirolamo, R., J. Liston, and J.R. Matches. 1970. Survival of virus in chilled, frozen, and processed oysters. *Appl. Microbiol.* 20:58–63.

Dominguez-Ortega, J., A. Alonso-Llamazares, I. Rodriguez, and M. Chamorro. 2001. Anaphylaxis due to hypersensitivity to *Anisakis simplex*. *International Archives of Allergy and Immunology* 125:86–88.

Drapchl, C.M., and D.E. Brune. 2000. The partition aquaculture system: impact of design and environmental parameters on algal productivity and photosynethic oxygen production. *Aquacultural Engineering* 21:151–168.

Dressel, D.M., and M.I. Snyder. 1991. Depuration—The regulatory perspective. In *Molluscan Shellfish Depuration*. W.S. Otwell, G.E. Rodrick, and R.E. Martin, eds. Boca Raton, FL: CRC Press, 19–23.

Durborrow, R.M., and R. Francis-Floyd. 1996. Medicated feed for food fish. *Southern Regional Aquaculture Center Publication* No. 473. 4 pp.

Durborrow, R.M. 1999. Health and safety concerns in fisheries and aquaculture. *Occupational Med.* 14:373–406.

Dyckman, L.J., K.W. Oleson, M.C. Gobin, R.D. Jones, L. Acosta, R. Pinero, A. Antonetti, F. Featherston, C. Herrnstadt-Shulman, and O. Easterwood. 2001. Food Safety: Federal Oversight of Seafood Does Not Sufficiently Protect Consumers. *Report to the Committee on Agriculture, Nutrition, and Forestry, U.S. Senate*. Government Accounting Office. GAO-01-204. 60 pp.

Eastburn, R.L., T.R. Fritsche, and C.A. Terhune. 1987. Human intestinal infection with *Nanophyetus salmincola* from salmonid fishes. *Am. J. Trop. Med. Hygiene* 36:586–591.

Egna, H.S., C.E. Boyd, and D.A. Burke. 1997. Introduction. In *Dynamics of Pond Aquaculture*. H.S. Egna and C.E. Boyd, eds. Boca Raton: CRC Press, 1–18.

Elliot, E.L., and R.R. Colwell. 1985. Indicator organisms for estuarine and marine waters. *FEMS Microbiol. Rev.* 32:61–79.

Endtz, H.P., J.S. Vliegenthart, P. Vandamme, H.W. Weverink, N.P. van den Braak, H.A. Verbrugh, and A. Belkum. 1997. Genotypic diversity of *Campylobacter lari* isolated from mussels and oysters in the Netherlands. *Intl. J. Food Microbiol.* 34:79–88.

Enriquez, R., G.G. Frosner, V. Hochstein-Mintzel, S. Riedemann, and G. Reinhardt. 1992. Accumulation and persistence of hepatitis A virus in mussels. *J. Med. Virol.* 37:174.

Ericesson, H., A. Eklow, M.L. Danielsson-Tham, S. Loncarevic, L.O. Mentzing, I. Persson, H. Unnerstad, and W. Tham. 1997. An outbreak of listeriosis suspected to have been caused by rainbow trout. *J. Clin. Microbiol.* 35:2904–2907.

Eskesen, A., E.A. Strand, S.N. Andersonen, A. Rosseland, K.B. Hellum, and O.A. Strand. 2001. Anisakiasis presenting as an obstructive duodenal tumor. A Scandinavian case. *Scandinavian J. Infect. Dis.* 33:75–76.

Fayer, R., T.K. Graczyk, E.J. Lewis, J.M. Trout, C.A. Farley. 1998. Survival of infectious *Cryptosporidium parvum* oocysts in seawater and eastern oysters (*Crassostrea virginica*) in the Chesapeake Bay. *Appl. Environ. Microbiol.* 64:1070–1074.

Fayer, R., E.J. Lewis, J.M. Trout, T.K. Graczyk, M.C. Jenkins, J. Higgins, L. Xiao, and A.A. Lal. 1999. *Cryptosporidium parvum* in oysters from commercial harvesting sites in the Chesapeake Bay. *Emerg. Infect. Dis.* 5:706–710.

Fell, G., O. Hamouda, R. Lindner, S. Rehmet, A. Liesegang, R. Prager, B. Gericke, and L. Peterson. 2000. An outbreak of *Salmonella* blockley infections following smoked eel consumption in Germany. *Epidemiol. and Infect.* 125:9–12.

Fenlon, D.R. 1999. *Listeria monocytogenes* in the natural environment. In *Listeria, Listeriosis, and Food Safety*, 2nd Edition. E.T. Ryser and E.H. Marth, eds. New York: Marcel Dekker Inc., 21–37.

Finelli, L., D. Swerdlow, K. Mertz, H. Ragazzoni, and K. Spitalny. 1992. Outbreak of cholera associated with crab brought from an area with epidemic disease. *J. Infect. Dis.* 166:1433–1435.

Food and Agricultural Organization. 1999. *Understanding The Codex Alimentarius*. Rome, Italy: Food and Agricultural Organization United Nations/World Health Organization.

Fossato, V.U., and W.J. Canzonier. 1976. Hydrocarbon uptake and loss by the mussel *Mytilus edulis*. *Marine Biol.* 36:243–250.

Freire-Santos, F., A.M. Oteiza-Lopez, C.A. Vergara-Castiblanco, E. Ares-Mazas, E. Alvarez-Suarez, and O. Garcia-Martin. 2000. Detection of *Cryptosporidium oocysts* in bivalve molluscs destined for human consumption. *J. Parasitol.* 86:853–854.

Freire-Santos, F., A.M. Oteiza-Lopez, J.A. Castro-Hermida, O. Garcia-Martin, and M.E. Ares-Mazas. 2001. Viability and infectivity of oocysts recovered from clams, *Rudtapes philipparum*, experimentally contaminated with *Crypotsporidium parvum*. *Parasitol. Res.* 87:428–430.

Gholami, P., S.Q. Lew, and K.C. Klontz. 1998. Raw shellfish consumption among renal disease patients. A risk factor for severe *Vibrio vulnificus*. *Am. J. Prev. Med.* 15:243–245.

Gillespie, I.A., G.K. Adak, S.J. O'Brien, M. M. Brett, and F.J. Bolton. 2001. General outbreaks of infectious intestinal disease associated with fish and shellfish, England and Wales, 1992–1999. *Communicable Dis. Publ. Health* 4:117–123.

Gobat, P.F., and T. Jemmi. 1993. Distribution of mesophilic *Aeromonas* species in raw and ready-to-eat fish and meat products in Switzerland. *Intl. J. Food Microbiol.* 20:117–120.

Goma Epidemiology Group. 1995. Public health impact of Rwandan refugee crisis: what happened in Goma, Zaire, in July, 1994? *Lancet* 345:339–344.

Gomez-Bautista, M., L.M. Ortega-Mora, E. Tabares, V. Lopez-Rodas, and E. Costas. 2000. Detection of infectious *Cryptosporidium parvum* oocysts in mussels (*Mytilus galloprovincialis*) and cockles (*Cerastoderma edule*). *Appl. Envion. Microbiol.* 66:1866–1870.

Goulletquer, P., and M. Heral. 1997. Marine molluscan production trends in France: from fisheries to aquaculture. *NOAA Technical Report National Marine Fisheries Service* 129:137–164.

Gonzalez, C.J., T.M. Lopez-Diaz, M.L. Garcia-Lopez, M. Prieto, and A. Otero. 1999. Bacterial microflora of wild brown trout (*Salmo trutta*), wild pike (*Esox lucius*), and aquacultured rainbow trout (*Oncorhynchus mykiss*). *J. Food Prot.* 62:1270–1277.

Gracia-Bara, M.T., V. Matheu, J.M Zubeldia, M. Rubio, M.P. Ordoqui Eopez-Saez, Z. Sierra, P. Tornero, and M. L. Baeza. 2001. Anisakis simplex-sensitized patients: should fish be excluded from their diet? *Annals of Allergy Asthma and Immunology* 86:679–685.

Graczyk, T.K., and B. Fried. 1998. Echinostomiasis: a common but forgotten food-borne disease. *Am. J. Trop. Med. Hygiene* 58:501–504.

Graczyk, T.K., R. Fayer, M.R. Cranfield, and D.B. Conn. 1998. Recovery of waterborne C*ryptosporidium* oocysts by freshwater benthic clams (*Corbicula fluminea*). *Appl. Envion. Microbiol.* 64:427–430.

Greenlees, K.J., J Machado, T. Bell, and S.F. Sundlof. 1998. Foodborne microbial pathogens of cultured aquatic species. *Vet. Clinics of N. America Food Animal Practice* 14:101–112.

Greenwood, M., G. Winnard, and B. Bagot. 1998. An outbreak of *Salmonella enteritidis* phage type 19 infection associated with cockles. *Communicable Dis. Publ. Health* 1:35–37.

Gudding, R., A. Lillehaug, and O. Evensen. 1999. Recent developments in fish vaccinology. *Vet. Immunol. Immunopathol.* 15:203–212.

Hackney, C.R., and M.E. Potter. 1994. Human-associated bacterial pathogens. In *Environmental Indicators and Shellfish Safety*. C.R. Hackney, and M.D. Pierson, eds. New York: Chapman and Hall, 154–171.

Hall, M.A., D.L. Alverson, and K.I. Metuzalis. 2000. Bycatch: Problems and Solutions. *Marine Pollution Bulletin* 41:204–219.

Halstead, B.W. 2001. Fish toxins. In *Foodborne Disease Handbook 2nd Edition*. Y.H Hui, D. Kitts, and P.S. Stanfield, eds. New York: Marcel Dekker Inc., 23–50.

Hanninen, M.L., P. Oivanen, and V. Hirvela-Koski. 1997. *Aeromonas* species in fish, fish-eggs, shrimp and freshwater. *Intl. J. Food Microbiol.* 34:17–26.

Hansen, G.H., and J. A. Olafson. 1999. Bacterial interactions in early life stages of marine cold water fish. *Microbial Ecol.* 38:1–26.

Harada, M. 1995. Minamata disease: methylmercury poisoning in Japan caused by environmental pollution. *Crit. Rev.Toxicol.* 25:1–24.

Harrell, L.W., and T.L. Deardorff. 1990. Human nanophyetiasis: transmission by handling naturally infected coho salmon (*Oncorhynchus kisutch*). *J. Infect. Dis.*161:146–148.

Håstein, T., B.J. Hill, F. Berthe, and D.V. Lightner. 2001. Traceability of aquatic animals. *Scientific and Technical Review Office of International Des Epizooties* 20:564–583.

Haya, K., L.E. Burridge, and B.D. Chang. 2001. Environmental impact of chemical waste produced by the salmon industry. *Fisheries and Oceans* 58:492–496.

Hielm, S., E. Hyytia, J. Ridell, and H. Korkeala. 1996. Detection of *Clostridium botulinum* in fish and environmental samples using polymerase chain reaction. *Intl. J. Food Microbiol.* 31:357–365.

Heinitz, M.L., J.M. Johnson. 1998. The incidence of *Listeria* spp., *Salmonella* spp., and *Clostridium botulinum* in smoked fish and shellfish. *J. Food Prot.* 61:318–323.

Heinitz, M.L., R.D. Ruble, D.E. Wagner, and S.R. Tatini. 2000. Incidence of *Salmonella* in fish and seafood. *J. Food Prot.* 63:579–592.

Hinshaw, J.M., T.M. Losordo. 1993. Aeration, oxygenation, and energy use in coldwater aquaculture. Raleigh: NC, Agricultural Extension Service, 6 pp.

Ho, B.S.W., and T.Y. Tam. 2000. Natural depuration of shellfish for human consumption: a note of caution. *Water Res.* 34:1401–1406.

Hoff, J.C., and R.C. Becker. 1969. The accumulation and elimination of crude and clarified poliovirus suspensions by shellfish. *Am. J. Epidemiol.* 90:53–61.

Hongpromyart, M., R. Keawvichit, K. Wongworapat, S. Suwanrangsi, M. Hongpromyart, K. Sukhawat, K. Tonguthai, and C.A. Lima dos Santos. 1997. Application of hazard analysis critical control point (HACCP) as a possible control measure for *Opisthorchis viverrini* infection in cultured carp (*Puntius gonionotus*) *The Southeast Asian J. Trop. Med. Publ. Health* Supplement 1:65–72.

Hughes, J.M., M.A. Horwitz, M.H. Merson, W.H. Barker Jr., and E.J. Gangarosa. 1977. Foodborne disease outbreaks of chemical etiology in the United States 1970–1974. *Am. J. Epidemiol.* 105:233–244.

Hussong, D., J.M. Damare, R.J. Limpert, W.J. Sladen, R.M. Weiner, and R.R. Colwell. 1979. Microbial impact of Canada geese (*Branta canadensis*) and whistling swans (*Cygnus columbianus* columbianus) on aquatic ecosystems. *Appl. Envion. Microbiol.* 37:14–20.

Hutchinson, J.W., J.W. Bass, D.M. Demers, and G.B. Myers. 1997. Diphyllobothriasis after eating raw salmon. *Hawaii Med. J.* 56:176–177.

Ishida, H., N. Muramatsu, H. Nukaya, T. Kosuge, and K. Tsuji. 1996. Study on neurotoxic shellfish poisoning involving the oyster, *Crassostrea gigas*, in New Zealand. *Toxicol.* 34:1050–1053.

James, K.J., A.G. Bishop, E.P. Carmody, and S.S. Kelly. 2000. Detection methods for okadaic acid and analogues. In *Seafood and Freshwater Toxins, Pharmacology, Physiology, and Detection*. L.M. Botana, ed. New York: Marcel Dekker Inc., 217–238.

Janda, J.M., and S.L. Abbott. 1999. Unusual food-borne pathogens, *Listeria monocytogenes, Aeromonas, Plesimonas*, and *Edwardsiella* species. *Clin. Lab. Med.* 19:553–582.

Juan, H.J.,R.B. Tang, T.C. Wu, and K.W. Yu. 2000. Isolation of *Aeromonas hydrophilia* in children wih diarrhea. *J. Microbiol. Immunol. and Infection* 33:115–117.

Jaykus, L., M.T. Hemard, and M.D. Sobsey. 1994. Human enteric pathogenic viruses. In *Environmental Indicators and Shellfish Safety*. C.R. Hackney and M.D. Pierson, eds. New York: Chapman and Hall, 92–153.

Jensen, G.L., and K.J. Greenless. 1997. Public health issues in aquaculture. *Scientific and Technical Review Office of International Des Epizooties* 16:641–651.

Jhingran, V.G., and V. Gopalakrishnan. 1974. Catalogue of cultivated aquatic organisms. *FAO Fisheries Technology Paper*, 130 pp.

Jinnerman, K.C., M.M. Wekell, and M.W. Eklund. 1999. Incidence and behavior of *Listeria monocytogenes* in fish and seafood. In *Listeria, Listeriosis, and Food Safety* 2nd ed. E.T. Ryser and E.H. Marth, eds. New York: Marcel Dekker Inc., 601–630.

Johnsen, P.B. 1991. Aquaculture product quality issues: market position opportunities under mandatory seafood inspection regulations. *J. Animal Sci.* 69:4209–4215.

Jung, S.H., J.W. Kim, I.G. Jeon, and Y.H. Lee. 2001. Formaldehyde residues in formalin-treated olive flounder (*Paralichtys olivaceus*), blackrockfish (*Sebastes schlegeli*), and seawater. *Aquaculture* 194:253–262.

Kaiser, M.J. 2001. Ecological effects of shellfish cultivation. Sustainability of aquaculture. In *Environmental Impacts of Aquaculture*. K.D. Black, ed. Sheffield, UK: Sheffield Academic Press Ltd. Boca Raton, FL: CRC Press, 51–75.

Kam, K.M., T.H. Leung, Y.Y. Ho, N.K. Ho, and T.A. Saw. 1995. Outbreak of *Vibrio cholerae* 01 in Hong Kong related to contaminated fish tank water. *Publ. Health* 109:389–395.

Kator, H., and M.W. Rhodes. 1991. Indicators and Alternate indicators of growing water quality. In *Microbiology of Marine Food Products*. D.R. Ward and C. Hackney, eds. New York: Van Nostrand Reinhold, 211–266.

Kearney, J.P., D.C. Cole, L.A. Ferron, and J. Weber. 1999. Blood PCB, *p,p′*-DDE, and Mirex levels in Great Lakes fish and waterfowl consumers in two Ontario communities. *Environ. Res.* 80:S138–S149.

Kelly, A.M., and C.C. Kohler. 1997. Climate, Site and Pond Design. In *Dynamics of Pond Aquaculture*. H.S. Egna and C.E. Boyd, eds. Boca Raton: CRC Press, 1–18.

Khamboonruang, C., R. Keawvichit , K. Wongworapat, S. Suwanrangsi, M. Hongpromyart, K. Sukhawat, K. Tonguthai, and C.A. Lima dos Santos. 1997. Application of hazard analysis critical control point (HACCP) as a possible control measure for *Opisthorchis viverrini* infection in culture carp (*Puntius gonionotus*). *The Southeast Asian J. Trop. Med. Publ. Health* 28:Supplement 1:65–72.

Kilgen, M.B, and M.T. Cole. 1991. Viruses in Seafoods. In *Microbiology of Marine Food Products*. D.R Ward and C. Hackney, eds. New York: Van Nostrand Reinhold, 450 pp.

Kino, H., H. Inaba, N. Van De, L. Van Chau, D.T. Son, H.T. Hao, N.D. Toan, L.D. Cong, and M. Sano. 1998. Epidemiology of clonorchiasis in Ninh Binh Province, Vietnam. *The Southeast Asian J. Trop. Med. Publ. Health* 29:250–254.

Kirkland, K.B., R.A. Meriwether, J.K. Leiss, and W.R. MacKenzie. 1996. Steaming oysters does not prevent Norwalk-like gastroenteritis. *Publ. Health Repts.* 111:527–530.

Klontz, K.C., R.V. Tauxe, W.L. Cook, W.H. Riley, and I.K. Wachsmuth. 1987. Cholera after the consumption of raw oysters. A case report. *Ann. Internal Med.* 107:846–848.

Kobayashi, J., B. Vannachone, Y. Sato, K. Manivong, S. Nambanya, and S. Inthakone. 2000. An epidemiological study on *Opisthorchis viverrini* infection in Lao villages. *The Southeast Asian J. Trop. Med. Publ. Health* 31:128–132.

Kodama, M. 2000. Ecobiology, Classification, and Origin. In *Seafood and Freshwater Toxins Pharmacology, Physiology and Detection*. L.M. Botana, ed. New York: Marcel Dekker Inc., 125–149.

Kosow, J.A., J. Bell, P. Virtue, and D.C. Smith. 1996. Fecundity and its variability in orange roughy: effects of population density, condition, egg size and senescence. *J. Fish Biol.* 47:1063–1080.

Krantz, G.E., and J.W. Chamberlin. 1978. Blue crab predation on cultchless oyster spat. *Proceedings of the National Shellfish Assoc.* 68:38–41.

L'Abee-Lund, T.M., and H. Sorum. 2001. Class 1 integrons mediate antibiotic resistance in the fish pathogen *Aeromonas salmonicida* worldwide. *Microbial Drug Resistance* 7:263–272.

Lagos, N.W., and D. Andrinolo. 2000. Paralytic shellfish poisoning (PSP): Toxicology and kinetics. In *Seafood and Freshwater Toxins, Pharmacology, Physiology, and Detection*. L.M. Botana, ed. New York: Marcel Dekker Inc., 203–215.

Lee, K.W., H.C. Suhk, K.S. Pai, H.J. Shin, S.Y. Jung, E.T. Han, and J.Y. Chai. 2001. *Diphyllobothrium latum* infection after eating domestic salmon flesh. *Korean J. Parasitol.* 39:319–321.

Lees, D. 2000. Viruses and bivalve shellfish. *Intl. J. Food Microbiol.* 25:81–116.

Le Guyander, F., L. Haugarreau, L. Miossec, E. Dubois, and M. Pommepuy. 2000. Three-year study to assess human enteric viruses in shellfish. *Appl. Envion. Microbiol.* 66:3241–3248.

Leung, C., Y. Huang, and O.C. Pancorbo. 1992. Bacterial pathogens and indicators in catfish and pond environments. *J. Food Prot.* 55:424–427.

Li, S., 1994. Introduction: Freshwater fish culture. In *Freshwater fish culture in China: principles and practice*. S. Li and J. Mathias, eds. Amsterdam: Elsevier, 1–25.

Lim-Quizon, M.C., R.M. Benabaye, F.M. White, M.M. Dayrit, and M.E. White. 1994. Cholera in metropolitan Manila: foodborne transmission via street vendors. *Bull. World Health Org.* 72:745–749.

Ling, B, T.X. Den, Z.P. Lu, L.W. Min, Z.X. Wang, and A.X. Yuan. 1993. Use of night soil in agriculture and fish farming. *World Health Forum* 14:67–70.

Llewellyn, L.E. 2001. Shellfish chemical poisoning. In *Foodborne Disease Handbook 2ⁿᵈ Edition, Volume 4: Seafood and Environmental Toxins*. Y.H. Hui, D. Kitts, and P.S. Stanfield, eds. New York: Marcel Dekker Inc., 77–108.

Losordo, T.M., M.P. Masser, and J. Rakocy. 1998. Recirculating Aquaculture Tank Production Systems: An overview of critical considerations. *Southern Regional Aquaculture Center* #451. 6 pp.

Lowery, C.J., P. Nugent, J.E. Moore, B.C. Millar, X. Xiru, J.S. Dooley. 2001. PCR-IMS detection and molecular typing of *Cryptosporidium parvum* recovered from a recreational river source and an associated mussel (*Mytilus edulis*) bed in Northern Ireland. *Epidemiol. and Infect.* 127:545–553.

MacLean, J.D., J.R. Arthur, B.J. Ward, T.W. Gyorkos, M.A. Curtis, and E. Kokoskin. 1996. Common-source outbreak of acute infection due to the North American liver fluke *Metorchis conjunctus*. *Lancet* 347:154–158.

Makino, S.I., T. Kii, H. Asakura, T. Shirahata, T. Ikeda, K. Takeshi, and K. Itoh. 2000. Does enterohemorrhagic *Escherichia coli* O157:H7 enter the viable but nonculturable state in salted salmon roe? *Appl. Envion. Microbiol.* 66:5536–5539.

Margolis, L. 1977. Public health aspects of "codworm" infection: a review. *J. Fish Res. Board Can.* 34:887–898.

Markestad, A., and K. Grave. 1997. Reduction of antibacterial drug use in Norwegian fish farming due to vaccination. In *Fish Vaccinology. Developments in Biological Standardization*. R. Gudding, A. Lillehaug, P.J. Midtlyng, and F. Brown, eds. Basel: Karger, 90:365–369.

Mead, P.S., L. Slustker, V. Dietz, L.F. McCaig, J.S. Bresee, C. Shapiro, P.M. Griffin, and R.V. Tauxe. 1999. Food-related illness and death in the United States. *Emerg. Infect. Dis.* 5:607–625.

Melnick, J.L., C.P. Gerba, and C. Wallis. 1978. Viruses in water. Update Lepoint. *Bull. World Health Org.* 6:499–508.

Meyers-Samuel, P. 1995. Developments and trends in fisheries processing: Value-added product development and total resource utilization. *Bull. Korean Fisheries Soc.* 27:839–846.

Miao, X.S., L.A. Woodward, C. Swenson, Q.X. Li. 2001. Comparative concentrations of metals in marine species from French Frigate Shoals, North Pacific Ocean. *Marine Pollution Bull.* 42:1049–1054.

Millemann, R.E., and S.E. Knapp. 1970. Biology of *Nanophyetus salmincola* and "salmon poisoning" disease. *Adv. Parasitol.* 8:1–41.

Morbidity Mortality Weekly Report. 1996. Invasive infection with *Streptococcus iniae*—Ontario 1995–1996. *M.M.W.R.* 45:650–653.

Morohashi, A., M. Satake, H. Naoiki, H.F. Kaspar, Y. Oshima, and T. Yasumoto. 1999. Brevetoxin B4 isolated from greenshell mussels *Perna canaliculus*, the major toxin involved in neurotoxic shellfish poisoning in New Zealand. *Natural Toxins* 7:45–48.

Motes, M.L. 1991. Incidence of *Listeria* spp. in shrimps, oysters, and estuarine waters. *J. Food Prot.* 54:170–173.

Muir, J.F., and C.G. Nugent. 1999. Kyoto Conference Outcome and Papers Presented. Aquaculture Trends: Perspectives for Food Security. International Conference on the Sustainable Contribution of Fisheries to Food Security. Kyoto, Japan.

Nam, H.S., and W.M. Sohn. 2000. Infection status with trematode metacercariae in pond smelts, *Hypomesus olidus*. *Korean J. Parasitol.* 200 38:37–39.

NASS 2001. Catfish and Trout Production. National Agricultural Statistics Service. United States Department of Agriculture. Aq 2 (2-01). 24 pp.

Nedoluha, P.C., and D. Westhoff. 1997. Microbiological analysis of striped bass (*Morone saxatilis*) grown in a recirculating system. *J. Food Prot.* 60:948–953.

Ng, S., and J. Gregory. 2000. An outbreak of ciguatera fish poisoning in Victoria. *Communicable Dis. Intelligence* 24:344–346.

Niijar, M.S., and S.S. Niijar. 2001. Ecobiology, clinical symptoms, and mode of action of domoic acid, an amnesic shellfish toxin. In *Seafood and Freshwater Toxins, Pharmacology, Physiology, And Detection.* L.M. Botana,ed. New York: Marcel Dekker Inc., 325–372.

Nishio, T.J., J. Nakamori, K. Miyazaki. 1981. Survival of *Salmonella typhi* in oysters. *Zentralblatt Fur Bakteriologie Mikrobiologie Und Hygiene* 172:415–426.

Nolan, D.T., P. Reilly, and E.E. Wendelear-Bonga. 1999. Infection with low numbers of the sea louse (*Lepeoph-theirus salmonis*) induces stress-related effects in post-molt Atlantic salmon (*Salmo salar*). *Can. J. Fisheries and Aquatic Sci.* 56:947–959.

Ogawa, K. 1996. Marine parasitology with special reference to Japanese fisheries and mariculture. *Vet. Parasitol.* 64:95–105.

Ohyama, T., S. Yoshizumi, H. Sawada, Y. Uchiyama, Y. Katoh, N. Hamaoka, and E. Utagawa. 1999. Detection and nucleotide sequence analysis of human calicivirus (HuCVs) from samples in non-bacterial gastroenteritis outbreaks in Hokkaido, Japan. *Microbiol. and Immunol.* 43:543–550.

Olsen, S.J., L.C. MacKinnon, J.S. Goulding, N.H. Bean, and L. Slutsker. 2000. Surveillance for food-borne disease outbreaks United States, 1993–1997. *M.M.W.R. CDC Surveillance Summaries* 49:1–62.

O'Reilly, A., and F. Kaferstein. 1997. Food safety hazards and the application of the principles of the hazard analysis and critical control point (HACCP) system for their control in aquaculture production. *Aquaculture Res.* 28:735–752.

Osterholm, M.T., J.C. Forfang, T.L. Ristinen, A.G. Dean, J.W. Washburn, J.R. Godes, R.A. Rude, and J. G. McCullough. 1981. An outbreak of foodborne giardiasis. *N. Engl. J. Med.* 304:24–28.

Paez-Osuna, F. 2001. The environmental impact of shrimp aquaculture: causes, effects, and mitigating alternatives. *Environ. Mgmt.* 28:131–140.

Pancorbo, O.C. 1992. Microbial pathogens and indicators in estuarine environments and shellfish. *J. Environ. Health* 54:57–63.

Perl, T., L. Bedard, R. Remis, T. Kosatsky, J. Hoey, and R. Masse. 1987. Intoxication following mussel ingestion in Montreal. *Can. Dis. Wkly Rept.* 49:224–225.

Peterson, D.A., L.G. Wolfe, E.P. Larkin, and F.W. Deinhardt. 1978. Thermal treatment and infectivity of hepatitis A virus in human feces. *J. Med. Virol.* 2:201–206.

Pillay, T.V.R. 1993. *Aquaculture Principles and Practices.* London: Fishing News Books. 575 pp.

Pollack, C.V., and J. Fuller. 1999. Update on emerging infections from the Centers for Disease Control and Prevention. Outbreak of *Vibrio parahaemolyticus* infection associated with eating raw oysters and clams harvested from Long Island Sound—Connecticut, New Jersey and New York, 1998. *Ann. Emergency Med.* 34:679–680.

Portnoy, B.L., P.A. Mackowiak, C.T. Caraway, J.A. Walker, T.W. McKinley, and C.A. Klein. 1975. Oyster-associated hepatitis—Failure of shellfish certification programs to prevent outbreaks. *J.A.M.A.* 233:1065–1068.

Potasman, I., A. Paz, and M. Odeh. 2002. Infectious outbreaks associated with bivalve shellfish consumption: a worldwide perspective. *Clin. Infect. Dis.* 35:921–928.

Pullela, S.,C.F. Fernandes, G.J. Flick, G.S. Libey, S.A. Smith, C.W. Coale. 1998. Indicative and pathogenic microbiological quality of aquacultured finfish grown in different production systems. *J. Food Prot.* 61:205–210.

Rahim, Z., and K.M. Aziz. 1994. Enterotoxigenicity, hemolytic activity and antibiotic resistance of *Aeromonas* spp. isolated from freshwater prawn mar-

keted in Dhaka, Bangladesh. *Microbiol. and Immunol.* 38:773–778.

Randal, J.E. 1980. A survey of ciguatera at Eniwetok and Bikini Marshall Islands, with notes on the systematics and food habits of ciguatoxic fish. *Fish Bulletin* 78:201–249.

Rausch, R.L., and A.M. Adams. 2000. Natural transfer of helminths of marine origin to freshwater fishes with observations on the development of *Diphyllobothrium alascense. J. Parasitol.* 86:319–327.

Reeve, G., D.L. Martin, J. Pappas, R.E. Thompson, and K.D. Green. 1989. An outbreak of shigellosis associated with the consumption of raw oysters. *N. Engl. J. Med.* 32:224–227.

Ringo, E., and R.E. Olsen. 1999. The effect of diet on aerobic bacterial flora associated with intestine of Arctic charr (*Salvelinus alpinus* L.) *J. Appl. Microbiol.* 86:22–28.

Ripabelli, G., M.L. Sammarco, G.M. Grasso, I. Fanelli, A. Caprioli, and I. Luzzi. 1999. Occurrence of *Vibrio* and other pathogenic bacteria in *Mytilus galloprovincialis* (mussels) harvested from Adriatic Sea, Italy. *Intl. J. Food Microbiol.* 49:43–48.

Rippey, S.R. 1994. Infectious diseases associated with molluscan shellfish consumption. *Clin. Microbiol. Rev.* 7:419–425.

Rocourt, J., C. Jacquet, and A. Reilly. 2000. Epidemiology of human listeriosis and seafoods. *Intl. J. Food Microbiol.* 62:197–209.

Roderick, G.E., K.R. Schneider, F.A. Steslow, and N.J. Blake. 1987. Uptake, fate, and elimination by shellfish in a laboratory depuration system. *Proceedings Oceans 87* 5:1752–1756.

Ross, T., P. Dalgaard, S. Tienungoon. 2000. Predictive modeling of the growth and survival of *Listeria* in fishery products. *Intl. J. Food Microbiol.* 20:231–245.

Ruiz, J., L. Capitano, L. Nunez, D. Castro, J.M. Sierra, M. Hatha, J.J. Borrego, and J. Vila. 1999. Mechanisms of resistance to ampicillin, chloramphenicol and quinolones in multi-resistant *Salmonella typhimurium* strains isolated from fish. *J. Antimicrobial Chemotherapy* 43:699–702.

Ruttenber, A.J., B.G. Weniger, F. Sorvillo, B.A. Murray, and S.L. Ford. 1984. Diphyllobothriasis associated with salmon consumption in Pacific Coast states. *Am. J. Trop. Med. Hygiene* 33:455–459.

Rylander, L., U. Stromberg, and L. Hagmer. 2000. Lowered birth weight among infants born to women with a high intake of fish contaminated with persistent organochlorine compounds. *Chemosphere* 2000 40:1255–1256.

Sahagun, A.M., M.T. Teran, J.J. Garcia, M. Sierra, N. Fernandez, and M.J. Diez. 1997. Organochlorine pesticide-residues in rainbow trout, *Oncorhynchus mykis,* taken from four fish farms in Leon, Spain. *Bull. Environ. Toxicol.* 58:779–786.

Sakanari, J.A., and J.H. McKerrow. 1989. Anisakiasis. *Clin. Microbiol.* 2:278–284.

Santerre, C.R., R. Ingram, G.W. Lewis, J.T. Davis, L.G. Lane, R.M. Grodner, C.I. Wei, D.H. Bush, J. Shelton, E.G. Alley, and J.M. Hinshaw. 2000. Organochlorines, organophosphates, and pyrethroids in channel catfish, rainbow trout, and red swamp crayfish from aquaculture facilities. *J. Food Sci.* 65:231–235.

Santerre, C.R., P.B. Bush, D.H. Xu, G.W. Lewis, J.T. Davis, R.M. Grodner, R. Ingram, C. Wei, and J.M. Hinshaw. 2001. Metal residues in farm-raised channel catfish, rainbow trout and red swamp crayfish from the southern U.S. *J. Food Sci.* 66:270–273.

Schmidt, A.S., M.S. Bruun, I. Dalsgaard, and J.L. Larsen. 2001. Incidence, distribution, and spread of tetracycline resistance determinants and integron-associated antibiotic resistance genes among motile aeromonids from a fish farming environment. *Appl. Envion. Microbiol.* 67:5675–5682.

Scholz, E., U. Heinricy, and B. Flehmig. 1989. Acid stability of hepatitis A virus. *J. Gen. Virol.* 70:2481–2485.

Sericano, J.L., E.L. Atlas, T.L. Wade, and J.M. Brooks. 1990. NOAA's status and trends mussel watch program: chlorinated pesticides and PCBs in oysters (*Crassostrea virginica*) and sediments from the Gulf of Mexico, 1986–1987. *Marine Environ. Res.* 29:161–203.

Shapiro, R.L., S. Altekruse, L. Hutwagner, R. Bishop, R. Hammond, S. Wilson, B. Ray, S. Thompson, R.V. Tauxe, and P.M. Griffin. 1998. The role of Gulf Coast oysters harvested in warmer months in *Vibrio vulnificus* infections in the United States, 1988–1996. *Vibrio* Working Group. *J. Infect. Dis.* 178:752–759.

Shieh, Y., S.S. Monroe, R.L. Fankhauser, G.W. Langlois, W. Burkhardt 3rd, and R.S. Baric. 2000. Detection of Norwalk-like virus in shellfish implicated in illness. *J. Infect. Dis.* 181 Supplement 2:360–366.

Shoemaker, C.A, P.H. Klesius, and J.J. Evans. 2001. Prevalence of *Streptococcus iniae* in tilapia, hybrid striped bass, and channel catfish on commercial fish farms in the United States. *Am. J. Vet. Res.* 62:174–177.

Shpigel, M.A. Neori, D.M. Popper, and H. Gordin. 1993. A proposed model for "environmentally clean" land-based culture of fish, bivalves and seaweeds. *Aquaculture* 117:115–128.

Slifko, T.R., H.V. Smith, and J.B. Rose. 2000. Emerging parasite zoonoses associated with water and food. *Intl. J. Parasitol.* 30:1379–1393.

Sobsey, M.D., A.L. Davis, and V.A. Rullman. 1987. Persistence of hepatitis A virus and other viruses in depurated Eastern oysters. *Proceedings Oceans 87* 5:1740–1745.

Stegeman, J.J., and J.M. Teal. 1973. Accumulation, release and retention of petroleum hydrocarbons by the oyster *Crassostrea virginica. Marine Biol.* 22:37–44.

Steidinger, K.A. 1993. Some taxonomic and biologic aspects of toxic dinoflagellates. In *Algal Toxins in Seafood and Drinking Water.* I.R. Falconer, ed. San Diego: Academic Press, 75–86.

Steinberg, E.B., K.D. Greene, C.A. Bopp, D.N. Cameron, J.G. Wells, and E.D. Mintz. 2001. Cholera in the United States 1995–2000: trends at the end of the twentieth century. *J. Infect. Dis.* 184:799–802.

Strand, H.K., and V. Oiestad. 1997. Growth and the effect of grading, of turbot in a shallow raceway system. *Aquaculture Intl.* 5:397–406.

Stoskopf, M. 1993. Clinical physiology. In *Fish medicine.* M. Stoskopf, ed. Philadelphia: WB Saunders Co., 48–57.

Stuart, K.R., A.G. Eversole, and D.E. Brune. 2001. Filtration of green algae and cyanobacteria by freshwater mussels in a partitioned aquaculture system. *J. World Aquaculture Soc.* 32:105–111.

Sukontason, K, S. Piangjai, Y. Muangyimpong, K. Sukontason, R. Methanitikorn, and U. Chaithong. 1999. Prevalence of trematode metacercariae in cyprinoid fish of Ban Pao District, Chiang Mai Province, northern Thailand. *The Southeast Asian J. Trop. Med. Publ. Health* 30:365–370.

Sumaya, C.V. 1991. Major infectious diseases causing excess morbidity in the Hispanic population. *Arch. Intern. Med.* 151:1513–1520.

Szabo, P, F. Pekar, and J. Olah. 1993. Physical and chemical environment in a fish-cum-duck culture system in Hungary. In *From Discovery to Commercialization.* M. Carrillo, L. Dahle, J. Morales, P. Sorgeloos, N. Svennevig, and J. Wyban, eds. Belgium: Ostend, EAS no. 19, 271 pp.

Tamburrini, A., and E. Pozio. 1999. Long-term survival of *Cryptosporidium parvum* oocysts in seawater and in experimentally infected mussels (*Mytilus galloprovincialis*). *Intl J. Parasitol.* 29:711–715.

Tangtrongchitr, A., and R.B. Monzon. 1991. Eating habits associated with *Echinostoma malayanum* infections in the Philippines. *The Southeast Asian J. Trop. Med. and Publ. Health* 22 Supplement:212–216.

Taylor, S.L., L.S. Guthertz, M. Leatherwood, and E.R. Lieber. 1979. Histamine production by *Klebsiella pneumoniae* and an incident of scromboid fish poisoning. *Appl. Envion. Microbiol.* 37:274–278.

Telzak, E.E., E.P. Bell, D.A. Kautter, L. Crowell, L.D. Budnick, D.L. Morse, and S. Schultz. 1990. An international outbreak of type E botulism due to uneviscerated fish. *J. Infect. Dis.* 161:340–342.

Tham, T., H. Ericsson, S. Loncarevic, H. Unnerstad, M.L. Danielsson-Tham. 2000. Lessons from and outbreak of listeriosis related to vacuum-packed gravad and cold-smoked fish. *Intl. J. Food Microbiol.* 62:173–175.

Timmons, M.B., and Losorodo T. 1994. Aquaculture water reuse systems: engineering design and management. In *Developments in Aquaculture and Fisheries Science.* Amsterdam: Elsevier, 333 pp.

Ting, J., and A. Brown. 2001. Ciguatera poisoning: a global issue with common management problems. *Europ. J. Emergency Med.* 8:295–300.

Todd, E.C. 1982. Foodborne and waterborne disease in Canada—1977 annual summary. *J. Food Prot.* 45:865–873.

Torne, J., R. Miralles, S. Tomas, and P. Saballs. 1988. Typhoid fever and acute non-A non-B hepatitis after shellfish consumption. *Europ. J. Clin. Microbiol. Infect. Dis.* 7:581–582.

Tripuraneni, J., A. Koutsouris, L. Pestic, P. De Lanerolle, and G. Hecht. 1997. The toxin of diarrheic shellfish poisoning, okadaic acid, increases intestinal epithelial paracellular permeability. *Gastronenterology* 112:100–108.

Truman, B.I., H.P. Madore, M.A. Memegus, J.L. Nitzkin, and R. Dolin. 1987. Snow Mountain agent gastroenteritis from clams. *Am. J. Epidemiol.* 126:516–525.

Tsai, T.H., Y.C. Lin, S.R. Wann, W.R. Lin, S.J., Lee, H.H. Lin, Y.S. Chen, M.Y. Yen, and C.M. Yen. 2001. An outbreak of meningitis caused by *Angiostrongylus cantonensis* in Kaohsiung. J. Microbiol. Immunol. Infect. 34:50–56.

Unger, M.E., T.T. Chen, C.M. Murphy, M.M. Vestling, C.C. Fenselau, and G. Roesijadi. 1991. Primary structure of molluscan metallothionein deduced from PCR-amplified cDNA and mass spectrometry of purified proteins. Biochimica et Biophysica Acta 1074:371–377.

United States Food and Drug Administration. 1994. Definition of Seafood. FDA Federal Register Proposed Rule: To establish procedures for safe processing and importing of fish and fishery products.

United States Trout Farmers Association. 1994. Trout producer quality assurance program. Shepherdstown, WV: United States Trout Farmers Association. 28 pp.

Vanderzant, C., E. Mroz, and R. Nickelson. 1970. Microbial flora of a Gulf of Mexico and pond shrimp. J. Milk and Food Techn. 33:346.

Van Dolah, F.M. 2000. Diversity of marine and freshwater algal toxins. In Seafood and Freshwater Toxins; Pharmacology, Physiology, and Detection. L.M. Botana, ed. New York: Marcel Dekker Inc., 19–43.

Van Wagner, L.R., 1989. FDA takes action to combat seafood contamination. Food Protection 5:8–12.

Waikagul, J. 1998. Opisthorchis viverrini metacercaria in Thai freshwater fish. The Southeast Asian J. Trop. Med. Publ. Health 29:324–326.

Wakabayashi, H., and S. Egusa. 1973. Edwardsiella tarda (Paracoobactrum anguillimortiferum) associated with pond-cultured eel disease. Bull. of the Japanese Society of Scientific Fisheries 39:931–936.

Wang, W. 1980. Fractionation of sediment oxygen demand. Water Res. 14:603–612.

Warren, H. 1998. Quality assurance and food safety. Vet. Human Toxicol. 40(Suppl 2)34–36.

Weber, J.T., E.D. Mintz, R. Canizares, A. Semiglia, I. Gomez, R. Sempertegui, A. Davilla, K.D. Greene, N.D. Puhr, and D.N. Cameron. 1994. Epidemic cholera in Ecuador: multidrug-resistance and transmission by water and seafood. Epidemiol. and Infect. 112:1–11.

Wijkstrom, U., A. Gurny, and R. Grainger. 2000. Status of World Fisheries and Aquaculture. Fisheries Department, Food and Agriculture Organization, United Nations. Rome, Italy.

Wilson, I.G., and J.E. Moore. 1996. Presence of Salmonella spp. and Campylobacter spp. in shellfish. Epidemiol. Infect. 116:147–153.

Wittner, M., J.W. Turner, G. Jacquette, L.R. Ash, M.P. Salgo, and H.B. Tanowitz. 1989. Eustrongylidiasis—a parasitic infection acquired by eating sushi. N. Engl. J. Med. 320:1124–1126.

Wongsawad, C., A. Rojanapaibul, N. Mhad-arehin, A. Pachanawan, T. Marayong, S. Suwattanacoupt, J. Rojtinnakorn, P. Wongsawad, K. Kumchoo, and A. Nichapu. 2000. Metacercaria from freshwater fishes of Mae Sa stream, Chiang Mai, Thailand. The Southeast Asian J. Trop. Med. Publ. Health 31 Supplement: 1:54–57.

Wooten, R., and J.W. Smith. 1979. The occurrence of plerocercoids of Diphyllobothrium spp. in wild and cultured salmonids from the Loch Awe Area. Scottish Fisheries Res. Repts. 13:1–8.

Wyatt E.E., R. Nickelson II, and C. Vanderzant. 1979. Edwardsiella tarda in freshwater catfish and their environment. Appl. Environ. Microbiol. 38:710–714.

Yam, W.C., C.Y. Chan, S.W. Ho Bella, T.Y. Tam, C. Kueh, and T. Lee. 1999. Abundance of clinical enteric bacterial pathogens in coastal waters and shellfish. Water Res. 34:51–56.

Yang, H. 1994. Integrated Fish Farming. In Freshwater Fish Culture in China: Principles and Practice. S. Li and J. Mathias, eds. Amsterdam: Elsevier, 219–270.

Yang, M.S., S.T. Chiu, and M.H. Wong. 1995. Uptake, depuration, and subcellular distribution of cadmium in various tissues of Perna viridis. Biomed. Environ. Sci. 8:176–185.

Yao, G. 1991. Clinical spectrum and natural history of viral hepatitis A in the 1988 Shanghai epidemic. In Viral Hepatitis and Liver Disease. F.B. Hollinger, S.M. Oemon, and H.S. Margolis, eds. Baltimore: Williams and Wilkins, 76 pp.

Yu, S.-H. 1989. Drinking water and primary liver cancer. In Primary Liver Cancer. Z.Y. Tang, M.C. Wu, and S.S. Xia, eds. Berlin: Springer-Verlag, 30–37.

Zapatka, F.A, and B. Bartolomeo. 1973. Microbiological evaluation of cold-water shrimp (Pandalus borealis). Appl. Microbiol. 25:858–861.

38

ANTIMICROBIAL RESIDUES AND RESIDUE DETECTION IN MILK AND DAIRY PRODUCTS

SHEILA M. ANDREW

INTRODUCTION. The production of high-quality animal-based foods for human consumption is a priority for agricultural producers and is essential for maintaining the safety of the human food supply. Avoiding antibiotic residues in milk and dairy products is a fundamental aspect of quality food production. Violative concentrations of antibiotics in commingled milk affect its marketability and will result in regulatory actions, which may include financial penalties and possible suspension of license for food production. In addition, there is the potential for antibiotic residues in animal-based foods to have a negative impact on human health.

Dairy and livestock producers need effective and reliable means to reduce the risk of antibiotic contamination of milk. Residue detection methodologies used by regulatory agencies are targeted at identifying residues in commingled milk at the processing level. Testing at this level provides assurances for maintaining human food safety, but does not directly address the initial source of residues from antibiotic-treated cows. Implementing on-farm management practices and effective monitoring for residues has been shown to reduce the risk of antibiotic residue contamination of commingled milk (McEwen et al. 1991). Therefore, it is imperative that quality control programs be established to promote food safety beginning at the farm and monitored on an individual animal basis.

ANTIBIOTIC USE IN FOOD-PRODUCING ANIMALS.
Antibiotics are used in food-producing animals to eliminate or control bacterial infections for all livestock species and to improve growth and feed efficiency in several classes of livestock. A comprehensive review of the history of antibiotic use in animal agriculture has been published (Mitchell et al. 1998). Briefly, animals have been treated with antibiotics since 1948, following the initial use of these antibiotics in humans. The classes of antibiotics used for livestock have evolved as new antibiotics have been identified and have followed the same patterns as antibiotics used in human health. From a food-animal practitioner survey, the most common classes of antibiotics used therapeutically are beta-lactams, including penicillin G, ceftiofur sodium, cloxacillin, cephapirin, and ampicillin (Sundlof et al. 1995). Although antibiotics are used to treat a variety of infections, mastitis is the most common reason for antibiotic use in dairy cattle (Sundlof et al. 1995).

The U.S. Food and Drug Administration (FDA), Center for Veterinary Medicine, evaluates and establishes the appropriate dosage and withholding time for milk discard and slaughter for a particular antibiotic. The submitted drug applications are reviewed and evaluated for safety and effectiveness of the new animal drug, effects of the drug on the environment, and manufacturing methodologies. A primary responsibility of the FDA is to evaluate the safety of new animal drugs because a result of using these drugs in animals is that they can contaminate foods for humans. Thus, the lowest effective dosage is approved in order to prevent residues that can negatively impact human health. The impact of an antibiotic residue on human health must be evaluated, and a no-effect level of antibiotic is established as a tolerance or safe concentration that is allowable in edible animal tissues and milk. Currently, there is a draft FDA document to include the assessment of the effect of antimicrobial drug residues from animal-based foods on the human gastrointestinal tract microflora prior to approval and to assess currently approved animal drugs (USFDA 2001a). Application of these guidelines will advance the knowledge of the impact of drug residues and may affect the classes of antibiotics that will be available for animal use in the future.

REGULATION OF ANTIBIOTIC RESIDUES AND RESIDUE VIOLATIONS IN MILK.
The Center for Food Safety and Applied Nutrition of the FDA is responsible for monitoring milk, seafood, and other foods for residues and pathogens. Monitoring and inspection begins at the initial commingling of farm milk, which is at the bulk tank level and continues through the retail sale of milk and dairy products. Milk safety regulations and enforcement are provided by the National Conference on Interstate Milk Shipments, a collaborative program of the State Public Health Service, the FDA, state, and industry representatives for certification of interstate milk shippers (Smucker 1997). From this program, the Pasteurized Milk Ordinance (PMO) document is produced and provides def-

initions of milk and milk products and methods to assure milk quality and human food safety. This guidance document is the basis for regulations that are enforced within each state. Within the PMO, Appendix N focuses on testing for beta-lactam residues, which are the most common class of antimicrobial agents used in the dairy industry (Smucker 1996). Under Appendix N, all bulk milk tankers are tested for beta-lactam residues using qualitative, rapid screening tests. In addition, the program includes random testing for other antimicrobial agents and contaminants.

A National Milk Drug Residue Data Base was established in 1993 and is an annual summary of the results of the mandatory and voluntary testing of milk products for antibiotic residues. A summary of this milk monitoring program identified that an average 0.09 percent of all tanker trucks tested were positive for beta-lactam residue from 1994–1996 (Smucker 1997). More recently, 0.08 percent of tanker loads were identified as positive for beta-lactam antibics in 2000 (USFDA 2001a). Although the rate of residue violations was very low, beta-lactams were the most common antibiotic residue class detected, at 95.1 percent of all antibiotic residues identified. Continued monitoring of milk for antibiotic residues is an important component of human food safety in providing assurances of high-quality milk production for human food consumption.

RISK FACTORS ASSOCIATED WITH THE OCCURRENCE OF ANTIBIOTIC RESIDUES IN MILK.
Several studies and residue-reporting programs have investigated the causes of violative antibiotic residues in milk. Understanding the risks associated with residues is an important step for residue avoidance.

Several studies have indicated that unintentional milking of an antibiotic-treated cow or a cow treated with a dry-cow intramammary antibiotic preparation at the end of lactation can cause violative residues in the bulk or tanker load of milk (Kaneene and Ahl 1987; Sischo et al. 1997). Antibiotics used in an extra-label manner also may increase the risk of residues unless extended withholding times are established. Reduced clearance rates by cows that may have reduced milk production because of infection may also increase risks. In contrast, there was a low risk of antibiotic residues in milk from cows treated with oxytetracycline externally for treatment of papillomatous digital dermatitis (Britt et al. 1999).

In a Michigan farm survey, insufficient knowledge of drug withholding times, lack of identifying treated animals, and errors caused by part-time help were all associated with increased risk of antibiotic residues (Kaneene and Ahl 1987). Using a risk assessment tool, Sischo and coworkers (1997) identified that a lack of appropriate treatment records was the primary risk factor for violative residues in milk. This was followed by a lack of understanding of how to use the antibiotic and a lack of an adequate client/patient relationship with the veterinarian.

Several management practices on dairy farms may increase the risk of antibiotic residues. There is a potential for antibiotic residues in calves if they are marketed within twenty-four days of consuming milk containing residues of penicillin G and amoxicillin (Musser et al. 2001). Method of administration of antibiotic can also affect the clearance rate. Residues of penicillin G in milk were detectable for 132 hours with a subcutaneous administration, compared to 96 hours for intramuscular administration (Dubreuil et al. 2001).

Elevated somatic cell counts (SCC) in bulk tank milk are indicative of intramammary infection within a herd (Deluyker et al. 1993). This increase in SCC has been correlated with an increase in intramammary antibiotic use on the farm and ultimately a greater risk for antibiotic residues in bulk milk. In a longitudinal study, an increase in antibiotic residues in bulk tanks from 10,568 farms in the United States and Canada was positively associated with elevated SCC (Saville et al. 2000). This finding suggests that compromised udder health may be related to an increased use of antibiotics and, therefore, increased residue risk. In agreement, the relative risk of antibiotic violations in bulk tank milk increased from 1.0 for SCC of less than 250,000/ml to 4.33 for SCC of 551,000 to 700,000/ml in bulk tanks sampled from Wisconsin dairy farms over a four-year period (Ruegg and Tabone 2000).

Implementing a residue-prevention program that includes maintaining appropriate records reduces the risk of antibiotics from animal-based foods (McEwen et al. 1991; Sischo et al. 1997). One aspect of reducing the risk of antibiotic contamination is screening for antibiotic residues at the farm. The use of antibiotic residue screening tests and implementation of good management practices on dairy farms have been positively correlated with reductions in the occurrence of antibiotic residues in milk (McEwen et al. 1991). Likewise, Sischo et al. (1997) reported that the use of antibiotic residue screening tests for evaluating individual cow's milk was associated with a reduction in the risk of residue violations.

THE IMPACT OF ANTIBIOTIC RESIDUES FROM ANIMAL-BASED FOODS ON HUMAN HEALTH.
Antibiotic residues from animal-based foods can potentially impact microflora in the human gastrointestinal tract. Research has indicated that several potential human health hazards are associated with antibiotic residues in animal-based foods (Finegold et al. 1983; Gorbach 1993). The primary areas of concern are alterations in microbial functions, populations, and antimicrobial resistance patterns that may develop when human intestinal microflora are exposed to antibiotic residues from consumed meat and milk (USFDA 2001b). Also, there is a possibility for a reduction of the barrier effect that prevents overgrowth of a single species of bacteria within the intestine

(USFDA 2001b). In the 1960s, therapeutic dosages of antibiotics were identified as risk factors for the growth of antibiotic-resistant bacteria (Goldberg et al. 1961). There is evidence that, depending on the dosage and duration of treatment, subtherapeutic concentrations of antibiotics can alter host microflora. However, the significance and long-term effect of these alterations on the functioning of the gastrointestinal tract is unknown.

Antibiotics present in subtherapeutic concentrations have been shown to alter microbial populations in both *in vitro* and *in vivo* systems. Using the human flora–associated rodent model (a rodent implanted with human fecal flora), low concentrations of antibiotics are shown to shift the patterns of gut microbes (Finegold et al. 1983). These changes in populations may also result in alterations in the metabolic activity within the lumen of the digestive tract (USFDA 1999). Biotransformations by microbes that normally detoxify compounds may be reduced in the presence of residue levels of antibiotics (USFDA 1999). However, no direct relationship has been established at this time.

It is known that therapeutic concentrations of orally administered antibiotics have temporarily increased the growth of resistant bacteria for several classes of antibiotics (Nord et al. 1984). Subtherapeutic dosages as low as 2 mg/l of oxytetracycline administered over seven days resulted in increased drug-resistant *Enterobacteriaceae* (Corpet et al. 1996). In an *in vitro* system, several antibiotics present in concentrations considered below the tolerance level for milk resulted in an increase in resistant *Staphylococcus aureus* with a fourteen-day exposure to the antibiotics (Brady et al. 1993). However, it is not known whether the same resistance factors would be conferred in an *in vivo* system. In contrast, Elder et al. (1993) compared the frequency of antibiotic-resistant bacteria in human stool microflora between vegetarians and nonvegetarians over a twelve-month period, and no treatment difference in prevalence of resistant organisms was found.

Another concern for antibiotic residues is the potential for an allergenic immune response by sensitive individuals to an antibiotic that may be present as a residue in consumed animal tissue and milk (Huber 1986). However, it is theorized that it is unlikely that antibiotic residues would be immunogenic, because of the low dose, oral route, and the rarity of reported cases duringtwenty-five years (Dowdney at al. 1991).

Although the research does not conclusively implicate a link between antibiotic residues in animal-based foods and antibiotic resistance in human microflora, continued vigilance is warranted. In addition, the goal of reducing antibiotic residues in the human food supply is vital to ensure high-quality food.

DETECTION OF RESIDUES IN MILK. Currently, rapid screening tests are available for the detection of antibiotics in milk. These are qualitative tests used by the dairy industry and government regulators to screen bulk tank/tanker truck milk for antibiotic residues. Several of these screening tests have application for use on the farm to test milk from individual cows for antibiotic residues. Testing milk from antibiotic-treated cows, following an appropriate milk-withholding period, allows the dairy producer to make informed decisions before adding a treated cow's milk to the bulk tank. The use of these rapid screening tests is an integral part of the 10-point Milk and Dairy Beef Quality Assurance Program (Boeckman and Carlson 2001), and the proper use of these tests can assist in maintaining the quality of milk entering the human food supply.

The value of these rapid screening tests depends on the accuracy of the tests to detect a truly positive, violative milk sample (that is, antibiotic present in milk above the tolerance or safe level) with a minimum of false-positive outcomes. However, the false-positive rate of screening tests used for milk from individual cows has ranged from 0 percent to 83 percent depending on the study and the screening test evaluated (Bishop et al. 1985; Cullor et al. 1994; Andrew et al. 1997; Hillerton et al. 1999; Andrew 2000, 2001). A false-positive outcome for a screening test erroneously indicates the presence of an antibiotic in milk at violative concentrations either from untreated cows or in milk that contains an antibiotic at concentrations below the tolerance/safe level. Therefore, this milk would be discarded even though it is legal to sell, which can be a significant loss to the producer.

It is necessary to understand the factors that may negatively impact screening-test performance in order to identify the most accurate and appropriate tests to use to ensure human food safety starting at the farm level.

PERFORMANCE OF RAPID SCREENING TESTS FOR MILK FROM INDIVIDUAL COWS. Although the screening tests were initially designed for use with commingled milk, several tests are used on the farm for individual cow milk. Also, milk processors and milk handlers may offer their customers the service of testing individual cow milk for antibiotic residues. Therefore, the majority of the tests used on the farm are the same tests that have been evaluated for bulk tank milk. This review of screening-test performance includes the beta-lactam screening tests that are commercially available and commonly used for individual cow milk (Table 38.1).

Table 38.2 shows the results of seven studies that evaluated the false-positive rate of beta-lactam screening tests using milk from individual cows. Specificity rates (defined as the rate of truly negative samples that are found by the assay to be negative) were calculated for each of the tests. The specificity rate was calculated as the number of truly negative test outcomes divided by the total number of samples analyzed. The specificity rate can be converted to the false positive rate by subtracting the specificity rate from 1.0.

In several studies, milk was sampled from cows with

Table 38.1—The Analytical Principles for Rapid Screening Tests for the Detection of Beta-lactam Residues in Milk

Screening test	Analytical Principle
Charm BsDa[1]	Bacterial growth inhibition
Delvotest-P	Bacterial growth inhibition
Charm Farm	Bacterial growth inhibition
LacTek	ELISA competitive enzyme immunoassay
Penzyme	Enzymatic
CITE Snap	Antibiotic-antigen capture system
Charm II	Competitive-binding penicillin displacement

[1]BsDa - Bacillus stearothermophilus Disc Assay.

Table 38.2—A Summary of Specificity Rates for Beta-lactam Screening Tests Using Milk from Untreated Cows or Cows Not Treated with an Antibiotic for At Least Twenty-One Days

Experiment[1]	1		2		3	4	5	6		7	
Trial[2]	a.	b.	a.	b.				a.	b.	a.	b.
n	148	239	51	95	131	90	144	134	144	25	25
					Specificity rate						
BsDa	.95	–	.88	.93	1.0	–	–	–	–	–	
Delvotest-P	.75	.85	.78	.94	.97	.99	.88	1.0	.89	.88	.92
Charm Farm	–	.34	–	–	.97	–	–	–	–		
LacTek	.97	1.0	1.0	.94	.98	–	–	–			
Penzyme	–	–	–	.91	.61	–	–	–	.60	1.0	
Cite Snap	–	–	–	–	.96	–	–	–	.16	.64	
Charm II	–	–	–	–1.0	.99	–	–	–	.28	.60	

[1]Experiment 1—Cullor et al. 1994.
  [2]a. Milk sampled at twenty-one days after antibiotic treatment for clinical mastitis.
  [2]b. Milk sampled from bucket milk from cows with visually normal milk.
[1]Experiment 2—Van Eenennaam et al. 1993.
  [2]a. Milk sampled at four days after oxytocin treatment for clinical mastitis.
  [2]b. Milk sampled at twenty-one days after antibiotic or oxytocin treatment for clinical mastitis.
[1]Experiment 3—Andrew et al. 1997.
[1]Experiment 4—Andrew 2000.
[1]Experiment 5—Angelidis et al. 1999.
[1]Experiment 6—Hillerton, et al. 1999.
  [2]a. Milk from mid-lactation cows.
  [2]b. Milk from cows at four to six days postpartum.
[1]Experiment 7—Andrew et al. 2001.
  [2]a. Colostrum
  [2]b. Transition milk
n Number of samples tested
–Screening test not evaluated or unknown sample size or detection level.

clinical mastitis and visibly abnormal milk prior to antibiotic treatment for evaluation of screening tests (Cullor et al. 1994; Van Eenennaam et al. 1993). Somatic cell counts in milk from cows with clinical mastitis are generally markedly elevated (Kitchen 1981) and, under field conditions, milk would not be tested for the presence of antibiotics prior to treatment. Also, it is assumed that milk added to the bulk tank would be from cows that have visibly normal milk from a visibly normal udder. Visibly normal milk is defined here as milk that is free from clots, flakes, strings, clumps, is not discolored from blood, or is not of a serum consistency (Andrew et al. 1997). This summary includes only those studies that evaluated screening tests using milk that would be representative of milk that would be tested for antibiotics by dairy producers, and does not include data from studies using milk from clinical mastitis cases prior to drug therapy and from visibly abnormal milk.

Specificity rates varied considerably across studies and antibiotic residue screening tests (Table 38.2). Overall, the specificity rates ranged from .16 to 1.0 (no false-positive outcomes). The lowest specificity rates were observed for screening tests based on enzymatic or binding assay formats. Also, reduced specificity rates were noted when colostrum and transition milk were tested. It is possible that other components in milk, colostrum, and transition milk may interfere with the assay, particularly if there is a cross reactivity of proteins in milk for active sites on the enzyme or binding protein. However, many of the screening tests based on receptor binding assay formats showed no false positives when testing visibly normal milk (Table 38.2).

The specificity rates of screening tests based on bacterial inhibition ranged from .75 to 1.0 with a specificity rate of greater than .90 in seven of the fifteen trials. These screening tests may be affected by any component that inhibits bacterial growth. Such a component could include inhibitory substances in milk (Carlsson et al. 1989) or sanitizing agents that may contaminate the test. Prior heat treatment and enzymatic treatment of milk samples can inactivate inhibitors other than antibiotics and can improve test performance for tests based on bacterial inhibition. Hillerton and coworkers (1999) determined that heat treatment and penicillinase enzyme treatment of presumptive positive milk samples improved the ability to identify milk with violative concentrations of antibiotics using the Delvotest.

The screening tests summarized represent a wide range in analytical principles. Except for results from a few trials, the antibiotic screening tests evaluated in this study can be used to detect residues without significant losses of saleable milk caused by false-positive outcomes.

FACTORS AFFECTING FALSE-POSITIVE OUTCOMES.

An examination of factors affecting the false-positive outcomes can define the limitations of the screening tests and provide data to support improvements of test performance. Factors affecting the outcome of the screening tests can be divided into two categories: those associated with the operation of the test and those associated with the particular milk tested.

The screening tests are designed to render a positive result if the test procedure fails for any reason. Test storage conditions, including improper environmental temperature, will result in false-positive outcomes. Not following directions, such as using an inadequate or excessive quantity of milk, failure to use all reagents, and changing incubation times will all result in a false-positive reading. These causes for test failure can be minimized through effective education. Training and information transfer concerning the proper use of the screening tests on the farm is the responsibility of the manufacturer of the test, the veterinarian, and extension personnel.

Factors Associated with the Milk to Be Tested. The milk that is sampled for analysis of antimicrobial residues can differ significantly in milk composition and SCC. Also, during the milking process, milk composition fluctuates from the initiation to the end of milking an individual cow. Therefore, these variations in milk composition and other factors affect the reliability of the antimicrobial residue screening tests.

SOMATIC CELLS. Mastitis is the most common reason for antibiotic use on the dairy farm; therefore, testing milk following mastitis treatment would be the most frequent use for the residue screening tests (Reneau

and Packard 1991). Somatic cell counts can increase to greater than 1,000,000/ml in milk during intramammary infection, and because of the inflammatory response, components of plasma are found in higher concentrations in milk from mastitic cows than in that of normal milk (Kitchen 1981). These components may remain elevated after treatment and, depending on the analytical principle of the screening test, these milk components may impact test performance (Eagan and Meaney 1984). The effect of SCC on false-positive rates was not conclusive, as is shown in the summary of studies in Table 38.3. In two studies where SCC were elevated (>600,000/ml), SCC did not affect the false-positive rate, except for the BsDa screening test, in which SCC average was greater than 2,500,000/ml. In contrast, elevated SCC were associated with increased false-positive rates for the CITE Snap test used to evaluate colostrum and transition milk in early lactation. There was an effect of SCC on the false positive rate for the Charm Farm and Delvotest-P, in which SCC averaged less than 200,000/ml (Sischo and Burns 1993). For these screening tests, an increase in SCC was associated with an increase in the rate of false-positive outcomes. In a study to evaluate the specificity of the Delvotest for use with milk from individual cows, SCC in milk from cows in mid-lactation did not affect test performance (Hillerton et al. 1999). However, sampling milk from cows in the fourth to sixth milking postpartum resulted in an 11.1 percent false-positive rate (Table 38.3). Thirteen of the sixteen false-positive results were from milk that contained greater than 4,000,000 SCC/ml. It was suggested that other inhibitors may be present in milk at this time, but heat treatment resulted in only one sample becoming negative. Thus, there may be other heat resistant factors present in milk from cows that have recently calved.

The relationship between SCC and false-positive rates is not clearly established. Carlsson et al. (1989) observed a positive association between increased false-positive outcomes for the Delvotest-P and an increase in lactoferrin and lysozyme in milk. These components are found in higher concentrations in milk from cows with mastitis (Kitchen 1981); therefore, the effect of somatic cells on the false-positive rates observed in Table 38.3 may be indirect. Further work is needed to establish whether a direct relationship exists between milk quality and test performance and, if so, whether the effects can be minimized in order to improve the accuracy of the rapid screening tests.

FAT AND PROTEIN CONTENT OF MILK. The fat and protein content of milk varies by breed, stage of lactation, and intramammary infection status (Laben 1963; Kitchen 1981). These components may influence the residue screening test outcomes when used with individual cow milk. Andrew (2000) determined that there was an increase in the probability of false-positive outcomes with an increase in milk fat content for the CITE Snap test. In the same study, for the Penzyme test, the rate of false-positive responses increased with

Table 38.3. Effect of Somatic Cell counts (SCC) on Rate of False-Positive Outcomes for Beta-Lactam Tests Used for Milk from Untreated Individual Cows or Cows Not Treated with an Antibiotic for at Least Twenty-One Days

Experiment[1]	Trial[2]	SCC (x1,000/ml) Geometric mean	Screening Test					
			Delvotest-P	LacTek	BsDa	Charm Farm	CITE Snap	Penzyme
			Number of false-positive outcomes/total number of milk samples tested					
1	a	≤600	2/28	2/28	2/28	–	–	–
	b	>600	3/53	2/53	2/53	–	–	–
2		>2,500	6/95	6/95	7/95*	–	–	–
3		235	NS	–	–	–	4/90*	NS
4		<2,819	13/199*	NS	NS	25/199*	–	–
5	a	–	0/134	–	–	–	–	–
	b	–	16/144*	–	–	–	–	–
6	a	2,458	3/25	–	–	–	10/25	21/25*
	b	866	2/25	–	–	–	0/25	9/25*

[1]Experiment 1—Cullor et al. 1994.
  [2]a. Milk sampled at twenty-one days after antibiotic treatment for clinical mastitis.
  [2]b. Milk sampled from bucket milk from cows with visually normal milk.
[1]Experiment 2—Van Eenennaam et al. 1993; Sampled at twenty-one days after antibiotic or oxytocin treatment for clinical mastitis.
[1]Experiment 3—Andrew 2000.
[1]Experiment 4—Sischo and Burns 1993.
[1]Experiment 5—Hillerton et al. 1999.
  [2]a. Milk sampled from mid-lactation cows.
  [2]b. Milk sampled from cows at four to six days postpartum.
[1]Experiment 6—Andrew 2001.
  [2]a. Colostrum.
  [2]b. Transition milk.
– = Screening test not evaluated.
* = Significant effect of SCC on false-positive outcomes (P<.05).
NS = No significant effect of SCC.

increased protein content in milk from Jersey and Holstein cows. Because the Penzyme test is an enzymatic reaction, proteins may interfere with the test performance. These tests should be evaluated with milk from cows that vary in milk protein and fat content in order to determine the extent of the effect of these components on test performance.

METHOD OF SAMPLING MILK. In many studies evaluating screening test performance, foremilk is sampled and analyzed (Cullor et al. 1994; Van Eenennaam et al. 1993). Generally, foremilk is higher in SCC compared to the total composite milk produced by the cow (Ostensson et al. 1988; Paape and Tucker 1966). Using foremilk for antibiotic residue screening may result in a higher false-positive rate than would be expected if total composite milk were tested (Norell et al. 1994). Norell et al. (1994) observed an increase in false-positive rates for several screening tests when foremilk was compared to total composite milk in cows recovering from clinical mastitis. Possibly, significant differences between foremilk and total composite milk are apparent only at higher levels of SCC. Likewise, Hillerton and coworkers (1999) observed a higher rate of false-positive outcomes for quarter milk compared to testing whole udder milk samples. Kang and Kondo (2001)

determined that thirteen of 321 quarter-milk samples and nine of 207 whole-udder samples resulted in false positives using the Delvotest SP assay and did not differ by sample analyzed.

OTHER FACTORS. Several other factors associated with milk can negatively impact test performance. Lactoferrin and lysozyme, present in milk, have antimicrobial properties that may affect the performance of antibiotic residue screening tests that are based on bacterial inhibition (Carlsson et al. 1989). These compounds are elevated in milk from cows with mastitis (Kitchen 1981). The addition of 1 mg/ml lactoferrin or 20–40 percent bovine plasma resulted in increased rates of false-positive outcomes when milk was tested using the Delvo-X-Press (Angelidis et al. 1999). These concentrations are greater than what would be expected in milk from infected mammary glands (Kitchen 1981). However, these results demonstrate the potential effects of other inhibitors present in milk on screening test performance.

Also, residue screening tests can be used for milk from heifers that were treated prepartum with an antibiotic (Oliver et al. 1992). Residues have been detected up to seven days postpartum. Also, elevated concentrations of SCC, immunoglobulins and protein that are typically found in colostrum were associated with false-positive outcomes for several screening tests (Andrew 2001).

Determination of the specific factors that affect false-positive outcomes for the antibiotic screening tests may provide information that can be used to improve test performance for screening milk from individual cows.

LIMITATIONS TO SCREENING MILK FROM INDIVIDUAL COWS.

The sensitivity and specificity of a test provide estimates of the ability of the test to identify milk containing violative concentrations of an antibiotic. However, the predictive value of a test can provide a method to compare screening tests and is the most useful indicator of test performance (Sischo 1996). A positive predictive value is the probability that the positive response of the screening test is associated with the presence of an antibiotic. However, this is difficult to measure because in addition to knowing the sensitivity and specificity of a test, the overall probability of the event in the population is needed. To improve the use of a test, a screening test is most appropriate to use after antibiotic treatment and an appropriate withholding time, which would maximize the predictive value (Sischo 1996).

Ideally, a quantitative measure of antibiotic in milk should accompany the qualitative screening test evaluation of false-positive outcomes (Anderson et al. 1996). This would provide data to support a more complete diagnostic evaluation of the screening tests. However, several screening tests have a lower level of detection than the quantitative assays, including HPLC analysis. Riediker et al. (2001) analyzed eighteen presumptively positive raw milk samples using quantitative techniques and found sixteen to contain at least trace amounts of antibiotic residues that were below the level of detection for the quantitative assay. Only five of the positives were at concentrations above the established safe or tolerance level. In addition, the presence of more than one residue in a sample may impact test accuracy. Even though significant variability in test performance was noted, 75 percent of the screening tests evaluated across the studies had specificity rates above .88; therefore, these screening tests can be useful in a program to reduce residue contamination of commingled milk.

OPTIMIZING THE USE OF SCREENING TESTS ON THE FARM.

Results of a survey indicate that dairy producers who utilize screening tests to evaluate milk had lower incidence of violative residues in milk (McEwen et al. 1991). Thus, it appears that the screening tests can be useful on the farm. Although many of the screening tests used on the farm may have some limitations because of false-positive responses, steps can be taken to improve the reliability of the screening tests used for milk from individual cows. The variability in the quality of milk sampled contributes to the inconsistent test results when screening tests are used on-farm to test milk from cows recovering from mastitis. One way to improve test performance is to screen milk that is visibly normal from cows recovering from mastitis (Sischo and Burns 1993; Norell et al. 1994; Andrew et al. 1997).

Adopting a well-planned mastitis control program is the most effective means of reducing antibiotic use and the potential for antibiotic residues in milk. An effective program includes developing a valid veterinarian-client-patient relationship to assure that antibiotic use is appropriate, particularly for drugs used in an extra-label manner (Boeckman and Carlson 2001).

FUTURE PLANS FOR REDUCING RISKS OF ANTIBIOTIC CONTAMINATION OF MEAT AND MILK.

The future of antibiotic residue detection is dependent on many factors. There will always be a need to test for antibiotic and other residues in food-animal tissues and milk. However, in the future, there may be less reliance on antibiotic therapy and prophylactic treatments because of improved management practices and possible human health concerns associated with antibiotic use in animal agriculture. With this goal in mind, the USDA is embarking on a Food Safety Initiative with the objective of maintaining human food safety from farm-to-table (USFDA 2000). The goal is to implement a program to reduce risks of microbial pathogens and antibiotic and other residues. One such program is the Hazard Analysis Critical Control Point (HACCP) program that has been in effect at processing facilities and is now being modified to use at the farm level (Sischo 1996). This program has provided an effective tool to identify and reduce risk factors at each step of the processing procedures. Implementing a program that focuses on risk factors for human health starting at the farm level will strengthen the ability to produce wholesome food and maintain consumer confidence in animal-based foods.

Another area of research is to develop novel nontraditional methods for treating infections. The antibiotic, nisin, has antibacterial activity against Gram-positive bacteria and has been used in the food industry (Breukink and de Kruijff 1999). It has the classification as "Generally Regarded As Safe" by the US FDA and, therefore is not subject to human health regulations. This product has potential to be used in treating Gram-positive infections in animals.

Improving the sensitivity of screening tests and developing multiresidue analytical detection abilities are presently being investigated (Baxter et al. 2001; Medina 1997). Biosensors are an example of one newer technology that is being developed and evaluated to use as a rapid and quantitative assay for several antibiotics simultaneously (Baxter et al. 2001). Biosensors are biomolecule analogs of the analytes to be detected. These analogs are covalently bound to an inert surface and, based on competitive binding, can quantitate an antibiotic present in very low concentrations. This new technology has the potential to improve the ability to detect violative residues and provide information for more appropriate regulatory actions.

SUMMARY AND CONCLUSIONS. The goal of agriculture is to provide a high-quality, safe food supply. The use of antibiotics to eliminate infections in food-producing animals is an integral component of animal agriculture; therefore, effective monitoring for antibiotic residues is necessary to ensure food safety. Rapid screening tests used for detection of violative levels of beta-lactam residues in milk reduce the risk of contamination of the milk supply. Testing individual cow milk after antibiotic treatment and an appropriate milk withholding time provide the dairy producer with additional information to support the decision to add a cow's milk to the bulk tank. However, there is the potential for false-positive outcomes when the tests are used according to label directions. Several factors, such as somatic cells, milk fat, and protein content have been associated with increased rates of false-positive outcomes. In many studies, the performance of tests commonly used on the farm was acceptable. However, the screening tests are qualitative, and results should be interpreted with care. Development of more accurate antibiotic residue screening tests combined with implementation of risk-reduction programs at the farm level will provide additional assurances for maintaining human food safety.

REFERENCES

Anderson, K.L., W.A. Moats, J.E. Rushing, D.P. Wesen, and M.G. Papich. 1996. Ampicillin and amoxicillin residue detection in milk, using microbial receptor assay and liquid chromatography methods, after extra-label administration of the drugs to lactating cows. *Am. J. Vet. Res.* 57:73–78.

Andrew, S.M. 2001. Effect of composition of colostrum and transition milk from Holstein heifers on specificity rates of antibiotic residue tests. *J. Dairy Sci.* 84:100–106.

Andrew, S.M. 2000. Effect of fat and protein content of milk from individual cows on the specificity rates of antibiotic residue screening tests. *J. Dairy Sci.* 83:2992–2997.

Andrew, S.M., R.A. Frobish, M.J. Paape, and L.J. Maturin. 1997. Evaluation of selected antibiotic residue screening tests for milk from individual cows and examination of factors that affect the probability of false positive outcomes. *J. Dairy Sci.* 80:3050–3057.

Angelidis, A.S., T.B. Farver, and J.S. Cullor. 1999. Evaluation of the Delvo-X-Press assay for detecting antibiotic residues in milk samples from individual cows. *J. Food Prot.* 62:1183–1190.

Baxter, G.A., J.P. Ferguson, M.C. O'Connor, and C.T. Elliot. 2001. Detection of streptomycin residues in whole milk using an optical immunobiosensor. *J. Agric. Food Chem.* 49:3204–3207.

Bishop, J.R., A.B. Bodine, G.D. O'Dell, and J.J. Janzen. 1985. Quantitative assay for antibiotics used commonly in treatment of bovine infections. *J. Dairy Sci.* 68:3031–3036.

Boeckman, S., and K.R. Carlson. 2001. *Milk and Dairy Beef Residue Prevention Protocol*. Stratford, IA: Agri-Education, Inc..

Brady, M.S., N. White, and S. Katz. 1993. Resistance development potential of antibiotic/antimicrobial residue levels designated as "safe levels". *J. Food Prot.* 56:229–233.

Britt, J.S., M.C. Carson, J.D. Von Bredow, and R.J. Condon. 1999. Antibiotic residues in milk samples obtained from cows after treatment for papillomatous digital dermatitis. *J.A.V.M.A.* 215:833–836.

Breukink, E., and B. de Kruijff. 1999. The lantobiotic nisin, a special case or not. *Biochim. et Biophys. Acta.* 1462:223–234.

Carlsson, A., L. Bjorck, and K. Persson. 1989. Lactoferrin and lysozyme in milk during acute mastitis and their inhibitory effect in Delvotest P. *J. Dairy Sci.* 72:3166–3175.

Corpet, D.E., and H.B. Brugere. 1996. Antimicrobial drug residues in Foods: Evaluation of the no effect on human microflora. Recent Res. Devel. *Antimicrob. Agents and Chemother.* 1:113–123.

Cullor, J.S., A. Van Eenennaam, I. Gardner, L. Perani, J. Dellinger, W.L. Smith, T. Thompson, M.A. Payne, L. Jensen, and W.M. Guterbock. 1994. Performance of various tests used to screen antibiotic residues in milk samples from individual animals. *J.A.O.A.C. Internatl.* 77:862–870.

Deluyker, H.A., J.M. Gay, and L.D. Weaver. 1993. Interrelations of somatic cell count, mastitis, and milk yield in a low somatic cell count herd. *J. Dairy Sci.* 76:3445–3452.

Dowdney, J.M., L. Maes, J.P. Raynaud, F. Blanc, J.P. Scheid, T. Jackson, S. Lens, C. Verschueren. 1991. Risk assessment of antibiotic residues of beta-lactams and macrolides in food products with regard to their immunoallergenic potential. *Food Chem. Toxicol.* 29:477–483.

Dubreuil, P., J. Daigneault, Y. Couture, P. Guay, and D. Landry. 2001. Penicillin concentrations in serum, milk, and urine following intramuscular and subcutaneous administration of increasing doses of procaine penicillin G in lactating dairy cows. *Can. J. Vet. Res.* 65:173–180.

Eagan, J., and W.J. Meaney. 1984. The inhibitory effect of mastitic milk and colostrum on test methods used for antibiotic detection. *J. Food Sci. Technol.* 8:115–120.

Elder, H.A., I. Roy, S. Lehman, R.L. Phillips, and E.H. Kass. 1993. Human studies to measure the effect of antibiotic residues. *Vet. Human Toxicol.* 35:31–36.

Finegold, S.M., G.E. Mathisen, and W. L. George. 1983. Changes in human intestinal flora related to the administration of antimicrobial agents. In *Human Intestinal Microflora in Health and Disease.* Kent, U.K.: Academic Press, Inc, 355–429.

Goldberg, H.S., R.N. Goodman, J.T. Logue, and F.P. Handler. 1961. Long-term low level antibiotics and the emergence of antibiotic-resistant bacteria in human volunteers. In *Antimicrobial Agents Chemotherapy*, 80–88.

Gorbach, S.L. 1993. Pertabation of intestinal microflora. *Vet. Human Toxicol.* 35(suppl. 1):15–23.

Hillerton, J.E., B.I. Halley, P. Neaves, and M.D. Rose. 1999. Detection of antimicrobial substances in individual cow and quarter milk samples using Delvotest microbial inhibitor tests. *J. Dairy Sci.* 82:704–711.

Huber, W.G. 1986. Allergenicity of antibacterial drug residues. In *Drug Residues in Animals.* A.G. Rico, ed. Toronto, Ontario: Academic Press,. 33–49.

Kaneene, J.B., and A.S. Ahl. 1987. Drug residues in the dairy cattle industry: Epidemiological evaluation of factors influencing their occurrence. *J. Dairy Sci.* 70:2176–2180.

Kang, J.H., and F. Kondo. 2001. Occurrence of false-positive results of inhibitor on milk samples using the Delotest SP assay. *J. Food Prot.* 64:1211–1215.

Katz, S.E., and M. Siewierski. 1995. Bacillus Stearothermophilus disc assay: a review. *J.A.O.A.C. Internatl.* 78:1408.

Kitchen, B.J. 1981. Review of the progress of Dairy Science. Bovine mastitis: milk compositional changes and related diagnostic tests. *J. Dairy Res.* 48:167–188.

Kosikowski, F.V., and M. O'Leary. 1963. Natural inhibitory characteristics of some Irish manufacturing milks. *J. Dairy Sci.* 46:89–94.

Laben, R.C. 1963. Factors responsible for variation in milk composition. *J. Dairy Sci.* 46:1293–1296.

McEwen, S.C., W.D. Black, and A.H. Meek. 1991. Antibiotic residue prevention methods, farm management, and occurrence of antibiotic residues in milk. *J. Dairy Sci.* 74:2128–2137.

Medina, M. B. 1997. Hygromycin B antibody production and characterization by surface plasmon resonance biosensor. *J. Agric. Food. Chem.* 45:389–394.

Mitchell, J.M., M.W. Griffiths, S.A. McEwen, W.B. McNab, and A.J. Yee. 1998. Antimicrobial drug residues in milk and meat: causes, concerns, prevalence, regulations, tests, and test performance. *J. Food Prot.* 61:742–756.

Musser, J.M.B., K.L. Anderson, J.E. Rushing, and W.A. Moats. 2001. Potential for milk containing penicillin G or amoxicillin to cause residues in calves. *J. Dairy Sci.* 84:126–133.

Nord, C.E., A. Heimdahl, and L. Kager. 1984. The impact of different antimicrobial agents on normal gastrointestinal microflora of humans. *Rev. Infect. Dis.* 6(suppl.1): S270–275.

Norell, R.J., J.H. Packham, and L.J. Fox. 1994. Effect of clinical mastitis on antibiotic residue tests. In *National Mastitis Council Annual Meeting Proceedings, Orlando, FL.* Madison, WI: National Mastitis Council, Inc., 377–378.

Oliver, S.P., M.J. Lewis, B.E. Gillespie, and H.H. Dowlen. 1992. Influence of prepartum antibiotic therapy on intramammary infections in primigravid heifers during early lactation. *J. Dairy Sci.* 75:406–414.

Ostensson, K., M. Hageltorn, and G. Astrom. 1988. Differential cell counting in fraction-collected milk from dairy cows. *Acta Vet. Scand.* 29:493–500.

Paape, M.J., and H.A. Tucker. 1966. Somatic cell content variation in fraction-collected milk. *J. Dairy Sci.* 49:265–267.

Reneau, J.K., and V.S. Packard. 1991. Monitoring mastitis, milk quality, and economic losses in dairy fields. *Dairy Food Environ. Sanit.* 11:4–8.

Riediker, S., J. Diserens, and R. H. Stadler. 2001. Analysis of B-lactam antibiotics in incurred raw milk by rapid test methods and liquid chromatography coupled with electrospray ionization tandem mass spectrometry. *J. Agric. Food Chem.* 49:4171–4176.

Ruegg, P.L., and T.J. Tabone. 2000. The relationship between antibiotic residue violations and somatic cell counts in Wisconsin dairy herds. *J. Dairy Sci.* 83:2805–2809.

Saville, W.J., T.E. Wittum, and K.L. Smith. 2000. Association between measures of milk quality and risk of violative antimicrobial residues in grade-A raw milk. *J.A.V.M.A.* 217:541–545.

Sischo, W.M., and C. M. Burns. 1993. Field trial of four cowside antibiotic-residue screening tests. *J.A.V.M.A.* 202:1249–1254.

Sischo, W.M. 1996. Quality milk and tests for antibiotic residues. *J. Dairy Sci.* 79:1065–1073.

Sischo, W.M., N.E. Kiernan, C.M. Burns, and L.I. Byler. 1997. Implementing a quality assurance program using a risk assessment tool on dairy operations. *J. Dairy Sci.* 80:777–787.

Smucker, J.M. 1996. Beta lactam test methods for use under Appendix N of the PMO. M-a-85. Revision number 5. July 31, 1996. Food and Drug Administration, Washington, D.C.

Smucker, J.M. 1997. Results of the national tanker monitoring program. In *National Mastitis Council Annual Meeting Proceedings, Albuquerque, NM.* Madison, WI: National Mastitis Council, Inc.,. 185–186.

Sundlof, S.F. 1989. Drug and chemical residues in livestock. *Vet. Clin. N. Am. Food Anim. Pract.* 5:411–449.

Sundlof, S.F., J.B. Kaneene, and R. Miller. 1995. National Survey on veterinarian-initiated drug use in lactating cows. *J.A.V.M.A.* 207:347–352.

U.S. Food and Drug Administration. 1999. Consideration of the human health impact of the microbial effects of microbial effects of antimicrobial new animal drugs intended for use in food-producing animals. Center for Veterinary Medicine, Guideline 78, December 13, 1999.

U.S. Food and Drug Administration. 2000. A description of the Food Safety System. U.S. Department of Agriculture. March 3, 2000.

U.S. Food and Drug Administration.2001a. Center for Food Safety and Applied Nutrition. National Milk Drug Residue Data Base, Fiscal Year 2000 Annual Report. June 1, 2001.

U.S. Food and Drug Administration. 2001b. Assessment of the effects of antimicrobial drug residues from food of animal origin on the human intestinal tract. USFDA, Center of Veterinary Medicine Draft Guidance Document #52, December 19, 2001.

Van Eenennaam, A.L., J.S. Cullor, L. Perani, I.A. Gardner, W.L. Smith, J. Dellinger, and W.M. Guterbock. 1993. Evaluation of milk antibiotic residue screening tests in cattle with naturally occurring clinical mastitis. *J. Dairy Sci.* 76:3041–3053.

INDEX